LONDON MATHEMATICAL SOCIETY LECTURE NOTE

Managing Editor: Professor Endre Süli, Mathematical Institute, Univer
Woodstock Road, Oxford OX2 6GG, United Kingdom

The titles below are available from booksellers, or from Cambridge Ur
www.cambridge.org/mathematics

375 Triangulated categories, T. HOLM, P. JØRGENSEN & R. ROUQUIER (eds)
376 Permutation patterns, S. LINTON, N. RUŠKUC & V. VATTER (eds)
377 An introduction to Galois cohomology and its applications, G. BERHUY
378 Probability and mathematical genetics, N. H. BINGHAM & C. M. GOLDIE (eds)
379 Finite and algorithmic model theory, J. ESPARZA, C. MICHAUX & C. STEINHORN (eds)
380 Real and complex singularities, M. MANOEL, M.C. ROMERO FUSTER & C.T.C WALL (eds)
381 Symmetries and integrability of difference equations, D. LEVI, P. OLVER, Z. THOMOVA &
    P. WINTERNITZ (eds)
382 Forcing with random variables and proof complexity, J. KRAJÍČEK
383 Motivic integration and its interactions with model theory and non-Archimedean geometry I, R. CLUCKERS,
    J. NICAISE & J. SEBAG (eds)
384 Motivic integration and its interactions with model theory and non-Archimedean geometry II, R. CLUCKERS,
    J. NICAISE & J. SEBAG (eds)
385 Entropy of hidden Markov processes and connections to dynamical systems, B. MARCUS, K. PETERSEN &
    T. WEISSMAN (eds)
386 Independence-friendly logic, A.L. MANN, G. SANDU & M. SEVENSTER
387 Groups St Andrews 2009 in Bath I, C.M. CAMPBELL et al (eds)
388 Groups St Andrews 2009 in Bath II, C.M. CAMPBELL et al (eds)
389 Random fields on the sphere, D. MARINUCCI & G. PECCATI
390 Localization in periodic potentials, D.E. PELINOVSKY
391 Fusion systems in algebra and topology, M. ASCHBACHER, R. KESSAR & B. OLIVER
392 Surveys in combinatorics 2011, R. CHAPMAN (ed)
393 Non-abelian fundamental groups and Iwasawa theory, J. COATES et al (eds)
394 Variational problems in differential geometry, R. BIELAWSKI, K. HOUSTON & M. SPEIGHT (eds)
395 How groups grow, A. MANN
396 Arithmetic differential operators over the p-adic integers, C.C. RALPH & S.R. SIMANCA
397 Hyperbolic geometry and applications in quantum chaos and cosmology, J. BOLTE & F. STEINER (eds)
398 Mathematical models in contact mechanics, M. SOFONEA & A. MATEI
399 Circuit double cover of graphs, C.-Q. ZHANG
400 Dense sphere packings: a blueprint for formal proofs, T. HALES
401 A double Hall algebra approach to affine quantum Schur–Weyl theory, B. DENG, J. DU & Q. FU
402 Mathematical aspects of fluid mechanics, J.C. ROBINSON, J.L. RODRIGO & W. SADOWSKI (eds)
403 Foundations of computational mathematics, Budapest 2011, F. CUCKER, T. KRICK, A. PINKUS &
    A. SZANTO (eds)
404 Operator methods for boundary value problems, S. HASSI, H.S.V. DE SNOO & F.H. SZAFRANIEC (eds)
405 Torsors, étale homotopy and applications to rational points, A.N. SKOROBOGATOV (ed)
406 Appalachian set theory, J. CUMMINGS & E. SCHIMMERLING (eds)
407 The maximal subgroups of the low-dimensional finite classical groups, J.N. BRAY, D.F. HOLT &
    C.M. RONEY-DOUGAL
408 Complexity science: the Warwick master's course, R. BALL, V. KOLOKOLTSOV & R.S. MACKAY (eds)
409 Surveys in combinatorics 2013, S.R. BLACKBURN, S. GERKE & M. WILDON (eds)
410 Representation theory and harmonic analysis of wreath products of finite groups,
    T. CECCHERINI-SILBERSTEIN, F. SCARABOTTI & F. TOLLI
411 Moduli spaces, L. BRAMBILA-PAZ, O. GARCÍA-PRADA, P. NEWSTEAD & R.P. THOMAS (eds)
412 Automorphisms and equivalence relations in topological dynamics, D.B. ELLIS & R. ELLIS
413 Optimal transportation, Y. OLLIVIER, H. PAJOT & C. VILLANI (eds)
414 Automorphic forms and Galois representations I, F. DIAMOND, P.L. KASSAEI & M. KIM (eds)
415 Automorphic forms and Galois representations II, F. DIAMOND, P.L. KASSAEI & M. KIM (eds)
416 Reversibility in dynamics and group theory, A.G. O'FARRELL & I. SHORT
417 Recent advances in algebraic geometry, C.D. HACON, M. MUSTAȚĂ & M. POPA (eds)
418 The Bloch–Kato conjecture for the Riemann zeta function, J. COATES, A. RAGHURAM, A. SAIKIA &
    R. SUJATHA (eds)
419 The Cauchy problem for non-Lipschitz semi-linear parabolic partial differential equations, J.C. MEYER &
    D.J. NEEDHAM
420 Arithmetic and geometry, L. DIEULEFAIT et al (eds)
421 O-minimality and Diophantine geometry, G.O. JONES & A.J. WILKIE (eds)
422 Groups St Andrews 2013, C.M. CAMPBELL et al (eds)
423 Inequalities for graph eigenvalues, Z. STANIĆ

London Mathematical Society Lecture Note Series: 485

The Calabi Problem for Fano Threefolds

CAROLINA ARAUJO
Institute for Pure and Applied Mathematics (IMPA), Rio de Janeiro

ANA-MARIA CASTRAVET
University of Versailles

IVAN CHELTSOV
University of Edinburgh

KENTO FUJITA
Osaka University, Japan

ANNE-SOPHIE KALOGHIROS
Brunel University, London

JESUS MARTINEZ-GARCIA
University of Essex

CONSTANTIN SHRAMOV
Steklov Mathematical Institute, Moscow

HENDRIK SÜß
University of Jena, Germany

NIVEDITA VISWANATHAN
Loughborough University

CAMBRIDGE
UNIVERSITY PRESS

CAMBRIDGE
UNIVERSITY PRESS

Shaftesbury Road, Cambridge CB2 8EA, United Kingdom

One Liberty Plaza, 20th Floor, New York, NY 10006, USA

477 Williamstown Road, Port Melbourne, VIC 3207, Australia

314–321, 3rd Floor, Plot 3, Splendor Forum, Jasola District Centre, New Delhi – 110025, India

103 Penang Road, #05–06/07, Visioncrest Commercial, Singapore 238467

Cambridge University Press is part of Cambridge University Press & Assessment, a department of the University of Cambridge.

We share the University's mission to contribute to society through the pursuit of education, learning and research at the highest international levels of excellence.

www.cambridge.org
Information on this title: www.cambridge.org/9781009193399
DOI: 10.1017/9781009193382

© Carolina Araujo, Ana-Maria Castravet, Ivan Cheltsov, Kento Fujita, Anne-Sophie Kaloghiros, Jesus Martinez-Garcia, Constantin Shramov, Hendrik Süß and Nivedita Viswanathan 2023

First published 2023

Printed in the United Kingdom by TJ Books Limited, Padstow Cornwall

A catalogue record for this publication is available from the British Library.

A Cataloging-in-Publication data record for this book is available from the Library of Congress.

ISBN 978-1-009-19339-9 Paperback

Contents

Introduction

The Kähler–Einstein K-stability correspondence for Fano varieties is one of the most important contributions achieved in the 21st century [71, 212, 78, 82, 59, 214]. It links together complex algebraic geometry and analytic geometry:

a smooth Fano variety admits a Kähler–Einstein metric $\Longleftrightarrow$ it is K-polystable.

However, the notion of K-stability is elusive and often difficult to check (see Chapter 1). On the other hand, for two-dimensional Fano varieties, Tian and Yau proved

Theorem ([215, 211]) *Let S be a smooth del Pezzo surface. Then S is K-polystable if and only if it is not a blow up of $\mathbb{P}^2$ in one or two points.*

Smooth Fano threefolds have been classified in [118, 119, 158, 159] into 105 families, which are labeled as №1.1, №1.2, №1.3, ..., №9.1, №10.1 (see the Big Table in Chapter 6). Threefolds in each of these 105 deformation families can be parametrized by a non-empty irreducible rational variety [161, 163]. We pose the following problem.

Calabi Problem *Find all K-polystable smooth Fano threefolds in each family.*

This problem has already been solved for many families, and partial results are known in many cases [2, 3, 7, 14, 39, 46, 47, 55, 69, 79, 93, 101, 117, 146, 165, 199, 202, 212, 219, 227]. In particular, it has been proved in [93] that all smooth threefolds in the 26 families

№2.23, №2.28, №2.30, №2.31, №2.33, №2.35, №2.36, №3.14,
№3.16, №3.18, №3.21, №3.22, №3.23, №3.24, №3.26, №3.28, №3.29,
№3.30, №3.31, №4.5, №4.8, №4.9, №4.10, №4.11, №4.12, №5.2

1

are divisorially unstable (see Definition 1.20), so that none of them is K-polystable.

We show that all smooth Fano threefolds №2.26 are not K-polystable, and prove

Main Theorem   *Let X be a general Fano threefold in the family №$\mathcal{N}$. Then*

$$X \text{ is K-polystable} \iff \mathcal{N} \neq 2.26 \text{ and } \mathcal{N} \notin \begin{cases} 2.23, 2.28, 2.30, 2.31, 2.33, \\ 2.35, 2.36, 3.14, 3.16, 3.18, \\ 3.21, 3.22, 3.23, 3.24, 3.26, \\ 3.28, 3.29, 3.30, 3.31, 4.5, \\ 4.8, 4.9, 4.10, 4.11, 4.12, 5.2 \end{cases}.$$

Corollary   *Let X be a general Fano threefold in the family №$\mathcal{N} \neq$ №2.26. Then*

X is K-polystable $\iff$ X is divisorially semistable $\iff$ X is K-semistable.

Note that K-stability is an open property [170, 80, 19, 147]. Therefore, to prove that a general element of a given deformation family is K-polystable, it is enough to produce at least one K-stable (possibly singular) threefold in this family. However, this approach does not always work because many deformation families contain only Fano threefolds with infinite automorphism groups [45], so that none of these threefolds are K-stable, but some of them a priori could be K-polystable.

Before we finished the proof of the Main Theorem, its assertion had been already known for 65 deformation families (see Chapter 3 and Section 4.1 for more details). These families are

№1.1, №1.2, №1.3, №1.4, №1.5, №1.6, №1.7, №1.8, №1.10, №1.11,
№1.12, №1.13, №1.14, №1.15, №1.16, №1.17, №2.4, №2.23, №2.28,
№2.6, №2.29, №2.30, №2.31, №2.32, №2.33, №2.34, №2.35, №2.36,
№3.1, №3.11, №3.14,
№3.16, №3.18, №3.19, №3.20, №3.21, №3.22, №3.23, №3.24, №3.26,
№3.27, №3.28, №3.29, №3.30, №3.31, №4.4, №4.5, №4.7, №4.8, №4.9,
№4.10, №4.11, №4.12, №5.2, №5.3, №6.1, №7.1, №8.1, №9.1, №10.1.

For some families, we solved the Calabi Problem for all smooth threefolds in the family. For details, see the proof of the Main Theorem and check the Big Table in Chapter 6.

Example (see Section 4.7)   Smooth Fano threefolds №2.24 are divisors in $\mathbb{P}^2 \times \mathbb{P}^2$ that have degree $(1, 2)$. For a suitable choice of coordinates $([x : y :$

z], [$u : v : w$]) on $\mathbb{P}^2 \times \mathbb{P}^2$, these smooth Fano threefolds can be described as follows.

(i) One parameter family that consists of threefolds given by

$$xu^2 + yv^2 + zw^2 + \mu(xvw + yuw + zuv) = 0, \qquad (\bigstar)$$

where $\mu \in \mathbb{C}$ such that $\mu^3 \neq -1$. All such threefolds are K-polystable.

(ii) One non-K-polystable threefold given by $(u^2+vw)x+(uw+v^2)y+w^2z = 0$,

(iii) One non-K-polystable threefold given by $(u^2 + vw)x + v^2y + w^2z = 0$.

If $\mu^3 = -1$ or $\mu = \infty$, then $(\bigstar)$ defines a singular K-polystable Fano threefold.

Smooth Fano threefolds with infinite automorphism groups have been described in [45]. We completely solve the Calabi Problem for all of them. To be precise, we proved

Theorem *Let X be a smooth Fano threefold in the family №$\mathcal{N}$ such that $\mathrm{Aut}^0(X) \neq 1$. Then X is K-polystable if and only if either*

$$\mathcal{N} \in \left\{ \begin{array}{l} 1.15, 1.16, 1.17, 2.20, 2.22, 2.27, 2.32, 2.34, 2.29, 3.5, 3.8, 3.9, 3.12, \\ 3.15, 3.17, 3.19, 3.20, 3.25, 3.27, 4.2, 4.3, 4.4, 4.6, 4.7, 4.13, 5.1, 5.3, \\ 6.1, 7.1, 8.1, 9.1, 10.1 \end{array} \right\}$$

or one of the following cases hold:

- $\mathcal{N} = 1.10$ *and* $\mathrm{Aut}^0(X) \cong \mathrm{PGL}_2(\mathbb{C})$ *or* $\mathrm{Aut}^0(X) \cong \mathbb{G}_m$;
- $\mathcal{N} = 2.21$ *and* $\mathrm{Aut}^0(X) \cong \mathrm{PGL}_2(\mathbb{C})$ *or* $\mathrm{Aut}^0(X) \cong \mathbb{G}_m$;
- $\mathcal{N} = 2.24$ *and* $\mathrm{Aut}^0(X) \cong \mathbb{G}_m^2$;
- $\mathcal{N} = 3.10$ *and either* $\mathrm{Aut}^0(X) \cong \mathbb{G}_m^2$, *or* $\mathrm{Aut}^0(X) \cong \mathbb{G}_m$ *and X can be obtained by blowing up the smooth quadric threefold in $\mathbb{P}^4$ given by*

$$w^2 + xy + zt + a(xt + yz) = 0$$

along two conics that are given by $w^2 + zt = x = y = 0$ and $w^2 + xy = z = t = 0$, where $a \in \mathbb{C}$ is such that $a \notin \{0, \pm 1\}$, and x, y, z, t, w are coordinates on $\mathbb{P}^4$;

- $\mathcal{N} = 3.13$ *and* $\mathrm{Aut}^0(X) \cong \mathrm{PGL}_2(\mathbb{C})$ *or* $\mathrm{Aut}^0(X) \cong \mathbb{G}_m$.

At present, the Calabi Problem is not yet completely solved for the following 34 families:

№1.9, №1.10, №2.1, №2.2, №2.3, №2.4, №2.5, №2.6,
№2.7, №2.8, №2.9, №2.10, №2.11, №2.12, №2.13, №2.14,
№2.15, №2.16, №2.17, №2.18, №2.19, №2.20, №2.21, №2.22, №3.2,
№3.3, №3.4, №3.5, №3.6, №3.7, №3.8, №3.11, №3.12, №4.1.

For 27 of these families, we expect the following to be true:

Conjecture　*All smooth Fano threefolds in the deformation families*

№1.9, №2.1, №2.2, №2.3, №2.4, №2.5, №2.6, №2.7, №2.8,
№2.9, №2.10, №2.11, №2.12, №2.13, №2.14, №2.15, №2.16, №2.17,
№2.18, №2.19, №3.2, №3.3, №3.4, №3.6, №3.7, №3.11, №4.1

are K-stable and, in particular, they are K-polystable.

The remaining seven families №1.10, №2.20, №2.21, №2.22, №3.5, №3.8, №3.12 contain non-K-polystable smooth Fano threefolds, but their general members are K-polystable. We present conjectural characterizations of their K-polystable members in Chapter 7.

Remark　After the original version of this book appeared in June 2021 as Preprint 2021-31 in the preprint series of the Max Planck Institute for Mathematics, our Conjecture has been confirmed for the families №2.8, №3.3 and №4.1 in [16, 34, 145], and our conjectural characterizations of the K-polystable members of the families №2.22 and №3.12 have been proved in [38, 68]. Note that it follows from [38, 68] that every smooth Fano threefold in the deformation families №2.22 and №3.12 is K-semistable.

In Chapter 1, we present some K-stability results used in the proof of the Main Theorem. In Chapter 2, we prove the Tian–Yau theorem and find δ-invariants of del Pezzo surfaces. In Chapters 3, 4, and 5, we prove the Main Theorem. In Chapter 6, we present the Big Table that summarizes our results. In the Appendix, we present technical results used in the book.

Notations and conventions.　Throughout this book, all varieties are assumed to be projective and defined over the field $\mathbb{C}$. For a variety X, we denote by $\overline{\mathrm{Eff}}(X)$, $\overline{\mathrm{NE}}(X)$ and $\mathrm{Nef}(X)$ the closure of the cone of effective divisors on X, the Mori cone of X, and the cone of nef divisors on X, respectively. For a subgroup $G \subset \mathrm{Aut}(X)$, we denote by $\mathrm{Cl}^G(X)$ and $\mathrm{Pic}^G(X)$ the subgroups in $\mathrm{Cl}(X)$ and $\mathrm{Pic}(X)$ consisting of Weil and Cartier divisors whose classes are G-invariant, respectively.

A subvariety $Y \subset X$ is said to be G-irreducible if Y is G-invariant and is not a union of two proper G-invariant subvarieties. We also denote by $\mathrm{Aut}(X, Y)$ the group consisting of automorphisms in $\mathrm{Aut}(X)$ that maps Y into itself.

We denote by $\mathbb{F}_n$ the Hirzebruch surface $\mathbb{P}(\mathcal{O}_{\mathbb{P}^1} \oplus \mathcal{O}_{\mathbb{P}^1}(n))$. In particular, $\mathbb{F}_0 \cong \mathbb{P}^1 \times \mathbb{P}^1$, and the surface $\mathbb{F}_1$ is the blow up of $\mathbb{P}^2$ at a point.

For a divisor D on $\mathbb{P} = \mathbb{P}^{n_1} \times \mathbb{P}^{n_2} \times \cdots \times \mathbb{P}^{n_k}$, we say that D has degree $(a_1, a_2, \ldots, a_k)$ if

$$D \sim \sum_{i=1}^{k} \mathrm{pr}_i^* \left(O_{\mathbb{P}^{n_i}}(a_i) \right),$$

where $\mathrm{pr}_i \colon \mathbb{P} \to \mathbb{P}^{n_i}$ is the projection to the ith factor. For a curve $C \subset \mathbb{P}$, we say that C has degree $(a_1, a_2, \ldots, a_k)$ if $\mathrm{pr}_i^*(O_{\mathbb{P}^{n_i}}(1)) \cdot C = a_i$ for every $i \in \{1, \ldots, k\}$.

We denote by μ_n the cyclic group of order n, we denote by D_{2n} the dihedral group of order $2n$, where $n \geqslant 2$ and $\mathrm{D}_4 = \mu_2^2$. Similarly, we denote by $\mathfrak{S}_n$ and $\mathfrak{A}_n$ the symmetric group and its alternating subgroup, respectively. We denote by $\mathbb{G}_a$ the one-dimensional unipotent additive group, and we denote by $\mathbb{G}_m$ the one-dimensional algebraic torus.

We denote by $\mathbb{G}_m \rtimes \mu_2$ the unique non-trivial semi-direct product of $\mathbb{G}_m$ and μ_2, we denote by $\mathbb{G}_m \rtimes \mathfrak{S}_3$ the unique non-trivial semi-direct product of $\mathbb{G}_m$ and $\mathfrak{S}_3$, and we denote by $\mathbb{G}_a \rtimes \mathbb{G}_m$ the semi-direct product such that $\mathbb{G}_m$ acts on $\mathbb{G}_a$ as $\mathbf{x} \mapsto t\mathbf{x}$.

For positive integers $n > k_1 > \cdots > k_r$, we denote by $\mathrm{PGL}_{n;k_1,\ldots,k_r}(\mathbb{C})$ the parabolic subgroup in $\mathrm{PGL}_n(\mathbb{C})$ that consists of images of matrices in $\mathrm{GL}_n(\mathbb{C})$ preserving a flag of subspaces of dimensions $k_1, \ldots, k_r$. For $n \geqslant 5$, we denote by $\mathrm{PSO}_{n;k}(\mathbb{C})$ the parabolic subgroup of $\mathrm{PSO}_n(\mathbb{C})$ preserving an isotropic linear subspace of dimension k. By $\mathrm{PGL}_{(2,2)}(\mathbb{C})$ we denote the image in $\mathrm{PGL}_4(\mathbb{C})$ of the group of block-diagonal matrices in $\mathrm{GL}_4(\mathbb{C})$ with two 2×2 blocks. This group acts on $\mathbb{P}^3$ preserving two skew lines. By $\mathrm{PGL}_{(2,2);1}(\mathbb{C})$ we denote the stabilizer in $\mathrm{PGL}_{(2,2)}(\mathbb{C})$ of a point on one of these lines.

Acknowledgments. We started this project in 2020 during a workshop on K-stability at the American Institute of Mathematics (San Jose, California, USA), which was organized by Mattias Jonsson and Chenyang Xu. We are very grateful to the Institute and the organizers for this workshop, which triggered our research.

Carolina Araujo was partially supported by CNPq and Faperj Research Fellowships. Ana-Maria Castravet was supported by the grant ANR-20-CE40-0023. Ivan Cheltsov was supported by EPSRC grant EP/V054597/1 and is also very grateful to the Max Planck Institute for Mathematics in Bonn for its hospitality. Kento Fujita has been partially supported by JSPS KAKENHI Grant Numbers 18K13388 and 22K03269. Anne-Sophie Kaloghiros was supported by an LMS Emmy Noether Fellowship and by EPSRC Grant EP/V056689/1. Jesus Martinez-Garcia was supported by EPSRC grant EP/V055399/1. Hendrik Süß was supported by EPSRC grants EP/V055445/1 and EP/V013270/1 and by the Carl Zeiss Foundation. Nivedita Viswanathan was partially supported by EPSRC Grant EP/V048619/1.

The authors would like to thank Hamid Abban (Ahmadinezhad), Harold Blum, Igor Dolgachev, Sir Simon Donaldson, Mattias Jonsson, Alexander Kuznetsov, Yuchen Liu, Yuji Odaka, Jihun Park, Andrea Petracci, Yuri Prokhorov, Sandro Verra, Chenyang Xu, Shing-Tung Yau and Ziquan Zhuang for many useful comments.

1

K-Stability

1.1 What is K-stability?

Let X be a Fano variety of dimension $n \geqslant 2$ that has Kawamata log terminal singularities. In most of the cases we consider, the variety X will be smooth. Set $L = -K_X$. A (normal) *test configuration* of the (polarized) pair $(X; L)$ consists of

- a normal variety $\mathcal{X}$ with a $\mathbb{G}_m$ action,
- a flat $\mathbb{G}_m$-equivariant morphism $p: \mathcal{X} \to \mathbb{P}^1$, where $\mathbb{G}_m$ acts naturally on $\mathbb{P}^1$ by

$$(t, [x : y]) \mapsto [tx : y],$$

- a $\mathbb{G}_m$-invariant p-ample $\mathbb{Q}$-line bundle $\mathcal{L} \to \mathcal{X}$ and a $\mathbb{G}_m$-equivariant isomorphism

$$\left(\mathcal{X} \backslash p^{-1}(0), \mathcal{L}|_{\mathcal{X} \backslash p^{-1}(0)}\right) \cong \left(X \times (\mathbb{P}^1 \backslash \{0\}), \mathrm{pr}_1^*(L)\right),$$

where pr_1 is the projection to the first factor, and $0 = [0 : 1]$.

For such a test configuration, we let

$$\mathrm{DF}(\mathcal{X}; \mathcal{L}) = \frac{1}{L^n}\left(\mathcal{L}^n \cdot K_{\mathcal{X}/\mathbb{P}^1} + \frac{n}{n+1}\mathcal{L}^{n+1}\right). \tag{1.1}$$

This number is called *Donaldson–Futaki invariant* of the test configuration $(\mathcal{X}, \mathcal{L})$.

Remark 1.1   Quite often, we will omit $\mathcal{L}$ in $\mathrm{DF}(\mathcal{X}; \mathcal{L})$ and write it as $\mathrm{DF}(\mathcal{X})$.

Denote the central fiber $p^{-1}(0)$ by $\mathcal{X}_0$, and denote the fiber at infinity $p^{-1}(\infty)$ by $\mathcal{X}_\infty$, where $\infty = [1 : 0]$. The test configuration $(\mathcal{X}, \mathcal{L})$ is said to be

- *trivial* if there is a $\mathbb{G}_m$-equivariant isomorphism

$$\left(\mathcal{X}\backslash\mathcal{X}_\infty, \mathcal{L}|_{\mathcal{X}\backslash\mathcal{X}_\infty}\right) \cong \left(X \times (\mathbb{P}^1\backslash\infty), \mathrm{pr}_1^*(L)\right),$$

- *product-type* if we have an isomorphism $\mathcal{X}\backslash\mathcal{X}_\infty \cong X \times (\mathbb{P}^1\backslash\infty)$,
- *special* if the fiber $\mathcal{X}_0$ is irreducible, reduced, and $(\mathcal{X}, \mathcal{X}_0)$ has purely log terminal singularities, so that $\mathcal{X}_0$ is a Fano variety with Kawamata log terminal singularities.

Definition 1.2   The Fano variety X is said to be K-semistable if for every test configuration $(\mathcal{X}, \mathcal{L})$, $\mathrm{DF}(\mathcal{X}; \mathcal{L}) \geqslant 0$. Similarly, the Fano variety X is said to be K-stable if for every non-trivial test configuration $(\mathcal{X}, \mathcal{L})$, $\mathrm{DF}(\mathcal{X}; \mathcal{L}) > 0$. Finally, the Fano variety X is said to be K-polystable if it is K-semistable and

$$\mathrm{DF}(\mathcal{X}; \mathcal{L}) = 0 \iff (\mathcal{X}, \mathcal{L}) \text{ is of product type.}$$

Thus, we have the following implications:

$$X \text{ is K-stable} \implies X \text{ is K-polystable} \implies X \text{ is K-semistable.}$$

If X is not K-semistable, we say that X is K-unstable. Similarly, if X is K-semistable, but the Fano variety X is not K-polystable, we say that X is strictly K-semistable.

Theorem 1.3 ([6, 155])   *If X is K-polystable, then $\mathrm{Aut}(X)$ is reductive.*

Theorem 1.4 ([21, Corollary 1.3])   *If X is K-stable, then $\mathrm{Aut}(X)$ is finite.*

Corollary 1.5   *If $\mathrm{Aut}(X)$ is finite, then X is K-stable if and only if it is K-polystable.*

By the Chen–Donaldson–Sun theorem, the product of smooth K-polystable Fano varieties is K-polystable. This can be proved purely algebraically:

Theorem 1.6 ([225])   *Let V and Y be Fano varieties with Kawamata log terminal singularities. Then $V \times Y$ is K-semistable (resp. K-polystable, K-stable) if and only if V and Y are both K-semistable (resp. K-polystable, K-stable).*

Let G be a reductive subgroup in $\mathrm{Aut}(X)$. A given test configuration $(\mathcal{X}, \mathcal{L})$ is said to be G-equivariant if the product $G \times \mathbb{G}_m$ acts on $(\mathcal{X}, \mathcal{L})$ such that

- $\{1\} \times \mathbb{G}_m$ acting on $(\mathcal{X}, \mathcal{L})$ is the original $\mathbb{G}_m$-action,
- the $\mathbb{G}_m$-equivariant isomorphism

$$\left(\mathcal{X}\backslash p^{-1}(0), \mathcal{L}|_{\mathcal{X}\backslash p^{-1}(0)}\right) \cong \left(X \times (\mathbb{P}^1\backslash\{0\}), \mathrm{pr}_1^*(L)\right)$$

is $G \times \mathbb{G}_m$-equivariant.

Definition 1.7 The Fano variety X is said to be G-equivariantly K-polystable if for every G-equivariant test configuration $(\mathcal{X}, \mathcal{L})$, $\mathrm{DF}(\mathcal{X}; \mathcal{L}) \geqslant 0$, and $\mathrm{DF}(\mathcal{X}; \mathcal{L}) = 0$ if and only if $(\mathcal{X}, \mathcal{L})$ is of product type.

Remark 1.8 It has been proved in [142, 89] that it is enough to consider only special test configurations in Definitions 1.2 and 1.7.

If X is K-polystable, then X is G-equivariantly K-polystable. Surprisingly, we have

Theorem 1.9 ([67, 140, 148, 226]) *Suppose that X is G-equivariantly K-polystable. Then X is K-polystable.*

Remark 1.10 One can naturally define K-polystability for Fano varieties defined over an arbitrary field $\mathbb{F}$ of characteristic 0. By [226, Corollary 4.11], if X is defined over $\mathbb{F}$, and G is a reductive subgroup in $\mathrm{Aut}_{\mathbb{F}}(X)$, then

X is G-equivariantly K-polystable over $\mathbb{F} \iff X$ is K-polystable over $\overline{\mathbb{F}}$,

where $\overline{\mathbb{F}}$ is the algebraic closure of the field $\mathbb{F}$.

Let us conclude this section by briefly explaining how K-stability behaves in families.

Theorem 1.11 ([6, 19, 20, 80, 147, 141, 170, 218]) *Let $\eta \colon \mathcal{X} \to Z$ be a projective flat morphism such that $\mathcal{X}$ is $\mathbb{Q}$-Gorenstein, Z is normal, and all fibers of η are Fano varieties with at most Kawamata log terminal singularities. For every closed point $P \in Z$, let X_P be the fiber of the morphism η over P. Then the set*

$$\left\{ P \in Z \,\middle|\, X_P \text{ is K-stable} \right\}$$

is a Zariski open subset of the variety Z. Similarly, the set

$$\left\{ P \in Z \,\middle|\, X_P \text{ is K-semistable} \right\}$$

is a Zariski open subset of the variety Z. Furthermore, the set

$$\left\{ P \in Z \,\middle|\, X_P \text{ is K-polystable} \right\}$$

is a constructible subset of the variety Z.

Thus, if X is a K-polystable smooth Fano threefold such that the group $\mathrm{Aut}(X)$ is finite, then X is K-stable by Corollary 1.5, so that general Fano threefolds in the deformation family of X are K-stable. We will use this observation often in the proof of the Main Theorem to prove that a general member of a given family is K-stable. Vice versa, to prove that a given Fano threefold is not K-polystable, we will use the following result (cf. [31, 170]).

Theorem 1.12 ([21, Theorem 1.1])   *Let $\eta\colon X \to Z$ and $\eta'\colon X' \to Z$ be projective surjective morphisms such that both X and X' are $\mathbb{Q}$-Gorenstein, Z is a smooth curve, and all fibers of η and η' are Fano varieties with at most Kawamata log terminal singularities. Let P be a point in Z, and let X_P and X'_P be the fibers of the morphisms η and η' over P, respectively. Suppose that there is an isomorphism $X \setminus X_P \cong X' \setminus X'_P$ that fits the following commutative diagram:*

$$
\begin{array}{ccc}
X \setminus X_P & \xrightarrow{\;\cong\;} & X' \setminus X'_P \\
{\scriptstyle \eta|_{X\setminus X_P}}\big\downarrow & & \big\downarrow{\scriptstyle \eta'|_{X'\setminus X'_P}} \\
Z \setminus P & =\!\!=\!\!= & Z \setminus P
\end{array}
$$

If both X_P and X'_P are K-polystable, then they are isomorphic.

Together with Theorem 1.11, this result gives

Corollary 1.13   *Let $p\colon X \to \mathbb{P}^1$ be a test configuration for the Fano variety X such that the fiber $p^{-1}(0)$ is a K-polystable Fano variety with at most Kawamata log terminal singularities that is not isomorphic to X. Then X is strictly K-semistable.*

In some cases, it is possible to prove that the general element of the deformation family of a K-polystable Fano threefold X is also K-polystable, even when X has infinite automorphism group. This is achieved by relating K-polystability and GIT polystability, an idea first investigated in [26, 206] in the analytic context. Suppose that X is a smooth K-polystable Fano variety of dimension n, and set $d = (-K_X)^n$. Let us briefly recall the setup of deformation theory; proofs and details can be found in [192, 153].

The infinitesimal deformation functor of the Fano variety X is denoted Def_X; recall that for an Artinian local $\mathbb{C}$-algebra A with residue field $\mathbb{C}$, $\mathrm{Def}_X(A)$ consists of isomorphism classes of commutative diagrams:

$$
\begin{array}{ccc}
X & \hookrightarrow & X_S \\
\big\downarrow & & \big\downarrow \\
\{0\}{=}\mathrm{Spec}(\mathbb{C}) & \hookrightarrow & S{=}\mathrm{Spec}(A)
\end{array}
$$

An element $\{X_S \to S\} \in \mathrm{Def}_X(A)$ is a deformation family of X over S. The tangent space of the deformation functor Def_X is $T^1_X = Ext^1(\Omega_X, \mathcal{O}_X)$ and $T^2_X = Ext^2(\Omega_X, \mathcal{O}_X)$ is an obstruction space for Def_X. As X is a smooth Fano, $T^1_X = H^1(X, \mathcal{T}_X)$ and $T^2_X = 0$ (deformations of X are unobstructed).

Let A be the noetherian complete local $\mathbb{C}$-algebra with residue field $\mathbb{C}$ which is the hull of the functor of deformations of X; in other words, the formal spectrum of A is the base of the miniversal deformation of X. By the above, denoting by $S = \mathrm{Spec}(A)$, $T_{S,0} \to T_X^1$ is an isomorphism and S is smooth (deformations are unobstructed), so we can identify S with an analytic neighborhood of the origin in the affine space T_X^1.

Recall that G is a reductive subgroup in $\mathrm{Aut}(X)$. For instance, we may let $G = \mathrm{Aut}(X)$, since $\mathrm{Aut}(X)$ is reductive by Theorem 1.3 because X is assumed to be K-polystable. The group G acts on A and the Luna étale slice theorem for algebraic stacks [5] gives in this case a cartesian square

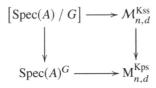

where $\mathcal{M}_{n,d}^{\mathrm{Kss}}$ is the stack that parametrizes n-dimensional K-semistable Fano varieties with at most Kawamata log terminal singularities that have anticanonical degree d [217], and $\mathrm{M}_{n,d}^{\mathrm{Kps}}$ is the algebraic space parametrizing n-dimensional K-polystable Fano varieties with Kawamata log terminal singularities that have anticanonical degree d. The horizontal arrows in this diagram are formally étale and map the closed point into the point corresponding to X.

Lemma 1.14 *Assume that the affine space T_X^1 contains a Zariski open subset consisting of GIT-polystable points with respect to the induced G-action. Then a general fiber $\mathcal{X}_t$ of the miniversal deformation $\mathcal{X} \to S$ of X is K-polystable.*

Proof An analytic formulation of this result is due to [26, 206]. Let $\mathcal{X}_t$ denote a general fiber in the miniversal deformation of X. Then $\mathcal{X}_t$ is K-semistable by Theorem 1.11. However, by the local description of K-moduli we can conclude that $\mathcal{X}_t$ is K-polystable.

Indeed, the general point in T_X^1 is GIT-polystable with respect to the G-action, so that it belongs to a closed G-orbit. But S coincides with a neighborhood of the origin in the tangent space T_X^1. By the Luna étale slice theorem for algebraic stacks, $\mathcal{X}_t$ gives rise to a closed point in $\mathcal{M}_{n,d}^{\mathrm{Kss}}$, so that $\mathcal{X}_t$ is K-polystable. $\square$

Corollary 1.15 *Assume that $\mathrm{Aut}^0(X) \cong \mathrm{PGL}_2(\mathbb{C})$, then the general fiber $\mathcal{X}_t$ of the miniversal deformation $\mathcal{X} \to S$ of X is K-polystable.*

Proof Here, T_X^1 is a sum of irreducible representations of $\mathrm{Aut}^0(X)$, which are odd-dimensional irreducible representations of $\mathrm{SL}_2(\mathbb{C})$. Thus, an orbit of a general vector in T_X^1 is closed by [181, Theorem 1] and the result follows from Lemma 1.14. $\qquad \square$

Corollary 1.16 *Assume that* $\mathrm{Aut}(X) \cong \mathbb{G}_m \rtimes G$, *where* G *is a finite group, some element of which acts on* $\mathbb{G}_m$ *by sending elements to their inverses. A general fiber* X_t *of the miniversal deformation* $\mathcal{X} \to S$ *of* X *is K-polystable.*

Proof The vector space T_X^1 is a linear representation of $\mathrm{Aut}(X) \cong \mathbb{G}_m \rtimes G$; it is entirely determined by the $\mathbb{G}_m$-weights for a chosen basis, and by the G-action on the basis elements. If all weights are 0, then the $\mathbb{G}_m$-action is trivial, and the result follows from Lemma 1.14. Now assume there is at least one non-zero weight $u \neq 0$. Then, the G-orbit of the corresponding basis element contains a basis element of weight $-u$, and every vector in T_X^1 with non-zero coordinates with respect to those two basis elements has a closed $\mathbb{G}_m$-orbit. Now, the result follows from Lemma 1.14. $\qquad \square$

Remark 1.17 If $\mathrm{Aut}^0(X) \cong \mathbb{G}_m^n$ for $n \geqslant 1$, but a general fiber X_t of the miniversal deformation $\mathcal{X} \to S$ of X has finite automorphism group, then $\mathcal{X} \to S$ contains strictly K-semistable smooth members. Indeed, every $\mathbb{G}_m$-fixed point in S lies in the closure of a maximal orbit, giving rise to a destabilizing test configuration for family members parametrized by these orbits.

1.2 Valuative criterion

Let X be a Fano variety with Kawamata log terminal singularities, let G be a reductive subgroup of $\mathrm{Aut}(X)$, let $f \colon \widetilde{X} \to X$ be a G-equivariant birational morphism, let E be a G-invariant prime divisor in $\widetilde{X}$, and let $n = \dim(X)$.

Definition 1.18 We say that E is a G-invariant prime divisor *over* the Fano variety X. If E is f-exceptional, we say that E is an exceptional G-invariant prime divisor *over* X. We will denote the subvariety $f(E)$ by $C_X(E)$. We say that E is *dreamy* if the $\mathbb{C}$-algebra

$$\bigoplus_{m,j \in \mathbb{Z}_{\geqslant 0}} H^0\big(\widetilde{X}, \mathcal{O}_{\widetilde{X}}\big(f^*(-mK_X) - jE\big)\big)$$

is finitely generated.

   Let

$$S_X(E) = \frac{1}{(-K_X)^n} \int_0^{\tau} \mathrm{vol}(f^*(-K_X) - xE)dx,$$

where $\tau = \tau(E)$ is the pseudo-effective threshold of E with respect to $-K_X$, i.e. we have

$$\tau(E) = \sup\{x \in \mathbb{Q}_{>0} \mid f^*(-K_X) - xE \text{ is big}\}.$$

Let $\beta(E) = A_X(E) - S_X(E)$, where $A_X(E)$ is the log discrepancy of the divisor E.

Theorem 1.19 ([95, 139, 21]) *The following assertions hold:*

- X *is* K-*stable* $\iff$ $\beta(F) > 0$ *for every prime divisor* F *over* X;
- X *is* K-*semistable* $\iff$ $\beta(F) \geqslant 0$ *for every prime divisor* F *over* X.

This criterion leads to the notion of *divisorial stability*, which is weaker than K-stability.

Definition 1.20 ([93, Definition 1.1]) The Fano variety X is said to be divisorially stable (respectively, semistable) if $\beta(F) > 0$ (respectively, $\beta(F) \geqslant 0$) for every prime divisor F on X. We say that X is divisorially unstable if it is not divisorially semistable.

For toric Fano varieties, divisorial semistability and K-polystability coincide by

Theorem 1.21 ([219, 93]) *Let X be a toric Fano variety, and let P be its associated polytope in $M \otimes_{\mathbb{Z}} \mathbb{R}$, where M is the character lattice of the torus. Then*

$$X \text{ is divisorially semistable} \iff X \text{ is } K\text{-polystable}$$
$$\iff \text{ the barycenter of } P \text{ is } 0.$$

To prove K-polystability, we can use the following handy criterion:

Theorem 1.22 ([226, Corollary 4.14]) *Suppose that $\beta(F) > 0$ for every G-invariant dreamy prime divisor F over X. Then X is K-polystable.*

Proof Let $(\mathcal{X}, \mathcal{L})$ be some G-equivariant special test configuration, so that $\mathcal{X}_0$ is integral. By Remark 1.8 and Theorem 1.9, it is enough to prove that $\mathrm{DF}(\mathcal{X}; \mathcal{L}) > 0$.

The fiber $\mathcal{X}_0$ defines a G-invariant prime divisor over $X \times \mathbb{A}^1$ since $\mathcal{X}$ is clearly birational to the product $X \times \mathbb{A}^1$. This gives us a divisorial valuation $\mathrm{ord}_{\mathcal{X}_0} : \mathbb{C}(X)(t)^* \to \mathbb{Z}$, so that we can consider the restricted valuation:

$$v_{\mathcal{X}_0} := \mathrm{ord}_{\mathcal{X}_0}|_{\mathbb{C}(X)^*} : \mathbb{C}(X)^* \to \mathbb{Z}.$$

This valuation is non-trivial and G-invariant by construction. Then, by [24, Lemma 4.5], there exists a G-invariant prime divisor F over X such that

$$v_{\mathcal{X}_0} = c \cdot \mathrm{ord}_F$$

for some integer $c > 0$. Moreover, it follows from [95, Theorem 5.1] that F is dreamy and

$$\mathrm{DF}(\mathcal{X}; \mathcal{L}) = A_X(F) - S_X(F),$$

so that $A_X(F) - S_X(F) > 0$ by our assumption. □

Remark 1.23 By [226, Corollary 4.14], Theorem 1.22 can be generalized for varieties defined over arbitrary fields as follows. If X is a Fano variety defined over an arbitrary field $\mathbb{F}$ of characteristic 0, and G is a reductive subgroup in $\mathrm{Aut}_{\mathbb{F}}(X)$ such that $\beta(F) > 0$ for every G-invariant geometrically irreducible divisor F over X, then X is K-polystable over the algebraic closure of the field $\mathbb{F}$.

In some cases, it is not easy to compute $S_X(E)$, but one can estimate it using basic properties of volumes. To explain this in detail, let V be an arbitrary n-dimensional normal projective variety, let L be a big and nef $\mathbb{Q}$-divisor on V, let $h: \widetilde{V} \to V$ be a birational morphism such that $\widetilde{V}$ is also a normal projective variety, and let F be prime divisor in $\widetilde{V}$. Abusing our previous notations, we let $\tau = \sup\{x \in \mathbb{Q}_{>0} \mid h^*(L) - xF \text{ is big}\}$. Fix $a \in (0, \tau)$. Then

$$\int_0^\tau \mathrm{vol}(h^*(L) - xF)dx \leqslant \int_0^a \mathrm{vol}(h^*(L) - xF)dx + (\tau - a)\mathrm{vol}(h^*(L) - aF)$$

because $\mathrm{vol}(h^*(L) - xF)$ is a decreasing function of x. This observation is very handy since the volume function $\mathrm{vol}(h^*(L) - xF)$ is often difficult to compute for large $x \in (0, \tau)$. Using log concavity of the volumes and the restricted volumes [138, 87], we can improve the latter inequality. Namely, arguing as in the proof of [96, Proposition 2.1], we get

$$\int_0^\tau \mathrm{vol}(h^*(L) - xF)dx \leqslant \int_0^a \mathrm{vol}(h^*(L) - xF)dx + \frac{n}{n+1}(\tau - a)\mathrm{vol}(h^*(L) - aF). \tag{1.2}$$

Furthermore, it follows from the proof of [97, Proposition 2] that

$$\mathrm{vol}(h^*(L) - xF) \leqslant \mathrm{vol}(h^*(L) - aF)\left(1 - \frac{(x-a)\phi(a)}{\mathrm{vol}(h^*(L) - aF)}\right)^n \tag{1.3}$$

for any $x \in (a, \tau)$, and $\tau \leqslant a + \frac{\mathrm{vol}(h^*(L) - aF)}{\phi(a)}$, where $\phi(x) = -\frac{1}{n}\frac{\partial}{\partial x}\mathrm{vol}(h^*(L) - xF)$.

1.3 Complexity one $\mathbb{T}$-varieties

Let X be a Fano variety with Kawamata log terminal singularities, let $\mathbb{T}$ be the maximal torus in $\mathrm{Aut}(X)$, and let $\mathbb{C}(X)^{\mathbb{T}}$ be the subfield in $\mathbb{C}(X)$ consisting of all $\mathbb{T}$-invariant rational functions.

Definition 1.24   The *complexity* of the $\mathbb{T}$-action on X is the number $\dim(X) - \dim(\mathbb{T})$.

First, we observe that the complexity of the $\mathbb{T}$-action is 0 if and only if X is toric. If the complexity of the $\mathbb{T}$-action is 1 then $\mathbb{C}(X)^{\mathbb{T}} = \mathbb{C}(Y)$ for some smooth curve Y, and the inclusion of fields $\mathbb{C}(Y) = \mathbb{C}(X)^{\mathbb{T}} \subset \mathbb{C}(X)$ gives the rational quotient map $\pi \colon X \dashrightarrow Y$. Moreover, we have $Y \cong \mathbb{P}^1$ since X is rationally connected [224].

Let M be the character lattice of $\mathbb{T}$, and let N be the dual lattice of one-parameter subgroups of the torus $\mathbb{T}$. We will denote by $\langle \cdot, \cdot \rangle$ the natural pairing between M and N.

Remark 1.25   Let w be an element in N. We write λ_w for the induced $\mathbb{G}_m$-action on X. We will consider N to be an additive group. But once we pass to λ_w, we will write the composition of two such $\mathbb{G}_m$-actions multiplicatively: $\lambda_{w+w'} = \lambda_w \lambda_{w'}$ for any $w' \in N$.

Denote by $\mathbb{C}(X)^{(\mathbb{T})}$ the multiplicative subgroup in $\mathbb{C}(X)^{\mathbb{T}}$ consisting of non-zero semi-invariant functions. We fix a group homomorphism $M \to \mathbb{C}(X)^{(\mathbb{T})}$ given by $u \mapsto \chi^u$, where χ^u is semi-invariant function in $\mathbb{C}(X)_u^{(\mathbb{T})}$ that has weight u. Given two semi-invariant functions f_u and g_u of the same weight u, their quotient f_u/g_u must be $\mathbb{T}$-invariant. Hence, every semi-invariant function can be expressed as $f\chi^u$ with $f \in \mathbb{C}(X)^{\mathbb{T}}$.

Let E be a $\mathbb{T}$-invariant prime divisor *over* X (see Definition 1.18).

Definition 1.26   The divisor E is said to be *vertical* if a maximal $\mathbb{T}$-orbit in E has the same dimension as the torus $\mathbb{T}$. Otherwise, the divisor E is said to be *horizontal*.

Remark 1.27   If X is toric, then all $\mathbb{T}$-invariant divisors in X are horizontal.

If E is horizontal, then the generic $\mathbb{T}$-orbit in the divisor E has dimension $\dim(\mathbb{T}) - 1$, so that the generic stabilizer must be a one-dimensional subtorus of the torus $\mathbb{T}$, which corresponds to rank-one sublattice, which we will denote by $N_E \subset N$.

Fix an integer $\ell \gg 0$ such that $-\ell K_X$ is an ample Cartier divisor. Let $L = \mathcal{O}_X(-\ell K_X)$, and let $l_k = \dim H^0(X, L^{\otimes k})$. For every $\mathbb{G}_m$-action λ on the threefold X and its canonical linearization for L, we set

$$w_k(\lambda) = \sum_m m \cdot \dim \left(H^0(X, L^{\otimes k})_m \right),$$

where $H^0(X, L^{\otimes k})_m$ is the subspace of the semi-invariant sections of λ-weight m.

Definition 1.28 The function

$$\mathrm{Fut}_X(\lambda) := -\lim_{k \to \infty} \frac{w_k(\lambda)}{k \cdot l_k \cdot \ell}$$

is called the *Futaki character* of the Fano variety X.

The following lemma summarizes properties of the Futaki character that we need

Lemma 1.29 *The following assertions hold:*

(i) *For two commuting $\mathbb{G}_m$-actions λ and λ' on the threefold X, we have*

$$\mathrm{Fut}_X(\lambda\lambda') = \mathrm{Fut}_X(\lambda) + \mathrm{Fut}_X(\lambda')$$

where $\lambda\lambda'$ stands for the composition $\lambda \circ \lambda'$.

(ii) *Let $(\mathcal{X}, \mathcal{L})$ be a special test configuration for (X, L), and let λ be the corresponding action of the group $\mathbb{G}_m$ on the variety $\mathcal{X}$. Then*

$$\mathrm{DF}(\mathcal{X}, \mathcal{L}) = \mathrm{Fut}_{X_0}(\lambda) \tag{1.4}$$

for the induced $\mathbb{G}_m$-action λ on the central fiber X_0. Moreover, we have

$$\mathrm{Fut}_X(\lambda') = \mathrm{Fut}_{X_0}(\lambda')$$

for a $\mathbb{G}_m$-action λ' on $(\mathcal{X}, \mathcal{L})$ that acts along the fibers of $p \colon \mathcal{X} \to \mathbb{P}^1$.

Proof The first assertion is obvious. The equality $\mathrm{DF}(\mathcal{X}, \mathcal{L}) = \mathrm{Fut}_{X_0}(\lambda)$ is the original definition of the Donaldson–Futaki invariant $\mathrm{DF}(\mathcal{X}, \mathcal{L})$ that is given in Tian's work [212]. The equality (1.4) is proved in [220], see also [142]. The final equality follows from the flatness of $\mathcal{L}$ over $\mathbb{P}^1$, which implies the flatness of its homogeneous components.   □

Now, we are ready to present a generalization of Definition 1.20.

Definition 1.30 We say that X is *divisorially polystable* if the following holds:

- $\beta(F) > 0$ for every vertical $\mathbb{T}$-invariant prime divisor F on the variety X,
- $\beta(F) = 0$ for every horizontal $\mathbb{T}$-invariant prime divisor F on the variety X.

By Lemma 1.29, if X is K-polystable, then $\mathrm{Fut}_X = 0$. This is Futaki's theorem [110]. If X is toric, it follows from Theorem 1.21, [196, Proposition 3.2] and [140, Theorem 1.4] that the Fano variety X is K-polystable $\iff$ it is divisorially polystable $\iff$ $\mathrm{Fut}_X = 0$. The aim of this section is to prove the following result:

Theorem 1.31 *Suppose that the complexity of the $\mathbb{T}$-action on X is 1. Then the Fano variety X is K-polystable $\iff$ it is divisorially polystable and $\mathrm{Fut}_X = 0$.*

Let us prove Theorem 1.31. Suppose that the complexity of the $\mathbb{T}$-action on X is 1. For our $\mathbb{T}$-invariant prime divisor E over X, let $v = \operatorname{ord}_E$ be the associated divisorial valuation. Consider the graded algebra

$$ R = \bigoplus_k R_k = \bigoplus_k H^0(X, L^k). $$

Recall from [217] that E induces test configurations via the filtration of R defined by

$$ \mathcal{F}_v^p R_k = \left\{ s \in R_k \,\middle|\, v(s) \geqslant p \right\}, $$

where $v(s) = v(f)$ with $s = f \cdot e$ for $f \in \mathbb{C}(X)$ and e being a local generator of the line bundle L at the generic point of the divisor E. Let

$$ \mathcal{R}_v = \bigoplus_k \bigoplus_p \mathcal{F}_v^p R_k \cdot \frac{1}{t^p}. $$

If the algebra $\mathcal{R}_v$ is finitely generated, then the Rees construction gives rise to a polarized family $\mathcal{X}_v \to \mathbb{A}^1 = \operatorname{Spec}(\mathbb{C}[t])$ with central fiber $(\mathcal{X}_v)_0$. In this case, $\mathcal{X}_v = \operatorname{Proj}_{\mathbb{A}^1}(\mathcal{R}_v)$, where the Proj is taken with respect to the k-grading. Here, we have $\mathcal{R}_v \subset R[t, t^{-1}]$ and

$$ \mathcal{X}_v \setminus (\mathcal{X}_v)_0 \cong X \times \mathbb{C}^*, $$

so that we can compactify the variety $\mathcal{X}_v$ by gluing it with $X \times \mathbb{P}^1 \setminus [1 : 0]$ along $X \times \mathbb{C}^*$. Let $\overline{\mathcal{X}}_v$ be the result of this gluing, and let $p \colon \overline{\mathcal{X}}_v \to \mathbb{P}^1$ be the corresponding projection. Then the $\mathbb{G}_m$-action λ_v on the variety $\overline{\mathcal{X}}_v$ is given by the p-grading.

Since E is $\mathbb{T}$-invariant, the filtration $\mathcal{F}_v^p R$ must respect the corresponding M-grading, so that $\overline{\mathcal{X}}_v$ admits a $\mathbb{T}$-action along the fibers of p that commutes with the $\mathbb{G}_m$-action λ_v. Then $p^{-1}(0)$ is given by the associated graded ring of the filtration: $(\mathcal{X}_v)_0 \cong \operatorname{Proj}(\operatorname{gr}\mathcal{F}_v)$. By construction, the variety $\overline{\mathcal{X}}_v$ is naturally equipped with a p-ample line bundle $\mathcal{L}_v$ such that the pair $(\overline{\mathcal{X}}_v, \mathcal{L}_v)$ is a test configuration for the pair (X, L), see [217] for details. Choosing $v = 0$ leads to the trivial test configuration $\overline{\mathcal{X}}_v = X \times \mathbb{P}^1$.

Remark 1.32   In the presented construction of the test configuration $\overline{\mathcal{X}}_v$, we can replace the valuation v with the valuation av for some $a \in \mathbb{Z}_{>0}$. Then we have $(\mathcal{X}_{av})_0 \cong (\mathcal{X}_v)_0$ because $\operatorname{gr}\mathcal{F}_{av}$ is the ath Veronese subring of $\operatorname{gr}\mathcal{F}_v$. Then $\operatorname{DF}(\overline{\mathcal{X}}_{av}, \mathcal{L}_{av}) = a \cdot \operatorname{DF}(\overline{\mathcal{X}}_v, \mathcal{L}_v)$ because the induced $\mathbb{G}_m$-actions would be λ_v^a.

The following lemma follows from [180, Proposition 3.14] or [216, Section 16].

Lemma 1.33 *The following assertions hold:*

(i) *the divisor E is horizontal $\Longleftrightarrow$ there exists $w \in N$ such that $v(f_u) = \langle w, u \rangle$ for every $f_u \in \mathbb{C}(X)^{(\mathrm{T})}$. Moreover, in this case, $w \in N_E$.*

(ii) *If E is vertical, there are $w \in N$ and $a \in \mathbb{Z}_{>0}$ such that $v(f\chi^u) = \langle w, u \rangle + a\,\mathrm{ord}_P(f)$ for every non-zero $f \in \mathbb{C}(X)^{\mathrm{T}}$, where $P = \pi(E) \in Y \cong \mathbb{P}^1$.*

Proof The restriction $v|_{\mathbb{C}(X)^{\mathrm{T}}}$ defines a discrete valuation on $\mathbb{C}(X)^{\mathrm{T}} = \mathbb{C}(Y)$, which we denote by $\widehat{v}$. Then either $\widehat{v} = 0$ or $\widehat{v}(f) = a\,\mathrm{ord}_P(f)$ for some $P \in Y$ and $a \in \mathbb{Z}_{>0}$. Since π corresponds to the field inclusion, it follows that $P = \pi(E)$ in the latter case. In either case, we get $v(f\chi^u) = \widehat{v}(f) + v(\chi^u)$, where $\chi \colon M \to \mathbb{C}(X)^{(\mathrm{T})}$ is a section fixed earlier. Since $v \colon \mathbb{C}(X)^* \to \mathbb{Z}$ is a homomorphism, the map $M \to \mathbb{Z}$ given by $u \mapsto v(\chi^u)$ must be linear, i.e. it is given by an element $w \in N = M^*$. Thus, it remains to show that the divisor E is horizontal exactly if $\widehat{v}$ is trivial and that $w \in N_E$.

First assume that $\widehat{v} = 0$. Consider the M-graded ideal sheaf $\mathcal{I}$ of E. Then the semi-invariant sections of $\mathcal{I}$ are those for which $v(f_u) > 0$. Hence, we have

$$\mathcal{I} = \bigoplus_{\langle w, u \rangle > 0} (\mathcal{O}_X)_u,$$

so that $(\mathcal{O}_X/\mathcal{I})_u \neq 0$ only if $\langle w, u \rangle = 0$. So, the $\mathbb{Z}$-grading on $\mathcal{O}_X/\mathcal{I}$ induced by w must be trivial. Therefore, we see that the corresponding $\mathbb{G}_m$-action on the divisor E is trivial, so that E is horizontal with $w \in N_E$.

Now we assume that E is horizontal and $v(f) \neq 0$ for some T-invariant function f. We may pick any $u \in M$ with both $\langle N_E, u \rangle \neq 0$ and $\langle w, u \rangle \neq 0$. Then $v(f^a\chi^{bu}) = 0$ for an appropriate choice of integers a and b. Hence, we have $(\mathcal{O}_X/\mathcal{I})_{au} \neq 0$ and therefore the $\mathbb{G}_m$-action on the divisor E induced by $N_E \subset N$ is not trivial. This is a contradiction. Hence, $v(f\chi^u) = \langle w, u \rangle$ and we have seen already that in this case $w \in N_E$.   □

Now we are ready to prove

Proposition 1.34 *Let E_1 and E_2 be two T-invariant prime divisors over the variety X. We let $v_1 = a_1 \cdot \mathrm{ord}_{E_1}$ and $v_2 = a_2 \cdot \mathrm{ord}_{E_2}$, where a_1 and a_2 are some positive integers. Then the following two conditions are equivalent:*

(i) *There exists $w \in N$ such that $v_1(f_u) = v_2(f_u) + \langle w, u \rangle$ for every $f_u \in \mathbb{C}(X)^{(\mathrm{T})}$.*

(ii) *There is an isomorphism $\varphi \colon \mathcal{R}_{v_1} \cong \mathcal{R}_{v_2}$ of $(M \times \mathbb{Z}^2)$-graded algebras with $\varphi(t) = t$ and inducing the identity on $\mathcal{R}_{v_1}/(t-1) = \mathcal{R}_{v_2}/(t-1)$.*

Moreover, if the two equivalent conditions hold, then there exists $\ell \in \mathbb{Z}$ such that φ sends homogeneous elements of weight (u, k, p) to elements of weight $(u, k, p + \langle w, u \rangle + k\ell)$.

Proof Assume that (1) holds. Consider the homomorphism $\mathcal{R}_{\nu_2} \to \mathcal{R}_{\nu_1}$ given by

$$s_{k,u} t^p \mapsto s_{k,u} t^{p+\langle w,u\rangle + k\ell}, \qquad (1.5)$$

where $s_{u,k}$ is a section in $H^0(X, L^k)$ of weight $u \in M$ and $\ell = \langle w, u_1 - u_2\rangle$ with u_i being the weight of a local generator of L at the center $C_X(E_i)$. Then $\nu_1(s_{u,k}) = \nu_2(s_{u,k}) + \langle w, u\rangle + k \cdot \ell$, so that $\mathcal{R}_{\nu_2} \to \mathcal{R}_{\nu_1}$ is the required isomorphism.

For the other direction, assume that we have an isomorphism $\varphi \colon \mathcal{R}_{\nu_1} \cong \mathcal{R}_{\nu_2}$ as in (2). The condition that φ induces the identity on $\mathcal{R}_{\nu_1}/(t-1) = \mathcal{R}_{\nu_2}/(t-1)$ implies that

$$\phi(s_{u,k} t^p) = s_{u,k} t^{p+m}.$$

Since ϕ is a graded isomorphism, we have $m = F(u,k,p)$ for some linear form $F(u,k,p)$. But the equality $\varphi(t) = t$ implies that $F(0,0,p) = 0$, so that $F(u,k,p) = \langle w,u\rangle + \ell \cdot k$ for some $w \in N$ and $\ell \in \mathbb{Z}$. Then φ is given by (1.5). Since φ is an isomorphism, we get

$$\mathcal{F}_{\nu_1}^{p+\langle w,u\rangle + k\ell} R_k = \mathcal{F}_{\nu_2}^p R_k$$

for any integers p and k. Then $\nu_1(s_{u,k}) = \nu_2(s_{u,k}) + \langle w,u\rangle + k\ell$. Now, for

$$f_u = \frac{s_{u+u',k}}{s_{u',k}} \in \mathbb{C}(X)^{(\mathbb{T})},$$

we have $\nu_1(f_u) = \nu_2(f_u) + \langle w,u\rangle$ as claimed. $\qquad\square$

Corollary 1.35 *In the notations and assumption of Proposition 1.34, suppose that there is $w \in N$ such that $\nu_1(f_u) = \nu_2(f_u) + \langle w,u\rangle$ for every $f_u \in \mathbb{C}(X)^{(\mathbb{T})}$. Then*

$$\mathrm{DF}(\overline{\mathcal{X}}_{\nu_1}) = \mathrm{DF}(\overline{\mathcal{X}}_{\nu_2}) + \mathrm{Fut}_X(\lambda_w).$$

Proof By Proposition 1.34, we have $X_0 := (\overline{\mathcal{X}}_{\nu_1})_0 \cong (\overline{\mathcal{X}}_{\nu_2})_0$ and $\lambda_{\nu_1} = \lambda_{\nu_2}\lambda_w$. Then

$$\mathrm{DF}(\overline{\mathcal{X}}_{\nu_1}) = \mathrm{Fut}_{X_0}(\lambda_{\nu_1}) = \mathrm{Fut}_{X_0}(\lambda_{\nu_2}\lambda_w)$$

by Lemma 1.29. Then, by Lemma 1.29, we obtain

$$\mathrm{DF}(\overline{\mathcal{X}}_{\nu_1}) = \mathrm{Fut}_{X_0}(\lambda_{\nu_2}) + \mathrm{Fut}_{X_0}(\lambda_w).$$

Now, applying Lemma 1.29 again, we conclude that $\mathrm{Fut}_{X_0}(\lambda_{\nu_2}) = \mathrm{DF}(\overline{\mathcal{X}}_{\nu_2})$, which implies that $\mathrm{Fut}_{X_0}(\lambda_w) = \mathrm{Fut}_X(\lambda_w)$ by Lemma 1.29. This gives us the desired result. $\qquad\square$

Corollary 1.36 *The test configuration $\overline{\mathcal{X}}_\nu$ is of product-type $\Longleftrightarrow E$ is horizontal. In this case, the corresponding $\mathbb{G}_m$-action on $(\overline{\mathcal{X}}_\nu)_0 \cong X$ is given by λ_w with $w \in N_E$.*

Proof By Lemma 1.33, the divisor E is horizontal $\iff v(f_u) = \langle w, u \rangle$ for $w \in N_E$. We have $\mathcal{X}_v \cong X \times \mathbb{A}^1$ if $v = 0$. Now, the claim follows from Proposition 1.34.   □

Corollary 1.37 *Suppose E is horizontal. Then $\beta(E) = \mathrm{Fut}_X(\lambda_w)$ for some $w \in N_E$.*

Proof Using Corollary 1.36, we see that the test configuration $\overline{\mathcal{X}}_v$ is of product-type. By Lemma 1.29, we have $\mathrm{DF}(\overline{\mathcal{X}}_v) = \mathrm{Fut}_X(\lambda_w)$ for some $w \in N_E$. On the other hand, it follows from [95, Theorem 5.1] that $\mathrm{DF}(\overline{\mathcal{X}}_v) = \beta(E)$.   □

Let G be a subgroup in $\mathrm{Aut}^{\mathbb{T}}(X)$ such that $\mathbb{T} \subset G$ and $G \cong \mathbb{T} \rtimes \mathbb{W}$ for some group $\mathbb{W}$. Note that $\mathrm{Aut}^{\mathbb{T}}(X)/\mathbb{T}$ is a finite group by [203, Lemma 2.9], so that $\mathbb{W}$ is finite as well. Then the quotient map $\pi \colon X \dashrightarrow Y$ is G-equivariant, so that $\mathbb{W}$ naturally acts on $Y \cong \mathbb{P}^1$. The following result is a reformulation of the main result of [117].

Proposition 1.38 *Suppose that the following two conditions hold:*

(i) $\mathrm{Fut}_X = 0$,
(ii) *for every point $P \in Y$ that is fixed by $\mathbb{W}$, there exists at least one irreducible component D of the fiber $\pi^{-1}(P)$ such that $\beta(D) > 0$.*

Then X is K-polystable.

Proof By Theorem 1.9, it is enough to consider G-equivariant special test configurations to check K-polystability. Moreover, given a special G-equivariant test configuration, it follows from [95, Theorem 5.1] that there is a G-invariant prime divisor F over X such that the test configuration is obtained as $\overline{\mathcal{X}}_{c \cdot \mathrm{ord}_F}$ for some $c \in \mathbb{Z}_{>0}$. If F is horizontal then

$$\mathrm{ord}_F\left(f \cdot \chi^u\right) = \langle w_F, u \rangle$$

and $\overline{\mathcal{X}}_{c \cdot \mathrm{ord}_F}$ is of product-type, so that its Donaldson–Futaki invariant is 0 by Lemma 1.29.

If F is vertical, then it follows from Lemma 1.33 that

$$\mathrm{ord}_F\left(f \cdot \chi^u\right) = \langle w, u \rangle + a\,\mathrm{ord}_P(f)$$

with $a > 0$ and $\pi(F) = P \in \mathbb{P}^1$. Note that P is $\mathbb{W}$-invariant since F is G-invariant. By assumption, there is an irreducible component D of the fiber $\pi^{-1}(P)$ with $\beta(D) > 0$. Then D is a $\mathbb{T}$-invariant prime divisor on X, so that Lemma 1.33 gives

$$\mathrm{ord}_D(f \cdot \chi^u) = \langle w', u \rangle + b \cdot \mathrm{ord}_P(f)$$

for some $w \in N$ and $b \in \mathbb{Z}_{\geqslant 0}$. Hence, we have

$$\text{bord}_F(f_u) = a\text{ord}_D(f_u) + \langle bw - aw', u \rangle \tag{1.6}$$

for a semi-invariant funcion f_u of weight $u \in M$. It follows by Corollary 1.35 that

$$
\begin{aligned}
b\text{DF}\big(\overline{\mathcal{X}}_{c \cdot \text{ord}_F}\big) &= \text{DF}\big(\overline{\mathcal{X}}_{a\text{ord}_D}\big) + \text{Fut}_X\big(\lambda_{bw-aw'}\big) && \text{by (1.6) and Corollary 1.35} \\
&= \text{DF}\big(\overline{\mathcal{X}}_{a\text{ord}_D}\big) + 0 && \\
&= a\text{DF}\big(\overline{\mathcal{X}}_{\text{ord}_D}\big) && \text{by Remark 1.32} \\
&= a \cdot \beta(D) > 0 && \text{by [95, Theorem 5.1].}
\end{aligned}
$$

This also shows that $\beta(D') > 0$ for every other component D' of the fiber $\pi^{-1}(P)$. $\qquad\square$

Corollary 1.39 *If* $\text{Fut}_X = 0$ *and* Y *has no* $\mathbb{W}$-*fixed points, then* X *is K-polystable.*

Corollary 1.40 *Suppose that* $\text{Fut}_X = 0$, *all G-invariant fibers of* π *are irreducible, and* $\beta(D) > 0$ *for one fiber* D *of the map* π. *Then* X *is K-polystable.*

Proof This follows from Proposition 1.38 since fibers of π are rationally equivalent. $\qquad\square$

Corollary 1.41 *Suppose that* $\text{Fut}_X = 0$, *not all G-invariant fibers of* π *are irreducible, and* $\beta(D) > 0$ *for at least one irreducible component* D *of every reducible G-invariant fiber of the map* π. *Then* X *is K-polystable.*

Proof Using Proposition 1.38, we see that to prove the required assertion it is enough to check that $\beta(F) > 0$ for an irreducible fiber F of the map π. Observe that $F \sim D + D'$, where D is an irreducible component of some reducible fiber of π such that $\beta(D) > 0$, and D' is an effective divisor on X. Then $\beta(F) \geqslant \beta(D) > 0$ as required. $\qquad\square$

If $\text{Fut}_X = 0$, then to check the K-polystability of the variety X, it is enough to check that $\beta(D) > 0$ for finitely many $\mathbb{T}$-invariant divisors D in X.

Proof of Theorem 1.31 First, we suppose that $\text{Fut}_X = 0$ and X is divisorially polystable. Then X is K-polystable by Proposition 1.38.

Now, we suppose that X is K-polystable. Then we must have $\mathrm{DF}(\mathcal{X}, \mathcal{L}) = 0$ for every test configuration $(\mathcal{X}, \mathcal{L})$ of product-type. By Lemma 1.29, this is equivalent to $\mathrm{Fut}_X = 0$. Moreover, we have

$$\beta(D) = \mathrm{DF}(\overline{\mathcal{X}}_{\mathrm{ord}_D}) > 0$$

for every $\mathbb{T}$-invariant prime divisor D on X such that $\overline{\mathcal{X}}_{\mathrm{ord}_D}$ is not of product-type. By Lemma 1.36, the latter condition is equivalent to D being vertical. □

1.4 Tian's criterion

Let X be a Fano variety with at most Kawamata log terminal singularities of dimension $n \geqslant 2$, and let G be a reductive subgroup in $\mathrm{Aut}(X)$. Then

$$\alpha_G(X) = \sup \left\{ \epsilon \in \mathbb{Q} \;\middle|\; \begin{array}{l} \text{the log pair } \left(X, \dfrac{\epsilon}{m}\mathcal{D}\right) \text{ is log canonical for any} \\ m \in \mathbb{Z}_{>0} \text{ and every } G\text{-invariant linear system} \\ \mathcal{D} \subset \left|-mK_X\right| \end{array} \right\}.$$

This number, also known as the global log canonical threshold (see [48, Definition 3.1]), has been defined by Tian in [210] in a very different way (see also [213, Appendix 2]). However, both definitions coincide by [46, Theorem A.3].

Lemma 1.42 *Suppose that $G = \mathbb{G}_m^r \rtimes B$ for some finite group B. Then*

$$\alpha_G(X) = \sup \left\{ \epsilon \in \mathbb{Q} \;\middle|\; \begin{array}{l} \text{the log pair } (X, \epsilon D) \text{ is log canonical for every} \\ G\text{-invariant effective } \mathbb{Q}\text{-divisor } D \sim_{\mathbb{Q}} -K_X \end{array} \right\}.$$

Proof Let D be an effective G-invariant $\mathbb{Q}$-divisor on X that satisfies $D \sim_{\mathbb{Q}} -K_X$. Take a positive integer r such that rD is a Cartier $\mathbb{Z}$-divisor. Then rD is a G-invariant zero-dimensional linear subsystem in $|-rK_X|$, and $\mathrm{lct}(X; D) = \frac{\mathrm{lct}(X; rD)}{r}$. This gives

$$\alpha_G(X) \leqslant \sup \left\{ \epsilon \in \mathbb{Q} \;\middle|\; \begin{array}{l} \text{the log pair } (X, \epsilon D) \text{ is log canonical for every} \\ G\text{-invariant effective } \mathbb{Q}\text{-divisor } D \sim_{\mathbb{Q}} -K_X \end{array} \right\}.$$

Thus, to complete the proof, we have to prove the opposite inequality.

Let m be large positive integer, let $\mathcal{D}$ be a G-invariant linear subsystem in $|-mK_X|$, and let $c = \mathrm{lct}(X; \frac{1}{m}\mathcal{D})$. Then $c \geqslant \alpha_G(X)$, and we can choose m and $\mathcal{D}$ in $|-mK_X|$ such that c is arbitrary close to $\alpha_G(X)$. On the other hand, the linear system $\mathcal{D}$ contains a $\mathbb{G}_m^r$-invariant divisor. Denote it by D. Then for every $g \in B$, we have $g^*(D) \in \mathcal{D}$ and the log pair $(X, \frac{c}{m}g^*(D))$ is not Kawamata log terminal (cf. [130, Theorem 4.8]). Let

$$\mathcal{D} = \frac{1}{m|B|} \sum_{g \in B} g^*(D).$$

Then $\mathcal{D}$ is an effective G-invariant divisor such that $\mathcal{D} \sim_{\mathbb{Q}} -K_X$. Moreover, it follows from the proof of [130, Theorem 4.8] that $(X, c\mathcal{D})$ is not Kawamata log terminal, so that

$$\alpha_G(X) \geqslant \sup \left\{ \epsilon \in \mathbb{Q} \; \middle| \; \begin{array}{l} \text{the log pair } (X, \epsilon D) \text{ is log canonical for every} \\ G\text{-invariant effective } \mathbb{Q}\text{-divisor } D \sim_{\mathbb{Q}} -K_X \end{array} \right\}$$

because c can be arbitrary close to $\alpha_G(X)$. $\qquad\square$

If G is a trivial group, we let $\alpha(X) = \alpha_G(X)$. By Lemma 1.42, we have

$$\alpha(X) = \inf \left\{ \mathrm{lct}(X, D) \; \middle| \; D \text{ is effective } \mathbb{Q}\text{-divisor such that } D \sim_{\mathbb{Q}} -K_X \right\}.$$

All possible values of the α-invariants of smooth del Pezzo surfaces are found in [30, 154], and we present them in Section 1.5. Similarly, α-invariants of del Pezzo surfaces with at most Du Val singularities have been computed in a series of papers [176, 31, 177, 178, 35]. For smooth Fano threefolds, we only know partial results about their α-invariants [46].

Observe that the invariant $\alpha(X)$ has a global nature. It measures the singularities of effective $\mathbb{Q}$-divisors on X that are $\mathbb{Q}$-linearly equivalent to the anticanonical divisor $-K_X$. We can also localize $\alpha(X)$ as follows. Let Z be a proper irreducible subvariety in X. Let

$$\alpha_Z(X) = \sup \left\{ \lambda \in \mathbb{Q} \; \middle| \; \begin{array}{l} \text{the log pair } (X, \lambda D) \text{ is log canonical at a general} \\ \text{point of } Z \text{ for every effective } \mathbb{Q}\text{-divisor } D \text{ on } X \\ \text{such that } D \sim_{\mathbb{Q}} -K_X \end{array} \right\}.$$

Clearly, we have

$$\alpha(X) = \inf_{P \in X} \alpha_P(X),$$

where the infimum is taken over all (closed) points in X. If the subvariety Z is G-invariant, we can also define the number $\alpha_{G,Z}(X)$ as follows:

$$\alpha_{G,Z}(X) = \sup \left\{ \lambda \in \mathbb{Q} \; \middle| \; \begin{array}{l} \text{the pair } (X, \lambda D) \text{ is log canonical at a general} \\ \text{point of } Z \text{ for any effective } G\text{-invariant } \mathbb{Q}\text{-divisor} \\ D \text{ on } X \text{ such that } D \sim_{\mathbb{Q}} -K_X \end{array} \right\}.$$

Then $\alpha_G(X) \leqslant \alpha_{G,Z}(X)$.

Remark 1.43 ([100, Lemma 2.5])   Let $f \colon \widetilde{X} \to X$ be an arbitrary G-equivariant birational morphism, let F be a G-invariant prime divisor in $\widetilde{X}$ such that $Z \subseteq f(F)$, and let

$$\tau(F) = \sup\left\{x \in \mathbb{Q}_{>0}\big|\ f^*(-K_X) - xF \text{ is big}\right\}.$$

Then $\frac{A_X(F)}{\tau(F)} \geqslant \alpha_{G,Z}(X)$. Indeed, fix any positive rational number $x < \tau(F)$, let $\mathcal{D}$ be the image on the variety X of the (non-empty) complete linear system $|M(f^*(-K_X) - xF)|$ for sufficiently large and divisible integer M. Then $\mathcal{D}$ is G-invariant. If F is f-exceptional, then the log pair $(X, \frac{A_X(F)}{xM}\mathcal{D})$ is not Kawamata log terminal along $f(F)$. Similarly, if the divisor F is not f-exceptional, then $(X, \frac{1}{xM}\mathcal{D} + f(F))$ is not Kawamata log terminal along $f(F)$, and $\frac{1}{xM}\mathcal{D} + f(F) \sim_{\mathbb{Q}} \frac{1}{x}(-K_X)$. Thus, in both cases $\alpha_{G,Z}(X) \leqslant \frac{A_X(F)}{x}$, which implies the required inequality since we can choose x to be as close to $\tau(F)$ as we wish.

Corollary 1.44   *In the notations and assumptions of Remark 1.43, we have*

$$\frac{A_X(F)}{S_X(F)} \geqslant \frac{n+1}{n}\alpha_{G,Z}(X).$$

Proof   By [18, Proposition 3.11], one has $\frac{1}{n+1}\tau(F) \leqslant S_X(F) \leqslant \frac{n}{n+1}\tau(F)$, so that the result follows from Remark 1.43.   □

In some cases, this inequality can be improved a little bit:

Lemma 1.45   *In the notations and assumptions of Remark 1.43, suppose in addition that X is smooth and $\dim(Z) \geqslant 1$. Then*

$$\frac{A_X(F)}{S_X(F)} > \frac{n+1}{n}\alpha_{G,Z}(X).$$

Proof   By [94, Proposition 3.2], we have $S_X(F) < \frac{n}{n+1}\tau(F)$, so that the required result follows from Remark 1.43.   □

We can also define α-invariants for

- log Fano varieties (see Section 1.5);
- weak Fano varieties (see Lemma 1.47, Example 4.10 and Section 4.1);
- Fano varieties defined over arbitrary fields (sec Theorem 1.52 and Section A.5).

To save space and to keep the exposition simple, we leave these definitions to the reader.

Lemma 1.46 *Suppose that G is finite. Let $Y = X/G$, let $\pi\colon X \to Y$ be the quotient morphism, and let Δ be the effective $\mathbb{Q}$-divisor on Y such that $\pi^*(K_Y + \Delta) = K_X$. Then the log pair (Y, Δ) has Kawamata log terminal singularities, $-(K_Y + \Delta)$ is ample, and $\alpha(Y, \Delta) = \alpha_G(X)$.*

Proof The required assertion is [48, Remark 3.2]. Let D_Y be an effective $\mathbb{Q}$-divisor on the variety Y such that $D_Y \sim_{\mathbb{Q}} -(K_Y + \Delta)$. Then $\pi^*(D_Y) \sim_{\mathbb{Q}} \pi^*(K_Y + \Delta) \sim_{\mathbb{Q}} -K_X$, the divisor $\pi^*(D_Y)$ is G-invariant, and $\mathrm{lct}(X; \pi^*(D_Y)) = \mathrm{lct}(Y, \Delta; D_Y)$ by [130, Proposition 3.16]. This immediately gives $\alpha_G(X) \geqslant \alpha(Y, \Delta)$. Vice versa, for every effective G-invariant divisor D on X such that $D \sim_{\mathbb{Q}} -K_X$, $D = \pi^*(D_Y)$ for some effective $\mathbb{Q}$-divisor D_Y on the variety Y such that D_Y satisfies $D_Y \sim_{\mathbb{Q}} -(K_Y + \Delta)$. As above, this gives $\alpha_G(X) \leqslant \alpha(Y, \Delta)$. $\quad\square$

Lemma 1.47 *Let $\pi\colon Y \to X$ be a G-equivariant birational morphism such that Y has Kawamata log terminal singularities, and $-K_Y \sim_{\mathbb{Q}} \pi^*(-K_X)$. Then $\alpha_G(Y) = \alpha_G(X)$.*

Proof The proof is similar to the proof of Lemma 1.46, so it is left to the reader. $\quad\square$

The α-invariants are important because of the following result:

Theorem 1.48 ([67, 148, 210, 226]) *The Fano variety X is K-semistable if*

$$\alpha_G(X) \geqslant \frac{n}{n+1}.$$

Moreover, if $\alpha_G(X) > \frac{n}{n+1}$, then X is K-polystable.

Remark 1.49 By [226, Corollary 4.15], Theorem 1.48 can be generalized for varieties defined over arbitrary fields of characteristic 0 as follows. If X is a Fano variety defined over an arbitrary field $\mathbb{F}$ of characteristic 0, and G is a reductive subgroup in $\mathrm{Aut}_{\mathbb{F}}(X)$ such that $\alpha_G(X) > \frac{n}{n+1}$, then X is K-polystable over the algebraic closure of the field $\mathbb{F}$.

Recall that $n = \dim(X) \geqslant 2$ by assumption. If G is trivial, we have the following result:

Theorem 1.50 ([94, 171]) *If X is smooth and $\alpha(X) \geqslant \frac{n}{n+1}$, then X is K-stable.*

Recall that we assume that the group G is reductive.

Theorem 1.51 *If X is smooth and $\alpha_G(X) \geqslant \frac{n}{n+1}$, then X is K-polystable.*

Proof Suppose that the Fano variety X is smooth and $\alpha_G(X) \geqslant \frac{n}{n+1}$. We must show that X is K-polystable. By Theorem 1.48, we may assume $\alpha_G(X) = \frac{n}{n+1}$.

Let $f \colon \widetilde{X} \to X$ be a G-equivariant birational morphism, and let E be a G-invariant prime divisor in $\widetilde{X}$. By Theorem 1.22, it is enough to show that $\beta(E) > 0$ provided that E is dreamy (see Section 1.1).

Suppose that E is dreamy. By Remark 1.43, we have

$$A_X(E) \geqslant \alpha_G(X)\tau(E) = \frac{n}{n+1}\tau(E).$$

If $A_X(E) > S_X(E)$, then we are done. Thus, we may assume that $A_X(E) \leqslant S_X(E)$. Since $A_X(E) \geqslant \frac{n}{n+1}\tau(E)$, we get $X \cong \mathbb{P}^n$ by [95, Theorem 1]. Then X is K-polystable. $\qquad\square$

To estimate $\alpha_G(X)$ in the case when X is a smooth (or mildly singular) Fano threefold, we will use the following result, which is a refinement of [165, Theorem 0.1] for threefolds.

Theorem 1.52 *Let X be a Fano threefold that has canonical Gorenstein singularities, let G be a reductive subgroup of $\mathrm{Aut}(X)$, and let μ be a positive number such that $\mu \leqslant 1$. Suppose that $\alpha_G(X) < \mu$. Then one of the following assertions holds:*

(1) *There exists a G-invariant irreducible normal surface S on X such that*

$$-K_X \sim_{\mathbb{Q}} \lambda S + \Delta,$$

   where Δ is an effective $\mathbb{Q}$-divisor, and $\lambda \in \mathbb{Q}$ is such that $\lambda > \frac{1}{\mu}$.
(2) *There exists a G-invariant point $P \in X$. Moreover, the following holds:*

   (2.1) *if there is a del Pezzo fibration $\pi \colon X \to \mathbb{P}^1$, and F is the scheme theoretic fiber that contains the point P, then*

   $$\alpha(F) \leqslant \alpha_\Gamma(F) < \mu,$$

   where Γ is the image in $\mathrm{Aut}(F)$ of the stabilizer of the fiber F in the group G, and we assume that $\alpha_\Gamma(F) = 0$ in the case when F is not a del Pezzo surface with Du Val singularities.
(3) *There exists a smooth rational G-invariant curve $C \subset X$ such that*

$$-K_X \cdot C \leqslant \frac{(-K_X)^3}{2} + 2.$$

   Moreover, in this case, the following additional assertions hold:
   (3.1) *if $\mu < 1$, then $-K_X \cdot C < \frac{2}{1-\mu}$, e.g. if $\mu = \frac{3}{4}$, then $-K_X \cdot C < 8$;*
   (3.2) *if there is a del Pezzo fibration $\pi \colon X \to \mathbb{P}^1$, then $F \cdot C \in \{0, 1\}$ and*

   $$\alpha(F) \leqslant \alpha_\Gamma(F) < \mu,$$

   where F is any fiber of the fibration π that intersects (or contains) *the curve C, and Γ is the image in* Aut(*F*) *of the stabilizer of F in the group G;*

(3.3) *if in* (3.2) *we have F · C* = 1, *then*

$$\alpha\big(F_\pi\big) \leqslant \alpha_\Gamma\big(F_\pi\big) < \mu,$$

where F_π is the (scheme) generic fiber of the fibration π, which is a del Pezzo surface with Du Val singularities defined over the function field of the line $\mathbb{P}^1$, and Γ is the image in Aut(F_π) *of the stabilizer of the fiber F_π in the group G.*

Proof By definition, there exists a *G*-invariant linear system $\mathcal{D} \subset |-nK_X|$ for some $n \geqslant 1$ such that the log pair $(X, \frac{\epsilon}{n}\mathcal{D})$ is strictly log canonical for some positive rational $\epsilon < \mu$. We write $\frac{\epsilon}{n}\mathcal{D} = B_X + \mathcal{M}_X$, where B_X is a *G*-invariant effective $\mathbb{Q}$-divisor on *X*, and $\mathcal{M}_X$ is a *G*-invariant mobile boundary (see Section A.3). Let *Z* be the *G*-orbit of its minimal log canonical center. Then, using Lemma A.28, we may assume that the only log canonical centers of the log pair $(X, \frac{\epsilon}{n}\mathcal{D})$ are the irreducible components of *Z*.

The irreducible components of *Z* cannot intersect by Lemma A.19. On the other hand, it follows from Corollary A.4 that the locus *Z* is connected, so that *Z* is an irreducible *G*-invariant subvariety of the threefold *X*.

If *Z* is a surface, then we get (1) since *Z* must be normal by Theorem A.20. Thus, we assume that *Z* is not a surface.

Suppose that *Z* is a point. Then we get (2). To prove (2.1), we suppose that there is a del Pezzo fibration $\pi \colon X \to \mathbb{P}^1$. Let *F* be its scheme fiber over $\pi(Z)$, and let Γ be the image in Aut(*F*) of the stabilizer of the fiber *F* in the group *G*. Suppose that *F* is an irreducible normal surface that has at most Du Val singularities. Write $B_X = aF + \Delta$ where *a* is a non-negative rational number, and Δ is an effective $\mathbb{Q}$-divisor, whose support does not contain the surface *F*. Then $a < 1$ because otherwise *F* would be a log canonical center of the log pair $(X, B_X + \mathcal{M}_X)$, which is not the case since *Z* is the unique log canonical center of this log pair. Then the pair $(X, F + \Delta + \mathcal{M}_X)$ is not log canonical at *Z*. Now using Theorem A.15, we see that $(F, \Delta|_F + \mathcal{M}_X|_F)$ is not log canonical at *Z*. On the other hand, $\Delta|_F + \mathcal{M}_X|_F \sim_{\mathbb{Q}} \epsilon(-K_F)$ and $\Delta|_F + \mathcal{M}_X|_F$ are Γ-invariant, so that we have $\alpha_\Gamma(F) < \epsilon$. Then $\alpha(F) \leqslant \alpha_\Gamma(F) < \mu$, which proves (2.1).

Thus, we may assume that *Z* is a curve, so that we let *C* = *Z*. Then the curve *C* is smooth and rational by Theorem A.20.

Let $\mathcal{I}_C$ be the ideal sheaf of the curve *C*. Then $h^1(\mathcal{I}_C \otimes \mathcal{O}_X(-K_X)) = 0$ by Theorem A.3. Thus, we have the following exact sequence of *G*-representations:

$$1 \longrightarrow H^0\big(\mathcal{I}_C \otimes \mathcal{O}_X(-K_X)\big) \longrightarrow H^0\big(\mathcal{O}_X(-K_X)\big) \longrightarrow H^0\big(\mathcal{O}_C \otimes \mathcal{O}_X(-K_X)\big) \longrightarrow 1,$$

which, in particular, gives

$$\frac{(-K_X)^3}{2} + 3 = h^0\big(O_X(-K_X)\big) \geqslant h^0\big(O_C \otimes O_X(-K_X)\big) = -K_X \cdot C + 1,$$

which gives $-K_X \cdot C \leqslant \frac{(-K_X)^3}{2} + 2$ as required in (3).

Observe that (3.1) follows from Corollary A.21.

To prove (3.2), we suppose (again) that there exists a del Pezzo fibration $\pi\colon X \to \mathbb{P}^1$. Let F be a fiber of this fibration such that $F \cap C \neq \varnothing$. Then either $C \subset F$ and $F \cdot C = 0$, or the intersection $F \cap C$ consists of finitely many points. Arguing as in the proof of (2.1), we see that $\alpha(F) \leqslant \alpha_\Gamma(F) \leqslant \epsilon < \mu$, where Γ is the image in $\mathrm{Aut}(F)$ of the stabilizer of the fiber F in the group G.

Let us show that $F \cdot C \in \{0, 1\}$. Suppose that $F \cdot C \neq 0$. Let us show that $F \cdot C = 1$. Let S be a general fiber of the fibration π. Then S is a del Pezzo surface with Du Val singularities, and $S \cap C$ consists of $F \cdot C \geqslant 1$ distinct points. On the other hand, the log pair $(S, B_X|_S + M_X|_S)$ is not Kawamata log terminal at any point of $S \cap C$, and is log canonical away from this set. Since $B_X|_S + M_X|_S \sim_{\mathbb{Q}} -\epsilon K_S$ and $\epsilon < 1$, it follows from Corollary A.4 that $S \cap C$ is connected, so that $F \cdot C = 1$. This proves (3.2).

Finally, to prove (3.3), let F_π be the generic fiber of the fibration π, let $\mathbb{F}$ be the function field of the line $\mathbb{P}^1$, and let Γ be a subgroup in G such that π is Γ-equivariant and Γ acts trivially on its base. Then Γ is the stabilizer of the fiber F_π in the group G, and we can identify Γ with a subgroup of $\mathrm{Aut}(F_\pi)$. Then F_π is a del Pezzo surface with Du Val singularities defined over $\mathbb{F}$, the curve C defines a Γ-invariant $\mathbb{F}$-point in F_π, the log pair $(F_\pi, B_X|_{F_\pi} + M_X|_{F_\pi})$ is not Kawamata log terminal at this point, and $B_X|_{F_\pi} + M_X|_{F_\pi}$ is Γ-invariant. This gives $\alpha_\Gamma(F_\pi) \leqslant \epsilon$, which proves (3.3). $\qquad\qquad\square$

Let us present several corollaries of Theorem 1.52, which are easier to apply.

Corollary 1.53 ([165, Corollary 4.1]) *Let X be a Fano threefold that has canonical Gorenstein singularities, and let G be a finite subgroup in $\mathrm{Aut}(X)$ such that X does not have G-orbits of length 1 or 2, and G does not admit any epimorphisms to any of the following groups $\mathfrak{A}_4$, $\mathfrak{S}_4$ or $\mathfrak{S}_5$. Suppose that X does not contain any G-invariant irreducible surface S such that $-K_X \sim_{\mathbb{Q}} \lambda S + \Delta$ for some positive rational number $\lambda > 1$ and an effective $\mathbb{Q}$-divisor Δ. Then $\alpha_G(X) \geqslant 1$.*

Proof Apply Theorem 1.52 and use the classification of finite subgroups in $\mathrm{PGL}_2(\mathbb{C})$. $\qquad\qquad\square$

Corollary 1.54 *Let X be a smooth Fano threefold, and let G be a finite simple non-abelian subgroup in $\mathrm{Aut}(X)$ such that $G \not\cong \mathfrak{A}_5$, $G \not\cong \mathrm{PSL}_2(\mathbb{F}_7)$.*

Suppose that X does not contain any G-invariant irreducible surface S such that $-K_X \sim_{\mathbb{Q}} \lambda S + \Delta$ for some positive rational number $\lambda > 1$ and an effective $\mathbb{Q}$-divisor Δ. Then $\alpha_G(X) \geq 1$.

Proof Apply Theorem 1.52. The condition Theorem 1.52(1) is not satisfied by assumption. Since $G \ncong \mathfrak{A}_5$ and $G \ncong \mathrm{PSL}_2(\mathbb{F}_7)$, the group G does not have faithful three-dimensional representations, so that the threefold X does not have G-invariant points by Lemma A.25. Thus, since $G \ncong \mathfrak{A}_5$, we see that X does not contain rational G-invariant curves. □

Corollary 1.55 *Let V be a weak Fano threefold that has canonical Gorenstein singularities, let G be a reductive subgroup of Aut(V), let $\pi\colon V \to \mathbb{P}^1$ be a G-equivariant weak del Pezzo fibration. Suppose that the following three conditions are satisfied:*

(i) *π does not have G-invariant fibers;*
(ii) *V does not contain G-invariant (irreducible) sections of π;*
(iii) *V does not contain G-irreducible surface S such that $-K_V \sim_{\mathbb{Q}} \lambda S + \Delta$ for some rational number $\lambda > 1$ and effective $\mathbb{Q}$-divisor Δ.*

Then $\alpha_G(V) \geq 1$.

Proof If V is a Fano threefold, then the required assertion follows from Theorem 1.52. In general, it follows from the proof of this theorem. Indeed, suppose $\alpha_G(V) < 1$. Then there are rational number $\mu < 1$ and G-invariant linear system $\mathcal{D} \subset |-nK_V|$ for some $n \geq 1$ such that $(V, \frac{\mu}{n}\mathcal{D})$ is strictly log canonical. Let us seek for a contradiction.

Let C be a center of log canonical singularities of the log pair $(V, \frac{\mu}{n}\mathcal{D})$ that has maximal dimension, and let Z be its G-orbit. Then Z is a G-irreducible subvariety of V, so that C is not a surface by (iii). In particular, the locus $\mathrm{Nklt}(V, \frac{\mu}{n}\mathcal{D})$ is at most one-dimensional.

If C is a point, then the locus $\mathrm{Nklt}(V, \frac{\mu}{n}\mathcal{D})$ is zero-dimensional. Since it is connected by Corollary A.4, we conclude that $Z = C$ must be a G-invariant point in this case, which is impossible by (i). Thus, we see that C is a curve.

Let F be a general fiber of π. If $F \cdot Z \neq 0$, then the log pair $(F, \frac{\mu}{n}\mathcal{D}|_F)$ is not Kawamata log terminal at every intersection point in $F \cap Z$, and $\mathrm{Nklt}(F, \frac{\mu}{n}\mathcal{D}|_F)$ is zero-dimensional, so that $F \cdot Z = 1$ and $Z = C$ by Corollary A.4. The latter is impossible by (ii). Hence, we see that $F \cdot Z = 0$.

Thus, the locus $\mathrm{Nklt}(V, \frac{\mu}{n}\mathcal{D})$ is one-dimensional, and each of its irreducible components is contained in a fiber of the G-equivariant fibration π. On the other hand, this locus is connected by Corollary A.4. This shows that $Z = C$ and C

is contained in a fiber of π, which must be G-invariant. The latter is impossible by (i).   □

Corollary 1.56 *Let V be a weak Fano threefold that has canonical Gorenstein singularities, let G be a reductive subgroup of $\mathrm{Aut}(V)$, let $\pi\colon V \to \mathbb{P}^1$ be a G-equivariant weak del Pezzo fibration, and let F_π be the (scheme-theoretic) generic fiber of the fibration π, which is a weak del Pezzo surface with Du Val singularities that is defined over the function field of $\mathbb{P}^1$. Suppose that π does not have G-invariant fibers. Then $\alpha_G(V) \geqslant \alpha(F_\pi)$.*

Proof The assertion follows from the proof of Theorem 1.52. Indeed, suppose that $\alpha_G(V) < \alpha(F_\pi)$. Then there is a G-invariant linear system $\mathcal{D} \subset |-nK_V|$ for some $n \geqslant 1$ such that $(V, \frac{\mu}{n}\mathcal{D})$ is strictly log canonical for some positive rational number $\mu < \alpha(F_\pi)$. Note that $\alpha(F_\pi) \leqslant 1$ because $|-K_{F_\pi}|$ is not empty.

Let $Z = \mathrm{Nklt}(V, \frac{\mu}{n}\mathcal{D})$. If an irreducible component of the locus Z is not contained in any fiber of the fibration π, then the log pair $(F_\pi, \frac{\mu}{n}\mathcal{D}|_{F_\pi})$ is not Kawamata log terminal and $\mathcal{D}|_{F_\pi} \subset |-nK_{F_\pi}|$, so that $\mu \geqslant \alpha(F_\pi)$, which is a contradiction. Therefore, we conclude that each irreducible component of the locus Z is contained in a fiber of the fibration π. But Z is connected by Corollary A.4. Hence, the whole locus Z is contained in one fiber of the fibration π, so that this fiber must be G-invariant, which is impossible since π does not have G-invariant fibers by assumption.   □

Corollary 1.57 *Let V be a weak Fano threefold with isolated canonical Gorenstein singularities, let G be a reductive subgroup of $\mathrm{Aut}(V)$, and let $\pi\colon X \to \mathbb{P}^1$ be a G-equivariant fibration whose general fiber is a smooth quintic del Pezzo surface. Suppose, in addition, that $\mathrm{rk}\,\mathrm{Cl}(V) = 2$, and π does not have G-invariant fibers. Then $\alpha_G(V) \geqslant \frac{4}{5}$.*

Proof Apply Corollary 1.56 and Lemma A.43.   □

Let us conclude this section by presenting one application of Corollary 1.53.

Example 1.58 Let x_0, x_1, x_2, x_3, x_4 be coordinates on $\mathbb{P}^4$, and for each $t \in \mathbb{C}$, let

$$X_t = \left\{ \sum_{i=0}^{4} x_i^4 + \left(\sum_{i=0}^{4} x_i \right)^4 = t \left(\sum_{i=0}^{4} x_i^2 + \left(\sum_{i=0}^{4} x_i \right)^2 \right)^2 \right\} \subset \mathbb{P}^4.$$

As explained in [112, Section 4], the threefold X_t admits a natural action of the symmetric group $\mathfrak{S}_6$. Moreover, it follows from [36, Lemma 3.4] that

$$\text{Aut}(X_t) \cong \begin{cases} \text{PSp}_4(\mathbb{F}_3) \text{ if } t = \dfrac{1}{2}, \\ \mathfrak{S}_6 \text{ if } t \neq \dfrac{1}{2}. \end{cases}$$

Further, the threefold X_t is singular. If $t = \frac{1}{4}$, it has canonical Gorenstein singularities. Moreover, if $t \neq \frac{1}{4}$, then X_t has isolated ordinary double points [112, Theorem 4.1]. Then

- the smallest $\mathfrak{S}_6$-orbit on X_t contains at least six points [52];
- the subgroup of $\mathfrak{S}_6$-invariant divisors in $\text{Cl}(X)$ is generated by $-K_X$, see [36].

By Corollary 1.53, $\alpha_{\mathfrak{S}_6}(X) \geqslant 1$, and X is K-stable by Theorem 1.48 and Corollary 1.5 because the automorphism group of X_t is finite.

1.5 Stability threshold

The paper [102] introduces a new invariant of Fano varieties, called the δ-invariant, that yields a criterion for K-stability. In this section, we will give a slightly simplified definition of the δ-invariant together with its equivariant counterpart, and we will also consider some applications, e.g. Proposition 1.66 and Corollary 1.76.

Let X be a normal projective variety of dimension n, let Δ be an effective $\mathbb{Q}$-divisor on it such that the log pair (X, Δ) has at most Kawamata log terminal singularities, and let L be an ample $\mathbb{Q}$-divisor on X. Let $f: Y \to X$ be a projective birational morphism with normal variety Y, and let E be a (not necessarily f-exceptional) prime divisor in Y. Then E is a divisor over X (see Definition 1.18). Let

$$A_{X,\Delta}(E) = 1 + \text{ord}_E\left(K_Y - f^*(K_X + \Delta)\right),$$

and we let

$$S_L(E) = \frac{1}{L^n} \int_0^\infty \text{vol}(L - xE)dx.$$

If (X, Δ) is a log Fano variety and $L = -(K_X + \Delta)$, we set $S_{X,\Delta}(E) = S_L(E)$ for simplicity. Note that this (infinite) integral is actually finite since $\text{vol}(L-xE) = 0$ for $x > \tau_L(E)$, where $\tau_L(E)$ is the pseudo-effective threshold:

$$\tau_L(E) = \sup\left\{\lambda \in \mathbb{R}_{>0} \,\middle|\, \text{vol}(L - \lambda E) > 0\right\}.$$

Following [18], let us define $\alpha(X, \Delta; L)$ and $\delta(X, \Delta; L)$ as follows:

$$\alpha(X, \Delta; L) = \inf_{E/X} \frac{A_{X,\Delta}(E)}{\tau_L(E)},$$

and

$$\delta(X, \Delta; L) = \inf_{E/X} \frac{A_{X,\Delta}(E)}{S_L(E)},$$

where both infima are taken over all prime divisors over X. Then

$$\alpha(X, \Delta; L) = \inf\left\{\operatorname{lct}(X, \Delta; D) \mid D \text{ is effective } \mathbb{Q}\text{-divisor such that } D \sim_{\mathbb{Q}} L\right\},$$

which can be shown arguing as in Remark 1.43. This equality can be restated as

$$\alpha(X, \Delta; L) = \sup\left\{\lambda \in \mathbb{Q} \;\middle|\; \begin{array}{l} \text{the log pair } (X, \Delta + \lambda D) \text{ is log canonical} \\ \text{for any effective } \mathbb{Q}\text{-divisor } D \sim_{\mathbb{Q}} -K_X \end{array}\right\}.$$

If (X, Δ) is a log Fano variety, we let

$$\alpha(X, \Delta) = \alpha\big(X, \Delta; -(K_X + \Delta)\big),$$
$$\delta(X, \Delta) = \delta\big(X, \Delta; -(K_X + \Delta)\big).$$

In this very important case, the number $\delta(X, \Delta)$ is also known as *the stability threshold* because of the following result (cf. Theorem 1.19).

Theorem 1.59 ([18, 63, 95, 102, 139, 147]) *If (X, Δ) is a log Fano variety, then*

- $\delta(X, \Delta) > 1 \iff (X, \Delta)$ *is K-stable;*
- $\delta(X, \Delta) \geqslant 1 \iff (X, \Delta)$ *is K-semistable.*

Actually, we did not define the K-stability and K-semistability for log Fano varieties. Both these notions can be defined similar to what we did for Fano varieties in Section 1.1. For details, we refer the reader to the excellent survey [217].

Theorem 1.60 ([225]) *Suppose that $X = X_1 \times X_2$, $\Delta = \Delta_1 \boxtimes \Delta_2$, $L = L_1 \boxtimes L_2$, where*

- X_1 *and X_2 are projective varieties,*
- Δ_1 *is an effective $\mathbb{Q}$-divisor on X_1 such that (X_1, Δ_1) is Kawamata log terminal,*
- Δ_2 *is an effective $\mathbb{Q}$-divisor on X_2 such that (X_2, Δ_2) is Kawamata log terminal,*
- L_1 *and L_2 are ample divisors on X_1 and X_2, respectively.*

Then $\delta(X, \Delta; L) = \min\{\delta(X_1, \Delta_1; L_1), \delta(X_2, \Delta_2; L_2)\}$.

We can also define local analogues of the numbers $\alpha(X, \Delta; L)$ and $\delta(X, \Delta; L)$ as follows. For a point $P \in X$, we let

$$\alpha_P(X, \Delta; L) = \inf_{\substack{E/X \\ P \in C_X(E)}} \frac{A_{X,\Delta}(E)}{\tau_L(E)},$$

and

$$\delta_P(X, \Delta; L) = \inf_{\substack{E/X \\ P \in C_X(E)}} \frac{A_{X,\Delta}(E)}{S_L(E)},$$

where the infima are taken over all prime divisors over X whose centers on X contain P. Then

$$\alpha(X, \Delta; L) = \inf_{P \in X} \alpha_P(X, \Delta; L),$$

$$\delta(X, \Delta; L) = \inf_{P \in X} \delta_P(X, \Delta; L).$$

By [18, Proposition 3.11], we have $\frac{1}{n+1}\tau_L(E) \leqslant S_L(E) \leqslant \frac{n}{n+1}\tau_L(E)$ for any prime divisor E over X. Thus, we have $\frac{n+1}{n}\alpha_P(X, \Delta; L) \leqslant \delta_P(X, \Delta; L) \leqslant (n+1)\alpha_P(X, \Delta; L)$, which implies that $\frac{n+1}{n}\alpha(X, \Delta; L) \leqslant \delta(X, \Delta; L) \leqslant (n+1)\alpha(X, \Delta; L)$.

Arguing as in Remark 1.43, one can show that

$$\alpha_P(X, \Delta; L) = \inf\left\{\mathrm{lct}_P\left(X, \Delta; D\right) \,\middle|\, D \text{ is an effective } \mathbb{Q}\text{-divisor such that } D \sim_{\mathbb{Q}} L\right\},$$

which is the original definition of $\alpha_P(X, \Delta; L)$. Note that it can be restated as

$$\alpha_P(X, \Delta; L) = \sup\left\{\lambda \in \mathbb{Q} \,\middle|\, \begin{array}{l} \text{the log pair } (X, \Delta + \lambda D) \text{ is log canonical at } P \\ \text{for every effective } \mathbb{Q}\text{-divisor } D \sim_{\mathbb{Q}} -K_X \end{array}\right\}.$$

It would be useful to have a similar alternative definition of $\delta_P(X, \Delta; L)$, using log canonical thresholds of some divisors on X. To give this alternative definition, we need

Definition 1.61 An effective $\mathbb{Q}$-divisor D such that $D \sim_{\mathbb{Q}} L$ is called *cool* if the inequality $\mathrm{ord}_E(D) \leqslant S_L(E)$ holds for every prime Weil divisor E over X.

The following result can be considered as an alternative definition of the δ-invariant.

Proposition 1.62 *Let P be a point in X. Then*

$$\delta_P(X, \Delta; L) = \inf\left\{\mathrm{lct}_P\left(X, \Delta; D\right) \,\middle|\, \begin{array}{l} D \text{ is a cool effective } \mathbb{Q}\text{-divisor such that} \\ D \sim_{\mathbb{Q}} L \end{array}\right\}.$$

We can restate the equality in this proposition as follows:

$$\delta_P(X, \Delta; L) = \sup \left\{ \lambda \in \mathbb{Q} \,\middle|\, \begin{array}{l} \text{the log pair } (X, \Delta + \lambda D) \text{ is log canonical at } P \\ \text{for any effective cool } \mathbb{Q}\text{-divisor } D \sim_{\mathbb{Q}} -K_X \end{array} \right\}.$$

Corollary 1.63

$$\delta(X, \Delta; L) = \inf \left\{ \text{lct}(X, \Delta; D) \,\middle|\, \begin{array}{l} D \text{ is a cool effective } \mathbb{Q}\text{-divisor such that} \\ D \sim_{\mathbb{Q}} L \end{array} \right\}.$$

To prove this result, we need the following auxiliary.

Lemma 1.64 *Fix any $\epsilon \in \mathbb{Q}_{>0}$. Then there exists an effective $\mathbb{Q}$-divisor $D \sim_{\mathbb{Q}} L$ such that $\text{ord}_E(D) \leqslant \epsilon \tau_L(E)$ for every divisor E over X.*

Proof Let $\pi \colon \widetilde{X} \to X$ be a log resolution of the pair (X, Δ), let N be a sufficiently divisible integer such that $N \geqslant \frac{n}{\epsilon}$, and let $D_1, D_2, \ldots, D_N$ be general divisors in the linear system $|NL|$. By Bertini's theorem, the divisor $\pi^*(D_1) + \cdots + \pi^*(D_N)$ has simple normal crossing singularities since we may assume that $|NL|$ is basepoint free. Let

$$D = \frac{1}{N^2} \sum_{i=1}^{N} D_i.$$

Then $D \sim_{\mathbb{Q}} L$. Let us show that D is the required divisor.

Let E be any prime divisor over X, and let $C_{\widetilde{X}}(E)$ be the center on $\widetilde{X}$ of the discrete valuation defined by E. Then $C_{\widetilde{X}}(E)$ is contained in the support of at most n divisors among $\pi^*(D_1), \ldots, \pi^*(D_N)$. If $C_{\widetilde{X}}(E) \not\subset \text{Supp}(\pi^*(D_i))$, then we get $\text{ord}_E(D_i) = 0$. On the other hand, if $C_{\widetilde{X}}(E) \subset \text{Supp}(\pi^*(D_i))$, then $\text{ord}_E(D_i) \leqslant N\tau_L(E)$. Thus, we have

$$\text{ord}_E(D) \leqslant \frac{n}{N^2} N\tau_L(E) = \frac{n}{N} \tau_L(E) \leqslant \epsilon \tau_L(E)$$

as required. $\square$

We also need the following lemma. For the definition of m-basis type divisors, see [102].

Lemma 1.65 ([18, Corollary 3.6]) *Fix $\epsilon > 0$. Then there exists $m_0(\epsilon) \in \mathbb{Z}_{>0}$ with the following property: for every integer $m \geqslant m_0(\epsilon)$ with mL a Cartier divisor and for every prime divisor E over X, $\text{ord}_E(D_m) \leqslant (1 + \epsilon)S_L(E)$ for any m-basis type divisor $D_m \sim_{\mathbb{Q}} L$.*

Let us now prove Corollary 1.63. The proof of Proposition 1.62 is almost identical, so we omit it.

Proof of Corollary 1.63 Let F be any prime divisor over X. We have to prove that

$$\delta(X, \Delta; L) = \inf_{F/X} \frac{A_{X,\Delta}(F)}{\sup \left\{ \mathrm{ord}_F(D) | D \text{ is a cool effective } \mathbb{Q}\text{-divisor with } D \sim_{\mathbb{Q}} L \right\}}.$$

To do this, it is enough to prove that the denominator in this formula is equal to $S_L(F)$. But this denominator does not exceed $S_L(F)$. Thus, we only have to prove that

$$\sup \left\{ \mathrm{ord}_F(D) \mid D \text{ is cool effective } \mathbb{Q}\text{-divisor such that } D \sim_{\mathbb{Q}} L \right\} \geq S_L(F).$$
(1.7)

Fix a prime divisor F over X and $\epsilon > 0$. Let $m_0(\epsilon)$ be the constant from Lemma 1.65. Take a sufficiently large and divisible integer $k \geq m_0(\epsilon)$ such that kL is a Cartier divisor, and $|kL|$ is not empty. It follows from [18, Corollary 3.6] that for each $m \in \mathbb{N}$ divisible by k, there exists an m-basis type divisor $D_m \sim_{\mathbb{Q}} L$ such that

$$\lim_{m \to \infty} \mathrm{ord}_F(D_m) = S_L(F).$$

But it follows from Lemma 1.64 that there is an effective $\mathbb{Q}$-divisor $D' \sim_{\mathbb{Q}} L$ such that

$$\mathrm{ord}_E(D') \leq \frac{1}{2(n+1)} \tau_L(E) \leq \frac{1}{2} S_L(E)$$

for every prime divisor E over X. Moreover, by construction of the divisor D', we may assume that $\mathrm{ord}_F(D') = 0$. Now for every positive m divisible by k, we let

$$D = \frac{1-\epsilon}{1+\epsilon} D_m + \frac{2\epsilon}{1+\epsilon} D'.$$

Then $D \sim_{\mathbb{Q}} L$. We claim that D is cool. Indeed, since $m \geq k \geq m_0(\epsilon)$, for every prime divisor E over X, we have

$$\mathrm{ord}_E(D) \leq \frac{1-\epsilon}{1+\epsilon}(1+\epsilon) S_L(E) + \frac{2\epsilon}{1+\epsilon} \times \frac{1}{2} S_L(E) < S_L(E)$$

by Lemma 1.65. On the other hand, we have

$$\mathrm{ord}_F(D) \geq \frac{1-\epsilon}{1+\epsilon} \mathrm{ord}_F(D_m),$$

where $\mathrm{ord}_F(D_m) \to S_L(F)$ as $m \to \infty$. This gives (1.7) as required. □

Let us use δ-invariants to prove the following generalization of [69, Theorem 1.1], which we obtained after a discussion with Ziquan Zhuang.

Proposition 1.66 *Let X be a Fano variety of dimension $n \geqslant 2$ that has Kawamata log terminal singularities. Suppose that there exists a cyclic cover $f \colon X \to Y$ of degree m such that Y is also a Fano variety that has at most Kawamata log terminal singularities, and f is branched along an effective reduced divisor $B \subset Y$ such that $B \sim_{\mathbb{Q}} b(-K_Y)$ for some positive rational number $b < \frac{m}{m-1}$. Suppose that one of the following holds:*

(i) *the log pair (Y, B) is log canonical and $\delta(Y) > m - (m-1)b$,*
(ii) *the log pair (Y, B) is log canonical, $\delta(Y) = m - (m-1)b$, and for every prime divisor F over Y such that $A_{Y,B}(F) = 0$, $\frac{A_Y(F)}{S_Y(F)} > m - (m-1)b$.*

Then X is K-stable.

Proof Let $L = -K_Y$. Let us show that the log pair $(Y, \frac{m-1}{m}B)$ is K-stable, which would imply the required result by [148, Proposition 3.4].

Let $(\mathcal{Y}, \mathcal{L})$ be any non-trivial test configuration of the pair (Y, L) over $\mathbb{P}^1$ such that its central fiber $\mathcal{Y}_0$ is reduced and irreducible, and let M_t be the non-Archimedean Mabuchi functional of $(\mathcal{Y}, \mathcal{L})$ defined in [24, Definition 7.13], where $t \in [0, 1]$. Then

$$M_t = \frac{1}{L^n} \mathcal{L}^n \cdot (K_{\mathcal{Y}/\mathbb{P}^1} + t\mathcal{B}) - \frac{n}{n+1} \frac{L^{n-1} \cdot (K_Y + tB)}{L^n} (\mathcal{L}^{n+1}),$$

where $\mathcal{B}$ is the closure of $B \times (\mathbb{P}^1 \setminus [0 : 1])$ in $\mathcal{Y}$. Moreover, it follows from [142] that to prove K-stability of $(Y, \frac{m-1}{m}B)$, it is enough to prove that $M_t > 0$ for $t = \frac{(m-1)}{m}$.

Let $v := \mathrm{ord}_{\mathcal{Y}_0}|_{\mathbb{C}(Y)^*} \colon \mathbb{C}(Y)^* \to \mathbb{Z}$ be the divisorial valuation given by $\mathcal{Y}$. Then there exists a prime divisor F over Y such that $v = c\,\mathrm{ord}_F$ for some $c \in \mathbb{Z}_{>0}$. Moreover, it follows from [95, Theorem 5.1] and [99, Theorem 3.2] that $M_t = A_{(Y,tB)}(F) - (1-tb)S_Y(F)$. Thus, in order to prove that $M_{\frac{m-1}{m}} > 0$, it is enough to prove that

$$A_{(Y, \frac{(m-1)B}{m})}(F) > \left(1 - \frac{(m-1)b}{m}\right)S_L(F).$$

Since (Y, B) is log canonical, we have $A_{(Y,B)}(F) \geqslant 0$, so that

$$A_{(Y, \frac{(m-1)B}{m})}(F) \geqslant \frac{1}{m} A_Y(F).$$

Moreover, we have $\delta(Y) \geqslant m - (m-1)b$, which gives $\frac{A_Y(F)}{S_L(F)} \geqslant m - (m-1)b$. Thus, using conditions (1) or (2), we see that

$$A_{(Y, \frac{(m-1)B}{m})}(F) \geqslant \frac{1}{m} A_Y(F) > \left(1 - \frac{(m-1)b}{m}\right)S_L(F)$$

or

$$A_{(Y,\frac{(m-1)B}{m})}(F) > \frac{1}{m}A_Y(F) = \left(1 - \frac{(m-1)b}{m}\right)S_L(F),$$

respectively. This proves the proposition. □

Using Proposition 1.66 and Theorem 1.59, we get

Corollary 1.67 ([69]) *Suppose that X is a smooth Fano variety of dimension $n \geqslant 2$, and there is a cyclic cover $f : X \to Y$ of degree m such that Y is a smooth Fano variety, and f is branched along an effective reduced divisor $B \subset Y$ with $B \sim_{\mathbb{Q}} b(-K_Y)$ for a rational number b such that $1 \leqslant b < \frac{m}{m-1}$. Note that the ramification divisor B is smooth. If the Fano variety Y is K-semistable, then X is K-stable.*

Let G be a reductive algebraic subgroup of $\mathrm{Aut}(X, \Delta)$ such that the class of the ample divisor L in the group $\mathrm{Pic}(X) \otimes \mathbb{Q}$ is G-invariant. As in Section 1.4, we can define

$$\alpha_G(X, \Delta; L) = \sup \left\{ \epsilon \in \mathbb{Q} \;\middle|\; \begin{array}{l} \left(X, \Delta + \dfrac{\epsilon}{m}\mathcal{D}\right) \text{ is log canonical for any} \\ m \in \mathbb{Z}_{>0} \text{ such that } mL \text{ is a } \mathbb{Z}\text{-divisor and any} \\ G\text{-invariant subsystem } \mathcal{D} \subset |mL| \end{array} \right\}.$$

Note that we can reformulate the definition of $\alpha_G(X, \Delta; L)$ as follows:

$$\alpha_G(X, \Delta; L) = \inf \left\{ \mathrm{lct}\left(X, \Delta; \dfrac{1}{m}\mathcal{D}\right) \;\middle|\; \begin{array}{l} m \text{ is a positive integer such that } mL \\ \text{is a } \mathbb{Z}\text{-divisor and } \mathcal{D} \text{ is a } G\text{-invariant} \\ \text{linear subsystem in } |mL| \end{array} \right\}.$$

Moreover, arguing as in Remark 1.43, one can show that

$$\alpha_G(X, \Delta; L) = \inf_{E/X} \frac{A_{X,\Delta}(E)}{\tau_L(E)}, \tag{1.8}$$

where the infimum is taken over all G-irreducible (not necessarily prime) divisors over X. If (X, Δ) is a log Fano variety, then we let $\alpha_G(X, \Delta) = \alpha_G(X, \Delta; -(K_X + \Delta))$.

Remark 1.68 In (1.8), we cannot take the infimum over all G-invariant prime divisors over X in general. But if (X, Δ) is a log Fano variety, $L = -(K_X + \Delta)$ and $\alpha_G(X, \Delta) < 1$, then we can assume that the infimum in (1.8) is taken over all G-invariant prime divisors over X. This follows from Corollary 1.53, Lemma A.28 and [226, Lemma 4.8].

Inspired by [148, Definition 2.5] and Theorem 1.22, we can define

$$\delta_G(X,\Delta;L) = \inf_{E/X} \frac{A_{X,\Delta}(E)}{S_L(E)},$$

where the infimum is taken over all possible G-invariant prime divisors over the variety X. If (X,Δ) is a log Fano variety, we also let

$$\delta_G(X,\Delta) = \delta_G(X,\Delta;-(K_X+\Delta)).$$

In this case, the strict inequality $\delta_G(X,\Delta) > 1$ implies that (X,Δ) is K-polystable [226]. Similarly, if X is a Fano variety, we let $\delta_G(X) = \delta_G(X,0;-K_X)$.

In the remaining part of this section, we will show another way to define $\delta_G(X,\Delta;L)$, which resembles the original definition of the δ-invariant given in [102].

First, we fix a positive integer m such that mL is a very ample Cartier divisor, and the action of G lifts to its linear representation in $H^0(X,mL)$. We let $N_m = h^0(X,mL)$. For every linear subspace $W \subseteq H^0(X,mL)$, we denote by $|W|$ the corresponding linear subsystem in $|mL|$. If the subspace W is G-invariant, we say that $|W|$ is G-invariant. Note that $H^0(X,mL)$ splits as a sum of irreducible G-representations [200, Section 4.6.6].

Definition 1.69 Fix positive integers $m_1,\ldots,m_t$ such that each divisor m_iL is Cartier, and take positive rational numbers $a_1,\ldots,a_t$ such that

$$\sum_{i=1}^{t} a_i m_i = 1.$$

Let $\mathcal{D}_1,\ldots,\mathcal{D}_t$ be linear subsystems in $|m_1 L|,\ldots,|m_t L|$, respectively. Then

$$\mathcal{D} = \sum_{i=1}^{t} a_i \mathcal{D}_i$$

is said to be a $\mathbb{Q}$-system of the ample $\mathbb{Q}$-divisor L. If each linear system $\mathcal{D}_i$ is G-invariant, then we say that $\mathcal{D}$ is G-invariant. Similarly, if no $\mathcal{D}_i$ has fixed components, we say that $\mathcal{D}$ is mobile [4, 29]. We say that $\mathcal{D}$ is m-decomposed if the following holds:

(i) $m_1 = \cdots = m_t = m$, so that each $\mathcal{D}_i$ is given by a subspace $W_i \subset H^0(X,mL)$,

(ii)

$$H^0(X,mL) = \bigoplus_{i=1}^{t} W_i,$$

(iii) for each $i \in \{1,\ldots,t\}$, $a_i = \frac{\dim(W_i)}{mN_m}$.

Note that $\frac{1}{m}|mL|$ is a G-invariant m-decomposed $\mathbb{Q}$-system of the divisor L. Likewise, if

$$D = \frac{1}{mN_m} \sum_{i=1}^{N_m} D_i$$

is an m-basis type $\mathbb{Q}$-divisor of L, then D is an m-decomposed $\mathbb{Q}$-system of the divisor L, where each linear system $\mathcal{D}_i$ consists of one divisor D_i.

Let $V_m = H^0(X, mL)$. Consider a G-invariant filtration $\mathcal{F}$ of the space V_m given by

$$H^0(X, mL) = \mathcal{F}^0 V_m \supseteq \mathcal{F}^1 V_m \supseteq \mathcal{F}^2 V_m \supseteq \cdots \supseteq \mathcal{F}^t V_m \supseteq \{0\},$$

where each $\mathcal{F}^j V_m$ is a G-invariant vector subspace of $H^0(X, mL)$. Since the group G is reductive by assumption, the vector space $H^0(X, mL)$ decomposes as a direct sum of G-subrepresentations $W_1 \oplus W_2 \oplus \cdots \oplus W_t$ such that

$$\mathcal{F}^j V_m = \bigoplus_{i=j}^{t} W_i.$$

Note that this decomposition is not necessarily unique. We set

$$\mathcal{D}_m^{\mathcal{F}} = \sum_{i=0}^{t} \frac{\dim(W_i)}{mN_m} |W_i|.$$

Then $\mathcal{D}_m^{\mathcal{F}}$ is a G-invariant m-decomposed $\mathbb{Q}$-system of the divisor L, which can depend on the decomposition of the vector space $H^0(X, mL)$ into a sum of G-subrepresentations.

Now, for a G-invariant prime divisor F over X, we consider a G-invariant filtration

$$H^0(X, mL) = \mathcal{F}_F^0 V_m \supseteq \mathcal{F}_F^1 V_m \supseteq \mathcal{F}_F^2 V_m \supseteq \cdots \supseteq \mathcal{F}_F^s V_m \supseteq \{0\},$$

where $\mathcal{F}_F^j V_m = H^0(X, mL - jF)$, and s is the largest integer such that $h^0(X, mL - sF) \neq 0$. We set $\mathcal{D}_m^F = \mathcal{D}_m^{\mathcal{F}_F}$. This means that the G-representation $H^0(X, mL)$ decomposes as a direct sum of G-subrepresentations $U_1 \oplus U_2 \oplus \cdots \oplus U_s$ such that

$$H^0(X, mL - jF) = \bigoplus_{i=j}^{s} U_i$$

and

$$\mathcal{D}_m^F = \sum_{i=0}^{s} \frac{\dim(U_i)}{mN_m} |U_i|. \tag{1.9}$$

Then

$$S_m(F) = \mathrm{ord}_F\left(\mathcal{D}_m^F\right) = \sum_{j=0}^{s} \frac{h^0(X, mL - jF) - h^0(X, mL - (j+1)F)}{mN_m}$$

$$= \sum_{i=1}^{\infty} \frac{h^0(X, mL - iF)}{mN_m}.$$

Lemma 1.70   *It holds that*

$S_m(F) =$

$\sup\left\{\mathrm{ord}_F\left(\mathcal{D}\right) \mid \mathcal{D} \text{ is a } G\text{-invariant } m\text{-decomposed } \mathbb{Q}\text{-system of the divisor } L\right\}.$

Proof   We only need to prove the $\geqslant$-part of the assertion because $S_m(F) = \mathrm{ord}_F(\mathcal{D}_m^F)$. Let $\mathcal{D}$ be a G-invariant m-decomposed $\mathbb{Q}$-system of the divisor L. Then

$$\mathcal{D} = \sum_{i=1}^{t} a_i \mathcal{D}_i,$$

where $\mathcal{D}_i = |W_i|$ and $a_i = \frac{\dim(W_i)}{mN_m}$ for some G-invariant linear subspace $W_i \subset H^0(X, mL)$, and $H^0(X, mL)$ decomposes as a direct sum of G-subrepresentations $W_1 \oplus W_2 \oplus \cdots \oplus W_t$. For every $j \in \{1, \ldots, t\}$,

$$W_j = \bigoplus_{i=1}^{s} (W_j \cap U_i),$$

where $W_j \cap U_i$ is G-invariant for every i and j. Therefore, we have the following G-invariant decomposition:

$$H^0(X, mL) = \bigoplus_{j=1}^{t}\bigoplus_{i=1}^{s} (W_j \cap U_i).$$

Therefore, we can set

$$\mathcal{D}' = \sum_{j=1}^{t}\sum_{i=1}^{s} \frac{\dim(W_j \cap U_i)}{mN_m}\left|W_j \cap U_i\right|.$$

We observe that $\mathcal{D}'$ is a G-invariant m-decomposed $\mathbb{Q}$-system of the ample $\mathbb{Q}$-divisor L. On the other hand, we have $\mathrm{ord}_F(|W_j \cap U_i|) = i$ for each i and j. Thus, we conclude that

$$\mathrm{ord}_F\left(\mathcal{D}'\right) = \frac{1}{mN_m}\sum_{j=1}^{t}\sum_{i=1}^{s} i \cdot \dim(W_j \cap U_i) = \frac{1}{mN_m}\sum_{i=1}^{s} i \cdot \dim(U_i) = S_m(F).$$

But $\mathrm{ord}_F(\mathcal{D}) \leqslant \mathrm{ord}_F(\mathcal{D}')$ by construction, which completes the proof of the lemma.   □

Now we define

$$\widehat{\delta}_{G,m}(X,\Delta;L)$$

$$= \inf\Big\{\mathrm{lct}(X,\Delta;\mathcal{D}) \mid \mathcal{D} \text{ is a } G\text{-invariant } m\text{-decomposed } \mathbb{Q}\text{-system of } L\Big\}$$

and

$$\widehat{\delta}_{G}(X,\Delta;L) = \limsup_{\substack{m\in\mathbb{Z}_{>0} \\ mL \text{ is Cartier}}} \widehat{\delta}_{G,m}(X,\Delta;L).$$

As above, if (X,Δ) is a log Fano variety, we simply let

$$\widehat{\delta}_{G}(X,\Delta) = \widehat{\delta}_{G}(X,\Delta;-(K_X+\Delta)).$$

Likewise, if X is a Fano variety, we let $\widehat{\delta}_{G}(X) = \widehat{\delta}_{G}(X,0;-K_X)$.

Note that the number $\widehat{\delta}_{G}(X,\Delta;L)$ differs from $\delta_{G}(X,\Delta;L)$, and $\widehat{\delta}_{G}(X,\Delta;L)$ also differs from the counterpart of the number $\delta_{G}(X,\Delta;L)$ defined in [89].

Proposition 1.71 (cf. [18, §4])

$$\widehat{\delta}_{G}(X,\Delta;L) \leqslant \delta_{G}(X,\Delta;L).$$

Proof Let m be a sufficiently large and divisible integer. Then

$$\widehat{\delta}_{G,m}(X,\Delta;L) = \inf_{\mathcal{D}} \inf_{E/X} \frac{A_{X,\Delta}(E)}{\mathrm{ord}_E(\mathcal{D})},$$

where the first infimum is taken over all G-invariant m-decomposed $\mathbb{Q}$-systems of L, and the second infimum is taken over all prime divisors over X. Using Lemma 1.70, we get

$$\widehat{\delta}_{G,m}(X,\Delta;L) \leqslant \inf_{E/X} \frac{A_{X,\Delta}(E)}{S_m(E)},$$

where the infimum now is taken over all G-invariant prime divisors over the variety X. Therefore, we conclude that

$$\widehat{\delta}_{G}(X,\Delta;L) \leqslant \limsup_{\substack{m\in\mathbb{Z}_{>0} \\ mL \text{ is Cartier}}} \inf_{E/X} \frac{A_{X,\Delta}(E)}{S_m(E)} \leqslant \inf_{E/X} \limsup_{\substack{m\in\mathbb{Z}_{>0} \\ mL \text{ is Cartier}}} \frac{A_{X,\Delta}(E)}{S_m(E)}$$

$$= \inf_{E/X} \frac{A_{X,\Delta}(E)}{S_L(E)},$$

where E runs through all G-invariant prime divisors over X. □

Thus, applying Theorem 1.22 and [226, Corollary 4.14], we get

Corollary 1.72 *Suppose that (X,Δ) is a log Fano variety such that $\widehat{\delta}_{G}(X,\Delta) > 1$. Then (X,Δ) is K-polystable.*

If (X, Δ) is a log Fano variety, $L = -(K_X + \Delta)$ and $\widehat{\delta}_G(X, \Delta; L) < 1$, then

$$\widehat{\delta}_G(X, \Delta; L) = \lim_{\substack{m \in \mathbb{Z}_{>0} \\ mL \text{ is Cartier}}} \widehat{\delta}_{G,m}(X, \Delta; L) = \delta_G(X, \Delta)$$

by [226, Lemma 4.7]. However, in general, $\widehat{\delta}_G(X, \Delta; L) \neq \delta_G(X, \Delta; L)$.

Example 1.73 Suppose that $X = \mathbb{P}^1$, $\Delta = 0$, $L = -K_X$ and that G is the infinite group generated by the transformations $[x : y] \mapsto [y : x]$ and $[x : y] \mapsto [x : \lambda y]$ for $\lambda \in \mathbb{C}^*$. Then $\widehat{\delta}_G(X, \Delta; L) = 2$. But $\delta_G(X, \Delta; L) = +\infty$ since X has no G-fixed point.

As in Definition 1.61, we say that a $\mathbb{Q}$-system $\mathcal{D}$ of L is *cool* if $\mathrm{ord}_E(\mathcal{D}) \leqslant S_L(E)$ for every prime Weil divisor E over X. Then, inspired by Proposition 1.62, we let

$$\widetilde{\delta}_G(X, \Delta; L) =$$
$$\inf\{\mathrm{lct}(X, \Delta; \mathcal{D}) \mid \mathcal{D} \text{ is a } G\text{-invariant cool } \mathbb{Q}\text{-system of the divisor } L\}.$$

If (X, Δ) is a log Fano variety, we let $\widetilde{\delta}_G(X, \Delta) = \widetilde{\delta}_G(X, \Delta; -(K_X + \Delta))$ for simplicity. Similarly, if X is a Fano variety, we simply let $\widetilde{\delta}_G(X) = \widetilde{\delta}_G(X, 0; -K_X)$.

Lemma 1.74 *Let F be a G-invariant prime divisor over X. Then*

$$\sup\{\mathrm{ord}_F(\mathcal{D}) \mid \mathcal{D} \text{ is a } G\text{-invariant cool } \mathbb{Q}\text{-system of the divisor } L\} = S_L(F).$$

Proof The inequality $\leqslant$ is trivial. Let us prove the inequality $\geqslant$. Take $\epsilon > 0$ very small. By Lemma 1.65, there exists a sufficiently divisible integer $m \gg 0$ such that mL is Cartier and $\mathrm{ord}_E(\mathcal{D}_m^F) \leqslant (1 + \epsilon)S_L(E)$, where $\mathcal{D}_m^F$ is defined in (1.9). Now, we let

$$\mathcal{D} = \frac{1}{1 + \epsilon} \mathcal{D}_m^F + \frac{\epsilon}{(1 + \epsilon)k} |kL|$$

for some sufficiently large positive integer k such that kL is a very ample Cartier divisor. Then $\mathcal{D}$ is a G-invariant cool $\mathbb{Q}$-system of the divisor L. On the other hand, we have

$$\mathrm{ord}_F(\mathcal{D}) = \frac{1}{1 + \epsilon} \mathrm{ord}_F(\mathcal{D}_m^F) = \frac{1}{1 + \epsilon} S_m(F),$$

which gives the required inequality since $S_m(F) \to S_L(F)$ when $m \to \infty$. $\square$

Now, arguing as in the proof of Proposition 1.71 and using our Proposition 1.74, we can prove that $\widetilde{\delta}_G(X, \Delta; L) \leqslant \delta_G(X, \Delta; L)$. In fact, we can say more.

Proposition 1.75 $\widetilde{\delta}_G(X, \Delta; L) \leqslant \widehat{\delta}_G(X, \Delta; L)$.

Proof Take any sufficiently small $\epsilon > 0$. By Lemmas 1.65 and 1.70, there is a positive integer m_0 such that $\mathrm{ord}_E(\mathcal{D}) \leqslant (1+\epsilon)S_L(E)$ for every m-decomposed $\mathbb{Q}$-system $\mathcal{D}$ of the divisor L, where m is any integer such that $m \geqslant m_0$ and mL a Cartier divisor. As in the proof of Proposition 1.74, we let

$$\mathcal{D}' = \frac{1}{1+\epsilon}\mathcal{D} + \frac{\epsilon}{(1+\epsilon)k}|kL|$$

for some sufficiently large positive integer k such that kL is a very ample Cartier divisor. Then $\mathcal{D}'$ is a cool $\mathbb{Q}$-system of the divisor L since $\mathrm{ord}_E(\mathcal{D}) = (1+\epsilon)\mathrm{ord}_E(\mathcal{D}')$. This inequality also implies that

$$\widehat{\delta}_{G,m}(X,\Delta;L) \geqslant \frac{1}{1+\epsilon}\widetilde{\delta}_G(X,\Delta;L).$$

Since ϵ can be chosen arbitrarily small, we get $\widetilde{\delta}_G(X,\Delta;L) \leqslant \widehat{\delta}_G(X,\Delta;L)$. $\square$

By Propositions 1.71 and 1.75, we have $\widetilde{\delta}_G(X,\Delta;L) \leqslant \widehat{\delta}_G(X,\Delta;L) \leqslant \delta_G(X,\Delta;L)$. Therefore, applying Theorem 1.22 and [226, Corollary 4.14], we get

Corollary 1.76 *Suppose that (X,Δ) is a log Fano variety such that $\widetilde{\delta}_G(X,\Delta) > 1$. Then (X,Δ) is K-polystable.*

Let us show how to apply this corollary.

Example 1.77 Let $G = \mathfrak{S}_5$. Consider the G-action on $\mathbb{P}^3$ that is given by the standard representation of the group G. Then $\mathbb{P}^3$ contains a unique G-invariant quadric surface, and a unique G-invariant cubic surface. Denote them by S_2 and S_3, respectively. We let $\mathscr{C} = S_2 \cap S_3$. Then $\mathscr{C}$ is a G-invariant smooth curve, known as the *Bring's curve*. Let $\pi\colon X \to \mathbb{P}^3$ be the blow up of the curve $\mathscr{C}$, let E be the π-exceptional divisor, and let Q be the proper transform of the surface S_2 on the threefold X. Then X is a smooth Fano threefold №2.15, and there exists the G-equivariant commutative diagram

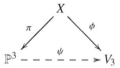

where V_3 is a cubic threefold with one singular point, which is an ordinary double point, the morphism ϕ is a contraction of the surface Q to the singular point of the cubic V_3, and ψ is given by the linear system of cubic surfaces that contain $\mathscr{C}$. Let

$$\varepsilon = \max\left\{ S_X(E), S_X(Q), S_X(H), \frac{4}{5}\right\}$$

where H is the proper transform on X of a plane from $\mathbb{P}^3$. Then $\varepsilon < 1$ by Theorem 3.17. We claim that $\widetilde{\delta}_G(X) \geqslant \frac{1}{\varepsilon}$, which would imply that X is K-polystable by Corollary 1.76. Namely, suppose that $\widetilde{\delta}_G(X) < \frac{1}{\varepsilon}$. Then there exists a G-invariant cool $\mathbb{Q}$-system $\mathcal{D}$ of the divisor $-K_X$ such that the log pair $(X, \lambda\mathcal{D})$ is strictly log canonical for some positive rational number $\lambda < \frac{1}{\varepsilon}$. Write $\mathcal{D} = aQ + \Delta$, where a is a non-negative rational number, and Δ is a $\mathbb{Q}$-system whose support does not contain Q. Then $a \leqslant S_X(Q) \leqslant \epsilon$ because the $\mathbb{Q}$-system $\mathcal{D}$ is cool. On the other hand, we have

$$\Delta|_Q \sim_{\mathbb{Q}} -K_X|_Q - aQ|_Q \sim -(1+a)Q|_Q,$$

so that $\Delta|_Q$ is a $\mathbb{Q}$-system on $Q \cong \mathbb{P}^1 \times \mathbb{P}^1$ of degree $(1+a, 1+a)$. But Q does not contain G-invariant curves of degree $(1,0)$, $(0,1)$, $(1,1)$, which implies that $\mathrm{Nklt}(Q, \lambda\Delta|_Q)$ is zero-dimensional, so that $\mathrm{Nklt}(Q, \lambda\Delta|_Q) = \varnothing$ by Corollary A.4 since Q has no G-fixed points. Then $\mathrm{Nklt}(X, \lambda\mathcal{D}) \cap Q = \varnothing$ by Theorem A.15. Note that $\mathrm{Nklt}(X, \lambda\mathcal{D})$ is at most one-dimensional. Indeed, if S is an irreducible surface in X, then $S \not\subset \mathrm{Nklt}(X, \lambda\mathcal{D})$ because we have $\mathrm{ord}_S(\mathcal{D}) \leqslant \varepsilon$ since one of the divisors $S - Q$, $S - E$ or $S - H$ is pseudo-effective. Denote by Z the union of all irreducible components of the locus $\mathrm{Nklt}(X, \lambda\mathcal{D})$ that have maximal dimension. Then Z is either a G-invariant curve or a union of G-orbits. Suppose that Z is a union of G-orbits in X. Since Z is disjoint from Q, we see that

$$\phi(Z) \subseteq \mathrm{Nklt}\big(V_3, \lambda\phi(\mathcal{D})\big) \subset \phi(Z) \cup \phi(Q),$$

so that the locus $\mathrm{Nklt}(V_3, \lambda\phi(\mathcal{D}))$ is a finite set that consists of at least $|Z| \geqslant 1$ points. Now, applying Corollary A.6 to the log pair $(V_3, \lambda\phi(\mathcal{D}))$, we immediately get $|Z| \leqslant 5$. In particular, we see that $Z \not\subset E$ since E does not contain G-orbits of length less than 24, because $\mathscr{C}$ does not contain G-orbits of length less than 24 by [53, Lemma 5.1.5]. Then

$$\pi(Z) \subseteq \mathrm{Nklt}\big(\mathbb{P}^3, \lambda\pi(\mathcal{D})\big) \subset \pi(Z) \cup \mathscr{C}$$

and $\pi(Z) \not\subset \mathscr{C}$. Now, applying Corollary A.6 to the log pair $(\mathbb{P}^3, \lambda\pi(\mathcal{D}))$, we get $|Z| \leqslant 4$. But $\mathbb{P}^3$ has no G-orbits of length less than 5. This shows that Z is a G-invariant curve. Suppose that $Z \not\subset E$. Let C be any G-irreducible component of Z such that $C \not\subset E$. Then $\pi(C)$ is a G-irreducible curve in $\mathbb{P}^3$. Let d be its degree. Since C is disjoint from Q, we have $0 = C \cdot Q = 2d - E \cdot C$, so that $d \geqslant 12$ because

$$24 \leqslant |\mathscr{C} \cap \pi(C)| \leqslant |E \cap C| \leqslant E \cdot C = 2d.$$

On the other hand, the log pair $(\mathbb{P}^3, \lambda\pi(\mathcal{D}))$ is not Kawamata log terminal along $\pi(C)$. Thus, applying Corollary A.8 to this pair with $\mathcal{S} = |\mathcal{O}_{\mathbb{P}^3}(1)|$, $S = \mathbb{P}^2$ and $L_S = \mathcal{O}_{\mathbb{P}^2}(2)$, we immediately get $d \leqslant 6$, which is a contradiction. Therefore, we conclude that $Z \subset E$. Now, let S be a general hyperplane section of the cubic threefold V_3. Then $\phi(\mathcal{D}) \sim_{\mathbb{Q}} 2S$. Then, applying Corollary A.8 with $\mathcal{S} = |S|$ and $L_S = O_{\mathbb{P}^4}(2)|_S$, we get $S \cdot \phi(Z) \leqslant 10$, so that $0 \neq \phi^*(S) \cdot Z = S \cdot \phi(Z) \leqslant 10$. Then

$$\pi^*\big(O_{\mathbb{P}^3}(1)\big) \cdot Z = \big(\phi^*(S) - Q\big) \cdot Z = \phi^*(S) \cdot Z \leqslant 10,$$

which implies that $\phi^*(S) \cdot Z = 6$ and Z is a section of the natural projection $E \to \mathscr{C}$. Thus, we see that Z is a smooth curve of genus 4 that is G-equivariantly isomorphic to $\mathscr{C}$. The locus $\mathrm{Nklt}(V_3, \lambda\phi(\mathcal{D}))$ consists of the curve $\phi(Z)$ and a (possibly empty) finite set. Observe also that the smooth curve $\phi(Z) \cong Z$ cannot be a minimal center of log canonical singularities of the log pair $(V_3, \lambda\phi(\mathcal{D}))$ because otherwise Corollary A.21 would give

$$6 = \phi^*(S) \cdot Z = S \cdot \phi(Z) > 12.$$

Thus, applying Lemma A.28, we get a G-invariant $\mathbb{Q}$-system D_{V_3} on V_3 together with a rational number $\mu < \frac{1}{\varepsilon}$ such that $D_{V_3} \sim_{\mathbb{Q}} -K_{V_3}$, the locus $\mathrm{Nklt}(V_3, \mu D_{V_3})$ is zero-dimensional, and the intersection $\mathrm{Nklt}(V_3, \mu D_{V_3}) \cap \phi(Z)$ contains a non-empty finite subset. Applying Corollary A.6, we see that $|\mathrm{Nklt}(V_3, \mu D_{V_3}) \cap \phi(Z)| \leqslant 5$, which is impossible because $\phi(Z) \cong \mathscr{C}$ contains no G-orbits of length less than 24. The obtained contradiction shows that $\tilde{\delta}(X) \geqslant \frac{1}{\varepsilon} > 1$. Thus, the threefold X is K-polystable.

Let us present localized versions of the invariants $\tilde{\delta}_G(X, \Delta; L)$, $\hat{\delta}_G(X, \Delta; L)$, $\delta_G(X, \Delta; L)$. Fix a proper closed subvariety $Z \subset X$. Let

$$\delta_{G,Z}(X, \Delta; L) = \inf_{\substack{E/X \\ Z \subseteq C_X(E)}} \frac{A_{X,\Delta}(E)}{S_L(E)},$$

where the infimum runs over all G-invariant prime divisors over X such that $Z \subseteq C_X(E)$. We also define

$$\hat{\delta}_{G,Z,m}(X, \Delta; L) =$$
$$\sup\left\{\lambda \in \mathbb{Q} \,\middle|\, \begin{array}{l} (X, \Delta; \lambda\mathcal{D}) \text{ is log canonical at a general point of } Z \\ \text{for every } G\text{-invariant } m\text{-decomposed } \mathbb{Q}\text{-system } \mathcal{D} \text{ of } L \end{array}\right\}$$

and

$$\hat{\delta}_{G,Z}(X, \Delta; L) = \limsup_{\substack{m \in \mathbb{Z}_{>0} \\ mL \text{ is Cartier}}} \hat{\delta}_{G,m}(X, \Delta; L).$$

Finally, we let

$$\tilde{\delta}_{G,Z}(X, \Delta; L) =$$
$$\sup\left\{\lambda \in \mathbb{Q} \,\middle|\, \begin{array}{l} (X, \Delta; \lambda\mathcal{D}) \text{ is log canonical at a general point of } Z \text{ for} \\ \text{every } G\text{-invariant cool } \mathbb{Q}\text{-system } \mathcal{D} \text{ of the divisor } L \end{array}\right\}.$$

Then, arguing as in the proof of Propositions 1.71 and 1.75, we obtain

$$\widetilde{\delta}_{G,Z}(X,\Delta;L) \leqslant \widehat{\delta}_{G,Z}(X,\Delta;L) \leqslant \delta_{G,Z}(X,\Delta;L). \qquad (1.10)$$

As above, if (X,Δ) is a log Fano variety, we let $\widetilde{\delta}_{G,Z}(X,\Delta) = \widetilde{\delta}_{G,Z}(X,\Delta; -(K_X + \Delta))$. Finally, if X is a Fano variety, we let $\widetilde{\delta}_{G,Z}(X) = \widetilde{\delta}_{G,Z}(X,0; -K_X)$.

1.6 Equivariant Stibitz–Zhuang theorem

Let us consider a log Fano variety (X,Δ), i.e. X is a normal variety, Δ is an effective $\mathbb{Q}$-divisor on X such that the log pair (X,Δ) has Kawamata log terminal singularities, and $-(K_X + \Delta)$ is an ample $\mathbb{Q}$-Cartier divisor. As in Section 1.5, let us fix a reductive algebraic subgroup $G \subseteq \mathrm{Aut}(X)$ such that the divisor Δ is G-invariant. Suppose, in addition, that we have $\mathrm{rk}\,\mathrm{Cl}^G(X) = 1$. This condition means the following: for every Weil divisor D on the variety X whose class in $\mathrm{Cl}(X)$ is G-invariant, $D \sim_{\mathbb{Q}} -\lambda(K_X + \Delta)$ for some $\lambda \in \mathbb{Q}$.

Remark 1.78   It should be noted that the condition $\mathrm{rk}\,\mathrm{Cl}^G(X) = 1$ is rather restrictive. For instance, if X is a smooth Fano threefold, then the condition $\mathrm{rk}\,\mathrm{Cl}^G(X) = 1$ implies that either $\mathrm{Cl}(X) \cong \mathbb{Z}$, or X is contained in one of the families №2.6, №2.12, №2.21, №2.32, №3.1, №3.13, №3.27, №4.1. See [185] for details. Note also that every smooth Fano threefold in these eight deformation families is fiber-like [113], i.e. it can appear as the fiber of a Mori fiber space.

Let us also assume that $\dim(X) \geqslant 2$. In this section, we prove the following result.

Theorem 1.79 (cf. [201, Theorem 1.2])   *Suppose that $\alpha_G(X,\Delta) \geqslant \frac{1}{2}$ and*

(★) *for any G-invariant mobile linear system $\mathcal{M}$ on X, the pair $(X,\Delta + \lambda\mathcal{M})$ has log canonical singularities for $\lambda \in \mathbb{Q}_{>0}$ such that $\lambda\mathcal{M} \sim_{\mathbb{Q}} -(K_X + \Delta)$.*

Then (X,Δ) is K-semistable. Moreover, (X,Δ) is K-polystable if $\alpha_G(X,\Delta) > \frac{1}{2}$ or

(♦) *for any G-invariant mobile linear system $\mathcal{M}$ on X, the pair $(X,\Delta + \lambda\mathcal{M})$ has Kawamata log terminal singularities for $\lambda \in \mathbb{Q}_{>0}$ such that $\lambda\mathcal{M} \sim_{\mathbb{Q}} -(K_X + \Delta)$.*

For the definition of $\alpha_G(X,\Delta)$, see Section 1.5. If $\Delta = 0$, we let $\alpha_G(X) = \alpha_G(X,\Delta)$.

Corollary 1.80 *Suppose that* $\Delta = 0$, $\alpha_G(X) \geqslant \frac{1}{2}$ *and*

($\heartsuit$) *for any G-invariant mobile linear system* $\mathcal{M}$ *on X, the pair* $(X, \lambda\mathcal{M})$ *has canonical singularities for* $\lambda \in \mathbb{Q}_{>0}$ *such that* $\lambda\mathcal{M} \sim_{\mathbb{Q}} -K_X$.

Then X is K-polystable.

The condition ($\heartsuit$) is equivalent to X being G-birationally super-rigid [53, §3.1.1]. Therefore, Corollary 1.80 can be restated as follows:

Corollary 1.81 (cf. [201, Theorem 1.2]) *Let V be a Fano variety with at most terminal singularities, let G be a reductive subgroup of the group* Aut(V) *such that* rk Cl$^{\mathbf{G}}(V) = 1$. *Suppose that X is G-birationally super-rigid and* $\alpha_{\mathbf{G}}(V) \geqslant \frac{1}{2}$. *Then V is K-polystable.*

This corollary naturally leads to the following

Conjecture 1.82 (cf. [128, Conjecture 1.1.1]) *Let V be a Fano variety with terminal singularities, and let G be a reductive subgroup of the group* Aut(V) *such that* rk Cl$^{\mathbf{G}}(V) = 1$. *Suppose that X is G-birationally super-rigid. Then V is K-polystable.*

This conjecture says that we can remove the condition $\alpha_G(V) \geqslant \frac{1}{2}$ from Corollary 1.81, which leads to the following

Question 1.83 (cf. [201, Question 1.5]) *Let V be a Fano variety with at most terminal singularities, let G be a reductive subgroup of the group* Aut(V) *such that* rk Cl$^{\mathbf{G}}(V) = 1$. *Is it always true that* $\alpha_{\mathbf{G}}(V) \geqslant \frac{1}{2}$?

Let us prove Theorem 1.79. First, we observe that to prove that (X, Δ) is K-semistable it is enough to show that $\beta(F) \geqslant 0$ for every G-invariant dreamy prime divisor F over X. This follows from [226, Theorem 4.14] and [20, Lemma 3.2]. Similarly, we have

Lemma 1.84 *To prove that* (X, Δ) *is K-polystable it is enough to show that* $\beta(F) > 0$ *for every G-invariant dreamy prime divisor F over X.*

Proof We may assume that (X, Δ) is K-semistable. If (X, Δ) is not K-polystable, then, arguing as in the proof of [226, Corollary 4.11], we see that there exists a G-equivariant special test configuration for (X, Δ) whose central fiber is a K-polystable log Fano pair. The Donaldson–Futaki invariant of this test configuration vanishes, so that there exists a G-invariant dreamy prime divisor F over X with $\beta(F) = 0$ by [95, Theorem 5.1]. $\qquad\square$

Suppose that $\alpha_G(X, \Delta) \geqslant \frac{1}{2}$ and ($\bigstar$) holds. Let us fix some G-invariant dreamy prime divisor F over X. To prove Theorem 1.79, it is enough to prove the following assertions:

(i) $\beta(F) \geqslant 0$;
(ii) if $\alpha_G(X, \Delta) > \frac{1}{2}$ or ($\blacklozenge$) holds, then $\beta(F) > 0$.

Since F is dreamy, there exists a G-equivariant birational morphism $\sigma : Y \to X$ such that Y is normal, and one of the following two possibilities holds:

- either σ is an identity map, and F is a G-invariant prime divisor on X;
- or the prime divisor F is the σ-exceptional locus, and $-F$ is σ-ample.

For simplicity, we set $n = \dim(X)$, $L = -(K_X + \Delta)$, $A = A_{X,\Delta}(F)$ and $S = S_{X,\Delta}(F)$. Let $\tau = \sup\{t \in \mathbb{R} \mid \sigma^*(L) - tF \text{ is pseudo-effective}\}$. Then $\beta(F) = A - S$ and

$$S = \frac{1}{L^n} \int_0^\tau \mathrm{vol}\big(\sigma^*(L) - tF\big)dt.$$

Note that $\tau > S$. Thus, to prove Theorem 1.79, we may assume that $\tau > A$.

Lemma 1.85 *Suppose that σ is an identity map. Then $\beta(F) > 0$.*

Proof $F \sim_{\mathbb{Q}} \lambda L$ for some $\lambda \in \mathbb{Q}_{>0}$. Then the pair $(X, \Delta + \frac{1}{2\lambda}F)$ is log canonical since $\alpha_G(X) \geqslant \frac{1}{2}$. In particular, we see that $\lambda \geqslant \frac{1}{2} > \frac{1}{n+1}$ because $n \geqslant 2$ by assumption. Now, applying [93, Lemma 9.2], we get $A > S$, so that $\beta(F) = A - S > 0$. $\square$

To proceed, we may assume that F is σ-exceptional. Take any $x \in (A, \tau) \cap \mathbb{Q}$.

Lemma 1.86 *There exists a G-irreducible effective Weil divisor D in X such that the inequality $\mathrm{ord}_F(\mu D) > x$ holds for $\mu \in \mathbb{Q}_{>0}$ such that $\mu D \sim_{\mathbb{Q}} L$. Moreover, such a divisor D is unique.*

Proof To prove the existence part, take a sufficiently large and divisible integer $m \gg 0$. Now we consider the G-invariant complete (non-empty) linear system $|m(\sigma^*(L) - xF)|$. Let $\mathcal{M}_Y$ be its mobile part, let $\mathcal{F}_Y$ be its fixed part, and let $\mathcal{M}$ and $\mathcal{F}$ be their proper transforms on X, respectively. Then $\mathcal{M} \neq \emptyset$, $\mathcal{M}$ and $\mathcal{F}$ are G-invariant, $\mathcal{M} + \mathcal{F} \sim_{\mathbb{Q}} mL$. But there exists $\epsilon \in \mathbb{Q}_{\geqslant 0}$ such that $\mathcal{F} \sim_{\mathbb{Q}} \epsilon L$. Then $\frac{1}{m-\epsilon}\mathcal{M} \sim_{\mathbb{Q}} L$ and $\epsilon < m$, so that the log pair $(X, \Delta + \frac{1}{m-\epsilon}\mathcal{M})$ is log canonical, which gives

$$0 \leqslant A - \frac{1}{m-\epsilon}\mathrm{ord}_F(\mathcal{M}) = A - \frac{1}{m-\epsilon}\Big(mx - \mathrm{ord}_F(\mathcal{F})\Big) = A + \frac{\mathrm{ord}_F(\mathcal{F}) - mx}{m-\epsilon},$$

which implies that $\epsilon \neq 0$ and $\mathrm{ord}_F(\mathcal{F}) > x\epsilon$ because $x > A$.

If $\mathcal{F}$ is G-irreducible, we let $D = \mathcal{F}$ and $\mu = \frac{1}{\epsilon}$. Otherwise, we have

$$\mathcal{F} = \sum_{i=1}^{r} a_i D_i,$$

where each D_i is a G-irreducible effective Weil divisor, and each a_i is a positive integer. For every $i \in \{1, \ldots, r\}$, there is $\mu_i \in \mathbb{Q}_{>0}$ such that $\mu_i D_i \sim_{\mathbb{Q}} L$, so that $\mathrm{ord}_F(\mu_j D_j) > x$ for some $j \in \{1, \ldots, r\}$. Thus, we let $D = D_j$ and $\mu = \mu_j$. This proves the existence part.

To prove the uniqueness part, suppose that D is not unique. Then there exists another G-irreducible effective Weil divisor D' on X with $\mathrm{ord}_F(\mu'D') > x$, where μ' is a positive rational number such that $\mu'D \sim_{\mathbb{Q}} L$. Then $aD \sim bD'$ for some positive integers a and b because $\mathrm{rk}\,\mathrm{Cl}^G(X) = 1$. Let $\mathcal{P}$ be the pencil $\langle aD, bD' \rangle$. Then $\mathcal{P}$ is mobile because both divisors D and D' are G-irreducible, and $D \neq D'$. But $\mathrm{ord}_F(\frac{\mu}{a}\mathcal{P}) > x$ and $\frac{\mu}{a}\mathcal{P} \sim_{\mathbb{Q}} L$. Since $x > A$, this implies that $(X, \Delta + \frac{\mu}{a}\mathcal{P})$ is not log canonical, which contradicts ($\bigstar$). This shows that D is unique. $\qquad\square$

Let D be the divisor constructed in Lemma 1.86, and let $\mu \in \mathbb{Q}_{>0}$ such that $\mu D \sim_{\mathbb{Q}} L$. By Lemma 1.86, the divisor D is unique, so that it does not depend on $x \in (A, \tau) \cap \mathbb{Q}$. But $\mathrm{ord}_F(\mu D) > x$ for every $x \in (A, \tau) \cap \mathbb{Q}$ by construction. This gives $\mathrm{ord}_F(\mu D) \geqslant \tau$. On the other hand, we have $\mathrm{ord}_F(\mu D) \leqslant \tau$ by the definition of τ, which implies

Corollary 1.87 $\mathrm{ord}_F(\mu D) = \tau$.

Let $\widetilde{D}$ be the proper transform of D on Y. Then $\mu\widetilde{D} \sim_{\mathbb{Q}} \sigma^*(L) - \tau F$ by Corollary 1.87.

Lemma 1.88 *For every* $t \in [A, \tau]$,

$$\mathrm{vol}\big(\sigma^*(L) - tF\big) = \Big(\frac{\tau - t}{\tau - A}\Big)^n \mathrm{vol}\big(\sigma^*(L) - A \cdot F\big).$$

Proof Since $\mathrm{vol}(\sigma^*(L) - tF)$ is a continuous function, we may assume that $t \in (A, \tau) \cap \mathbb{Q}$. Suppose that $\sigma^*(L) - tF \sim_{\mathbb{Q}} R + a\mu\widetilde{D}$ for some effective $\mathbb{Q}$-divisor R on the variety Y and some $a \in \mathbb{Q}_{\geqslant 0}$. Then the class of the divisor R in $\mathrm{Cl}(Y) \otimes \mathbb{Q}$ is G-invariant, $a < 1$ and

$$R \sim_{\mathbb{Q}} \sigma^*(L) - tF - a\mu\widetilde{D} \sim_{\mathbb{Q}} \sigma^*(L) - tF - a\big(\sigma^*(L) - \tau F\big) \sim_{\mathbb{Q}}$$

$$\sim_{\mathbb{Q}} (1-a)\sigma^*(L) - (t - a\tau)F \sim_{\mathbb{Q}} (1-a)\Big(\sigma^*(L) - \frac{t - a\tau}{1 - a}F\Big).$$

Thus, arguing as in the proof of Lemma 1.86 and using the uniqueness of the divisor D, we see that $\mathrm{Supp}(R)$ contains $\widetilde{D}$ provided that $\frac{t - a\tau}{1 - a} > A$. But

$\frac{t-a\tau}{1-a} > A \iff a < \frac{t-A}{\tau-A}$. Thus, arguing as in the proof of [100, Proposition 3.2], we obtain

$$\mathrm{vol}\big(\sigma^*(L) - tF\big) = \mathrm{vol}\Big(\sigma^*(L) - tF - \frac{t-A}{\tau-A}\mu\widetilde{D}\Big).$$

On the other hand, we have

$$\sigma^*(L) - tF - \frac{t-A}{\tau-A}\mu\widetilde{D} \sim_{\mathbb{Q}} \frac{\tau-t}{t-A}\big(\sigma^*(L) - AF\big),$$

so that

$$\mathrm{vol}\big(\sigma^*(L) - tF\big) = \mathrm{vol}\Big(\frac{\tau-t}{t-A}\big(\sigma^*(L) - AF\big)\Big) = \Big(\frac{\tau-t}{t-A}\Big)^n \mathrm{vol}\big(\sigma^*(L) - AF\big)$$

as required.   □

Now, using Lemma 1.88 and [100, Proposition 3.1], we conclude that $S \leqslant \frac{(n-1)A+\tau}{n+1}$. On the other hand, it follows from (1.8) that $A \geqslant \frac{\tau}{2}$ because $\alpha_G(X,\Delta) \geqslant \frac{1}{2}$, so that

$$S \leqslant \frac{(n-1)A+\tau}{n+1} \leqslant \frac{(n-1)A+2A}{n+1} = A,$$

so that $\beta(F) = A - S \geqslant 0$. Similarly, if we have $\alpha_G(X,\Delta) > \frac{1}{2}$, then (1.8) gives $A > \frac{\tau}{2}$, which implies that $S < A$, so that $\beta(F) = A - S > 0$ as required. Finally, we get

Lemma 1.89 *Suppose that (♦) is satisfied. Then $\beta(F) = A - S > 0$.*

Proof We already know that $A \geqslant S$. Suppose that $S = A$. Let us seek for a contradiction. Arguing as above, we conclude that $A = \frac{\tau}{2}$ and $S = \frac{(n-1)A+\tau}{n+1}$.

Recall that $\mathrm{vol}(\sigma^*(L) - tF)$ is a differentiable function for $t \in [0, \tau)$. Set

$$f(t) = -\frac{1}{n}\frac{\partial}{\partial t}\mathrm{vol}\big(\sigma^*(L) - tF\big).$$

Arguing as in the proof of [100, Proposition 3.1] and using Lemma 1.88, we get

$$f(t) = \begin{cases} \dfrac{f(A)t^{n-1}}{A^{n-1}} & \text{if } 0 \leqslant t \leqslant A, \\[2mm] \dfrac{f(A)(\tau-t)^{n-1}}{(\tau-A)^{n-1}} & \text{if } A \leqslant t < \tau, \end{cases}$$

and $L^n = \tau f(A)$. Thus, we have

$$\mathrm{vol}\big(\sigma^*(L) - tF\big) = L^n - n\int_0^t f(\xi)d\xi = \begin{cases} L^n\Big(1 - \dfrac{t^n}{2A^n}\Big) & \text{if } 0 \leqslant t \leqslant A, \\[2mm] L^n\dfrac{(\tau-t)^n}{2(\tau-A)^n} & \text{if } A \leqslant t < \tau. \end{cases}$$

If t is sufficiently small, then $\sigma^*(L) - tF$ is ample, so that $\mathrm{vol}(\sigma^*(L) - tF) = (\sigma^*(L) - tF)^n$. Now, using [94, Claim 3.3], we see that $\sigma(F)$ must be a point and $(-F|_F)^{n-1} = \frac{L^n}{2A^n}$. Then $\sigma^*(L) - tF$ is nef for $t \in [0, A]$ by [144, Lemma 10] (cf. [24, Proposition 1.12]), so that the divisor $\sigma^*(L) - AF$ is semiample, because F is dreamy.

Take sufficiently divisible $m \gg 0$. Then $|m(\sigma^*(L) - AF)|$ is a G-invariant basepoint free linear system. Let $\mathcal{M}$ be its proper transform on X. Then $\mathcal{M}$ is a G-invariant mobile linear system such that $\frac{1}{m}\mathcal{M} \sim_{\mathbb{Q}} L$, so that the log pair $(X, \Delta + \frac{1}{m}\mathcal{M})$ has Kawamata log terminal singularities. Then $A = \mathrm{ord}_F(\mathcal{M})/m > A_{X,\Delta} = A$, which is absurd. $\qquad\qquad\square$

Therefore, Theorem 1.79 is completely proved. Now, let us present its applications. We start with the following result, which also follows from Proposition 1.66.

Theorem 1.90 *Let $\pi\colon X \to \mathbb{P}^3$ be a double cover such that π is branched over a sextic surface S_6 that has isolated ordinary double points. Then X is K-stable.*

Proof Let G be the subgroup of $\mathrm{Aut}(X)$ generated by the Galois involution of the double cover π. Then $\mathrm{Cl}^G(X)$ is generated by $-K_X$ since the quotient X/G is isomorphic to $\mathbb{P}^3$. This implies that the Fano threefold X is G-birationally superrigid. Indeed, the required assertion follows from the proof of [37, Theorem A]. The only difference is that one should use Theorem A.24 instead of the standard Noether–Fano inequality.

We claim that $\alpha_G(X) \geqslant \frac{1}{2}$. In fact, this follows from the proof of [39, Proposition 3.2]. Indeed, suppose that $\alpha_G(X) < \frac{1}{2}$. Then there is a G-invariant divisor $D \in |-nK_X|$ such that the pair $(X, \frac{1}{2n}D)$ is not log canonical at some point $P \in X$. Using Corollary A.33, we may assume that the divisor D is G-irreducible. Let us seek for a contradiction.

Since D is G-invariant, either $n = 3$ and D is the preimage of the sextic surface S_6, or $D = \pi^*(F)$ for some irreducible surface $F \subset \mathbb{P}^3$ of degree $n \geqslant 2$. In the former case, the singularities of the log pair $(X, \frac{1}{6}D)$ are log canonical because S_6 has at most isolated ordinary double points. Thus, we are in the latter case. Then $(\mathbb{P}^3, \frac{1}{2}S_6 + \frac{1}{2n}F)$ is not log canonical at $\pi(P)$ by [130, Proposition 8.12]. Then $\pi(P) \in \mathrm{Sing}(S_6)$ by Lemma A.1, so that P is a singular point of the threefold X.

Let $\eta\colon \widetilde{X} \to X$ be the blow up of the point P, let E be the η-exceptional surface, and let $\widetilde{D}$ be the proper transform on $\widetilde{X}$ of the divisor D. Then $\widetilde{D} \sim \eta^*(-nK_X) - mE$ for some integer $m \geqslant 0$. Let $\widetilde{S}_1$ and $\widetilde{S}_2$ be proper transforms of two sufficiently general surfaces in $|-K_X|$ that pass through P. Then $\widetilde{S}_1 \sim \widetilde{S}_2 \sim \eta^*(-K_X) - E$, which gives

$$0 \leqslant \tilde{S}_1 \cdot \tilde{S}_2 \cdot \tilde{D} = \left(\eta^*(-K_X) - E\right)^2 \cdot \left(\eta^*(-nK_X) - mE\right) = 2n - 2m,$$

so that $m \leqslant n$. But this inequality contradicts [61, Theorem 3.10] or [29, Theorem 1.7.20]. The obtained contradiction completes the proof of the theorem.   □

Example 1.91 ([11]) Let S_6 be the sextic surface in $\mathbb{P}^3$ that is given by

$$4\left(\tau^2 x^2 - y^2\right)\left(\tau^2 y^2 - z^2\right)\left(\tau^2 z^2 - x^2\right) = (1 + 2\tau)w^2\left(x^2 + y^2 + z^2 - w^2\right)^2,$$

where $\tau = \frac{1+\sqrt{5}}{2}$, and x, y, z and w are coordinates on $\mathbb{P}^3$. Then S_6 has 65 singular points, and all these points are ordinary double points. This surface is called the *Barth sextic*. Let $\pi\colon X \to \mathbb{P}^3$ be a double cover that is ramified over S_6. Then the threefold X is rational by [44, Proposition 3.6], and X is K-stable by Theorem 1.90.

Let us present more applications of Theorem 1.79.

Example 1.92 ([114, 43, 9]) Let us identify $\mathbb{P}^3$ with the hyperplane in $\mathbb{P}^4$ given by the equation $x_0 + x_1 + x_2 + x_3 + x_4 = 0$, where x_0, x_1, x_2, x_3 and x_4 are coordinates on $\mathbb{P}^4$. Let $S_\lambda = \{x_0^4 + x_1^4 + x_2^4 + x_3^4 + x_4^4 = \lambda(x_0^2 + x_1^2 + x_2^2 + x_3^2 + x_4^2)^2\} \subset \mathbb{P}^3$ for some number λ. Let $\mathfrak{S}_5$ be the symmetric subgroup in $\mathrm{Aut}(\mathbb{P}^3)$ that acts by permutation of the coordinates. Then the surface S_λ is $\mathfrak{S}_5$-invariant, and S_λ has at most isolated ordinary double points. To describe its singularities, let Σ_5, Σ_{10}, Σ'_{10}, Σ_{15} be the orbits of the points

$$[-4:1:1:1:1], \quad [0:0:0:-1:1], \quad [-2:-2:-2:3:3], \quad [0:-1:-1:1:1],$$

respectively. Then $|\Sigma_5| = 5$, $|\Sigma_{10}| = |\Sigma'_{10}| = 10$ and $|\Sigma_{15}| = 15$. Moreover, one has

$$\mathrm{Sing}(S_\lambda) = \begin{cases} \Sigma_5 & \text{if } \lambda = \dfrac{13}{20}, \\ \Sigma_{10} & \text{if } \lambda = \dfrac{1}{2}, \\ \Sigma'_{10} & \text{if } \lambda = \dfrac{7}{30}, \\ \Sigma_{15} & \text{if } \lambda = \dfrac{1}{4}, \\ \varnothing & \text{otherwise.} \end{cases}$$

Let $\pi\colon X \to \mathbb{P}^3$ be the double cover branched over S_λ. Then X is a Fano threefold №1.12. Note that the $\mathfrak{S}_5$-action lifts to X, so that we can identify $\mathfrak{S}_5$ with a subgroup in $\mathrm{Aut}(X)$. Let G be the subgroup in $\mathrm{Aut}(X)$ generated by $\mathfrak{S}_5$ and the involution of the cover π. Then $G \cong \mathfrak{S}_5 \times \mu_2$, and X has no G-fixed points, so that $\alpha_G(X) \geqslant \frac{1}{2}$ by Theorem 1.52. If $\lambda \neq \frac{13}{20}$, then X is G-birationally super-rigid [9], so that X is K-polystable.

Example 1.93 ([51, 8, 146])   Now we let X be the Segre cubic hypersurface in $\mathbb{P}^4$. Then X has 10 ordinary double points, it admits a faithful action of the group $G = \mathfrak{S}_6$, and X is G-birationally super-rigid [51]. Arguing as in Example 1.92, we get $\alpha_G(X) \geqslant \frac{1}{2}$. Thus, the threefold X is K-polystable by Theorem 1.79.

Example 1.94   Let X be the smooth Fano threefold №3.13 with

$$\mathrm{Aut}^0(X) \cong \mathrm{PGL}_2(\mathbb{C}).$$

Such a threefold exists and is unique [185, 45]. For an explicit description, see Section 5.19. Let W be a smooth divisor in $\mathbb{P}^2 \times \mathbb{P}^2$ of degree $(1, 1)$. Then there is a $\mathrm{PGL}_2(\mathbb{C})$-equivariant commutative diagram

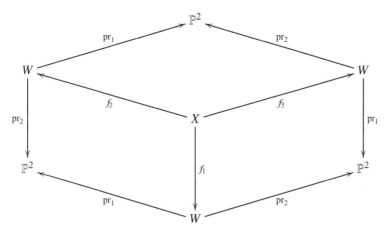

where each morphism f_i is a blow up of a smooth curve of degree $(2,2)$. Let $G = \mathrm{Aut}(X)$. One can show that $G \cong \mathrm{PGL}_2(\mathbb{C}) \times \mathfrak{S}_3$ and $\mathrm{Pic}^G(X) = \mathbb{Z}[-K_X]$, see [185] or Section 5.19. Let E_1, E_2 and E_3 be the exceptional surfaces of the birational morphisms f_1, f_2 and f_3, respectively. Then $E_1 + E_2 + E_3 \sim -K_X$, and $E_1 \cap E_2 \cap E_3$ is an irreducible smooth curve, so that $\alpha_G(X) \leqslant \frac{2}{3}$. If $\alpha_G(X) < \frac{2}{3}$, then applying Theorem 1.52 with $\mu = \frac{2}{3}$, we see that there exists a G-invariant irreducible curve C such that $-K_X \cdot C \leqslant 5$, which gives

$$5 \geqslant -K_X \cdot C = \sum_{i=1}^{3} (\mathrm{pr}_i \circ f_i)^* \left(\mathcal{O}_{\mathbb{P}^2}(1) \right) \cdot C = 3 (\mathrm{pr}_1 \circ f_1)^* \left(\mathcal{O}_{\mathbb{P}^2}(1) \right) \cdot C,$$

so that $\mathrm{pr}_1 \circ f_1(C)$ must be a $\mathrm{PGL}_2(\mathbb{C})$-invariant line in the plane $\mathbb{P}^2$, which does not exist. Therefore, we see that $\alpha_G(X) = \frac{2}{3}$. We claim that X is G-birationally super-rigid because otherwise X contains a G-invariant mobile linear system

M such that $(X, \lambda M)$ has canonical singularities, where λ is a rational number such that $\lambda M \sim_{\mathbb{Q}} -K_X$. Then $(X, \lambda M)$ is not canonical along $E_1 \cap E_2 \cap E_3$ because $E_1 \cap E_2 \cap E_3$ is the unique G-invariant curve in X, and X does not contain G-invariant finite subsets. This gives

$$\frac{1}{\lambda} = M \cdot \ell \geqslant \mathrm{mult}_C(M) > \frac{1}{\lambda},$$

where M is a general surface in $\mathcal{M}$, and ℓ is a general fiber of the restriction $f_1|_{E_1}$. The obtained contradiction shows that X is G-birationally super-rigid. Thus, we see that the Fano threefold X is K-polystable by Corollary 1.81. This can also be proved using the technique described in the next section (see the proof of Lemma 4.18).

1.7 Abban–Zhuang theory

Let X be a normal variety of dimension n that has at most Kawamata log terminal singularities, let $Z \subseteq X$ be an irreducible subvariety, let L be some big line bundle on X, and let $M(L)$ be the set consisting of all positive integers m such that $h^0(X, \mathcal{O}_X(mL)) \neq 0$. The δ-invariant $\delta_Z(X; L)$ along Z is defined by

$$\delta_Z(X; L) = \inf_{\substack{E/X \\ Z \subseteq C_X(E)}} \frac{A_X(E)}{S_L(E)},$$

where the infimum runs over all prime divisors E over the variety X such that $Z \subseteq C_X(E)$. In the case when X is a Fano variety and $L = -K_X$, we let

$$\delta_Z(X) = \delta_Z(X; L).$$

In this section, we explain how to estimate $\delta_Z(X; L)$ using the technique developed in [2].

Let Y be a Cartier prime divisor in X such that $Z \subset Y$, and Y is not contained in the support of the negative part of the σ-decomposition of L, see [166, Definition III.1.12]. The latter condition always holds if L is nef. Then [2, Theorem 3.2] implies the following.

Theorem 1.95 *Let $\delta_Z(Y; W^Y_{\bullet,\bullet})$ be the number defined in (1.12). Then*

$$\delta_Z(X, L) \geqslant \min\left\{\frac{1}{S_L(Y)}, \delta_Z\left(Y; W^Y_{\bullet,\bullet}\right)\right\}.$$

To define $\delta_Z(Y; W^Y_{\bullet,\bullet})$, let us present notations from [2] that will be used throughout this section and occasionally in other sections of this book. Let $L_Y = L|_Y$ and $M = -Y|_Y$. For every $m \in \mathbb{Z}_{\geqslant 0}$, we let $V_m = H^0(X, mL)$. Put

$$V_\bullet^X = \bigoplus_{m \geqslant 0} V_m.$$

The refinement of $V_\bullet^X$ by the prime divisor Y is the $\mathbb{Z}_{\geqslant 0}^2$-graded linear series

$$W_{\bullet,\bullet}^Y = \bigoplus_{m,j \geqslant 0} W_{m,j}^Y$$

such that

$$W_{m,j}^Y = \mathrm{Im}\Big(H^0\big(X, mL - jY\big) \to H^0\big(Y, mL_Y + jM\big)\Big)$$

where $\to$ is the restriction map. Observe that $W_{m,j}^Y \subseteq H^0(Y, mL_Y + jM)$ for all m and j. Moreover, the refinement $W_{\bullet,\bullet}^Y$ satisfies the following two conditions:

(i) there exists $\tau \in \mathbb{R}_{\geqslant 0}$ such that $W_{m,j}^Y = 0$ whenever $j/m > \tau$;

(ii) there is (m_0, j_0) and a decomposition

$$m_0 L_Y + j_0 M \sim_{\mathbb{Q}} A + E,$$

where A is an ample $\mathbb{Q}$-Cartier divisor on Y, and E is an effective $\mathbb{Q}$-divisor on Y, such that $H^0(Y, mA) \subseteq W_{mm_0, mj_0}^Y$ for all sufficiently divisible $m \in \mathbb{Z}_{>0}$.

In the language of [2, Definition 2.9], these conditions mean that $W_{\bullet,\bullet}^Y$ has bounded support, and $W_{\bullet,\bullet}^Y$ contains an ample linear series. Recall from [2, Definition 2.9] that

$$\mathrm{vol}\big(V_\bullet^X\big) = \mathrm{vol}(L) = \lim_{m \to \infty} \frac{\dim(V_m)}{m^n/n!}$$

and

$$\mathrm{vol}\big(W_{\bullet,\bullet}^Y\big) = \lim_{m \to \infty} \frac{\sum_{j \geqslant 0} \dim\big(W_{m,j}^Y\big)}{m^n/n!}.$$

Similarly, one can define volumes of any $\mathbb{Z}_{\geqslant 0}$-graded linear series and $\mathbb{Z}_{\geqslant 0}^2$-graded linear series with bounded support (see [2, Definition 2.9] for details).

Lemma 1.96 $\mathrm{vol}(W_{\bullet,\bullet}^Y) = \mathrm{vol}(V_\bullet^X)$.

Proof For all non-negative integers m and j, we have an isomorphism of vector spaces

$$W_{m,j}^Y \cong \frac{V_m \cap H^0(X, mL - jY)}{V_m \cap H^0(X, mL - (j+1)Y)},$$

so that $\sum_{j \geqslant 0} \dim(W_{m,j}^Y) = \dim(V_m)$ and the equality $\mathrm{vol}(W_{\bullet,\bullet}^Y) = \mathrm{vol}(V_\bullet^X)$ follows. $\qquad\square$

For every prime divisor F over Y with $Z \subseteq C_Y(F)$, we let

$$S(W^Y_{\bullet,\bullet}; F) = \frac{1}{\operatorname{vol}(W^Y_{\bullet,\bullet})} \int_0^\infty \operatorname{vol}(\mathcal{F}_F^t W^Y_{\bullet,\bullet}) \, dt, \qquad (1.11)$$

where for any $t \in \mathbb{R}_{\geqslant 0}$ we define the $\mathbb{Z}^2_{\geqslant 0}$-graded linear series $\mathcal{F}_F^t W^Y_{\bullet,\bullet}$ on Y by

$$\mathcal{F}_F^t W^Y_{\bullet,\bullet} = \bigoplus_{m,j \geqslant 0} \mathcal{F}_F^{mt} W^Y_{m,j}$$

with

$$\mathcal{F}_F^{mt} W^Y_{m,j} = \left\{ s \in W^Y_{m,j} \mid \operatorname{ord}_F(s) \geqslant mt \right\}.$$

Now, following [2, Lemma 2.9] and [2, Corollary 2.22], we are ready to define

$$\delta_Z(Y; W^Y_{\bullet,\bullet}) = \inf_{\substack{F/Y, \\ Z \subseteq C_Y(F)}} \frac{A_Y(F)}{S(W^Y_{\bullet,\bullet}; F)}, \qquad (1.12)$$

where the infimum is taken over prime divisors F over the variety Y with $Z \subseteq C_Y(F)$.

Remark 1.97   One can generalize $S(W^Y_{\bullet,\bullet}; F)$ and $\delta_Z(Y; W^Y_{\bullet,\bullet})$ for any $\mathbb{Z}^2_{\geqslant 0}$-graded linear series with bounded support that contains an ample linear series (see [2] for details).

Remark 1.98   Let D be a $\mathbb{Q}$-Cartier divisor on Y, let $g: \widehat{Y} \to Y$ be a birational morphism, let F be a prime divisor in $\widehat{Y}$, and let u be a real number. To make the exposition simpler, we will abuse notations and write $D - uF$ for the divisor $g^*(D) - uF$ on the variety $\widehat{Y}$. In particular, $\operatorname{vol}(D - uF)$ would mean the volume $\operatorname{vol}(g^*(D) - uF)$.

As in Section 1.2, we let

$$\tau = \sup\left\{ v \in \mathbb{R}_{>0} \mid L - vY \text{ is pseudo-effective} \right\}.$$

Similarly, for a prime divisor F over the variety Y with $Z \subseteq C_Y(F)$, we let

$$\tau' = \sup\left\{ v \in \mathbb{R}_{>0} \mid L_Y + uM - vF \text{ is pseudo-effective for some } u \in [0, \tau] \right\}.$$

We let

$$S_m(W^Y_{\bullet,\bullet}; F) = \frac{1}{mN^{W^Y_{\bullet,\bullet}}_m} \sum_{j=0}^{m\tau} \sum_{k \geqslant 0} \dim(\mathcal{F}_F^k W^Y_{m,j}),$$

where $N^{W^Y_{\bullet,\bullet}}_m = \sum_{j \geqslant 0} \dim(W^Y_{m,j})$. Note that $\mathcal{F}_F^k W^Y_{m,j} = 0$ for $k > m\tau'$.

Lemma 1.99 ([2, Lemma 2.9])   *It holds that*

$$S(W^Y_{\bullet,\bullet}; F) = \lim_{m \to \infty} S_m(W^Y_{\bullet,\bullet}; F).$$

Proof Let $h(t) = \text{vol}(\mathcal{F}_F^t W_{\bullet,\bullet}^Y)$. Then

$$S(W_{\bullet,\bullet}^Y; F) = \frac{1}{\text{vol}(W_{\bullet,\bullet}^Y)} \int_0^{\tau'} h(t)dt.$$

Let

$$h_m(t) = \frac{n!}{m^n} \sum_{j=0}^{m\tau} \dim(\mathcal{F}^{\lceil mt \rceil} W_{m,j}^Y).$$

Then $h(t) = \lim_{m \to \infty} h_m(t) = f(t)$. On the other hand, we have

$$\lim_{m \to \infty} S_m(W_{\bullet,\bullet}^Y; F) = \frac{1}{\text{vol}(W_{\bullet,\bullet}^Y)} \lim_{m \to \infty} \frac{n!}{m^{n+1}} \sum_{j=0}^{m\tau} \sum_{k=0}^{m\tau'} \dim(\mathcal{F}_F^k W_{m,j}^Y),$$

where

$$\frac{n!}{m^{n+1}} \sum_{j=0}^{m\tau} \sum_{k=0}^{m\tau'} \dim(\mathcal{F}_F^k W_{m,j}^Y) = \frac{1}{m} \sum_{k=0}^{m\tau'} f_m(k/m).$$

Thus, we have

$$\lim_{m \to \infty} \frac{1}{m} \sum_{k=0}^{m\tau'} f_m(k/m) = \int_0^{\tau'} f(t)dt,$$

which implies the required equality. $\qquad\square$

Remark 1.100 Lemma 1.99 holds for any $\mathbb{Z}_{\geqslant 0}^2$-graded linear series with bounded support that contains an ample linear series, where $S_m(W_{\bullet,\bullet}^Y; F)$ and $S(W_{\bullet,\bullet}^Y; F)$ should be replaced by their counterparts. We state this lemma for $W_{\bullet,\bullet}^Y$ to simplify the exposition.

In this book, we will often use Theorem 1.95, which is a corollary of [2, Theorem 3.2]. Occasionally, we will use another (similar but more technical) corollary of this theorem. To state it, suppose (temporarily) that there is a birational morphism $\pi \colon \widehat{X} \to X$ such that

- the π-exceptional locus consists of a single prime divisor E_Z such that $\pi(E_Z) = Z$,
- the divisor $-E_Z$ is $\mathbb{Q}$-Cartier and is π-ample,
- the log pair $(\widehat{X}, E_Z)$ has purely log terminal singularities [130].

The birational map π is known as a *plt blow up* of the subvariety Z. Write

$$K_{E_Z} + \Delta_{E_Z} = (K_{\widehat{X}} + E_Z)\big|_{E_Z},$$

where Δ_{E_Z} is an effective $\mathbb{Q}$-divisor on E_Z known as the different of the log pair $(\widehat{X}, E_Z)$. Note that the log pair (E_Z, Δ_{E_Z}) has at most Kawamata log terminal singularities, and the divisor $-(K_{E_Z} + \Delta_{E_Z})$ is $\pi|_{E_Z}$-ample. Similar to the refinement $W^Y_{\bullet,\bullet}$, we can define the refinement $W^{E_Z}_{\bullet,\bullet}$ of the linear series $V^X_\bullet$ by the prime divisor E_Z. Namely, it is enough to replace $W^Y_{m,j}$ in the definition of $W^Y_{\bullet,\bullet}$ by

$$W^{E_Z}_{m,j} = \mathrm{Im}\left(H^0\left(\widehat{X}, m\pi^*(L) - jE_Z\right) \to H^0\left(E_Z, m\pi^*(L)\big|_{E_Z} - jE_Z\big|_{E_Z}\right)\right).$$

Then, for every prime divisor F over E_Z, we can define $S(W^{E_Z}_{\bullet,\bullet}; F)$ similar to (1.11). The following result is a special (but slightly different) case of [2, Theorem 3.2].

Theorem 1.101 *Let $\widehat{Z}$ be an irreducible subvariety in E_Z, and let*

$$\delta_{\widehat{Z}}(X, L) = \inf_{\substack{E/X, \\ \widehat{Z} \subseteq C_Y(E)}} \frac{A_X(E)}{S_L(E)},$$

where the infimum is taken over prime divisors E over X such that $\widehat{Z} \subseteq C_{\widehat{X}}(E)$. Then

$$\delta_{\widehat{Z}}(X, L) \geqslant \min\left\{\frac{A_X(E_Z)}{S_L(E_Z)}, \delta_{\widehat{Z}}\left(E_Z, \Delta_{E_Z}; W^{E_Z}_{\bullet,\bullet}\right)\right\}, \qquad (1.13)$$

where

$$\delta_{\widehat{Z}}\left(E_Z, \Delta_{E_Z}; W^{E_Z}_{\bullet,\bullet}\right) = \inf_{\substack{F/E_Z, \\ \widehat{Z} \subseteq C_{E_Z}(F)}} \frac{A_{E_Z, \Delta_{E_Z}}(F)}{S\left(W^{E_Z}_{\bullet,\bullet}; F\right)},$$

where the infimum is taken over all prime divisors F over E_Z such that $\widehat{Z} \subseteq C_{E_Z}(F)$. Moreover, if the inequality (1.13) is an equality and there exists a prime divisor E over the variety X such that $\widehat{Z} \subseteq C_{\widehat{X}}(E) \subseteq E_Z$ and $\delta_{\widehat{Z}}(X, L) = \frac{A_X(E)}{S_L(E)}$, then $\delta_{\widehat{Z}}(X, L) = \frac{A_X(E_Z)}{S_L(E_Z)}$.

Proof The required assertion follows from the proof of [2, Theorem 3.2]. For the reader's convenience, we present its proof here. Let $\mathcal{F}$ be a filtration on $V^X_\bullet$. For $m \in M(L)$, let

$$\delta_{\widehat{Z}, m}\left(V^X_\bullet, \mathcal{F}\right) =$$
$$\sup\left\{\lambda \in \mathbb{Q} \;\middle|\; \begin{array}{l} \left(\widehat{X}, (1 - A_X(E_Z))E_Z + \lambda\pi^*(D)\right) \text{ is log canonical} \\ \text{at a general point of the variety } \widehat{Z} \text{ for any } m\text{-basis type} \\ \mathbb{Q}\text{-divisor } D \sim_{\mathbb{Q}} L \text{ that is compatible with } \mathcal{F} \end{array}\right\}.$$

See [2, Definition 1.5] for the definition of compatibility. We have

$$\delta_{\widehat{Z}, m}\left(V^X_\bullet, \mathcal{F}\right) = \inf_D \inf_{\substack{E/X, \\ \widehat{Z} \subseteq C_{\widehat{X}}(E)}} \frac{A_{\widehat{X}, (1 - A_X(E_Z))E_Z}(E)}{\mathrm{ord}_E(D)},$$

where the first infimum is taken over all m-basis type divisors of the line bundle L that are compatible with $\mathcal{F}$, and the second infimum is taken over prime divisors E over X such that $\widehat{Z} \subseteq C_{\widehat{X}}(E)$. Swapping these infima and using [2, Proposition 3.1], we get

$$\delta_{\widehat{Z},m}\big(V_{\bullet}^{X}, \mathcal{F}\big) = \inf_{\substack{E/X, \\ \widehat{Z} \subseteq C_{\widehat{X}}(E)}} \frac{A_{\widehat{X},(1-A_X(E_Z))E_Z}(E)}{\sup_D \operatorname{ord}_E(D)} = \inf_{\substack{E/X, \\ \widehat{Z} \subseteq C_{\widehat{X}}(E)}} \frac{A_X(E)}{S_m\big(V_{\bullet}^{X}; E\big)},$$

so that $\delta_{\widehat{Z},m}(V_{\bullet}^{X}, \mathcal{F})$ does not depend on the choice of the filtration $\mathcal{F}$. We set

$$\delta_{\widehat{Z}}\big(V_{\bullet}^{X}, \mathcal{F}\big) = \limsup_{m \in M(L)} \delta_{\widehat{Z},m}\big(V_{\bullet}^{X}, \mathcal{F}\big).$$

Then $\delta_{\widehat{Z}}(V_{\bullet}^{X}, \mathcal{F}) \leqslant \delta_{\widehat{Z}}(X, L)$. Moreover, it follows from the proof of [2, Lemma 2.9] that for every $\epsilon > 0$ there exists a positive integer $m_0(\epsilon)$ such that

$$S_m\big(V_{\bullet}^{X}; E'\big) \leqslant (1 + \epsilon)S\big(V_{\bullet}^{X}; E'\big)$$

for every prime divisor E' over X and for every $m \in M(L)$ with $m > m_0(\epsilon)$. Thus, we get

$$\inf_{\substack{E/X, \\ \widehat{Z} \subseteq C_{\widehat{X}}(E)}} \frac{A_X(E)}{S\big(V_{\bullet}^{X}; E\big)} \leqslant (1 + \epsilon) \limsup_{m \in M(L)} \inf_{\substack{E/X, \\ \widehat{Z} \subseteq C_{\widehat{X}}(E)}} \frac{A_X(E)}{S_m\big(V_{\bullet}^{X}; E\big)} = (1 + \epsilon)\delta_{\widehat{Z}}\big(V_{\bullet}^{X}, \mathcal{F}\big).$$

Therefore, we conclude that $\delta_{\widehat{Z}}(V_{\bullet}^{X}, \mathcal{F}) = \delta_{\widehat{Z}}(X, L)$.

Let $M(W_{\bullet,\bullet}^{E_Z})$ be the set consisting of all positive integers m such that $W_{m,j}^{E_Z} \neq 0$, and let θ be the right hand side of (1.14). For every $m \in M(L) \cap M(W_{\bullet,\bullet}^{E_Z})$, we set

$$\theta_m = \min\left\{\frac{A_X(E_Z)}{S_m\big(V_{\bullet}^{X}; E_Z\big)}, \delta_{\widehat{Z},m}\Big(E_Z, \Delta_Z; W_{\bullet,\bullet}^{E_Z}\Big)\right\},$$

where $\delta_{\widehat{Z},m}(E_Z, \Delta_Z; W_{\bullet,\bullet}^{E_Z})$ is defined similar to $\delta_{\widehat{Z},m}(V_{\bullet}^{X}, \mathcal{F})$. Then $\theta_m \to \theta$ as $m \to \infty$.

Now, let us show that $\delta_{\widehat{Z},m}(V_{\bullet}^{X}, \mathcal{F}) \geqslant \theta_m$. Since $\delta_{\widehat{Z},m}(V_{\bullet}^{X}, \mathcal{F})$ does not depend on the choice of the filtration $\mathcal{F}$, we may assume that $\mathcal{F}$ is the filtration induced by E_Z. Arguing as in [2, §3.1], we see that for every m-basis type divisor D of $V_{\bullet}^{X}$ compatible with $\mathcal{F}$,

$$\pi^*(D) = S_m\big(V_{\bullet}^{X}; E_Z\big)E_Z + \Gamma,$$

where Γ is an effective $\mathbb{Q}$-divisor such that $E_Z \not\subset \operatorname{Supp}(\Gamma)$, and $\Gamma|_{E_Z}$ is an m-basis type divisor of $W_{\bullet,\bullet}^{E_Z}$. Note that

$$\pi^*\big(K_X + \theta_m D\big) = K_{\widehat{X}} + a_m E_Z + \theta_m \Gamma$$

with $a_m = 1 - A_X(E_Z) + \theta_m S_m(V_\bullet^X; E_Z) \leqslant 1$. Since $(E_Z, \Delta_{E_Z} + \theta_m \Gamma|_{E_Z})$ is log canonical in a neighborhood of the subvariety $\widehat{Z}$, we see that $(\widehat{X}, E_Z + \theta_m \Gamma)$ is also log canonical in a neighborhood of the subvariety $\widehat{Z}$ by Theorem A.15, so that $(\widehat{X}, a_m E_Z + \theta_m \Gamma)$ is log canonical in a neighborhood of $\widehat{Z}$ as well. This shows that $\delta_{\widehat{Z},m}(V_\bullet^X, \mathcal{F}) \geqslant \theta_m$.

Moreover, since $(\widehat{X}, E_Z + \theta_m \Gamma)$ is log canonical in a neighborhood of the subvariety $\widehat{Z}$, for every prime divisor E over X such that $\widehat{Z} \subseteq C_{\widehat{X}}(E)$, we have

$$A_X(E) \geqslant \theta_m \mathrm{ord}_E(D) + (1 - a_m)\mathrm{ord}_E(E_Z)$$

for every m-basis type divisor of $V_\bullet^X$ that is compatible with $\mathcal{F}$. This gives

$$A_X(E) \geqslant \theta_m S_m(V_\bullet^X; E) + \left(A_X(E_Z) - \theta_m S(V_\bullet^X; E_Z)\right)\mathrm{ord}_E(E_Z).$$

Hence, taking the limit when $m \to \infty$, we get

$$A_X(E) \geqslant \theta S(V_\bullet^X; E) + \left(A_X(E_Z) - \theta S(V_\bullet^X; E_Z)\right)\mathrm{ord}_E(E_Z), \qquad (1.14)$$

where $A_X(E_Z) - \theta S(V_\bullet^X; E_Z) \geqslant 0$. This proves (1.14).

Finally, if there exists a prime divisor E over the variety X such that $\widehat{Z} \subseteq C_{\widehat{X}}(E) \subseteq E_Z$ and $A_X(E) = \theta S(V_\bullet^X; E)$, then $A_X(E_Z) = \theta S(V_\bullet^X; E_Z)$ by (1.14) since $\mathrm{ord}_E(E_Z) > 0$. This completes the proof of Theorem 1.101.   □

Theorem 1.101 implies the following corollary of [2, Theorem 3.2].

Corollary 1.102

$$\delta_Z(X, L) \geqslant \min\left\{\frac{A_X(E_Z)}{S_L(E_Z)}, \inf_{\widehat{Z} \subset E_Z} \delta_{\widehat{Z}}\left(E_Z, \Delta_{E_Z}; W_{\bullet,\bullet}^{E_Z}\right)\right\},$$

where the infimum is taken over all irreducible subvarieties $\widehat{Z} \subset E_Z$ such that $\pi(\widehat{Z}) = Z$, and $\delta_{\widehat{Z}}(E_Z, \Delta_{E_Z}; W_{\bullet,\bullet}^{E_Z})$ is defined in Theorem 1.101.

Now, we give a simple formula for $S(W_{\bullet,\bullet}^Y; F)$ when X is a Mori Dream Space [127, 173]. This formula is especially simple when X is a Mori Dream Space with $\mathrm{Nef}(X) = \overline{\mathrm{Mov}}(X)$. In this book, we will mostly apply this formula in the following situation:

- X is a smooth Fano threefold,
- $L = -K_X$,
- Y is a smooth (explicitly described) surface in X,
- Z is an irreducible curve in Y, which is often also explicitly described.

Our formula is given in Theorem 1.106. Before presenting it, we consider one inspirational example, which is redone in Section 4.4 using Theorem 1.106.

Example 1.103 (cf. Lemma 4.41) Suppose that X is a smooth Fano threefold
№2.15. Then there exists a blow up $\pi\colon X \to \mathbb{P}^3$ of a smooth curve $\mathscr{C}$ of degree
6 and genus 4. Observe that $\mathscr{C}$ is contained in a unique quadric surface in $\mathbb{P}^3$,
which we denote by S_2. Suppose that the quadric S_2 is smooth. Then $\mathscr{C}$ is a
curve of degree $(3,3)$ in $S_2 \cong \mathbb{P}^1 \times \mathbb{P}^1$. Let E be the π-exceptional divisor, let
Q be the proper transform on X of the quadric S_2, let H be a hyperplane in $\mathbb{P}^3$,
and let $C = E \cap Q$. Then $\overline{\mathrm{Eff}}(X) = \mathbb{R}_{\geqslant 0}[E] + \mathbb{R}_{\geqslant 0}[Q]$ and

$$\mathrm{Nef}(X) = \overline{\mathrm{Mov}}(X) = \mathbb{R}_{\geqslant 0}\big[\pi^*(H)\big] + \mathbb{R}_{\geqslant 0}\big[3\pi^*(H) - E\big].$$

We suppose that $L = -K_X$, $Y = Q$ and Z is an irreducible curve in Q. Then
$L^3 = 22$. We claim that $S_X(Q) = \frac{37}{44}$. Indeed, take $u \in \mathbb{R}_{\geqslant 0}$ and observe that

$$-K_X - uQ \sim_{\mathbb{Q}} (4 - 2u)\pi^*(H) - (1 - u)E.$$

Let $P(u)$ be the positive (nef) part of the Zariski decomposition of the divisor
$-K_X - xQ$, and let $N(u)$ be its negative part. Then

$$P(u) = \begin{cases} -K_X - uQ & \text{if } 0 \leqslant u \leqslant 1, \\ (4 - 2u)\pi^*(H) & \text{if } 1 \leqslant u \leqslant 2, \end{cases}$$

and

$$N(u) = \begin{cases} 0 & \text{if } 0 \leqslant u \leqslant 1, \\ (u - 1)E & \text{if } 1 \leqslant u \leqslant 2. \end{cases}$$

Note that $-K_X - uQ$ is not pseudo-effective for $u > 2$, so that $\tau = 2$. Then

$$S_X(Q) = \frac{1}{22} \int_0^1 \big(-K_X - uQ\big)^3 du + \frac{1}{22} \int_1^2 (4 - 2u)^3 du = \frac{37}{44}.$$

Let us show that $S(W_{\bullet,\bullet}^Q; Z) < \frac{37}{44}$. Let M be a divisor on Q of degree $(1,1)$.
Then

$$W_{m,j}^Q = \begin{cases} H^0\big(Q, (m + j)M\big) & \text{if } 0 \leqslant j \leqslant m, \\ (j - m)C + H^0\big(Q, (4m - 2j)M\big) & \text{if } m < j \leqslant 2m, \\ 0 & \text{otherwise.} \end{cases}$$

This follows from Kawamata–Viehweg vanishing or Theorem A.3. Then

$$\mathrm{vol}\big(W_{\bullet,\bullet}^Q\big) = \lim_{m \to \infty} \frac{\sum_{j \geqslant 0} \dim(W_{m,j})}{m^3/3!}$$

$$= \lim_{m \to \infty} \left(\sum_{j=0}^{m} (m + j + 1)^2 + \sum_{j=m+1}^{2m} (4m - 2j + 1)^2 \right)$$

$$= 3!\left(\int_0^1 (1 + x)^2 dx + \int_1^2 (4 - 2x)^2 dx \right) = 22.$$

Meanwhile we have

$$S(W_{\bullet,\bullet}^Q; Z) = \frac{1}{\text{vol}(W_{\bullet,\bullet}^Q)} \int_0^\infty \text{vol}(\mathcal{F}_Z^t W_{\bullet,\bullet}^Z) dt.$$

First, let us compute $S(W_{\bullet,\bullet}^Q; Z)$ in the case when $Z = C$. If $0 \leqslant j \leqslant m$, then

$$\mathcal{F}_C^{mt} W_{m,j}^Q = \begin{cases} \lceil mt \rceil C + H^0((m+j)M - \lceil mt \rceil C) & \text{if } m + j \geqslant 3mt, \\ 0 & \text{otherwise.} \end{cases}$$

Similarly, if $m < j \leqslant 2m$, then

$$\mathcal{F}_C^{mt} W_{m,j}^Q = \begin{cases} (j-m)C + H^0((4m-2j)M) & \text{if } j - m \geqslant mt, \\ \lceil mt \rceil C + H^0((m+j)M - \lceil mt \rceil C) & \text{if } j - m < mt, m + j \geqslant 3mt, \\ 0 & \text{otherwise.} \end{cases}$$

We now summarize this as follows. If $0 \leqslant t < \frac{1}{3}$, we have

$$\mathcal{F}_C^{mt} W_{m,j}^Q = \begin{cases} \lceil mt \rceil C + H^0((m+j)M - \lceil mt \rceil C) & \text{if } 0 \leqslant j < m(t+1), \\ (j-m)C + H^0((4m-2j)M) & \text{if } m(t+1) \leqslant j \leqslant 2m, \\ 0 & \text{otherwise.} \end{cases}$$

Similarly, if $\frac{1}{3} \leqslant t \leqslant 1$, then

$$\mathcal{F}_C^{mt} W_{m,j}^Q = \begin{cases} 0 & \text{if } 0 \leqslant j < m(3t-1), \\ \lceil mt \rceil C + H^0((m+j)M - \lceil mt \rceil C) & \text{if } m(3t-1) \leqslant j < m(t+1), \\ (j-m)C + H^0((4m-2j)M) & \text{if } m(t+1) \leqslant j \leqslant 2m, \\ 0 & \text{otherwise.} \end{cases}$$

Finally, if $t > 1$, then $\mathcal{F}_C^{mt} W_{m,j}^Q = 0$ for all $j, m \in \mathbb{Z}_{\geqslant 0}^2$. Thus, if $0 \leqslant t < \frac{1}{3}$, then

$$\text{vol}(\mathcal{F}_C^t W_{\bullet,\bullet}^Q) = 3! \left(\int_0^{t+1} (1 - 3t + x)^2 dx + \int_{t+1}^2 (4 - 2x)^2 dx \right)$$
$$= 2(15t^3 + 9t^2 - 27t + 11).$$

Similarly, if $\frac{1}{3} \leqslant t \leqslant 1$, then

$$\text{vol}(\mathcal{F}_C^t W_{\bullet,\bullet}^Q) = 3! \left(\int_{3t-1}^{t+1} (1 - 3t + x)^2 dx + \int_{t+1}^2 (4 - 2x)^2 dx \right) = 24(1 - t)^3.$$

Hence we have

$$S(W_{\bullet,\bullet}^Q; C) = \frac{1}{22} \int_0^{\frac{1}{3}} 2(15t^3 + 9t^2 - 27t + 11)dt + \frac{1}{22} \int_{\frac{1}{3}}^1 24(1-t)^3 dt$$

$$= \frac{35}{132} < \frac{37}{44}.$$

Now we consider the case when $Z \neq C$. We may assume that Z is a curve on $Q = \mathbb{P}^1 \times \mathbb{P}^1$ of degree (b_1, b_2) with $0 \leqslant b_1 \leqslant b_2 \neq 0$. If $0 \leqslant j \leqslant m$, then

$$\mathcal{F}_Z^{mt} W_{m,j}^Q = \begin{cases} \lceil mt \rceil Z + H^0((m+j)M - \lceil mt \rceil Z) & \text{if } mt \leqslant \frac{m+j}{b_2}, \\ 0 & \text{otherwise.} \end{cases}$$

Similarly, if $m < j \leqslant 2m$,

$$\mathcal{F}_Z^{mt} W_{m,j}^Q = \begin{cases} (j-m)C + \lceil mt \rceil Z + H^0((4m-2j)M - \lceil mt \rceil Z) & \text{if } mt \leqslant \frac{4m-2j}{b_2}, \\ 0 & \text{otherwise.} \end{cases}$$

We summarize this as follows. If $0 \leqslant t < \frac{1}{b_2}$, then

$$\mathcal{F}_Z^{mt} W_{m,j}^Q =$$
$$\begin{cases} \lceil mt \rceil Z + H^0((m+j)M - \lceil mt \rceil Z) & \text{if } 0 \leqslant j \leqslant m, \\ (j-m)C + \lceil mt \rceil Z + H^0((4m-2j)M - \lceil mt \rceil Z) & \text{if } m < j \leqslant m(2 - \frac{1}{2}b_2 t), \\ 0 & \text{otherwise.} \end{cases}$$

Similarly, if $\frac{1}{b_2} \leqslant t \leqslant \frac{2}{b_2}$, then

$$\mathcal{F}_Z^{mt} W_{m,j}^Q =$$
$$\begin{cases} 0 & \text{if } 0 \leqslant j < m(b_2 t - 1), \\ \lceil mt \rceil Z + H^0((m+j)M - \lceil mt \rceil Z) & \text{if } m(b_2 t - 1) \leqslant j \leqslant m, \\ (j-m)C + \lceil mt \rceil Z + H^0((4m-2j)M - \lceil mt \rceil Z) & \text{if } m < j \leqslant m(2 - \frac{1}{2}b_2 t), \\ 0 & \text{otherwise.} \end{cases}$$

Finally, if $t > \frac{2}{b_2}$, then $\mathcal{F}_Z^{mt} W_{m,j}^Q = 0$ for all j and m. Thus, if $0 \leqslant t < \frac{1}{b_2}$, then

$$\text{vol}(\mathcal{F}_Z^t W_{\bullet,\bullet}^Q) = 3! \left(\int_0^1 (1 - b_1 t + x)(1 - b_2 t + x)dx \right.$$
$$\left. + \int_1^{2-\frac{1}{2}b_2 t} (4 - b_1 t - 2x)(4 - b_2 t - 2x)dx \right)$$
$$= \frac{1}{2}(44 - 30b_1 t - 30b_2 t + 24b_1 b_2 t^2 - 3b_1 b_2^2 t^3 + b_2^3 t^3).$$

Likewise, if $\frac{1}{b_2} \leqslant t < \frac{2}{b_2}$, then

$$
\begin{aligned}
\mathrm{vol}\big(\mathcal{F}_Z^t W_{\bullet,\bullet}^{\mathcal{Q}}\big) = 3! \bigg(& \int_{b_2 t - 1}^{1} (1 - b_1 t + x)(1 - b_2 t + x)\,dx \\
& + \int_{1}^{2 - \frac{1}{2} b_2 t} (4 - b_1 t - 2x)(4 - b_2 t - 2x)\,dx \bigg) \\
= & \frac{3}{2}(4 - 3b_1 t + b_2 t)(b_2 t - 2)^2.
\end{aligned}
$$

Hence, if $Z \neq C$, then

$$
\begin{aligned}
S(W_{\bullet,\bullet}^{\mathcal{Q}}, Z) &= \frac{1}{22} \int_{0}^{\frac{1}{b_2}} \mathrm{vol}\big(\mathcal{F}_Z^t W_{\bullet,\bullet}^{\mathcal{Q}}\big)\,dt + \frac{1}{22} \int_{\frac{1}{b_2}}^{\frac{2}{b_2}} \mathrm{vol}\big(\mathcal{F}_Z^t W_{\bullet,\bullet}^{\mathcal{Q}}\big)\,dt \\
&= \frac{23}{88}\left(\frac{3b_2 - b_1}{b_2^2}\right) \leqslant \frac{69}{88} < \frac{37}{44}
\end{aligned}
$$

because $\frac{3b_2 - b_1}{b_2^2} \leqslant 3$. Therefore, we see that $S(W_{\bullet,\bullet}^{\mathcal{Q}}; Z) < \frac{37}{44}$, so that

$$
\delta_Z\big(\mathcal{Q}; W_{\bullet,\bullet}^{\mathcal{Q}}\big) = \frac{A_{\mathcal{Q}}(Z)}{S(W_{\bullet,\bullet}^{\mathcal{Q}}; Z)} = \frac{1}{S(W_{\bullet,\bullet}^{\mathcal{Q}}; Z)} > \frac{44}{37}.
$$

Thus it follows from Theorem 1.95 that $\delta_Z(X) \geqslant \frac{44}{37}$.

From now until the end of this section, we assume that X is $\mathbb{Q}$-factorial, and that X is a Mori Dream Space. Consider the diagram

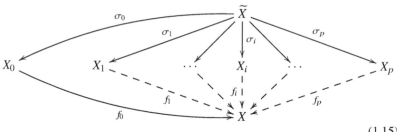

$$(1.15)$$

where $X_0 = X$ and $f_0 = \mathrm{Id}_X$, every X_i is a $\mathbb{Q}$-factorial variety, every f_i is a small birational modification, $\widetilde{X}$ is some smooth variety, and every σ_i is a birational morphism. Let $\sigma = \sigma_0$. By [173, Proposition 2.13], we may assume that the following holds:

($\bigstar$) For any pseudo-effective $\mathbb{R}$-divisor D on X, there is an $i \in \{0, \ldots, p\}$ such that

$$
f_i^*(D) \sim_{\mathbb{R}} P_i(D) + N_i(D),
$$

where $P_i(D)$ is a semiample divisor on X_i, and $N_i(D)$ is an effective $\mathbb{R}$-divisor whose support consists of exceptional divisors of the birational morphism $X_i \to Y_i$ that corresponds to $P_i(D)$. This is a Zariski decomposition of the divisor D in the sense of [173, Definition 2.11], which also gives a Zariski decomposition

$$\sigma^*(D) \sim_{\mathbb{R}} \sigma_i^*\big(P_i(D)\big) + N\big(\sigma_i^*(D)\big) \tag{1.16}$$

with $(\sigma_i)_*(N(\sigma^*(D))) = N_i(D)$ and $\sigma_i^*(P_i(D))$ being semiample on $\widetilde{X}$. Moreover, if mD is a $\mathbb{Z}$-divisor for a positive integer m, then

$$H^0\big(X, mD\big) = H^0\big(X_i, f_i^*(mD)\big) = H^0\Big(X_i, \big\lfloor mP_i(D) \big\rfloor\Big)$$
$$= H^0\Big(\widetilde{X}, \big\lfloor m\sigma_i^*(P_i(D)) \big\rfloor\Big).$$

Remark 1.104   The decomposition (1.16) is Nakayama's σ-decomposition from [166]. Indeed, for a pseudo-effective $\mathbb{R}$-divisor D on X, Nakayama's σ-decomposition is given by

$$D \sim_{\mathbb{R}} N_\sigma(D) + P_\sigma(D)$$

for $P_\sigma(D) = D - N_\sigma(D)$ and

$$N_\sigma(D) = \sum_E \sigma_E(D)E,$$

where the sum runs over all prime divisors E in X, and $\sigma_E(D)$ is defined as

$$\sigma_E(D) = \inf\left\{ \mathrm{ord}_E\left(D'\right) \,\middle|\, \begin{array}{l} D' \text{ is a pseudo-effective } \mathbb{R}\text{-divisor} \\ \text{on } X \text{ such that } D' \sim_{\mathbb{R}} D \end{array} \right\}$$

in the case when D is big [166, Definition III.1.1]. If D is not big but pseudo-effective, the value $\sigma_E(D)$ is the limit of $\sigma_E(D + \varepsilon A)$ as $\varepsilon \searrow 0$ for some ample divisor $A \in \mathrm{Pic}(X)$. Note that $N_\sigma(xD) = xN_\sigma(D)$ for all $x \in \mathbb{R}_{\geqslant 0}$. Similarly, we have

$$N_\sigma\big(D_1 + D_2\big) \leqslant N_\sigma\big(D_1\big) + N_\sigma\big(D_2\big)$$

for any pseudo-effective divisors D_1 and D_2 on the variety X. If the divisor $P_\sigma(D)$ is nef, then the σ-decomposition is called *Zariski decomposition* [166, Definition III.1.12].

Let $\widetilde{Y}$ be the proper transform of the divisor Y on the variety $\widetilde{X}$. For every $u \in [0, \tau]$, consider the Zariski decomposition of $\sigma^*(L - uY)$ described above:

$$\sigma^*(L - uY) \sim_{\mathbb{R}} \widetilde{P}(u) + \widetilde{N}(u),$$

where $\widetilde{P}(u)$ is a semiample $\mathbb{R}$-divisor (the positive part), and $\widetilde{N}(u)$ is its negative part. We also consider the Nakayama–Zariski decomposition

$$L - uY \sim_{\mathbb{R}} P(u) + N(u),$$

where $P(u) = P_\sigma(L - uY)$ and $N(u) = N_\sigma(L - uY)$ as described earlier in Remark 1.104. Recall that Y is not contained in the support of the divisor $N_\sigma(L)$ by assumption, so that the divisor $\widetilde{Y}$ is not contained in $\mathrm{Supp}(\widetilde{N}(u))$ by [166, Corollary 1.9].

Let $L_{\widetilde{Y}} = (\sigma|_{\widetilde{Y}})^*(L_Y)$ and let $M_{\widetilde{Y}} = (\sigma|_{\widetilde{Y}})^*(M)$. Using the identification

$$H^0(X, mL - jY) = H^0(\widetilde{X}, \sigma^*(mL - jY)),$$

we can identify the image of the restriction map

$$H^0(\widetilde{X}, \sigma^*(mL - jY)) \to H^0(\widetilde{Y}, mL_{\widetilde{Y}} + jM_{\widetilde{Y}})$$

with the vector space $W^Y_{m,j}$. In particular, the linear series $W^Y_{\bullet,\bullet}$ can be seen as a linear series on $\widetilde{Y}$ associated with $L_{\widetilde{Y}}$ and $M_{\widetilde{Y}}$. Let $V^{\widetilde{Y}}_{\bullet,\bullet}$ be the linear series on $\widetilde{Y}$ defined by

$$V^{\widetilde{Y}}_{\bullet,\bullet} = \bigoplus_{m,j} V^{\widetilde{Y}}_{m,j},$$

where

$$V^{\widetilde{Y}}_{m,j} = \lceil m\widetilde{N}(j/m)\rceil\big|_{\widetilde{Y}} + H^0\Big(\widetilde{Y}, \lfloor m\widetilde{P}(j/m)\rfloor\big|_{\widetilde{Y}}\Big)$$

for all $(m, j) \in \mathbb{Z}^2_{\geqslant 0}$ such that $0 \leqslant \frac{j}{m} \leqslant \tau$, and $V^{\widetilde{Y}}_{m,j} = 0$ otherwise.

Lemma 1.105 *The linear series $V^{\widetilde{Y}}_{\bullet,\bullet}$ is $\mathbb{Z}^2_{\geqslant 0}$-graded, it has bounded support and it contains an ample linear series.*

Proof Take (j_1, m_1) and (j_2, m_2) in $\mathbb{Z}^2_{\geqslant 0}$. Then the canonical map

$$H^0(\widetilde{Y}, m_1 L_{\widetilde{Y}} + j_1 M_{\widetilde{Y}}) \otimes H^0(\widetilde{Y}, m_2 L_{\widetilde{Y}} + j_2 M_{\widetilde{Y}}) \to H^0(\widetilde{Y}, (m_1+m_2)L_{\widetilde{Y}} + (j_1+j_2)M_{\widetilde{Y}})$$

maps $V^{\widetilde{Y}}_{m_1,j_1} \otimes V^{\widetilde{Y}}_{m_2,j_2}$ into $V^{\widetilde{Y}}_{m_1+m_2,j_1+j_2}$. Therefore, in order to show that $V^{\widetilde{Y}}_{\bullet,\bullet}$ is $\mathbb{Z}^2_{\geqslant 0}$-graded, it suffices to check that

$$\lfloor m_1 \widetilde{P}(j_1/m_1)\rfloor + \lfloor m_2 \widetilde{P}(j_2/m_2)\rfloor \leqslant \Big\lfloor (m_1 + m_2)\widetilde{P}\Big(\frac{j_1 + j_2}{m_1 + m_2}\Big)\Big\rfloor,$$

or equivalently that

$$\lceil m_1 \widetilde{N}(j_1/m_1)\rceil + \lceil m_2 \widetilde{N}(j_2/m_2)\rceil \geqslant \Big\lceil (m_1 + m_2)\widetilde{N}\Big(\frac{j_1 + j_2}{m_1 + m_2}\Big)\Big\rceil.$$

The latter follows from $N_\sigma(D_1 + D_2) \leqslant N_\sigma(D_1) + N_\sigma(D_2)$ applied to the divisors

$$D_1 = N_\sigma(m_1 L_{\widetilde{Y}} - j_1\widetilde{Y}) = m_1 \widetilde{N}(j_1/m_1),$$
$$D_2 = N_\sigma(m_2 L_{\widetilde{Y}} - j_2\widetilde{Y}) = m_2 \widetilde{N}(j_2/m_2).$$

Clearly, the linear series $V^{\widetilde{Y}}_{\bullet,\bullet}$ has bounded support. Moreover, it contains an ample linear series because $V^{\widetilde{Y}}_{\bullet,\bullet}$ contains $W^Y_{\bullet,\bullet}$, and $W^Y_{\bullet,\bullet}$ contains an ample linear series. □

Let $U_{\bullet,\bullet}$ be the $\mathbb{Z}^2_{\geqslant 0}$-graded complete linear series on $\widetilde{Y}$ associated to $L_{\widetilde{Y}}$ and $M_{\widetilde{Y}}$, i.e.

$$U_{\bullet,\bullet} = \bigoplus_{m,j} U_{m,j},$$

where $U_{m,j} = H^0(\widetilde{Y}, mL_{\widetilde{Y}} + jM_{\widetilde{Y}})$. Recall that $\sigma^*(mL - jY) \sim_{\mathbb{Q}} m\widetilde{N}(j/m) + m\widetilde{P}(j/m)$. Note that $H^0(\widetilde{X}, \sigma^*(mL - jY)) = H^0(\widetilde{X}, \lfloor m\widetilde{P}(j/m) \rfloor)$. It follows that for all $(m,j) \in \mathbb{Z}^2_{\geqslant 0}$, we have $W^Y_{m,j} \subseteq V^{\widetilde{Y}}_{m,j} \subseteq U_{m,j}$, as we can identify

$$W^Y_{m,j} = \lceil m\widetilde{N}(j/m) \rceil|_{\widetilde{Y}} + H^0\big(\widetilde{X}, \lfloor m\widetilde{P}(j/m) \rfloor\big)\big|_{\widetilde{Y}}.$$

Therefore, for all non-negative integers m and j, there are injective maps:

$$V^{\widetilde{Y}}_{m,j}/W_{m,j} \hookrightarrow H^1\big(\widetilde{X}, \lfloor m\widetilde{P}(j/m) \rfloor - \widetilde{Y}\big). \tag{1.17}$$

Theorem 1.106 *The following assertions hold:*

(i)

$$\mathrm{vol}\big(W^Y_{\bullet,\bullet}\big) = \mathrm{vol}\big(V^{\widetilde{Y}}_{\bullet,\bullet}\big) = \mathrm{vol}(L) = n\int_0^\tau \big(\widetilde{P}(u)^{n-1} \cdot \widetilde{Y}\big)\,du.$$

(ii) *For every prime divisor F over Y,*

$$S\big(W^Y_{\bullet,\bullet}; F\big) = S\big(V^{\widetilde{Y}}_{\bullet,\bullet}; F\big) = \frac{n}{\mathrm{vol}(L)}\int_0^\tau h(u)\,du,$$

where

$$h(u) = \big(\widetilde{P}(u)^{n-1} \cdot \widetilde{Y}\big) \cdot \mathrm{ord}_F\big(\widetilde{N}(u)\big|_Y\big) + \int_0^\infty \mathrm{vol}\big(\widetilde{P}(u)\big|_{\widetilde{Y}} - vF\big)\,dv.$$

To prove Theorem 1.106, we need the following auxiliary result.

Lemma 1.107 *There are rational numbers $0 = \tau_0 < \tau_1 < \cdots < \tau_l = \tau$ such that for every $i \in \{1,\ldots,l\}$ and every $u \in [\tau_{i-1}, \tau_i]$, the Nakayama–Zariski decomposition*

$$\sigma^*(L - uY) = \widetilde{P}(u) + \widetilde{N}(u)$$

satisfies

$$\widetilde{N}(u) = \frac{\tau_i - u}{\tau_i - \tau_{i-1}}\widetilde{N}(\tau_{i-1}) + \frac{u - \tau_{i-1}}{\tau_i - \tau_{i-1}}\widetilde{N}(\tau_i).$$

Proof This follows from [173, Proposition 2.13]. The half line $L - uY$ given by $u \geq 0$ intersects finitely many walls of the Mori chamber decomposition of $\overline{\mathrm{Eff}}(X)$ at finitely many rational values. If L is in the interior of a chamber, we denote these values by

$$0 < \tau_1 < \cdots < \tau_l = \tau$$

and set $\tau_0 = 0$. If L is on a wall, set $\tau_0 = 0$ and denote the next values by $\tau_1 < \cdots < \tau_l = \tau$.

By [173, Proposition 2.13], the Zariski decomposition is linear within each chamber. In particular, $\widetilde{N}(u)$ is an affine function of u on the interval $[\tau_{i-1}, \tau_i]$, i.e. we have

$$\widetilde{N}(u) = uD_1 + D_2$$

for some pseudo-effective $\mathbb{R}$-divisors D_1 and D_2 on the variety $\widetilde{X}$. Hence, the required formula follows by setting $\widetilde{N}(\tau_i) = \tau_i D_1 + D_2$ and $\widetilde{N}(\tau_{i-1}) = \tau_{i-1}D_1 + D_2$.   □

For every $i \in \{1, \dots, l\}$, we set

$$C_i = \mathbb{R}_{\geq 0}(1, \tau_{i-1}) + \mathbb{R}_{\geq 0}(1, \tau_i) \subseteq \mathbb{R}_{\geq 0}^2.$$

We also let $C = \cup_{i=1}^l C_i$. Then C is the region of $\mathbb{R}_{\geq 0}^2$ that contains pairs (m, j) such that the divisor $mL - jY$ is pseudo-effective. Now, we choose $n_0 \in \mathbb{Z}_{>0}$ such that

(i) all numbers $n_0\tau_1, \dots, n_0\tau_l$ are positive integers,
(ii) the number $\frac{n_0}{\tau_i - \tau_{i-1}}$ is an integer for every $i \in \{1, \dots, l\}$,
(iii) for any $(m, j) \in C \cap \mathbb{Z}_{\geq 0}^2$, both $n_0 m\widetilde{N}(j/m)$ and $n_0 m\widetilde{P}(j/m)$ are $\mathbb{Z}$-divisors.

Such n_0 does exist. Indeed, we have $\frac{j}{m} \in [\tau_{i-1}, \tau_i]$ for some $i \in \{1, \dots, l\}$, so that

$$\widetilde{N}(j/m) = \frac{\tau_i - \frac{j}{m}}{\tau_i - \tau_{i-1}}\widetilde{N}(\tau_{i-1}) + \frac{\frac{j}{m} - \tau_{i-1}}{\tau_i - \tau_{i-1}}\widetilde{N}(\tau_i)$$

by Lemma 1.107. Hence, we can choose n_0 to clear denominators appearing in all $\widetilde{N}(\tau_i)$, as well as the denominators of $\tau_i - \tau_{i-1}$ for every $i \in \{1, \dots, l\}$.

Proof of Theorem 1.106 Let $m_0 = n_0^4$, $\overline{W}_{\bullet,\bullet}^Y = m_0 W_{\bullet,\bullet}^Y$, $\overline{V}_{\bullet,\bullet}^{\widetilde{Y}} = m_0 V_{\bullet,\bullet}^{\widetilde{Y}}$ and $\overline{U}_{\bullet,\bullet} = m_0 U_{\bullet,\bullet}$. Then $\overline{W}_{m,j}^Y = W_{m_0 m, m_0 j}^Y$, $\overline{V}_{m,j}^{\widetilde{Y}} = V_{m_0 m, m_0 j}^{\widetilde{Y}}$ and $\overline{U}_{m,j} = U_{m_0 m, m_0 j}$ for every m and j in $\mathbb{Z}_{\geq 0}$. For every $t \in \mathbb{R}_{\geq 0}$, consider the $\mathbb{Z}_{\geq 0}^2$-graded linear series $\mathcal{F}_F^t \overline{U}_{\bullet,\bullet} \subseteq \overline{U}_{\bullet,\bullet}$ defined by

$$\mathcal{F}_F^t \overline{U}_{\bullet,\bullet} = \bigoplus_{m,j \in \mathbb{Z}_{\geq 0}} \mathcal{F}_F^{mt} \overline{U}_{m,j},$$

where

$$\mathcal{F}_F^{mt}\overline{U}_{m,j} = \left\{ s \in \overline{U}_{m,j} \mid \mathrm{ord}_F(s) \geqslant mt \right\}.$$

Then $\mathcal{F}_F^t\overline{U}_{\bullet,\bullet}$ is a filtration on $\overline{U}_{\bullet,\bullet}$ in the sense of [2, Definition 2.10], which also induces the filtrations $\mathcal{F}_F^t\overline{W}_{\bullet,\bullet}^Y$ and $\mathcal{F}_F^t\overline{V}_{\bullet,\bullet}^{\widetilde{Y}}$ on $\overline{W}_{\bullet,\bullet}^Y$ and $\overline{V}_{\bullet,\bullet}^{\widetilde{Y}}$, respectively. Namely, we have

$$\mathcal{F}_F^{mt}\overline{W}_{m,j}^Y = \overline{W}_{m,j}^Y \cap \mathcal{F}_F^{mt}\overline{U}_{m,j}$$

and

$$\mathcal{F}_F^{mt}\overline{V}_{m,j}^{\widetilde{Y}} := \overline{V}_{m,j}^{\widetilde{Y}} \cap \mathcal{F}_F^{mt}\overline{U}_{m,j}.$$

It follows from (1.17) that for all $t \in \mathbb{R}_{\geqslant 0}$ and $(m,j) \in C \cap \mathbb{Z}_{\geqslant 0}^2$ there are injective maps

$$\mathcal{F}^{mt}\overline{V}_{m,j}^{\widetilde{Y}} / \mathcal{F}^{mt}\overline{W}_{m,j} \hookrightarrow H^1\left(\widetilde{X}, m_0 m\widetilde{P}(j/m) - \widetilde{Y}\right). \tag{1.18}$$

In particular, when $t = 0$ we recover the usual inclusion (1.17). For $m \in \mathbb{Z}_{\geqslant 0}$, we have

$$\sum_{j \geqslant 0} \dim\left(\overline{V}_{m,j}^{\widetilde{Y}}\right) - \sum_{j \geqslant 0} \dim\left(\overline{W}_{m,j}^Y\right) = \sum_{j=0}^{m\tau} \dim\left(\overline{V}_{m,j}^{\widetilde{Y}} / \overline{W}_{m,j}^Y\right)$$

$$\leqslant \sum_{j=0}^{m\tau} h^1\left(\widetilde{X}, m_0 m\widetilde{P}(j/m) - \widetilde{Y}\right).$$

Let us first prove that $\mathrm{vol}\left(\overline{W}_{\bullet,\bullet}^Y\right) = \mathrm{vol}(\overline{V}_{\bullet,\bullet}^{\widetilde{Y}})$. For this, it suffices to prove that

$$\sum_{j=0}^{m\tau} h^1\left(\widetilde{X}, m_0 m\widetilde{P}(j/m) - \widetilde{Y}\right) \leqslant O(m^{n-1}). \tag{1.19}$$

Further dividing the sum, it suffices to prove that for every $i \in \{1, \ldots, l\}$, we have

$$\sum_{j=m\tau_{i-1}}^{m\tau_i} h^1\left(\widetilde{X}, m_0 m\widetilde{P}(j/m) - \widetilde{Y}\right) \leqslant O(m^{n-1}). \tag{1.20}$$

Assume that $\tau_{i-1} \leqslant \frac{j}{m} \leqslant \tau_i$. By Lemma 1.107, we have

$$\widetilde{N}(u) = \frac{\tau_i - u}{\tau_i - \tau_{i-1}}\widetilde{N}(\tau_{i-1}) + \frac{x - \tau_{i-1}}{\tau_i - \tau_{i-1}}\widetilde{N}(\tau_i)$$

for any $u \in [\tau_{i-1}, \tau_i]$. In particular, we have

$$\widetilde{N}(j/m) = \frac{\tau_i - \frac{j}{m}}{\tau_i - \tau_{i-1}}\widetilde{N}(\tau_{i-1}) + \frac{\frac{j}{m} - \tau_{i-1}}{\tau_i - \tau_{i-1}}\widetilde{N}(\tau_i).$$

Since $\widetilde{P}(x) = \sigma^*(L - xY) - \widetilde{N}(x)$, we obtain

$$\widetilde{P}(j/m) = \frac{\tau_i - \frac{j}{m}}{\tau_i - \tau_{i-1}}\widetilde{P}(\tau_{i-1}) + \frac{\frac{j}{m} - \tau_{i-1}}{\tau_i - \tau_{i-1}}\widetilde{P}(\tau_i)$$

since $n_0^2\widetilde{P}(\tau_{i-1})$ and $n_0^2\widetilde{P}(\tau_i)$ are $\mathbb{Z}$-divisors. Hence for $m_0 = n_0^4$, we can write

$$m_0 m\widetilde{P}(j/m) = mA + kB$$

for $k = n_0(j - m\tau_{i-1})$, $A = m_0\widetilde{P}(\tau_{i-1})$ and

$$B = \frac{n_0^3}{\tau_i - \tau_{i-1}}\Big(\widetilde{P}(\tau_i) - \widetilde{P}(\tau_{i-1})\Big).$$

We also let $a = n_0(\tau_i - \tau_{i-1})$. Then a and k are positive integers such that $0 \leqslant k \leqslant ma$. Furthermore, both A and B are $\mathbb{Z}$-divisors. But $A = m_0\widetilde{P}(\tau_{i-1})$ and $A + aB = m_0\widetilde{P}(\tau_i)$. Then A and $A + aB$ are semiample. But the divisor $\widetilde{P}(\tau_{i-1})$ is big, so that A is nef and big. Then (1.20) follows from Lemma A.55 applied to A, B and $D = \widetilde{Y}$.

Observe that $\mathrm{vol}(\overline{V}^{\widetilde{Y}}_{\bullet,\bullet}) = m_0^{n-1} \cdot \mathrm{vol}(V^{\widetilde{Y}}_{\bullet,\bullet})$ by the asymptotic Riemann–Roch theorem. Thus, to finish the proof of part (i), it suffices to prove that

$$\mathrm{vol}(\overline{V}^{\widetilde{Y}}_{\bullet,\bullet}) = m_0^{n-1} \cdot n \int_0^\tau \big(\widetilde{P}(u)^{n-1} \cdot \widetilde{Y}\big)du.$$

By definition, we have

$$\mathrm{vol}(\overline{V}^{\widetilde{Y}}_{\bullet,\bullet}) = \lim_{m\to\infty} \frac{\sum_{j\geqslant 0}\dim(\overline{V}^{\widetilde{Y}}_{m,j})}{m^n/n!} = \lim_{m\to\infty}\frac{n!}{m^n}\cdot\sum_{j=0}^{m\tau}h^0\big(\widetilde{Y}, mm_0\widetilde{P}(j/m)|_{\widetilde{Y}}\big).$$

The result now follows from the asymptotic Riemann–Roch theorem [137, Corollary 1.4.41] because the divisor $\widetilde{P}(j/m)|_{\widetilde{Y}}$ is nef. Namely, we get

$$h^0\big(\widetilde{Y}, mm_0\widetilde{P}(j/m)|_{\widetilde{Y}}\big) = \frac{m^{n-1}}{(n-1)!}\big(m_0\widetilde{P}(j/m)|_{\widetilde{Y}}\big)^{n-1} + O(m^{n-2}).$$

Then

$$\sum_{j=m\tau_{i-1}}^{m\tau_i}h^0\big(\widetilde{Y}, mm_0\widetilde{P}(j/m)|_{\widetilde{Y}}\big) = \sum_{j=m\tau_{i-1}}^{m\tau_i}\frac{m^{n-1}}{(n-1)!}\big(m_0\widetilde{P}(j/m)|_{\widetilde{Y}}\big)^{n-1} + O(m^{n-1}),$$

and hence we have

$$\lim_{m\to\infty}\frac{n!}{m^n}\cdot\sum_{j=m\tau_{i-1}}^{m\tau_i}h^0\big(\widetilde{Y}, mm_0\widetilde{P}(j/m)|_{\widetilde{Y}}\big) = \lim_{m\to\infty}\frac{n}{m}\cdot\sum_{j=m\tau_{i-1}}^{m\tau_i}m_0^{n-1}\big(\widetilde{P}(j/m)^{n-1} \cdot \widetilde{Y}\big)$$

$$= nm_0^{n-1}\cdot\int_{\tau_{i-1}}^{\tau_i}\big(\widetilde{P}(u)^{n-1} \cdot \widetilde{Y}\big)du.$$

This equality can be deduced from Lemma A.55 similarly to how we proved (1.20). Together with Lemma 1.96, this finishes the proof of part (i).

Let us prove part (ii). First, we prove that $S(V_{\bullet,\bullet}^{Y}; F) = S(W_{\bullet,\bullet}^{Y}; F)$. This follows from

$$S\big(\widetilde{V}_{\bullet,\bullet}^{\widetilde{Y}}; F\big) = S\big(\overline{W}_{\bullet,\bullet}^{Y}; F\big)$$

since $S(\widetilde{V}_{\bullet,\bullet}^{\widetilde{Y}}; F) = m_0 \cdot S(V_{\bullet,\bullet}^{\widetilde{Y}}; F)$ and $S(\overline{W}_{\bullet,\bullet}^{Y}; F) = m_0 S(W_{\bullet,\bullet}^{Y}; F)$ by [2, Lemma 2.24]. To prove the equality $S(V_{\bullet,\bullet}^{\widetilde{Y}}; F) = S(\overline{W}_{\bullet,\bullet}^{Y}; F)$, observe that part (i) gives

$$\lim_{m \to \infty} \frac{N_m^{\widetilde{V}^{\widetilde{Y}}}}{m^n/n!} = \mathrm{vol}\big(\widetilde{V}_{\bullet,\bullet}^{\widetilde{Y}}\big) = \mathrm{vol}\big(\overline{W}_{\bullet,\bullet}^{Y}\big) = \lim_{m \to \infty} \frac{N_m^{\overline{W}^{Y}}}{m^n/n!},$$

where $N_m^{\widetilde{V}^{\widetilde{Y}}} = \sum_{j \geqslant 0} \dim(\widetilde{V}_{m,j}^{\widetilde{Y}})$ and $N_m^{\overline{W}^{Y}} = \sum_{j \geqslant 0} \dim(\overline{W}_{m,j}^{Y})$. These limits are non-zero because $\overline{W}_{\bullet,\bullet}^{Y}$ contains an ample linear series. By Lemma 1.99 and Remark 1.100, we have

$$S\big(\widetilde{V}_{\bullet,\bullet}^{\widetilde{Y}}; F\big) = \lim_{m \to \infty} S_m\big(\widetilde{V}_{\bullet,\bullet}^{\widetilde{Y}}; F\big),$$

$$S\big(\overline{W}_{\bullet,\bullet}^{Y}; F\big) = \lim_{m \to \infty} S_m\big(\overline{W}_{\bullet,\bullet}^{Y}; F\big),$$

where

$$S_m\big(\widetilde{V}_{\bullet,\bullet}^{\widetilde{Y}}; F\big) = \frac{1}{mN_m^{\widetilde{V}^{\widetilde{Y}}}} \sum_{j=0}^{m\tau} \sum_{k \geqslant 0} \mathcal{F}_F^k \widetilde{V}_{m,j}^{\widetilde{Y}},$$

and

$$S_m\big(\overline{W}_{\bullet,\bullet}^{Y}; F\big) = \frac{1}{mN_m^{\overline{W}^{Y}}} \sum_{j=0}^{m\tau} \sum_{k \geqslant 0} \mathcal{F}_F^k \overline{W}_{m,j}^{Y}.$$

Thus, to prove the equality $S(\widetilde{V}_{\bullet,\bullet}^{\widetilde{Y}}; F) = S(\overline{W}_{\bullet,\bullet}^{Y}; F)$, it is enough to prove that

$$\lim_{m \to \infty} \frac{1}{m^{n+1}} \Big(S_m\big(\widetilde{V}_{\bullet,\bullet}^{\widetilde{Y}}; F\big) - S_m\big(\overline{W}_{\bullet,\bullet}^{Y}; F\big)\Big) = 0.$$

This limit equals

$$\lim_{m \to \infty} \frac{1}{m^{n+1}} \sum_{j=0}^{m\tau} \sum_{k=0}^{m_0 m\tau'} \Big(\dim\big(\mathcal{F}_F^k \widetilde{V}_{m,j}^{\widetilde{Y}}\big) - \dim\big(\mathcal{F}_F^k \overline{W}_{m,j}^{Y}\big)\Big),$$

which is non-negative. Moreover, by (1.18), it is bounded from above by

$$\lim_{m \to \infty} \frac{1}{m^{n+1}} \sum_{j=0}^{m\tau} \sum_{k=0}^{m_0 m\tau'} h^1\left(\widetilde{X}, m_0 m \widetilde{P}(j/m) - \widetilde{Y}\right)$$

$$= \lim_{m \to \infty} \frac{m_0 \tau'}{m^n} \sum_{j=0}^{m\tau} h^1\left(\widetilde{X}, m_0 m \widetilde{P}(j/m) - \widetilde{Y}\right).$$

The latter limit equals 0 by (1.19). Hence, $S(V_{\bullet,\bullet}^{\widetilde{Y}}; F) = S(W_{\bullet,\bullet}^{Y}; F)$.

To finish the proof of part ii, by using part i, it suffices to prove that

$$\mathrm{vol}\left(\overline{V}_{\bullet,\bullet}^{\widetilde{Y}}\right) \cdot \lim_{m \to \infty} S_m\left(\overline{V}_{\bullet,\bullet}^{\widetilde{Y}}; F\right) = \lim_{m \to \infty} \frac{n!}{m^{n+1}} \sum_{j=0}^{m\tau} \sum_{k \geqslant 0} \mathcal{F}_F^k \overline{W}_{m,j}$$

$$= n m_0^n \cdot \int_0^\tau h(u) du.$$

The first equality is clear. To prove the second, recall that

$$\overline{V}_{m,j}^{\widetilde{Y}} = m m_0 \widetilde{N}(j/m)\big|_{\widetilde{Y}} + H^0\left(\widetilde{Y}, m m_0 \widetilde{P}(j/m)\big|_{\widetilde{Y}}\right)$$

since both $m m_0 \widetilde{N}(j/m)$ and $m m_0 \widetilde{P}(j/m)$ are $\mathbb{Z}$-divisors. For $m \geqslant 0$, let

$$\phi_{j,m} = \mathrm{ord}_F\left(\widetilde{N}(j/m)\big|_{\widetilde{Y}}\right).$$

Then $m m_0 \phi_{j,m}$ is an integer, and we have

$$\dim\left(\mathcal{F}_F^k \overline{V}_{m,j}^{\widetilde{Y}}\right) = \begin{cases} h^0\left(\widetilde{Y}, m m_0 \widetilde{P}(j/m)\big|_{\widetilde{Y}}\right) & \text{if } 0 \leqslant k \leqslant m m_0 \phi_{j,m}, \\ h^0\left(\widetilde{Y}, m m_0 \widetilde{P}(j/m)\big|_{\widetilde{Y}} - (k - m m_0 \phi_{j,m}) F\right) & \text{if } m m_0 \phi_{j,m} \leqslant k. \end{cases}$$

Therefore, we have

$$S_m\left(\overline{V}_{\bullet,\bullet}^{\widetilde{Y}}; F\right) = \frac{\Sigma_1 + \Sigma_2}{m N_m^{\overline{V}^{\widetilde{Y}}}},$$

where

$$\Sigma_1 = \sum_{j=0}^{m\tau} m m_0 \phi_{j,m} \cdot h^0\left(\widetilde{Y}, m m_0 \widetilde{P}(j/m)\big|_{\widetilde{Y}}\right),$$

and

$$\Sigma_2 = \sum_{j=0}^{m\tau} \sum_{s=0}^{m m_0 \tau'} h^0\left(\widetilde{Y}, m m_0 \widetilde{P}(j/m)\big|_{\widetilde{Y}} - sF\right).$$

Since $\widetilde{P}(j/m)|_{\widetilde{Y}}$ is nef, using the asymptotic Riemann–Roch theorem, we get

$$\lim_{m\to\infty} \frac{n!}{m^{n+1}} \Sigma_1 = \lim_{m\to\infty} \frac{n}{m} \sum_{j=0}^{m\tau} m_0 \phi_{j,m} \cdot \frac{h^0\left(\widetilde{Y}, m m_0 \widetilde{P}(j/m)|_{\widetilde{Y}}\right)}{m^{n-1}/(n-1)!}$$

$$= \lim_{m\to\infty} \frac{n}{m} \sum_{j=0}^{m\tau} m_0 \phi_{j,m} \cdot \left(m_0 \widetilde{P}(j/m)|_{\widetilde{Y}}\right)^{n-1}$$

$$= m_0^n \cdot n \int_0^\tau \mathrm{ord}_F\left(\widetilde{N}(u)|_{\widetilde{Y}}\right)\left(\widetilde{P}(u)^{n-1} \cdot \widetilde{Y}\right) du.$$

Furthermore, for the second sum Σ_2, we have

$$\lim_{m\to\infty} \frac{n!}{m^{n+1}} \Sigma_2 = \lim_{m\to\infty} \frac{n}{m^2} \sum_{j=0}^{m\tau} \sum_{s=0}^{m m_0 \tau'} \frac{h^0\left(\widetilde{Y}, m\left(m_0 \widetilde{P}(j/m)|_{\widetilde{Y}} - \frac{s}{m} F\right)\right)}{m^{n-1}/(n-1)!}$$

$$= n \int_0^\tau \int_0^{m_0 \tau'} \mathrm{vol}\left(m_0 \widetilde{P}(u)|_{\widetilde{Y}} - x F\right) dx\, du$$

$$= m_0^n \cdot n \int_0^\tau \int_0^\infty \mathrm{vol}\left(\widetilde{P}(u)|_{\widetilde{Y}} - v F\right) dv\, du.$$

Hence, it follows that

$$\lim_{m\to\infty} \frac{n!}{m^{n+1}} \left(\Sigma_1 + \Sigma_2\right) = m_0^n \cdot n \cdot \int_0^\tau h(u)\, du,$$

which completes the proof of Theorem 1.106. □

If $\mathrm{Nef}(X) = \overline{\mathrm{Mov}}(X)$, then all varieties $X_0, X_1, \ldots, X_p$ in (1.15) are isomorphic to X, and we can also take $\widetilde{X} = X$ and $\sigma = \mathrm{Id}_X$. Therefore, Theorem 1.106 implies

Corollary 1.108 *Suppose that* $\mathrm{Nef}(X) = \overline{\mathrm{Mov}}(X)$. *For every* $u \in [0, \tau]$, *write*

$$L - uY \sim_{\mathbb{R}} P(u) + N(u),$$

where $P(u)$ *is the positive (nef) part of the Zariski decomposition of the divisor* $L - uY$, *and* $N(u)$ *is its negative part. Then for every prime divisor* F *over* Y, *we have*

$$S\left(W_{\bullet,\bullet}^Y; F\right) = \frac{n}{\mathrm{vol}(L)} \int_0^\tau h(u)\, du,$$

where

$$h(u) = \left(P(u)^{n-1} \cdot Y\right) \cdot \mathrm{ord}_F\left(N(u)|_Y\right) + \int_0^\infty \mathrm{vol}\left(P(u)|_Y - v F\right) dv.$$

If Y *is normal, and* Z *is a prime divisor on* Y, *then* (1.12) *simplifies as*

$$\delta_Z\left(Y; W_{\bullet,\bullet}^Y\right) = \frac{1}{S(W_{\bullet,\bullet}^Y; Z)}.$$

Corollary 1.109 *In the assumption and notations of Corollary 1.108, suppose that the variety Y is normal, and Z is a prime divisor on Y. Then $\delta_Z(Y; W^Y_{\bullet,\bullet}) = \frac{1}{S(W^Y_{\bullet,\bullet};Z)}$ and*

$$S(W^Y_{\bullet,\bullet}; Z) = \frac{n}{\mathrm{vol}(L)} \int_0^\tau h(u)du,$$

where

$$h(u) = \left(P(u)^{n-1} \cdot Y\right) \cdot \mathrm{ord}_Z\left(N(u)\big|_Y\right) + \int_0^\infty \mathrm{vol}_Y\left(P(u)\big|_Y - vZ\right)dv.$$

Examples of varieties that satisfy the condition $\mathrm{Nef}(X) = \overline{\mathrm{Mov}}(X)$ are the following:

- two-dimensional Mori Dream Spaces [209],
- smooth Fano threefolds [157].

For smooth Fano threefolds, Corollary 1.109 and [2, Theorem 3.3] give the following very handy corollary that will be often used in the proof of the Main Theorem.

Corollary 1.110 *Let X be a smooth Fano threefold, let Y be an irreducible normal surface in the threefold X, let Z be an irreducible curve in Y, and let E be a prime divisor over the threefold X such that $C_X(E) = Z$. Then*

$$\frac{A_X(E)}{S_X(E)} \geq \min\left\{\frac{1}{S_X(Y)}, \frac{1}{S(W^Y_{\bullet,\bullet}; Z)}\right\} \tag{1.21}$$

and

$$S(W^Y_{\bullet,\bullet}; Z) = \frac{3}{(-K_X)^3} \int_0^\tau \left(P(u)^2 \cdot Y\right) \cdot \mathrm{ord}_Z\left(N(u)\big|_Y\right)du$$
$$+ \frac{3}{(-K_X)^3} \int_0^\tau \int_0^\infty \mathrm{vol}\left(P(u)\big|_Y - vZ\right)dvdu,$$

where P(u) is the positive part of the Zariski decomposition of the divisor $-K_X - uY$, and N(u) is its negative part. Moreover, if the equality holds in (1.21), then

$$\frac{A_X(E)}{S_X(E)} = \frac{1}{S_X(Y)}.$$

Remark 1.111 Observe that the assertion of Corollary 1.110 remains valid in the case when X is a smooth Fano threefold, Y is a possibly non-normal irreducible surface in X, and Z is an irreducible curve Y such that $Z \not\subset \mathrm{Sing}(Y)$. In this case, we should replace both $P(u)|_Y$ and $N(u)|_Y$ by their pull backs on the normalization of the surface Y.

Let us conclude this chapter by proving one very useful generalization of Corollary 1.110. To state it, we fix the following assumptions:

- X is a smooth Fano threefold, so that $\text{Nef}(X) = \text{Mov}(X)$;
- Y is an irreducible normal surface in X that has at most Du Val singularities;
- Z is an irreducible smooth curve in Y such that the log pair (Y, Z) has purely log terminal singularities, e.g. Z is contained in the smooth locus of the surface Y;
- Δ_Z is the different of the log pair (Y, Z), i.e. Δ_Z is an effective $\mathbb{Q}$-divisor on the curve Z such that $\text{Supp}(\Delta_Z) = \text{Sing}(Y) \cap Z$ and $K_Z + \Delta_Z = (K_Y + Z)|_Z$.

As usual, we denote by τ the largest $u \in \mathbb{Q}_{\geqslant 0}$ such that $-K_X - uY$ is pseudo-effective. For $u \in [0, \tau]$, let $P(u)$ be the positive part of the Zariski decomposition of this divisor, and let $N(u)$ be its negative part. Then

(i) $Y \not\subset \text{Supp}(N(u))$ for every $u \in [0, \tau]$;
(ii) $N(u)$ is continuous at every point $u \in [0, \tau]$;
(iii) $N(u)$ is a $\mathbb{Q}$-divisor for $u \in [0, \tau] \cap \mathbb{Q}$;
(iv) $N(u)$ is *convex* [166] in the following sense: for every u and $u' \in [0, \tau]$,

$$N\big((1 - s)u + su'\big) \leqslant (1 - s)N\big(u\big) + sN\big(u'\big)$$

for every $s \in [0, 1]$;
(v) the restriction $P(u)|_Y$ is nef and big for every $u \in [0, \tau)$.

Therefore, for every $u \in [0, \tau]$, we can define the effective $\mathbb{R}$-divisor

$$N(u)\big|_Y = d(u)Z + N'_Y(u), \qquad (1.22)$$

where $N'_Y(u)$ is an effective divisor such that $Z \not\subset \text{Supp}(N'_Y(u))$, and $d(u) = \text{ord}_Z(N(u)|_Y)$. This gives the function $d \colon [0, \tau] \to \mathbb{R}_{\geqslant 0}$ given by $u \mapsto d(u)$, which is continuous and convex. Now, for every $u \in [0, \tau]$, we define the pseudo-effective threshold $t(u) \in \mathbb{R}_{\geqslant 0}$ as follows:

$$t(u) = \max \left\{ v \in \mathbb{R}_{\geqslant 0} \,\middle|\, P(u)|_Y - vZ \text{ is pseudo-effective} \right\}.$$

For $v \in [0, t(u)]$, the divisor $P(u)|_Y - vZ$ is pseudo-effective. Let $P(u, v)$ be the positive part of the Zariski decomposition of this divisor, and let $N(u, v)$ be its negative part. Then the following assertions hold:

(i) $N(u, 0) = 0$ for every $u \in [0, \tau]$ because $P(u)|_Y$ is nef for $u \in [0, \tau]$;
(ii) $P(u, v) \cdot Z > 0$ and $Z \not\subset \text{Supp}\big(N(u, v)\big)$ for every $u \in [0, \tau)$ and $v \in (0, t(u))$;
(iii) $P(u, v)$ and $N(u, v)$ are $\mathbb{Q}$-divisors if both $u, v \in \mathbb{Q}$.

Let $V^Y_{\bullet,\bullet}$ be the $\mathbb{Z}^2_{\geq 0}$-graded linear series on Y defined by

$$V^Y_{\bullet,\bullet} = \bigoplus_{m,j} V^Y_{m,j},$$

where

$$V^Y_{m,j} = \begin{cases} \big\lceil mN(j/m)\big\rceil\big|_Y + H^0\Big(Y, \lfloor mP(j/m)\rfloor\big|_Y\Big) & \text{if } 0 \leqslant \dfrac{j}{m} \leqslant \tau, \\ 0 & \text{otherwise.} \end{cases}$$

Denote by $W^{Y,Z}_{\bullet,\bullet,\bullet}$ the refinement of $V^Y_{\bullet,\bullet}$ by the curve Z in the sense of [2, Example 2.7]. For every point $P \in Z$, we also define

$$F_P\big(W^{Y,Z}_{\bullet,\bullet,\bullet}\big) := \frac{6}{(-K_X)^3} \int_0^\tau \int_0^{t(u)} \big(P(u,v)\cdot Z\big)\cdot \mathrm{ord}_P\big(N'_Y(u)|_Z + N(u,v)|_Z\big)\,dv\,du.$$

Theorem 1.112   *Let P be a point in the curve Z. Then*

$$\delta_P(X) \geqslant \min\left\{\frac{1-\mathrm{ord}_P(\Delta_Z)}{S(W^{Y,Z}_{\bullet,\bullet,\bullet}; P)}, \frac{1}{S(V^Y_{\bullet,\bullet}; Z)}, \frac{1}{S_X(Y)}\right\}, \tag{1.23}$$

where

$$S(W^{Y,Z}_{\bullet,\bullet,\bullet}; P) = \frac{3}{(-K_X)^3} \int_0^\tau \int_0^{t(u)} \big(P(u,v)\cdot Z\big)^2 dv\,du + F_P\big(W^{Y,Z}_{\bullet,\bullet,\bullet}\big).$$

Moreover, if the inequality (1.23) *is an equality and there exists a prime divisor E over the threefold X such that $C_X(E) = P$ and $\delta_P(X) = \frac{A_X(E)}{S_X(E)}$, then $\delta_P(X) = \frac{1}{S_X(Y)}$.*

Proof   By [2, Theorem 3.2] and Theorem 1.106, we have

$$\delta_P(X) \geqslant \min\left\{\frac{1}{S_X(Y)}, \delta_P\big(Y; W^Y_{\bullet,\bullet}\big)\right\} = \min\left\{\frac{1}{S_X(Y)}, \delta_P\big(Y; V^Y_{\bullet,\bullet}\big)\right\}$$

$$\geqslant \min\left\{\frac{1}{S_X(Y)}, \frac{1}{S(V^Y_{\bullet,\bullet}; Z)}, \frac{1-\mathrm{ord}_P(\Delta_Z)}{S(W^{Y,Z}_{\bullet,\bullet,\bullet}; P)}\right\}.$$

Moreover, if we have equality here, then [2, Theorem 3.2] gives $\delta_P(X) = \frac{1}{S_X(Y)}$ provided that there exists a prime divisor E over X such that $C_X(E) = P$ and $\delta_P(X) = \frac{A_X(E)}{S_X(E)}$.

Now, we set

$$\Delta^{Y,Z} = \Big\{(u,v) \in \mathbb{R}^2_{\geq 0} \mid u \in [0,\tau], v \in [d(u), d(u)+t(u)]\Big\}.$$

The subset $\Delta^{Y,Z}$ is closed and convex since $d+t: [0,\tau] \to \mathbb{R}_{\geq 0}$ is continuous and concave. Then, as in [2, Corollary 2.15], we set

$$\Delta^{\mathrm{Supp}} = \mathrm{Supp}\big(W^{Y,Z}_{\bullet,\bullet,\bullet}\big) \cap \Big(\{1\} \times \mathbb{R}^2_{\geq 0}\Big).$$

We claim that $\Delta^{\text{Supp}} = \Delta^{Y,Z}$. Indeed, take any $(u,v) \in \mathbb{R}^2_{\geqslant 0} \setminus \Delta^{Y,Z}$ such that $(u,v) \in \mathbb{Q}^2$. If $u > \tau$, then $V^Y_{m,mu} = 0$, which gives $W^{Y,Z}_{m,mu,mv} = 0$ for all sufficiently divisible $m \in \mathbb{Z}_{>0}$. Similarly, if $0 \leqslant u \leqslant \tau$ and $v > d(u) + t(u)$, then $W^{Y,Z}_{m,mu,mv} = 0$ as well for all sufficiently divisible $m \in \mathbb{Z}_{>0}$ because

$$\text{ord}_Z\left(m\big(N(u)\big|_Y\big)\right) = md(u)$$

and $m(P(u)|_Y - (v - d(u))Z)$ does not have global sections since it is not pseudo-effective. This shows that $\Delta^{\text{Supp}} \subseteq \Delta^{Y,Z}$.

Similarly, to show that $\Delta^{Y,Z} \subseteq \Delta^{\text{Supp}}$, we take $(u,v) \in \text{Int}(\Delta^{Y,Z})$ such that $(u,v) \in \mathbb{Q}^2$. If m is a sufficiently divisible integer, then $W^{Y,Z}_{m,mu,mv}$ is the image of the restriction map

$$m\left(N'_Y(u) + N(u, v - d(u))\right) + H^0\left(Y, m\big(P(u, v - d(u))\big)\right) \xrightarrow{\text{rest}}$$
$$\xrightarrow{\text{rest}} m\left(N'_Y(u)\big|_Z + N(u, v - d(u))\big|_Z\right) + H^0\left(Z, m\big(P(u, v - d(u))\big)\big|_Z\right).$$

The cokernel of this map lives in $H^1(Y, mP(u, v - d(u)) - Z)$, whose dimension is bounded when m goes to infinity by [106, Corollary 7] since $P(u, v - d(u))$ is nef and big. Then

$$\text{vol}_{W^{Y,Z}_{\bullet,\bullet,\bullet}}(u,v) = \lim_{m \to \infty} \frac{\dim\left(W^{Y,Z}_{m,mu,mv}\right)}{m} = P\big(u, v - d(u)\big) \cdot Z > 0,$$

where the limit is taken over sufficiently divisible m. In particular, we have $(u,v) \in \Delta^{\text{Supp}}$, which proves that $\Delta^{Y,Z} \subseteq \Delta^{\text{Supp}}$. Thus, we see that $\Delta^{\text{Supp}} = \Delta^{Y,Z}$ as claimed.

Let us prove the formula for $S(W^{Y,Z}_{\bullet,\bullet,\bullet}; P)$. For $c \in \mathbb{Z}_{\geqslant 0}$, let $W^{Y,Z}_{(m,mu,mv),c} = W^{Y,Z}_{cm,cmu,cmv}$. Let $\Delta = \Delta(W^{Y,Z}_{\bullet,\bullet,\bullet})$ be the Okounkov body of $W^{Y,Z}_{\bullet,\bullet,\bullet}$ that is associated to the flag $\{P\} \subset Z$. Then $\Delta \subset \mathbb{R}^3_{\geqslant 0}$. Let $p \colon \Delta \twoheadrightarrow \Delta^{\text{Supp}} \subset \mathbb{R}^2$ be the projection to the first two coordinates. By [138, Theorem 4.21], for any $(u,v) \in \text{Int}(\Delta^{\text{Supp}}) \cap \mathbb{Q}^2_{\geqslant 0}$, the preimage $p^{-1}(u,v) \subset \mathbb{R}_{\geqslant 0}$ is the Okounkov body of $W^{Y,Z}_{(1,u,v),\bullet}$ that is associated to the same admissible flag $\{P\} \subset Z$. To be fully precise, the preimage $p^{-1}(u,v)$ is $(1/m)$th of the Okounkov body of $W^{Y,Z}_{(m,mu,mv),\bullet}$, where m is sufficiently divisible. On the other hand, we have $p^{-1}(\Delta) = [a,b]$, where

$$\begin{cases} a = \text{ord}_P\big(\big(N'_Y(u) + N(u, v - d(u))\big)\big|_Z, \\ b = \text{ord}_P\big(\big(N'_Y(u) + N(u, v - d(u))\big)\big|_Z\big) + P(u, v - d(u)) \cdot Z. \end{cases}$$

The prime divisor $P \in Z$ gives a filtration $\mathcal{F} = \mathcal{F}_P$ on $W^{Y,Z}_{\bullet,\bullet,\bullet}$ (see [2, Example 2.2]). For each $t \in \mathbb{R}_{\geqslant 0}$, let $W^{Y,Z,t}_{\bullet,\bullet,\bullet}$ be the induced linear series defined by

$$W^{Y,Z,t}_{m,j,k} = \mathcal{F}^{mt} W^{Y,Z}_{m,j,k},$$

and let $\Delta^t = \Delta(W_{\bullet,\bullet,\bullet}^{Y,Z,t}) \subset \Delta$ be the associated Okounkov body (cf. [23, §1.2], [2, §2.6]). For all $(u,v,x) \in \Delta$, we let

$$G(u,v,x) = \sup \left\{ t \in \mathbb{R}_{\geqslant 0} \mid (u,v,x) \in \Delta^t \right\}.$$

Observe that $\mathrm{vol}(\Delta) = \frac{1}{3!}\mathrm{vol}(W_{\bullet,\bullet,\bullet}^{Y,Z}) = \frac{1}{3!}\mathrm{vol}(V_{\bullet,\bullet}^{Y}) = \frac{1}{3!}(-K_X)^3$ by [2, Remark 2.3]. Therefore, arguing as in the proof of [2, Lemma 2.9], we get

$$S(W_{\bullet,\bullet,\bullet}^{Y,Z}; P) = \frac{1}{\mathrm{vol}(\Delta)} \int_\Delta G d\rho = \frac{6}{(-K_X)^3} \int_\Delta G d\rho,$$

where ρ is the Lebesgue measure on $\mathrm{Int}(\Delta)$. Now, we let

$$\Gamma(W_{\bullet,\bullet,\bullet}^{Y,Z}) = \left\{ \left(m,j,k; \mathrm{ord}_P(s)\right) \mid s \in W_{m,j,k}^{Y,Z} \setminus \{0\} \right\} \subset \mathbb{R}_{\geqslant 0}^4$$

and let $\Sigma(W_{\bullet,\bullet,\bullet}^{Y,Z})$ be the closure of the cone spanned by $\Gamma(W_{\bullet,\bullet,\bullet}^{Y,Z})$. Then

$$\Delta(W_{\bullet,\bullet,\bullet}^{Y,Z}) = \Delta = \Sigma(W_{\bullet,\bullet,\bullet}^{Y,Z}) \cap \left(1 \times \mathbb{R}_{\geqslant 0}^3\right).$$

For every $(u,v,x) \in \mathrm{Int}(\Delta) \cap \mathbb{Q}_{\geqslant 0}^3$, we have $G((u,v,x)) = x$. Indeed, for every sufficiently divisible $m \gg 0$, it follows from [22, Lemma 1.13] that

$$\left(m, m(u,v,x)\right) \in \Gamma\left(W_{\bullet,\bullet,\bullet}^{Y,Z}\right),$$

so there exists $s \in W_{m,mu,mv}$ such that $\mathrm{ord}_P(s) \geqslant mx$. Hence $G((u,v,x)) \geqslant x$. Vice versa, if $G((u,v,x)) \geqslant x' > x$ for some $x' \in \mathbb{Q}$, then $\mathrm{ord}_P(s) \geqslant mx'$ for every sufficiently divisible $m \gg 0$ and every $s \in W_{m,mu,mv}^{Y,Z} \setminus \{0\}$, so that

$$\left(m, m(u,v,x)\right) \notin \Gamma\left(W_{\bullet,\bullet,\bullet}^{Y,Z}\right),$$

which is a contradiction. Therefore, we see that $G((u,v,x)) = x$.

Observe that the function $G \colon \Delta \to \mathbb{R}_{\geqslant 0}$ is concave on the interior $\mathrm{Int}(\Delta)$, which implies that its restriction $G|_{\mathrm{Int}(\Delta)} \colon \mathrm{Int}(\Delta) \to \mathbb{R}_{\geqslant 0}$ is just the projection to the third factor. Now, since $\Delta^{\mathrm{Supp}} = \Delta^{Y,Z} \subset \mathbb{R}_{\geqslant 0}^2$, we obtain

$$S(W_{\bullet,\bullet,\bullet}^{Y,Z}; P) = \frac{6}{(-K_X)^3} \int_\Delta G d\rho = \frac{6}{(-K_X)^3} \int_{(u,v)\in\Delta^{\mathrm{Supp}}} \left(\int_{x\in p^{-1}(\Delta)} x\, dx \right) du dv$$

$$= \frac{6}{(-K_X)^3} \int_{u=0}^\tau \int_{v=d(u)}^{d(u)+t(u)} \int_a^b x dx du dv$$

$$= \frac{6}{(-K_X)^3} \int_{u=0}^\tau \int_{v=d(u)}^{d(u)+t(u)} \frac{b^2 - a^2}{2} du dv$$

$$= \frac{6}{(-K_X)^3} \int_{u=0}^{\tau} \int_{v=0}^{t(u)} \left(\mathrm{ord}_P\big((N'_{\widetilde{Y}}(u) + N(u,v))\big|_Z\big) \cdot \big(P(u,v) \cdot Z\big) \right.$$
$$\left. + \frac{1}{2}\big(P(u,v) \cdot Z\big)^2 \right) dv\,du$$
$$= \frac{3}{(-K_X)^3} \int_{u=0}^{\tau} \int_{v=0}^{t(u)} \big(P(u,v) \cdot Z\big)^2 dv\,du + F_P\big(W^{Y,Z}_{\bullet,\bullet,\bullet}\big),$$

which is exactly what we want. $\qquad\square$

In this book, we will always apply Theorem 1.112 to a smooth surface Y, so that the different Δ_Z will always be zero in all our applications.

Remark 1.113 Let Q be a point in Y, let $\varepsilon\colon \widetilde{Y} \to Y$ be the *plt blow up* of the point Q, and let $\widetilde{Z}$ be the ε-exceptional curve. Then $(\widetilde{Y},\widetilde{Z})$ has purely log terminal singularities, so that $K_{\widetilde{Z}} + \Delta_{\widetilde{Z}} \sim_{\mathbb{Q}} (K_{\widetilde{Y}} + \widetilde{Z})|_{\widetilde{Z}}$, where $\Delta_{\widetilde{Z}}$ is the different of the log pair $(\widetilde{Y},\widetilde{Z})$. The formula in Theorem 1.112 remains valid if we replace (Z,Δ_Z) by $(\widetilde{Z},\Delta_{\widetilde{Z}})$ after appropriate modifications. Let us state this more precisely. For every $u \in [0,\tau]$, we let

$$\widetilde{t}(u) = \max\left\{ v \in \mathbb{R}_{\geqslant 0} \;\middle|\; \varepsilon^*\big(P(u)|_Y\big) - v\widetilde{Z} \text{ is pseudo-effective}\right\}.$$

For every $v \in [0,\widetilde{t}(u)]$, let us denote by $\widetilde{P}(u,v)$ the positive part of the Zariski decomposition of the divisor $\varepsilon^*(P(u)|_Y) - v\widetilde{Z}$, and let us denote by $\widetilde{N}(u,v)$ its negative part. Let $W^{Y,\widetilde{Z}}_{\bullet,\bullet,\bullet}$ be the refinement of $V^Y_{\bullet,\bullet}$ by the curve $\widetilde{Z}$. Finally, let $N'_{\widetilde{Y}}(u)$ be the proper transform on $\widetilde{Y}$ of the divisor $N(u)$. Then

$$\delta_Q(X) \geqslant \min\left\{ \min_{P\in\widetilde{Z}} \frac{1 - \mathrm{ord}_P(\Delta_{\widetilde{Z}})}{S(W^{Y,\widetilde{Z}}_{\bullet,\bullet,\bullet};P)}, \frac{A_Y(\widetilde{Z})}{S(V^Y_{\bullet,\bullet};\widetilde{Z})}, \frac{1}{S_X(Y)} \right\}, \qquad (1.24)$$

where for every $P \in \widetilde{Z}$ we have

$$S(W^{Y,\widetilde{Z}}_{\bullet,\bullet,\bullet};P) = \frac{3}{(-K_X)^3} \int_0^{\tau} \int_0^{\widetilde{t}(u)} \left(\big(\widetilde{P}(u,v) \cdot \widetilde{Z}\big)\right)^2 dv\,du + F_P\big(W^{Y,\widetilde{Z}}_{\bullet,\bullet,\bullet}\big)$$

and

$$F_P\big(W^{Y,\widetilde{Z}}_{\bullet,\bullet,\bullet}\big) = \frac{6}{(-K_X)^3} \int_0^{\tau} \int_0^{\widetilde{t}(u)} \big(\widetilde{P}(u,v)\cdot\widetilde{Z}\big) \times \mathrm{ord}_P\Big(N'_{\widetilde{Y}}(u)\big|_{\widetilde{Z}} + \widetilde{N}(u,v)\big|_{\widetilde{Z}}\Big) dv\,du.$$

Moreover, if the inequality (1.24) is an equality and there exists a prime divisor E over the threefold X such that $C_X(E) = Q$ and $\delta_Q(X) = \frac{A_X(E)}{S_X(E)}$, then we have $\delta_Q(X) = \frac{1}{S_X(Y)}$. The proof of this assertion is essentially the same as the proof of Theorem 1.112.

2

Warm-Up: Smooth del Pezzo Surfaces

Let S be a smooth del Pezzo surface. Then $1 \leqslant K_S^2 \leqslant 9$, and the surface S can be described as follows:

- if $K_S^2 \in \{6,7,8,9\}$, then S is toric and one of the following cases holds:
    - $K_S^2 = 9$ and $S = \mathbb{P}^2$;
    - $K_S^2 = 8$ and $S = \mathbb{P}^1 \times \mathbb{P}^1$;
    - $K_S^2 = 8$ and S is a blow up of $\mathbb{P}^2$ in one point;
    - $K_S^2 = 7$ and S is a blow up of $\mathbb{P}^2$ in two points;
    - $K_S^2 = 6$ and S is a divisor in $\mathbb{P}^1 \times \mathbb{P}^1 \times \mathbb{P}^1$ of degree $(1,1,1)$;
- if $K_S^2 = 5$, then the surface S is unique up to isomorphism. It can be obtained as a section of the Grassmannian $\mathrm{Gr}(2,5) \subset \mathbb{P}^9$ in its Plücker embedding by a linear space of dimension 5;
- if $K_S^2 = 4$, then S is a complete intersection of two quadrics in $\mathbb{P}^4$;
- if $K_S^2 = 3$, then S is a cubic surface in $\mathbb{P}^3$;
- if $K_S^2 = 2$, then S is a quartic hypersurface in $\mathbb{P}(1,1,1,2)$;
- if $K_S^2 = 1$, then S is a sextic hypersurface in $\mathbb{P}(1,1,2,3)$.

In [215, 211], Tian and Yau proved that S is K-polystable $\iff$ it is not a blow up of $\mathbb{P}^2$ in one or two points. Let us illustrate the methods described in Chapter 1 by giving a short proof of this theorem. We will split the proof into ten lemmas, which show several ways to prove or disprove the K-polystability of the corresponding surfaces.

Lemma 2.1 *Suppose that $S = \mathbb{P}^2$. Then S is K-polystable.*

Proof Let G be a finite subgroup in $\mathrm{Aut}(S)$ such that S does not have G-fixed points, e.g. $G = \mathfrak{A}_5$ or $G = \mathrm{PSL}_2(\mathbb{F}_7)$. Then, arguing as in the proof of [48, Theorem 3.21], we see that $\alpha_G(S) \geqslant \frac{2}{3}$. Indeed, if $\alpha_G(S) < \frac{2}{3}$, then S contains a G-invariant effective $\mathbb{Q}$-divisor $D \sim_{\mathbb{Q}} -K_S$ such that the log pair $(S, \lambda D)$

is not log canonical for some positive rational number $\lambda < \frac{2}{3}$. Since S does not contain G-invariant lines, the locus $\mathrm{Nklt}(S, \lambda D)$ is zero-dimensional. Now, using Corollary A.4, we see that the locus $\mathrm{Nklt}(S, \lambda D)$ consists of a G-fixed point. Thus, $\alpha_G(S) \geqslant \frac{2}{3}$, so that S is K-polystable by Theorem 1.51.

Alternatively, we can use Theorem 1.22 to show that the surface S is K-polystable. Indeed, suppose that S is not K-polystable. By Theorem 1.22, there exists a G-invariant prime divisor F over S such that $\beta(F) = A_S(F) - S_S(F) \leqslant 0$. Let $Z = c_S(F)$. Then Z is a curve, since S does not have G-fixed points by assumption. By Corollary 1.44, we have

$$\alpha_{G,Z}(S) \leqslant \frac{2}{3} \frac{A_S(F)}{S_S(F)} \leqslant \frac{2}{3}.$$

Since S does not contain G-invariant lines, this immediately implies that Z is a G-invariant conic, which would lead to a contradiction if $G = \mathrm{PSL}_2(\mathbb{F}_7)$. In fact, using Lemma 1.45, we conclude that $\alpha_{G,Z}(X) < \frac{2}{3}$, which implies that Z is a line, which is a contradiction since S does not contain G-invariant lines. Hence, we see that S is K-polystable. $\qquad\square$

Lemma 2.2 *Suppose that* $S = \mathbb{P}^1 \times \mathbb{P}^1$. *Then S is K-polystable.*

Proof Let G be a finite subgroup in $\mathrm{Aut}(S)$ such that the following conditions hold:

(i) S does not have G-fixed points,
(ii) S does not contain G-invariant curves of degree $(1,0)$ or $(0,1)$,
(iii) S does not contain G-invariant curves of degree $(1,1)$.

For instance, if $G = \mathfrak{A}_4 \times \mathfrak{A}_4$ or $G = \mathfrak{A}_5 \times \mathfrak{A}_5$, then these three conditions hold (cf. [57]). Now, arguing as in the proof of Lemma 2.1, we see that S is K-polystable.

Alternatively, let S_2 be a quadric in $\mathbb{P}^3_{\mathbb{R}}$ that is given by $x^2 + y^2 + z^2 + t^2 = 0$, where x, y, z and t are coordinates on $\mathbb{P}^3_{\mathbb{R}}$. Then S_2 is defined over $\mathbb{R}$, it does not contain real points, and $\mathrm{Pic}_{\mathbb{R}}(S) \cong \mathbb{Z}$. This implies that $\beta(F) > 0$ for every geometrically irreducible divisor F over the surface S_2, so that S_2 is K-polystable over $\mathbb{C}$ by Remark 1.23, which implies that S is K-polystable since $S \cong S_2$ over $\mathbb{C}$. $\qquad\square$

Lemma 2.3 *Suppose that* $S = \mathbb{F}_1$. *Then S is not K-semistable.*

Proof Observe that $\mathrm{Aut}(S) \cong (\mathbb{B}_2 \times \mathbb{B}_2) \rtimes \mu_2$, where $\mathbb{B}_2$ is the Borel subgroup of $\mathrm{PGL}_2(\mathbb{C})$. Since $\mathrm{Aut}(S)$ is not reductive, the surface S is not K-polystable by Theorem 1.3.

To show that S is not K-semistable, let E be the unique (-1)-curve in the surface S, and let L be a fiber of the natural projection $S \to \mathbb{P}^1$. Then $-K_X \sim 3L + 2E$, so that

$$\beta(E) = 1 - \frac{1}{8} \int_0^2 \left(3L + (2-x)E\right)^2 dx = 1 - \frac{1}{8} \int_0^2 (8 - 2x - x^2) dx = -\frac{1}{6},$$

which implies that S is not K-semistable by Theorem 1.19. □

Lemma 2.4 *Suppose that $K_S^2 = 7$. Then S is not K-semistable.*

Proof First, we observe that $\mathrm{Aut}(S) \cong \mathbb{G}_a^2 \rtimes \mathrm{PGL}_2(\mathbb{C})$. Since this group is not reductive, we conclude that the surface S is not K-polystable by Theorem 1.3.

To show that S is not K-semistable, let E_1, E_2 and E be (-1)-curves in S such that we have $E_1 \cdot E_2 = 0$, $E_1 \cdot E = 1$ and $E_2 \cdot E = 1$. Then $-K_S \sim 3E + 2E_1 + 2E_2$.

Let us compute $\beta(E)$. Take $x \in \mathbb{R}_{>0}$. If $x \leqslant 1$, then $-K_S - xE$ is nef, so that

$$\mathrm{vol}\left(-K_S - xE\right) = \left(-K_S - xE\right)^2 = 7 - 2x - x^2.$$

Similarly, if $1 < x \leqslant 3$, then the Zariski decomposition of the divisor $-K_S - xE$ is

$$-K_S - xE \sim_{\mathbb{R}} \underbrace{(3-x)\left(E + E_1 + E_2\right)}_{\text{positive part}} + \underbrace{(x-1)\left(E_1 + E_2\right)}_{\text{negative part}},$$

which gives $\mathrm{vol}(-K_S - xE) = (3-x)^2$. If $x > 3$, then $-K_S - xE$ is not pseudo-effective. Integrating, we get $\beta(E) = -\frac{4}{21}$, so S is not K-semistable by Theorem 1.19. □

Lemma 2.5 *Suppose that $K_S^2 = 6$. Then S is K-polystable.*

Proof It is well known (see [77]) that there exists the following exact sequence of groups:

$$1 \longrightarrow \mathbb{G}_m^2 \longrightarrow \mathrm{Aut}(S) \longrightarrow \mathfrak{S}_3 \times \mu_2.$$

This implies that $\mathrm{Aut}(S)$ contains a finite subgroup G such that S has no G-fixed points and $\mathrm{Pic}^G(S) = \mathbb{Z}[-K_S]$. Now, the proof of Lemma 2.1 implies the required assertion. □

Lemma 2.6 *Suppose that $K_S^2 = 5$. Then S is K-stable.*

Proof Recall from [77] that $\mathrm{Aut}(S) \cong \mathfrak{S}_5$. Let G be a subgroup in $\mathrm{Aut}(S)$. Then

$$\mathrm{Pic}^G(S) = \mathbb{Z}[-K_S] \iff G \text{ is one of the following groups: } \mu_5, \mathrm{D}_{10}, \mu_5 \rtimes \mu_4,$$
$$\mathfrak{A}_5, \mathfrak{S}_5.$$

Moreover, if $G \cong \mu_5$, then $\alpha_G(S) = \frac{4}{5}$ by [30, Lemma 5.8]. Thus, if $\mathrm{Pic}^G(S) = \mathbb{Z}[-K_S]$, then $\alpha_G(S) \geq \frac{4}{5}$, cf. Lemma A.43. Hence, we see that S is K-stable by Theorem 1.48.

We can also prove the assertion using Remark 1.49. Namely, let $f(t)$ be an irreducible quintic polynomial in $\mathbb{Q}[t]$, and let $\xi_1, \xi_2, \xi_3, \xi_4, \xi_5$ be its roots in $\mathbb{C}$. Then

$$\mathrm{Gal}\big(\mathbb{Q}(\xi_1, \xi_2, \xi_3, \xi_4, \xi_5), \mathbb{Q}\big) \text{ is one of the following groups: } \mu_5, \mathrm{D}_{10}, \mu_5 \rtimes \mu_4,$$
$$\mathfrak{A}_5, \mathfrak{S}_5.$$

Let Σ be the reduced subscheme of the plane $\mathbb{P}^2_{\mathbb{Q}}$ that consists of the points $[\xi_1 : \xi_1^2 : 1]$, $[\xi_2 : \xi_2^2 : 1]$, $[\xi_3 : \xi_3^2 : 1]$, $[\xi_4 : \xi_4^2 : 1]$, $[\xi_5 : \xi_5^2 : 1]$, and let C be the conic $\{yz = x^2\} \subset \mathbb{P}^2_{\mathbb{Q}}$, where x, y, z are coordinates on $\mathbb{P}^2_{\mathbb{Q}}$. Then C contains Σ, and we have the diagram

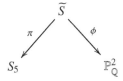

where ϕ is a blow up of the subscheme Σ, and π is a birational contraction of the proper transform of the conic C. Then S_5 is a smooth del Pezzo surface, which is defined over $\mathbb{Q}$. Then we have $\mathrm{Pic}_{\mathbb{Q}}(S_5) = \mathbb{Z}[-K_{S_5}]$ by construction, so that $\alpha(S_5) \geq \frac{4}{5}$ by Lemma A.43. Then S_5 is K-stable over $\mathbb{C}$ by Remark 1.49 and Corollary 1.5, which implies that the surface S is K-stable since $S \cong S_5$ over $\mathbb{C}$. □

Lemma 2.7 *Suppose that $K_S^2 = 4$. Then S is K-stable.*

Proof It follows from [187, Proposition 2.1] that S can be given by

$$S = \big\{x_0^2 + x_1^2 + x_2^2 + x_3^2 + x_4^2 = 0, \lambda_0 x_0^2 + \lambda_1 x_1^2 + \lambda_2 x_2^2 + \lambda_3 x_3^2 = 0\big\} \subset \mathbb{P}^4$$

for some non-zero numbers λ_0, λ_1, λ_2 and λ_3, where x_0, x_1, x_2, x_3, x_4 are coordinates on $\mathbb{P}^4$. Let G be a subgroup in $\mathrm{Aut}(S)$ that is generated by

$$[x_0 : x_1 : x_2 : x_3 : x_4] \mapsto \big[x_0 : (-1)^a x_1 : (-1)^b x_2 : (-1)^c x_3 : (-1)^d x_4\big]$$

for all possible a, b, c and d in $\{0, 1\}$. Then $G \cong \mu_2^4$, the surface S has no G-fixed points, and $\mathrm{Pic}^G(S) = \mathbb{Z}[-K_S]$. Hence, all G-invariant prime divisors over S are curves in S. On the other hand, if C is a G-invariant curve in S, then $C \sim m(-K_S)$ for some $m \in \mathbb{N}$, which implies that $\beta(C) = 1 - \frac{1}{3m} > 0$, so that S is K-polystable by Theorem 1.22. Then S is K-stable by Corollary 1.5 because the group $\mathrm{Aut}(S)$ is finite.

Arguing as in the proof of Lemma 2.1, we can also show that $\alpha_G(S) \geqslant 1$, cf. [150, 30]. This would imply that S is K-stable by Theorem 1.48 and Corollary 1.5.

Alternatively, we can prove the K-stability of the surface S using Corollary 1.67. Indeed, the projection $\mathbb{P}^4 \dashrightarrow \mathbb{P}^3$ given by $[x_0 : x_1 : x_2 : x_3 : x_4] \mapsto [x_0 : x_1 : x_2 : x_3]$ induces a double cover $S \to S_2$, where S_2 is a smooth quadric in $\mathbb{P}^3$. This double cover is branched over a smooth anticanonical elliptic curve in S_2, so that S is K-stable by Corollary 1.67 because the quadric surface S_2 is K-polystable by Lemma 2.2.   □

Lemma 2.8 *Suppose that $K_S^2 = 3$. Then S is K-stable.*

Proof We claim that $\alpha(S) \geqslant \frac{2}{3}$. Indeed, suppose that $\alpha(S) < \frac{2}{3}$. Then there is an effective $\mathbb{Q}$-divisor D on the surface S such that $D \sim_{\mathbb{Q}} -K_X$, and the log pair $(S, \lambda D)$ is not log canonical for some positive rational number $\lambda < \frac{2}{3}$. Let us seek for a contradiction.

We claim that the locus $\mathrm{Nklt}(S, \lambda D)$ does not contain curves. Indeed, if it does, then the surface S contains an irreducible curve C such that $D = aC + \Delta$ for some $a \geqslant \frac{1}{\lambda} > \frac{3}{2}$, where Δ is an effective $\mathbb{Q}$-divisor such that $C \not\subset \mathrm{Supp}(\Delta)$. Then

$$3 = -K_S \cdot D = a(-K_S) \cdot C - K_S \cdot \Delta \geqslant a(-K_S) \cdot C > \frac{3}{2}(-K_S) \cdot C,$$

which gives $-K_S \cdot C < 2$. Then $-K_S \cdot C = 1$, so that C is a (-1)-curve in the surface S. Since S is a cubic surface in $\mathbb{P}^3$, we see that C is a line. Let H be a general hyperplane section of the surface S that contains C. Then $H = C + Z$, where Z is an irreducible conic such that the intersection $Z \cap C$ consists of two points. Moreover, the generality in the choice of the hyperplane H implies that $Z \not\subset \mathrm{Supp}(D)$. Therefore, we have

$$2 = -K_S \cdot Z = D \cdot Z = (aC + \Delta) \cdot Z \geqslant aC \cdot Z + \Delta \cdot Z \geqslant aC \cdot Z = 2a,$$

so that $a \leqslant 1$. The obtained contradiction shows that $\mathrm{Nklt}(S, \lambda D)$ contains no curves.

Using Corollary A.4, we see that the locus $\mathrm{Nklt}(S, \lambda D)$ consists of a single point O. Since O is contained in at most three (-1)-curves, S contains 6 disjoint (-1)-curves that do not contain O. Let $\pi \colon S \to \mathbb{P}^2$ be the birational contraction of these six (-1)-curves, and let L be a line in $\mathbb{P}^2$ that does not contain $\pi(O)$. Then $L \cup O \subseteq \mathrm{Nklt}(\mathbb{P}^2, L + \lambda\pi(D))$, but $\mathrm{Nklt}(\mathbb{P}^2, L + \lambda\pi(D))$ contains no curves except L. This contradicts Corollary A.4. Then $\alpha(S) \geqslant \frac{2}{3}$, so that S is K-stable by Theorem 1.50.   □

Lemma 2.9 *Suppose that $K_S^2 = 2$. Then S is K-stable.*

Proof In this case S is a double cover of $\mathbb{P}^2$ that is branched over a smooth quartic curve, so that S is K-stable by Corollary 1.67. Alternatively, we can prove that S is K-stable arguing as in the proof of Lemma 2.8. □

Lemma 2.10 *Suppose that* $K_S^2 = 1$. *Then* S *is K-stable.*

Proof We claim that $\alpha(S) \geqslant \frac{5}{6}$. Indeed, suppose that $\alpha(S) < \frac{5}{6}$. Then there is an effective $\mathbb{Q}$-divisor D on the surface S such that $D \sim_{\mathbb{Q}} -K_X$, and $(S, \lambda D)$ is not log canonical at some point $P \in S$ for some $\lambda \in \mathbb{Q} \cap (0, \frac{5}{6})$. Let C be a curve in $|-K_S|$ that contains P. Then C is irreducible, and the log pair $(S, \lambda C)$ is log canonical. Thus, using Lemma A.34, we may assume that $C \not\subset \mathrm{Supp}(D)$. Then

$$1 = K_S^2 = -K_S \cdot D = C \cdot D \geqslant \mathrm{mult}_P(D) > \frac{1}{\lambda} > \frac{6}{5},$$

which is absurd. This shows that $\alpha(S) \geqslant \frac{5}{6}$. Then S is K-stable by Theorem 1.48.

Alternatively, one can also show that the surface S is K-stable using Proposition 1.66. Indeed, the surface S is a double cover of $\mathbb{P}(1, 1, 2)$ branched over a smooth sextic curve. Then S is K-stable by Proposition 1.66 since $\delta(\mathbb{P}(1, 1, 2)) = \frac{3}{4}$ by [18, Corollary 7.7]. □

All possible values of the number $\alpha(S)$ have been found in [30, 154], cf. Section A.5. In particular, we have $\alpha(S) \geqslant \frac{2}{3} \iff K_S^2 = 4$. Moreover, as we mentioned in Section 1.5, $\frac{3\alpha(S)}{2} \leqslant \delta(S) \leqslant 3\alpha(S)$, which gives certain estimates for $\delta(S)$. These estimates have been improved in [179, 58]. If $K_S^2 \geqslant 6$ or $K_S^2 = 3$, all possible values of the number $\delta(S)$ have been found in [18, 2]. In the remaining part of this section, we show how to compute $\delta(S)$ for $K_S^2 \in \{1, 2, 3, 4, 5\}$. These results are summarized in Table 2.1.

In particular, we observe that $\delta(S) > 1 \iff K_S^2 \leqslant 5$. This gives another proof that the surface S is K-stable $\iff K_S^2 \leqslant 5$, which follows from Lemmas 2.6, 2.7, 2.8, 2.9, 2.10.

In the proof of the following five lemmas, we will use notations introduced in Section 1.7, which include notations used in Theorem 1.95 and Corollaries 1.102, 1.108, 1.109.

Lemma 2.11 *Suppose that* $K_S^2 = 5$. *Then* $\delta(S) = \frac{15}{13}$.

Proof The surface S can be obtained by blowing up $\mathbb{P}^2$ at four points P_1, P_2, P_3, P_4 no three of which lie on a line. Let E_1, E_2, E_3 and E_4 be the exceptional curves of this blow up that are mapped to the points P_1, P_2, P_3 and P_4, respectively. For every $1 \leqslant i < j \leqslant 4$, we denote by L_{ij} the proper transform on S of the line in $\mathbb{P}^2$ that passes through the points P_i and P_j. Then E_1, E_2, E_3, E_4, L_{12}, L_{13}, L_{14}, L_{23}, L_{24}, L_{34} are all (-1)-curves in the surface S, so that they generate the Mori cone of the surface S. Moreover, the group

Table 2.1 *Computing* $\delta(S)$ *for* $K_S^2 \in \{1,2,3,4,5\}$.

Smooth del Pezzo surface S	K_S^2	$\alpha(S)$	$\delta(S)$
$\mathbb{P}^2$	9	$\frac{1}{3}$	1
$\mathbb{P}^1 \times \mathbb{P}^1$	8	$\frac{1}{2}$	1
the blow up of $\mathbb{P}^2$ in one point	8	$\frac{1}{3}$	$\frac{6}{7}$
the blow up of $\mathbb{P}^2$ in two points	7	$\frac{1}{3}$	$\frac{21}{25}$
a divisor in $\mathbb{P}^1 \times \mathbb{P}^1 \times \mathbb{P}^1$ of degree $(1,1,1)$	6	$\frac{1}{2}$	1
a section of the Grassmannian $\mathrm{Gr}(2,5) \subset \mathbb{P}^9$ in its Plücker embedding by a linear space of codimension 4	5	$\frac{1}{2}$	$\frac{15}{13}$
a complete intersection of two quadrics in $\mathbb{P}^4$	4	$\frac{2}{3}$	$\frac{4}{3}$
a cubic surface in $\mathbb{P}^3$ with an Eckardt point	3	$\frac{2}{3}$	$\frac{3}{2}$
a cubic surface in $\mathbb{P}^3$ without an Eckardt point	3	$\frac{3}{4}$	$\frac{27}{17}$
a quartic surface in $\mathbb{P}(1,1,1,2)$ such that the linear system $\lvert -K_S \rvert$ contains a tacnodal curve	2	$\frac{3}{4}$	$\frac{9}{5}$
a quartic surface in $\mathbb{P}(1,1,1,2)$ such that the linear system $\lvert -K_S \rvert$ does not contain tacnodal curves	2	$\frac{5}{6}$	$\frac{15}{8}$
a sextic surface in $\mathbb{P}(1,1,2,3)$ such that the linear system $\lvert -K_S \rvert$ contains a cuspidal curve	1	$\frac{5}{6}$	$\frac{15}{7}$
a sextic surface in $\mathbb{P}(1,1,2,3)$ such that the linear system $\lvert -K_S \rvert$ does not contain cuspidal curves	1	1	$\frac{12}{5}$

$\mathrm{Aut}(S) \cong \mathfrak{S}_5$ acts transitively on the set of these ten curves. Thus, for every irreducible curve $C \subset S$, we have $S_S(C) \leqslant S_S(E_1)$.

Let us compute $S_S(E_1)$. Take $u \in \mathbb{R}_{\geqslant 0}$. Then $-K_S - uE_1 \sim_{\mathbb{R}} (2-u)E_1 + L_{12} + L_{13} + L_{14}$, so that the divisor $-K_S - uE_1$ is pseudo-effective $\iff u \leqslant 2$. Moreover, if $u \in [0,1]$, then $-K_S - uE_1$ is nef. Furthermore, if $u \in [1,2]$, then its Zariski decomposition is

$$-K_S - uE_1 \sim_{\mathbb{R}} \underbrace{(2-u)\big(E_1 + L_{12} + L_{13} + L_{14}\big)}_{\text{positive part}} + \underbrace{(u-1)\big(L_{12} + L_{13} + L_{14}\big)}_{\text{negative part}}.$$

Thus, in the notations of Corollary 1.108 with $X = S$, $Y = C$, $L = -K_S$, we have

$$P(u) = \begin{cases} (2-u)E_1 + L_{12} + L_{13} + L_{14} & \text{if } 0 \leqslant u \leqslant 1, \\ (2-u)\big(E_1 + L_{12} + L_{13} + L_{14}\big) & \text{if } 1 \leqslant u \leqslant 2, \end{cases}$$

and

$$N(u) = \begin{cases} 0 & \text{if } 0 \leqslant u \leqslant 1, \\ (u-1)(L_{12} + L_{13} + L_{14}) & \text{if } 1 \leqslant u \leqslant 2. \end{cases}$$

Therefore, we have

$$\text{vol}(-K_S - uE_1) = \begin{cases} 5 - 2u - u^2 & \text{if } 0 \leqslant u \leqslant 1, \\ 2(2-u)^2 & \text{if } 1 \leqslant u \leqslant 2. \end{cases}$$

Thus, integrating, we get $S_S(E_1) = \frac{13}{15}$. In particular, we have

$$\delta(S) = \inf_{E/S} \frac{A_S(E)}{S_S(E)} \leqslant \frac{A_S(E_1)}{S_S(E_1)} = \frac{15}{13},$$

where the infimum is taken over all prime divisors over S.

Let us show that $\delta(S) \geqslant \frac{15}{13}$. Suppose that this is not true. Then there exists a prime divisor E over S such that $\frac{A_S(E)}{S_S(E)} < \frac{15}{13}$. If E is a curve in S, then $S_S(E) \leqslant S_S(E_1) = \frac{13}{15}$, which is impossible. Thus, we see that $C_S(E)$ is a point. Let $P = C_S(E)$.

Let C be an irreducible smooth curve in the surface S that passes through the point P. By Theorem 1.95 and Corollary 1.109, we have

$$\frac{15}{13} > \frac{A_S(E)}{S_S(E)} \geqslant \min\left\{ \frac{1}{S_S(C)}, \frac{1}{S(W^C_{\bullet,\bullet}; P)} \right\},$$

where we use the notation of Corollary 1.109 with $X = S$, $Y = C$, $L = -K_S$ and $Z = P$. On the other hand, we have $S_S(C) \leqslant S_S(E_1) \leqslant \frac{13}{15}$. Therefore, we have $S(W^C_{\bullet,\bullet}; P) > \frac{13}{15}$. Moreover, it follows from Corollary 1.109 that

$$S(W^C_{\bullet,\bullet}; P) = \frac{2}{K_S^2} \int_0^\tau h(u)du = \frac{2}{5} \int_0^\tau h(u)du,$$

where τ is the largest real number such that $-K_S - uC$ is pseudo-effective, and

$$h(u) = (P(u) \cdot C) \times \text{ord}_P(N(u)|_C) + \int_0^\infty \text{vol}_C(P(u)|_C - vP)dv$$

$$= (P(u) \cdot C) \times (N(u) \cdot C)_P + \int_0^{P(u) \cdot C} (P(u) \cdot C - v)dv$$

$$= (P(u) \cdot C) \times (N(u) \cdot C)_P + \frac{(P(u) \cdot C)^2}{2}.$$

Suppose that $P \in E_1$. In this case, it is natural to let $C = E_1$. Then we have $\tau = 2$, and both $\mathbb{R}$-divisors $P(u)$ and $N(u)$ have been already computed earlier in the proof. In particular, if $P \notin L_{12} \cup L_{13} \cup L_{14}$, then $P \notin \mathrm{Supp}(N(u))$ for every $u \in [0,2]$, so that

$$h(u) = \left(P(u) \cdot E_1\right) \times \left(N(u) \cdot E_1\right)_P + \frac{\left(P(u) \cdot E_1\right)^2}{2}$$

$$= \frac{\left(P(u) \cdot E_1\right)^2}{2} = \begin{cases} (1+u)^2/2 & \text{if } 0 \leqslant u \leqslant 1, \\ 2(2-u)^2 & \text{if } 1 \leqslant u \leqslant 2, \end{cases}$$

which gives $S(W_{\bullet,\bullet}^C; P) = \frac{11}{15}$. Similarly, if $P \in L_{12} \cup L_{13} \cup L_{34}$, then

$$h(u) = \begin{cases} \dfrac{\left(P(u) \cdot E_1\right)^2}{2} & \text{if } 0 \leqslant u \leqslant 1, \\ (u-1)\left(P(u) \cdot E_1\right) + \dfrac{\left(P(u) \cdot E_1\right)^2}{2} & \text{if } 1 \leqslant u \leqslant 2, \end{cases}$$

so that

$$h(u) = \begin{cases} \dfrac{(1+u)^2}{2} & \text{if } 0 \leqslant u \leqslant 1, \\ 2(u-1)(2-u) + 2(2-u)^2 & \text{if } 1 \leqslant u \leqslant 2, \end{cases}$$

which gives us $S(W_{\bullet,\bullet}^C; P) = \frac{13}{15}$. Since we know that $S(W_{\bullet,\bullet}^C; P) > \frac{13}{15}$, we get $P \notin E_1$. Similarly, we see that P is not contained in any (-1)-curve in S.

Let Z_0, Z_1, Z_2, Z_3, Z_4 be the curves in the pencils $|L_{12} + L_{34}|$, $|L_{13} + E_3|$, $|L_{24} + E_4|$, $|L_{13} + E_1|$, $|L_{24} + E_2|$, respectively, that contain P. Then Z_0, Z_1, Z_2, Z_3, Z_4 are smooth and irreducible because P is not contained in any (-1)-curve in S. In fact, these five curves are all (0)-curves in S that pass through the point P, see the proof of Lemma A.43.

Let $\sigma \colon \widetilde{S} \to S$ be the blow up of the point P, let E_P be the σ-exceptional curve, and let $\widetilde{Z}_0, \widetilde{Z}_1, \widetilde{Z}_2, \widetilde{Z}_3, \widetilde{Z}_4$ be the proper transforms on $\widetilde{S}$ of the curves Z_0, Z_1, Z_2, Z_3, Z_4, respectively. Then $A_S(E_P) = 2$. Let us compute $S_S(E_P)$. To do this, we observe that

$$\sigma^*(-K_S) - uE_P \sim_{\mathbb{R}} \left(\frac{5}{2} - u\right)E_P + \frac{1}{2}\left(\widetilde{Z}_0 + \widetilde{Z}_1 + \widetilde{Z}_2 + \widetilde{Z}_3 + \widetilde{Z}_4\right).$$

Abusing our previous notations, we denote by $P(u)$ and $N(u)$ the positive and the negative parts of the Zariski decomposition of the divisor $\sigma^*(-K_S) - uE_P$, respectively. Then

$$P(u) = \begin{cases} \sigma^*(-K_S) - uE_P & \text{if } 0 \leqslant u \leqslant 2, \\ \left(\dfrac{5}{2} - u\right)\left(E_P + \widetilde{Z}_0 + \widetilde{Z}_1 + \widetilde{Z}_2 + \widetilde{Z}_3 + \widetilde{Z}_4\right) & \text{if } 2 \leqslant u \leqslant \dfrac{5}{2}, \end{cases}$$

and

$$N(u) = \begin{cases} 0 & \text{if } 0 \leqslant u \leqslant 2, \\ (u-2)(\tilde{Z}_0 + \tilde{Z}_1 + \tilde{Z}_2 + \tilde{Z}_3 + \tilde{Z}_4) & \text{if } 2 \leqslant u \leqslant \dfrac{5}{2}, \end{cases}$$

so that

$$\text{vol}\big(\sigma^*(-K_S) - uE_P\big) = P(u) \cdot P(u) = \begin{cases} 5 - u^2 & \text{if } 0 \leqslant u \leqslant 2, \\ (5 - 2u)^2 & \text{if } 2 \leqslant u \leqslant \dfrac{5}{2}. \end{cases}$$

Integrating, we get $S_S(E_P) = \frac{3}{2}$, so that $\frac{A_S(E_P)}{S_S(E_P)} = \frac{4}{3} > \frac{15}{13}$, which implies that $C_{\tilde{S}}(E) \neq E_P$. On the one hand, it follows from Corollary 1.102 that

$$\frac{15}{13} > \frac{A_S(E)}{S_S(E)} \geqslant \delta_P(S) \geqslant \min\left\{ \frac{A_S(E_P)}{S_S(E_P)}, \inf_{O \in E_P} \delta_O\big(E_P; W^{E_P}_{\bullet,\bullet}\big) \right\}.$$

Thus, there is a point $O \in E_P$ such that $\delta_O(E_P; W^{E_P}_{\bullet,\bullet}) < \frac{15}{13}$. Recall from (1.12) that

$$\delta_O\big(E_P; W^{E_P}_{\bullet,\bullet}\big) = \frac{1}{S(W^{E_P}_{\bullet,\bullet}; O)},$$

so that $S(W^{E_P}_{\bullet,\bullet}; O) > \frac{13}{15}$. But $\tilde{S}$ is a smooth (quartic) del Pezzo surface by construction. Hence, we can apply Corollary 1.109 to compute $S(W^{E_P}_{\bullet,\bullet}; O)$. This gives

$$S\big(W^{E_P}_{\bullet,\bullet}; O\big)$$

$$= \frac{2}{5} \int_0^{\frac{5}{2}} \left((P(u) \cdot E_P) \text{ord}_O\big(N(u)\big|_{E_P}\big) + \int_0^\infty \text{vol}_{E_P}\big(P(u)\big|_{E_P} - vO\big) dv \right) du$$

$$= \frac{2}{5} \int_0^{\frac{5}{2}} \left((P(u) \cdot E_P)(N(u) \cdot E_P)_O + \int_0^{P(u) \cdot E_P} (P(u) \cdot E_P - v) dv \right) du$$

$$= \frac{2}{5} \int_0^{\frac{5}{2}} \left((P(u) \cdot E_P)(N(u) \cdot E_P)_O + \frac{(P(u) \cdot E_P)^2}{2} \right) du.$$

Now, using the description of $P(u)$ and $N(u)$ obtained earlier, we see that

$$S\big(W^{E_P}_{\bullet,\bullet}; O\big) = \frac{2}{5} \int_2^{\frac{5}{2}} 2(5 - 2u)(u - 2) du \times \big((\tilde{Z}_0 + \tilde{Z}_1 + \tilde{Z}_2 + \tilde{Z}_3 + \tilde{Z}_4) \cdot E_P\big)_O$$

$$+ \frac{2}{5} \int_0^2 \frac{u^2}{2} du + \frac{2}{5} \int_2^{\frac{5}{2}} 2(5 - 2u)^2 du$$

$$= \frac{1}{30} \big((\tilde{Z}_0 + \tilde{Z}_1 + \tilde{Z}_2 + \tilde{Z}_3 + \tilde{Z}_4) \cdot E_P\big)_O + \frac{2}{3}.$$

Thus, if $O \in \tilde{Z}_0 \cup \tilde{Z}_1 \cup \tilde{Z}_2 \cup \tilde{Z}_3 \cup \tilde{Z}_4$, then $S(W^{E_P}_{\bullet,\bullet}; O) = \frac{7}{10}$. Otherwise, $S(W^{E_P}_{\bullet,\bullet}; O) = \frac{2}{3}$. Then $S(W^{E_P}_{\bullet,\bullet}; O) < \frac{13}{15}$, which is a contradiction. $\square$

Lemma 2.12   *Suppose that $K_S^2 = 4$. Then $\delta(S) = \frac{4}{3}$.*

Proof   There exists a birational morphism $\pi \colon S \to \mathbb{P}^2$ that blows up five (general) points. Let E_1, E_2, E_3, E_4, E_5 be the exceptional curves of the morphism π, let C be the proper transform on S of the conic in $\mathbb{P}^2$ that passes through $\pi(E_1)$, $\pi(E_2)$, $\pi(E_3)$, $\pi(E_4)$, $\pi(E_5)$, and let L_{ij} be the proper transform on S of the line that passes through $\pi(E_i)$ and $\pi(E_j)$, where $1 \leqslant i < j \leqslant 5$. Then the curves E_1, E_2, E_3, E_4, E_5, C, L_{12}, L_{13}, L_{14}, L_{15}, L_{23}, L_{24}, L_{25}, L_{34}, L_{35}, L_{45} are all (-1)-curves in the del Pezzo surface S. Moreover, arguing as in the proof of Lemma 2.11, we see that $S_S(Z) = \frac{17}{24}$ for any (-1)-curve Z in the surface S.

Let $\sigma \colon \widetilde{S} \to S$ be the blow up of the point $E_1 \cap C$, let E be the σ-exceptional curve, let $\widetilde{E}_1$ and $\widetilde{C}$ be the proper transforms on $\widetilde{S}$ of the (-1)-curves E_1 and C, respectively, and let $\widetilde{L}$ be the proper transform on $\widetilde{S}$ of the line in $\mathbb{P}^2$ that is tangent to $\pi(C)$ at $\pi(E_1)$. Then $\sigma^*(-K_S) - uE \sim_{\mathbb{R}} (3 - u)E + \widetilde{E}_1 + \widetilde{C} + \widetilde{L}$, where u is a non-negative real number. Moreover, the curves $\widetilde{E}_1$, $\widetilde{C}$ and $\widetilde{L}$ are disjoint, and we have $\widetilde{E}_1^2 = \widetilde{C}^2 = -2$ and $\widetilde{L}^2 = -1$. Therefore, we conclude that the divisor $\sigma^*(-K_S) - uE$ is pseudo-effective $\iff u \leqslant 3$. Denote by $P(u)$ and $N(u)$ the positive and the negative parts of its Zariski decomposition, respectively. Then

$$
P(u) = \begin{cases} (3 - u)E + \widetilde{E}_1 + \widetilde{C} + \widetilde{L} & \text{if } 0 \leqslant u \leqslant 1, \\[2mm] (3 - u)\left(E + \dfrac{1}{2}\widetilde{E}_1 + \dfrac{1}{2}\widetilde{C}\right) + \widetilde{L} & \text{if } 1 \leqslant u \leqslant 2, \\[2mm] (3 - u)\left(E + \dfrac{1}{2}\widetilde{E}_1 + \dfrac{1}{2}\widetilde{C} + \widetilde{L}\right) & \text{if } 2 \leqslant u \leqslant 3, \end{cases}
$$

and

$$
N(u) = \begin{cases} 0 & \text{if } 0 \leqslant u \leqslant 1, \\[2mm] \dfrac{u - 1}{2}\left(\widetilde{E}_1 + \widetilde{C}\right) & \text{if } 1 \leqslant u \leqslant 2, \\[2mm] \dfrac{u - 1}{2}\left(\widetilde{E}_1 + \widetilde{C}\right) + (u - 2)\widetilde{L} & \text{if } 2 \leqslant u \leqslant 3. \end{cases}
$$

Now, integrating $\mathrm{vol}(\sigma^*(-K_S) - uE_P) = P(u) \cdot P(u)$ from $u = 0$ to $u = 3$, we get

$$
S_S(E) = \frac{1}{4}\int_0^1 (4 - u^2)\,du + \frac{1}{4}\int_1^2 (5 - 2u)\,du + \frac{1}{4}\int_2^3 (u - 3)^2\,du = \frac{3}{2},
$$

so that $\delta(S) \leqslant \frac{A_S(E)}{S_S(E)} = \frac{4}{3}$. Moreover, if $P = E_1 \cap C$, then Corollary 1.102 and (1.12) give

$$
\frac{4}{3} \geqslant \delta_P(S) \geqslant \min\left\{\frac{A_S(E)}{S_S(E)}, \inf_{O \in E} \delta_O\left(E; W_{\bullet,\bullet}^E\right)\right\} = \min\left\{\frac{4}{3}, \inf_{O \in E} \frac{1}{S(W_{\bullet,\bullet}^E; O)}\right\}.
$$

But $\widetilde{S}$ is a weak del Pezzo surface, so that Corollary 1.109 gives

$$S\left(W_{\bullet,\bullet}^E; O\right) = \frac{1}{2}\int_0^3\left((P(u)\cdot E)(N(u)\cdot E)_O + \frac{(P(u)\cdot E)^2}{2}\right)du \leqslant \frac{17}{24}$$

for every point $O \in E$. If P is the intersection point $E_1 \cap C$, then $\delta_P(S) = \frac{4}{3}$. Likewise, we see that $\delta_P(S) = \frac{4}{3}$ if P is an intersection point of any two (-1)-curves in S.

Now, let us show that $\delta_P(S) \geqslant \frac{4}{3}$ for every point $P \in C$. Take $u \in \mathbb{R}_{\geqslant 0}$. Then

$$-K_S - uC \sim_{\mathbb{R}} \left(\frac{3}{2} - u\right)C + \frac{1}{2}(E_1 + E_2 + E_3 + E_4 + E_5),$$

so that the divisor $-K_S - uC$ is pseudo-effective $\iff u \leqslant \frac{3}{2}$, cf. the proof of Lemma 2.11. Abusing our previous notations, denote by $P(u)$ and $N(u)$ the positive and the negative parts of the Zariski decomposition of the divisor $-K_S - uC$, respectively. Then

$$P(u) = \begin{cases} -K_S - uC & \text{if } 0 \leqslant u \leqslant 1, \\ (3 - 2u)\pi^*\left(O_{\mathbb{P}^2}(1)\right) & \text{if } 1 \leqslant u \leqslant \frac{3}{2}, \end{cases}$$

and

$$N(u) = \begin{cases} 0 & \text{if } 0 \leqslant u \leqslant 1, \\ (u - 1)(E_1 + E_2 + E_3 + E_4 + E_5) & \text{if } 1 \leqslant u \leqslant \frac{3}{2}, \end{cases}$$

so that

$$\text{vol}\left(-K_S - uC\right) = \begin{cases} 4 - 2u - u^2 & \text{if } 0 \leqslant u \leqslant 1, \\ (3 - 2u)^2 & \text{if } 1 \leqslant u \leqslant \frac{3}{2}, \end{cases}$$

giving $S_S(C) = \frac{17}{24}$, as already mentioned. Now, using Corollary 1.109, we get

$$S(W_{\bullet,\bullet}^C; P) = \frac{1}{2}\int_0^{\frac{3}{2}}\left((P(u)\cdot C)(N(u)\cdot C)_P + \frac{(P(u)\cdot C)^2}{2}\right)du$$

$$= \frac{1}{2}\int_0^{\frac{3}{2}}\frac{(P(u)\cdot C)^2}{2}du$$

$$= \frac{1}{2}\int_0^1\frac{(1 + u)^2}{2}du + \frac{1}{2}\int_1^{\frac{3}{2}}\frac{(6 - 4u)^2}{2}du = \frac{3}{4}$$

for every point $P \in C \setminus (E_1 \cup E_2 \cup E_3 \cup E_4 \cup E_5)$. Hence, it follows from Theorem 1.95 that $\delta_P(C) \geqslant \frac{4}{3}$ for every point $P \in C$. Similarly, we see that the same inequality holds for every point of the surface S that is contained in a (-1)-curve.

Let P be a point in S that is not contained in any (-1)-curve. To complete the proof, it is enough to show that $\delta_P(S) \geqslant \frac{4}{3}$. We will do this arguing as in the proof of Lemma 2.11.

Let $\upsilon\colon \widehat{S} \to S$ be the blow up of the point P, let E_P be the υ-exceptional divisor, let $\widehat{L}_P$ be the proper transform on $\widehat{S}$ of the line in $\mathbb{P}^2$ that passes through $\pi(P)$ and $\pi(E_1)$, let $\widehat{Z}$ be the proper transform of the conic that contains $\pi(P)$, $\pi(E_2)$, $\pi(E_3)$, $\pi(E_4)$, $\pi(E_5)$, and let u be a non-negative real number. Then $\upsilon^*(-K_S)-uE_P \sim_{\mathbb{R}} (2-u)E_P + \widehat{Z}+\widehat{L}_P$, the curves $\widehat{Z}$ and $\widehat{L}_P$ meet transversally in one point, and $\widehat{Z}^2 = \widehat{L}_P^2 = -1$. Using this, we conclude that the divisor $\upsilon^*(-K_S) - uE_P$ is pseudo-effective $\iff \upsilon^*(-K_S) - uE_P$ is nef $\iff u \leqslant 2$. This implies that $S_S(E_P) = \frac{4}{3}$, so that $\frac{A_S(E_P)}{S_S(E_P)} = \frac{3}{2}$. Now, using Corollary 1.102, we get

$$\delta_P(S) \geqslant \min\left\{\frac{3}{2}, \inf_{O\in E_P} \delta_O\left(E_P; W^{E_P}_{\bullet,\bullet}\right)\right\}.$$

But $\widehat{S}$ is a smooth cubic surface. Hence, using Corollary 1.109, we get

$$\frac{1}{\delta_O\left(E_P; W^{E_P}_{\bullet,\bullet}\right)} = S\left(W^{E_P}_{\bullet,\bullet}; O\right) = \frac{1}{2}\int_0^2 \frac{(P(u)\cdot E_P)^2}{2}\,du = \frac{1}{4}\int_0^2 u^2\,du = \frac{2}{3}$$

for every point $O \in E_P$. This shows that $\delta_P(S) \geqslant \frac{3}{2} > \frac{4}{3}$, which completes the proof. $\square$

The next lemma has been proved in [2]. We present its (slightly simplified) proof.

Lemma 2.13 *Suppose that S is a smooth cubic surface in $\mathbb{P}^3$. Then*

$$\delta(S) = \begin{cases} \dfrac{3}{2} & \text{if } S \text{ contains an Eckardt point,} \\ \dfrac{27}{17} & \text{if } S \text{ contains no Eckardt point.} \end{cases}$$

Proof Let P be a point in S, and let T be the hyperplane section of the surface S such that the curve T is singular at the point P. Then we have the following cases:

(i) T is a union of 3 lines that pass through P, i.e. P is an Eckardt point;
(ii) T is a union of a line and a conic that intersect transversally at P;
(iii) T is a union of 3 lines such that not all of them pass through P;
(iv) T is a union of a line and a conic that are tangent at P;

(v) T is an irreducible curve that has a cuspidal singularity at P;

(vi) T is an irreducible curve that has a nodal singularity at P.

It is well known that a general cubic surface in $\mathbb{P}^3$ does not contain Eckardt points. Moreover, if S does not contain Eckardt points, there is a hyperplane section of the cubic surface S that consists of a line and an irreducible conic that are tangent at some point. Thus, to prove the required assertion, it is enough to prove the following assertions:

- $\delta_P(S) = \frac{3}{2}$ if P is an Eckardt point;
- $\delta_P(S) = \frac{27}{17}$ if T is a union of a line and a conic that are tangent at P;
- $\delta_P(S) \geqslant \frac{27}{17}$ in all remaining cases.

We will do this case by case. But first, let us unify the notations that we will use.

Let $\sigma \colon \widetilde{S} \to S$ be the blow up of the point P, let E_P be the σ-exceptional divisor, let u be a non-negative real number, and let τ be the largest real number such that the divisor $\sigma^*(-K_S) - uE_P$ is pseudo-effective. For every number u such that $0 \leqslant u \leqslant \tau$, we will denote by $P(u)$ the positive part of the Zariski decomposition of $\sigma^*(-K_S) - uE_P$, and we will denote its negative part by $N(u)$. For every irreducible curve $Z \subset S$, we will denote by $\widetilde{Z}$ its proper transform on $\widetilde{S}$. Observe also that $\widetilde{S}$ is a weak del Pezzo surface, so that it is a Mori Dream Space [209].

Case 1. Suppose that $T = L_1 + L_2 + L_3$, where L_1, L_2 and L_3 are lines containing P. Then $\sigma^*(-K_S) - uE_P \sim_{\mathbb{Q}} (3-u)E_P + \widetilde{L}_1 + \widetilde{L}_2 + \widetilde{L}_3$, the curves $\widetilde{L}_1, \widetilde{L}_2, \widetilde{L}_2$ are disjoint, and $\widetilde{L}_1^2 = \widetilde{L}_2^2 = \widetilde{L}_3^2 = -2$. This implies that $\tau = 3$ and

$$
P(u) = \begin{cases} (3-u)E_P + \widetilde{L}_1 + \widetilde{L}_2 + \widetilde{L}_3 & \text{if } 0 \leqslant u \leqslant 1, \\ (3-u)\left(E_P + \frac{1}{2}\widetilde{L}_1 + \frac{1}{2}\widetilde{L}_2 + \frac{1}{2}\widetilde{L}_3\right) & \text{if } 1 \leqslant u \leqslant 3, \end{cases}
$$

and

$$
N(u) = \begin{cases} 0 & \text{if } 0 \leqslant u \leqslant 1, \\ \dfrac{u-1}{2}(\widetilde{L}_1 + \widetilde{L}_2 + \widetilde{L}_3) & \text{if } 1 \leqslant u \leqslant 3, \end{cases}
$$

so that

$$
\text{vol}\left(\sigma^*(-K_S) - uE_P\right) = P(u) \cdot P(u) = \begin{cases} 3 - u^2 & \text{if } 0 \leqslant u \leqslant 1, \\ \dfrac{(u-3)^2}{2} & \text{if } 1 \leqslant u \leqslant 3, \end{cases}
$$

which gives $S_S(E_P) = \frac{4}{3}$. Then $\delta(S) \leqslant \delta_P(S) \leqslant \frac{A_S(E_P)}{S_S(E_P)} = \frac{3}{2}$. For every $O \in E_P$, we get

$$S\left(W^{E_P}_{\bullet,\bullet};O\right) = \frac{2}{3}\int_0^3 \left((P(u)\cdot E_P)(N(u)\cdot E_P)_O + \frac{(P(u)\cdot E_P)^2}{2}\right)du$$

$$= \frac{2}{3}\int_1^3 \frac{(u-1)(3-u)}{4}du \times \mathrm{ord}_O\left(\widetilde{L}_1 + \widetilde{L}_2 + \widetilde{L}_3\right) + \frac{2}{3}\int_0^1 \frac{u^2}{2}du$$

$$+ \frac{2}{3}\int_1^3 \frac{(3-u)^2}{8}du$$

by Corollary 1.109, so that

$$S\left(W^{E_P}_{\bullet,\bullet};O\right) \leqslant \frac{2}{3}\int_1^3 \frac{(u-1)(3-u)}{4}du + \frac{2}{3}\int_0^1 \frac{u^2}{2}du + \frac{2}{3}\int_1^3 \frac{(3-u)^2}{8}du = \frac{5}{9}.$$

Recall from (1.12) that $\delta_O\left(E_P;W^{E_P}_{\bullet,\bullet}\right) = \frac{1}{S\left(W^{E_P}_{\bullet,\bullet};O\right)}$. Using Corollary 1.102, we then get

$$\delta_P(S) \geqslant \min\left\{\frac{A_S(E_P)}{S_S(E_P)}, \inf_{O\in E_P}\delta_O\left(E_P;W^{E_P}_{\bullet,\bullet}\right)\right\} = \min\left\{\frac{3}{2},\frac{9}{5}\right\} = \frac{3}{2}.$$

This shows that $\delta_P(S) = \frac{3}{2}$ as required.

Case 2. Suppose $T = C + L$, where C is a smooth irreducible conic, and L is a line that intersects C transversally at P. Let $\rho\colon \widehat{S} \to \widetilde{S}$ be the blow up of the point $\widetilde{L} \cap E_P$, let F be the exceptional curve of the blow up ρ, let $\widehat{L}$, $\widehat{C}$ and $\widehat{E}_P$ be the proper transforms on $\widehat{S}$ of the curves $\widetilde{L}$, $\widetilde{C}$ and E_P, respectively. Then $(\sigma\circ\rho)^*(-K_S) \sim \widehat{L} + \widehat{C} + 2\widehat{E}_P + 3F$.

Let $\phi\colon \widehat{S} \to \overline{S}$ be the contraction of the curve $\widehat{E}_P$, let $\overline{L} = \phi(\widehat{L})$, $\overline{C} = \phi(\widehat{C})$, $\overline{F} = \phi(F)$. Then $\phi(\widehat{E}_P) = \overline{C} \cap \overline{F}$ is an isolated ordinary double singular point of the surface $\overline{S}$, and the intersections of the curves $\overline{L}$, $\overline{C}$ and $\overline{F}$ are contained in the following table:

	$\overline{L}$	$\overline{C}$	$\overline{F}$
$\overline{L}$	-3	1	1
$\overline{C}$	1	$-\frac{1}{2}$	$\frac{1}{2}$
$\overline{F}$	1	$\frac{1}{2}$	$-\frac{1}{2}$

Observe that the divisor $-K_{\widehat{S}}$ is big. Then $\widehat{S}$ is a Mori Dream Space by [209, Theorem 1], so that $\overline{S}$ is also a Mori Dream Space. Moreover, we have a commutative diagram

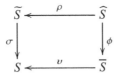

where υ is a weighted blow up of P with weights $(1,2)$, and the υ-exceptional curve is $\overline{F}$. Then $\upsilon^*(-K_S) - u\overline{F} \sim_{\mathbb{R}} \overline{L} + \overline{C} + (3-u)\overline{F}$. Using this equivalence, we conclude that the divisor $\upsilon^*(-K_S) - u\overline{F}$ is pseudo-effective $\iff u \in [0,3]$. For $u \in [0,3]$, the Zariski decomposition of this divisor can be described as follows. If $0 \leqslant u \leqslant 1$, then $\upsilon^*(-K_S) - u\overline{F}$ is nef. If $1 \leqslant u \leqslant \frac{14}{5}$, then

$$\upsilon^*(-K_S) - u\overline{F} \sim_{\mathbb{R}} \underbrace{\overline{C} + \frac{4-u}{3}\overline{L} + (3-u)\overline{F}}_{\text{positive part}} + \underbrace{\frac{u-1}{3}\overline{L}}_{\text{negative part}} .$$

Finally, if $\frac{14}{5} \leqslant u \leqslant 3$, then

$$\upsilon^*(-K_S) - u\overline{F} \sim_{\mathbb{R}} \underbrace{(3-u)(5\overline{C} + 2\overline{L} + \overline{F})}_{\text{positive part}} + \underbrace{(2u-5)\overline{L} + (5u-14)\overline{C}}_{\text{negative part}} .$$

Therefore, we have

$$\text{vol}(\upsilon^*(-K_S) - u\overline{F}) = \begin{cases} 3 - \dfrac{u^2}{2} & \text{if } 0 \leqslant u \leqslant 1, \\[2mm] 3 - \dfrac{u^2}{2} + \dfrac{(u-1)^2}{3} & \text{if } 1 \leqslant u \leqslant \dfrac{14}{5}, \\[2mm] 4(3-u)^2 & \text{if } \dfrac{14}{5} \leqslant u \leqslant 3. \end{cases}$$

Integrating this function, we get $S_S(\overline{F}) = \frac{9}{5}$, so that $\delta_P(S) \leqslant \frac{A_S(\overline{F})}{S_S(\overline{F})} = \frac{5}{3}$ since $A_S(\overline{F}) = 3$. On the one hand, it follows from Corollary 1.102 that

$$\delta_P(S) \geqslant \min\left\{ \frac{A_S(\overline{F})}{S_S(\overline{F})}, \inf_{O \in \overline{F}} \delta_O(\overline{F}, \Delta_{\overline{F}}; W^{\overline{F}}_{\bullet,\bullet}) \right\},$$

where $\Delta_{\overline{F}}$ is an effective $\mathbb{Q}$-divisor on $\overline{F} \cong \mathbb{P}^1$ known as the "different", which is defined via the subadjunction formula $(K_{\overline{S}} + \overline{F})|_{\overline{F}} = K_{\overline{F}} + \Delta_{\overline{F}}$. In our case, we have $\Delta_{\overline{F}} = \frac{1}{2}(\overline{C} \cap \overline{F})$. Let O be a point in the curve $\overline{F}$. Recall from [2] that

$$\delta_O(\overline{F}, \Delta_{\overline{F}}; W^{\overline{F}}_{\bullet,\bullet}) = \frac{A_{\overline{F}, \Delta_{\overline{F}}}(O)}{S(W^{\overline{F}}_{\bullet,\bullet}; O)},$$

where $A_{\overline{F}, \Delta_{\overline{F}}}(O) = \frac{1}{2}$ if $O = \overline{C} \cap \overline{F}$, and $A_{\overline{F}, \Delta_{\overline{F}}}(O) = 1$ otherwise. On the other hand, using Corollary 1.109, we get

$$S(W^{\overline{F}}_{\bullet,\bullet}; O) = \frac{2}{3}\left(\epsilon_O + \int_0^1 \frac{u^2}{8}du + \int_1^{\frac{14}{5}} \frac{(u+2)^2}{72}du + \int_{\frac{14}{5}}^3 \frac{(12-4u)^2}{2}du\right)$$

$$= \frac{2}{3}\epsilon_O + \frac{3}{10},$$

where

$$\epsilon_O = \begin{cases} \dfrac{131}{300} & \text{if } O = \overline{L} \cap \overline{F}, \\[2mm] \dfrac{1}{75} & \text{if } O = \overline{C} \cap \overline{F}, \\[2mm] 0 & \text{otherwise.} \end{cases}$$

Using this we get that

$$\frac{A_{\overline{F},\Delta_{\overline{F}}}(O)}{S(W^{\overline{F}}_{\bullet,\bullet}; O)} = \begin{cases} \dfrac{225}{133} & \text{if } O = \overline{L} \cap \overline{F}, \\[2mm] \dfrac{225}{139} & \text{if } O = \overline{C} \cap \overline{F}, \\[2mm] \dfrac{10}{3} & \text{otherwise.} \end{cases}$$

Combining our inequalities, we get $1.666 \approx \frac{5}{3} \geqslant \delta_P(S) \geqslant \frac{225}{139} \approx 1.612$, so that $\delta_P(S) > \frac{27}{17}$. In fact, it follows from [2] that $\delta_P(S) = \frac{225}{241} + \frac{72}{241}\sqrt{6} \approx 1.665$.

Case 3. Suppose that $T = L_1 + L_2 + L_3$ for lines L_1, L_2, L_3 such that $L_1 \cap L_2 = P \notin L_3$. Then $\sigma^*(-K_S) - uE_P \sim_{\mathbb{R}} (2-u)E_P + \widetilde{L}_1 + \widetilde{L}_2 + \widetilde{L}_3$, so that $\tau = 2$. Moreover, we have

$$P(u) = \begin{cases} (2-u)E_P + \widetilde{L}_1 + \widetilde{L}_2 + \widetilde{L}_3 & \text{if } 0 \leqslant u \leqslant 1, \\[2mm] (2-u)E_P + \dfrac{3-u}{2}\left(\widetilde{L}_1 + \widetilde{L}_2\right) + \widetilde{L}_3 & \text{if } 1 \leqslant u \leqslant 2, \end{cases}$$

and

$$N(u) = \begin{cases} 0 & \text{if } 0 \leqslant u \leqslant 1, \\[2mm] \dfrac{u-1}{2}\left(\widetilde{L}_1 + \widetilde{L}_2\right) & \text{if } 1 \leqslant u \leqslant 2, \end{cases}$$

so that

$$\text{vol}\left(\sigma^*(-K_S) - uE_P\right) = \begin{cases} 3 - u^2 & \text{if } 0 \leqslant u \leqslant 1, \\ 4 - 2u & \text{if } 1 \leqslant u \leqslant 2. \end{cases}$$

Integrating this function, we get $S_S(E_P) = \frac{11}{9}$, hence $\delta_P(S) \leqslant \frac{A_S(E_P)}{S_S(E_P)} = \frac{18}{11}$. On the other hand, it follows from Corollary 1.109 that

$$S(W^{E_P}_{\bullet,\bullet}; O) \leqslant \frac{2}{3}\left(\int_1^2 \frac{u-1}{2}du + \int_0^1 \frac{u^2}{2}du + \int_1^2 \frac{1}{2}du\right) = \frac{11}{18}$$

for every point $O \in E_P$. Now, using Corollary 1.102, we get $\delta_P(S) = \frac{18}{11}$.

Case 4. Suppose that $T = C + L$, where C is a smooth conic, and L is a line that is tangent to the conic C at the point P. Let $\rho\colon \widehat{S} \to \widetilde{S}$ be the blow up of the point $\widetilde{L} \cap \widetilde{C} \cap E_P$, let F be the exceptional curve of the blow up ρ, and let $\widehat{L}$, $\widehat{C}$, $\widehat{E}_P$ be the proper transforms on $\widehat{S}$ of the curves $\widetilde{L}$, $\widetilde{C}$, E_P, respectively. Then $(\sigma \circ \rho)^*(-K_S) \sim \widehat{L} + \widehat{C} + 2\widehat{E}_P + 4F$.

Let $\phi\colon \widehat{S} \to \overline{S}$ be the contraction of the curve $\widehat{E}_P$, let $\overline{L} = \phi(\widehat{L})$, $\overline{C} = \phi(\widehat{C})$, $\overline{F} = \phi(F)$. Then $\phi(\widehat{E}_P)$ is an ordinary double point of the surface $\overline{S}$. But $\phi(\widehat{E}_P) \notin \overline{L}$ and $\phi(\widehat{E}_P) \notin \overline{C}$. The intersections of the curves $\overline{L}$, $\overline{C}$ and $\overline{F}$ are contained in the following table:

	$\overline{L}$	$\overline{C}$	$\overline{F}$
$\overline{L}$	-3	0	1
$\overline{C}$	0	-2	1
$\overline{F}$	1	1	$-\frac{1}{2}$

Observe that the divisor $-K_{\widehat{S}}$ is big. Then $\widehat{S}$ is a Mori Dream Space by [209, Theorem 1], so that $\overline{S}$ is also a Mori Dream Space. Moreover, we have commutative diagram

where υ is the contraction of the curve $\overline{F}$. Observe that $\upsilon^*(-K_S) - u\overline{F} \sim_{\mathbb{R}} \overline{L} + \overline{C} + (4 - u)\overline{F}$. Using this, we conclude that the divisor $\upsilon^*(-K_S) - u\overline{F}$ is pseudo-effective $\iff u \in [0,4]$. For $u \in [0,4]$, the Zariski decomposition of the divisor $\upsilon^*(-K_S) - u\overline{F}$ can be described as follows. If $0 \leqslant u \leqslant 1$, then $\upsilon^*(-K_S) - u\overline{F}$ is nef. If $1 \leqslant u \leqslant 2$, then

$$\upsilon^*(-K_S) - u\overline{F} \sim_{\mathbb{R}} \underbrace{\overline{C} + \frac{4-u}{3}\overline{L} + (4-u)\overline{F}}_{\text{positive part}} + \underbrace{\frac{u-1}{3}\overline{L}}_{\text{negative part}}.$$

Finally, if $2 \leqslant u \leqslant 4$, then

$$\upsilon^*(-K_S) - u\overline{F} \sim_{\mathbb{R}} \underbrace{(4-u)\left(\frac{1}{2}\overline{C} + \frac{1}{3}\overline{L} + \overline{F}\right)}_{\text{positive part}} + \underbrace{\frac{u-1}{3}\overline{L} + \frac{u-2}{2}\overline{C}}_{\text{negative part}}.$$

Therefore, we have

$$\mathrm{vol}\big(\upsilon^*(-K_S) - u\overline{F}\big) = \begin{cases} 3 - \dfrac{u^2}{2} & \text{if } 0 \leqslant u \leqslant 1, \\[2mm] 3 - \dfrac{u^2}{2} + \dfrac{(u-1)^2}{3} & \text{if } 1 \leqslant u \leqslant 2, \\[2mm] \dfrac{(4-u)^2}{3} & \text{if } 2 \leqslant u \leqslant 4. \end{cases}$$

Integrating this volume function, we see that $S_S(\overline{F}) = \frac{17}{9}$, so that $\delta_P(S) \leqslant \frac{A_S(\overline{F})}{S_S(\overline{F})} = \frac{27}{17}$. Now, using Corollary 1.102, we see that

$$\delta_P(S) \geqslant \min\left\{\frac{27}{17}, \inf_{O \in \overline{F}} \frac{A_{\overline{F}, \Delta_{\overline{F}}}(O)}{S(W^{\overline{F}}_{\bullet,\bullet}; O)}\right\},$$

where $\Delta_{\overline{F}} = \frac{1}{2}\phi(\widehat{E}_P)$. Let O be a point in the curve $\overline{F}$. Then Corollary 1.109 gives

$$S\big(W^{\overline{F}}_{\bullet,\bullet}; O\big) = \begin{cases} \dfrac{5}{9} & \text{if } O = \overline{L} \cap \overline{F}, \\[2mm] \dfrac{7}{18} & \text{if } O = \overline{C} \cap \overline{F}, \\[2mm] \dfrac{13}{54} & \text{otherwise,} \end{cases}$$

so that

$$\frac{27}{17} < \frac{A_{\overline{F}, \Delta_{\overline{F}}}(O)}{S(W^{\overline{F}}_{\bullet,\bullet}; O)} = \begin{cases} \dfrac{9}{5} & \text{if } O = \overline{L} \cap \overline{F}, \\[2mm] \dfrac{18}{7} & \text{if } O = \overline{C} \cap \overline{F}, \\[2mm] \dfrac{27}{13} & \text{if } O = \phi(\widehat{E}_P), \\[2mm] \dfrac{54}{13} & \text{otherwise.} \end{cases}$$

Therefore, we see that $\delta_P(S) = \frac{27}{17}$ as required.

Case 5. Suppose that T has a cusp. Let $\rho \colon \widehat{S} \to \widetilde{S}$ be the blow up of the point $\widetilde{T} \cap E_P$, let F be the exceptional curve of the blow up ρ, let $\widehat{T}$ and $\widehat{E}_P$ be the proper transforms on the surface $\widehat{S}$ of the curves $\widetilde{T}$ and E_P, respectively. Then $(\sigma \circ \rho)^*(-K_S) \sim \widehat{T} + 2\widehat{E}_P + 3F$. Let $\eta \colon \overline{S} \to \widehat{S}$ be the blow up of the point $\widehat{T} \cap \widehat{E}_P \cap F$, let G be the η-exceptional curve, and let $\overline{T}, \overline{E}_P, \overline{F}$ be the proper transforms on $\overline{S}$ of the curves $\widehat{T}, \widehat{E}_P, F$, respectively. Then $(\sigma \circ \rho \circ \eta)^*(-K_S) \sim \overline{T} + 2\overline{E}_P + 3\overline{F} + 6G$.

Let $\phi \colon \overline{S} \to \mathscr{S}$ be the contraction of the curves $\overline{E}_P$ and $\overline{F}$, let $\mathscr{T} = \upsilon(\overline{T})$ and $\mathscr{G} = \phi(G)$. Then $\phi(\overline{F})$ is an ordinary double point of the surface $\mathscr{S}$, and

$\phi(\overline{E}_P)$ is its quotient singular point of type $\frac{1}{3}(1,1)$. Note that these singular points are not contained in the curve $\mathscr{T}$. Note also that $\mathscr{T}^2 = -3$, $\mathscr{G}^2 = -\frac{1}{6}$ and $\mathscr{T} \cdot \mathscr{G} = 1$. Observe that

$$-K_{\overline{S}} \sim_{\mathbb{Q}} (\sigma \circ \rho \circ \eta)^* \left(-\frac{1}{3} K_S \right) + \frac{2}{3}\overline{T} + \frac{1}{3}\overline{E}_P,$$

so that $-K_{\overline{S}}$ is big. Then $\overline{S}$ is a Mori Dream Space by [209, Theorem 1], which implies that $\mathscr{S}$ is also a Mori Dream Space. Moreover, we have commutative diagram

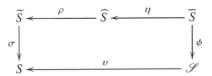

where υ is a weighted blow up of P with weights $(2,3)$, and the υ-exceptional curve is $\mathscr{G}$. Since $\upsilon^*(-K_S) - u\mathscr{G} \sim_{\mathbb{R}} (6-u)\mathscr{G} + \mathscr{T}$, $\upsilon^*(-K_S) - u\mathscr{G}$ is pseudo-effective $\iff u \leqslant 6$, and this divisor is nef $\iff u \leqslant 3$. If $3 \leqslant u \leqslant 6$, then the positive part of its Zariski decomposition is $(6-u)\mathscr{G} + \frac{6-u}{3}\mathscr{T}$, and the negative part is $\frac{u-3}{3}\mathscr{T}$. This gives

$$\mathrm{vol}\big(\upsilon^*(-K_S) - u\mathscr{G}\big) = \begin{cases} 3 - \dfrac{u^2}{6} & \text{if } 0 \leqslant u \leqslant 3, \\[2mm] \dfrac{(6-u)^2}{6} & \text{if } 3 \leqslant u \leqslant 6. \end{cases}$$

Integrating this function, we get $S_S(\mathscr{G}) = 3$, so that $\delta_P(S) \leqslant \frac{A_S(\mathscr{G})}{S_S(\mathscr{G})} = \frac{5}{3}$ since $A_S(\mathscr{G}) = 5$. Now, to get a lower bound for $\delta_P(S)$, we use Corollary 1.102 that gives

$$\delta_P(S) \geqslant \min\left\{ \frac{5}{3}, \inf_{O \in \mathscr{G}} \frac{A_{\mathscr{G},\Delta_{\mathscr{G}}}(O)}{S(W^{\mathscr{G}}_{\bullet,\bullet}; O)} \right\},$$

where $\Delta_{\mathscr{G}} = \frac{2}{3}\phi(\overline{E}_P) + \frac{1}{2}\phi(\overline{F})$. On the other hand, if O is a point in $\mathscr{G}$, then

$$S(W^{\mathscr{G}}_{\bullet,\bullet}; O) = \begin{cases} \dfrac{1}{3} & \text{if } O = \mathscr{G} \cap \mathscr{T}, \\[2mm] \dfrac{1}{6} & \text{otherwise}, \end{cases}$$

by Corollary 1.109, so that

$$\frac{A_{\mathscr{G},\Delta_{\mathscr{G}}}(O)}{S(W^{\mathscr{G}}_{\bullet,\bullet};O)} = \begin{cases} 3 & \text{if } O = \mathscr{G} \cap \mathscr{T}, \\ 3 & \text{if } O = \phi(\overline{E}_P), \\ 2 & \text{if } O = \phi(\overline{F}), \\ 6 & \text{otherwise.} \end{cases}$$

This gives $\delta_P(S) \geqslant \frac{5}{3} > \frac{27}{17}$.

Case 6. Finally, we suppose that T is an irreducible cubic curve that has a node at P. Then $\sigma^*(-K_S) - uE_P \sim_{\mathbb{R}} (2-u)E_P + \widetilde{T}$ and $\widetilde{T}^2 = -1$, so that $\tau = 2$. Moreover, we have

$$P(u) = \begin{cases} (2-u)E_P + \widetilde{T} & \text{if } 0 \leqslant u \leqslant \frac{3}{2}, \\ (2-u)\left(E_P + 2\widetilde{T}\right) & \text{if } \frac{3}{2} \leqslant u \leqslant 2, \end{cases}$$

and

$$N(u) = \begin{cases} 0 & \text{if } 0 \leqslant u \leqslant \frac{3}{2}, \\ (2u-3)\widetilde{T} & \text{if } \frac{3}{2} \leqslant u \leqslant 2, \end{cases}$$

so that

$$\mathrm{vol}\left(\sigma^*(-K_S) - uE_P\right) = \begin{cases} 3 - u^2 & \text{if } 0 \leqslant u \leqslant \frac{3}{2}, \\ 3(2-u)^2 & \text{if } \frac{3}{2} \leqslant u \leqslant 2. \end{cases}$$

Integrating this function, we get $S_S(E_P) = \frac{7}{6}$. Hence, we see that $\delta_P(S) \leqslant \frac{A_S(E_P)}{S_S(E_P)} = \frac{12}{7}$. Now, using Corollary 1.109, we conclude that $S(W^{E_P}_{\bullet,\bullet}; O) \leqslant \frac{7}{12}$ for every point $O \in E_P$. Hence, it follows from Corollary 1.102 that $\delta_P(S) = \frac{12}{7}$. This completes the proof. $\qquad\square$

To compute δ-invariants of smooth del Pezzo surfaces of degree 2, we have to recall a few basic facts about (-1)-curves on these surfaces. These are collected in the next remark.

Remark 2.14 Suppose that $K_S^2 = 2$. Then there exists a ramified double cover $\pi\colon S \to \mathbb{P}^2$, which is branched over a smooth curve of degree four [74]. Let us denote this curve by R. The double cover π induces an involution $\tau \in \mathrm{Aut}(S)$ that is known as a Geiser involution. For any (-1)-curve L in the surface S, the curve $\tau(L)$ is a (-1)-curve, $L \cdot \tau(L) = 2$ and

$$L + \tau(L) \sim -K_S,$$

so that $\pi(L) = (\pi \circ \tau)(L)$ is a line in $\mathbb{P}^2$, which is a bi-tangent (or four-tangent) of R. Therefore, we see that (-1)-curves in S always come in pairs. There are 28 such pairs, which correspond to 28 bi-tangents of the quartic curve R. This gives 56 (-1)-curves. For every line $\ell \subset \mathbb{P}^2$, its preimage on S via π is a reduced curve $C \subset |-K_S|$ such that exactly one of the following possibilities holds:

(i) if ℓ intersects R transversally, then C is a smooth elliptic curve;

(ii) if ℓ is tangent to R at one point that is not an inflection point, then C is an irreducible curve of arithmetic genus 1 that has one node;

(iii) if ℓ is tangent to R at an ordinary inflection point (not a hyperinflection point), then C is an irreducible curve of arithmetic genus 1 that has one cusp;

(iv) if ℓ is tangent to R at two distinct points, then $C = L + \tau(L)$ for a (-1)-curve L such that the intersection $L \cap \tau(L)$ consists of two points, so that C is nodal;

(v) if ℓ is tangent to R at a hyperinflection point, then $\ell \cap R$ consists of one point, and $C = L + \tau(L)$ for a (-1)-curve L such that the curves L and $\tau(L)$ are tangent, so that the anticanonical curve C has a tacnodal singularity.

The inflection points of the curve R are precisely the intersection points of this curve with its Hessian sextic curve, which intersects the quartic curve R transversally at ordinary inflection points and meets R at hyperinflection points (undulations) with multiplicity 2. In particular, we see that the quartic curve R always has at least one inflection point. However, if the curve R is general, then it has no hyperinflection point. Surprisingly, it may happen that R has no ordinary inflection point, see [84, 133] and [41, §6.1]. In fact, there are exactly two such curves: the Fermat quartic curve, and the curve given by

$$x^4 + y^4 + z^4 + 3(x^2y^2 + y^2z^2 + z^2x^2) = 0,$$

where x, y, z are coordinates on $\mathbb{P}^2$. Moreover, if L and L' are two distinct (-1)-curves in the surface S, then $1 = L' \cdot (-K_S) = L' \cdot (L + \tau(L)) = L' \cdot L + L' \cdot \tau(L)$, so that we have one of the following three mutually excluding possibilities:

(i) $L' \cap L = \varnothing$, $L' \cdot L = 0$ and $L' \cdot \tau(L) = 1$;

(ii) $L' \cap \tau(L) = \varnothing$, $L' \cdot L = 1$ and $L' \cdot \tau(L) = 0$;

(iii) $L' = \tau(L)$ and $L' \cdot L = 2$.

For any point $P \in S$ such that $\pi(P) \in R$, there exists a unique curve $C \subset |-K_S|$ such that C is singular at the point P, and every (-1)-curve in S that contains P must be an irreducible component of the curve C since $C \cdot L = 1$ for every (-1)-curve $L \subset S$. If P is a point in S such that $\pi(P) \notin R$, then $|-K_S|$ contains no curve singular at P. In this case, the point P is contained in at most four

(-1)-curves in S by [191, Lemma 5], which can easily be derived directly from the intersection graph of all (-1)-curves in S. This can also be proved as follows: if S has five (-1)-curves L_1, L_2, L_3, L_4, L_5 that have a common point, then contracting $\tau(L_5)$ we obtain a smooth cubic surface that contains four lines sharing a common point, which is impossible. Moreover, if P is a point in S such that $\pi(P) \notin R$, then it follows from [74, Exercise 6.17] that the point P is contained in four (-1)-curves $\iff \pi(P) = [1 : 0 : 0]$ and R is given by $x^4 + f_2(y,z)x^2 + f_4(y,z) = 0$ for an appropriate choice of coordinates x, y, z, where $f_2(y,z)$ and $f_4(y,z)$ are quadratic and quartic forms, respectively. Intersections of four (-1)-curves on the surface S are called *generalized Eckardt points* [191].

Now, we are ready to prove

Lemma 2.15 *Suppose that* $K_S^2 = 2$. *Then*

$$\delta(S) = \begin{cases} \dfrac{9}{5} & \text{if } |-K_S| \text{ contains a tacnodal curve,} \\[2mm] \dfrac{15}{8} & \text{if } |-K_S| \text{ contains no tacnodal curve.} \end{cases}$$

Proof Let us use the notation introduced in Remark 2.14. Fix some point $P \in S$. Let $\sigma \colon \widetilde{S} \to S$ be the blow up of the point P, and let E_P be the σ-exceptional divisor. Then $\widetilde{S}$ is a weak del Pezzo surface, $K_{\widetilde{S}}^2 = 1$, and $|-2K_{\widetilde{S}}|$ gives a morphism $\widetilde{S} \to \mathbb{P}(1,1,2)$, which has the Stein factorization

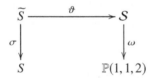

where ϑ is a contraction of all (-2)-curves in the surface $\widetilde{S}$ (if any), and ω is a double cover branched over the union of a sextic curve in $\mathbb{P}(1,1,2)$ and the singular point of $\mathbb{P}(1,1,2)$. Observe that S is a del Pezzo surface of degree 1 with at most Du Val singularities, and the morphism ϑ is an isomorphism $\iff -K_{\widetilde{S}}$ is ample $\iff$ the surface S is smooth. Moreover, if the divisor $-K_{\widetilde{S}}$ is not ample, then $\widetilde{S}$ contains at most four (-2)-curves. Furthermore, if Z is a (-2)-curve in $\widetilde{S}$, then either $\pi(P) \in R$ and $\sigma(Z)$ is the curve in the linear system $|-K_S|$ that is singular at P, or $\sigma(Z)$ is a (-1)-curve that contains P. The double cover $S \to \mathbb{P}(1,1,2)$ induces an involution $\iota \in \mathrm{Aut}(\widetilde{S})$ such that

$$\iota(E_P) = E_P \iff \pi(P) \in R \text{ or } P \text{ is a generalized Eckardt point.}$$

The involution ι is known as a Bertini involution. It gives the involution $\sigma \circ$ $\iota \circ \sigma^{-1} \in \mathrm{Bir}(S)$, which is biregular $\iff \iota(E_P) = E_P$. Let $\varsigma = \sigma \circ \iota$. Then $\varsigma \colon \widetilde{S} \to S$ contracts $\iota(E_P)$. Thus, we have the following commutative diagram:

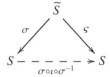

If $-K_{\widetilde{S}}$ is ample, then $E_P + \iota(E_P) \sim -2K_{\widetilde{S}}$. Similarly, if $\pi(P) \notin R$, then

$$E_P + \iota(E_P) + \big(\text{the sum of all } (-2)\text{-curves in } \widetilde{S}\big) \sim -2K_{\widetilde{S}}.$$

In particular, if $-K_{\widetilde{S}}$ is ample, then $E_P \cdot \iota(E_P) = 3$, so that $\varsigma(E_P)$ is an irreducible curve in the linear system $|-2K_S|$ that has a singular point of multiplicity 3 at the point P.

If $\pi(P) \in R$, let us denote by C_P the unique curve in $|-K_S|$ that is singular at P. Then we have the following cases:

(i) the divisor $-K_{\widetilde{S}}$ is ample, so that $\widetilde{S}$ is a del Pezzo surface;
(ii) $\pi(P) \in R$, and C_P is an irreducible nodal curve;
(iii) $\pi(P) \in R$, and C_P is an irreducible cuspidal curve;
(iv) $\pi(P) \in R$, and C_P is a union of two (-1)-curves that meet transversally;
(v) $\pi(P) \in R$, and C_P is a union of two (-1)-curves that are tangent at P;
(vi) $\pi(P) \notin R$, and P is contained in exactly one (-1)-curve;
(vii) $\pi(P) \notin R$, and P is contained in exactly two (-1)-curves;
(viii) $\pi(P) \notin R$, and P is contained in exactly three (-1)-curves;
(ix) the point P is a generalized Eckardt point.

It follows from Remark 2.14 that to prove the required assertion, it is enough to prove the following three assertions:

- $\delta_P(S) = \frac{9}{5}$ if $\pi(P) \in R$, and C_P is a tacnodal curve;
- $\delta_P(S) = \frac{15}{8}$ if $\pi(P) \in R$, and C_P is a cuspidal curve;
- $\delta_P(S) \geqslant \frac{15}{8}$ in all remaining seven cases.

We will do this case by case similarly to what we did in the proof of Lemma 2.13.

Take $u \in \mathbb{R}_{\geqslant 0}$. Let τ be the largest number such that $\sigma^*(-K_S) - uE_P$ is pseudo-effective. For every real number $u \in [0, \tau]$, let us denote by $P(u)$ and $N(u)$ the positive and the negative parts of the Zariski decomposition of the divisor $\sigma^*(-K_S) - uE_P$, respectively. For every irreducible $Z \subset S$, let us denote by $\widetilde{Z}$ its proper transform on the surface $\widetilde{S}$. For instance, if $\pi(P) \in R$, then $\widetilde{C}_P$ is the proper transform on $\widetilde{S}$ of the curve C_P.

Case 1. Suppose that $-K_{\widetilde{S}}$ is ample. Then $E_P + \iota(E_P) \sim \sigma^*(-2K_S) - 2E_P$, so that

$$\sigma^*(-K_S) - uE_P \sim_{\mathbb{R}} \frac{3 - 2u}{2}E_P + \frac{\iota(E_P)}{2},$$

which immediately implies that $\tau = \frac{3}{2}$. Moreover, we have

$$P(u) = \begin{cases} \dfrac{3 - 2u}{2}E_P + \dfrac{\iota(E_P)}{2} & \text{if } 0 \leqslant u \leqslant \dfrac{4}{3}, \\[2ex] \dfrac{3 - 2u}{2}(E_P + 3\iota(E_P)) & \text{if } \dfrac{4}{3} \leqslant u \leqslant \dfrac{3}{2}, \end{cases}$$

and

$$N(u) = \begin{cases} 0 & \text{if } 0 \leqslant u \leqslant \dfrac{4}{3}, \\[2ex] (3u - 4)\iota(E_P) & \text{if } \dfrac{4}{3} \leqslant u \leqslant \dfrac{3}{2}, \end{cases}$$

so that

$$\mathrm{vol}\big(\sigma^*(-K_S) - uE_P\big) = \begin{cases} 2 - u^2 & \text{if } 0 \leqslant u \leqslant \dfrac{4}{3}, \\[2ex] 2(3 - 2u)^2 & \text{if } \dfrac{4}{3} \leqslant u \leqslant \dfrac{3}{2}. \end{cases}$$

Integrating, we get $S_S(E_P) = \frac{17}{18}$, so that $\delta_P(S) \leqslant \frac{A_S(E_P)}{S_S(E_P)} = \frac{36}{17}$. For every $O \in E_P$, we get

$$S\big(W_{\bullet,\bullet}^{E_P}; O\big) = \int_0^{\frac{3}{2}} \left((P(u) \cdot E_P)(N(u) \cdot E_P)_O + \frac{(P(u) \cdot E_P)^2}{2} \right) du$$

$$= \int_{\frac{4}{3}}^{\frac{3}{2}} (3u - 4)(12 - 8u)du \times (\iota(E_P) \cdot E_P)_O + \int_0^{\frac{4}{3}} \frac{u^2}{2}du$$

$$+ \int_{\frac{4}{3}}^{\frac{3}{2}} \frac{(12 - 8u)^2}{2}du$$

$$= \frac{(\iota(E_P) \cdot E_P)_O}{54} + \frac{4}{9} \leqslant \frac{\iota(E_P) \cdot E_P}{54} + \frac{4}{9} = \frac{3}{54} + \frac{4}{9} = \frac{1}{2},$$

by Corollary 1.109. Now, using Corollary 1.102, we obtain

$$\delta_P(S) \geqslant \min\left\{ \frac{36}{17}, \inf_{O \in E_P} \frac{1}{S(W_{\bullet,\bullet}^{E_P}; O)} \right\} \geqslant 2,$$

so that $\frac{36}{17} \geqslant \delta_P(S) \geqslant \frac{15}{8}$ as required. In fact, we proved that $\delta_P(S) = \frac{36}{17}$ if $|\iota(E_P) \cap E_P| \geqslant 2$.

Case 2. Suppose that $\pi(P) \in R$, and C_P is an irreducible nodal curve. We have $\tau = 2$ because $\sigma^*(-K_S) - uE_P \sim_{\mathbb{R}} (2 - u)E_P + \widetilde{C}_P$. Moreover, we have

$$P(u) = \begin{cases} (2 - u)E_P + \widetilde{C}_P & \text{if } 0 \leqslant u \leqslant 1, \\ (2 - u)(E_P + \widetilde{C}_P) & \text{if } 1 \leqslant u \leqslant 2, \end{cases}$$

and

$$N(u) = \begin{cases} 0 & \text{if } 0 \leqslant u \leqslant 1, \\ (u - 1)\widetilde{C}_P & \text{if } 1 \leqslant u \leqslant 2, \end{cases}$$

so that

$$\mathrm{vol}\big(\sigma^*(-K_S) - uE_P\big) = \begin{cases} 2 - u^2 & \text{if } 0 \leqslant u \leqslant 1, \\ (2 - u)^2 & \text{if } 1 \leqslant u \leqslant 2. \end{cases}$$

Integrating, we get $S_S(E_P) = 1$, so that $\delta_P(S) \leqslant \frac{A_S(E_P)}{S_S(E_P)} = 2$. For every $O \in E_P$, we obtain

$$S\big(W_{\bullet,\bullet}^{E_P}; O\big) = \int_0^2 \left(\big(P(u) \cdot E_P\big)\big(N(u) \cdot E_P\big)_O + \frac{\big(P(u) \cdot E_P\big)^2}{2} \right) du$$

$$= \int_1^2 (u - 1)(2 - u) du \times \big(\widetilde{C}_P \cdot E_P\big)_O + \int_0^1 \frac{u^2}{2} du + \int_1^2 \frac{(2 - u)^2}{2} du$$

$$= \frac{\big(\widetilde{C}_P \cdot E_P\big)_O}{6} + \frac{1}{3} \leqslant \frac{1}{2}$$

by Corollary 1.109. Now, using Corollary 1.102, we get

$$\delta_P(S) \geqslant \min\left\{ \frac{A_S(E_P)}{S_S(E_P)}, \inf_{O \in E_P} \frac{1}{S(W_{\bullet,\bullet}^{E_P}; O)} \right\} \geqslant 2,$$

so that $\delta_P(S) = 2 > \frac{15}{8}$ as required.

Case 3. Suppose that $\pi(P) \in R$, and C_P is an irreducible curve that has a cusp at P. Let $\rho \colon \widetilde{S} \to \widehat{S}$ be the blow up of the point $\widetilde{C}_P \cap E_P$, let F be the ρ-exceptional curve, let $\widehat{C}_P$ and $\widehat{E}_P$ be the proper transforms on $\widehat{S}$ via ρ of the curves $\widetilde{C}_P$ and E_P, respectively. Then there exists commutative diagram

where η is the blow up of the point $\widehat{C}_P \cap \widehat{E}_P$, the map ϕ is the contraction of the proper transforms of the curves $\widehat{E}_P$ and F, and υ is the birational contraction of the proper transform of the η-exceptional curve. It is not hard to see that the divisor $-K_{\widetilde{S}}$ is big. Then $\widetilde{S}$ is a Mori Dream Space [209], so that $\mathscr{S}$ is a Mori Dream Space as well.

Let $\mathscr{G}$ be the υ-exceptional curve, and let $\mathscr{C}_P$ be the proper transform of the curve C_P on the surface $\mathscr{S}$. Then $\mathscr{G}$ contains two singular points Q_1 and Q_2 such that Q_1 is an ordinary double point, and Q_2 is a quotient singular point of type $\frac{1}{3}(1,1)$. However, these singular points are not contained in the curve $\mathscr{C}_P$. Note also that $\mathscr{C}^2 = -4, \mathscr{G}^2 = -\frac{1}{6}, \mathscr{C} \cdot \mathscr{G} = 1$. Since $\upsilon^*(-K_S) - u\mathscr{G} \sim_{\mathbb{R}} (6-u)\mathscr{G} + \mathscr{C}_P$, the divisor $\upsilon^*(-K_S) - u\mathscr{G}$ is nef $\iff u \leqslant 2$, and the divisor $\upsilon^*(-K_S) - u\mathscr{G}$ is pseudo-effective $\iff u \leqslant 6$. If $2 \leqslant u \leqslant 6$, then the positive part of its Zariski decomposition is $(6-u)(\mathscr{G} + \frac{1}{4}\mathscr{C}_P)$, and its negative part is $\frac{u-2}{4}\mathscr{C}_P$. Then

$$\mathrm{vol}(\upsilon^*(-K_S) - u\mathscr{G}) = \begin{cases} 2 - \dfrac{u^2}{6} & \text{if } 0 \leqslant u \leqslant 2, \\[2mm] \dfrac{(6-u)^2}{12} & \text{if } 1 \leqslant u \leqslant 6. \end{cases}$$

Integrating this function, we obtain $S_S(\mathscr{G}) = \frac{8}{3}$, which implies that $\delta_P(S) \leqslant \frac{A_S(\mathscr{G})}{S_S(\mathscr{G})} = \frac{15}{8}$. On the one hand, it follows from Corollary 1.102 that

$$\delta_P(S) \geqslant \min\left\{\frac{15}{8}, \inf_{O \in \mathscr{G}} \frac{A_{\mathscr{G},\Delta_{\mathscr{G}}}(O)}{S(W_{\bullet,\bullet}^{\mathscr{G}};O)}\right\},$$

where $\Delta_{\mathscr{G}} = \frac{1}{2}Q_1 + \frac{2}{3}Q_2$. On the other hand, if O is a point in $\mathscr{G}$, then

$$S(W_{\bullet,\bullet}^{\mathscr{G}};O) = \frac{1}{9} + \frac{2}{9}\mathrm{ord}_O(\mathscr{C}_P|_{\mathscr{G}}) = \begin{cases} \dfrac{1}{3} & \text{if } O = \mathscr{G} \cap \mathscr{C}, \\[2mm] \dfrac{1}{9} & \text{otherwise,} \end{cases}$$

by Corollary 1.109, so that

$$\frac{A_{\mathscr{G},\Delta_{\mathscr{G}}}(O)}{S(W_{\bullet,\bullet}^{\mathscr{G}};O)} = \begin{cases} 3 & \text{if } O = \mathscr{G} \cap \mathscr{C}_P, \\[1mm] \dfrac{9}{2} & \text{if } O = Q_1, \\[1mm] 3 & \text{if } O = Q_2, \\[1mm] 9 & \text{otherwise.} \end{cases}$$

This gives $\delta_P(S) = \frac{15}{8}$ as required.

Case 4. Suppose that $\pi(P) \in R$, and $C_P = L_1 + L_2$, where L_1 and L_2 are (-1)-curves that intersect transversally at the point P. Then $\sigma^*(-K_S) - uE_P \sim_{\mathbb{R}} (2-u)E_P + \widetilde{L}_1 + \widetilde{L}_2$. This gives $\tau = 2$. Moreover, we have

$$P(u) = \begin{cases} (2-u)E_P + \widetilde{L}_1 + \widetilde{L}_2 & \text{if } 0 \leqslant u \leqslant 1, \\[1mm] (2-u)(E_P + \widetilde{L}_1 + \widetilde{L}_2) & \text{if } 1 \leqslant u \leqslant 2, \end{cases}$$

and

$$N(u) = \begin{cases} 0 & \text{if } 0 \leqslant u \leqslant 1, \\ (u-1)(\widetilde{L}_1 + \widetilde{L}_2) & \text{if } 1 \leqslant u \leqslant 2, \end{cases}$$

so that

$$\text{vol}\big(\sigma^*(-K_S) - uE_P\big) = \begin{cases} 2 - u^2 & \text{if } 0 \leqslant u \leqslant 1, \\ (2-u)^2 & \text{if } 1 \leqslant u \leqslant 2. \end{cases}$$

Integrating, we get $S_S(E_P) = 1$, so that $\delta_P(S) \leqslant \frac{A_S(E_P)}{S_S(E_P)} = 2$. For every $O \in E_P$, we get

$$S(W_{\bullet,\bullet}^{E_P}; O)$$
$$= \int_0^2 \left((P(u) \cdot E_P)(N(u) \cdot E_P)_O + \frac{(P(u) \cdot E_P)^2}{2} \right) du$$
$$= \int_1^2 (u-1)(2-u)du \times \big((\widetilde{L}_1 + \widetilde{L}_2) \cdot E_P\big)_O + \int_0^1 \frac{u^2}{2} du + \int_1^2 \frac{(2-u)^2}{2} du$$
$$= \frac{\big((\widetilde{L}_1 + \widetilde{L}_2) \cdot E_P\big)_O}{6} + \frac{1}{3}$$

by Corollary 1.109, so that $S(W_{\bullet,\bullet}^{E_P}; O) \leqslant \frac{1}{2}$. Then $\delta_P(S) = 2 > \frac{15}{8}$ by Corollary 1.102.

Case 5. Suppose that $\pi(P) \in R$, and $C_P = L_1 + L_2$, where L_1 and L_2 are (-1)-curves that are tangent at the point P. Let $\rho \colon \widehat{S} \to \widetilde{S}$ be the blow up of the point $\widetilde{L}_1 \cap \widetilde{L}_2 \cap E_P$, let F be the ρ-exceptional curve, and let $\widehat{L}_1, \widehat{L}_2, \widehat{E}_P$ be the proper transforms on $\widehat{S}$ of the curves $\widetilde{L}_1, \widetilde{L}_2, E_P$, respectively. Then $(\sigma \circ \rho)^*(-K_S) \sim \widehat{L}_1 + \widehat{L}_1 + 2\widehat{E}_P + 4F$.

Let $\phi \colon \widehat{S} \to \overline{S}$ be the contraction of $\widehat{E}_P$. Set $\overline{L}_1 = \phi(\widehat{L}_1)$, $\overline{L}_2 = \phi(\widehat{L}_2)$ and $\overline{F} = \phi(F)$. Then $\phi(\widehat{E}_P)$ is an ordinary double point of the surface $\overline{S}$, $\phi(\widehat{E}_P) \notin \overline{L}_1$ and $\phi(\widehat{E}_P) \notin \overline{L}_2$. The intersections of the curves $\overline{L}_1, \overline{L}_2$ and $\overline{F}$ are contained in following table:

	$\overline{L}_1$	$\overline{L}_2$	$\overline{F}$
$\overline{L}_1$	-3	0	1
$\overline{L}_2$	0	-3	1
$\overline{F}$	1	1	$-\frac{1}{2}$

Observe that $-K_{\widetilde{S}} \sim_{\mathbb{Q}} \widetilde{L}_1 + \widehat{L}_1 + \widehat{E}_P + 2F$, which implies that the divisor $-K_{\widehat{S}}$ is big. Then $\widehat{S}$ is a Mori Dream Space by [209, Theorem 1], so that $\widetilde{S}$ is a Mori Dream Space. Moreover, we have commutative diagram

$$
\begin{array}{ccc}
\widetilde{S} & \xleftarrow{\ \rho\ } & \widehat{S} \\
{\scriptstyle\sigma}\downarrow & & \downarrow{\scriptstyle\phi} \\
S & \xleftarrow{\ \upsilon\ } & \overline{S}
\end{array}
$$

where υ is a contraction of the curve $\overline{F}$. Then $\upsilon^*(-K_S) - u\overline{F} \sim_{\mathbb{R}} \overline{L}_1 + \overline{L}_2 + (4-u)\overline{F}$, Using this, we conclude that the divisor $\upsilon^*(-K_S) - u\overline{F}$ is pseudo-effective $\iff$ $u \in [0,4]$. Moreover, if $0 \leqslant u \leqslant 1$, then the divisor $\upsilon^*(-K_S) - u\overline{F}$ is nef. Furthermore, if $u \in [1,4]$, then the Zariski decomposition of this divisor can be described as follows:

$$
\upsilon^*(-K_S) - u\overline{F} \sim_{\mathbb{R}} \underbrace{\frac{4-u}{3}\left(\overline{L}_1 + \overline{L}_2 + 3\overline{F}\right)}_{\text{positive part}} + \underbrace{\frac{u-1}{3}\left(\overline{L}_1 + \overline{L}_2\right)}_{\text{negative part}}.
$$

Therefore, we have

$$
\mathrm{vol}\left(\upsilon^*(-K_S) - u\overline{F}\right) =
\begin{cases}
2 - \dfrac{u^2}{2} & \text{if } 0 \leqslant u \leqslant 1, \\[2mm]
\dfrac{(4-u)^2}{6} & \text{if } 1 \leqslant u \leqslant 4.
\end{cases}
$$

Integrating this function, we get $S_S(\overline{F}) = \frac{5}{3}$, so that $\delta_P(S) \leqslant \frac{A_S(\overline{F})}{S_S(\overline{F})} = \frac{9}{5}$ since $A_S(\overline{F}) = 3$. Now, using Corollary 1.102 and writing $\Delta_{\overline{F}} = \frac{1}{2}\phi(\widehat{E}_P)$, we see that

$$
\delta_P(S) \geqslant \min\left\{ \frac{9}{5}, \inf_{O \in \overline{F}} \frac{A_{\overline{F}, \Delta_{\overline{F}}}(O)}{S(W_{\bullet,\bullet}^{\overline{F}}; O)} \right\}.
$$

Let O be a point in the curve $\overline{F}$. Then Corollary 1.109 gives

$$
\begin{aligned}
S\left(W_{\bullet,\bullet}^{\overline{F}}; O\right) \\
= \int_0^4 \left((P(u) \cdot \overline{F}) \times \mathrm{ord}_O\left(N(u)\big|_{\overline{F}}\right) + \frac{(P(u) \cdot \overline{F})^2}{2} \right) du \\
= \int_1^4 \frac{(u-1)(4-u)}{18} du \times \left((\overline{L}_1 + \overline{L}_2) \cdot \overline{F} \right)_O + \int_0^1 \frac{u^2}{8} du + \int_1^4 \frac{(4-u)^2}{72} du \\
= \frac{\left((\overline{L}_1 + \overline{L}_2) \cdot \overline{F} \right)_O}{4} + \frac{1}{6},
\end{aligned}
$$

so that

$$S\left(W_{\bullet,\bullet}^{\overline{F}}; O\right) = \begin{cases} \dfrac{5}{12} & \text{if } O = \overline{L}_1 \cap \overline{F}, \\[2mm] \dfrac{5}{12} & \text{if } O = \overline{L}_2 \cap \overline{F}, \\[2mm] \dfrac{1}{6} & \text{otherwise,} \end{cases}$$

so that

$$\frac{A_{\overline{F}, \Delta_{\overline{F}}}(O)}{S\left(W_{\bullet,\bullet}^{\overline{F}}; O\right)} = \begin{cases} \dfrac{12}{5} & \text{if } O = \overline{L}_1 \cap \overline{F}, \\[2mm] \dfrac{12}{5} & \text{if } O = \overline{L}_2 \cap \overline{F}, \\[2mm] 3 & \text{if } O = \phi\left(\widehat{E}_P\right), \\[2mm] 6 & \text{otherwise.} \end{cases}$$

This gives $\delta_P(S) = \frac{9}{5}$ as required.

Case 6. Suppose that $\pi(P) \notin R$, and P is contained in exactly one (-1)-curve L. Then $E_P + \iota(E_P) + \widetilde{L} \sim -2K_{\widetilde{S}}$, so that $\sigma^*(-K_S) - uE_P \sim_{\mathbb{R}} \frac{3-2u}{2}E_P + \frac{1}{2}(\iota(E_P) + \widetilde{L})$, which gives $\tau = \frac{3}{2}$ because the intersection form of the curves $\iota(E_P)$ and $\widetilde{L}$ is negative definite. Similarly, we see that

$$P(u) = \begin{cases} \dfrac{3-2u}{2}E_P + \dfrac{1}{2}\left(\iota(E_P) + \widetilde{L}\right) & \text{if } 0 \leqslant u \leqslant 1, \\[3mm] \dfrac{3-2u}{2}E_P + \dfrac{1}{2}\iota(E_P) + \dfrac{2-u}{2}\widetilde{L} & \text{if } 1 \leqslant u \leqslant \dfrac{7}{5}, \\[3mm] \dfrac{3-2u}{2}\left(E_P + 5\iota(E_P) + 3\widetilde{L}\right) & \text{if } \dfrac{7}{5} \leqslant u \leqslant \dfrac{3}{2}, \end{cases}$$

and

$$N(u) = \begin{cases} 0 & \text{if } 0 \leqslant u \leqslant 1, \\[3mm] \dfrac{u-1}{2}\widetilde{L} & \text{if } 1 \leqslant u \leqslant \dfrac{7}{5}, \\[3mm] (5u-7)\iota(E_P) + (3u-4)\widetilde{L} & \text{if } \dfrac{7}{5} \leqslant u \leqslant \dfrac{3}{2}. \end{cases}$$

Note that $E_P \cdot \iota(E_P) = 2$. Computing $P(u) \cdot P(u)$ for $u \in [0, \frac{3}{2}]$, we get

$$\mathrm{vol}\left(\sigma^*(-K_S) - uE_P\right) = \begin{cases} 2 - u^2 & \text{if } 0 \leqslant u \leqslant 1, \\[3mm] \dfrac{5 - u^2 - 2u}{2} & \text{if } 1 \leqslant u \leqslant \dfrac{7}{5}, \\[3mm] 3(3-2u)^2 & \text{if } \dfrac{7}{5} \leqslant u \leqslant \dfrac{3}{2}. \end{cases}$$

Integrating, we get $S_S(E_P) = \frac{19}{20}$, so that $\delta_P(S) \leqslant \frac{A_S(E_P)}{S_S(E_P)} = \frac{40}{19}$. For every $O \in E_P$, we get

$$
S(W_{\bullet,\bullet}^{E_P}; O) = \int_0^{\frac{3}{2}} \left(\big(P(u) \cdot E_P\big) \times \big(N(u) \cdot E_P\big)_O + \frac{\big(P(u) \cdot E_P\big)^2}{2} \right) du
$$

$$
= \int_0^{\frac{3}{2}} \big(P(u) \cdot E_P\big) \times \big(N(u) \cdot E_P\big)_O du + \int_0^1 \frac{u^2}{2} du
$$

$$
+ \int_1^{\frac{7}{5}} \frac{(1+u)^2}{8} du + \int_{\frac{7}{5}}^{\frac{3}{2}} \frac{(18-12u)^2}{2} du
$$

$$
= \left(\int_1^{\frac{7}{5}} \frac{(u-1)(1+u)}{4} du + \int_{\frac{7}{5}}^{\frac{3}{2}} (3u-4)(18-12u) du \right) \times \big(\widetilde{L} \cdot E_P\big)_O
$$

$$
+ \int_{\frac{7}{5}}^{\frac{3}{2}} (5u-7)(18-12u) du \times \big(\iota(E_P) \cdot E_P\big)_O + \frac{13}{30}
$$

$$
= \frac{19}{300}\big(\widetilde{L} \cdot E_P\big)_O + \frac{\big(\iota(E_P) \cdot E_P\big)_O}{100} + \frac{13}{30}
$$

$$
\leqslant \frac{19}{300}\widetilde{L} \cdot E_P + \frac{\iota(E_P) \cdot E_P}{100} + \frac{13}{30} = \frac{31}{60}
$$

by Corollary 1.109, Now, using Corollary 1.102, we get

$$
\frac{40}{19} \geqslant \delta_P(S) \geqslant \min\left\{ \frac{40}{19}, \inf_{O \in E_P} \frac{1}{S(W_{\bullet,\bullet}^{E_P}; O)} \right\} \geqslant \frac{60}{31},
$$

so that $\frac{40}{19} \geqslant \delta_P(S) \geqslant \frac{60}{31}$. In particular, we see that $\delta_P(S) > \frac{15}{8}$ as required.

Case 7. Suppose that $\pi(P) \notin R$, and P is contained in two (-1)-curves L_1 and L_2. Then $E_P + \iota(E_P) + \widetilde{L}_1 + \widetilde{L}_2 \sim \sigma^*(-2K_S) - 2E_P$. This gives

$$
\sigma^*(-K_S) - uE_P \sim_{\mathbb{R}} \frac{3-2u}{2}E_P + \frac{1}{2}\big(\iota(E_P) + \widetilde{L}_1 + \widetilde{L}_2\big),
$$

so that $\tau = \frac{3}{2}$, because the intersection form of the curves $\iota(E_P)$, $\widetilde{L}_1$, $\widetilde{L}_2$ is semi-negative definite. Moreover, we have

$$
P(u) = \begin{cases} \dfrac{3-2u}{2}E_P + \dfrac{1}{2}\big(\iota(E_P) + \widetilde{L}_1 + \widetilde{L}_2\big) & \text{if } 0 \leqslant u \leqslant 1, \\[2ex] \dfrac{3-2u}{2}E_P + \dfrac{1}{2}\iota(E_P) + \dfrac{2-u}{2}\big(\widetilde{L}_1 + \widetilde{L}_2\big) & \text{if } 1 \leqslant u \leqslant \dfrac{3}{2}, \end{cases}
$$

and

$$
N(u) = \begin{cases} 0 & \text{if } 0 \leqslant u \leqslant 1, \\[2ex] \dfrac{u-1}{2}\big(\widetilde{L}_1 + \widetilde{L}_2\big) & \text{if } 1 \leqslant u \leqslant \dfrac{3}{2}, \end{cases}
$$

so that

$$\text{vol}\left(\sigma^*(-K_S) - uE_P\right) = \begin{cases} 2 - u^2 & \text{if } 0 \leqslant u \leqslant 1, \\ 3 - 2u & \text{if } 1 \leqslant u \leqslant \dfrac{3}{2}. \end{cases}$$

Integrating, we get $S_S(E_P) = \frac{23}{24}$, so that $\delta_P(S) \leqslant \frac{A_S(E_P)}{S_S(E_P)} = \frac{48}{23}$. For every $O \in E_P$, we get

$$S\left(W_{\bullet,\bullet}^{E_P}; O\right) = \frac{\left((\widetilde{L}_1 + \widetilde{L}_2) \cdot E_P\right)_O}{16} + \frac{5}{12} \leqslant \frac{23}{48}$$

by Corollary 1.109. Now, using Corollary 1.102, we get $\delta_P(S) = \frac{48}{23} > \frac{15}{8}$.

Case 8. Suppose that $\pi(P) \notin R$, and P is contained in three (-1)-curves L_1, L_2, L_3. Then $E_P + \iota(E_P) + \widetilde{L}_1 + \widetilde{L}_2 + \widetilde{L}_3 \sim -K_{\widetilde{S}}$ and $E_P \cdot \iota(E_P) = 0$, so that the (-1)-curves E_P and $\iota(E_P)$ are disjoint. Then $\sigma \circ \iota(E_P) = \varsigma(E_P)$ is a (-1)-curve in S that does not pass through P. We have $\varsigma(E_P) + L_1 + L_2 + L_3 \sim -2K_S$, which implies that $\varsigma(E_P) \cdot L_1 = 1$, $\varsigma(E_P) \cdot L_2 = 1$ and $\varsigma(E_P) \cdot L_3 = 1$. Let $B = \tau(\varsigma(E_P))$. Then B is a (-1)-curve such that $B + \varsigma(E_P) \sim -K_S$, that gives $B \cdot L_1 = B \cdot L_2 = B \cdot L_3 = 0$. Thus, we see that B is disjoint from L_1, L_2, L_3. In particular, it does not contain P.

Now, we denote by $\widetilde{B}$ the proper transform of the (-1)-curve B on the surface $\widetilde{S}$. Then $\widetilde{B}$ is a (-1)-curve that is disjoint from $E_P, \widetilde{L}_1, \widetilde{L}_2, \widetilde{L}_3$. Let $\widetilde{Z} = \iota(\widetilde{B})$ and $Z = \sigma(\widetilde{Z})$. We have $\widetilde{Z} + \widetilde{B} \sim -2K_{\widetilde{S}}$. This gives $\widetilde{Z} \cdot \widetilde{L}_1 = \widetilde{Z} \cdot \widetilde{L}_2 = \widetilde{Z} \cdot \widetilde{L}_3 = 0$ and $\widetilde{Z} \cdot E_P = 2$. Thus, the curves $\widetilde{Z}, \widetilde{L}_1, \widetilde{L}_2, \widetilde{L}_3$ are disjoint, $Z + B \sim -2K_S$ and $\text{mult}_P(Z) = \widetilde{Z} \cdot E_P = 2$. Summarizing, we get $Z + B \sim -2K_S$, $B + \varsigma(E_P) \sim -K_S$ and $\varsigma(E_P) + L_1 + L_2 + L_3 \sim -2K_S$. Using these rational equivalences, we get $L_1 + L_2 + L_3 + Z \sim -3K_S$. Then

$$\sigma^*(-K_S) - uE_P \sim_{\mathbb{R}} \frac{5 - 3u}{3}E_P + \frac{1}{3}\left(\widetilde{Z} + \widetilde{L}_1 + \widetilde{L}_2 + \widetilde{L}_3\right).$$

This gives $\tau = \frac{5}{3}$ because the curves $\widetilde{Z}, \widetilde{L}_1, \widetilde{L}_2, \widetilde{L}_3$ are disjoint and all of them have negative self-intersections. Similarly, we see that

$$P(u) = \begin{cases} \dfrac{5 - 3u}{3}E_P + \dfrac{1}{3}\left(\widetilde{Z} + \widetilde{L}_1 + \widetilde{L}_2 + \widetilde{L}_3\right) & \text{if } 0 \leqslant u \leqslant 1, \\[3mm] \dfrac{5 - 3u}{6}\left(2E_P + \widetilde{L}_1 + \widetilde{L}_2 + \widetilde{L}_3\right) + \dfrac{1}{3}\widetilde{Z} & \text{if } 1 \leqslant u \leqslant \dfrac{3}{2}, \\[3mm] \dfrac{5 - 3u}{6}\left(2E_P + 4\widetilde{Z} + \widetilde{L}_1 + \widetilde{L}_2 + \widetilde{L}_3\right) & \text{if } \dfrac{3}{2} \leqslant u \leqslant \dfrac{5}{3}, \end{cases}$$

and

$$N(u) = \begin{cases} 0 & \text{if } 0 \leqslant u \leqslant 1, \\ \dfrac{u-1}{2}\left(\widetilde{L}_1 + \widetilde{L}_2 + \widetilde{L}_3\right) & \text{if } 1 \leqslant u \leqslant \dfrac{3}{2}, \\ \dfrac{u-1}{2}\left(\widetilde{L}_1 + \widetilde{L}_2 + \widetilde{L}_3\right) + (2u-3)\widetilde{Z} & \text{if } \dfrac{3}{2} \leqslant u \leqslant \dfrac{5}{3}. \end{cases}$$

Now, computing $P(u) \cdot P(u)$ for $u \in [0, \frac{5}{3}]$, we obtain

$$\mathrm{vol}\left(\sigma^*(-K_S) - uE_P\right) = \begin{cases} 2 - u^2 & \text{if } 0 \leqslant u \leqslant 1, \\ \dfrac{u^2 - 6u + 7}{2} & \text{if } 1 \leqslant u \leqslant \dfrac{3}{2}, \\ \dfrac{(5-3u)^2}{2} & \text{if } \dfrac{3}{2} \leqslant u \leqslant \dfrac{5}{3}. \end{cases}$$

Integrating, we get $S_S(E_P) = \frac{35}{36}$, so that $\delta_P(S) \leqslant \frac{A_S(E_P)}{S_S(E_P)} = \frac{72}{35}$. For every $O \in E_P$, we get

$S\left(W_{\bullet,\bullet}^{E_P}; O\right)$

$= \displaystyle\int_0^{\frac{5}{3}} \left((P(u) \cdot E_P) \times (N(u) \cdot E_P)_O + \frac{(P(u) \cdot E_P)^2}{2}\right) du$

$= \displaystyle\int_0^{\frac{5}{3}} (P(u) \cdot E_P) \times (N(u) \cdot E_P)_O \, du + \int_0^1 \frac{u^2}{2} du + \int_1^{\frac{3}{2}} \frac{(3-u)^2}{8} du$

$\quad + \displaystyle\int_{\frac{3}{2}}^{\frac{5}{3}} \frac{9(4-3u)^2}{8} du$

$= \left(\displaystyle\int_1^{\frac{3}{2}} \frac{(u-1)(3-u)}{4} du + \int_{\frac{3}{2}}^{\frac{5}{3}} \frac{3(u-1)(5-3u)}{4} du\right) \times \left((\widetilde{L}_1 + \widetilde{L}_2 + \widetilde{L}_3) \cdot E_P\right)_O$

$\quad + \displaystyle\int_{\frac{3}{2}}^{\frac{5}{3}} \frac{3(2u-3)(5-3u)}{2} du \times \left(\widetilde{Z} \cdot E_P\right)_O + \frac{3}{8}$

$= \dfrac{5}{72}\left((\widetilde{L}_1 + \widetilde{L}_2 + \widetilde{L}_3) \cdot E_P\right)_O + \dfrac{\left(\widetilde{Z} \cdot E_P\right)_O}{144} + \dfrac{3}{8}$

by Corollary 1.109, which gives

$$S\left(W_{\bullet,\bullet}^{E_P}; O\right) = \begin{cases} \dfrac{4}{9} & \text{if } O \in \widetilde{L}_1 \cup \widetilde{L}_2 \cup \widetilde{L}_3, \\ \dfrac{3}{8} + \dfrac{\left(\widetilde{Z} \cdot E_P\right)_O}{144} & \text{if } O \in \widetilde{Z}, \\ \dfrac{3}{8} & \text{if } O \notin \widetilde{L}_1 \cup \widetilde{L}_2 \cup \widetilde{L}_3 \cup \widetilde{Z}, \end{cases}$$

so $S(W_{\bullet,\bullet}^{E_P}; O) \leqslant \frac{4}{9}$ since $(\widetilde{Z} \cdot E_P)_O \leqslant \widetilde{Z} \cdot E_P = 2$. Then $\delta_P(S) = \frac{72}{35}$ by Corollary 1.102.

Case 9. Finally, we suppose that $\pi(P) \notin R$, and P is contained in four (-1)-curves. Denote them by L_1, L_2, L_3 and L_4. Then $L_1 + L_2 + L_3 + L_4 \sim -2K_S$, so that

$$\sigma^*(-K_S) - uE_P \sim_\mathbb{R} (2-u)E_P + \frac{1}{2}(\widetilde{L}_1 + \widetilde{L}_2 + \widetilde{L}_3 + \widetilde{L}_4).$$

This gives $\tau = 2$ because the (-2)-curves $\widetilde{L}_1, \widetilde{L}_2, \widetilde{L}_3, \widetilde{L}_4$ are disjoint. Moreover, we have

$$P(u) = \begin{cases} (2-u)E_P + \frac{1}{2}(\widetilde{L}_1 + \widetilde{L}_2 + \widetilde{L}_3 + \widetilde{L}_4) & \text{if } 0 \leqslant u \leqslant 1, \\ (2-u)\left(E_P + \frac{1}{2}(\widetilde{L}_1 + \widetilde{L}_2 + \widetilde{L}_3 + \widetilde{L}_4)\right) & \text{if } 1 \leqslant u \leqslant 2, \end{cases}$$

and

$$N(u) = \begin{cases} 0 & \text{if } 0 \leqslant u \leqslant 1, \\ \frac{u-1}{2}(\widetilde{L}_1 + \widetilde{L}_2 + \widetilde{L}_3 + \widetilde{L}_4) & \text{if } 1 \leqslant u \leqslant 2, \end{cases}$$

so that

$$\text{vol}\big(\sigma^*(-K_S) - uE_P\big) = \begin{cases} 2 - u^2 & \text{if } 0 \leqslant u \leqslant 1, \\ (2-u)^2 & \text{if } 1 \leqslant u \leqslant 2. \end{cases}$$

Integrating, we get $S_S(E_P) = 1$, so that $\delta_P(S) \leqslant \frac{A_S(E_P)}{S_S(E_P)} = 2$. For every $O \in E_P$, we get

$$S(W_{\bullet,\bullet}^{E_P}; O) = \frac{\big((\widetilde{L}_1 + \widetilde{L}_2 + \widetilde{L}_3 + \widetilde{L}_4) \cdot E_P\big)_O}{12} + \frac{1}{3} \leqslant \frac{5}{12}$$

by Corollary 1.109. Now, using Corollary 1.102, we get $\delta_P(S) = 2 > \frac{15}{8}$. This completes the proof of the lemma. $\square$

Let us conclude this chapter by proving the following lemma:

Lemma 2.16 *Suppose that $K_S^2 = 1$. Then*

$$\delta(S) = \begin{cases} \dfrac{15}{7} & \text{if } |-K_S| \text{ contains a cuspidal curve,} \\ \dfrac{12}{5} & \text{if } |-K_S| \text{ contains no cuspidal curve.} \end{cases}$$

Proof Let P be a point in S, and let C be a curve in the pencil $|-K_S|$ that contains P. Then C is an irreducible curve of arithmetic genus 1, so that either C is smooth at P, or the curve C has a node at P, or the curve C has a cusp at P. Note that the pencil $|-K_S|$ always contains singular curves. Observe also that $S_S(C) = \frac{1}{3}$. Moreover, if the curve C is smooth at P, then we have $S(W_{\bullet,\bullet}^C; P) = \frac{1}{3}$ by Corollary 1.109, so that $\delta_P(C) \geqslant 3$ by Theorem 1.95. Thus, to complete the proof, we must prove that

$$
\delta_P(S) = \begin{cases} \dfrac{15}{7} & \text{if } C \text{ has a cusp at } P, \\[2mm] \dfrac{12}{5} & \text{if } C \text{ has a node at } P. \end{cases}
$$

Therefore, we may assume that our curve C is singular at the point P.

Let $\sigma \colon \widetilde{S} \to S$ be the blow up of the point P, let E_P be the σ-exceptional divisor, let $\widetilde{C}$ be the proper transform on $\widetilde{S}$ of the curve C, and let u be a non-negative number. Then $\widetilde{C}$ is a smooth curve such that $\widetilde{C}^2 = -3$ and $\sigma^*(-K_S) - uE_P \sim_{\mathbb{R}} (2-u)E_P + \widetilde{C}$. Then $\sigma^*(-K_S) - uE_P$ is pseudo-effective $\iff u \leqslant 2$. This divisor is nef $\iff u \leqslant \frac{1}{2}$. Furthermore, if $\frac{1}{2} \leqslant u \leqslant 2$, then its Zariski decomposition can be described as follows:

$$
\sigma^*(-K_S) - uE_P \sim_{\mathbb{R}} \underbrace{(2-u)\left(E_P + \frac{2}{3}\widetilde{C}\right)}_{\text{positive part}} + \underbrace{\frac{2u-1}{3}\widetilde{C}}_{\text{negative part}},
$$

so that

$$
\mathrm{vol}\left(\sigma^*(-K_S) - uE_P\right) = \begin{cases} 1 - u^2 & \text{if } 0 \leqslant u \leqslant \dfrac{1}{2}, \\[3mm] \dfrac{(2-u)^2}{3} & \text{if } \dfrac{1}{2} \leqslant u \leqslant 2, \end{cases}
$$

which gives $S_S(E_P) = \frac{5}{6}$. Thus, we have $\delta_P(S) \leqslant \frac{A_S(E_P)}{S_S(E_P)} = \frac{12}{5}$.

Note that the divisor $-K_{\widetilde{S}}$ is big. Then $\widetilde{S}$ is a Mori Dream Space by [209, Theorem 1]. Therefore, we can apply Corollary 1.109 to compute $S(W_{\bullet,\bullet}^{E_P}; O)$ for every point $O \in E_P$. To be precise, if O is a point in E_P, then Corollary 1.109 gives

$$
S\left(W_{\bullet,\bullet}^{E_P}; O\right) = \frac{1}{6} + \frac{\left(E_P \cdot \widetilde{C}\right)_O}{4},
$$

which implies that $S(W_{\bullet,\bullet}^{E_P}; O) \leqslant \frac{5}{12}$ in the case when C has a nodal singularity at P. Thus, if C has a node at P, then $\delta_P(S) = \frac{12}{5}$ by Corollary 1.102. Hence, we may assume that the curve C has a cusp at P. Then the intersection $\widetilde{C} \cap E_P$ consists of one point, and $\widetilde{C}$ is tangent to E_P at this point.

Let $\rho \colon \widehat{S} \to \widetilde{S}$ be the blow up of the point $\widetilde{C} \cap E_P$, let F be the ρ-exceptional curve, let $\widehat{C}$ and $\widehat{E}_P$ be the proper transforms on $\widehat{S}$ via ρ of the curves $\widetilde{C}$ and E_P, respectively. Then there exists a commutative diagram

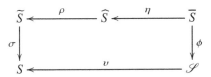

where η is the blow up of $\widehat{C} \cap \widehat{E}_P$, ϕ is the contraction of the proper transforms of the curves $\widehat{E}_P$ and F, and υ is the contraction of the proper transform of the η-exceptional curve.

Let $\mathscr{G}$ be the υ-exceptional curve, and let $\mathscr{C}$ be the proper transform of the curve C on the surface $\mathscr{S}$. Then $\mathscr{G}$ contains two singular points Q_1 and Q_2 such that Q_1 is an ordinary double point, and Q_2 is a quotient singular point of type $\frac{1}{3}(1,1)$. However, these singular points are not contained in the curve $\mathscr{C}$. Note also that $\mathscr{C}^2 = -5$, $\mathscr{G}^2 = -\frac{1}{6}$, $\mathscr{C} \cdot \mathscr{G} = 1$, and $\mathscr{S}$ is a Mori Dream Space by [209, Theorem 1] since $-K_{\overline{S}}$ is big.

Since $\upsilon^*(-K_S) - u\mathscr{G} \sim_{\mathbb{R}} (6 - u)\mathscr{G} + \mathscr{C}$, the divisor $\upsilon^*(-K_S) - u\mathscr{G}$ is pseudo-effective if and only if $u \leqslant 6$. Moreover, this divisor is nef $\iff u \leqslant 1$. Furthermore, if $1 \leqslant u \leqslant 6$, then the positive part of its Zariski decomposition is $(6 - u)\mathscr{G} + \frac{6-u}{5}\mathscr{C}$, and the negative part of its Zariski decomposition is $\frac{u-1}{5}\mathscr{C}$. This gives

$$\mathrm{vol}\big(\upsilon^*(-K_S) - u\mathscr{G}\big) = \begin{cases} 1 - \dfrac{u^2}{6} & \text{if } 0 \leqslant u \leqslant 1, \\ \dfrac{(6 - u)^2}{30} & \text{if } 1 \leqslant u \leqslant 6. \end{cases}$$

Integrating this function, we obtain $S_S(\mathscr{G}) = \frac{7}{3}$, which implies that $\delta_P(S) \leqslant \frac{A_S(\mathscr{G})}{S_S(\mathscr{G})} = \frac{15}{7}$. On the one hand, it follows from Corollary 1.102 that

$$\delta_P(S) \geqslant \min\left\{\frac{15}{7}, \inf_{O \in \mathscr{G}} \frac{A_{\mathscr{G}, \Delta_{\mathscr{G}}}(O)}{S\big(W_{\bullet, \bullet}^{\mathscr{G}}; O\big)}\right\},$$

where $\Delta_{\mathscr{G}} = \frac{1}{2}Q_1 + \frac{2}{3}Q_2$. On the other hand, if O is a point in $\mathscr{G}$, then

$$S\big(W_{\bullet, \bullet}^{\mathscr{G}}; O\big) = \begin{cases} \dfrac{1}{3} & \text{if } O = \mathscr{G} \cap \mathscr{C}, \\ \dfrac{1}{18} & \text{otherwise,} \end{cases}$$

by Corollary 1.109, so that

$$\frac{A_{\mathscr{G},\Delta_{\mathscr{G}}}(O)}{S(W^{\mathscr{G}}_{\bullet,\bullet};O)} = \begin{cases} 3 & \text{if } O = \mathscr{G} \cap \mathscr{C}, \\ 9 & \text{if } O = Q_1, \\ 6 & \text{if } O = Q_2, \\ 18 & \text{otherwise.} \end{cases}$$

This gives $\delta_P(S) = \frac{15}{7}$ as required. This completes the proof of the lemma. □

Let P be a point of the surface S. All possible values of $\alpha_P(S)$ have been found in [40]. It would be interesting to find all values of $\delta_P(S)$. For $K_S^2 = 3$, this is done in [2].

3

Proof of Main Theorem: Known Cases

3.1 Direct products

Let S be a smooth del Pezzo surface. If S is not a blow up of $\mathbb{P}^2$ in one or two points, then S is K-polystable (see Chapter 2), so that $\mathbb{P}^1 \times S$ is also K-polystable by Theorem 1.6. Therefore, every smooth Fano threefold in the families №2.34, №3.27, №5.3, №6.1, №7.1, №8.1, №9.1, №10.1 is K-polystable. On the other hand, blow ups of the plane $\mathbb{P}^2$ in one or two points are K-unstable by Lemmas 2.3 and 2.4, so that the smooth Fano threefolds №3.28 and №4.10 are K-unstable by Theorem 1.6.

3.2 Homogeneous spaces

The following smooth Fano threefolds are homogeneous spaces under actions of reductive groups: a smooth quadric threefold in $\mathbb{P}^4$ (family №1.16), $\mathbb{P}^3$ (family №1.17), a smooth divisor in $\mathbb{P}^2 \times \mathbb{P}^2$ of degree $(1,1)$ (family №2.32), $\mathbb{P}^1 \times \mathbb{P}^2$ (family №2.34), and $\mathbb{P}^1 \times \mathbb{P}^1 \times \mathbb{P}^1$ (family №3.27). All of them are K-polystable by Theorem 1.22 or Theorem 1.48.

3.3 Fano threefolds with torus action

There are exactly 18 smooth toric Fano threefolds [12, 222], and each such threefold is the unique member of the corresponding deformation family. Theorem 1.21 tells us which of these threefolds are K-polystable [14, 219]. These results are summarized in Table 3.1, where S_6 and S_7 are the smooth del Pezzo surfaces of degree 6 and 7, respectively. Smooth non-toric Fano threefolds admitting a faithful action of the two-dimensional torus $\mathbb{G}_m^2$ have been classified in [203], see also [45]. These are smooth Fano threefolds №1.16, №2.29,

Table 3.1

Family	Short description	K-polystable
№1.17	$\mathbb{P}^3$	Yes
№2.33	blow up of $\mathbb{P}^3$ in a line	No
№2.34	$\mathbb{P}^1 \times \mathbb{P}^2$	Yes
№2.35	V_7 =blow up of $\mathbb{P}^3$ in a point	No
№2.36	$\mathbb{P}(O_{\mathbb{P}^2} \oplus O_{\mathbb{P}^2}(2))$	No
№3.25	blow up of $\mathbb{P}^3$ in two disjoint lines	Yes
№3.26	blow up of $\mathbb{P}^3$ in a line and a point	No
№3.27	$\mathbb{P}^1 \times \mathbb{P}^1 \times \mathbb{P}^1$	Yes
№3.28	$\mathbb{P}^1 \times \mathbb{F}_1$	No
№3.29	blow up of V_7 in a line in the exceptional divisor of the blow up $V_7 \to \mathbb{P}^3$	No
№3.30	blow up of V_7 in a fiber of the $\mathbb{P}^1$-bundle $V_7 \to \mathbb{P}^2$	No
№3.31	blow up of the quadric cone in its vertex	No
№4.9	blow up of the smooth Fano threefold №3.25 in a curve contracted by the birational morphism to $\mathbb{P}^3$	No
№4.10	$\mathbb{P}^1 \times S_7$	No
№4.11	blow up of $\mathbb{P}^1 \times \mathbb{F}_1$ in a curve that is a (-1)-curve of a fiber of the projection $\mathbb{P}^1 \times \mathbb{F}_1 \to \mathbb{P}^1$	No
№4.12	blow up of the smooth Fano threefold №2.33 in two curves contracted by the birational morphism to $\mathbb{P}^3$	No
№5.2	blow up of the smooth Fano threefold №3.25 in two curves contracted by the birational morphism to $\mathbb{P}^3$ which are both contained in one exceptional surface	No
№5.3	$\mathbb{P}^1 \times S_6$	Yes

№2.30, №2.31, №2.32, №3.18, №3.19, №3.20, №3.21, №3.22, №3.23, №3.24, №4.4, №4.5, №4.7, №4.8, and two special threefolds in the families №2.24 and №3.10. Their Futaki invariants have been found in [203]. This allowed to solve the Calabi Problem for all of them [202, 203, 117]. We summarize these results in Table 3.2. Let us illustrate these results:

Lemma 3.1 ([117, Theorem 6.1]) *Let X be the unique smooth Fano threefold №3.19. Then X is K-polystable.*

Proof Let Q be the smooth quadric hypersurface in $\mathbb{P}^4$ given by $x_0^2 + x_1 x_2 - x_3 x_4 = 0$, where x_0, x_1, x_2, x_3, x_4 are coordinates on $\mathbb{P}^4$. Then Q admits a

Table 3.2

Fano threefold	Short description	Futaki invariant	K-poly-stable
№1.16	Q =smooth quadric threefold in $\mathbb{P}^4$	zero	Yes
№2.24 (special)	divisor in $\mathbb{P}^2 \times \mathbb{P}^2$ of degree $(1,2)$	zero	Yes
№2.29	blow up of Q in a conic	zero	Yes
№2.30	blow up of Q in a point	non-zero	No
№2.31	blow up of Q in a line	non-zero	No
№2.32	W =divisor in $\mathbb{P}^2 \times \mathbb{P}^2$ of degree $(1,1)$	zero	Yes
№3.10 (special)	blow up of Q in two disjoint conics	zero	Yes
№3.18	blow up of Q in a point and a conic	non-zero	No
№3.19	blow up of Q in two points	zero	Yes
№3.20	blow up of Q in two disjoint lines	zero	Yes
№3.21	blow up of $\mathbb{P}^1 \times \mathbb{P}^2$ in a curve of degree $(2,1)$	non-zero	No
№3.22	blow up of $\mathbb{P}^1 \times \mathbb{P}^2$ in a curve of degree $(0,2)$	non-zero	No
№3.23	blow up of Q in a point and the strict transform of a line passing through this point	non-zero	No
№3.24	blow up of W in a curve of degree $(0,1)$	non-zero	No
№4.4	blow up of Q in two non-collinear points and the strict transform of a conic passing through both of these points	zero	Yes
№4.5	blow up of $\mathbb{P}^1 \times \mathbb{P}^2$ in a disjoint union of a curve of degree $(2,1)$ and a curve of degree $(1,0)$	non-zero	No
№4.7	blow up of W in a disjoint union of a curve of degree $(0,1)$ and a curve of degree $(1,0)$	zero	Yes
№4.8	blow up of $(\mathbb{P}^1)^3$ in a curve of degree $(0,1,1)$	non-zero	No

$\mathbb{G}_m^2$-action that is generated by two commuting $\mathbb{G}_m$-actions λ_1 and λ_2 defined as follows:

$$\lambda_1(t).[x_0 : x_1 : x_2 : x_3 : x_4] = [x_0 : tx_1 : x_2/t : x_3 : x_4],$$
$$\lambda_2(s).[x_0 : x_1 : x_2 : x_3 : x_4] = [x_0 : x_1 : x_2 : sx_3 : x_4/s],$$

where $t \in \mathbb{G}_m$ and $s \in \mathbb{G}_m$. Now, we let $P_1 = [0 : 0 : 0 : 0 : 1]$ and $P_2 = [0 : 0 : 0 : 1 : 0]$. Then P_1 and P_2 are $\mathbb{G}_m^2$-fixed. We may assume that

X is a blow up of Q at these points. Denote this blow up by ϕ, and denote the exceptional divisors by E_1 and E_2, respectively. Let σ be the involution in $\mathrm{Aut}(Q)$ given by $[x_0 : x_1 : x_2 : x_3 : x_4] \mapsto [x_0 : x_2 : x_1 : x_4 : x_3]$. Then σ swaps P_1 and P_2, so that both the $\mathbb{G}_m^2$-action and σ lifts to the threefold X. Therefore, we may consider $\mathbb{G}_m^2$ and $\langle \sigma \rangle$ as subgroups in $\mathrm{Aut}(X)$.

Let us apply results of Section 1.3 to X and $\mathbb{T} = \mathbb{G}_m^2$. We will use notations introduced in this section. First, we observe that σ acts on the $\mathbb{G}_m$-actions λ_1 and λ_2 by conjugation and sends λ_1 to λ_1^{-1} and λ_2 to λ_2^{-1}. Then by Lemma 1.29 we must have $\mathrm{Fut}_X = 0$.

Let $\pi \colon X \dashrightarrow \mathbb{P}^1$ be the quotient map. Then $\pi \circ \phi^{-1}$ is given by

$$[x_0 : x_1 : x_2 : x_3 : x_4] \mapsto [x_1 x_2 : x_3 x_4]$$

because the field of $\mathbb{G}_m^2$-invariant rational functions on Q is generated by $x_1 x_2 / x_3 x_4$. Moreover, both divisors E_1 and E_2 are horizontal because λ_2 induces a trivial action on both of them. Furthermore, the reducible fibres of the map π can be described as follows:

$$\pi^{-1}([0 : 1]) = D_1 \cup D_2,$$
$$\pi^{-1}([1 : 0]) = D_3 \cup D_4,$$
$$\pi^{-1}([1 : 1]) = 2D_0,$$

where each D_i is the proper transform on X of the hyperplane section of Q that are cut out by $x_i = 0$. Therefore, using Proposition 1.38, we see that to complete the proof, it is sufficient to show that $\beta(D_1) > 0$, $\beta(D_3) > 0$ and $\beta(D_0) > 0$.

Let $H = \phi^*(\mathcal{O}_{\mathbb{P}^4}(1)|_Q)$. Then we have $H - E_1 - E_2 \sim D_0 \sim D_1 \sim D_3 - E_1$, which implies that $\beta(D_0) = \beta(D_1) \leqslant \beta(D_3)$. Hence, it is actually sufficient to check that $\beta(D_0) > 0$. This is easy. Since $-K_X \sim 3H - 2E_1 - 2E_2$, we have

$S_X(D_0)$

$$= \frac{1}{38} \int_0^2 \left((3-x)H - (2-x)E_1 - (2-x)E_1 \right)^3 dx + \frac{1}{38} \int_2^3 \left((3-x)H \right)^3 dx$$
$$= \frac{1}{38} \int_0^2 \left(2(3-x)^3 - 2(2-x)^3 \right) dx + \frac{1}{38} \int_2^3 2(3-x)^3 dx = \frac{65}{76},$$

so that $\beta(D_0) = 1 - \frac{65}{76} = \frac{13}{76} > 0$. This implies that X is K-polystable. $\qquad \square$

3.4 Del Pezzo threefolds

Let V_d be a smooth Fano threefold such that $-K_{V_d} \sim 2H$ for some $H \in \mathrm{Pic}(V_d)$ such that $d = H^3$. Then V_d is a del Pezzo threefold of degree d. One can show that a general surface in $|H|$ is a smooth del Pezzo surface of degree d. Moreover, it follows from [103, 104, 105, 107, 108] that we have the following possibilities:

- $d = 1$ and V_1 is a sextic hypersurface in $\mathbb{P}(1, 1, 1, 2, 3)$;
- $d = 2$ and V_2 is quartic hypersurface in $\mathbb{P}(1, 1, 1, 1, 2)$;
- $d = 3$ and V_3 is a cubic hypersurface in $\mathbb{P}^4$;
- $d = 4$ and V_4 is an intersection of two quadrics in $\mathbb{P}^5$;
- $d = 5$ and V_5 is the quintic del Pezzo threefold (see Example 3.2 below);
- $d = 6$ and V_6 is a divisor in $\mathbb{P}^2 \times \mathbb{P}^2$ of degree $(1, 1)$;
- $d = 6$ and $V_6 = \mathbb{P}^1 \times \mathbb{P}^1 \times \mathbb{P}^1$;
- $d = 7$ and V_7 is a blow up of $\mathbb{P}^3$ at a point;
- $d = 8$ and $V_8 = \mathbb{P}^3$.

Hence, V_d belongs to the family №1.11, №1.12, №1.13, №1.14, №1.15, №2.32, №3.27, №2.35, respectively. From Sections 3.1, 3.2, 3.3, we know that V_6 and V_8 are K-polystable, but V_7 is not K-polystable, which will also be discussed later in Sections 3.6 and 3.7. The family №1.15 contains a unique smooth Fano threefold, and it is K-polystable:

Example 3.2 Let V_5 be a smooth intersection of the Grassmannian $\mathrm{Gr}(2, 5) \subset \mathbb{P}^9$ in its Plücker embedding with a linear subspace of dimension 5. Then V_5 is the unique smooth Fano threefold №1.15. By [47, Theorem 1.17], $\alpha_G(V_5) = \frac{5}{6}$ for $G = \mathrm{Aut}(V_5)$, where $\mathrm{Aut}(V_5) \cong \mathrm{PGL}_2(\mathbb{C})$, see, for example, [53, Proposition 7.1.10]. Thus, the smooth Fano threefold V_5 is K-polystable by Theorem 1.48.

Let us present K-stable smooth Fano threefolds in each of the remaining deformation families №1.11, №1.12, №1.13, №1.14.

Example 3.3 Let V_1 be a smooth sextic hypersurface in $\mathbb{P}(1, 1, 1, 2, 3)$ that is given by

$$w^2 = t^3 + x^6 + y^6 + z^6,$$

where x, y, z are coordinates of weight 1, t and w are coordinates of weights 2 and 3, respectively. Then V_1 is a smooth Fano threefold №1.11, the group $\mathrm{Aut}(V_1)$ is finite [45], and $\alpha_G(V_1) \geqslant 1$ by [47, Theorem 1.18], so that V_1 is K-stable by Theorem 1.48.

Example 3.4 Let V_2 be the quartic hypersurface in $\mathbb{P}(1, 1, 1, 1, 2)$ given by

$$w^2 = t^3 y + 6txyz + tz^3 + 2x^4 + y^3 z,$$

where x, y, z and t are coordinates of weight one, and w is a coordinate of weight two. Then V_2 is a smooth Fano threefold №1.12, and it follows from [151] that

Aut(V_2) contains a finite subgroup G such that $G \cong \mu_2 \times \mathrm{PSL}_2(\mathbb{F}_7)$ and G acts on V_2 without fixed points. Using Theorem 1.52, we see that $\alpha_G(V_2) \geqslant 1$. Then V_2 is K-stable by Theorem 1.48 since Aut(V_2) is a finite group [45].

Example 3.5 Let V_3 be the Klein smooth cubic threefold in $\mathbb{P}^4$ that is given by

$$x_0 x_1^2 + x_1 x_2^2 + x_2 x_3^2 + x_3 x_4^2 + x_4 x_0^2 = 0,$$

and let $G = \mathrm{Aut}(V_3)$. It follows from [1] that $G \cong \mathrm{PSL}_2(\mathbb{F}_{11})$, and the cubic V_3 does not contain G-invariant hyperplane sections. Thus, $\alpha_G(V_3) \geqslant 1$ by Corollary 1.54, so that the cubic threefold V_3 is K-stable by Theorem 1.48.

Example 3.6 ([165, §6.2]) Let V_4 be a smooth complete intersection of two quadrics in $\mathbb{P}^5$. Then V_4 is a Fano threefold in the family №1.14, and $G = \mathrm{Aut}(V_4)$ is finite [45]. It follows from [187, Proposition 2.1] that

$$V_4 \cong \left\{ x_0^2 + x_1^2 + x_2^2 + x_3^2 + x_4^2 + x_5^2 \right.$$
$$\left. = \lambda_0 x_0^2 + \lambda_1 x_1^2 + \lambda_2 x_2^2 + \lambda_3 x_3^2 + \lambda_4 x_4^2 + \lambda_5 x_5^2 = 0 \right\} \subset \mathbb{P}^5$$

for some (pairwise distinct) numbers $\lambda_0, \ldots, \lambda_5$, where $x_0, \ldots, x_5$ are coordinates on $\mathbb{P}^5$. If $\lambda_i = \omega^i$ for a primitive sixth root of unity ω, then $\alpha_G(V_4) \geqslant 1$ by Corollary 1.53, so that V_4 is K-stable by Theorem 1.48.

Thus, using Theorem 1.11, we conclude that general smooth Fano threefolds in the families №1.11, №1.12, №1.13, №1.14 are K-stable. In fact, all smooth Fano threefolds in these families are K-polystable [7, 69, 199, 146, 2].

3.5 K-stable cyclic covers

Some smooth Fano threefolds are cyclic covers of other smooth Fano threefolds. Therefore, it is tempting to apply Proposition 1.66 to these threefolds to prove their K-stability. Let us present a few examples that show how to apply Proposition 1.66 or its Corollary 1.67 to smooth Fano threefolds.

Example 3.7 ([7, Theorem 3.2]) Let X be any smooth Fano threefold №1.1. Then X is a smooth sextic hypersurface in $\mathbb{P}(1,1,1,1,3)$, so that X is a double cover of $\mathbb{P}^3$ branched over a smooth sextic surface. Then X is K-stable by Corollary 1.67.

Example 3.8 ([69, Theorem 1.1]) Let X be a smooth Fano threefold №1.2. Then X can be obtained as a complete intersection in $\mathbb{P}(1,1,1,1,1,2)$ given by

$$\begin{cases} \lambda w + x^2 + y^2 + z^2 + t^2 + u^2 = 0, \\ w^2 = f(x,y,z,t,u) \end{cases}$$

for some $\lambda \in \mathbb{C}$ and a quartic polynomial f, where x, y, z, t, u are coordinates of weight 1, and w is a coordinate of weight 2. If $\lambda \neq 0$, then X is isomorphic to a smooth quartic threefold in $\mathbb{P}^4$ that is given by $(x^2 + y^2 + z^2 + t^2 + u^2)^2 = \lambda^2 f(x,y,z,t,u)$. If $\lambda = 0$, then X is a double cover of the quadric in $\mathbb{P}^4$ given by $x^2 + y^2 + z^2 + t^2 + u^2 = 0$, which is branched over the smooth surface cut out on the quadric by $f(x,y,z,t,u) = 0$, where we consider x, y, z, t, u as coordinate on $\mathbb{P}^4$. Applying Corollary 1.67, we see that the threefold X is K-stable if $\lambda = 0$. Now, using Theorem 1.11, we deduce that general quartic threefolds in $\mathbb{P}^4$ are also K-stable.

Example 3.9 ([69, Theorem 1.1]) Let X be the complete intersection

$$\left\{ x^2 + y^2 + z^2 + t^2 + u^2 = 0, w^3 = f(x,y,z,t,u) \right\} \subset \mathbb{P}^5,$$

where $f(x,y,z,t,u)$ is a cubic polynomial, and x, y, z, t, u, w are coordinates on $\mathbb{P}^5$. Then X is a smooth Fano threefold №1.3, which is a triple cover of the quadric threefold in $\mathbb{P}^4$ branched over a smooth anticanonical surface. Applying Corollary 1.67, we see that the threefold X is K-stable. Thus, general Fano threefolds №1.3 are also K-stable.

Example 3.10 Let X be a smooth Fano threefold №1.4. Suppose that

$$X = \left\{ \sum_{i=0}^{5} x_i^2 = 0, \sum_{i=0}^{5} \lambda_i x_i^2 = 0, x_6^2 = f(x_0, x_1, x_2, x_3, x_4, x_5) \right\} \subset \mathbb{P}^6$$

for some pairwise different numbers $\lambda_1, \dots, \lambda_5$ and some quadratic polynomial f that does not depend on x_6, where $x_0, \dots, x_6$ are coordinates on $\mathbb{P}^6$. The projection to the first 6 coordinates gives a double cover $\varpi \colon X \to Y$, where Y is the smooth complete intersection of two quadrics in $\mathbb{P}^4$ described in Example 3.6. Since Y is K-stable by Example 3.6, and the ramification divisor of ϖ is a smooth surface in $|-K_Y|$, we see that X is K-stable by Corollary 1.67, so that a general Fano threefold №1.4 is K-stable.

Example 3.11 ([69, Theorem 1.1]) Let X be a smooth Fano threefold №1.5. Then X can be obtained as an intersection of the cone $V \subset \mathbb{P}^{10}$ over the Grassmannian $\mathrm{Gr}(2,5)$ in its Plücker embedding in $\mathbb{P}^9$ with a quadric hypersurface and a linear subspace Λ of codimension 3. If Λ does not contain the vertex of the cone Y, the threefold X is isomorphic to an intersection of the Grassmannian $\mathrm{Gr}(2,5) \subset \mathbb{P}^9$ with a quadric hypersurface and a linear subspace of codimension 2. If Λ contains the vertex of the cone, then X admits a double cover of the unique smooth Fano threefold №1.15 that is branched in a smooth anticanonical surface. In this (special) case, X is K-stable by Corollary 1.67 because the unique smooth Fano threefold №1.15 is K-polystable (see Example 3.2). Therefore, a general Fano threefold in the family №1.5 is K-stable by Theorem 1.11.

Example 3.12 ([2, Corollary 4.9(5)], cf. Example 3.3)   Let X be any smooth Fano threefold №1.11. Then X is a sextic hypersurface in $\mathbb{P}(1, 1, 1, 2, 3)$. Let $Y = \mathbb{P}(1, 1, 1, 2)$. There is a double cover $\varpi : X \to Y$ such that ϖ is branched over a smooth surface B such that $B \sim_{\mathbb{Q}} \frac{6}{5}(-K_Y)$. Then $\delta(Y) = \frac{4}{5}$ by [18, Corollary 7.7]. Since B does not contain the only singular point of the threefold Y, the log pair (Y, B) is purely log terminal. Thus, the only prime divisor F over Y such that $A_{Y,B}(F) = 0$ is B. But $S_Y(B) = \frac{5}{12}$, which implies that $\frac{A_Y(B)}{S_Y(B)} = \frac{12}{5} > \frac{4}{5}$. Then X is K-stable by Proposition 1.66.

Example 3.13 ([69, Example 4.2], cf. Example 3.4)   Let X be a smooth Fano threefold in the family №1.12. Then X is a double cover of $\mathbb{P}^3$ branched over a smooth quartic surface, so that the threefold X is K-stable by Corollary 1.67.

Example 3.14 ([69, Example 4.4])   Let $\mathbb{P}^2 \times \mathbb{P}^2 \hookrightarrow \mathbb{P}^8$ be the Segre embedding, let V be the projective cone in $\mathbb{P}^9$ over its image, let H be a hyperplane in $\mathbb{P}^9$, let Q be a quadric in $\mathbb{P}^9$ such that $X = V \cap H \cap Q$ is a smooth threefold. Then X is a Fano threefold in the family №2.6, and every smooth Fano threefold in this family can be obtained in this way. If $\mathrm{Sing}(V) \notin H$, then X is isomorphic to a divisor in $\mathbb{P}^2 \times \mathbb{P}^2$ of degree $(2, 2)$. If $\mathrm{Sing}(V) \in H$, then there is a double cover $\varpi : X \to W$ such that W is a smooth divisor in $\mathbb{P}^2 \times \mathbb{P}^2$ of degree $(1, 1)$, and ϖ is branched over a surface in $|-K_W|$. In this (special) case, X is K-stable by Corollary 1.67 since W is K-polystable (see Section 3.2). By Theorem 1.11, general smooth Fano threefolds №2.6 are K-stable.

Example 3.15 ([69, Example 4.4])   Let X be a smooth threefold in the family №3.1. Then X is a double cover of $\mathbb{P}^1 \times \mathbb{P}^1 \times \mathbb{P}^1$ branched over a smooth surface of degree $(2, 2, 2)$. By Corollary 1.67, the threefold X is K-stable because $\mathbb{P}^1 \times \mathbb{P}^1 \times \mathbb{P}^1$ is K-polystable.

3.6 Threefolds with infinite automorphism groups

Recall the following result:

Theorem 3.16 ([45, Theorem 1.2])   *Every smooth Fano threefold that has an infinite automorphism group is contained in one of the following 63 deformation families:*

№1.10, №1.15, №1.16, №1.17, №2.20, №2.21, №2.22, №2.24, №2.26, №2.27,

№2.28, №2.29, №2.30, №2.31, №2.32, №2.33, №2.34, №2.35, №2.36, №3.5,

№3.8, №3.9, №3.10, №3.12, №3.13, №3.14, №3.15, №3.16, №3.17, №3.18, №3.19,

№3.20, №3.21, №3.22, №3.23, №3.24, №3.25, №3.26, №3.27, №3.28, №3.29,

№3.30, №3.31, №4.2, №4.3, №4.4, №4.5, №4.6, №4.7, №4.8, №4.9, №4.10,

№4.11, №4.12, №4.13, №5.1, №5.2, №5.3, №6.1, №7.1, №8.1, №9.1, №10.1.

Each smooth threefold in the following 53 families has an infinite automorphism group:

№1.15, №1.16, №1.17, №2.26, №2.27, №2.28, №2.29, №2.30, №2.31, №2.32,

№2.33, №2.34, №2.35, №2.36, №3.9, №3.13, №3.14, №3.15, №3.16, №3.17,

№3.18, №3.19, №3.20, №3.21, №3.22, №3.23, №3.24, №3.25, №3.26, №3.27,

№3.28, №3.29, №3.30, №3.31, №4.2, №4.3, №4.4, №4.5, №4.6, №4.7, №4.8, №4.9,

№4.10, №4.11, №4.12, №5.1, №5.2, №5.3, №6.1, №7.1, №8.1, №9.1, №10.1.

Each of the following 10 deformation families has at least one smooth member that has an infinite automorphism group:
№1.10, №2.20, №2.21, №2.22, №2.24, №3.5, №3.8, №3.10, №3.12, №4.13, *while their general members have finite automorphism groups.*

It follows from [45] that every smooth Fano threefold in the following 22 deformation families has a non-reductive automorphism group:

№2.28, №2.30, №2.31, №2.33, №2.35, №2.36, №3.16, №3.18, №3.21, №3.22, №3.23, №3.24, №3.26, №3.28, №3.29, №3.30, №3.31, №4.8, №4.9, №4.10, №4.11, №4.12.

Thus, smooth Fano threefolds in these families are not K-polystable by Theorem 1.3. We will see in Section 3.7 that they are K-unstable. We know from Sections 3.1, 3.2, 3.3 and Example 3.2 that smooth Fano threefolds

№1.15, №1.16, №1.17, №2.32, №2.34, №2.29, №3.19, №3.20, №3.25, №3.27, №4.4, №4.7, №5.3, №6.1, №7.1, №8.1, №9.1, №10.1

are K-polystable. We know from Section 3.3 that both smooth Fano threefolds №4.5 and №5.2 are K-unstable. For the remaining smooth Fano threefolds that have infinite automorphism groups, the Calabi Problem is solved in Sections 3.7, 4.2, 4.4, 4.6, 4.7, 5.8, 5.9, 5.10, 5.14, 5.16, 5.17, 5.19, 5.20, 5.21, 5.22, and 5.23. We present a summary of these results in Table 3.3.

Table 3.3

Family	$\mathrm{Aut}^0(X)$	K-polystable	K-semistable	References
	$\mathbb{G}_a$	No	Yes	[55, Example 1.4]
№1.10	$\mathbb{G}_m$	Yes	Yes	Example 4.11
	$\mathrm{PGL}_2(\mathbb{C})$	Yes	Yes	Example 4.11
№2.20	$\mathbb{G}_m$	Yes	Yes	Section 5.8
	$\mathbb{G}_a$	No	Yes	Remark 5.51
№2.21	$\mathbb{G}_m$	Yes	Yes	Section 5.9
	$\mathrm{PGL}_2(\mathbb{C})$	Yes	Yes	Lemma 4.15
№2.22	$\mathbb{G}_m$	Yes	Yes	Section 4.4
№2.24	$\mathbb{G}_m$	No	Yes	Corollary 4.71
	$\mathbb{G}_m^2$	Yes	Yes	Lemma 4.70
№2.26	$\mathbb{G}_a \rtimes \mathbb{G}_m$	No	No	Lemma 5.56
	$\mathbb{G}_m$	No	Yes	Corollary 5.58
№2.27	$\mathrm{PGL}_2(\mathbb{C})$	Yes	Yes	Lemma 4.17
№3.5	$\mathbb{G}_m$	Yes	Yes	Section 5.14
№3.8	$\mathbb{G}_m$	Yes	Yes	Section 5.16
№3.9	$\mathbb{G}_m$	Yes	Yes	Section 4.6
№3.10	$\mathbb{G}_m$	Yes/No	Yes	Lemma 5.81, Corollary 5.84
	$\mathbb{G}_m^2$	Yes	Yes	Sections 3.3, Lemma 5.80
№3.12	$\mathbb{G}_m$	Yes	Yes	Section 5.18
	$\mathbb{G}_a$	No	Yes	Lemma 5.98
№3.13	$\mathbb{G}_m$	Yes	Yes	Section 5.19
	$\mathrm{PGL}_2(\mathbb{C})$	Yes	Yes	Example 1.94, Lemma 4.18
№3.14	$\mathbb{G}_m$	No	No	Section 3.7
№3.15	$\mathbb{G}_m$	Yes	Yes	Section 5.20
№3.17	$\mathrm{PGL}_2(\mathbb{C})$	Yes	Yes	Lemma 4.19
№4.2	$\mathbb{G}_m$	Yes	Yes	Section 4.6
№4.3	$\mathbb{G}_m$	Yes	Yes	Section 5.21
№4.6	$\mathrm{PGL}_2(\mathbb{C})$	Yes	Yes	Lemma 4.14
№4.13	$\mathbb{G}_m$	Yes	Yes	Section 5.22
№5.1	$\mathbb{G}_m$	Yes	Yes	Section 5.23

3.7 Divisorially unstable threefolds

Let X be an arbitrary smooth Fano threefold. By [93, Theorem 10.1], the threefold X is divisorially unstable if and only if X is contained in one of the following 26 deformation families:

№2.23, №2.28, №2.30, №2.31, №2.33, №2.35, №2.36, №3.14,
№3.16, №3.18, №3.21, №3.22, №3.23, №3.24, №3.26, №3.28, №3.29,
№3.30, №3.31, №4.5, №4.8, №4.9, №4.10, №4.11, №4.12, №5.2.

Recall from Theorem 1.19 that X is K-unstable if it is divisorially unstable.

In the proof of the Main Theorem, we will often use the following relevant result:

Theorem 3.17 ([93, Theorem 10.1]) *Let X be any smooth Fano threefold that is not contained in the following 41 deformation families:*

*№1.17, №2.23, №2.26, №2.28, №2.30, №2.31, №2.33, №2.34, №2.35, №2.36,
№3.9, №3.14, №3.16, №3.18, №3.19, №3.21, №3.22, №3.23, №3.24, №3.25,
№3.26, №3.28, №3.29, №3.30, №3.31, №4.2, №4.4, №4.5, №4.7, №4.8, №4.9,
№4.10, №4.11, №4.12, №5.2, №5.3, №6.1, №7.1, №8.1, №9.1, №10.1.*

Then $S_X(E) < 1$ for every prime Weil divisor $E \subset X$, i.e. X is divisorially stable.

In the remaining part of this section, we recall the proof of the fact that all smooth Fano threefolds in the 26 deformation families listed above are divisorially unstable. To do this, it is enough to present an irreducible surface $S \subset X$ such that $\beta(S) < 0$. As in Chapter 1, we let

$$\tau(S) = \sup\{u \in \mathbb{R} \mid \text{the divisor } -K_X - uS \text{ is pseudo-effective}\}.$$

For every $u \in [0, \tau(S)]$, we denote by $P(-K_X - uS)$ and $N(-K_X - uS)$ the positive and the negative parts of the Zariski decomposition of the divisor $-K_X - uS$, respectively.

Lemma 3.18 *Suppose that X is contained in one of the following 13 families: №2.33, №2.35, №2.36, №3.26, №3.28, №3.29, №3.30, №3.31, №4.9, №4.10, №4.11, №4.12, №5.2. Then X is divisorially unstable.*

Proof From Section 3.3, we know that the smooth Fano threefold X is toric, so that the required assertion follows from Theorem 1.21. □

Example 3.19 (cf. Section 3.1) Let $X = \mathbb{P}^1 \times \mathbb{F}_1$, let **s** be the (-1)-curve in $\mathbb{F}_1$, let **f** be a fiber of the projection $\mathbb{F}_1 \to \mathbb{P}^1$, and let S and F be the preimages of these curves in X, respectively. Then $-K_X \sim 2S + 3F + 2R$, where R is a fiber of the projection to the first factor $X \to \mathbb{P}^1$. This implies that $\tau(S) = 2$, and the divisor $-K_X - uS$ is nef for every $u \in [0, 2]$, which gives

$$\beta(S) = 1 - \frac{1}{-K_X^3} \int_0^2 \left(-K_X - uS\right)^3 du = 1 - \frac{1}{48} \int_0^2 6(2-u)(u+4)du = -\frac{1}{6} < 0,$$

so that X is divisorially unstable (cf. Lemma 2.3).

To deal with the families №2.23, №2.28 and №2.30, we need the following lemma.

Lemma 3.20 ([93, Lemma 9.9])   *Let Y be a smooth Fano threefold such that $-K_Y \sim rH$ for an ample divisor $H \in \mathrm{Pic}(Y)$ and an integer $r \geqslant 2$, let S_1 and S_2 be two irreducible surfaces in Y such that $S_1 \sim d_1 H$ and $D_2 \sim d_2 H$ for some positive integers $d_1 \leqslant d_2 < r$. Suppose, in addition, that the scheme-theoretic intersection $C = S_1 \cap S_2$ is a smooth curve. Let $\pi\colon X \to Y$ be the blow up of the curve C, and let $\widetilde{S}_1$ be the proper transform on X of the surface S_1. Suppose that X is a Fano threefold. Then*

$$\beta(\widetilde{S}_1) = \frac{2d_1^3 d_2 + 3d_1^2 d_2^2 - 8d_1^2 d_2 r + 4d_1 r^3 - r^4}{4d_1(d_1^2 d_2 + d_1 d_2^2 - 3d_1 d_2 r + r^3)}.$$

Proof   Let E be the π-exceptional surface. Then

$$-K_X - u\widetilde{S}_1 \sim_{\mathbb{R}} \left(\frac{r}{d_1} - u\right)\widetilde{S}_1 + \left(\frac{r}{d_1} - 1\right)E,$$

which implies that $\tau(\widetilde{S}_1) = \frac{r}{d_1}$. We have

$$\left(-K_X - u\widetilde{S}_1\right)\big|_{\widetilde{S}_1} \sim_{\mathbb{R}} (r - d_2 - u(d_1 - d_2))\pi^*(H)\big|_{\widetilde{S}_1}$$

and

$$\left(-K_X - u\widetilde{S}_1\right)\big|_E \sim_{\mathbb{R}} (r - d_2 + u(d_2 - d_1))\pi^*(H)\big|_E + (1 - u)\widetilde{S}_2\big|_E,$$

where $\widetilde{S}_2$ is the proper transform on X of the surface S_2. Note that $\widetilde{S}_2|_E \sim \widetilde{S}_1|_E + (d_2 - d_1)\pi^*(H)|_E$, so that the divisor $\widetilde{S}_2|_E$ is nef. Therefore, we conclude that $-K_X - u\widetilde{S}_1$ is also nef for $u \in [0,1]$. If $1 \leqslant u \leqslant \frac{r}{d_1}$, then $P(-K_X - u\widetilde{S}_1) = (r - ud_1)\pi^*(H)$ and $N(-K_X - u\widetilde{S}_1) = (u - 1)E$.

We have $(\pi^*(H))^2 \cdot E = 0$ and $(\pi^*(H)) \cdot E^2 = -H \cdot C = -d_1 d_2 H^3$. Note also that

$$\mathcal{N}_{C/Y} \cong \mathcal{O}_C\left(d_1 H|_C\right) \oplus \mathcal{O}_C\left(d_2 H|_E\right),$$

so that $E^3 = -c_1(\mathcal{N}_{C/Y}) = -(d_1 + d_2)H \cdot C = -d_1 d_2(d_1 + d_2)H^3$. Thus, if $u \in [0,1]$, then

$$\mathrm{vol}\left(-K_X - u\widetilde{S}_1\right) = \left((r - ud_1)^3 - 3d_1 d_2(r - ud_1)(1-u)^2 + d_1 d_2(d_1 + d_2)(1-u)^3\right)H^3.$$

Likewise, if $1 \leqslant u \leqslant \frac{r}{d_1}$, then $\mathrm{vol}(-K_X - u\widetilde{S}_1) = (r - ud_1)^3 H^3$. Now, integrating, we get the required formula for $\beta(\widetilde{S}_1)$.   □

Lemma 3.21 *Suppose that X is contained in one of the families №2.23, №2.28, №2.30. Then X is divisorially unstable.*

Proof A smooth Fano threefold №2.23 is a blow up of a smooth quadric in $\mathbb{P}^3$ along its section by a hyperplane and another quadric. Likewise, any smooth Fano threefold №2.28 can be obtained by blowing up $\mathbb{P}^3$ along an intersection of a plane and a cubic surface. Finally, a smooth Fano threefold №2.30 is a blow up of $\mathbb{P}^3$ along an intersection of a plane and a quadric surface. Thus, we can apply Lemma 3.20 with

- $r = 3$, $d_1 = 1$, $d_2 = 2$ if X is contained in the family №2.23,
- $r = 4$, $d_1 = 1$, $d_2 = 3$ if X is contained in the family №2.28,
- $r = 4$, $d_1 = 1$, $d_2 = 2$ if X is contained in the family №2.30.

This gives a surface $S \subset X$ with $\beta(S) = -\frac{1}{12}$, $\beta(S) = -\frac{63}{160}$, $\beta(S) = -\frac{6}{23}$, respectively. □

In the remaining part of the section, we will deal with smooth Fano threefolds in the families №2.31, №3.14, №3.16, №3.18, №3.21, №3.22, №3.23, №3.24, №4.5, №4.8.

Lemma 3.22 *Suppose X is a smooth Fano threefold in the deformation family №2.31. Then X is divisorially unstable.*

Proof Let Q be a smooth quadric hypersurface in $\mathbb{P}^4$, and let L be a line in the quadric Q. Then we have the commutative diagram

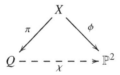

where π is the blow up of the line L, the map χ is a projection from L, and ϕ is a $\mathbb{P}^1$-bundle.

Let E be the π-exceptional surface, let $H_Q = \pi^*(O_{\mathbb{P}^4}(1)|_Q)$, and let $H_{\mathbb{P}^2} = \phi^*(O_{\mathbb{P}^2}(1))$. Then $-K_X \sim 3H_{\mathbb{P}^2} + 2E$, so that $\tau(E) = 2$, and $-K_X - uE$ is nef for $u \in [0, 2]$, so that

$$\beta(E) = 1 - \frac{1}{46}\int_0^2 (-K_X - uE)^3 du = 1 - \frac{1}{46}\int_0^2 (2 - u)(23 - u^2 + 4u)du$$

$$= -\frac{2}{23} < 0.$$

Therefore, X is divisorially unstable. □

Lemma 3.23 *Suppose X is a smooth Fano threefold in the deformation family №3.14. Then X is divisorially unstable.*

Proof Let $\mathscr{C}$ be a smooth plane cubic curve in $\mathbb{P}^3$, let Π be the plane in $\mathbb{P}^3$ that contains $\mathscr{C}$, let P be a point in $\mathbb{P}^3$ such that $P \notin \Pi$, let $\phi\colon V_7 \to \mathbb{P}^3$ be the blow up of this point, and let C be the proper transform on V_7 of the cubic curve $\mathscr{C}$. Then the threefold X can be obtained as a blow up $\pi\colon X \to V_7$ along the curve C.

Let E_C be the π-exceptional surface, and let E_P, H_C, F be the proper transforms on the threefold X of the ϕ-exceptional surface, the plane Π, and the cubic cone in $\mathbb{P}^3$ over the curve $\mathscr{C}$ with vertex P, respectively. We have the commutative diagram

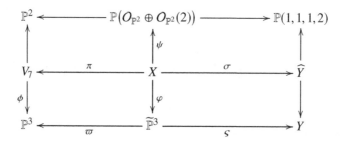

where ϖ is the blow up of the curve $\mathscr{C}$, φ is the contraction of the surface E_P, σ and ψ are the contractions of the surfaces H_C and F, respectively, ς is the contraction of the surface $\varphi(H_C)$, Y is a Fano threefold that has a singular point of type $\frac{1}{2}(1,1,1)$, the morphism $\widehat{Y} \to Y$ is a blow up of a smooth point of the threefold Y, both $V_7 \to \mathbb{P}^2$ and $\mathbb{P}(O_{\mathbb{P}^2} \oplus O_{\mathbb{P}^2}(2)) \to \mathbb{P}^2$ are $\mathbb{P}^1$-bundles, the morphism $\mathbb{P}(O_{\mathbb{P}^2} \oplus O_{\mathbb{P}^2}(2)) \to \mathbb{P}(1,1,1,2)$ is the contraction of the surface $\psi(H_C)$, and $\widehat{Y} \to \mathbb{P}(1,1,1,2)$ is the contraction of $\sigma(F)$.

Observe that $-K_X \sim 2H_C + 2H_P + E_C$, where H_P is the proper transform on X of a general plane in $\mathbb{P}^3$ containing P. Then $\tau(H_C) = 2$. For $u \in [0,2]$, we get

$$P\big(-K_X - uH_C\big) = \begin{cases} (2-u)H_C + 2H_P + E_C & \text{if } u \in [0,1], \\ (2-u)(H_C + E_C) + 2H_P & \text{if } u \in [1,2], \end{cases}$$

and

$$N\big(-K_X - uH_C\big) = \begin{cases} 0 & \text{if } u \in [0,1], \\ (u-1)E_C & \text{if } u \in [1,2]. \end{cases}$$

Hence we obtain

$$\beta(H_C)$$

$$= 1 - \frac{1}{32} \int_0^1 \left((2-u)H_C + 2H_P + E_C\right)^3 du$$

$$- \frac{1}{32} \int_1^2 \left((2-u)(H_C + E_C) + 2H_P\right)^3 du$$

$$= 1 - \frac{1}{32} \int_0^1 \left(32 - 3u - 6u^2 - 4u^3\right)du - \frac{1}{32} \int_1^2 \left(56 - 48u + 12u^2 - u^3\right)du$$

$$= -\frac{15}{128} < 0.$$

Therefore the threefold X is divisorially unstable. $\qquad\square$

Lemma 3.24 *Suppose X is a smooth Fano threefold in the deformation family №3.16. Then X is divisorially unstable.*

Proof Let $\mathscr{C}$ be a twisted cubic curve in the space $\mathbb{P}^3$, let P be a point in the curve $\mathscr{C}$, let $\phi\colon V_7 \to \mathbb{P}^3$ be the blow up of this point, and let C be the proper transform of the cubic curve $\mathscr{C}$ on the threefold V_7. Then X can be obtained as a blow up $\pi\colon X \to V_7$ along the curve C. One can see that X fits into the commutative diagram

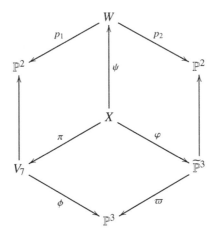

where W is a smooth divisor of degree $(1,1)$ in $\mathbb{P}^2 \times \mathbb{P}^2$, both p_1 and p_2 are $\mathbb{P}^1$-bundles, the morphism ϖ is the blow up of $\mathbb{P}^3$ along $\mathscr{C}$, the morphism $\widetilde{\mathbb{P}}^3 \to \mathbb{P}^2$ is the $\mathbb{P}^1$-bundle whose fibers are proper transforms of the secant lines in $\mathbb{P}^3$ of the twisted cubic curve $\mathscr{C}$, the morphism $V_7 \to \mathbb{P}^2$ is the $\mathbb{P}^1$-bundle whose fibers are proper transforms of the lines in the space $\mathbb{P}^3$ that pass through P, and φ is the blow up of the fiber of ϖ over P.

We denote by E_C the π-exceptional surface, we denote by E_P the φ-exceptional surface, and we denote by F the ψ-exceptional surface. Then $E_C \cong \mathbb{P}^1 \times \mathbb{P}^1$, $E_P \cong \mathbb{F}_1$ and $F \cong \mathbb{F}_2$ since $\phi \circ \pi(F)$ is the unique quadric cone in $\mathbb{P}^3$ with vertex P that contains the curve $\mathscr{C}$. Let us compute $\beta(E_P)$. First, we observe that $\tau(E_P) = 2$ since $-K_X \sim 2E_P + 2F + E_C$.

Denote by $\mathbf{s}$ the (-1)-curve in E_P, denote by $\mathbf{f}$ a fiber of the projection $E_P \to \mathbb{P}^1$, and denote by ℓ the proper transform on X of a ruling of the cone $\phi \circ \pi(F)$. Then $\mathbf{s} = E_C|_{E_P}$, and the curves $\mathbf{s}$, $\mathbf{f}$ and ℓ generate the Mori cone $\overline{NE}(X)$. Moreover,

$$\left(- K_X - uE_P \right) \cdot \mathbf{s} = 1,$$
$$\left(- K_X - uE_P \right) \cdot \mathbf{f} = 1 + u,$$
$$\left(- K_X - uE_P \right) \cdot \ell = 1 - u,$$

so that $-K_X - uE_P$ is nef for $u \in [0,1]$. If $u \in [1,2]$, then $N(-K_X - uE_P) = (u-1)F$ and

$$P\left(- K_X - uE_P \right) \sim_{\mathbb{R}} (\phi \circ \pi)^* \left(\mathcal{O}_{\mathbb{P}^3}(6 - 2u) \right) - (4-u)E_P - (2-u)E_C.$$

Therefore we see that

$$\begin{aligned}
\beta(D) &= 1 - \frac{1}{34} \int_0^1 \left(- K_X - uE_P \right)^3 du \\
&\quad - \frac{1}{34} \int_1^2 \left(- K_X - uE_P + (1-u)F \right)^3 du \\
&= 1 - \frac{1}{34} \int_0^1 \left(34 - 9u - 6u^2 - u^3 \right) du \\
&\quad - \frac{1}{34} \int_1^2 \left(48 - 36u + 6u^2 \right) du \\
&= -\frac{5}{136} < 0,
\end{aligned}$$

so that X is divisorially unstable. $\square$

Lemma 3.25 *Suppose X is a smooth Fano threefold in the deformation family №3.18. Then X is divisorially unstable.*

Proof The Fano threefold X can be obtained as a blow up $\pi\colon X \to \mathbb{P}^3$ along a disjoint union of a smooth conic C and a line L. There is a commutative diagram

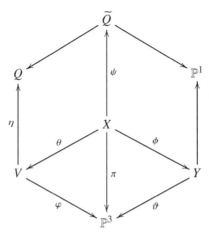

where ϑ is the blow up of the line L, the morphism φ is the blow up of the conic C, the morphisms θ and ϕ are the blow ups of the proper transforms of the curves L and C, respectively, Q is a smooth quadric in $\mathbb{P}^4$, the morphism η is a blow up of a point in Q, the morphism $\widetilde{Q} \to Q$ is the blow up of a conic (the proper transform of the line L), the morphism $Y \to \mathbb{P}^1$ is a $\mathbb{P}^2$-bundle, the morphism $\widetilde{Q} \to \mathbb{P}^1$ is a fibration into quadric surfaces, and ψ is the contraction of the proper transform of the plane in $\mathbb{P}^3$ containing C.

Let E_C and E_L be the π-exceptional surfaces that are mapped to C and L, respectively, let H_C be the proper transform on the threefold X of the plane in $\mathbb{P}^3$ that contains C, and let H_L be the proper transform on X of a general plane in $\mathbb{P}^3$ that passes through L. Then $-K_X \sim 3H_C + 2E_C + H_L$, which implies that $\tau(H_C) = 3$. Let us compute $\beta(H_C)$.

First, we observe that $H_C \cong \mathbb{F}_1$. Denote by $\mathbf{s}$ the unique (-1)-curve in the surface H_C, denote by $\mathbf{f}$ and ℓ fibers of the natural projections $H_C \to \mathbb{P}^1$ and $E_C \to C$, respectively. Then the curves $\mathbf{s}$, $\mathbf{f}$ and ℓ generate the Mori cone $\overline{\mathrm{NE}}(X)$, and the corresponding extremal contractions are ϕ, ψ and θ, respectively. Note also that $H_C|_{H_C} \sim -\mathbf{s} - \mathbf{f}$ and $H_C \cdot \ell = 1$. Therefore, for $u \in [0,3]$, we have $(-K_X - uH_C) \cdot \mathbf{s} = 1$, $(-K_X - uH_C) \cdot \mathbf{f} = 1+u$ and $(-K_X - uH_C) \cdot \ell = 1-u$, so that $-K_X - uH_C$ is nef for $u \in [0,1]$. If $u \in [1,3]$, then $N(-K_X - uE_P) = (u-1)E_C$, so that $P(-K_X - uH_C) \sim_\mathbb{R} (4-u)\pi^*(O_{\mathbb{P}^3}(4)) - E_L$. Then

$$\beta(H_C) = 1 - \frac{1}{36}\int_0^1 \left(36 - 9u - 6u^2 - u^3\right)du$$

$$- \frac{1}{36}\int_1^3 \left(54 - 45u + 12u^2 - u^3\right)du = -\frac{7}{48} < 0,$$

so that X is divisorially unstable. $\qquad\square$

Lemma 3.26 *Suppose X is a smooth Fano threefold in the deformation family №3.21. Then X is divisorially unstable.*

Proof Note that there is a blow up $\pi \colon X \to \mathbb{P}^1 \times \mathbb{P}^2$ of a smooth curve C of degree $(2,1)$. Let S be the proper transform on X of the surface in $\mathbb{P}^1 \times \mathbb{P}^2$ of degree $(0,1)$ that passes through the curve C, let ℓ_1 and ℓ_2 be the rulings of the surface $S \cong \mathbb{P}^1 \times \mathbb{P}^1$ such that the curves $\pi(\ell_1)$ and $\pi(\ell_2)$ are curves in $\mathbb{P}^1 \times \mathbb{P}^2$ of degree $(1,0)$ and $(0,1)$, respectively, let E be the π-exceptional surface, and let ℓ_3 be a fiber of the natural projection $E \to C$. Then $S|_S \sim -\ell_1 - \ell_2$, the curves ℓ_1, ℓ_2, ℓ_3 generate the Mori cone $\overline{\mathrm{NE}}(X)$, and the extremal rays $\mathbb{R}_{\geqslant 0}[\ell_1]$ and $\mathbb{R}_{\geqslant 0}[\ell_2]$ give birational contractions $X \to U_1$ and $X \to U_2$, respectively.

It follows from the proof of [46, Lemma 8.22] that there is a commutative diagram

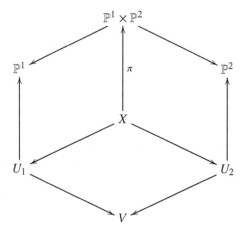

where the morphism $U_1 \to \mathbb{P}^1$ is a quadric bundle, the morphism $U_2 \to \mathbb{P}^2$ is a $\mathbb{P}^1$-bundle, the map $U_1 \dashrightarrow U_2$ is a flop, and V is a Fano threefold №1.15 with one isolated ordinary double singularity. For details, we refer the reader to the case (2.3.2) in [208, Theorem 2.3].

We have $\tau(S) = 3$ since $-K_X \sim 3S + 2E + (\mathrm{pr}_1 \circ \pi)^*(\mathcal{O}_{\mathbb{P}^1}(2))$, where $\mathrm{pr}_1 \colon \mathbb{P}^1 \times \mathbb{P}^2 \to \mathbb{P}^1$ is the projection to the first factor. If $u \in [1,3]$, then $N(-K_X - uS) = (u-1)E$ and $P(-K_X - uS) \sim_{\mathbb{R}} (3-u)(\mathrm{pr}_2 \circ \pi)^*(\mathcal{O}_{\mathbb{P}^2}(1)) + (\mathrm{pr}_1 \circ \pi)^*(\mathcal{O}_{\mathbb{P}^1}(2))$, where $\mathrm{pr}_2 \colon \mathbb{P}^1 \times \mathbb{P}^2 \to \mathbb{P}^2$ is the projection to the second factor. Integrating, we get $\beta(S) = -\frac{17}{76} < 0$. $\square$

Lemma 3.27 *Suppose X is a smooth Fano threefold in the deformation family №3.22. Then X is divisorially unstable.*

Proof Let $\mathrm{pr}_1 \colon \mathbb{P}^1 \times \mathbb{P}^2 \to \mathbb{P}^1$ and $\mathrm{pr}_2 \colon \mathbb{P}^1 \times \mathbb{P}^2 \to \mathbb{P}^2$ be the projections to the first and the second factors, respectively, let H_1 be a fiber of the map pr_1, let $H_2 = \mathrm{pr}_2^*(\mathcal{O}_{\mathbb{P}^2}(1))$, and let C be a conic in $H_1 \cong \mathbb{P}^2$. Then there is a blow up $\psi\colon X \to \mathbb{P}^1 \times \mathbb{P}^2$ along C.

Let E_C be the ψ-exceptional surface, let $\widetilde{H}_1$ be the proper transform of the surface H_1 on the threefold X, let F be the surface in $|H_2|$ that contains C, and let $\widetilde{F}$ be the proper transform of this surface on X. We have the commutative diagram

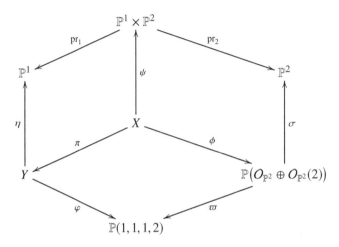

where π and ϕ are the contraction of the surfaces $\widetilde{H}_1 \cong \mathbb{P}^2$ and $\widetilde{F} \cong \mathbb{P}^1 \times \mathbb{P}^1$, respectively, the morphisms ϖ and φ are the contractions of the surfaces $\phi(\widetilde{H}_1)$ and $\pi(\widetilde{F})$, respectively, the morphism σ is a $\mathbb{P}^1$-bundle, and η is a fibration into del Pezzo surfaces such that all its fibers except $\pi(\widetilde{F})$ are isomorphic to $\mathbb{P}^2$, while $\pi(\widetilde{F}) \cong \mathbb{P}(1,1,4)$. Note that the Mori cone $\overline{\mathrm{NE}}(X)$ is generated by the extremal rays contracted by π, ϕ and ψ.

Let us compute $\beta(\widetilde{H}_1)$. Take $u \in \mathbb{R}_{\geqslant 0}$. Then

$$-K_X - u\widetilde{H}_1 \sim_{\mathbb{R}} (2-u)\widetilde{H}_1 + \frac{3}{2}\widetilde{F} + \frac{5}{2}E_C \sim_{\mathbb{R}} (2-u)\widetilde{H}_1 + E_C + \psi^*(3H_2),$$

so that $-K_X - u\widetilde{H}_1$ is pseudo-effective $\Longleftrightarrow u \leqslant 2$. Moreover, if $u \in [0,2]$, we have

$$P\big(-K_X - u\widetilde{H}_1\big) = \begin{cases} (2-u)\widetilde{H}_1 + E_C + \psi^*(3H_2) & \text{if } u \in [0,1], \\ (2-u)(\widetilde{H}_1 + E_C) + \psi^*(3H_2) & \text{if } u \in [1,2], \end{cases}$$

and

$$N\big(-K_X - u\widetilde{H}_1\big) = \begin{cases} 0 & \text{if } u \in [0,1], \\ (u-1)E_C & \text{if } u \in [1,2]. \end{cases}$$

Hence we see that $\beta(H_C)$ is equal to

$$1 - \frac{1}{40} \int_0^1 \left((2-u)\tilde{H}_1 + E_C + \psi^*(3H_2) \right)^3 du$$

$$- \frac{1}{40} \int_1^2 \left((2-u)(\tilde{H}_1 + E_C) + \psi^*(3H_2) \right)^3 du$$

$$= 1 - \frac{1}{40} \int_0^1 \left(40 - 3u - 6u^2 - 4u^3 \right) du - \frac{1}{40} \int_1^2 \left(54 - 27u \right) du$$

$$= -\frac{9}{40} < 0.$$

Therefore the threefold X is divisorially unstable.   □

Lemma 3.28 *Suppose X is a smooth Fano threefold in the deformation family №3.23. Then X is divisorially unstable.*

Proof   Let $\mathscr{C}$ be a smooth conic in the space $\mathbb{P}^3$, let P be an arbitrary point in the conic $\mathscr{C}$, let $\phi\colon V_7 \to \mathbb{P}^3$ be the blow up of the point P, and let C be the proper transform on the threefold V_7 of the conic $\mathscr{C}$. Then X can be obtained as a blow up $\pi\colon X \to V_7$ along the curve C. One can see that X fits into the commutative diagram

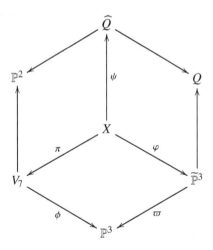

where Q is a smooth quadric threefold in $\mathbb{P}^4$, the morphism ϖ is the blow up of the conic $\mathscr{C}$, the morphism $\overline{\mathbb{P}}^3 \to Q$ is the contraction of the proper transform of the plane in $\mathbb{P}^3$ that contains $\mathscr{C}$ to a point, φ is a blow up of the fiber of the morphism ϖ over the point P, the morphism $\widehat{Q} \to Q$ is the blow up of a line on Q that passes through the latter point, and $\widehat{Q} \to \mathbb{P}^2$ is a $\mathbb{P}^1$-bundle.

Table 3.4

	H_C	E_P	E_C	$-K_X$
s	0	-1	1	1
f	-1	1	1	1
ℓ	1	0	-1	1

Denote by E_C the π-exceptional surface, denote by E_P the φ-exceptional surface, and denote by H_C the proper transform on X of the plane in $\mathbb{P}^3$ that contains C. Then

$$-K_X \sim \pi^*\big(\mathcal{O}_{\mathbb{P}^3}(4)\big) - 2E_P - E_C \sim 4H_C + 2E_P + 3E_C$$

since $H_C \sim (\phi \circ \pi)^*(\mathcal{O}_{\mathbb{P}^3}(1)) - E_P - E_C$. In particular, we have $\tau(H_C) = 4$.

Observe that $H_C \cong \mathbb{F}_1$. Let **s** be the (-1)-curve in H_C, and let **f** be a fiber of the natural projection $H_C \to \mathbb{P}^1$. Denote by ℓ a fiber of the projection $E_C \to C$ that is induced by π. Then the curves **s**, **f**, ℓ generate the cone $\overline{\mathrm{NE}}(X)$, the contractions of the corresponding extremal rays are φ, ψ, π, respectively, and the intersections of the curves **s**, **f**, ℓ with the divisors H_C, E_P, E_C, $-K_X$ are contained in Table 3.4.

Let $u \in [0,4]$. Since $-K_X - uH_C \sim_{\mathbb{R}} (4-u)H_C + 2E_P + 3E_C$, we obtain

$$P\big(-K_X - uH_C\big) = \begin{cases} (4-u)H_C + 2E_P + 3E_C & \text{if } u \in [0,1], \\ (4-u)\big(H_C + E_C\big) + 2E_P & \text{if } u \in [1,2], \\ (4-u)\big(H_C + E_C + E_P\big) & \text{if } u \in [2,4], \end{cases}$$

and

$$N\big(-K_X - uH_C\big) = \begin{cases} 0 & \text{if } u \in [0,1], \\ (u-1)E_C & \text{if } u \in [1,2], \\ (u-1)H_C + (u-2)E_P & \text{if } u \in [2,4]. \end{cases}$$

Hence, we obtain

$$\beta(H_C) = 1 - \frac{1}{42}\int_0^1 \big(42 - 9u - 6u^2 - u^3\big)du - \frac{1}{42}\int_1^2 \big(56 - 36u + 6u^2\big)du$$
$$- \frac{1}{42}\int_2^4 (4-u)^3 du,$$

so that $\beta(H_C) = -\frac{53}{168} < 0$. Therefore, the threefold X is divisorially unstable. $\quad\square$

Lemma 3.29 *Suppose X is a smooth Fano threefold in the deformation family №3.24. Then X is divisorially unstable.*

Proof Note that there is a blow up $\pi\colon X \to \mathbb{P}^1 \times \mathbb{P}^2$ of a smooth curve C of degree $(1, 1)$. Let E be the π-exceptional surface, and let S be the proper transform on X of the surface in $\mathbb{P}^1 \times \mathbb{P}^2$ of degree $(0, 1)$ that contains C. Then, arguing as in the proof of Lemma 3.26, we see that $\tau(S) = 3$. Similarly, we see that

$$P\big(-K_X - uS\big) = \begin{cases} -K_X - uS & \text{if } u \in [0, 1], \\ -K_X - uS - (u-1)E & \text{if } u \in [1, 3], \end{cases}$$

and

$$N\big(-K_X - uS\big) = \begin{cases} 0 & \text{if } u \in [0, 1], \\ (u-1)E & \text{if } u \in [1, 3]. \end{cases}$$

Integrating, we get $\beta(S) = -\frac{1}{7} < 0$. Therefore, we see that X is divisorially unstable. □

Lemma 3.30 *Suppose X is a smooth Fano threefold in the deformation family №4.5. Then X is divisorially unstable.*

Proof There is a birational morphism $\pi\colon X \to \mathbb{P}^1 \times \mathbb{P}^2$ such that π is a blow up along a disjoint union of a smooth curve C of degree $(2, 1)$ and a smooth curve L of degree $(1, 0)$. Denote by E_C and E_L the π-exceptional surfaces such that $\pi(E_C) = C$ and $\pi(E_L) = L$. We also let $H_1 = (\mathrm{pr}_1 \circ \pi)^*(\mathcal{O}_{\mathbb{P}^1}(1))$ and $H_2 = (\mathrm{pr}_2 \circ \pi)^*(\mathcal{O}_{\mathbb{P}^2}(1))$, where $\mathrm{pr}_1\colon \mathbb{P}^1 \times \mathbb{P}^2 \to \mathbb{P}^1$ and $\mathrm{pr}_2\colon \mathbb{P}^1 \times \mathbb{P}^2 \to \mathbb{P}^2$ are projections to the first and the second factors, respectively. Then $-K_X \sim 2H_1 + 3H_2 - E_C - E_L$.

Let S be the proper transform on X of the surface in $\mathbb{P}^1 \times \mathbb{P}^2$ of degree $(0, 1)$ that contains C, and let H_L be a general surface in the pencil $|H_2 - E_L|$. Then $S \sim H_2 - E_C$, so that $-K_X \sim 2S + E_C + H_L + 2H_1$, cf. [93, p. 577]. We have

$$P\big(-K_X - uS\big) = \begin{cases} -K_X - uS & \text{if } u \in [0, 1], \\ -K_X - uS - (u-1)E & \text{if } u \in [1, 2], \end{cases}$$

and

$$N\big(-K_X - uS\big) = \begin{cases} 0 & \text{if } u \in [0, 1], \\ (u-1)E & \text{if } u \in [1, 2]. \end{cases}$$

Therefore, we see that

$$\beta(S) = 1 - \frac{1}{32} \int_0^1 (32 - 2u^3 - 6u^2 - 6u)\,du - \frac{1}{32} \int_1^3 6(2-u)(4-u)\,du = -\frac{5}{64} < 0.$$

Therefore, we see that X is divisorially unstable. $\qquad\square$

Lemma 3.31 *Suppose X is a smooth Fano threefold in the deformation family №4.8. Then X is divisorially unstable.*

Proof For $i \in \{1,2,3\}$, let $\mathrm{pr}_i \colon \mathbb{P}^1 \times \mathbb{P}^1 \times \mathbb{P}^1 \to \mathbb{P}^1$ be the projection to the ith factor, and let H_i be a fiber of this projection. Let C be a curve of degree $(1,1)$ in $H_1 \cong \mathbb{P}^1 \times \mathbb{P}^1$. Then there exists a blow up $\pi \colon X \to \mathbb{P}^1 \times \mathbb{P}^1 \times \mathbb{P}^1$ along the curve C.

Let E be the exceptional surface of the birational morphism π, and let $\widetilde{H}_1$ be the proper transform of the surface H_1 on the threefold X. Then $\widetilde{H}_1 \sim \pi^*(H_1) - E$, so that

$$-K_X \sim \pi^*\big(2H_1 + 2H_2 + 2H_3\big) - E \sim 2\widetilde{H}_1 + E + \pi^*\big(2H_2 + 2H_3\big).$$

Let us find $\beta(\widetilde{H}_1)$. Take $u \in \mathbb{R}_{\geqslant 0}$. Then

$$-K_X - u\widetilde{H}_1 \sim_{\mathbb{R}} (2-u)\widetilde{H}_1 + E + \pi^*(2H_2 + 2H_3),$$

so that $-K_X - u\widetilde{H}_1$ is pseudo-effective $\iff u \leqslant 2$. Moreover, if $u \in [0,2]$, we have

$$P\big(-K_X - u\widetilde{H}_1\big) = \begin{cases} (2-u)\widetilde{H}_1 + E + \pi^*\big(2H_2 + 2H_3\big) & \text{if } u \in [0,1], \\ (2-u)\pi^*\big(H_1\big) + \pi^*\big(2H_2 + 2H_3\big) & \text{if } u \in [1,2], \end{cases}$$

and

$$N\big(-K_X - u\widetilde{H}_1\big) = \begin{cases} 0 & \text{if } u \in [0,1], \\ (u-1)E_C & \text{if } u \in [1,2]. \end{cases}$$

Integrating, we get $\beta(H_C) = -\frac{13}{76} < 0$, so that X is divisorially unstable. $\qquad\square$

4

Proof of Main Theorem: Special Cases

4.1 Prime Fano threefolds

A smooth Fano variety is *prime* if its Picard group is $\mathbb{Z}$. Smooth prime Fano threefolds were classified by Iskovskikh in [118, 119]. Smooth prime Fano threefolds whose Picard groups are generated by their anticanonical divisors form ten deformation families, which we denote by: №1.1, №1.2, №1.3, №1.4, №1.5, №1.6, №1.7, №1.8, №1.9, №1.10. We will present (at least one) K-stable Fano threefold in each family. Thus, a general smooth prime Fano threefold is K-stable by Theorem 1.11. With the exception of family №1.9, this is already known [3].

Example 4.1 Let X be a smooth Fano threefold №1.1. Then X is a double cover of $\mathbb{P}^3$ branched along a smooth surface $B \sim O_{\mathbb{P}^3}(6)$, so that X is K-stable by Corollary 1.67. Alternatively,

$$\alpha(X) \in \left\{ \frac{5}{6}, \frac{43}{50}, \frac{13}{15}, \frac{33}{38}, \frac{7}{8}, \frac{8}{9}, \frac{9}{10}, \frac{11}{12}, \frac{13}{14}, \frac{15}{16}, \frac{17}{18}, \frac{19}{20}, \frac{21}{22}, \frac{29}{30}, 1 \right\}$$

by [39, Proposition 3.7], so that X is K-stable by Theorem 1.48.

Example 4.2 A smooth Fano threefold X in this family is

(1.2^a) either a quartic threefold $X \subset \mathbb{P}^4$,

(1.2^b) or a double cover of a smooth quadric threefold in $\mathbb{P}^4$ branched along a smooth surface of degree 8.

In the latter case, the K-stability of the Fano threefold X follows from Corollary 1.67. In the former case, $\alpha(X) \geqslant \frac{3}{4}$ by [28, Theorem 1.3] and X is K-stable by Theorem 1.48. In the case where X is a Fermat hypersurface, the K-stability of X is proved in [210, 165].

Example 4.3 ([165, 6.3]) Let X be the complete intersection

$$\left\{ \sum_{i=0}^{6} x_i = \sum_{i=0}^{6} x_i^2 = \sum_{i=0}^{6} x_i^3 = 0 \right\} \subset \mathbb{P}^6.$$

Then X is a smooth Fano threefold №1.3, and X admits a faithful action of the symmetric group $\mathfrak{S}_7$. By Corollary 1.53, $\alpha_{\mathfrak{S}_7}(X) \geqslant 1$, and X is K-stable by Theorem 1.48.

Example 4.4 ([165, 6.1]) Let $a_0, \ldots, a_6$ and $b_0, \ldots, b_6$ be complex numbers such that every 3×3 minor of the matrix

$$\begin{pmatrix} 1 & 1 & 1 & 1 & 1 & 1 & 1 \\ a_0 & a_1 & a_2 & a_3 & a_4 & a_5 & a_6 \\ b_0 & b_1 & b_2 & b_3 & b_4 & b_5 & b_6 \end{pmatrix}$$

is invertible (this holds generically). Let X be the complete intersection

$$\left\{ \sum_{i=0}^{6} x_i^2 = \sum_{i=0}^{6} a_i x_i^2 = \sum_{i=0}^{6} b_i x_i^2 = 0 \right\} \subset \mathbb{P}^6,$$

and let $G = \mathrm{Aut}(X)$. Then X is a smooth Fano threefold №1.4, the group G is finite [45], and $\alpha_G(X) \geqslant 1$ by Corollary 1.53, so that X is K-stable by Theorem 1.48.

Example 4.5 Now, we give another argument for the K-stability of a Fano threefold in the family №1.5^b from the one outlined in Example 3.11. Let V_5 be the smooth Fano threefold №1.15. Then $\mathrm{Aut}(V_5) \cong \mathrm{PGL}_2(\mathbb{C})$. Fix a subgroup $\mathfrak{A}_5 \subset \mathrm{Aut}(V_5)$. It follows from [53, Theorem 8.2.1] that there is a pencil of $\mathfrak{A}_5$-invariant anticanonical surfaces, whose general member is smooth. Let $\pi \colon X \to V_5$ be the double cover of V_5 branched over a general $\mathfrak{A}_5$-invariant anticanonical surface B. Then X is a smooth Fano threefold that belongs to the deformation family №1.5^b. Moreover, the threefold X is endowed with a faithful action of the group $G = \mathfrak{A}_5 \times \mu_2$ and $\alpha_G(X) \geqslant 1$. Indeed, assume this is not the case, i.e. that $\alpha_G(X) < 1$. Applying Theorem 1.52 to X with $\mu = 1$, we obtain a contradiction. First, there can be no G-invariant surface as in Theorem 1.52(1) because the Picard group of X is generated by $-K_X$. Second, there are no G-fixed points on X because there are no $\mathfrak{A}_5$-fixed points on V_5 by [53, Theorem 7.3.5], and Theorem 1.52(2) doesn't hold. Last, we show that X does not contain smooth G-invariant rational curves of anticanonical degree less than 16, so that Theorem 1.52(3) fails as well. Let C be such a curve. Since G does not act faithfully on $\mathbb{P}^1$ and V_5 does not have $\mathfrak{A}_5$-fixed points, the action of the subgroup $\mathfrak{A}_5$ on C is faithful, and the action of the Galois involution of the double cover π on C is trivial, so that C lies on the ramification divisor. Therefore $\pi(C) \subset B$ is an irreducible $\mathfrak{A}_5$-invariant curve in V_5 of degree less than 16. There is no such curve by [53, Theorem 13.6.1] and [53, Corollary 8.1.9], so that $\alpha_G(X) \geqslant 1$, and X is K-stable by Theorem 1.48.

Example 4.6   Let X be the smooth Fano threefold constructed in [184, Example 2.11]. Then X belongs to the family №1.6 and $\mathrm{Aut}(X) \cong \mathrm{SL}_2(\mathbf{F}_8)$, which is a simple group [64]. Let $G = \mathrm{Aut}(X)$. Then $\alpha_G(X) \geqslant 1$ by Corollary 1.54. Therefore, we conclude that the threefold X is K-stable by Theorem 1.48.

Example 4.7   Let X be the smooth Fano threefold constructed in [184, Example 2.9]. Then X belongs to family №1.7, and it admits a non-trivial action of $G \cong \mathrm{PSL}_2(\mathbf{F}_{11})$. Since G is simple, $\alpha_G(X) \geqslant 1$ by Corollary 1.54 (see also the proof of [51, Theorem A.5]). Thus, the threefold X is K-stable by Theorem 1.48.

Example 4.8   Let C_λ be the quartic curve $\{x^4+y^4+z^4+\lambda(x^2y^2+y^2z^2+z^2x^2) = 0\} \subset \mathbb{P}^2$, where $\lambda \in \mathbb{C} \setminus \{-1,\pm 2\}$. Then C_λ is smooth, and $\mathrm{Aut}(C_\lambda)$ contains a subgroup isomorphic to $\mathfrak{S}_4$. In fact, by [74, Theorem 6.5.2], $\mathrm{Aut}(C_\lambda) \cong \mathfrak{S}_4$ when $\lambda \neq 0$ and $\lambda^2 + 3\lambda + 18 \neq 0$ (and is strictly larger otherwise). The action of $\mathrm{Aut}(C_\lambda)$ on the curve C_λ is induced by its linear action on the plane $\mathbb{P}^2$. Let $G \cong \mathfrak{S}_3$ be a subgroup in $\mathrm{Aut}(C_\lambda)$ that acts on $\mathbb{P}^2$ by permuting the coordinates x, y and z. Set

$$P_1 = [1:s:s], P_2 = [s:1:s], P_3 = [s:s:1],$$
$$P_4 = [1:\omega:\omega^2], P_5 = [1:\omega^2:\omega], P_6 = [1:1:1],$$

where ω is a primitive cube root of unity, and $s \in \mathbb{C}$ such that $(\lambda+2)s^4 + 2\lambda s^2 + 1 = 0$, so that $s \neq 0$, $s \neq 1$ and $s \neq \frac{1}{2}$. One can check that $\{P_1, P_2, P_3\}$ is a G-orbit of length 3 contained in C_λ, $\{P_4, P_5\}$ is a G-orbit of length 2 contained in C_λ, $P_6 \notin C_\lambda$, and P_6 is the only G-invariant point in $\mathbb{P}^2$. Moreover, no three points among P_1, P_2, P_3, P_4, P_5 and P_6 are collinear, and the points P_1, P_2, P_3, P_4, P_5 and P_6 are not contained in a conic. Let $\pi\colon S \to \mathbb{P}^2$ be the blow up of the points P_1, P_2, P_3, P_4, P_5, P_6. Then S is a smooth cubic surface in $\mathbb{P}^3$. By construction, G acts on S, and its action is induced by the linear action on $\mathbb{P}^3$. Let Γ_λ be the proper transform of C_λ on S; Γ_λ is a G-invariant smooth non-hyperelliptic curve of genus 3 and degree 7 in $\mathbb{P}^3$. By [119, 6.1] (see also the construction in [17]), we have a G-equivariant Sarkisov link:

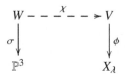

where σ is the blow up of Γ_λ, χ is the composition of five Atiyah flops, and ϕ contracts the proper transform of S to a smooth curve ℓ with $-K_X \cdot \ell = 1$, and X_λ is a smooth Fano in the family №1.8. As X_λ has no G-fixed points, we conjecture that $\alpha_G(X_\lambda) \geqslant \frac{3}{4}$, which would imply that X_λ is K-stable. Unfortunately, we were unable to show that $\alpha_G(X_\lambda) \geqslant \frac{3}{4}$. Nevertheless, we know from [3] that X_λ is K-stable.

Example 4.9 Let $W = \mathbb{P}(O_{\mathbb{P}^1} \oplus O_{\mathbb{P}^1} \oplus O_{\mathbb{P}^1}(1) \oplus O_{\mathbb{P}^1}(1))$, and let $\pi \colon W \to \mathbb{P}^1$ be the natural projection. Denote by H the tautological line bundle and by F a fiber of π. Write t_0, t_1 for the coordinates on $\mathbb{P}^1$, so that $|F| = < t_0, t_1 >$ and x, y, z, t are coordinates on the fiber with x, y sections of H and w, z sections of $H - F$. Let V be the divisor in $|2H + F|$ defined by

$$\left\{ t_1 x^2 + t_0 y^2 + t_0^2 xz + t_1^2 yw + t_1(t_1^2 - 4t_0^2)z^2 + t_0(t_0^2 - 4t_1^2)w^2 = 0 \right\}.$$

Denote by C the curve $\{z = w = 0\} \subset V$, and by $\phi \colon V \to \mathbb{P}^1$ the restriction of π to V. Then V is (2.3.8) in [208]: V is a Picard rank 2 weak Fano threefold with anticanonical degree $(-K_V)^3 = 16$. Further, the anticanonical map of V is small, and C is the only curve with trivial intersection $-K_V \cdot C = 0$. The curve C is a smooth rational curve that is a bisection of ϕ and $N_{C/V} \cong O_{\mathbb{P}^1}(-1) \oplus O_{\mathbb{P}^1}(-1)$. The morphism ϕ is a quadric fibration, so that V is a Mori fiber space since $\mathrm{Pic}(V) = \mathbb{Z}[H|_V] \oplus \mathbb{Z}[F|_V]$. Moreover, it follows from [208] that there is a Sarkisov link

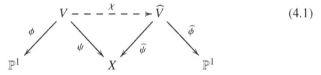

$$\tag{4.1}$$

where the anticanonical map ψ contracts C to an ordinary double point of X, χ is an Atiyah flop in C, $\widehat{\psi}$ is a birational morphism, and $\widehat{\phi}$ is a del Pezzo fibration of degree 4. Note that the map $\widehat{\phi} \circ \chi$ is given by $|(H - F)|_V|$, all surfaces in this pencil are singular along C, and its general surface is smooth away from this curve. Let S be the surface in the pencil $|(H - F)|_V|$ that is cut out by $w = \lambda z$, where λ is one of the 16 roots of

$$75759616\lambda^{16} - 303812608\lambda^{12} - 759031797\lambda^8 - 303812608\lambda^4 + 75759616 = 0.$$

Then S is singular along C, and it is also singular at the point in W with coordinates:

$$x = 1695122086341510483148,$$

$$y = -7749319224279144144456832\lambda^{15} + 28761189370681284199710721\lambda^{11}$$

$$+ 8753709667519885555073664\lambda^7 + 5089346293183564791988224\lambda^3,$$

$$z = 10795745280\lambda^{12} + 49800119500\lambda^8 - 5599436221125\lambda^4$$
$$- 5780173209600,$$

$$t_1 = 150338377728,$$

$$t_0 = -300841435136\lambda^{14} + 1487659560960\lambda^{10} + 1673023786335\lambda^6$$
$$- 2621369509012\lambda^2.$$

Thus, the pencil $|(H-F)|_V|$ contains at least 16 surfaces that are singular away from C. This implies that the group $\mathrm{Aut}(V)$ is finite. Indeed, $\mathrm{Aut}(V) \cong \mathrm{Aut}(\widehat{V}) \cong \mathrm{Aut}(X)$, and the link (4.1) is $\mathrm{Aut}(V)$-equivariant, which gives an exact sequence of groups

$$1 \longrightarrow \Gamma \longrightarrow \mathrm{Aut}(\widehat{V}) \xrightarrow{\nu} \mathrm{PGL}_2(\mathbb{C}),$$

where ν is given by the induced $\mathrm{Aut}(\widehat{V})$-action on $\mathbb{P}^1$ and Γ acts trivially on $\mathbb{P}^1$ in (4.1). It follows that Γ is finite since it acts faithfully on a general fiber of the fibration $\widehat{\phi}$. Since $\mathrm{im}(\nu)$ permutes points in $\mathbb{P}^1$ that correspond to the surfaces in $|(H-F)|_V|$ that are singular away from C, we see that $\mathrm{im}(\nu)$ is finite. This shows that $\mathrm{Aut}(V)$ is finite, which can also be proved using [13, 124]. Let G be the subgroup in $\mathrm{Aut}(V)$ generated by

$$A_1 : (x, y, z, w, t_0, t_1) \mapsto (y, x, w, z, t_1, t_0),$$
$$A_2 : (x, y, z, w, t_0, t_1) \mapsto (x, -y, z, -w, t_0, t_1),$$
$$A_3 : (x, y, z, w, t_0, t_1) \mapsto (ix, y, -iz, w, t_0, -t_1).$$

Observe that V is G-invariant, and G acts faithfully on V, so that we can identify G with a subgroup in $\mathrm{Aut}(V)$. Then ϕ is G-equivariant and the following assertions hold:

(i) V contains no G-invariant points,
(ii) $|F|_V|$ and $|(H-F)_V|$ do not contain G-invariant surfaces,
(iii) V contains no G-invariant irreducible curve C such that $C \cdot F \leqslant 1$,
(iv) V contains no G-invariant irreducible surface S such that $-K_V \sim_{\mathbb{Q}}$ $\lambda S + \Delta$ for some rational number $\lambda > 1$ and effective $\mathbb{Q}$-divisor Δ on the threefold V, because the cone of effective divisors on V is generated by $F|_V$ and $(H-F)|_V$.

Then $\alpha_G(V) \geqslant 1$ by Corollary 1.55. But we have $\alpha_G(X) = \alpha_G(V)$ by Lemma 1.47, where we identify G with a subgroup of $\mathrm{Aut}(X)$ using the fact that ψ is G-equivariant. Thus, the singular Fano threefold X is K-polystable by Theorem 1.48 and hence K-stable by Corollary 1.5. By [168, Theorem 11] and [122, Theorem 1.4], X has a smoothing to a member of the family №1.8. Now, using Theorem 1.11, we conclude that a general smooth Fano threefold №1.8 is K-stable.

Example 4.10 Let Y be the smooth complete intersection in $\mathbb{P}^5$ given by

$$\begin{cases} x_0x_2 - x_1^2 + x_4(x_1 + x_3) + x_5(x_0 + x_2) + x_4^2 = 0, \\ x_1x_3 - x_2^2 + x_5(x_2 + x_0) + x_4(x_3 + x_1) + x_5^2 = 0. \end{cases}$$

Let G be the subgroup in $\mathrm{Aut}(\mathbb{P}^5)$ that is generated by the involutions

$$[x_0 : x_1 : x_2 : x_3 : x_4 : x_5] \mapsto [x_3 : x_2 : x_1 : x_0 : x_5 : x_4],$$
$$[x_0 : x_1 : x_2 : x_3 : x_4 : x_5] \mapsto [x_0 : -x_1 : x_2 : -x_3 : -x_4 : x_5].$$

Then $G \cong \mu_2 \times \mu_2$, the threefold Y is G-invariant, G acts on Y faithfully, and G preserves the three-dimensional subspace $\Lambda = \{x_4 = x_5 = 0\} \subset \mathbb{P}^5$. Then $\Lambda \cap Y$ is given by

$$x_0x_2 - x_1^2 = x_1x_3 - x_2^2 = 0.$$

It consists of a twisted cubic curve C and its secant line L that is cut out by $x_1 = x_2 = 0$. Both C and L are G-invariant. Let $H = \mathcal{O}_{\mathbb{P}^5}(1)|_Y$, and let $\mathcal{H}$ be the pencil in $|H|$ consisting of surfaces passing through C. Then $\mathcal{H}$ is cut out by $\lambda x_4 + \mu x_5 = 0$, where $[\lambda : \mu] \in \mathbb{P}^1$. This pencil has no G-invariant surfaces. Let $\alpha \colon \widetilde{Y} \to Y$ be the blow up of the curve C, and let $\widetilde{L}$ be the proper transform on $\widetilde{Y}$ of the line L. Then $-K_{\widetilde{Y}}^3 = 18$, the divisor $-K_{\widetilde{Y}}$ is nef, and $\widetilde{L}$ is the only irreducible curve in $\widetilde{Y}$ that has trivial intersection with the divisor $-K_{\widetilde{Y}}$. Moreover, there is a G-equivariant commutative diagram (see [208, (2.13.3)])

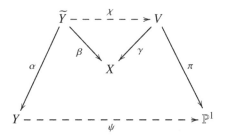

where χ is a flop of the curve $\widetilde{L}$, the morphism β is the contraction of the curve $\widetilde{L}$, the morphism γ is a flopping contraction, π is a fibration into quintic del Pezzo surfaces, and ψ is the map given by $\mathcal{H}$. Then V is a smooth weak Fano threefold, X is a Fano threefold with one Gorenstein terminal singular point such that $-K_X^3 = 18$ and $\mathrm{Pic}(X) \cong \mathbb{Z}[-K_X]$, and the group $\mathrm{Aut}(X)$ is finite since $\mathrm{Aut}(Y)$ is finite [45]. Thus, applying Corollary 1.57, we see that $\alpha_G(V) \geqslant \frac{4}{5}$. But $\alpha_G(X) = \alpha_G(V)$ by Lemma 1.47, so that X is K-stable by Theorem 1.48 and Corollary 1.5. By [168, Theorem 11] and [122, Theorem 1.4], it has a smoothing to a Fano threefold №1.9. Thus, a general Fano threefold in the family №1.9 is K-stable by Theorem 1.11.

Example 4.11 ([79, 81, 70, 135, 55, 100])   Fix $u \in \mathbb{C} \setminus \{0,1\}$, and let Q_u be the smooth quadric in $\mathbb{P}^4$ given by $u(xw - z^2) + (z^2 - yt) = 0$, and let G be the subgroup in $\mathrm{PGL}_4(\mathbb{C})$ generated by the involution $[x:y:z:t:w] \mapsto [w:t:z:y:x]$ and the transformations

$$[x:y:z:t:w] \mapsto [x : \lambda y : \lambda^3 z : \lambda^5 t : \lambda^6 w],$$

where $\lambda \in \mathbb{C}^*$. Then $G \cong \mathbb{G}_m \rtimes \mu_2$ and Q_u is G-invariant, so that G is naturally identified with a subgroup in $\mathrm{Aut}(Q_u)$. Let $S = \{xw - z^2 = z^2 - yt = 0\} \subset \mathbb{P}^4$, and let Γ be the sextic curve in $\mathbb{P}^4$ that is the locus $[s_0^6 : s_0^5 s_1 : s_0^3 s_1^3 : s_0 s_1^5 : s_1^6]$, where $[s_0 : s_1] \in \mathbb{P}^1$. Then S and Γ are G-invariant, $\Gamma \subset S \subset Q_u$, and there is a G-equivariant diagram

where V_u is a smooth Fano threefold in the family №1.10, π is the blow up of the curve Γ, the morphism ϕ is the blow up of the threefold V_u along a (unique) G-invariant smooth rational curve C_2 with $-K_{V_u} \cdot C_2 = 2$, and χ is the flop of two smooth rational curves. Every smooth Fano threefold №1.10 that admits an effective $\mathbb{G}_m$-action can be obtained in this way. We can identify G with a subgroup in $\mathrm{Aut}(V_u)$. Then

$$\alpha_G(V_u) = \begin{cases} \dfrac{4}{5} & \text{if } u \neq \dfrac{3}{4} \text{ and } u \neq 2, \\ \dfrac{3}{4} & \text{if } u = \dfrac{3}{4}, \\ \dfrac{2}{3} & \text{if } u = 2. \end{cases}$$

Hence, if $u \neq 2$ then V_u is K-polystable by Theorem 1.51. Moreover, it was proved in [100] that the threefold V_2 is also K-polystable. If $u \neq -\frac{1}{4}$, then $\mathrm{Aut}(V_u) = G$. Vice versa, if $u = -\frac{1}{4}$, then $\mathrm{Aut}(V_u) \cong \mathrm{PGL}_2(\mathbb{C})$, and V_u is the unique smooth threefold in the deformation family №1.10 with automorphism group $\mathrm{PGL}_2(\mathbb{C})$; this threefold is known as the Mukai–Umemura threefold [164].

It was proved in [136] that there exists a unique smooth Fano threefold №1.10 whose automorphism group is $\mathbb{G}_a \rtimes \mu_4$. By Theorem 1.3, this threefold is not K-polystable. This Fano threefold and the Fano threefolds described in Example 4.11 are the only smooth Fano threefolds in the family №1.10 that have infinite automorphism groups. In particular, the threefold in the following example has finite automorphism group.

Example 4.12 There is a unique smooth Fano threefold №1.10 such that Aut(X) contains a subgroup $G \cong \mathrm{PSL}_2(\mathbf{F}_7)$, and X has no G-fixed points [50]. Then $\alpha_G(X) \geqslant 1$ by Corollary 1.53, so that X is K-stable by Theorem 1.48.

Thus, a general Fano threefold in the family №1.10 is K-stable by Theorem 1.11. Using Corollaries 1.15 or 1.16 instead, we can also deduce this from Example 4.11. Similarly, other examples presented in this section show that the general members of the families №1.1, №1.2, №1.3, №1.4, №1.5, №1.6, №1.7, №1.8, №1.9 are also K-stable. In fact, a much stronger assertion holds:

Theorem 4.13 ([3]) *Every smooth Fano threefold in the families №1.1, №1.2, №1.3, №1.4, №1.5, №1.6, №1.7, №1.8 is K-stable.*

We expect that every smooth Fano threefold №1.9 is also K-stable.

4.2 Fano threefolds with PGL$_2$(ℂ)-action

Let X be a smooth Fano threefold such that $\mathrm{Aut}^0(X) \cong \mathrm{PGL}_2(\mathbb{C})$. By [45], X is one of:

(1) the Mukai–Umemura threefold (Example 4.11),
(2) the del Pezzo threefold V_5 (Example 3.2),
(3) the unique member of the family №2.21 with $\mathrm{Aut}^0(X) \cong \mathrm{PGL}_2(\mathbb{C})$,
(4) the unique member of family №2.27,
(5) the unique member of family №3.13 with $\mathrm{Aut}^0(X) \cong \mathrm{PGL}_2(\mathbb{C})$,
(6) the unique member of family №3.17,
(7) the unique member of family №4.6,
(8) $\mathbb{P}^1 \times S$, where S is a smooth del Pezzo surface of degree $K_S^2 \leqslant 5$.

We know from Section 3.1 and Examples 4.11 and 3.2 that the threefolds (1), (2), (8) are K-polystable. We now show that X is K-polystable in the remaining five cases. By Corollaries 1.15 and 1.5, this implies that a general member of family №2.21 is K-stable, and that a general member of the family №3.13 is K-polystable. We start with the simplest case:

Lemma 4.14 *The unique smooth Fano threefold in family №4.6 is K-polystable.*

Proof Let $\mathbb{V}$ be the vector space of 2×2-matrices

$$\begin{pmatrix} z_0 & z_1 \\ z_2 & z_3 \end{pmatrix}$$

and let $\mathbb{P}^3 = \mathbb{P}(V)$. Consider the $GL_2(\mathbb{C})$-action on V given by left (matrix) multiplication:

$$\text{for all } g \in GL_2(\mathbb{C}), \quad \left(g, \begin{pmatrix} z_0 & z_1 \\ z_2 & z_3 \end{pmatrix}\right) \mapsto g \cdot \begin{pmatrix} z_0 & z_1 \\ z_2 & z_3 \end{pmatrix}.$$

This induces a faithful $PGL_2(\mathbb{C})$-action on $\mathbb{P}^3$. The locus of invertible matrices in $\mathbb{P}^3$ is an open $PGL_2(\mathbb{C})$-orbit, and its complement is the $PGL_2(\mathbb{C})$-invariant quadric

$$S = \left\{ \det \begin{pmatrix} z_0 & z_1 \\ z_2 & z_3 \end{pmatrix} = 0 \right\} \subset \mathbb{P}^3.$$

For each $[a : b] \in \mathbb{P}^1$, define a line

$$\ell_{a,b} = \left\{ \begin{bmatrix} a\lambda & b\lambda \\ a\mu & b\mu \end{bmatrix} \in \mathbb{P}^3 \,\middle|\, [\lambda : \mu] \in \mathbb{P}^1 \right\}$$

and note that $\ell_{a,b}$ lies on S and is a $PGL_2(\mathbb{C})$-orbit.

Consider the subgroup of $GL_2(\mathbb{C})$ generated by

$$G = \left\langle \begin{pmatrix} 0 & 1 \\ -1 & 0 \end{pmatrix}, \begin{pmatrix} 1 & 1 \\ -1 & 0 \end{pmatrix} \right\rangle.$$

Then $G \cong \mathfrak{S}_3$. As above, consider the G-action on V given by left matrix multiplication. The G-action on $\mathbb{P}^3$ defined in this way is faithful and commutes with the $PGL_2(\mathbb{C})$-action, no $PGL_2(\mathbb{C})$-invariant line is fixed by G, and G acts freely on $\{\ell_{1,0}, \ell_{0,1}, \ell_{1,1}\}$.

Since (up to change of coordinates) X is the blow up of $\mathbb{P}^3$ in $\ell_{1,0} \cup \ell_{0,1} \cup \ell_{1,1}$, both the $PGL_2(\mathbb{C})$-action and G-action lift to X, so that $PGL_2(\mathbb{C})$ and G are identified with subgroups of $\mathrm{Aut}(X)$. One can show that $\mathrm{Aut}(X) = \langle PGL_2(\mathbb{C}), G \rangle \cong PGL_2(\mathbb{C}) \times \mathfrak{S}_3$. The strict transform $\mathscr{S}$ of the quadric S on X is the unique proper $\mathrm{Aut}(X)$-invariant irreducible subvariety of X. Furthermore, it follows from Theorem 3.17 that $\beta(\mathscr{S}) > 0$, so that X is K-polystable by Theorem 1.22. $\square$

Now, we consider the member of family №2.21 with an effective $PGL_2(\mathbb{C})$-action. Its K-polystability could be proved using Theorem 1.79, but we give an alternative proof.

Lemma 4.15 *Let X be the smooth Fano threefold №2.21 such that $\mathrm{Aut}^0(X) \cong PGL_2(\mathbb{C})$. Then X is K-polystable.*

Proof The smooth Fano threefold X can be constructed as follows. Let V be the standard representation of $GL_2(\mathbb{C})$, denote by $\mathbb{P}^4 = \mathbb{P}(\mathrm{Sym}^4(V))$, and let $C \subset \mathbb{P}^4$ be the image of the 4th Veronese embedding of $\mathbb{P}(\mathbb{V})$. Then the GL$_2(\mathbb{C})$-action on $\mathbb{V}$ induces an action of the group PGL$_2(\mathbb{C})$ on $\mathbb{P}^4$ and C is PGL$_2(\mathbb{C})$-invariant. The representation of GL$_2(\mathbb{C})$ on Sym$^4(\mathbb{V})$ is irreducible, and there is a smooth invariant quadric $Q \subset \mathbb{P}^4$ that contains C. Then X can be obtained as a blow up of Q along C.

Let $f_1 \colon X \to Q$ be the blow up of the curve C, and let E_1 be its exceptional divisor. Then f_1 is PGL$_2(\mathbb{C})$-equivariant and the PGL$_2(\mathbb{C})$-action lifts to X. The threefold X has a second PGL$_2(\mathbb{C})$-equivariant contraction $f_2 \colon X \to Q$. The f_2-exceptional divisor E_2 is the proper transform of the surface $\overline{E}_2 \subset Q$ that is cut out on Q by the secant variety of the curve C, which is a singular cubic hypersurface. We thus have a PGL$_2(\mathbb{C})$-equivariant commutative diagram:

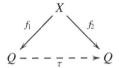

where f_2 is the contraction of the surface E_2 to the curve C, and τ is a birational involution that is given by the linear subsystem in $|O_{\mathbb{P}^4}(2)|_Q|$ consisting of surfaces that contain C. The birational action of τ lifts to a biregular action on X that swaps E_1 and E_2, and τ commutes with the PGL$_2(\mathbb{C})$-action on X (this is [185, Example 2.4.1]). Thus, we have Aut$(X) \cong$ PGL$_2(\mathbb{C}) \times \mu_2$, where the factor μ_2 is generated by τ.

Let $G = $ Aut(X) and denote by C the smooth irreducible G-invariant curve $C = E_1 \cap E_2$. By Lemma A.47, we have $E_1 \cong E_2 \cong \mathbb{P}^1 \times \mathbb{P}^1$, and C is the diagonal in both E_1 and E_2. Note that C is the only G-invariant irreducible proper subvariety of X, and that E_1 and E_2 are tangent along C, so that $E_1 \cdot E_2 = 2C$. Since $E_1 + E_2 \sim -K_X$ and $E_1 + E_2$ is G-invariant, this implies $\alpha_G(X) \leqslant \frac{3}{4}$ because $(X, \frac{3}{4}(E_1 + E_2))$ is strictly log canonical.

We claim that $\alpha_G(X) = \frac{3}{4}$. Indeed, suppose $\alpha_G(X) < \frac{3}{4}$. Then there is a G-invariant linear system $\mathcal{D} \subset |-nK_X|$ such that the singularities of the log pair $(X, \frac{3}{4n}\mathcal{D})$ are not log canonical. Write $\frac{1}{n}\mathcal{D} = a(E_1 + E_2) + b\mathcal{M}$, where a and b are some non-negative numbers, and $\mathcal{M}$ is the mobile part of the linear system $\mathcal{D}$. Then $a \leqslant 1$ since

$$a(E_1 + E_2) + b\mathcal{M} \sim_Q -K_X \sim E_1 + E_2.$$

Furthermore, since $(X, \frac{3}{4}(E_1 + E_2))$ is log canonical, we have $a < 1$.

Using Corollary A.32, we may assume that $a = 0$. Indeed, let $\mu = \frac{a}{1-a}$ and

$$D = (1 + \mu)\Big(aE_1 + aE_2 + b\mathcal{M}\Big) - \mu(E_1 + E_2).$$

Then $D \sim_{\mathbb{Q}} -K_X$ and $D = \frac{b}{1-a} M$. On the other hand, we have

$$a(E_1 + E_2) + bM = \frac{1}{1+\mu} D + \frac{\mu}{1+\mu}(E_1 + E_2),$$

so that $(X, \frac{3}{4}D)$ is also not log canonical. Therefore, replacing $a(E_1 + E_2) + bM$ by $\frac{b}{1-a} M$, we may assume that $a = 0$, so that $\mathcal{D} = \mathcal{M}$.

Since $\mathcal{M}$ is mobile, $(X, \frac{3}{4n}\mathcal{M})$ is not log canonical, and X does not have G-invariant zero-dimensional subschemes, and since C is the only G-invariant curve in X, C is a center of non-log canonical singularities of $(X, \frac{3}{4n}\mathcal{M})$. Let M be a general surface in $\mathcal{M}$, and let ℓ be a general fiber of the projection $E \to C$. Then $\ell \not\subset M$ and $n = M \cdot \ell \geq \mathrm{mult}_C(M) > \frac{4n}{3}$, which is absurd. Then $\alpha_G(X) = \frac{3}{4}$, so that X is K-polystable by Theorem 1.51.   □

In Section 5.9, we will give another proof of Lemma 4.15 that relies on the more general statement that all smooth Fano threefolds №2.21 with infinite reductive automorphism group are K-polystable. Using Lemma 4.15, Corollaries 1.5 and 1.15, we obtain

Corollary 4.16 (cf. Remark 5.116)   *A general member of family №2.21 is K-stable.*

Now we consider the unique smooth Fano threefold in family №2.27.

Lemma 4.17   *The smooth Fano threefold №2.27 is K-polystable.*

Proof   Let C_3 be a twisted cubic curve in $\mathbb{P}^3$, and let $\pi \colon X \to \mathbb{P}^3$ be its blow up. Since $\mathrm{Aut}(C_3) \cong \mathrm{PGL}_2(\mathbb{C})$, $\mathrm{Aut}(X) \cong \mathrm{PGL}_2(\mathbb{C})$ as well, and by [205, §2], there exists a $\mathrm{PGL}_2(\mathbb{C})$-equivariant commutative diagram

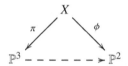

where ϕ is a conic bundle and $\mathbb{P}^3 \dashrightarrow \mathbb{P}^2$ is the map defined by the net of quadrics containing C_3. The group $G = \mathrm{PGL}_2(\mathbb{C})$ acts faithfully on $\mathbb{P}^2$, and $\mathbb{P}^2$ contains a unique G-invariant conic C_2, which is also the smooth conic of jumping lines of the bundle ϕ.

Let E be the exceptional divisor of π, and let R be the preimage of the conic C_2 in X. The restriction of ϕ is a double cover $E \to \mathbb{P}^2$ branched over C_2. Let C be the intersection $R \cap E$ taken with reduced structure; then R and E are tangent along C and $R \cdot E = 2C$. Moreover, the surface $\pi(R)$ is the non-normal quartic surface that has an ordinary cusp along the curve C_3. This surface is spanned by the lines in $\mathbb{P}^3$ that are tangent to C_3. Then $C \sqcup (R \setminus C) \sqcup (E \setminus C) \sqcup (X \setminus (R \cup E))$ is the decomposition of X into G-orbits.

Let $\upsilon\colon V \to X$ be the blow up of the curve C, and let F be the υ-exceptional surface. Denote by $\widetilde{R}$ and $\widetilde{E}$ the proper transforms on V of the surfaces R and E, respectively. Then $F \cong \mathbb{P}^1 \times \mathbb{P}^1$ and the intersection $F \cap \widetilde{R} \cap \widetilde{E}$ is a smooth rational curve $\widetilde{C}$, which is a divisor of degree $(1,1)$ on the surface F. Since C is G-invariant, the G-action lifts to V, but $\mathrm{Aut}(V)$ is larger than G. Indeed, it follows from [174, Section 2] or [75, Example 3.4.4] that there is a biregular involution $\tau \in \mathrm{Aut}(V)$ that swaps F and $\widetilde{R}$ and leaves $\widetilde{E}$ invariant. Thus, we can write the G-equivariant diagram

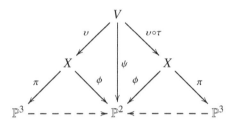

where ψ is a conic bundle. The involution $\upsilon \circ \tau \circ \upsilon^{-1}$ induced by the involution τ is an elementary birational transformation of the $\mathbb{P}^1$-bundle ϕ. Note that τ induces a Cremona transformation $\mathbb{P}^3 \dashrightarrow \mathbb{P}^3$, which (in appropriate coordinates) is given by the four partial derivatives of the defining quartic polynomial of the surface $\pi(R)$.

We claim that $\alpha_G(X) = \frac{3}{4}$. Indeed, observe that both divisors E and R are G-invariant and $-K_X \sim E + R$, so that $\alpha_G(X) \leqslant \frac{3}{4}$ because $(X, \frac{3}{4}(E + R))$ is strictly log canonical.

Suppose that $\alpha_G(X) < \frac{3}{4}$. Then there is a G-invariant linear system $\mathcal{D} \subset |-nK_X|$ such that the singularities of the pair $(X, \frac{3}{4n}\mathcal{D})$ are not log canonical. Write $\frac{1}{n}\mathcal{D} = aE + bR + c\mathcal{M}$ where $\mathcal{M}$ is the mobile part of the linear system $\mathcal{D}$, and a, b and $c = \frac{1}{n}$ are non-negative rational numbers. Then $a \leqslant 1$ and $b \leqslant 1$ since $aE + bR + c\mathcal{M} \sim_{\mathbb{Q}} R + E$. Furthermore, since the pair $(X, \frac{3}{4}(E + R))$ is log canonical, we have $a < 1$ or $b < 1$. Moreover, it follows from Lemma A.34 that we may assume that either $a = 0$ or $b = 0$.

If $a = 0$, then $1 = D \cdot \ell \geqslant \mathrm{mult}_C(D) > \frac{4}{3}$ by Lemma A.1, where ℓ is a general fiber of the natural projection $E \to C_3$. Thus, $a > 0$ so that $b = 0$.

Let $\widetilde{D}$ be the proper transform of D on the threefold V, and let $\widetilde{L}$ be a general fiber of the natural projection $\widetilde{R} \to C_2$. Then $\mathrm{mult}_C(D) \leqslant 2$ because $0 \leqslant \widetilde{D} \cdot \widetilde{L} = 2 - \mathrm{mult}_C(D)$. Now, using Lemma A.27, we see that F contains a G-invariant section Z of the natural projection $F \to C$ such that $\mathrm{mult}_C(D) + \mathrm{mult}_Z(\widetilde{D}) > \frac{8}{3}$. On the other hand, it follows from Lemma A.47 that $\widetilde{C}$ is the only G-invariant curve in F, so that $Z = \widetilde{C}$, which gives

$$\text{mult}_Z(\widetilde{D}) \leqslant \widetilde{D} \cdot \widetilde{L} = \left(v^*(-K_X) - \text{mult}_C(D)F\right) \cdot \widetilde{L} = 2 - \text{mult}_C(D),$$

where $\widetilde{L}$ is a general fiber of the natural projection $\widetilde{R} \to C_2$. The obtained contradiction shows that $\alpha_G(X) = \frac{3}{4}$, so that X is K-polystable by Theorem 1.51.

$\square$

Now, we deal with deformation family №3.13. The K-polystability of this threefold has been already shown in Example 1.94. Let us prove this using a different approach.

Lemma 4.18 *Let X be the smooth Fano threefold №3.13 with $\text{Aut}^0(X) \cong \text{PGL}_2(\mathbb{C})$. Then X is K-polystable.*

Proof The Fano threefold X can be described as follows. Take any smooth conic $C \subset \mathbb{P}^2$, and consider the $\text{PGL}_2(\mathbb{C})$-action on $\mathbb{P}^2$ that leaves C invariant. This defines the diagonal action of the group $\text{PGL}_2(\mathbb{C})$ on $\mathbb{P}^2 \times \mathbb{P}^2$, and there exists a smooth $\text{PGL}_2(\mathbb{C})$-invariant divisor $W \subset \mathbb{P}^2 \times \mathbb{P}^2$ of degree $(1,1)$. Then all $\text{PGL}_2(\mathbb{C})$-invariant irreducible closed subvarieties in the threefold W are the surfaces $\overline{E}_2 = \text{pr}_1^{-1}(C)$ and $\overline{E}_3 = \text{pr}_2^{-1}(C)$, and the smooth irreducible rational curve $\overline{C} = \overline{E}_2 \cap \overline{E}_3$. The threefold X can be obtained by blowing up W along the curve $\overline{C}$ (cf. [45]).

Let $f_1 \colon X \to W$ be the blow up of the curve $\overline{C}$. Then the $\text{PGL}_2(\mathbb{C})$-action lifts to X. Denote by E_1 the f_1-exceptional surface, and denote by E_2 and E_3 the proper transforms on the threefold X of the surfaces $\overline{E}_2$ and $\overline{E}_3$. Then there exists a $\text{PGL}_2(\mathbb{C})$-equivariant commutative diagram:

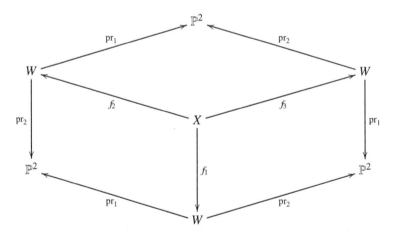

where f_2 and f_3 are contractions of the surfaces E_2 and E_3 to curves of degree $(2,2)$, Moreover, it follows from [185] that $\text{Aut}(X) \cong \text{PGL}_2(\mathbb{C}) \times \mathfrak{S}_3$ (see also

Section 5.19), and that the direct factor $\mathfrak{S}_3$ permutes the surfaces E_1, E_2 and E_3 transitively.

We let $G = \mathrm{Aut}(X)$. Then $E_1 \cap E_2 \cap E_3 = E_1 \cap E_2 = E_2 \cap E_3 = E_1 \cap E_3$ is a smooth irreducible G-invariant curve, which we denote by C. Then C is the only G-invariant proper irreducible closed subvariety in X.

Let $\varphi\colon Y \to X$ be the blow up of the curve C, and let E be the φ-exceptional surface, and $\widetilde{E}_1, \widetilde{E}_2, \widetilde{E}_3$ the proper transforms of E_1, E_2, E_3. Then $\widetilde{E}_1, \widetilde{E}_2, \widetilde{E}_3$ are pairwise disjoint, so that $\widetilde{E}_1|_E, \widetilde{E}_2|_E$, and $\widetilde{E}_3|_E$ are three pairwise disjoint sections of the projection $E \to C$. This is only possible if $E \cong \mathbb{P}^1 \times \mathbb{P}^1$.

The G-action lifts to Y, and E is G-invariant. Applying Lemma A.47, we see that PGL$_2(\mathbb{C})$ acts trivially on one factor of $E \cong \mathbb{P}^1 \times \mathbb{P}^1$, so that the sections of $E \to C$ are PGL$_2(\mathbb{C})$-orbits contained in E. On the other hand, the group $\mathfrak{S}_3$ permutes $\widetilde{E}_1|_E, \widetilde{E}_2|_E$, and $\widetilde{E}_3|_E$ transitively. This immediately implies that no section of $E \to C$ is G-invariant, so that E contains no proper closed G-invariant subvarieties. Therefore, the surface E is the only G-invariant prime divisor over X, and by Theorem 1.22, X is K-polystable if and only if $\beta(E) > 0$.

We claim that $\beta(E) = \frac{9}{10}$. Let $t \in \mathbb{R}_{\geqslant 0}$, then since $-K_X \sim E_1 + E_2 + E_3$, we have

$$\varphi^*(-K_X) - tE \sim \widetilde{E}_1 + \widetilde{E}_2 + \widetilde{E}_3 + 3E - tE = \widetilde{E}_1 + \widetilde{E}_2 + \widetilde{E}_3 + (3-t)E,$$

which implies that $\varphi^*(-K_X) - tE$ is pseudo-effective if and only if $t \leqslant 3$. Moreover, the divisor $\varphi^*(-K_X) - tE$ is nef precisely when $t \leqslant 1$. When $1 < t < 3$, the Zariski decomposition of $\varphi^*(-K_X) - tE$ is

$$\varphi^*(-K_X) - tE \sim_{\mathbb{R}} \underbrace{\frac{3-t}{2}(\widetilde{E}_1 + \widetilde{E}_2 + \widetilde{E}_3 + 2E)}_{\text{positive part}} + \underbrace{\frac{t-1}{2}(\widetilde{E}_1 + \widetilde{E}_2 + \widetilde{E}_3)}_{\text{negative part}}.$$

Hence, we calculate

$$
\begin{aligned}
S_X(E) &= \frac{1}{30} \int_0^1 \left(\varphi^*(-K_X) - tE \right)^3 dt \\
&\quad + \frac{1}{30} \int_1^3 \left(\varphi^*(-K_X) - tE - \frac{t-1}{2}(\widetilde{E}_1 + \widetilde{E}_2 + \widetilde{E}_3) \right)^3 dt \\
&= \frac{1}{30} \int_0^1 (30 - 18t^2 + 4t^3)dt + \frac{1}{30} \int_1^3 2(3-t)^3 dt = \frac{11}{10},
\end{aligned}
$$

which gives $\beta(E) = A_X(E) - S_X(E) = 2 - \frac{11}{10} = \frac{9}{10}$, so that X is K-polystable. $\square$

Therefore, a general member of the family №3.13 is K-polystable by Corollary 1.15. In fact, with a single exception, all smooth Fano threefolds in this deformation family are K-polystable, see Section 5.19 for details. Let us conclude this section by proving

Lemma 4.19 (cf. [56])   *The unique smooth Fano threefold №3.17 is K-polystable.*

Proof   Let X be the unique smooth Fano threefold №3.17. Then X is a smooth divisor in $\mathbb{P}^1 \times \mathbb{P}^1 \times \mathbb{P}^2$ that has degree $(1, 1, 1)$. Moreover, one can choose appropriate homogeneous coordinates $([x_0 : x_1], [y_0 : y_1], [z_0 : z_1 : z_2])$ on $\mathbb{P}^1 \times \mathbb{P}^1 \times \mathbb{P}^2$ such that X is given by

$$x_0 y_0 z_2 + x_1 y_1 z_0 = x_0 y_1 z_1 + x_1 y_0 z_1.$$

Let $G = \mathrm{Aut}(X)$. Then it follows from [45, Corollary 8.8] that $\mathrm{Aut}(X) \cong \mathrm{PGL}_2(\mathbb{C}) \rtimes \mu_2$, where μ_2 is generated by an involution ι that acts as

$$([x_0 : x_1], [y_0 : y_1], [z_0 : z_1 : z_2]) \mapsto ([y_0 : y_1], [x_0 : x_1], [z_0 : z_1 : z_2]),$$

and $\mathrm{PGL}_2(\mathbb{C})$ acts as follows:

$$([x_0 : x_1], [y_0 : y_1], [z_0 : z_1 : z_2])$$
$$\mapsto ([ax_0 + cx_1 : bx_0 + dx_1], [ay_0 + cy_1 : by_0 + dy_1],$$
$$[a^2 z_0 + 2ac z_1 + c^2 z_2 : abz_0 + (ad + bc)z_1 + cd z_2 : b^2 z_0 + 2bd z_1 + d^2 z_2]),$$

where $\left(\begin{smallmatrix} a & b \\ c & d \end{smallmatrix}\right) \in \mathrm{PGL}_2(\mathbb{C})$.

There are birational contractions $\pi_1 \colon X \to \mathbb{P}^1 \times \mathbb{P}^2$ and $\pi_2 \colon X \to \mathbb{P}^1 \times \mathbb{P}^2$ that contract smooth irreducible surfaces E_1 and E_2 to smooth curves C_1 and C_2 of degrees $(1, 2)$. Moreover, there is $\mathrm{PGL}_2(\mathbb{C})$-equivariant commutative diagram

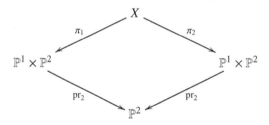

where pr_2 is the projection to the second factor, the $\mathrm{PGL}_2(\mathbb{C})$-action on $\mathbb{P}^2$ is faithful, and $\mathrm{pr}_2(C_1) = \mathrm{pr}_2(C_2)$ is the unique $\mathrm{PGL}_2(\mathbb{C})$-invariant conic.

Let $\mathrm{pr}_1 \colon \mathbb{P}^1 \times \mathbb{P}^2 \to \mathbb{P}^1$ be the projection to the first factor. Using $\mathrm{pr}_1 \circ \pi_1$ and $\mathrm{pr}_1 \circ \pi_2$, we obtain a $\mathrm{PGL}_2(\mathbb{C})$-equivariant $\mathbb{P}^1$-bundle $\phi \colon X \to \mathbb{P}^1 \times \mathbb{P}^1$, where the $\mathrm{PGL}_2(\mathbb{C})$-action on the surface $\mathbb{P}^1 \times \mathbb{P}^1$ is diagonal. Let $C = E_1 \cap E_2$. Then $\phi(C)$ is a diagonal curve. Denote its preimage on X by R. Then $C = R \cap E_1 \cap E_2$. Moreover, the curve C and the surface R are the only proper $\mathrm{Aut}(X)$-invariant proper irreducible subvarieties in X.

Observe that $\alpha_G(X) \leqslant \frac{2}{3}$ because $-K_X \sim E_1 + E_2 + R$ and $E_1 + E_2 + R$ is G-invariant. Therefore, we cannot apply Theorem 1.51 to prove that X is K-polystable.

Suppose that X is not K-polystable. By Theorem 1.22, there is a G-invariant prime divisor F over X such that $\beta(F) \leqslant 0$. Let $Z = C_X(F)$. Then $Z \neq R$ by Theorem 3.17, so that $Z = C$. Let us apply Corollary 1.110 with $Y = E_1$ to show that $Z \neq C$.

Let $H_1 = (\mathrm{pr}_1 \circ \pi_1)^*(\mathcal{O}_{\mathbb{P}^1}(1))$, let $H_2 = (\mathrm{pr}_1 \circ \pi_2)^*(\mathcal{O}_{\mathbb{P}^1}(1))$, let $H_3 = (\mathrm{pr}_2 \circ \pi_2)^*(\mathcal{O}_{\mathbb{P}^2}(1))$; and note that these generate the nef cone of X. The descriptions of X, π_1 and π_2, imply

$$-K_X \sim H_1 + H_2 + 2H_3 \sim 2H_1 + 3H_3 - E_1 \sim 2H_2 + 3H_3 - E_2,$$

so that $E_1 + E_2 \sim 2H_3$. Note also that $R \sim H_1 + H_2$. Writing

$$-K_X \sim 2H_2 + 3H_3 - E_2 \sim 2H_2 + \frac{3}{2}(E_1 + E_2) - E_2 \sim 2H_2 + \frac{3}{2}E_1 + \frac{1}{2}E_2,$$

we get

$$-K_X - uE_1 \sim_{\mathbb{R}} 2H_2 + \left(\frac{3}{2} - u\right)E_1 + \frac{1}{2}E_2,$$

where u is a non-negative real number. Hence, the divisor $-K_X - uE_1$ is nef for $u \leqslant 1$, and it is not pseudo-effective for $u > \frac{3}{2}$. Moreover, if $1 < u \leqslant \frac{3}{2}$, its Zariski decomposition is

$$-K_X - uE_1 \sim_{\mathbb{R}} \underbrace{2H_2 + \left(\frac{3}{2} - u\right)(E_1 + E_2)}_{\text{positive part}} + \underbrace{(u - 1)E_2}_{\text{negative part}},$$

noting that the positive part $2H_2 + \left(\frac{3}{2} - u\right)(E_1 + E_2) \sim_{\mathbb{R}} 2H_2 + (3 - 2u)H_3$ is nef. Thus, in the notations of Corollary 1.110, we have

$$P(u) = \begin{cases} -K_X - uE_1 & \text{if } 0 \leqslant u \leqslant 1, \\ 2H_2 + (3 - 2u)H_3 & \text{if } 1 \leqslant u \leqslant \dfrac{3}{2}, \end{cases}$$

and

$$N(u) = \begin{cases} 0 & \text{if } 0 \leqslant u \leqslant 1, \\ (u - 1)E_2 & \text{if } 1 \leqslant u \leqslant \dfrac{3}{2}. \end{cases}$$

Using this, one can easily check that $S_X(E_1) < 1$, which also follows from Theorem 3.17. Therefore, we have $S(W_{\bullet,\bullet}^{E_1}; C) \geqslant 1$ by Corollary 1.110.

Let us compute $S(W_{\bullet,\bullet}^{E_1}; C)$. Recall that $E_1 \cong \mathbb{P}^1 \times \mathbb{P}^1$. Let $\mathbf{f}$ be a fiber of the natural projection $E_1 \to C_1$, and let $\mathbf{s}$ be the section of this projection such that

$\mathbf{s}^2 = 0$ on E_1. Then $E_1|_{E_1} \sim -\mathbf{s} + 3\mathbf{f}$ and $C \sim \mathbf{s} + \mathbf{f}$. Therefore, for any $v \in \mathbb{R}$, we have

$$P(u)\big|_{E_1} - vC \sim_{\mathbb{R}} \begin{cases} (u + 1 - v)\mathbf{s} + (5 - 3u - v)\mathbf{f} & \text{if } 0 \leqslant u \leqslant 1, \\[2mm] (2 - v)\mathbf{s} + (6 - 4u - v)\mathbf{f} & \text{if } 1 \leqslant u \leqslant \dfrac{3}{2}. \end{cases}$$

Hence, using Corollary 1.110, we get

$$S(W^{E_1}_{\bullet,\bullet}; C)$$

$$= \frac{3}{(-K_X)^3} \int_0^{\frac{3}{2}} \left(P(u)^2 \cdot E_1 \right) \cdot \mathrm{ord}_C\left(N(u)\big|_{E_1} \right) du$$

$$+ \frac{3}{(-K_X)^3} \int_0^{\frac{3}{2}} \int_0^\infty \mathrm{vol}\left(P(u)\big|_{E_1} - vC \right) dv\, du$$

$$= \frac{1}{12} \int_1^{\frac{3}{2}} (u - 1)\left(P(u)^2 \cdot E_1 \right) du + \frac{1}{12} \int_0^{\frac{3}{2}} \int_0^\infty \mathrm{vol}\left(P(u)\big|_{E_1} - vC \right) dv\, du$$

$$= \frac{1}{12} \int_1^{\frac{3}{2}} 4(u - 1)(6 - 4u)du + \frac{1}{12} \int_0^1 \int_0^{u+1} 2(u + 1 - v)(5 - 3u - v)dv\, du$$

$$= \frac{1}{12} \int_1^{\frac{3}{2}} \int_0^{6-4u} 2(2 - v)(6 - 4u - v)dv\, du = \frac{5}{8} < 1,$$

which is a contradiction. $\square$

4.3 Blow ups of del Pezzo threefolds in elliptic curves

Let V_d be a smooth threefold such that $-K_{V_d} \sim 2H$ for an ample Cartier divisor H on the threefold V_d such that $d = H^3$, let H_1 and H_2 be two distinct surfaces in $|H|$ such that $\mathscr{C} = H_1 \cap H_2$ is a smooth curve, let $\mathcal{P}$ be the pencil generated by H_1 and H_2, let $\pi \colon X \to V_d$ be the blow up of the curve $\mathscr{C}$. Then $\mathscr{C}$ is an elliptic curve, X is a Fano threefold, and there exists a commutative diagram

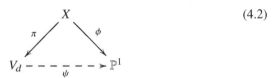

$$(4.2)$$

where ψ is the map given by $\mathcal{P}$, and ϕ is a fibration into del Pezzo surfaces of degree d. Let E be the π-exceptional surface, let F be a sufficiently general fiber of the morphism ϕ, and let $\widetilde{H}_1$ and $\widetilde{H}_2$ be proper transforms on X of the surfaces H_1 and H_2, respectively. Then $E \cong \mathscr{C} \times \mathbb{P}^1$, and $F \sim \widetilde{H}_1 \sim \widetilde{H}_2$ on the threefold X.

Recall from Section 3.4 that V_d is a smooth del Pezzo threefold of degree d, and we have the following nine possibilities:

(i) $d = 1$, V_1 is a Fano threefold №1.11, and X is a Fano threefold №2.1;

(ii) $d = 2$, V_2 is a Fano threefold №1.12, and X is a Fano threefold №2.3;

(iii) $d = 3$, V_3 is a Fano threefold №1.13, and X is a Fano threefold №2.5;

(iv) $d = 4$, V_4 is a Fano threefold №1.14, and X is a Fano threefold №2.10;

(v) $d = 5$, V_5 is a Fano threefold №1.15, and X is a Fano threefold №2.14;

(vi) $d = 6$, V_6 is a divisor in $\mathbb{P}^2 \times \mathbb{P}^2$ of degree $(1, 1)$, and X is a Fano threefold №3.7;

(vii) $d = 6$, $V_6 = \mathbb{P}^1 \times \mathbb{P}^1 \times \mathbb{P}^1$, and X is a Fano threefold №4.1;

(viii) $d = 7$, $V_7 = \mathbb{P}(O_{\mathbb{P}^2} \oplus O_{\mathbb{P}^2}(1))$, and X is a Fano threefold №3.11;

(ix) $d = 8$, $V_8 = \mathbb{P}^3$, $H = O_{\mathbb{P}^3}(2)$, and X is a Fano threefold №2.25.

Smooth Fano threefolds in the family №4.1 have an alternative description – they are divisors in $\mathbb{P}^1 \times \mathbb{P}^1 \times \mathbb{P}^1 \times \mathbb{P}^1$ of degree $(1, 1, 1, 1)$. This family contains one K-polystable singular member – the toric Gorenstein terminal Fano threefold №625 in [27], so that the general smooth member of the family №4.1 is also K-semistable by Theorem 1.11. Let us present one very special smooth Fano threefold №4.1 that is K-stable:

Lemma 4.20 *Let X be the divisor of $(\mathbb{P}^1)^4$ defined by*

$$\{x_1 x_2 x_3 x_4 + y_1 y_2 y_3 y_4 = 2(x_1 x_2 y_3 y_4 + y_1 y_2 x_3 x_4 + x_1 y_2 x_3 y_4 + x_1 y_2 y_3 x_4$$
$$+ y_1 x_2 x_3 y_4 + y_1 x_2 y_3 x_4)\},$$

where $[x_i : y_i]$ are coordinates on the ith factor of $(\mathbb{P}^1)^4$. Then X is smooth and K-stable.

Proof The smoothness of the threefold X is easy to check. To prove its K-stability, observe that $\mathrm{Aut}(X)$ contains a subgroup $G \cong \mu_2^2 \times \mathfrak{S}_4$, where $\sigma \in \mathfrak{S}_4$ acts by

$$([x_1 : y_1], [x_2 : y_2], [x_3 : y_3], [x_4 : y_4])$$
$$\mapsto ([x_{\sigma(1)} : y_{\sigma(1)}], [x_{\sigma(2)} : y_{\sigma(2)}], [x_{\sigma(3)} : y_{\sigma(3)}], [x_{\sigma(4)} : y_{\sigma(4)}]),$$

while the generator τ of the first factor of μ_2^2 acts by

$$([x_1 : y_1], [x_2 : y_2], [x_3 : y_3], [x_4 : y_4])$$
$$\mapsto ([y_1 : x_1], [y_2 : x_2], [y_3 : x_3], [y_4 : x_4]),$$

and the generator ι of the second factor of μ_2^2 acts by

$$([x_1 : y_1], [x_2 : y_2], [x_3 : y_3], [x_4 : y_4])$$
$$\mapsto ([x_1 : -y_1], [x_2 : -y_2], [x_3 : -y_3], [x_4 : -y_4]).$$

We claim that $\alpha_G(X) \geqslant 1$, so that X is K-stable by Theorem 1.48 since $\mathrm{Aut}(X)$ is finite. Indeed, suppose that $\alpha_G(X) < 1$. Let us seek for a contradiction.

First, we observe that $\mathrm{Pic}^G(X) = \mathbb{Z}[-K_X]$, and X does not contain G-fixed points. Thus, applying Theorem 1.52 with $\mu = 1$, we see that X contains a smooth G-invariant curve C such that $C \cdot S = 1$ for any fiber S of any of four (natural) projections $X \to \mathbb{P}^1$. Hence, the curve C is a curve of degree $(1, 1, 1, 1)$.

Let Γ be the stabilizer in G of the surface S. If S is given by $x_4 = y_4$, then $\Gamma \cong \mu_2 \times \mathfrak{S}_3$, where the group $\mathfrak{S}_3$ acts by simultaneous permutations of coordinates x_i and y_i for $i \neq 4$, and $\mu_2 = \langle \tau \rangle$. Then $P_1 = ([1 : -1], [1 : -1], [1 : -1], [1 : 1])$ is the only Γ-invariant point in the surface S, so that $P_1 = S \cap C$. Similarly, letting S be the surfaces $x_4 + y_4 = 0$, $x_4 = 0$ and $y_4 = 0$, we see that C contains the points

$$P_2 = ([1 : 1], [1 : 1], [1 : 1], [1 : -1]),$$
$$P_3 = ([1 : 0], [1 : 0], [1 : 0], [0 : 1]),$$
$$P_4 = ([0 : 1], [0 : 1], [0 : 1], [1 : 0]).$$

Let $\mathrm{pr}_{12} \colon X \to \mathbb{P}^1 \times \mathbb{P}^1$ be the projection to the first two factors of $(\mathbb{P}^1)^4$. The $\mathrm{pr}_{12}(C)$ is an irreducible curve of degree $(1, 1)$. Observe that the projection pr_{12} is equivariant with respect to the subgroup $\Xi \cong \mu_2^3$ of the group G generated by τ, ι and the involution

$$([x_1 : y_1], [x_2 : y_2], [x_3 : y_3], [x_4 : y_4])$$
$$\mapsto ([x_2 : y_2], [x_1 : y_1], [x_3 : y_3], [x_4 : y_4]).$$

Therefore, the curve $\mathrm{pr}_{12}(C)$ is Ξ-invariant, so that C is contained in one of the following four surfaces: $x_1 x_2 + y_1 y_2 = 0$, $x_1 x_2 = y_1 y_2$, $x_1 y_2 + y_1 x_2 = 0$, $x_1 y_2 = y_1 x_2$. Among them, only the surface $x_1 y_2 = y_1 x_2$ contains all points P_1, P_2, P_3, P_4. Hence, this surface must contain C. Since C is G-invariant, we see that C is contained in the subset given by

$$\big\{ x_1 y_2 = y_1 x_2, x_1 y_3 = y_1 x_3, x_1 y_4 = y_1 x_4, x_2 y_3 = y_2 x_3,$$
$$x_2 y_4 = y_2 x_4, x_3 y_4 = y_3 x_4 \big\} \subset (\mathbb{P}^1)^4.$$

This system of equations defines the diagonal, which is not contained in X. The obtained contradiction completes the proof. □

Therefore, we see that a general Fano threefold №4.1 is K-stable by Theorem 1.11. In the remaining part of this section, we will present examples of K-stable smooth Fano threefolds in the following families: №2.1, №2.3, №2.5, №2.10, №2.14, №2.25, №3.7, №3.11. This implies that general threefolds in these families are also K-stable. In fact, we will also prove that all smooth Fano threefolds in the deformation family №2.25 are K-stable. The K-stability of a general member of the family №2.10 has been proved in [123].

Setup for the rest of the section. Let G be some finite subgroup in $\mathrm{Aut}(V_d)$ such that the curve $\mathscr{C}$ is G-invariant. Since (4.2) is G-equivariant, we can identify G with a subgroup in $\mathrm{Aut}(X)$. Since ϕ is G-equivariant, it gives a homomorphism $\upsilon\colon G \to \mathrm{Aut}(\mathbb{P}^1)$, so that we have the following exact sequence of groups:

$$1 \longrightarrow \Theta \longrightarrow G \longrightarrow \Gamma \longrightarrow 1, \tag{4.3}$$

where $\Gamma = \mathrm{im}(\upsilon)$ is a finite subgroup in $\mathrm{Aut}(\mathbb{P}^1)$, and $\Theta = \ker(\upsilon)$ is the largest subgroup in the group G such that every surface in the pencil $\mathcal{P}$ is Θ-invariant.

Example 4.21 Suppose that $d = 7$, and let $\vartheta\colon V_7 \to \mathbb{P}^3$ be the blow up of a point P. Without loss of generality, we may assume that $P = [0 : 0 : 1 : 0]$. Let Q_1 be the smooth quadric surface $\{x^2 + y^2 + zt = 0\} \subset \mathbb{P}^3$, and let Q_2 be the quadric $\{yz + t^2 = 0\} \subset \mathbb{P}^3$, where x, y, z, t are coordinates on $\mathbb{P}^3$. Set $C = Q_1 \cap Q_2$. Then C is smooth and $P \in C$. Now, we let H_1 and H_2 be the proper transforms on X of the surfaces Q_1 and Q_2, respectively, let $\mathscr{C}$ be the proper transform on V_7 of the curve C, and let $G = \mathrm{Aut}(V_7; \mathscr{C})$. Then $G \cong \mu_6$. Indeed, the group $\mathrm{Aut}(\mathbb{P}^3; C)$ contains the involution $[x : y : z : t] \mapsto [-x : y : z : t]$ and also the automorphism of order three $[x : y : z : t] \mapsto [x : y : \omega z : \omega^2 t]$, where ω is a primitive cube root of unity. Since they fix P, their actions lift to V_7, and they generate a subgroup in G isomorphic to μ_6. But G cannot be larger than μ_6 since this group acts faithfully on the curve $\mathscr{C}$ and it preserves a point in this elliptic curve. In this case, the subgroup Θ is trivial and $\Gamma \cong \mu_6$, where Θ and Γ are defined in (4.3). Arguing as in the proof of [46, Lemma 8.12], we see that $\alpha_G(X) = \frac{1}{2}$. But X is K-stable [101].

For all remaining families, we will present an example consisting of a threefold V_d, a smooth elliptic curve $\mathscr{C}$, and a finite subgroup $G \subset \mathrm{Aut}(V_d; \mathscr{C})$ such that $\alpha_G(X) > \frac{3}{4}$, so that X is K-stable by Theorem 1.48 and Corollary 1.5 because $\mathrm{Aut}(X)$ is finite [45]. To proceed, we need one very easy auxiliary result.

Lemma 4.22 *Suppose that* $\mathrm{Pic}^G(V_d) = \mathbb{Z}[H]$, *and* $\mathcal{P}$ *contains no G-invariant surfaces. Then X does not have G-fixed points. Moreover, let S be a G-irreducible surface in X such that* $-K_X \sim_{\mathbb{Q}} \lambda S + \Delta$ *for some* $\lambda \in \mathbb{Q}$ *and effective $\mathbb{Q}$-divisor Δ on X. Then* $\lambda \leqslant 1$.

Proof Since $\mathcal{P}$ contains no G-invariant surface, $\mathbb{P}^1$ has no Γ-invariant point, which implies that X has no G-invariant point.

Now let us show that $\lambda \leqslant 1$. If $S = E$, then $\Delta|_F \sim_{\mathbb{Q}} (1 - \lambda)E|_F$, which gives $\lambda \leqslant 1$. Thus, we may assume that $S \neq E$. Then $S \sim \pi^*(nH) - mE$ for some integers n and m such that $n \geqslant 1$ and $m \geqslant 0$. If $\lambda > 1$, then $n = 1$. Further, restricting S to the surface F, we see that $m \leqslant 1$ in this case.

The case $n = 1$ and $m = 1$ is impossible since $\mathcal{P}$ does not contain G-invariant surfaces. If $n = 1$ and $m = 0$, then $\Delta|_F \sim_{\mathbb{Q}} (1 - \lambda)H|_F$, which gives $\lambda \leqslant 1$. $\square$

Now we are ready to present explicit examples of K-stable smooth Fano threefolds in the families №2.1, №2.3, №2.5, №2.10, №2.14 and №3.7.

Example 4.23 Suppose that $d = 1$, and V_1 is the smooth hypersurface in $\mathbb{P}(1, 1, 1, 2, 3)$ that is given by $x_0^6 + x_1^6 + x_2^6 + x_3^3 + x_4^2 = 0$, where x_0, x_1, x_2, x_3, x_4 are coordinates of weights $1, 1, 1, 2, 3$, respectively. Suppose that H_1 and H_2 are cut out by $x_0 = 0$, $x_1 = 0$, respectively. Observe that the curve $\mathscr{C}$ is smooth, so that X is a smooth Fano threefold in the family №2.1. Let G be the subgroup in $\mathrm{Aut}(V_1)$ that is generated by two involutions:

$$[x_0 : x_1 : x_2 : x_3 : x_4] \mapsto [x_0 : -x_1 : x_2 : x_3 : x_4]$$

and

$$[x_0 : x_1 : x_2 : x_3 : x_4] \mapsto [x_1 : x_0 : x_2 : x_3 : x_4].$$

Then $G \cong \mu_2^2$, the curve $\mathscr{C}$ is G-invariant, $\mathcal{P}$ does not contain G-invariant surfaces, and it follows from Lemma A.40 that the α-invariant of a general fiber of ϕ is at least $\frac{5}{6}$. Therefore, applying Lemma 4.22 and Corollary 1.56, we conclude that $\alpha_G(X) \geqslant \frac{5}{6}$.

Example 4.24 Suppose that $d = 2$. Let V_2 be the hypersurface in $\mathbb{P}(1, 1, 1, 1, 2)$ given by the equation $x_0^4 + x_1^4 + x_2^4 + x_3^4 + x_4^2 = 0$, where x_0, x_1, x_2, x_3 are coordinates of weight 1, and x_4 is a coordinate of weight 2. Suppose that $H_1 = \{x_0 = 0\}$ and $H_2 = \{x_1 = 0\}$. Then the curve $\mathscr{C}$ is smooth, so that X is a smooth Fano threefold in the family №2.3. Now, let G be the subgroup of the group $\mathrm{Aut}(V_2)$ such that $G \cong \mu_2 \times (\mu_4^3 \rtimes \mu_2)$, where the generator of the ith factor of μ_4^3 acts by multiplying the coordinate x_i by $\sqrt{-1}$, the generator of the non-normal subgroup $\mu_2 \subset \mu_4^3 \rtimes \mu_2$ acts by

$$[x_0 : x_1 : x_2 : x_3 : x_4] \mapsto [x_1 : x_0 : x_2 : x_3 : x_4],$$

and the generator of the factor μ_2 acts as $[x_0 : x_1 : x_2 : x_3 : x_4] \mapsto [x_0 : x_1 : x_2 : x_3 : -x_4]$. Then $\mathscr{C}$ is G-invariant, $\Gamma \cong D_8$, $\Theta \cong \mu_2 \times \mu_4^2$, and $\mathbb{P}^1$ does not contain Γ-fixed points. Further, X does not contain G-invariant rational curves. Indeed, let C be such a curve. Since V_2 does not contain G-invariant points, $\pi(C)$ is a rational curve. Since the largest quotients of μ_4^3 that admit a faithful action on $\mathbb{P}^1$ are μ_4 and μ_2^2, the curve $\pi(C)$ must have a trivial action of some non-cyclic subgroup in $\mu_4^3 \subset G$, which is impossible since the fixed points in V_2 of every non-cyclic subgroup of μ_4^3 are isolated. The obtained contradiction shows that the smooth Fano threefold X does not contain G-invariant rational curves. Now, applying Corollary 1.55 and Lemma 4.22, we see that $\alpha_G(X) \geqslant 1$.

Example 4.25 Now, suppose that $d = 3$. Let $V_3 = \{x_0^3 + x_1^3 + x_2^3 + x_3^3 + x_4^3 = 0\} \subset \mathbb{P}^4$, where x_0, x_1, x_2, x_3 and x_4 are coordinates on $\mathbb{P}^4$. Let $H_1 = \{x_0 = 0\}$ and $H_2 = \{x_1 = 0\}$. Then $\mathscr{C}$ is smooth, so that X is a smooth Fano threefold №2.5. Let G be the subgroup defined by $G = \mu_3^4 \rtimes \mu_2$, where the generator of the ith factor of μ_3^4 acts by multiplying x_i by a primitive cube root of unity, while μ_2 acts by

$$[x_0 : x_1 : x_2 : x_3 : x_4] \mapsto [x_1 : x_0 : x_2 : x_3 : x_4].$$

Then $\mathscr{C}$ is G-invariant, and $\mathcal{P}$ does not contain G-invariant surfaces. Then $\alpha_G(X) \geqslant 1$. Indeed, if $\alpha_G(X) < 1$, then Theorem 1.52 and Lemma 4.22 implies that X contains a G-invariant curve C such that $\widetilde{H}_1 \cdot C = 1$, so that $H_1 \cap \pi(C)$ is a point that is fixed by the subgroup $\mu_3^4 \subset G$, which is impossible since this subgroup has no fixed points in V_3.

Example 4.26 Suppose that $d = 4$, and that V_4 is the complete intersection of two smooth quadric hypersurfaces in $\mathbb{P}^5$ given by

$$\begin{cases} x_0^2 + x_1^2 + x_2^2 + x_3^2 + x_4^2 + x_5^2 = 0, \\ x_0^2 - x_1^2 + 2x_2^2 - 2x_3^2 + 3x_4^2 - 3x_5^2 = 0, \end{cases}$$

where x_0, x_1, x_2, x_3, x_4 and x_5 are coordinates on $\mathbb{P}^5$. Suppose that H_1 and H_2 are cut out by the equations $x_0 = 0$ and $x_1 = 0$, respectively. Then $\mathscr{C}$ is a smooth elliptic curve, and X is a smooth Fano threefold №2.10. Let G be a subgroup such that $G = \mu_2^5 \rtimes \mu_2$, where the generator of the ith factor of μ_2^5 acts by changing the sign of the coordinate x_i, while the generator of the non-normal subgroup μ_2 acts by

$$[x_0 : x_1 : x_2 : x_3 : x_4 : x_5] \mapsto [x_1 : x_0 : x_3 : x_2 : x_5 : x_4].$$

Then $\mathscr{C}$ is G-invariant, $\mathcal{P}$ has no G-invariant surfaces, and the subgroup $\mu_2^5 \subset G$ does not have fixed points in V_4. Thus, arguing as in Example 4.25, we see that $\alpha_G(X) \geqslant 1$.

Example 4.27 Suppose that $d = 5$ and V_5 is the unique smooth Fano threefold №1.15. Then $\mathrm{Aut}(V_5) \cong \mathrm{PGL}_2(\mathbb{C})$, see [53, Proposition 7.1.10]. Fix a subgroup $\mathfrak{A}_5 \subset \mathrm{Aut}(V_5)$, and let G be its subgroup such that $G \cong \mathbb{D}_{10}$. Then the actions of these groups lift to their linear action on $H^0(O_{V_5}(H))$. By [53, Lemma 7.1.6], we have $H^0(O_{V_5}(H)) \cong W_3 \oplus W_4$, where W_3 and W_4 are irreducible $\mathfrak{A}_5$-representations of dimensions 3 and 4, respectively. As G-representation, the representations W_3 and W_4 split as follows:

- W_3 is a sum of one-dimensional and irreducible two-dimensional representations;
- W_4 is a sum of two (different) irreducible two-dimensional representations.

Let us denote by $\mathcal{M}$ the two-dimensional linear subsystem in $|H|$ that corresponds to W_3. By [53, Lemma 7.5.8], its base locus is a $\mathfrak{A}_5$-orbit of length 5, which we denote by Σ_5. By [53, Lemma 7.3.4], this orbit is the unique $\mathfrak{A}_5$-orbit in V_5 consisting of at most 5 points. Without loss of generality, we may assume that H is the unique G-invariant surface in $\mathcal{M}$. Let $\mathcal{P}$ be the pencil in $\mathcal{M}$ that is given by the two-dimensional G-subrepresentation in W_3, let H_1 and H_2 be two distinct surfaces in $\mathcal{P}$, and let $\mathscr{C} = H_1 \cap H_2$. Then $H \cap \mathscr{C} = \Sigma_5$, so that $\mathscr{C}$ is reduced. We claim that it is smooth. Indeed, suppose that $\mathscr{C}$ is not smooth. Then it is reducible because otherwise it would have one singular point, but V_5 does not have G-fixed points. Since G acts transitively on Σ_5, we conclude that $\mathscr{C}$ is G-irreducible. Then $\mathscr{C}$ is a union of 5 lines, which are disjoint away from Σ_5 by [53, Corollary 9.1.10], so that $\mathscr{C}$ is not connected, which is absurd since it is an intersection of two ample divisors. Therefore, we conclude that $\mathscr{C}$ is smooth, so that X is a smooth Fano threefold №2.14. Using Corollary 1.57 and Lemma 4.22, we get $\alpha_G(X) \geqslant \frac{4}{5}$.

Example 4.28 Suppose that $d = 6$, and that $V_6 = \{x_0y_0 + x_1y_1 + x_2y_2 = 0\} \subset \mathbb{P}^2 \times \mathbb{P}^2$, where $[x_0 : x_1 : x_2]$ and $[y_0 : y_1 : y_2]$ are homogeneous coordinates on the first and the second factors of $\mathbb{P}^2 \times \mathbb{P}^2$, respectively. Suppose also that H_1 and H_2 are given by

$$x_0y_1 + \omega x_1y_2 + \omega^2 x_2y_0 = 0,$$
$$x_0y_2 + \omega x_1y_0 + \omega^2 x_2y_1 = 0,$$

respectively, where ω is a non-trivial cube root of unity. One can check that $\mathscr{C}$ is smooth. Then X is a smooth Fano threefold №3.7. Let G be a subgroup such that $G \cong \mu_3^2 \rtimes \mu_2$, the generator of the first factor μ_3 acts by

$$([x_0 : x_1 : x_2], [y_0 : y_1 : y_2]) \mapsto ([x_2 : x_0 : x_1], [y_2 : y_0 : y_1]),$$

the generator of the second factor μ_3 acts by

$$([x_0 : x_1 : x_2], [y_0 : y_1 : y_2]) \mapsto ([x_0 : \omega x_1 : \omega^2 x_2], [y_0 : \omega^2 y_1 : \omega y_2]),$$

and the generator of μ_2 acts by $([x_0 : x_1 : x_2], [y_0 : y_1 : y_2]) \mapsto ([y_0 : y_1 : y_2], [x_0 : x_1 : x_2])$. Then V_6 and $\mathscr{C}$ are G-invariant. We claim that

(i) $\mathbb{P}^2 \times \mathbb{P}^2$ does not have μ_3^2-invariant points,
(ii) $\mathbb{P}^2 \times \mathbb{P}^2$ does not contain $\mu_3^2 \rtimes \mu_2$-invariant rational curves.

Indeed, let $\pi_1 \colon V_6 \to \mathbb{P}^2$ and $\pi_2 \colon V_6 \to \mathbb{P}^2$ be the projections to the first and the second factors of $\mathbb{P}^2 \times \mathbb{P}^2$, respectively. Then π_1 and π_2 are μ_3^2-equivariant. Observe that

(i) the action of μ_3^2 on $\mathbb{P}^2$ has no fixed points,
(ii) no rational curve in $\mathbb{P}^2$ is μ_3^2-invariant since $\mathbb{P}^1$ admits no faithful μ_3^2-action.

Thus, if a point $P \in V_6$ is fixed by μ_3^2, then $\pi_1(P)$ is fixed by μ_3^2, which is impossible. Likewise, if C is a μ_3^2-invariant rational curve in V_6, then $\pi_1(C)$ or $\pi_2(C)$ is a μ_3^2-invariant rational curve, which is impossible. Then $\alpha_G(X) \geqslant 1$ by Lemma 4.22 and Theorem 1.52.

Now let us show that all smooth Fano threefolds in the family №2.25 are K-stable. From now on and until the end of this section, we assume that $d = 8$. Recall that $V_8 = \mathbb{P}^3$, and $\pi \colon X \to V_d$ is the blow up of a smooth elliptic curve curve $\mathscr{C}$, which is an intersection of two quadric surfaces H_1 and H_2. Note that Lemma 4.22 is not applicable in this case. Because of this, we need the following similar but more specific result.

Lemma 4.29 *Suppose that G is a finite group of order 2^r such that $r \geqslant 2$ and $\Gamma \cong \mu_2^2$. Then the following assertions hold:*

(i) *X contains no G-invariant points,*
(ii) *the pencil $\mathcal{P}$ contains no G-invariant surface,*
(iii) *$\mathbb{P}^3$ contains neither G-invariant points nor G-invariant planes,*
(iv) *X contains no G-invariant irreducible curve C such that $F \cdot C \leqslant 1$,*
(v) *X contains no G-invariant irreducible normal surface S such that $-K_X \sim_{\mathbb{Q}} \lambda S + \Delta$ for some rational number $\lambda > 1$ and effective $\mathbb{Q}$-divisor Δ on the threefold X.*

Proof Observe that Γ has no fixed points in $\mathbb{P}^1$, so that $\mathcal{P}$ contains no G-invariant surfaces, and X has no G-invariant points. This proves (i) and (ii).

Since G is not cyclic, the curve $\mathscr{C}$ does not have G-invariant points, so that $\mathbb{P}^3$ does not have G-invariant points by (i). Hence, the G-action on $\mathbb{P}^3$ is given by a four-dimensional representation of a central extension of the group G

that does not have one-dimensional subrepresentations, which implies that this representation does not have three-dimensional subrepresentations, so that $\mathbb{P}^3$ contains no G-invariant planes. This proves (iii).

To prove (iv), suppose that $F \cdot C \leqslant 1$ for some G-invariant irreducible curve $C \subset X$. Then $F \cdot C = 1$ because $\mathbb{P}^1$ contains no G-invariant points. Then $C \cong \mathbb{P}^1$ and

$$1 = F \cdot C = \left(\pi^*(H) - E\right) \cdot C = \pi^*(H) \cdot C - E \cdot C,$$

so that $E \cdot C$ is odd, because $H = O_{\mathbb{P}^3}(2)$. But C does not contain G-orbits of odd length because $|G| = 2^r$. Then $C \subset E$, so that $\pi(C) = \mathscr{C}$ since $\mathbb{P}^3$ has no G-invariant points. But $\pi(C) \neq \mathscr{C}$ because $\mathscr{C}$ is not rational. The obtained contradiction proves (iv).

Finally, to prove (v), we suppose that the threefold X contains a G-invariant irreducible normal surface S such that $2F + E \sim -K_X \sim_{\mathbb{Q}} \lambda S + \Delta$ for some rational number $\lambda > 1$ and effective $\mathbb{Q}$-divisor Δ on the threefold X. Then

$$\frac{2}{\lambda}F + \frac{1}{\lambda}E - \frac{1}{\lambda}\Delta \sim_{\mathbb{Q}} S \sim_{\mathbb{Q}} aF + bE$$

for some non-negative rational numbers a and b since F and E generates the cone $\mathrm{Eff}(X)$. Since $\lambda > 1$, we have $a < 2$ and $b < 1$. On the other hand, we have $2a \in \mathbb{Z}$ and $b - a \in \mathbb{Z}$ since $\mathrm{Pic}(X)$ is generated by $\pi^*(O_{\mathbb{P}^3}(1))$ and E. It follows that $2a, 2b \in \mathbb{Z}$ and we have the following possibilities: $(2a, 2b) \in \{(1,1), (2,0), (3,1)\}$. If $(2a, 2b) = (1,1)$, then $\pi(S)$ is a G-invariant plane in $V_8 = \mathbb{P}^3$, which is impossible by (iii). Similarly, if $(a, b) = (2, 0)$, then S is a G-invariant surface in $\mathcal{P}$, which contradicts (ii). Thus, we get $(2a, 2b) = (3, 1)$, so that $\pi(S)$ is a G-invariant cubic surface in $\mathbb{P}^3$ that contains $\mathscr{C}$.

Let $S_3 = \pi(S)$. Then S_3 has isolated singularities because S is normal by assumption, and S_3 is not singular along the curve $\mathscr{C}$ since $2b = 1$. Note that G acts faithfully on S_3, and this action lifts to the linear action of the group G on $H^0(O_{S_3}(-K_{S_3})) \cong H^0(O_{\mathbb{P}^3}(1))$. Then G is not abelian since $\mathbb{P}^3$ has no G-fixed points. Thus, we have $r \geqslant 3$.

Suppose that S_3 is smooth. Then, looking on the list of automorphism groups of smooth cubic surfaces [77, Table 4], we see that $|G| = 8$. Now, looking on the list of automorphism groups of smooth cubic surfaces again, we conclude that G must have a fixed point in $\mathbb{P}^3$, which is a contradiction. Thus, we conclude that S_3 is singular.

Singular cubic surfaces have been classified in [25]. Note that $|\mathrm{Sing}(S_3)| \leqslant 4$. Further, if S_3 has 4 singular points, then we have $\mathrm{Aut}(S_3) \cong \mathfrak{S}_4$, and S_3 can be given by

$$x_0 x_1 x_2 + x_0 x_1 x_3 + x_0 x_2 x_3 + x_1 x_2 x_3 = 0.$$

In this case, $\mathbb{P}^3$ has a G-fixed point, which contradicts (iii). Similarly, we see that S_3 cannot have three singular points because $\mathbb{P}^3$ does not have G-invariant points, and $\mathbb{P}^3$ does not have G-orbits of length 3. Thus, we conclude that S_3 has two singular points.

Let L be the line in $\mathbb{P}^3$ such that L contains both singular points of the surface S_3. Then $L \subset S_3$, and there is a unique plane $\Pi \subset \mathbb{P}^3$ that is tangent to S_3 along the line L. Since L is G-invariant, Π is G-invariant, which contradicts (iii). This proves (v). □

Applying Theorem 1.52 and Lemma 4.29, we get

Corollary 4.30 *Suppose that* $\Theta \cong \mu_2^3$ *and* $\Gamma \cong \mu_2^2$. *Then* $\alpha_G(X) \geqslant 1$.

To apply this corollary, take $\lambda \in \mathbb{C}^*$ such that $\lambda^4 \neq 1$. Suppose that H_1 is given by

$$x_0^2 + x_1^2 + \lambda(x_2^2 + x_3^2) = 0,$$

and suppose that H_2 is given by

$$\lambda(x_0^2 - x_1^2) + x_2^2 - x_3^2 = 0.$$

Then $\mathscr{C}$ is a smooth quartic elliptic curve, so that X is smooth Fano threefold №2.25. Moreover, every smooth Fano threefold №2.25 can be obtained in this way [83].

Recall that every surface in the pencil $\mathcal{P}$ is Θ-invariant. Using this, one can show that the group Θ is contained in the subgroup in $\mathrm{Aut}(\mathbb{P}^3)$ that is generated by

$$[x_0 : x_1 : x_2 : x_3] \mapsto \left[x_0 : (-1)^a x_1 : (-1)^b x_2 : (-1)^c x_3\right] \quad (4.4)$$

for all a, b, c in $\{0, 1\}$. Note that these automorphisms generate a group isomorphic to μ_2^3.

Lemma 4.31 *There exists a subgroup* $G \subset \mathrm{Aut}(\mathbb{P}^3; \mathscr{C})$ *such that* $\Theta \cong \mu_2^3$ *and* $\Gamma \cong \mu_2^2$.

Proof Let Σ be the subset in $\mathscr{C}$ consisting of the 16 points of the intersection of this curve with the tetrahedron $x_0 x_1 x_2 x_3 = 0$. Fix a point $O \in \Sigma$, and equip $\mathscr{C}$ with the group law such that O is the identity element. As it was noticed in [86, §2], the embedding $\mathscr{C} \hookrightarrow \mathbb{P}^3$ is given by the linear system $|4O|$, and $\Sigma \setminus O$ consists of all points of order 4.

Let G be the subgroup in $\mathrm{Aut}(\mathscr{C})$ generated by the translation by points in Σ and the involution $P \mapsto -P$. Then $|G| = 32$, and the embedding $\mathscr{C} \hookrightarrow \mathbb{P}^3$ is G-equivariant, so that we can identify G with a subgroup in $\mathrm{Aut}(\mathbb{P}^3; \mathscr{C})$.

We claim that G is the required group. Indeed, since Θ contains no elements of order 4, the group Γ is one the following groups: μ_2, μ_2^2, or μ_4. Using this, we see that $\Gamma \cong \mu_2 \times \mu_2$, and Θ is generated by translations by elements of order 2 and the involution $P \mapsto -P$. Then $\Theta \cong \mu_2^3$ as required. $\qquad \square$

Corollary 4.32 *All smooth Fano threefolds №2.25 are K-stable.*

One can describe the group constructed in the proof of Lemma 4.31 in coordinates. Namely, let ι be the involution in $\mathrm{Aut}(\mathbb{P}^3)$ given by

$$[x_0 : x_1 : x_2 : x_3] \mapsto [x_1 : x_0 : x_3 : x_2], \qquad (4.5)$$

and let τ be the automorphism of order 4 in $\mathrm{Aut}(\mathbb{P}^3)$ that is given by

$$[x_0 : x_1 : x_2 : x_3] \mapsto [x_2 : ix_3 : x_0 : ix_1], \qquad (4.6)$$

where $i = \sqrt{-1}$. Then $\mathscr{C}$ is ι-invariant and τ-invariant. Then the group constructed in the proof of Lemma 4.31 is the group generated by ι, τ and all automorphisms (4.4).

4.4 Blow up of $\mathbb{P}^3$ in curves lying on quadric surfaces

Let S_2 be a smooth quadric surface in $\mathbb{P}^3$, let $\mathscr{C}$ be a smooth curve in $S_2 \cong \mathbb{P}^1 \times \mathbb{P}^1$ of degree (a, b) with $a \leqslant b$, and let $\pi \colon X \to \mathbb{P}^3$ be the blow up of the curve $\mathscr{C}$. Then $\mathscr{C}$ has degree $a + b$ and genus $(a-1)(b-1)$, and X is a Fano threefold if and only if $b \leqslant 3$ by [17, Proposition 3.1]. This gives us the following possibilities:

- $(a, b) = (3, 3)$, and X is a smooth Fano threefold №2.15,
- $(a, b) = (2, 3)$, and X is a smooth Fano threefold №2.19,
- $(a, b) = (1, 3)$, and X is a smooth Fano threefold №2.22,
- $(a, b) = (2, 2)$, and X is a smooth Fano threefold №2.25,
- $(a, b) = (1, 2)$, and X is a smooth Fano threefold №2.27,
- $(a, b) = (1, 1)$, and X is a smooth Fano threefold №2.30,
- $(a, b) = (0, 1)$, and X is a smooth Fano threefold №2.33.

Both smooth Fano threefolds №2.30 and №2.33 are K-unstable (see Sections 3.3, 3.6, 3.7). In Section 4.2, we proved that the unique smooth Fano threefold №2.27 is K-polystable. In Section 4.3, we proved that all smooth Fano threefolds in the family №2.25 are K-stable. The goal of this section is to prove the following result:

Proposition 4.33 *A general Fano threefold in the deformation families №2.15, №2.19 and №2.22 is K-stable.*

Thus, we assume that one of the following three cases holds:

№2.15	$\mathscr{C}$ is a curve of degree $(3,3)$, and its genus is 4,
№2.19	$\mathscr{C}$ is a curve of degree $(2,3)$, it is hyperelliptic, and its genus is 2,
№2.22	$\mathscr{C}$ is a curve of degree $(1,3)$, and it is a smooth rational quartic curve.

In the first two cases, the group $\mathrm{Aut}(X)$ is finite [45]. In the third case, the automorphism group $\mathrm{Aut}(X)$ is also finite with a single exception, which is described in

Example 4.34 Let S_2 be the smooth quadric surface in $\mathbb{P}^3$ that is given by $x_0 x_3 = x_1 x_2$. Fix the isomorphism $S_2 \cong \mathbb{P}^1 \times \mathbb{P}^1$ that is given by

$$\left([s_0 : s_1], [t_0 : t_1] \right) \mapsto [s_0 t_0 : s_0 t_1 : s_1 t_0 : s_1 t_1].$$

Let $\mathscr{C}$ be the curve of degree $(1,3)$ in S_2 given by $s_0^3 t_0 = s_1^3 t_1$. Then its image in $\mathbb{P}^3$ is the rational quartic curve given by $[s_0 : s_1] \mapsto [s_0 s_1^3 : s_0^4 : s_1^4 : s_1 s_0^3]$, and X is a smooth Fano threefold in the family №2.22. Let G be the subgroup in $\mathrm{Aut}(\mathbb{P}^3)$ that is generated by the involution $[x_0 : x_1 : x_2 : x_3] \mapsto [x_3 : x_2 : x_1 : x_0]$, and automorphisms

$$\left[x_0 : x_1 : x_2 : x_3 \right] \mapsto \left[\lambda^3 x_0 : x_1 : \lambda^4 x_2 : \lambda x_3 \right],$$

where $\lambda \in \mathbb{G}_m$. Then $G \cong \mathbb{G}_m \rtimes \boldsymbol{\mu}_2$, and the curve $\mathscr{C}$ is G-invariant. Thus, the action of the group G lifts to the threefold X. Then X is the unique smooth Fano threefold №2.22 that has an infinite automorphism group [45].

Denote by Q the proper transform on X of the quadric S_2. As shown in [17], there exists the commutative diagram

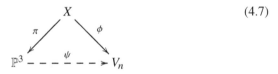

$$(4.7)$$

where ϕ is a contraction of the surface Q, ψ is a rational map given by the system of all cubic surfaces that contain $\mathscr{C}$, and V_n is a del Pezzo threefold in $\mathbb{P}^{n+1}$ of degree n such that we have the following three possibilities:

№2.15	$n = 3$, V_3 is a singular cubic threefold in $\mathbb{P}^4$ that has one ordinary double point, and ϕ is a blow up of this point,
№2.19	$n = 4$, V_4 is a smooth complete intersection of two quadric hypersurfaces in $\mathbb{P}^5$, and ϕ is the blow up of a line,
№2.22	$n = 5$, V_5 is described in Example 3.2, and ϕ is a blow up of a smooth conic.

Let G be a subgroup in $\mathrm{Aut}(\mathbb{P}^3, \mathscr{C})$. Then the diagram (4.7) is G-equivariant, so that we can also identify G with a subgroup of $\mathrm{Aut}(X)$. Let E be the π-exceptional surface. Then $-K_X \sim 2Q + E$, and both Q and E are G-invariant, so that $\alpha_G(X) \leqslant \frac{1}{2}$.

Let us prove that each of the three families №2.15, №2.19 and №2.22 contains a special threefold that is K-stable, so that Proposition 4.33 would follow from Theorem 1.11. To describe these special Fano threefolds, we have to specify the curve $\mathscr{C}$ and the group G. Let us do this in the next three examples.

Example 4.35 Let G be the symmetric group $\mathfrak{S}_5$, consider the G-action on $\mathbb{P}^4$ that permutes the coordinates $x_0, \ldots, x_4$, and identify $\mathbb{P}^3$ with the G-invariant hyperplane in $\mathbb{P}^4$. Then $\mathbb{P}^3 = \mathbb{P}(V)$, where V is the irreducible 4-dimensional representation of the group G, so that $\mathbb{P}^3$ does not contain G-fixed points, G-invariant lines and also G-invariant planes. Note that the same assertion holds for the alternating subgroup $\mathfrak{A}_5$ of the group G. Let S_2 be the smooth quadric surface in $\mathbb{P}^3$ that is given by its intersection with

$$x_0^2 + x_1^2 + x_2^2 + x_3^2 + x_4^2 = 0,$$

let S_3 be the cubic surface in $\mathbb{P}^3$ given by its intersection with $x_0^3 + x_1^3 + x_2^3 + x_3^3 + x_4^3 = 0$, and let $\mathscr{C} = S_2 \cap S_3$. Then $\mathscr{C}$ is a smooth curve of genus 4 and degree 6, which is canonically embedded in $\mathbb{P}^3$. Clearly, both S_2 and S_3 are G-invariant, so that $\mathscr{C}$ is also G-invariant. The curve $\mathscr{C}$ is known as the *Bring's curve*. It is the unique smooth curve of genus 4 that admits a faithful action of the group $\mathfrak{S}_5$. Therefore, the threefold X is the unique smooth Fano threefold in the family №2.15 that admits a faithful action of the group $\mathfrak{S}_5$.

Example 4.36 Recall the isomorphism $S_2 \cong \mathbb{P}^1 \times \mathbb{P}^1$ from Example 4.34. Let $\mathscr{C} \subset S_2$ be the curve of degree $(2,3)$ that is given by

$$(s_0^2 + s_1^2)(t_0^3 + t_1^3) + \varepsilon(s_0^2 - s_1^2)(t_0^3 - t_1^3) = 0,$$

where ε is a general number. Then $\mathscr{C}$ is smooth. In particular, it is smooth for $\varepsilon = 5$. Let $\tau \colon S_2 \to S_2$ be the involution that is given by

$$([s_0 : s_1], [t_0 : t_1]) \mapsto ([s_0 : -s_1], [t_0 : t_1]),$$

let $\iota \colon S_2 \to S_2$ be the involution that is given by

$$([s_0 : s_1], [t_0 : t_1]) \mapsto ([s_1 : s_0], [t_1 : t_0]),$$

and let $\gamma \colon S_2 \to S_2$ be the automorphism of order 3 that is given by

$$([s_0 : s_1], [t_0 : t_1]) \mapsto ([s_0 : s_1], [t_0 : \omega t_1]),$$

where ω is a primitive cube root of unity. Let $G = \langle \tau, \iota, \gamma \rangle \subset \mathrm{Aut}(S_2)$. Then $G \cong D_{12}$, and the curve $\mathscr{C}$ is G-invariant. Observe that the G-action extends to $\mathbb{P}^3$ as follows:

$$\tau\big([x_0 : x_1 : x_2 : x_3]\big) = [-x_0 : -x_1 : x_2 : x_3],$$
$$\iota\big([x_0 : x_1 : x_2 : x_3]\big) = [x_3 : x_2 : x_1 : x_0],$$
$$\gamma\big([x_0 : x_1 : x_2 : x_3]\big) = [x_0 : \omega x_1 : x_2 : \omega x_3].$$

Then $\mathbb{P}^3 = \mathbb{P}(\mathbb{V})$, where $\mathbb{V}$ is a four-dimensional representation of the group G which splits as a direct sum of two non-isomorphic irreducible two-dimensional representations. In particular, we conclude that $\mathbb{P}^3$ does not contain G-fixed points and G-invariant planes. Moreover, S_2 contains no G-invariant curves of degree $(1,0)$, $(0,1)$, $(1,1)$, $(1,2)$ and $(2,1)$. The threefold X is a smooth Fano threefold in the family №2.19.

Example 4.37 As in Example 4.34, identify the quadric S_2 with $\mathbb{P}^1 \times \mathbb{P}^1$. Let $G = \mathfrak{A}_4$. Fix a faithful G-action on $\mathbb{P}^1$, and consider the corresponding diagonal G-action on S_2. This action extends to $\mathbb{P}^3$ such that $\mathbb{P}^3 = \mathbb{P}(\mathbb{V})$, where $\mathbb{V}$ is the reducible four-dimensional permutation representation of the group G. Then $\mathbb{P}^3$ does not contain G-invariant lines, $\mathbb{P}^3$ contains one G-fixed point and one G-invariant plane, the G-fixed point in $\mathbb{P}^3$ is not contained in S_2, and the G-invariant plane in $\mathbb{P}^3$ intersects S_2 by the diagonal curve Δ. Let $\mathscr{C}$ be a smooth G-invariant curve in S_2 of degree $(1,3)$, which exists by Lemma A.53. Then X is a smooth Fano threefold №2.22 on which the group $G = \mathfrak{A}_4$ acts faithfully. Moreover, arguing as in the proof of [45, Lemma 6.13], we see that $\mathrm{Aut}(X) \cong \mathrm{Aut}(S_2, \mathscr{C})$. On the other hand, the group $\mathrm{Aut}(S_2, \mathscr{C})$ is finite by Lemma A.53.

In the remaining part of this section, we will assume that $\mathscr{C}$ is one of the curves described in Examples 4.34, 4.35, 4.36, 4.37, so that X is a smooth Fano threefold in the families №2.22, №2.15, №2.19, №2.22, and G is one of the groups $\mathbb{G}_m \rtimes \mu_2$, $\mathfrak{S}_5$, D_{12}, $\mathfrak{A}_4$, respectively. We will refer to these cases as $(2.22.D_\infty)$, $(2.15.\mathfrak{S}_5)$, $(2.19.D_{12})$, $(2.22.\mathfrak{A}_4)$, respectively. In the remaining part of this section, we will prove that X is K-polystable in each case, so that X is K-stable in the cases $(2.15.\mathfrak{S}_5)$, $(2.19.D_{12})$ and $(2.22.\mathfrak{A}_4)$ by Corollary 1.5. This would imply Proposition 4.33 by Theorem 1.11.

Lemma 4.38 *The following assertions hold:*

(i) *If we are in the case* $(2.15.\mathfrak{S}_5)$, *then* $\mathbb{P}^3$ *does not contain G-fixed points,* $\mathbb{P}^3$ *does not contain G-invariant planes,* S_2 *is the only G-invariant quadric in* $\mathbb{P}^3$, *and* $\mathbb{P}^3$ *does not contain G-invariant irreducible rational curves.*

(ii) *If we are in the case* (2.19.D_{12}), *then* $\mathbb{P}^3$ *does not contain G-fixed points,* $\mathbb{P}^3$ *does not contain G-invariant planes,* $\mathbb{P}^3$ *does not contain G-invariant conics and cubics, the only G-invariant lines in* $\mathbb{P}^3$ *are the lines* $x_0 = x_3 = 0$ *and* $x_1 = x_2 = 0$, *which are not contained in* S_2 *and do not intersect* $\mathscr{C}$.

(iii) *If we are in the case* (2.22.$\mathfrak{A}_4$), *then* $\mathbb{P}^3$ *contains a unique G-fixed point, which is not contained in* S_2, $\mathbb{P}^3$ *contains a unique G-invariant plane, which intersects* S_2 *by the diagonal* Δ, *and* $\mathbb{P}^3$ *does not contain G-invariant lines.*

(iv) *If we are in the case* (2.22.D_∞), *then* $\mathbb{P}^3$ *does not contain G-fixed points,* $\mathbb{P}^3$ *does not contain G-invariant planes, and the only G-invariant lines in* $\mathbb{P}^3$ *are the lines* $\{x_0 = x_3 = 0\}$ *and* $\{x_1 = x_2 = 0\}$. *Moreover,*

$$\{x_0 = x_3 = 0\} \cap S_2 = \{x_0 = x_3 = 0\} \cap \mathscr{C} = [0:1:0:0] \cup [0:0:1:0],$$

but

$$\{x_1 = x_2 = 0\} \cap S_2 = [1:0:0:0] \cup [0:0:0:1]$$

and

$$\{x_1 = x_2 = 0\} \cap \mathscr{C} = \varnothing.$$

Proof The assertions (i), (ii) and (iii) immediately follow from Example 4.35. To prove assertion (iv), observe that $\mathfrak{S}_5$ cannot faithfully act on a rational curve because $\mathrm{PGL}_2(\mathbb{P}^1)$ does not contain a subgroup isomorphic to $\mathfrak{S}_5$. On the other hand, the group G acts faithfully on any irreducible G-invariant curve in $\mathbb{P}^3$ in the case (2.15.$\mathfrak{S}_5$), because no such curve can be contained in a hyperplane since $\mathbb{P}^3 = \mathbb{P}(V)$ for the standard irreducible four-dimensional representation V of the group G. Thus, we see that $\mathbb{P}^3$ does not contain G-invariant irreducible rational curves.

Assertions (iii) and (iv) easily follow from Examples 4.37 and 4.34, respectively. So we leave their proofs to the reader. Let us only prove assertion (ii).

Suppose that we are in the case (2.19.D_{12}). Then, as we mentioned in Example 4.36, the G-action on $\mathbb{P}^3$ lifts to its linear action on $H^0(\mathbb{P}^3, O_{\mathbb{P}^3}(1))$, which splits as a sum of two irreducible two-dimensional representations of the group G. In particular, the projective space $\mathbb{P}^3$ does not contain G-fixed points and G-invariant planes, so that it does not contain G-invariant conics and G-invariant plane cubic curves.

Observe that $H^0(\mathbb{P}^3, O_{\mathbb{P}^3}(1))$ splits as a direct sum of two non-isomorphic two-dimensional irreducible G-representations. Thus $\mathbb{P}^3$ contains exactly two G-invariant lines. One can check that the lines $x_0 = x_3 = 0$ and $x_1 = x_2 = 0$ are indeed G-invariant, so that these are the only G-invariant lines in $\mathbb{P}^3$. They are not contained in S_2 because its defining equation is $x_0 x_3 = x_1 x_2$. In fact, the line $x_0 = x_3 = 0$ intersects $S_2 = \mathbb{P}^1 \times \mathbb{P}^1$ transversally by the points $([0:1],[1:0])$ and

([1 : 0], [0 : 1]), and the line $x_1 = x_2 = 0$ intersects $S_2 = \mathbb{P}^1 \times \mathbb{P}^1$ transversally by the points ([0 : 1], [0 : 1]) and ([1 : 0], [1 : 0]). But none of these four points is contained in the curve $\mathscr{C}$.

Finally, let us show that $\mathbb{P}^3$ does not contain G-invariant twisted cubic curves. Suppose that $\mathbb{P}^3$ contains a G-invariant twisted cubic curve C_3. Then the G-action on C_3 is faithful and $C_3 \cong \mathbb{P}^1$. Let G' be the subgroup in G generated by ι and γ. Then $G' \cong \mathfrak{S}_3$, so that $\mathbb{P}^1$ must contain G'-orbit of length 3, which is not contained in one line since C_3 is an intersection of quadrics. Thus, $\mathbb{P}^3$ contains a G'-invariant plane, which is impossible since $H^0(\mathbb{P}^3, O_{\mathbb{P}^3}(1))$ splits as a sum of two isomorphic two-dimensional irreducible representations of G'. This shows that $\mathbb{P}^3$ contains no G-invariant twisted cubics. □

Corollary 4.39 *If X contains a G-fixed point, then we are in the case* (2.22.$\mathfrak{A}_4$)*, such point is unique, and it is not contained in the surface Q.*

Corollary 4.40 *Suppose that we are in the case* (2.19.D_{12})*. Then V_4 contains neither G-fixed points nor G-invariant hyperplane sections.*

Proof The threefold V_4 does not have G-fixed points away from $\phi(Q)$ because X does not have G-fixed points. Moreover, the conic $\phi(Q)$ does not contain G-fixed points either, since curves contracted by $\phi|_Q : Q \to \phi(Q)$ are mapped to lines in S_2. By Lemma 4.38, none of such lines are G-invariant, so that V_4 does not contain G-fixed points.

To prove the final assertion, recall that ψ in (4.7) is given by the linear system of cubic surfaces that pass through $\mathscr{C}$. Thus, if there exists a G-invariant hyperplane section of the threefold V_4, then there exists a G-invariant surface S_3 in $\mathbb{P}^3$ that contains the curve $\mathscr{C}$. If $S_3 = S_2 + H$ for some hyperplane H in $\mathbb{P}^3$, then H is G-invariant, which contradicts Lemma 4.38. Hence, $S_2 \not\subset S_3$ and $S_3|_{S_2}$ is a curve of degree $(3, 3)$ that contains $\mathscr{C}$, which implies that $S_3|_{S_2} = \mathscr{C} + \ell$ for a G-invariant line ℓ. This contradicts Lemma 4.38. □

Now, we are ready to give a proof of the K-polystability of the threefold X that works in all cases (2.15.$\mathfrak{S}_5$), (2.19.D_{12}), (2.22.$\mathfrak{A}_4$), (2.22.D_{∞}). Suppose that X is not K-polystable. By Theorem 1.22, there exists a G-invariant prime divisor F over X such that $\beta(F) \leqslant 0$. Let us seek for a contradiction.

Let $Z = C_X(F)$. By Theorem 3.17, we know that Z is not a surface, so that Z is either a G-invariant irreducible curve or a G-fixed point. Moreover, if Z is a G-fixed point, then it follows from Corollary 4.39 that we are in the case (2.22.$\mathfrak{A}_4$), and Z is the unique G-fixed point in X, which is not contained in the surface Q.

Lemma 4.41 *The G-invariant center $Z = C_X(F)$ does not lie on Q.*

Proof We have seen that Z is a point or a curve and that if Z is a point, $Z \notin Q$, so that we only need to show that if Z is a curve, $Z \not\subset Q$. Let us compute $S_X(Q)$. Let H be a hyperplane in $\mathbb{P}^3$, and let u be a non-negative real number. Observe that $-K_X - uQ \sim_{\mathbb{R}} (4 - 2u)\pi^*(H) + (u - 1)E \sim_{\mathbb{R}} (1 - u)Q + 2\pi^*(H)$. Thus, the divisor $-K_X - uQ$ is nef for $u \in [0, 1]$, and it is not pseudo-effective for $u > 2$. Moreover, in the notations of Section 1.7, we have

$$P\big(-K_X - uQ\big) = \begin{cases} -K_X - uQ & \text{if } 0 \leqslant u \leqslant 1, \\ (4 - 2u)\pi^*(H) & \text{if } 1 \leqslant u \leqslant 2, \end{cases}$$

and $N(-K_X - uQ) = (u - 1)E$ for $u \in [1, 2]$. Note that $S_X(Q) < 1$ by Theorem 3.17.

Now, we suppose that $Z \subset Q$. Then Z is a curve. Using Corollary 1.110, we conclude that $S(W^Q_{\bullet,\bullet}; Z) \geqslant 1$. Let us show that $S(W^Q_{\bullet,\bullet}; Z) < 1$.

Let $P(u) = P(-K_X - uQ)$ and $N(u) = N(-K_X - uQ)$. Note that $Q \cong \mathbb{S}_2 \cong \mathbb{P}^1 \times \mathbb{P}^1$. Set

$$n = \begin{cases} 3 & \text{if we are in the case } (2.15.\mathfrak{S}_5), \\ 4 & \text{if we are in the case } (2.19.\mathrm{D}_{12}), \\ 5 & \text{if we are in the cases } (2.22.\mathfrak{A}_4) \text{ or } (2.22.\mathrm{D}_\infty). \end{cases}$$

Then $(-K_X^3) = 10 + 4n$, $E|_Q \sim O_Q(3, 6 - n)$ and

$$P(u)|_Q \sim_{\mathbb{R}} \begin{cases} O_Q(u + 1, 4u + n(1 - u) - 2) & \text{if } 0 \leqslant u \leqslant 1, \\ O_Q(4 - 2u, 4 - 2u) & \text{if } 1 \leqslant u \leqslant 2. \end{cases}$$

If $Z = E|_Q$, then Corollary 1.110 gives

$$S\big(W^Q_{\bullet,\bullet}; Z\big)$$

$$= \frac{3}{10 + 4n} \int_0^1 \int_0^\infty \mathrm{vol}\Big(O_Q(u + 1 - 3v, 4u + n(1 - u) - 2 - (6 - n)v)\Big) dv du$$

$$+ \frac{3}{10 + 4n} \int_1^2 2(4 - 2u)^2 (u - 1) du$$

$$\frac{3}{10 + 4n} \int_1^2 \int_0^\infty \mathrm{vol}\Big(O_Q(4 - 2u - 3v, 4 - 2u - (6 - n)v)\Big) dv du$$

$$= \frac{3}{10 + 4n} \int_0^1 \int_0^{\frac{u+1}{3}} 2(u + 1 - 3v)(4u + n(1 - u) - 2 - (6 - n)v) dv du$$

$$+ \frac{2}{10 + 4n} + \frac{3}{10 + 4n} \int_1^2 \int_0^{\frac{4-2u}{3}} 2(4 - 2u - 3v)(4 - 2u - (6 - n)v) dv du$$

$$= \frac{7(3 + 4n)}{36(5 + 2n)} < 1.$$

Hence we may assume that $Z \neq E|_Q$. It then follows from Lemma 4.38 that $|Z - \Delta| \neq \varnothing$, where Δ is the diagonal curve in $Q \cong S_2 \cong \mathbb{P}^1 \times \mathbb{P}^1$. Therefore, writing down explicitly the formulas for $S(W_{\bullet,\bullet}^Q; Z)$ and $S(W_{\bullet,\bullet}^Q; \Delta)$ in Corollary 1.110, we see that

$$
S(W_{\bullet,\bullet}^Q; Z) \leqslant S(W_{\bullet,\bullet}^Q; \Delta)
$$

$$
= \frac{3}{10 + 4n} \int_0^1 \int_0^\infty \mathrm{vol}\Big(O_Q(u + 1 - v, 4u + n(1 - u) - 2 - v)\Big) dv\, du
$$

$$
+ \frac{3}{10 + 4n} \int_1^2 \int_0^\infty \mathrm{vol}\Big(O_Q(4 - 2u - v, 4 - 2u - v)\Big) dv\, du
$$

$$
= \frac{3}{10 + 4n} \int_0^1 \int_0^{u+1} 2(u + 1 - v)(4u + n(1 - u) - 2 - v) dv\, du
$$

$$
+ \frac{3}{10 + 4n} \int_1^2 \int_0^{4 - 2u} 2(4 - 2u - v)^2 dv\, du
$$

$$
= \frac{13 + 11n}{40 + 16n} < 1.
$$

The obtained contradiction completes the proof of the lemma. $\qquad\square$

To deal with the case $(2.22.\mathfrak{A}_4)$, we also need the following result:

Lemma 4.42 *Suppose that we are in the case* $(2.22.\mathfrak{A}_4)$, *and let* H *be the* G*-invariant hyperplane in* $\mathbb{P}^3$. *Then* $\pi(Z) \not\subset H$.

Proof Suppose that $\pi(Z) \subset H$. Then, using Lemma 4.38, we see that Z is not a point, so that Z is a G-invariant irreducible curve in H. Let us seek for a contradiction.

Observe that H intersects the curve $\mathscr{C}$ transversally by 4 distinct points since $H \cdot \mathscr{C} = 4$, and the curve $\mathscr{C}$ does not contain G-orbits of length less than 4 (recall that $\mathscr{C} \cong \mathbb{P}^1$). Note also that the action of the group G on the surface H is faithful.

Let S be the proper transform on X of the surface H, let $\varpi \colon S \to H$ be birational morphism induced by π and let $C = Q \cap S$. Then S is a smooth del Pezzo surface of degree 5, the morphism φ is a G-equivariant blow up of the four intersection points $H \cap \mathscr{C}$, the curve C is a G-invariant irreducible smooth curve such that

$$
C \sim 2\ell - e_1 - e_2 - e_3 - e_4,
$$

where ℓ is the proper transform on S of a general line in H, and e_1, e_2, e_3 and e_4 are φ-exceptional curves. Moreover, the group $\mathrm{Pic}^G(S)$ is generated by the divisor classes ℓ and $e_1 + e_2 + e_3 + e_4$. Furthermore, the cone of effective G-invariant divisors on S is generated by C and $e_1 + e_2 + e_3 + e_4$ since $C^2 = 0$. Thus, since Z is irreducible, we have

$$Z \sim a\ell - b(e_1 + e_2 + e_3 + e_4)$$

for some integers a and $b \leqslant \frac{a}{2}$. Since H does not have G-invariant lines by Lemma 4.38, the linear system $|\ell|$ does not have G-invariant curves. Hence, we see that $a \geqslant 2$, so that the linear system $|Z - C|$ is not empty. Observe also that $|C|$ is a basepoint free pencil that contains two G-invariant smooth curves [53, Lemma 6.2.2]. One of these curves is C. Denote the other curve by C'.

As in the proof of Lemma 4.41, let us compute $S_X(S)$. Take $u \in \mathbb{R}_{\geqslant 0}$. Then

$$-K_X - uS \sim_{\mathbb{R}} (4-u)\pi^*(H) - E \sim_{\mathbb{R}} Q + (2-u)\pi^*(H)$$
$$\sim_{\mathbb{R}} (u-1)Q + (2-u)(\pi^*(3H) - E),$$

and the restriction $(-K_X - uS)|_Q$ is a divisor on $Q \cong \mathbb{P}^1 \times \mathbb{P}^1$ of degree $(3-u, 1-u)$. Let $P(u) = P(-K_X - uS)$ and $N(u) = N(-K_X - uS)$. Then

$$P\big(-K_X - uS\big) = \begin{cases} -K_X - uS & \text{if } 0 \leqslant u \leqslant 1, \\ (2-u)(\pi^*(3H) - E) & \text{if } 1 \leqslant u \leqslant 2, \end{cases}$$

and we have

$$N\big(-K_X - uS\big) = \begin{cases} 0 & \text{if } 0 \leqslant u \leqslant 1, \\ (u-1)S & \text{if } 1 \leqslant u \leqslant 2, \end{cases}$$

where we have used notations from Section 1.7. Note that $S_X(S) < 1$ by Theorem 3.17.

Since $S_X(S) < 1$, $S(W^S_{\bullet,\bullet}; Z) \geqslant 1$ by Corollary 1.110. Let us show that $S(W^S_{\bullet,\bullet}; Z) < 1$. It is enough to do this in the cases $Z = C$ and $Z = C'$. Indeed, the case $Z = C$ is special because we have $N(u)|_S = (u-1)C$ for every $u \in [1,2]$. Moreover, if $Z \neq C$, then $S(W^S_{\bullet,\bullet}; Z) \leqslant S(W^S_{\bullet,\bullet}; C')$ because $|Z - C'| \neq \emptyset$. Observe also that Corollary 1.110 gives

$$S(W^S_{\bullet,\bullet}; C)$$
$$= \frac{1}{10} \int_0^1 \int_0^\infty \mathrm{vol}\Big((-K_X - uS)\big|_S - vC\Big) dv\, du + \frac{1}{10} \int_1^2 5(u-1)(2-u)^2 du$$
$$+ \frac{1}{10} \int_1^2 \int_0^\infty \mathrm{vol}\Big((2-u)(\pi^*(3H) - E)\big|_S - vC\Big) dv\, du$$
$$= \frac{5}{120} + S(W^S_{\bullet,\bullet}, C') \geqslant S(W^S_{\bullet,\bullet}; C')$$

because $(\pi^*(3H) - E)^2 \cdot S = 5$ and $C \sim C'$. Thus it is enough to show that $S(W^S_{\bullet,\bullet}; C) < 1$.

For any $u \in [0,1]$, observe that

$$\big(-K_X - uS\big)\big|_S - vC \sim_{\mathbb{R}} \frac{4-u-2v}{2}C + \frac{2-u}{2}(e_1 + e_2 + e_3 + e_4).$$

If $0 \leqslant v \leqslant 1$, this divisor is nef, and its volume is equal to $(u-2)(u+4v-6)$. Similarly, if $1 \leqslant v \leqslant \frac{4-u}{2}$, then its Zariski decomposition is

$$\left(-K_X - uS\right)\big|_S - vC \sim_{\mathbb{R}} \underbrace{(4-u-2v)\ell}_{\text{positive part}} + \underbrace{(v-1)(e_1 + e_2 + e_3 + e_4)}_{\text{negative part}},$$

so that its volume is $(4-u-2v)^2$. For $v > \frac{4-u}{2}$, this divisor is not pseudo-effective, so that its volume is zero. Thus, we have

$$\int_0^1 \int_0^\infty \text{vol}\left(\left(-K_X - uS\right)\big|_S - vC\right) dv\, du$$

$$= \int_0^1 \int_0^{\frac{4-u}{2}} \text{vol}\left(\left(-K_X - uS\right)\big|_S - vC\right) dv\, du$$

$$= \int_0^1 \int_0^1 (u-2)(u+4v-6)dv\, du + \int_0^1 \int_1^{\frac{4-u}{2}} (4-u-2v)^2 dv\, du$$

$$= \frac{143}{24}.$$

Similarly, if $u \in [1,2]$, then, using $(\pi^*(3H) - E)|_S \sim 3\ell - e_1 - e_2 - e_3 - e_4$, we get

$$(2-u)\left(\pi^*(3H) - E\right)\big|_S - vC \sim_{\mathbb{R}} \frac{6-3u-2v}{2}C + \frac{2-u}{2}(e_1 + e_2 + e_3 + e_4).$$

Hence, if $0 \leqslant v \leqslant 2-u$, this divisor is nef, with volume $(u-2)(5u+4v-10)$. Likewise, if $2 - u \leqslant v \leqslant \frac{6-3u}{2}$, then its Zariski decomposition is

$$\left(-K_X - uS\right)\big|_S - vC \sim_{\mathbb{R}} \underbrace{(6-3u-2v)\ell}_{\text{positive part}} + \underbrace{(v-2+u)(e_1 + e_2 + e_3 + e_4)}_{\text{negative part}},$$

and its volume is $(6-3u-2v)^2$. For $v > \frac{6-3u}{2}$, this divisor is not pseudo-effective. Then

$$\int_1^2 \int_0^\infty \text{vol}\left((2-u)\left(\pi^*(3H) - E\right)\big|_S - vC\right) dv\, du$$

$$= \int_1^2 \int_0^{2-u} (u-2)(5u+4v-10)dv\, du + \int_1^2 \int_{2-u}^{\frac{6-3u}{2}} (6-3u-2v)^2 dv\, du = \frac{19}{24}.$$

Therefore we see that $S(W^S_{\bullet,\bullet}; C) = \frac{5}{120} + \frac{1}{10}\left(\frac{143}{24} + \frac{19}{24}\right) = \frac{43}{60} < 1$. The obtained contradiction completes the proof of the lemma. $\qquad\square$

Using Lemma 1.45, we see that $\alpha_{G,Z}(X) < \frac{3}{4}$. Now, using Lemma 1.42, we see that there are a G-invariant effective $\mathbb{Q}$-divisor D on the threefold X and a positive rational number $\lambda < \frac{3}{4}$ such that $D \sim_{\mathbb{Q}} -K_X$ and Z is contained in the locus $\text{Nklt}(X, \lambda D)$.

Lemma 4.43 *Suppose that the locus* $\mathrm{Nklt}(X, \lambda D)$ *contains a G-irreducible surface. Then either* $S = Q$ *or* $\pi(S)$ *is a G-invariant hyperplane in* $\mathbb{P}^3$.

Proof By assumption, we have $D = \gamma S + \Delta$, where γ is a rational number such that $\gamma \geqslant \frac{1}{\lambda}$, and Δ is an effective $\mathbb{Q}$-divisor on X whose support does not contain S. If $S = E$, then

$$2Q + E \sim \pi^*(4H) - E \sim_{\mathbb{Q}} \gamma E + \Delta,$$

which implies that $2Q - (\gamma - 1)E$ is pseudo-effective. The latter is not the case because the cone $\overline{\mathrm{Eff}}(X)$ is generated by Q and E. Then $S \neq E$, so that $S \sim \pi^*(\mathcal{O}_{\mathbb{P}^3}(a)) - bE$ for some positive integer a and some non-negative integer $b \leqslant \frac{a}{2}$. Moreover, we have $\gamma a \leqslant 4$ because $\mathcal{O}_{\mathbb{P}^3}(4) \sim_{\mathbb{Q}} \gamma \pi(S) + \pi(\Delta)$. Thus, either $a = 1$ or $a = 2$ since $\gamma > \frac{4}{3}$.

If $a = 2$ and $b = 0$, then we immediately obtain a contradiction as in the case $S = E$. If $a = 2$ and $b = 1$, then $S = Q$ because S_2 is the only quadric surface in $\mathbb{P}^3$ that contains the curve $\mathscr{C}$. We conclude that $a = 1$ and $b = 0$, so that S is a G-invariant plane. □

If we are not in the case $(2.22.\mathfrak{A}_4)$, then Q is the only surface (a priori) that can be contained in the locus $\mathrm{Nklt}(X, \lambda D)$ because $\mathbb{P}^3$ does not contain G-invariant hyperplanes in the cases $(2.15.\mathfrak{S}_5)$, $(2.19.\mathrm{D}_{12})$ and $(2.22.\mathrm{D}_\infty)$ by Lemma 4.38.

Lemma 4.44 *The subvariety* Z *is not a point.*

Proof Suppose that Z is a point. Then, by Lemma 4.38, we are in the case $(2.22.\mathfrak{A}_4)$, and Z is the unique G-invariant point in the threefold X. For transparency, let $P = Z$. Let H be the unique G-invariant plane in $\mathbb{P}^3$. Then $P \notin H$, so that $\mathrm{Nklt}(X, \lambda D)$ does not contain any G-invariant surface that passes through P by Lemma 4.43, which implies that the locus $\mathrm{Nklt}(X, \lambda D)$ does not contain surfaces that pass through P.

Now, we observe that the action of the group G on the plane H is given by the standard irreducible three-dimensional representation of the group $G \cong \mathfrak{A}_4$. The second symmetric power of this representation is a sum of all irreducible representations of the group G. This can be verified using the following GAP script:

```
G:=SmallGroup(12,3);
T:=CharacterTable(G);
Ir:=Irr(T);
V:=Ir[4];
S:=SymmetricParts(T,[V],2);
MatScalarProducts(Ir,S);
```

Geometrically, this means that H contains exactly three G-invariant irreducible conics. Let us denote by S_2', S_2'' and S_2''' the quadric cones in $\mathbb{P}^3$ over these conics with vertex P, and let us also denote by $\widetilde{S}_2'$, $\widetilde{S}_2''$ and $\widetilde{S}_2'''$ their proper transforms on X, respectively. We will use these surfaces a bit later.

Second, we observe that $\mathrm{mult}_P(D) \leqslant 4$. This follows from the fact that $\pi(D) \sim_{\mathbb{Q}} 4H$.

Let $f \colon \widehat{X} \to X$ be the blow up of the point P. Denote by F the f-exceptional surface. Let $\widehat{D}$ be the proper transform on $\widehat{X}$ of the divisor D, let $\widehat{S}_2'$, $\widehat{S}_2''$ and $\widehat{S}_2'''$ be the proper transforms on X of the surfaces $\widetilde{S}_2'$, $\widetilde{S}_2''$ and $\widetilde{S}_2'''$, respectively. Then

$$K_{\widehat{X}} + \lambda\widehat{D} + \big(\lambda\,\mathrm{mult}_P(D) - 2\big)F \sim_{\mathbb{Q}} f^*\big(K_X + \lambda D\big),$$

so that $(\widehat{X}, \lambda\widehat{D} + (\lambda\,\mathrm{mult}_P(D) - 2)F)$ is not Kawamata log terminal at some point in F. Since $\lambda\,\mathrm{mult}_P(D) - 2 \leqslant 4\lambda - 2 < 1$, we conclude that the log pair $(\widehat{X}, \lambda\widehat{D} + F)$ is also not log canonical at some point in F. Then, using Theorem A.15, we conclude that the log pair $(F, \lambda\widehat{D}|_F)$ is not log canonical.

Now, we identify $F = \mathbb{P}^2$. Since $\lambda\,\mathrm{mult}_P(D) < 3$, the divisor $-(K_F + \lambda\widehat{D}|_F)$ is ample, so that $\mathrm{Nklt}(F, \lambda\widehat{D}|_F)$ is connected by Corollary A.4. But the G-action on F is given by its irreducible three-dimensional representation, so that F does not contain G-fixed points and G-invariant lines. This implies that $\mathrm{Nklt}(F, \lambda\widehat{D}|_F)$ is a G-invariant irreducible conic. But F contains exactly three G-invariant conics – the conics $F \cap \widehat{S}_2'$, $F \cap \widehat{S}_2''$, $F \cap \widehat{S}_2'''$. Thus, without loss of generality, we may assume that $\mathrm{Nklt}(F, \lambda\widehat{D}|_F) = F \cap \widehat{S}_2'$.

Let $C = F \cap \widehat{S}_2'$. We proved that $C = \mathrm{Nklt}(F, \lambda\widehat{D}|_F)$. In fact, our proof implies that

- $\lambda\,\mathrm{mult}_P(D) > 2$, so that the divisor $\lambda\widehat{D} + (\lambda\,\mathrm{mult}_P(D) - 2)F$ is effective,
- $\mathrm{Nklt}(\widehat{X}, \lambda\widehat{D} + (\lambda\,\mathrm{mult}_P(D) - 2)F) \cap F \subset C$.

Applying [132, Corollary 5.49] to $(\widehat{X}, \lambda\widehat{D} + (\lambda\,\mathrm{mult}_P(D)-2)F)$ and the morphism f, we see that $\mathrm{Nklt}\big(\widehat{X}, \lambda\widehat{D} + (\lambda\,\mathrm{mult}_P(D)-2)F)\big) \cap F = C$ since F has no G-fixed points.

Write $D = a\widetilde{S}_2' + \Delta$, where Δ is an effective $\mathbb{Q}$-divisor whose support does not contain the surface $\widetilde{S}_2'$, and a is a non-negative rational number. Then $\lambda a \leqslant 1$ by Lemma 4.43. Let $\widehat{\Delta}$ be the proper transform of the divisor Δ on the threefold $\widehat{X}$. Then

$$C \subset \mathrm{Nklt}\big(\widehat{X}, \lambda a\widehat{S}_2' + \lambda\widehat{\Delta} + (2\lambda a + \lambda\,\mathrm{mult}_P(\Delta) - 2)F\big).$$

Hence, using Theorem A.15 again, we get

$$C \subset \mathrm{Nklt}(\widehat{S}_2', \lambda\widehat{\Delta}|_{\widehat{S}_2'} + (2\lambda a + \lambda\,\mathrm{mult}_P(\Delta) - 2)C).$$

This simply means that $\widehat{\Delta}\big|_{\widehat{S}_2'} = bC + \Omega$, where b is a non-negative rational number such that $\lambda b + 2\lambda a + \lambda \mathrm{mult}_P(\Delta) - 2 \geqslant 1$, and Ω is an effective $\mathbb{Q}$-divisor on $\widehat{S}_2'$ whose support does not contain the curve C. Thus, we see that $b \geqslant \frac{3}{\lambda} - 2a - \mathrm{mult}_P(\Delta) > 4 - 2a - \mathrm{mult}_P(\Delta)$. Now, we let $\widehat{\ell}$ be the proper transform on $\widehat{X}$ of a general ruling of the cone S_2'. Then

$$\widehat{\ell} \cdot \widehat{\Delta} = \widehat{\ell} \cdot \left((\pi \circ f)^* \left(-K_{\mathbb{P}^3} - aS_2' \right) - f^*(E) - \mathrm{mult}_P(\Delta)F \right) = 4 - 2a - \mathrm{mult}_P(\Delta).$$

Then $4 - 2a - \mathrm{mult}_P(\Delta) = \widehat{\ell} \cdot \widehat{\Delta} = b + \widehat{\ell} \cdot \Omega \geqslant b > 4 - 2a + \mathrm{mult}_P(\Delta)$, which is absurd. This completes the proof of the lemma. $\qquad\square$

Therefore, we see that Z is a G-invariant irreducible curve.

Lemma 4.45 *The curve Z is rational.*

Proof Let $\overline{D} = \phi(D)$ and $\overline{Z} = \phi(Z)$, where ϕ is the contraction of Q in (4.7). Since $Z \not\subset Q$, we see that $\overline{Z}$ is a G-invariant irreducible curve, the induced map $\phi|_Z \colon Z \to \overline{Z}$ is birational, and $\overline{Z} \subset \mathrm{Nklt}(V_d, \lambda\overline{D})$. If $\mathrm{Nklt}(V_d, \lambda\overline{D})$ does not have two-dimensional components, then $\overline{Z}$ is a smooth rational curve by Corollary A.14.

To complete the proof, we may assume that $\mathrm{Nklt}(V_d, \lambda\overline{D})$ contains a G-irreducible surface $\overline{S}$. Let S be its proper transform on X. Then $S \subseteq \mathrm{Nklt}(X, \lambda D)$ and $S \neq Q$, so that $\pi(S)$ is a hyperplane in $\mathbb{P}^3$ by Lemma 4.43. Then we must be in the case $(2.22.\mathfrak{A}_4)$, so that $d = 5$, and the surface $\overline{S}$ is a hyperplane section of the threefold $V_5 \subset \mathbb{P}^5$.

By Lemma 4.42, the curve $\pi(Z)$ is not contained in $\pi(S)$, so that $\overline{Z} \not\subset \overline{S}$.

Write $\overline{D} = \gamma\overline{S} + \overline{\Delta}$, where γ is a rational number such that $\gamma \geqslant \frac{1}{\lambda}$, and $\overline{\Delta}$ is an effective $\mathbb{Q}$-divisor such that $\overline{S} \not\subset \mathrm{Supp}(\overline{\Delta})$. Then $\overline{Z} \subset \mathrm{Nklt}(V_5, \lambda\overline{\Delta})$. But $\overline{\Delta} \sim_{\mathbb{Q}} -(1 - \frac{\gamma}{2})K_{V_5}$, so that $\overline{Z}$ is rational by Corollary A.14 since $\mathrm{Nklt}(V_5, \lambda\overline{\Delta})$ does not contain surfaces. $\qquad\square$

Corollary 4.46 *If we are in one of the cases $(2.15.\mathfrak{S}_5)$ or $(2.19.\mathrm{D}_{12})$, then $Z \not\subset E$.*

Proof By Corollary 4.45, the curve Z is rational. But $\pi(Z)$ is not a point since $\mathscr{C}$ does not contain G-fixed points by Lemma 4.38. If $Z \subset E$, then $\pi(Z) = \mathscr{C}$, which implies that $\mathscr{C}$ is also rational. But $\mathscr{C}$ is irrational in the cases $(2.15.\mathfrak{S}_5)$ or $(2.19.\mathrm{D}_{12})$. $\qquad\square$

Using Corollary 4.46, we conclude that $\pi(Z)$ must be a G-invariant rational curve in the case $(2.15.\mathfrak{S}_5)$, which contradicts Lemma 4.38. Thus, the case $(2.15.\mathfrak{S}_5)$ is impossible, which we already know from Example 1.77. In the remaining part of the section, we will show that the cases $(2.19.\mathrm{D}_{12})$, $(2.22.\mathfrak{A}_4)$ and $(2.22.\mathrm{D}_\infty)$ are also impossible.

Lemma 4.47 $Z \not\subset E$.

Proof Suppose that $Z \subset E$. Let us seek for a contradiction. Using Corollary 4.46, we see that we are in one of the cases (2.22.$\mathfrak{A}_4$) or (2.22.D_∞). Then $\mathscr{C}$ is a smooth rational quartic curve, so that $E \cong \mathbb{F}_n$ for some $n \in \mathbb{Z}_{\geqslant 0}$. Let us show that $E \cong \mathbb{F}_2$ or $E \cong \mathbb{P}^1 \times \mathbb{P}^1$.

Let **s** be a section of the projection $E \to \mathscr{C}$ such that $\mathbf{s}^2 = -n$, and let **l** be its fiber. Then $-E|_E \sim \mathbf{s} + k\mathbf{l}$ for some integer k. Then $-n + 2k = E^3 = -c_1(\mathcal{N}_{\mathscr{C}/\mathbb{P}^3}) = -14$, so that $k = \frac{n-14}{2}$. Then

$$Q|_E \sim \left(\pi^*(\mathcal{O}_{\mathbb{P}^3}(2)) - E\right)\big|_E \sim \mathbf{s} + (k+8)\mathbf{l} = \mathbf{s} + \frac{n+2}{2}\mathbf{l},$$

which implies that $Q|_E \not\sim \mathbf{s}$. Moreover, we know that $Q|_E$ is a smooth irreducible curve since the quadric surface S_2 is smooth. Thus, since $Q|_E \neq \mathbf{s}$, we have

$$0 \leqslant Q|_E \cdot \mathbf{s} = \left(\mathbf{s} + \frac{n+2}{2}\mathbf{l}\right) \cdot \mathbf{s} = -n + \frac{n+2}{2} = \frac{2-n}{2},$$

so that $n = 0$ or $n = 2$. Note that $n = 0$ in the case (2.22.D_∞) by [66, Theorem 3.2].

Now, we can obtain a contradiction arguing exactly as in the proof of Lemma 4.41. But there is a simpler way to do this. Write $D = aE + \Delta$, where Δ is an effective $\mathbb{Q}$-divisor whose support does not contain the surface E, and a is a non-negative rational number. Then $\lambda a \leqslant 1$ by Lemma 4.43.

Note that $2Q + E \sim -K_X$ and $Z \not\subset \text{Nklt}(X, \lambda 2Q + \lambda E)$ since $Z \not\subset Q$ by Lemma 4.41. Thus, using Lemma A.34, we can replace D by an effective $\mathbb{Q}$-divisor $D' \sim_{\mathbb{Q}} D$ such that

- $Z \subset \text{Nklt}(X, \lambda D')$,
- the support of the divisor D' does not contain either Q or E (or both of them).

Therefore, we may assume that $\text{Supp}(D)$ does not contain Q or E. In particular, if $a > 0$, then $\Delta|_Q$ is an effective $\mathbb{Q}$-divisor on the surface $Q \cong \mathbb{P}^1 \times \mathbb{P}^1$ of degree $(3 - a, 1 - 3a)$. This shows that we always have the inequality $a \leqslant \frac{1}{3}$.

Using Theorem A.15, we get $Z \subset \text{Nklt}(E, \lambda \Delta|_E)$. This means that $\Delta|_E = bZ + \Omega$, where b is a rational number such that $b \geqslant \frac{1}{\lambda} > \frac{4}{3}$, and Ω is an effective $\mathbb{Q}$-divisor whose support does not contain the curve Z. On the other hand, if $E \cong \mathbb{P}^1 \times \mathbb{P}^1$, then

$$bZ + \Omega = \Delta|_E \sim_{\mathbb{Q}} -K_X|_E - aE|_E \sim_{\mathbb{Q}} \mathbf{s} + 9\mathbf{l} + a(\mathbf{s} - 7\mathbf{l}) = (1+a)\mathbf{s} + (9 - 7a)\mathbf{l}$$

because $-E|_E \sim \mathbf{s} - 7\mathbf{l}$ in this case. Similarly, if $E \cong \mathbb{F}_2$, then

$$bZ + \Omega = \Delta|_E \sim_{\mathbb{Q}} -K_X|_E - aE|_E \sim_{\mathbb{Q}} \mathbf{s} + 10\mathbf{l} + a(\mathbf{s} - 6\mathbf{l}) = (1+a)\mathbf{s} + (10 - 6a)\mathbf{l}$$

because $-E|_E \sim \mathbf{s} - 6\mathbf{l}$ in this case. In both cases, we immediately obtain a contradiction:

$$\frac{4}{3} < \frac{1}{\lambda} \leqslant b \leqslant bZ \cdot \mathbf{l} \leqslant bZ \cdot \mathbf{l} + \Omega \cdot \mathbf{l} = (bZ + \Omega) \cdot \mathbf{l} = 1 + a \leqslant \frac{4}{3}$$

because $Z \cdot \mathbf{l} \neq 0$, since $\pi(Z)$ is not a point by Lemma 4.38.   □

Thus we see that $\pi(Z)$ is a G-invariant rational curve in $\mathbb{P}^3$ such that $\pi(Z) \not\subset S_2$. Moreover, if we are in the case $(2.22.\mathfrak{A}_4)$, then $\pi(Z)$ is not contained in the G-invariant hyperplane in $\mathbb{P}^3$ by Lemma 4.42.

Lemma 4.48 *The curve $\pi(Z)$ is a G-invariant line in $\mathbb{P}^3$.*

Proof Let $\widehat{D} = \pi(D)$ and $\widehat{Z} = \pi(Z)$. Then $\widehat{Z} \subset \mathrm{Nklt}(\mathbb{P}^3, \lambda \widehat{D})$, and $\widehat{Z}$ is not contained in any surface contained in $\mathrm{Nklt}(\mathbb{P}^3, \lambda \widehat{D})$ by Lemma 4.43. Now, apply Corollary A.10.   □

Thus, using Lemma 4.38, we conclude that we are in the case $(2.19.D_{12})$ or $(2.22.D_\infty)$. Then $\mathbb{P}^3$ contains two G-invariant lines by Lemma 4.38. These G-invariant lines are the lines $x_0 = x_3 = 0$ and $x_1 = x_2 = 0$. For simplicity, let us call them L_∞ and L_0, respectively. We know that either $\pi(Z) = L_\infty$ or $Z = \pi(L_0)$.

Let H be a a general hyperplane in $\mathbb{P}^3$ that contains $\pi(Z)$, and let S be its proper transform on X. Then S is smooth. Moreover, one of the following possibilities holds:

- if we are in the case $(2.19.D_{12})$, then S is a smooth del Pezzo surface of degree 4,
- if we are in the case $(2.22.D_\infty)$, then S is a smooth del Pezzo surface of degree 5.

Let u be a non-negative real number. Observe that $-K_X - uS \sim_{\mathbb{R}} (2-u)\pi^*(H) + Q$. This implies that $-K_X - uS$ is nef for every $u \in [0,1]$, it is not pseudo-effective for $u > 2$. Moreover, in the notations of Section 1.7, we have

$$P(-K_X - uS) = \begin{cases} (4-u)\pi^*(H) - E & \text{if } 0 \leqslant u \leqslant 1, \\ (2-u)(3\pi^*(H) - E) & \text{if } 1 \leqslant u \leqslant 2, \end{cases}$$

and we have

$$N(-K_X - uS) = \begin{cases} 0 & \text{if } 0 \leqslant u \leqslant 1, \\ (1-u)Q & \text{if } 1 \leqslant u \leqslant 2, \end{cases}$$

so that $Z \not\subset N(-K_X - uS)$. Moreover, we have $S_X(S) < 1$ by Theorem 3.17. Then $S(W^S_{\bullet,\bullet}; Z) \geqslant 1$ by Corollary 1.110. Let us compute $S(W^S_{\bullet,\bullet}; Z)$.

Let $\varpi \colon S \to H$ be the birational morphism induced by π. Then φ contracts d disjoint smooth curves, where d is the degree of the curve $\mathscr{C}$. Denote them by $e_1, e_2, \ldots, e_d$. Then $E|_S = e_1 + e_2 + \cdots + e_d$. Let $C = Q|_S$, and let ℓ be the proper transform of a general line in H on the surface S. Then $\varphi(C)$ is the conic $H \cap S_2$, and

$$C \sim 2\ell - \sum_{i=1}^{9-d} e_i,$$

so that $C^2 = 4 - d \leqslant 0$.

Lemma 4.49 *Suppose that $\pi(Z) \cap \mathscr{C} = \varnothing$. Then $S(W_{\bullet,\bullet}^S; Z) < 1$.*

Proof By Lemma 4.48, we have $Z \sim \ell$. Thus, if $u \in [0,1]$ and $v \in \mathbb{R}_{\geqslant 0}$, then

$$P(-K_X - uS)|_S - vZ \sim_{\mathbb{R}} (4 - u - v)\ell - \sum_{i=1}^{9-d} e_i \sim_{\mathbb{R}} (2 - u - v)\ell + C,$$

which implies the following assertions:

- the divisor $P(-K_X - uS)|_S - vZ$ is not pseudo-effective for $v > 2 - u$,
- if $d = 4$, then $P(-K_X - uS)|_S - vZ$ is nef $\iff v \leqslant 2 - u$,
- if $d = 5$, then $P(-K_X - uS)|_S - vZ$ is nef $\iff v \leqslant \frac{3-2u}{2}$,
- if $d = 5$ and $\frac{3-2u}{2} \leqslant v \leqslant 2 - u$, then the Zariski decomposition of $P(-K_X - uS)|_S - vZ$ is

$$\underbrace{(2 - u - v)(5\ell - 2e_1 - 2e_2 - 2e_3 - 2e_4 - 2e_5)}_{\text{positive part}} + \underbrace{(2u + 2v - 3)C}_{\text{negative part}}.$$

Thus, if $u \in [0,1]$, $0 \leqslant v \leqslant 2 - u$ and $d = 4$, then $\mathrm{vol}(P(-K_X - uS)|_S - vZ) = (4 - u - v)^2 - 4$. Likewise, if $u \in [0,1]$, $0 \leqslant v \leqslant 2 - u$ and $d = 5$, then

$$\mathrm{vol}(P(-K_X - uS)|_S - vZ) = \begin{cases} (4 - u - v)^2 - 5 & \text{if } v \leqslant \dfrac{3 - 2u}{2}, \\[2mm] 5(2 - u - v)^2 & \text{if } v \geqslant \dfrac{3 - 2u}{2}. \end{cases}$$

Similarly, if $u \in [1,2]$ and $v \in \mathbb{R}_{\geqslant 0}$, then

$$P(-K_X - uS)|_S - vZ \sim_{\mathbb{R}} (2 - u - v)\ell + (2 - u)C,$$

which implies the following assertions:

- the divisor $P(-K_X - uS)|_S - vZ$ is not pseudo-effective for $v > 2 - u$,
- if $d = 4$, then $P(-K_X - uS)|_S - vZ$ is nef $\iff 0 \leqslant v \leqslant 2 - u$,
- if $d = 5$, then $P(-K_X - uS)|_S - vZ$ is nef $\iff 0 \leqslant v \leqslant \frac{2-u}{2}$,
- if $d = 5$ and $\frac{2-u}{2} \leqslant v \leqslant 2 - u$, the Zariski decomposition of $P(-K_X - uS)|_S - vZ$ is

$$P(-K_X - uS)\big|_S - vZ$$
$$\sim_{\mathbb{R}} \underbrace{(2 - u - v)(5\ell - 2e_1 - 2e_2 - 2e_3 - 2e_4 - 2e_5)}_{\text{positive part}} + \underbrace{(2u + v - 3)C}_{\text{negative part}}.$$

If $u \in [1, 2]$, $0 \leqslant v \leqslant 2 - u$ and $d = 4$, it follows that

$$\mathrm{vol}\big(P(-K_X - uS)\big|_S - vZ\big) = (6 - 3u - v)^2 - 4(2 - u)^2.$$

Likewise, if $u \in [1, 2]$, $0 \leqslant v \leqslant 2 - u$ and $d = 5$, then

$$\mathrm{vol}\big(P(-K_X - uS)\big|_S - vZ\big) = \begin{cases} (6 - 3u - v)^2 - 5(2 - u)^2 & \text{if } v \leqslant \dfrac{2 - u}{2}, \\[2mm] 5(2 - u - v)^2 & \text{if } v \geqslant \dfrac{2 - u}{2}. \end{cases}$$

Now we are ready to compute $S(W^S_{\bullet,\bullet}; Z)$. If $d = 4$, then Corollary 1.110 gives

$$S(W^S_{\bullet,\bullet}; Z) = \frac{3}{26} \int_0^1 \int_0^{2-u} \big((4 - u - v)^2 - 4\big) du\, dv$$
$$+ \frac{3}{26} \int_1^2 \int_0^{2-u} \big((6 - 3u - v)^2 - 4(2 - u)^2\big) du\, dv,$$

so that $S(W^S_{\bullet,\bullet}; Z) = \frac{39}{54} < 1$. Similarly, if $d = 5$, then

$$S(W^S_{\bullet,\bullet}; Z) = \frac{1}{10} \int_0^1 \int_0^{\frac{3-2u}{2}} \big((4 - u - v)^2 - 5\big) du\, dv$$
$$+ \frac{1}{10} \int_0^1 \int_{\frac{3-2u}{2}}^{2-u} 5(2 - u - v)^2 du\, dv$$
$$+ \frac{1}{10} \int_1^2 \int_0^{\frac{2-u}{2}} \big((6 - 3u - v)^2 - 5(2 - u)^2\big) du\, dv$$
$$+ \frac{1}{10} \int_1^2 \int_{\frac{2-u}{2}}^{2-u} 5(2 - u - v)^2 du\, dv = \frac{119}{240} < 1.$$

This completes the proof of the lemma.   □

By Lemma 4.38, the lines L_0 and L_∞ are disjoint from $\mathscr{C}$ in the case $(2.19.\mathrm{D}_{12})$. Therefore, we are in the case $(2.22.\mathrm{D}_\infty)$, so that X is the three-fold from Example 4.34, and $\pi(Z)$ is a line such that $\pi(Z) \cap \mathscr{C} \neq \varnothing$.

Using Lemma 4.38, we see that $\pi(Z) = L_\infty$ and

$$L_\infty \cap \mathscr{C} = [0 : 1 : 0 : 0] \cup [0 : 0 : 1 : 0].$$

We may assume that $\varpi(e_1) = [0 : 1 : 0 : 0]$ and $\varpi(e_2) = [0 : 0 : 1 : 0]$. Then $Z \sim \ell - e_1 - e_2$, so that Z is a (-1)-curve on the surface S that is disjoint from the (-1)-curves e_3 and e_4. Let L_{34} be the proper transform on S of the line in H that contains $\varpi(e_3)$ and $\varpi(e_4)$.

If $u \in [0,1]$ and $v \in \mathbb{R}_{\geqslant 0}$, then

$$P(-K_X - uS)\big|_S - vZ \sim_{\mathbb{R}} (3 - u - v)L + (2 - u)(e_1 + e_2) + L_{34},$$

which implies the following assertions:

- the divisor $P(-K_X - uS)\big|_S - vZ$ is not pseudo-effective for $v > 3 - u$,
- if $0 \leqslant v \leqslant 1$, then $P(-K_X - uS)\big|_S - vZ$ is nef,
- if $1 \leqslant v \leqslant 2 - u$, the Zariski decomposition of $P(-K_X - uS)\big|_S - vZ$ is

$$P(-K_X - uS)\big|_S - vZ \sim_{\mathbb{R}} \underbrace{(3 - u - v)(L + e_1 + e_2) + L_{34}}_{\text{positive part}} + \underbrace{(v - 1)(e_1 + e_2)}_{\text{negative part}},$$

- if $2 - u \leqslant v \leqslant 3 - u$, the Zariski decomposition of $P(-K_X - uS)\big|_S - vZ$ is

$$P(-K_X - uS)\big|_S - vZ$$
$$\sim_{\mathbb{R}} \underbrace{(3 - u - v)(L + e_1 + e_2 + L_{34})}_{\text{positive part}} + \underbrace{(v - 1)(e_1 + e_2) + (v + u - 2)L_{34}}_{\text{negative part}}.$$

If $u \in [0,1]$ and $0 \leqslant v \leqslant 3 - u$, we have

$$\mathrm{vol}\big(P(-K_X - uS)\big|_S - vZ\big)$$
$$= \begin{cases} (4 - u - v)^2 - 2(v - 1)^2 - 2 & \text{if } v \leqslant 1, \\ (4 - u - v)^2 - 2 & \text{if } 1 \leqslant v \leqslant 2 - u, \\ 2(3 - u - v)^2 & \text{if } 2 - u \leqslant v \leqslant 3 - u. \end{cases}$$

If $u \in [1,2]$ and $v \in \mathbb{R}_{\geqslant 0}$, then

$$P(-K_X - uS)\big|_S - vZ \sim_{\mathbb{R}} (4 - 2u - v)L + (2 - u)(e_1 + e_2 + L_{34}),$$

which implies the following assertions:

- the divisor $P(-K_X - uS)\big|_S - vZ$ is not pseudo-effective for $v > 4 - 2u$,
- if $0 \leqslant v \leqslant 2 - u$, then $P(-K_X - uS)\big|_S - vZ$ is nef,
- if $2 - u \leqslant v \leqslant 4 - 2u$, the Zariski decomposition of $P(-K_X - uS)\big|_S - vZ$ is

$$P(-K_X - uS)\big|_S - vZ$$
$$\sim_{\mathbb{R}} \underbrace{(4 - 2u - v)(L + e_1 + e_2 + L_{34})}_{\text{positive part}} + \underbrace{(v + u - 2)(e_1 + e_2 + L_{34})}_{\text{negative part}}.$$

Hence, if $u \in [0,2]$ and $0 \leqslant v \leqslant 4 - 2u$, then

$$\mathrm{vol}\big(P(-K_X - uS)\big|_S - vZ\big)$$
$$= \begin{cases} (6 - 3u - v)^2 - 2(u + v - 2)^2 - 2(2 - u)^2 & \text{if } 1 \leqslant v \leqslant 2 - u, \\ 2(4 - 2u - v)^2 & \text{if } 2 - u \leqslant v \leqslant 4 - 2u. \end{cases}$$

Now, using Corollary 1.110, we can compute $S(W^S_{\bullet,\bullet}; Z)$ as follows:

$$S(W^S_{\bullet,\bullet}; Z)$$
$$= \frac{1}{10} \int_0^1 \int_0^1 \big((4 - u - v)^2 - 2(v - 1)^2 - 2\big) du dv$$
$$+ \frac{1}{10} \int_0^1 \int_1^{2-u} \big((4 - u - v)^2 - 2\big) du dv + \frac{1}{10} \int_0^1 \int_{2-u}^{3-u} 2(3 - u - v)^2 du dv$$
$$+ \frac{1}{10} \int_1^2 \int_0^{2-u} \big((6 - 3u - v)^2 - 2(u + v - 2)^2 - 2(2 - u)^2\big) dv du$$
$$+ \frac{1}{10} \int_1^2 \int_{2-u}^{4-2u} 2(4 - 2u - v)^2 dv du = 1.$$

Thus, using Corollary 1.110 again, we conclude that $S_X(F) = 1$ as well, which is not the case by Theorem 3.17. The obtained contradiction proves that X is K-polystable.

4.5 Threefolds fibered into del Pezzo surfaces

Many smooth Fano threefolds admit a surjective morphism to $\mathbb{P}^1$ whose general fiber is a smooth del Pezzo surface. However, if we want this del Pezzo fibration to be a Mori fibered space, the threefold belongs to one of the families №2.1, №2.2, №2.3, №2.4, №2.5, №2.7, №2.10, №2.14, №2.18, №2.25, №2.33, №2.34. In Section 4.3, we already proved that general Fano threefolds in the families №2.1, №2.3, №2.5, №2.10, №2.14 are K-stable, and we also proved that every smooth Fano threefold in the family №2.25 is K-stable. On the other hand, the unique smooth Fano threefold №2.33 is K-unstable by Theorem 3.17. The family №2.34 contains a unique smooth threefold: $\mathbb{P}^1 \times \mathbb{P}^2$, and it is K-polystable. The goal of this section is to show that general members of the families №2.2, №2.4, №2.7, №2.18 are K-stable.

First, we show that general members of the family №2.2 are K-stable. Every smooth member of this family is a double cover of $\mathbb{P}^1 \times \mathbb{P}^2$ branched over a surface of degree $(2,4)$, so that the projection to $\mathbb{P}^1$ gives a fibration into del Pezzo surfaces of degree 2.

Lemma 4.50 *Let X be a smooth Fano threefold in the family №2.2 such that X satisfies the following generality condition: for every fiber S of the natural projection $X \to \mathbb{P}^1$, the surface S has at most Du Val singularities and $\alpha(S) \geqslant \frac{3}{4}$. Then $\alpha(X) \geqslant \frac{3}{4}$.*

Proof The assertion follows from Theorem 1.52, since Theorem 1.52(1) cannot hold because $-K_X \sim S + H_L$, where H_L is a pull back of a line via the conic bundle $X \to \mathbb{P}^2$. □

If applicable, this lemma implies that a general member of the family №2.2 is K-stable by Theorem 1.50 because all smooth Fano threefolds №2.2 have finite automorphism groups [45]. Therefore, we have to show that smooth Fano threefolds №2.2 that satisfy the generality condition of Lemma 4.50 do exist. This is done in the following example:

Example 4.51 Let X be a double cover of $\mathbb{P}^1 \times \mathbb{P}^2$ that is branched over a divisor of degree $(2, 4)$ that is given by

$$u^2(z^3 x + yx^3 - y^3 z) + v^2(x^3 z - xy^3 - z^4) = 0,$$

where $([u : v], [x : y : z])$ are coordinates on $\mathbb{P}^1 \times \mathbb{P}^2$. Then the Fano threefold X is smooth. Let S be a fiber of the projection $X \to \mathbb{P}^1$ over a point $P \in \mathbb{P}^1$. Then S has at most Du Val singularities and $\alpha(S) \geqslant \frac{3}{4}$. Indeed, if S is smooth, then $\alpha(S) \geqslant \frac{3}{4}$ by Lemma A.40. Therefore, we may assume that S is singular. In particular, $P \neq [0 : 1]$ and $P \neq [1 : 0]$. Let $t = \frac{v^2}{u^2}$. Then S is a double cover of $\mathbb{P}^2$, which is branched over the quartic curve

$$C_4 = \left\{ z^3 x + yx^3 - y^3 z + t(x^3 z - xy^3 - z^4) = 0 \right\}.$$

Note that C_4 must be singular since S is singular. On the other hand, one can show that the curve C_4 is singular if and only if t is a root of the following polynomial:

$$14348907t^{27} + 43046721t^{25} + 47298249t^{24} + 73279809t^{23} + 88219206t^{22}$$
$$+ 160219620t^{21} + 136305504t^{20} + 141235569t^{19} + 230867372t^{18} + 180568521t^{17}$$
$$+ 91887093t^{16} + 200311947t^{15} + 129699756t^{14} + 50748768t^{13} - 18457896t^{12}$$
$$+ 103837464t^{11} - 60378876t^{10} - 55596213t^9 - 32802534t^8 - 6278553t^7$$
$$- 53247369t^6 - 13308057t^5 - 1577457t^4 - 12252303t^3 - 1058841t^2 - 823543 = 0.$$

This polynomial is irreducible over $\mathbb{Q}$. The singular locus of C_4 consists of one ordinary double point, so that we can apply Lemma A.36 to find $\alpha(S)$. Let O be the singular point of the curve C_4. Then there are two lines L and L' in $\mathbb{P}^2$ such that $(L \cdot C_4)_O \geqslant 3$ and $(L' \cdot C_4)_O \geqslant 3$. Then Lemma A.36 gives

$$\alpha(S) = \begin{cases} \dfrac{2}{3} & \text{if } (L \cdot C_4)_O = 4 \text{ or } (L' \cdot C_4)_O = 4, \\[2mm] \dfrac{3}{4} & \text{if } (L \cdot C_4)_O = (L' \cdot C_4)_O = 3. \end{cases}$$

Note that $L + L'$ is defined over $\mathbb{Q}(t)$. Taking an appropriate change of coordinates, we can assume that $O = [0 : 0 : 1]$, and C_4 is given by

$$z^2 q_2(x, y) + z q_3(x, y) + q_4(x, y) = 0,$$

where $q_2(x, y)$, $q_3(x, y)$ and $q_4(x, y)$ are polynomials of degrees 2, 3 and 4, respectively. The quadratic form $q_2(x, y)$ is not degenerate (this is how we check that S has an ordinary double point at O), and $q_2(x, y) = 0$ define $L + L'$. Then $(L \cdot C_4)_O = (L' \cdot C_4)_O = 3$ if and only if the forms $q_2(x, y)$ and $q_3(x, y)$ are coprime. One can check that this is indeed the case, so that $\alpha(S) = \frac{3}{4}$ by Lemma A.36.

By Lemma 4.50 and Theorem 1.50, a general member of the family №2.2 is K-stable. By Theorem 1.11, this also follows from Theorem 1.48 and the following

Example 4.52 Let ω be a primitive cubic root of unity, and let $\pi\colon X \to \mathbb{P}^1 \times \mathbb{P}^2$ be a double cover branched over a smooth surface of degree $(2, 4)$ given by

$$u^2\left(x^4 + y^4 + z^4\right) + v^2\left(x^4 + \omega y^4 + \omega^2 z^4\right) = 0,$$

where $([u : v], [x : y : z])$ are coordinates on $\mathbb{P}^1 \times \mathbb{P}^2$. Then X is a smooth Fano threefold in the family №2.2. It admits a faithful action of the group $G = \mu_2^2 \times (\mu_4^2 \rtimes \mu_3)$, where the generator of one of the copies of μ_2 is the Galois involution of the cover π, the generator of another copy of μ_2 acts by changing the sign of u and preserves all other coordinates, generators of the two copies of μ_4 multiply x (respectively, y) by $\sqrt{-1}$ and preserve all other coordinates, and a generator of μ_3 acts by

$$u \mapsto u, \ v \mapsto \omega v, \ x \mapsto z, \ y \mapsto x, \ z \mapsto y.$$

The natural projection $X \to \mathbb{P}^2$ is G-equivariant, so that it gives a homomorphism of groups $G \to \mathrm{Aut}(\mathbb{P}^2)$. Denote its image by Γ. Then $\Gamma \cong \mu_4^2 \rtimes \mu_3$. Observe that $\mathbb{P}^2$ does not contain Γ-invariant lines, which implies that it does not contain Γ-invariant rational curves because $\mathrm{Aut}(\mathbb{P}^1)$ does not have a subgroup isomorphic to Γ. Then X contains neither G-invariant points nor G-invariant rational curves. Therefore, applying Theorem 1.52 with $\mu = 1$, we see that $\alpha_G(X) \geqslant 1$ because condition Theorem 1.52(1) cannot hold (see the proof of Lemma 4.50).

Remark 4.53 An anonymous referee suggested an alternative way to show that a general smooth Fano threefold in the deformation family №2.2 is K-polystable.

Let us describe it. Let P_1 and P_2 be two distinct points in $\mathbb{P}^1$, let C be a smooth quartic curve in $\mathbb{P}^2$, and let F_1, F_2, S be smooth surfaces in $\mathbb{P}^1 \times \mathbb{P}^2$ that are defined as follows:

$$F_1 = \mathrm{pr}_1^*(P_1), F_2 = \mathrm{pr}_1^*(P_2), S = \mathrm{pr}_2^*(C),$$

where $\mathrm{pr}_1\colon \mathbb{P}^1 \times \mathbb{P}^2 \to \mathbb{P}^1$ and $\mathrm{pr}_2\colon \mathbb{P}^1 \times \mathbb{P}^2 \to \mathbb{P}^2$ are natural projections. Let X be a double cover of $\mathbb{P}^1 \times \mathbb{P}^2$ that is branched over $F_1 + F_2 + S$. Then X is a singular Fano threefold, which is contained in the deformation family №2.2. Note that

$$\mathrm{Aut}(X) \cong \mathrm{Aut}\big(\mathbb{P}^1, P_1 + P_2\big) \times \mathrm{Aut}(C) \times \mu_2 \cong \big(\mathbb{G}_m \rtimes \mu_2\big) \times \mathrm{Aut}(C) \times \mu_2.$$

Applying [148, Proposition 3.4] and [225, Proposition 4.1], we see that X is K-polystable because the pair $(\mathbb{P}^1, \frac{1}{2}(P_1 + P_2))$ is K-polystable (apply Theorem 1.48 or Theorem 1.9), and it follows from [225, Proposition 4.1] and Lemma 2.9 that the pair $(\mathbb{P}^2, \frac{1}{2}C)$ is K-stable. Now, one can generalize Corollary 1.16 for singular varieties and use this generalization to show that a general member of the family №2.2 is K-polystable.

Every smooth Fano threefold in the family №2.4 is a blow up of $\mathbb{P}^3$ in a smooth curve that is the complete intersection of two cubic surfaces, so that admits a fibration into cubic surfaces. Using this observation, one can prove the following

Lemma 4.54 ([46, Lemma 7.2]) *Let X be a general enough smooth Fano threefold №2.4. Then $\alpha(X) \geqslant \frac{3}{4}$.*

Since smooth Fano threefolds №2.4 have finite automorphism groups [45], Lemma 4.54 and Theorem 1.50 imply that general smooth Fano threefolds №2.4 must be K-stable. By Theorem 1.11, this also follows from Theorem 1.48 and the following

Example 4.55 Let $\mathscr{C}$ be the curve in $\mathbb{P}^3$ that is given by

$$\begin{cases} x_0^3 + x_1^3 + \lambda(x_2^3 + x_3^3) = 0, \\ \lambda(x_0^3 - x_1^3) + x_2^3 - x_3^3 = 0, \end{cases}$$

where $\lambda \in \mathbb{C} \setminus \{0, \pm 1, \pm i\}$. Then $\mathscr{C}$ is a smooth curve. Let $X \to \mathbb{P}^3$ be a blow up of this curve. Then X is a smooth Fano threefold №2.4. Observe that $\mathrm{Aut}(X) \cong \mathrm{Aut}(\mathbb{P}^3, \mathscr{C})$, so that we can identify these two groups. Let $G = \mathrm{Aut}(\mathbb{P}^3, \mathscr{C})$. Then we have the G-equivariant commutative diagram

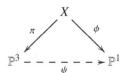

where ψ is the rational map that is given by the pencil of cubic surfaces on $\mathbb{P}^3$ which is generated by $\{x_0^3 + x_1^3 + \lambda(x_2^3 + x_3^3) = 0\}$ and $\{\lambda(x_0^3 - x_1^3) + x_2^3 - x_3^3 = 0\}$, and ϕ is a fibration into cubic surfaces. Let E be the π-exceptional surface, and let F be a sufficiently general fiber of the morphism ϕ. Then

$$-K_X \sim_{\mathbb{Q}} \frac{4}{3}F + \frac{1}{3}E,$$

and the cone $\mathrm{Eff}(X)$ is generated by E and F. This gives $\alpha(X) \leqslant \frac{3}{4}$, cf. [46, Lemma 7.2]. Observe that the group G contains transformations

$$[x_0 : x_1 : x_2 : x_3] \mapsto [x_0 : \omega^a x_1 : \omega^b x_2 : \omega^c x_3]$$

for all a, b, c in $\{0, 1, 2\}$, where ω is a primitive cube root of unity. These automorphisms generate a subgroup $\Gamma \subset G$ such that $\Gamma \cong \mu_3^3$. Note also that G contains two involutions

$$[x_0 : x_1 : x_2 : x_3] \mapsto [x_1 : x_0 : x_3 : x_2]$$

and

$$[x_0 : x_1 : x_2 : x_3] \mapsto [x_2 : -x_3 : x_0 : -x_1],$$

which generate a subgroup isomorphic to the Klein four group μ_2^2. Thus, we constructed a subgroup in G that is isomorphic to $\mu_3^3 \rtimes \mu_2^2$. On the other hand, the group G must permute the points $[1 : 0 : 0 : 0]$, $[0 : 1 : 0 : 0]$, $[0 : 0 : 1 : 0]$, $[0 : 0 : 0 : 1]$, because these are the vertices of all cubic cones that are contained in the pencil generated by the cubic surfaces $\{x_0^3 + x_1^3 + \lambda(x_2^3 + x_3^3) = 0\}$ and $\{\lambda(x_0^3 - x_1^3) + x_2^3 - x_3^3 = 0\}$. Using this, one can show that $G \cong \mu_3^3 \rtimes \mu_2^2$ when λ is general enough. In fact, one can show that

$$G = \begin{cases} \mu_3^3 \rtimes \mathfrak{A}_4 & \text{if } \lambda^4 - 2\lambda^3 + 2\lambda^2 + 2\lambda + 1 = 0, \\ \mu_3^3 \rtimes D_8 & \text{if } \lambda^4 + 6\lambda^2 + 1 = 0, \\ \mu_3^3 \rtimes \mu_2^2 & \text{otherwise.} \end{cases}$$

Arguing as in the proof of Lemma 4.29 and using Theorem 1.52, we get $\alpha_G(X) \geqslant 1$. Namely, suppose that we have $\alpha_G(X) < 1$, and let us apply Theorem 1.52 with $\mu = 1$. Since the pencil $|F|$ does not contain G-invariant surfaces, we immediately conclude that both conditions (i) and (ii) of Theorem 1.52 do not hold. Hence, it follows from Theorem 1.52 that there is an irreducible G-invariant curve $C \subset X$ with $F \cdot C \leqslant 1$. If $F \cdot C = 0$, then $|F|$ has a unique surface that passes through C, which is impossible since the pencil $|F|$ does not contain G-invariant surfaces. Thus, we see that $F \cdot C = 1$, so that the intersection $F \cap C$ consists of a single point. This point must be Γ-invariant since F is Γ-invariant.

On the other hand, one can check that Γ does not fix points in F. This shows that $\alpha_G(X) \geqslant 1$, so X is K-stable by Theorem 1.48 and Corollary 1.5.

Now, let us show that general smooth Fano threefolds №2.7 are K-stable. To do this, let Q, Q_1 and Q_2 be quadric hypersurfaces in $\mathbb{P}^4$ that are given by the equations

$$x_0^2 + x_1^2 + x_2^2 + x_3^2 + x_4^2 = 0,$$
$$x_0^2 + \xi_5 x_1^2 + \xi_5^2 x_2^2 + \xi_5^3 x_3^2 + \xi_5^4 x_4^2 = 0,$$
$$\xi_5^4 x_0^2 + \xi_5^3 x_1^2 + \xi_5^2 x_2^2 + \xi_5 x_3^2 + x_4^2 = 0,$$

respectively, where ξ_5 is a primitive fifth root of unity. Let $C = Q \cap Q_1 \cap Q_2$. Then C is a smooth curve of genus 5. Let σ be the automorphism of $\mathbb{P}^4$ of order 5 that acts by

$$\left[x_0 : x_1 : x_2 : x_3 : x_4\right] \mapsto \left[x_1 : x_2 : x_3 : x_4 : x_0\right],$$

let τ be the involution of $\mathbb{P}^4$ that acts as

$$\left[x_0 : x_1 : x_2 : x_3 : x_4\right] \mapsto \left[x_4 : x_3 : x_2 : x_1 : x_0\right],$$

let $\Gamma \subset \mathrm{Aut}(\mathbb{P}^4)$ be the subgroup such that $\Gamma \cong \mu_2^4$, and the generator of its ith factor acts by multiplying the coordinate x_i by -1. Set $G = \langle \sigma, \tau, \Gamma \rangle \subset \mathrm{Aut}(\mathbb{P}^4)$. Then $G \cong \mu_2^4 \rtimes \mathrm{D}_{10}$, and C is G-invariant, so that there exists a monomorphism $G \hookrightarrow \mathrm{Aut}(C)$. One can show that it is an isomorphism [134]. Observe that Q is also G-invariant, so that we may identify G with a subgroup in $\mathrm{Aut}(Q)$.

Lemma 4.56 *Let $\pi \colon X \to Q$ be the blow up along C. Then X is a Fano threefold №2.7, and we may identify G with a subgroup in $\mathrm{Aut}(X)$ because there exists a G-equivariant commutative diagram*

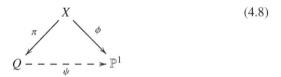

$$(4.8)$$

where ϕ is a fibration into del Pezzo surfaces of degree 4, and ψ is the map given by the pencil generated by the surfaces $Q_1|_Q$ and $Q_2|_Q$. Moreover, we have $\alpha_G(X) \geqslant 1$.

Proof All required assertions are clear except for $\alpha_G(X) \geqslant 1$. To show that $\alpha_G(X) \geqslant 1$, let us apply Theorem 1.52 with $\mu = 1$. First, we observe that the diagram (4.8) gives a homomorphism of groups $\nu \colon G \to \mathrm{Aut}(\mathbb{P}^1)$ such that $\ker(\nu) = \Gamma$ and $\mathrm{im}(\nu) \cong \mathrm{D}_{10}$. Observe that $\mathbb{P}^1$ has no $\nu(G)$-fixed points, so that X has no G-fixed points.

Let F be the proper transform on X of the surface $Q_1|_Q$, and let C be a G-invariant irreducible curve in X. We claim that $F \cdot C \notin \{0, 1\}$. Indeed, if $F \cdot C = 0$, then $\phi(C)$ must be a $\nu(G)$-fixed point in $\mathbb{P}^1$, which is impossible. Similarly, if $F \cdot C = 1$, then $F \cap C$ is a Γ-fixed point, so that $Q_1|_Q$ contains a Γ-fixed point, which is not the case. This shows that $F \cdot C \notin \{0, 1\}$.

Applying Theorem 1.52 with $\mu = 1$, we see that $\alpha_G(X) \geqslant 1$ provided that X does not contain a G-irreducible surface S such that $-K_X \sim_{\mathbb{Q}} \lambda S + \Delta$ for some $\lambda > 1$ and some effective $\mathbb{Q}$-divisor Δ. Suppose that such surface S exists. Then

$$\frac{3}{2}F + \frac{1}{2}E \sim -K_X \sim_{\mathbb{Q}} \lambda S + \Delta. \tag{4.9}$$

Let us seek for a contradiction. If $S = E$, then (4.9) gives $\Delta|_F \sim_{\mathbb{Q}} (1-2\lambda)(-K_F)$, which is a contradiction since Δ is effective. Thus, we have $S \sim \pi^*(dH) - mE$ for some integers $d \geqslant 1$ and $m \geqslant 0$, where H is a hyperplane section of the quadric Q. Then (4.9) gives $\lambda(\pi^*(dH) - mE) + \Delta \sim_{\mathbb{Q}} \pi^*(3H) - E$, so that either $d = 1$ or $d = 2$. Moreover, we have $S \sim_{\mathbb{Q}} \frac{d}{2}F + (\frac{d}{2} - m)E$, which gives $S|_F \sim_{\mathbb{Q}} (\frac{d}{2} - m)E|_F$. This shows that $m \leqslant \frac{d}{2}$.

If $d = 1$, then $\pi(S)$ is a G-invariant hyperplane section of the quadric Q, which is impossible since $\mathbb{P}^4$ does not have G-invariant hyperplanes. Then $d = 2$ and $m \in \{0, 1\}$.

If $m = 1$, then $S \sim F$, so that $\phi(S)$ is a $\nu(G)$-fixed point in $\mathbb{P}^1$, which is impossible. Then $d = 2$ and $m = 0$, so that $S \sim_{\mathbb{Q}} F + E$. Now, (4.9) gives $\Delta|_F \sim_{\mathbb{Q}} (1 - 2\lambda)(-K_F)$, which is absurd. This shows that $\alpha_G(X) \geqslant 1$. $\qquad\square$

Since all smooth Fano threefolds №2.7 have finite automorphism groups [45], we see that the Fano threefold in Lemma 4.56 is K-stable by Theorem 1.48 and Corollary 1.5. Hence, a general smooth Fano threefold in the family №2.7 is K-stable by Theorem 1.11.

Now let us present a K-stable smooth Fano threefold in the deformation family №2.18. Let B be the surface

$$\{x_0^2(y_0^2 + \omega y_1^2 + \omega^2 y_2^2) + x_1^2(y_0^2 + \omega^2 y_1^2 + \omega y_2^2) = 0\} \subset \mathbb{P}^1 \times \mathbb{P}^2,$$

where ω is a primitive cubic root of unity, x_0 and x_1 are homogeneous coordinates on $\mathbb{P}^1$, and y_0, y_1, y_2 are coordinates on $\mathbb{P}^2$. Then B is smooth. Let $\pi \colon X \to \mathbb{P}^1 \times \mathbb{P}^2$ be a double cover branched over the surface B. Then X is a smooth Fano threefold №2.18, and we have the commutative diagram

(4.10)

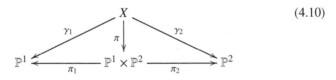

where π_1 and π_2 are natural projections, γ_1 is a fibration into quadric surfaces, and γ_2 is a (standard) conic bundle. Let ι_1 be the involution in $\mathrm{Aut}(\mathbb{P}^1 \times \mathbb{P}^2)$ that is given by

$$\big([x_0 : x_1], [y_0 : y_1 : y_2]\big) \mapsto \big([x_0 : x_1], [y_0 : -y_1 : y_2]\big),$$

let ι_2 be the involution that is given by

$$\big([x_0 : x_1], [y_0 : y_1 : y_2]\big) \mapsto \big([x_0 : x_1], [y_0 : y_1 : -y_2]\big),$$

let σ be the involution that is given by

$$\big([x_0 : x_1], [y_0 : y_1 : y_2]\big) \mapsto \big([x_0 : -x_1], [y_0 : y_1 : y_2]\big),$$

let ς be the involution that is given by

$$\big([x_0 : x_1], [y_0 : y_1 : y_2]\big) \mapsto \big([x_1 : x_0], [y_0 : y_2 : y_1]\big),$$

and let θ be the automorphism of order 3 that is given by

$$\big([x_0 : x_1], [y_0 : y_1 : y_2]\big) \mapsto \big([x_0 : \omega x_1], [y_2 : y_0 : y_1]\big).$$

These automorphisms leave the surface B invariant, and their actions on $\mathbb{P}^1 \times \mathbb{P}^2$ lift to the double cover X, so that we can consider them as automorphisms of the threefold X. Let τ be the Galois involution of the double cover π, and let $G = \langle \iota_1, \iota_2, \sigma, \varsigma, \theta, \tau \rangle$. Then G is a finite group because the whole automorphism group $\mathrm{Aut}(X)$ is finite [45]. Observe also that the commutative diagram (4.10) is G-equivariant.

Lemma 4.57 $\alpha_G(X) \geqslant 1$.

Proof Let us apply Theorem 1.52 with $\mu = 1$. Since $\mathbb{P}^2$ does not have G-fixed points, the threefold X has no G-fixed points, so Theorem 1.52(2) does not hold.

Let F be the fiber of γ_1 over $(0 : 1)$, and let $\widehat{G}$ be its stabilizer in G. Then $F \cong \mathbb{P}^1 \times \mathbb{P}^1$ and $\widehat{G} = \langle \iota_1, \iota_2, \sigma, \theta, \tau \rangle$. If X contains an irreducible G-invariant curve C such that $0 \leqslant F \cdot C \leqslant 1$, then either $\gamma_1(C)$ is a point or $F \cap C$ consists of one point. The former case is impossible since $\mathbb{P}^1$ does not have G-fixed points. The latter case is also impossible because F does not have $\widehat{G}$-fixed points. So, Theorem 1.52(3) does not hold.

Finally, suppose that X contains a G-irreducible surface S such that $-K_X \sim_{\mathbb{Q}} \lambda S + \Delta$, where Δ is effective $\mathbb{Q}$-divisor, and $\lambda \in \mathbb{Q}$ such that $\lambda > 1$. Then $S \in |\gamma_2^*(\mathcal{O}_{\mathbb{P}^2}(1))|$ since

$$\lambda S + \Delta \sim_{\mathbb{Q}} -K_X \sim \gamma_1^*\big(\mathcal{O}_{\mathbb{P}^1}(1)\big) + \gamma_2^*\big(\mathcal{O}_{\mathbb{P}^2}(2)\big).$$

This is impossible because $\mathbb{P}^2$ does not contain G-invariant lines. Therefore, we see that Theorem 1.52(1) does not hold, so that $\alpha_G(X) \geqslant 1$.   □

Thus, the smooth Fano threefold X is K-stable by Theorem 1.48 and Corollary 1.5, so that the general Fano threefold in the family №2.18 is also K-stable by Theorem 1.11.

In the remaining part of this section, we will use our construction of the threefold X to present one smooth K-stable Fano threefold in the family №3.4, which would imply that a general Fano threefold in this family is also K-stable.

Let O be the point $[1 : 0 : 0] \in \mathbb{P}^2$. Then the fiber of the conic bundle γ_2 over O is smooth. Let $\alpha: V \to X$ be the blow up of this fiber. Then V is a smooth Fano threefold in the family №3.4, and (4.10) can be extended to the commutative diagram

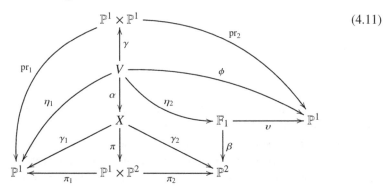

$$(4.11)$$

where β is a blow up of the point O, υ is the natural projection, η_1 is a fibration into del Pezzo surfaces of degree 6, η_2 and γ are conic bundles, ϕ is a fibration into del Pezzo surfaces of degree 4, and pr_1 and pr_2 are projections to the first and the second factors, respectively. Let $\Gamma = \langle \iota_1, \iota_2, \sigma, \varsigma, \tau \rangle$. Then the fiber of γ_2 over O is Γ-invariant, so that the Γ-action lifts to V. Therefore, we can identify Γ with a subgroup in $\mathrm{Aut}(V)$.

Lemma 4.58 (cf. the proof of Lemma 4.57) $\alpha_\Gamma(V) \geqslant 1$.

Proof Let us apply Theorem 1.52 with $\mu = 1$. Since both $\mathbb{P}^1$ in (4.11) have no Γ-fixed points, V does not have Γ-fixed points, so that Theorem 1.52(2) does not hold.

Let F be a fiber of η_1, let S be a fibers of ϕ, and let E be the exceptional divisor of α. Then F, S and E generate the cone $\mathrm{Eff}(V)$ (see [156]), and $-K_X \sim F + 2S + E$, so that Theorem 1.52(1) cannot hold, since $|S|$ does not have Γ-invariant surfaces.

Finally, suppose that V contains a Γ-irreducible curve C such that $0 \leqslant F \cdot C \leqslant 1$ and $0 \leqslant S \cdot C \leqslant 1$. Since both $\mathbb{P}^1$ in (4.11) do not have Γ-fixed points, we get $F \cdot C = S \cdot C = 1$. Then $\gamma(C)$ is a curve in $\mathbb{P}^1 \times \mathbb{P}^1$ of degree $(1, 1)$, which is impossible since $\mathbb{P}^1 \times \mathbb{P}^1$ does not have Γ-invariant curves of degree $(1, 1)$. Hence, we see that Theorem 1.52(3) does not hold either, so that $\alpha_\Gamma(V) \geqslant 1$. □

Thus, the threefold V is K-stable by Theorem 1.48 and Corollary 1.5 because its automorphism group is finite [45], so that general Fano threefolds №3.4 are also K-stable.

4.6 Blow ups of Veronese and quadric cones

In this section, we will prove that all smooth Fano threefolds in the families №3.9 and №4.2 are K-polystable.

Let $\mathscr{S}$ be one of the following surfaces: $\mathbb{P}^2$ or $\mathbb{P}^1 \times \mathbb{P}^1$. Then we fix a smooth irreducible curve $\mathscr{C}$ in the surface $\mathscr{S}$ such that

- if $\mathscr{S} = \mathbb{P}^2$, then $\mathscr{C}$ is a quartic curve, so that $\mathscr{C}$ has genus 3,
- if $\mathscr{S} = \mathbb{P}^1 \times \mathbb{P}^1$, then $\mathscr{C}$ is a curve of degree $(2, 2)$, so that $\mathscr{C}$ has genus 1.

Let pr_1 and pr_2 be projections of $\mathbb{P}^1 \times \mathscr{S}$ to the first and the second factors, respectively. Put $\mathscr{B} = \mathrm{pr}_2^*(\mathscr{C})$, put $\mathscr{E} = \mathrm{pr}_1^*([1 : 0])$ and put $\mathscr{E}' = \mathrm{pr}_1^*([0 : 1])$. Then $\mathscr{B} \cong \mathbb{P}^1 \times \mathscr{C}$, and

$$\mathrm{Aut}\big(\mathbb{P}^1 \times \mathscr{S}; \mathscr{E} + \mathscr{E}' + \mathscr{B}\big) \cong \mathrm{Aut}\big(\mathscr{S}; \mathscr{C}\big) \times \big(\mathbb{G}_m \rtimes \mu_2\big),$$

where $\mu_2 = \langle \iota \rangle$ for the involution $\iota \in \mathrm{Aut}(\mathbb{P}^1 \times \mathscr{S})$ that acts as $([u : v], P) \mapsto ([v : u], P)$, so that ι swaps $\mathscr{E}$ and $\mathscr{E}'$. Let $\eta \colon W \to \mathbb{P}^1 \times \mathscr{S}$ be a double cover branched over $\mathscr{E} + \mathscr{E}' + \mathscr{B}$. Denote by $\overline{E}$, $\overline{E}'$ and $\overline{B}$ the preimages on W of the surfaces $\mathscr{E}$, $\mathscr{E}'$ and $\mathscr{B}$, respectively. Then W is singular along the curves $\overline{E} \cap \overline{B}$ and $\overline{E}' \cap \overline{B}$, the composition $\mathrm{pr}_1 \circ \eta$ is a fibration into del Pezzo surfaces of degree 2 (when $\mathscr{S} = \mathbb{P}^2$) or 4 (when $\mathscr{S} = \mathbb{P}^1 \times \mathbb{P}^1$).

$$\mathrm{Aut}\big(W\big) \cong \mathrm{Aut}\big(\mathscr{S}; \mathscr{C}\big) \times \big(\mathbb{G}_m \rtimes \mu_2\big).$$

Then η gives an epimorphism $\mathrm{Aut}(W) \to \mathrm{Aut}(\mathbb{P}^1 \times \mathscr{S}, \mathscr{E} + \mathscr{E}' + \mathscr{B})$, whose kernel is generated by the Galois involution τ of the double cover η, which is contained in the torus $\mathbb{G}_m$.

Let $\alpha \colon \widehat{X} \to W$ be the blow up of the curves $\overline{E} \cap \overline{B}$ and $\overline{E}' \cap \overline{B}$, and let $\widehat{S}$ and $\widehat{S}'$ be the exceptional surfaces of this blow up that are mapped to $\overline{E} \cap \overline{B}$ and $\overline{E}' \cap \overline{B}$, respectively, let $\widehat{E}$, $\widehat{E}'$ and $\widehat{B}$ be the proper transforms on $\widehat{X}$ of the surfaces $\overline{E}$, $\overline{E}'$ and $\overline{B}$, respectively. Then $\widehat{X}$ is smooth. Note that $\widehat{B} \cong \mathbb{P}^1 \times \mathscr{C}$, $\widehat{E} \cap \widehat{B} = \varnothing$, $\widehat{E}' \cap \widehat{B} = \varnothing$, and $\widehat{E} \cong \widehat{E}' \cong \mathscr{S}$. If $\mathscr{S} = \mathbb{P}^2$, then the normal bundles of the surfaces $\widehat{E}$ and $\widehat{E}'$ are isomorphic to $\mathcal{O}_{\mathbb{P}^2}(-2)$. If $\mathscr{S} = \mathbb{P}^1 \times \mathbb{P}^1$, then their normal bundles are line bundles of degree $(-1, -1)$.

There is a birational morphism $\psi\colon \widehat{X} \to X$ contracting $\widehat{B}$ to a curve isomorphic to $\mathscr{C}$. Let E, E', S and S' be proper transforms on X of the surfaces $\widehat{E}$, $\widehat{E}'$, $\widehat{S}$ and $\widehat{S}'$, respectively. Then X, E, E', S and S' are smooth, ψ is a blow up of the curve $S \cap S'$, and there exists the commutative diagram

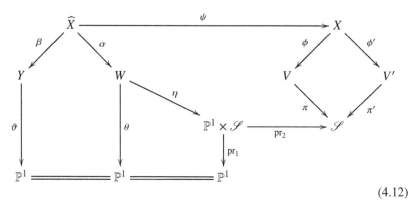

$$(4.12)$$

where β is a birational morphism contracting $\widehat{E}$ and $\widehat{E}'$ to isolated terminal singular points, both θ and ϑ are fibrations into del Pezzo surfaces, ϕ and ϕ' are birational morphisms that contract S and S' to smooth curves, respectively, and both π and π' are $\mathbb{P}^1$-bundles. Note that $\mathrm{Aut}(X) \cong \mathrm{Aut}(\widehat{X}) \cong \mathrm{Aut}(Y) \cong \mathrm{Aut}(W)$. Moreover, we have $V \cong V'$ and these threefolds can be described as follows:

- if $\mathscr{S} = \mathbb{P}^2$, then $V \cong V' \cong \mathbb{P}(O_{\mathbb{P}^2} \oplus O_{\mathbb{P}^2}(2))$, i.e. the smooth Fano threefold №2.36,
- if $\mathscr{S} = \mathbb{P}^1 \times \mathbb{P}^1$, then $V \cong V'$ is a blow up of the quadric cone in $\mathbb{P}^4$ in its vertex.

All birational morphisms in (4.12) are $\mathrm{Aut}(X)$-equivariant except for π, π', ϕ and ϕ'. The involution ι swaps S and S', so that it does not act on V and V' biregularly.

Let E_V and E'_V be the proper transforms on V of the surfaces E and E', respectively, let $E_{V'}$ and $E'_{V'}$ be the proper transforms on V' of the surfaces E and E', respectively. Then E_V and E'_V are disjoint sections of the $\mathbb{P}^1$-bundle π, while $E_{V'}$ and $E'_{V'}$ are disjoint sections of the $\mathbb{P}^1$-bundle π', so that $E_V \cong E'_V \cong E_{V'} \cong E'_{V'} \cong \mathscr{S}$. Moreover, if $\mathscr{S} = \mathbb{P}^2$, then $E_V|_{E_V} \cong E'_{V'}|_{E'_{V'}} \cong O_{\mathbb{P}^2}(2)$ and $E'_V|_{E'_V} \cong E_{V'}|_{E_{V'}} \cong O_{\mathbb{P}^2}(-2)$. Similarly, if $\mathscr{S} = \mathbb{P}^1 \times \mathbb{P}^1$, then both $E_V|_{E_V}$ and $E'_{V'}|_{E'_{V'}}$ are line bundles of degree $(1,1)$, but $E'_V|_{E'_V}$ and $E_{V'}|_{E_{V'}}$ are line bundles of degree $(-1,-1)$.

Let S'_V and $S_{V'}$ be the transforms on V and V' of the surfaces S' and S, respectively. Put $C = S'_V \cap E_V$ and $C' = S_{V'} \cap E'_{V'}$. Then $S'_V = \pi^*(\mathscr{C})$,

$S_{V'} = (\pi')^*(\mathscr{C})$, $C \cong C' \cong \mathscr{C}$. Note that ϕ and ϕ' are blow ups of the curves C and C', respectively.

If $\mathscr{S} = \mathbb{P}^2$, then X is a Fano threefold №3.9, and all smooth Fano threefolds №3.9 can be obtained in this way. Similarly, if $\mathscr{S} = \mathbb{P}^1 \times \mathbb{P}^1$, then X is a smooth Fano threefold №4.2, and all smooth Fano threefolds №4.2 can be obtained in this way.

Let $G = \mathrm{Aut}(X)$ and $C = S \cap S' = \psi(\widehat{B})$. Then C consists of all G-fixed points in X, and every G-invariant irreducible curve in X is either C or a smooth fiber of $\pi \circ \phi$. Thus, using Theorem 1.52, it is not hard to deduce the following.

Corollary 4.59 *If $\mathscr{S} = \mathbb{P}^2$ and $\mathscr{S}$ does not contain $\mathrm{Aut}(\mathscr{S};\mathscr{C})$-invariant lines and conics, then $\alpha_G(X) \geqslant 1$. If $\mathscr{S} = \mathbb{P}^1 \times \mathbb{P}^1$ and $\mathscr{S}$ does not contain $\mathrm{Aut}(\mathscr{S};\mathscr{C})$-invariant curves of degree $(1,0)$, $(0,1)$, $(1,1)$, then $\alpha_G(X) \geqslant 1$.*

Example 4.60 Suppose that $\mathscr{S} = \mathbb{P}^2$ and $\mathscr{C} = \{xy^3 + yz^3 + zx^3 = 0\}$, where x, y, z are coordinates on $\mathbb{P}^2$. Then $\mathrm{Aut}(\mathscr{S};\mathscr{C}) \cong \mathrm{PSL}_2(\mathbb{F}_7)$, so that $\alpha_G(X) = 1$ by Corollary 4.59. Then X is K-polystable smooth Fano threefold №3.9 by Theorem 1.48.

Example 4.61 Suppose that $\mathscr{S} = \mathbb{P}^1 \times \mathbb{P}^1$ and $\mathscr{C} = \{x_0^2 y_0^2 - x_0^2 y_1^2 - x_1^2 y_0^2 - x_1^2 y_1^2 = 0\}$, where $[x_0 : x_1]$ and $[y_0 : y_1]$ are coordinates on the first and the second factors of $\mathscr{S}$, respectively. Then $\mathscr{C}$ is a smooth curve, and $\mathrm{Aut}(\mathscr{S};\mathscr{C})$ contains the transformations

$$([x_0 : x_1],[y_0 : y_1]) \mapsto ([y_0 : y_1],[x_0 : x_1])$$

and $([x_0 : x_1],[y_0 : y_1]) \mapsto ([x_1, x_0],[y_0 : iy_1])$. Then $\alpha_G(X) = 1$ by Corollary 4.59, so that the threefold X is K-polystable by Theorem 1.48.

Let us use Corollary 1.110 and Theorem 1.112, to prove that X is always K-polystable. Suppose that X is not K-polystable. By Theorem 1.22, there exists G-invariant prime divisor F over X such that $\beta(F) \leqslant 0$. Let $Z = C_X(F)$.

Remark 4.62 It follows from [93, Section 10] that $\beta(D) > 0$ for every irreducible surface $D \subset X$ such that $D \neq E$ and $D \neq E'$. On the other hand, E and E' are not G-invariant.

Therefore, either $Z = C$, or Z is a smooth fiber of $\pi \circ \phi$, or Z is a point in C.

Lemma 4.63 $Z \neq C$.

Proof Suppose that $Z = C$. Then $Z \subset S$. Let us apply results of Section 1.7 to S and Z. As usual, we will use notations introduced in this section. Take $x \in \mathbb{R}_{\geqslant 0}$. Let

$$a = \begin{cases} 4 & \text{if } \mathscr{S} = \mathbb{P}^2, \\ 2 & \text{if } \mathscr{S} = \mathbb{P}^1 \times \mathbb{P}^1. \end{cases}$$

Let $P(x) = P(-K_X - xS)$ and $N(x) = N(-K_X - xS)$. Since $-K_X - xS$ is $\mathbb{R}$-rationally equivalent to $(\frac{a+1}{a} - x)S + \frac{1}{a}S' + 2E$, $-K_X - xS$ is not pseudo-effective for $x > \frac{a+1}{a}$. Then

$$P(x) = \begin{cases} \left(\frac{a+1}{a} - x\right)S + \frac{1}{a}S' + 2E & \text{if } 0 \leqslant x \leqslant \frac{1}{a}, \\ \left(\frac{a+1}{a} - x\right)(S + 2E) + \frac{1}{a}S' & \text{if } \frac{1}{a} \leqslant x \leqslant 1, \\ \left(\frac{a+1}{a} - x\right)(S + S' + 2E) & \text{if } 1 \leqslant x \leqslant \frac{a+1}{a} \end{cases}$$

and

$$N(x) = \begin{cases} 0 & \text{if } 0 \leqslant x \leqslant \frac{1}{a}, \\ \left(2x - \frac{2}{a}\right)E & \text{if } \frac{1}{a} \leqslant x \leqslant 1, \\ \left(2x - \frac{2}{a}\right)E + (x-1)S' & \text{if } 1 \leqslant x \leqslant \frac{a+1}{a}. \end{cases}$$

Let $e = E|_S$, let $s' = S'|_S$, and let ℓ be any fiber of the natural projection $S \to \mathscr{C}$. Then $Z = s' \equiv 2a\ell + e$ and $S|_S \equiv 2a\ell - e$ on the surface S, so that $-K_X|_S \equiv (2a + 4)\ell + e$. Therefore, we have

$P(x)|_S - yZ$

$$\equiv \begin{cases} (4 + 2a(1 - x - y))\ell + (x - y + 1)e & \text{if } 0 \leqslant x \leqslant \frac{1}{a}, \\ (4 - 2a(x + y - 1))\ell + \dfrac{2 - a(x + y - 1)}{a}e & \text{if } \frac{1}{a} \leqslant x \leqslant 1, \\ (4 - 2a(2x + y - 2))\ell + \dfrac{2 - a(2x + y - 2)}{a}e & \text{if } 1 \leqslant x \leqslant \frac{a+1}{a}. \end{cases}$$

Moreover, on the surface S, we have $e^2 = -2a$, $e \cdot \ell$ and $\ell^2 = 0$. Note that Z is contained in the support of $N(x)|_S$ only if $1 \leqslant x \leqslant \frac{a+1}{a}$. In this case, we have $N(x)|_S = (x - 1)Z$. Therefore, using Corollary 1.110, we conclude that

$S(W^S_{\bullet,\bullet}; Z)$

$$= \frac{3}{-K_X^3} \int_1^{\frac{a+1}{a}} (x - 1)P(x) \cdot P(x) \cdot S dx$$

$$+ \frac{3}{-K_X^3} \int_0^{\frac{1}{a}} \int_0^\infty \mathrm{vol}\big((4 + 2a(1 - x - y))\ell + (x - y + 1)e\big) dy dx$$

$$+ \frac{3}{-K_X^3} \int_{\frac{1}{a}}^1 \int_0^\infty \mathrm{vol}\Big((4 - 2a(x + y - 1))\ell + \frac{2 - a(x + y - 1)}{a}e\Big) dy dx$$

$$+ \frac{3}{-K_X^3} \int_1^{\frac{a+1}{a}} \int_0^\infty \mathrm{vol}\big((4 - 2a(2x + y - 2))\ell + \frac{2 - a(2x + y - 2)}{a}\mathbf{e}\big)dydx$$

$$= \frac{3}{-K_X^3} \int_1^{\frac{a+1}{a}} \frac{8}{a}(x - 1)(ax - a - 1)^2 dx$$

$$+ \frac{3}{-K_X^3} \int_0^{\frac{1}{a}} \int_0^{1+x} 2(x - y + 1)(4 + a - 3ax - ay)dydx$$

$$+ \frac{3}{-K_X^3} \int_{\frac{1}{a}}^1 \int_0^{1+\frac{2}{a}-x} \frac{2}{a}(ax + ay - a - 2)^2 dydx$$

$$+ \frac{3}{-K_X^3} \int_1^{\frac{a+1}{a}} \int_0^{2+\frac{2}{a}-2x} \frac{2}{a}(2ax + ay - 2a - 2)^2 dydx$$

$$= \frac{a^3 + 8a^2 + 24a + 16}{6a^2}.$$

Thus, if $S = \mathbb{P}^2$ then $S(W_{\bullet,\bullet}^S; Z) = \frac{19}{52}$. Similarly, if $S = \mathbb{P}^1 \times \mathbb{P}^1$, then $S(W_{\bullet,\bullet}^S; Z) = \frac{13}{28}$. Using Remark 4.62 and Corollary 1.110, we get $\beta(F) > 0$, which is a contradiction. $\qquad\square$

Hence we see that either Z is a smooth fiber of $\pi \circ \phi$, or Z is a point in C.

Lemma 4.64 *Suppose that $\mathscr{S} = \mathbb{P}^2$. Then Z is a point in C.*

Proof Suppose Z is a smooth fiber of the morphism $\pi \circ \phi$. Let us seek for a contradiction. Let H be a general surface in $|(\pi \circ \phi)^*(\mathcal{O}_{\mathbb{P}^2}(1))|$ such that H contains Z. Then H is smooth. Let us apply results of Section 1.7 to H and Z (using notations introduced therein).

Take $x \in \mathbb{R}_{\geqslant 0}$. If $0 \leqslant x \leqslant 1$, then $-K_X - xH$ is nef. If $1 \leqslant x \leqslant 3$, then

$$P\big(-K_X - xH\big) = -K_X - xH - \frac{x-1}{2}\big(E + E'\big) \sim_{\mathbb{R}} \frac{3-x}{2}\big(S + 2E\big)$$

and $N(-K_X - xH) = \frac{x-1}{2}(E + E')$. If $x > 3$, then $-K_X - xH$ is not pseudo-effective.

Let $\mathbf{e} = E|_H$, $\mathbf{e}' = E'|_H$ and $\ell = H|_H$. Then $Z \sim \ell$, $-K_X|_H \sim 3\ell + \mathbf{e} + \mathbf{e}'$ and

$$\mathbf{e}^2 = -2, (\mathbf{e}')^2 = -2, \mathbf{e} \cdot \mathbf{e}' = 0, \mathbf{e} \cdot \ell = \mathbf{e}' \cdot \ell = 1, \ell^2 = 0.$$

Thus, since $Z \not\subset \mathrm{Supp}(N(-K_X - xH)|_H)$, it follows from Corollary 1.110 that

$$S\big(W_{\bullet,\bullet}^H; Z\big) = \frac{3}{26} \int_0^1 \int_0^\infty \mathrm{vol}\big((3 - x - y)\ell + \mathbf{e} + \mathbf{e}'\big)dydx$$

$$+ \frac{3}{26} \int_1^3 \int_0^\infty \mathrm{vol}\Big((3 - x - y)\ell + \frac{3-x}{2}(\mathbf{e} + \mathbf{e}')\Big)dydx$$

$$= \frac{3}{26} \int_0^1 \int_0^{1-x} \left((3-x-y)\ell + \mathbf{e} + \mathbf{e}' \right)^2 dy dx$$

$$+ \frac{3}{26} \int_{1-x}^{3-x} \int_0^{1-x} \frac{(3-x-y)^2}{4} \left(2\ell + \mathbf{e} + \mathbf{e}' \right)^2 dy dx$$

$$+ \frac{3}{26} \int_1^3 \int_0^{3-x} \left((3-x-y)\ell + \frac{3-x}{2}(\mathbf{e} + \mathbf{e}') \right)^2 dy dx$$

$$= \frac{10}{13} < 1.$$

Therefore, since $S_X(H) < 1$ by Remark 4.62, Corollary 1.110 also gives $\beta(F) > 0$, which contradicts our assumption. $\qquad \square$

Similarly, we prove the following.

Lemma 4.65 *Suppose that $\mathscr{S} = \mathbb{P}^1 \times \mathbb{P}^1$. Then Z is a point in C.*

Proof Let ℓ_1 and ℓ_2 be different rulings of $\mathscr{S}$, let $H_1 = (\pi \circ \phi)^*(\ell_1)$ and $H_2 = (\pi \circ \phi)^*(\ell_2)$. Then $S \sim H_1 + H_2 - E + E'$, $S' \sim H_1 + H_2 + E - E'$ and $-K_X \sim 2H_1 + 2H_2 + E + E'$. Moreover, it follows from [93, Section 10] that $\mathrm{Pic}(X) = \mathbb{Z}[H_1] \oplus \mathbb{Z}[H_2] \oplus \mathbb{Z}[E] \oplus \mathbb{Z}[E']$,

$$\mathrm{Nef}(X) = \mathbb{R}_{\geqslant 0}[H_1] \oplus \mathbb{R}_{\geqslant 0}[H_2] \oplus \mathbb{R}_{\geqslant 0}[H_1 + H_2 + E]$$
$$\oplus \mathbb{R}_{\geqslant 0}[H_1 + H_2 + E'] \oplus \mathbb{R}_{\geqslant 0}[H_1 + H_2 + E + E'],$$

and

$$\overline{\mathrm{Eff}}(X) = \mathbb{R}_{\geqslant 0}[H_1] \oplus \mathbb{R}_{\geqslant 0}[H_2] \oplus \mathbb{R}_{\geqslant 0}[E] \oplus \mathbb{R}_{\geqslant 0}[E'] \oplus \mathbb{R}_{\geqslant 0}[S] \oplus \mathbb{R}_{\geqslant 0}[S'].$$

Let l_1 and l_2 be general fibers of the projections $S \to \mathscr{C}$ and $S' \to \mathscr{C}$, respectively, let l_3 and l_4 be the rulings of E mapped by $\pi \circ \phi$ to the rulings ℓ_1 and ℓ_2, respectively, let l_5 and l_6 be the rulings of E' mapped by $\pi \circ \phi$ to the rulings ℓ_1 and ℓ_2, respectively. Then

$$\overline{\mathrm{NE}}(X) = \mathbb{R}_{\geqslant 0}[l_1] + \mathbb{R}_{\geqslant 0}[l_2] + \mathbb{R}_{\geqslant 0}[l_3] + \mathbb{R}_{\geqslant 0}[l_4] + \mathbb{R}_{\geqslant 0}[l_5] + \mathbb{R}_{\geqslant 0}[l_6].$$

See [156] and [93, Section 10].

Suppose that Z is a smooth fiber of the morphism $\pi \circ \phi$. Let us seek for a contradiction. Let Y be the unique surface in $|H_1|$ that contains Z. Then Y is irreducible and normal. Note that Y is smooth along the curve Z. Let us apply results of Section 1.7 to Y and Z. As usual, we will use notations introduced in this section.

Let $\mathbf{e} = E|_Y$, $\mathbf{e}' = E'|_Y$, and let ℓ be a general fiber of the morphism $\pi \circ \phi|_Y \to \pi \circ \phi(Y)$. Then $-K_X|_Y \sim 2\ell + \mathbf{e} + \mathbf{e}'$ and $Z \sim \ell$. On the surface Y, we have $\mathbf{e}^2 = -1$, $(\mathbf{e}')^2 = -1$, $\mathbf{e} \cdot \mathbf{e}' = 0$, $\mathbf{e} \cdot \ell = \mathbf{e}' \cdot \ell = 1$, $\ell^2 = 0$. Note that the surface Y is smooth in the case when the ruling $\pi \circ \phi(Y)$ intersects $\mathscr{C}$

transversally. If the ruling $\pi \circ \phi(Y)$ is tangent to $\mathscr{C}$, then it follows from [167, §2] that Y has one isolated ordinary double point that is mapped to the point $\pi \circ \phi(Y) \cap \mathscr{C}$ by the conic bundle $\pi \circ \phi$. In both cases, Y is a del Pezzo surface of degree 6 that is smooth along Z, $\mathbf{e}$ and $\mathbf{e}'$.

Fix $x \in \mathbb{R}_{\geqslant 0}$. Then $-K_X - xY$ is pseudo-effective $\iff x \in [0, 2]$. Moreover, if $x \in [0, 1]$, then this divisor is nef. If $x \in [1, 2]$, then $P(-K_X - xY) = -K_X - xY - (x - 1)(E + E')$ and $N(-K_X - xY) = (x - 1)(E + E')$. If $0 \leqslant x \leqslant 1$, then $P(-K_X - xY)|_Y \sim_{\mathbb{R}} 2\ell + \mathbf{e} + \mathbf{e}'$. Similarly, if $1 \leqslant x \leqslant 2$, then we have $P(-K_X - xY)|_Y \sim_{\mathbb{R}} 2\ell + (2 - x)(\mathbf{e} + \mathbf{e}')$ and $N(-K_X - xY)|_Y = (x - 1)(\mathbf{e} + \mathbf{e}')$. Thus, by Corollary 1.110, $S(W_{\bullet,\bullet}^Y; Z)$ is equal to

$$\frac{3}{28} \int_0^1 \int_0^\infty \mathrm{vol}\big((2 - y)\ell + \mathbf{e} + \mathbf{e}'\big) dy dx$$

$$+ \frac{3}{28} \int_1^2 \int_0^\infty \mathrm{vol}\big((2 - y)\ell + (2 - x)(\mathbf{e} + \mathbf{e}')\big) dy dx$$

$$= \frac{3}{28} \int_0^1 \int_0^1 \big((2 - y)\ell + \mathbf{e} + \mathbf{e}'\big)^2 dy dx + \frac{3}{28} \int_0^1 \int_1^2 (2 - y)^2 \big(\ell + \mathbf{e} + \mathbf{e}'\big)^2 dy dx$$

$$+ \frac{3}{28} \int_1^2 \int_0^x \big((2 - y)\ell + (2 - x)(\mathbf{e} + \mathbf{e}')\big) dy dx$$

$$+ \frac{3}{28} \int_1^2 \int_x^2 (2 - y)^2 \big(\ell + \mathbf{e} + \mathbf{e}'\big)^2 dy dx$$

$$= \frac{3}{28} \int_0^1 \int_0^1 (6 - 4y) dy dx + \frac{3}{28} \int_0^1 \int_1^2 2(2 - y)^2 dy dx$$

$$+ \frac{3}{28} \int_1^2 \int_0^x 2(2 - x)(2 + x - 2y) dy dx + \frac{3}{28} \int_1^2 \int_x^2 2(2 - y)^2 dy dx$$

$$= \frac{45}{56} < 1.$$

Therefore, as in the proof of Lemma 4.64, we get $\beta(F) > 0$ by Corollary 1.110, which contradicts our assumption. $\square$

Hence, we see that Z is a point in C. Let us exclude the case $\mathscr{S} = \mathbb{P}^2$.

Lemma 4.66 $\mathscr{S} = \mathbb{P}^1 \times \mathbb{P}^1$.

Proof Suppose that $\mathscr{S} = \mathbb{P}^2$. Let H be a general surface in $|(\pi \circ \phi)^*(\mathcal{O}_{\mathbb{P}^2}(1))|$ containing Z. Then H is smooth. Let $\mathbf{e} = E|_H$ and $\mathbf{e}' = E'|_H$. Then $\mathbf{e}$ and $\mathbf{e}'$ are disjoint (-2)-curves. Moreover, we have $S|_H = \mathbf{f}_1 + \mathbf{f}_2 + \mathbf{f}_3 + \mathbf{f}_4$, where $\mathbf{f}_1, \mathbf{f}_2, \mathbf{f}_3$ and $\mathbf{f}_4$ are disjoint (-1)-curves that intersect transversally the curve $\mathbf{e}$, and do not intersect the curve $\mathbf{e}'$. Similarly, we have $S'|_H = \mathbf{f}_1' + \mathbf{f}_2' + \mathbf{f}_3' + \mathbf{f}_4'$, where $\mathbf{f}_1'$, $\mathbf{f}_2', \mathbf{f}_3'$ and $\mathbf{f}_4'$ are disjoint (-1)-curves such that

$$\mathbf{f}_i \cdot \mathbf{f}'_j = \begin{cases} 1 & \text{if } i = j, \\ 0 & \text{if } i \neq j. \end{cases}$$

The curves $\mathbf{f}'_1, \mathbf{f}'_2, \mathbf{f}'_3, \mathbf{f}'_4$ intersect transversally e', and they do not intersect the curve e. Then Z is one of the four points $\mathbf{f}_1 \cap \mathbf{f}'_1$, $\mathbf{f}_2 \cap \mathbf{f}'_2$, $\mathbf{f}_3 \cap \mathbf{f}'_3$, $\mathbf{f}_4 \cap \mathbf{f}'_4$. Without loss of generality, we may assume that $Z = \mathbf{f}_1 \cap \mathbf{f}'_1$. Now, we will apply results of Section 1.7 to H, $\mathbf{f}_1$ and Z. We will use notations introduced in this section.

Let x be some real number, let $P(x) = P(-K_X - xH)$, let $N(x) = N(-K_X - xH)$, and let ℓ be a general fiber of the conic bundle $\pi \circ \phi|_H : H \to \pi \circ \phi(H)$. Then $\ell \sim \mathbf{f}_1 + \mathbf{f}'_1$. As in the proof of Lemma 4.64, we see that $-K_X - xH$ is not pseudo-effective for $x > 3$. Similarly, if $0 \leqslant x \leqslant 3$, we have

$$P(x)\big|_H = \begin{cases} (3-x)\ell + e + e' & \text{if } 0 \leqslant x \leqslant 1, \\ (3-x)\ell + \dfrac{3-x}{2}(e+e') & \text{if } 1 \leqslant x \leqslant 3, \end{cases}$$

and

$$N(x)\big|_X = \begin{cases} 0 & \text{if } 0 \leqslant x \leqslant 1, \\ \dfrac{x-1}{2}(e+e') & \text{if } 1 \leqslant x \leqslant 3. \end{cases}$$

Recall from Remark 4.62 that $S_X(H) < 1$.

Let us compute $S(W^H_{\bullet,\bullet}; \mathbf{f}_1)$. Take a non-negative real number y. If $0 \leqslant x \leqslant 1$, then

$$P(x)\big|_H - y\mathbf{f}_1 \sim_{\mathbb{R}} (3-x)\ell + e + e' - y\mathbf{f}_1 \sim_{\mathbb{R}} (3-x-y)\mathbf{f}_1 + (3-x)\mathbf{f}'_1 + e + e'.$$

If $0 \leqslant x \leqslant 1$, then the divisor $P(x)|_H - y\mathbf{f}_1$ is pseudo-effective $\iff y \leqslant 3-x$. If $0 \leqslant x \leqslant 1$ and $0 \leqslant y \leqslant 3 - x$, its Zariski decomposition can be described as follows:

- if $0 \leqslant y \leqslant 1 - x$, then $P(x)|_H - y\mathbf{f}_1$ is nef,
- if $1 - x \leqslant y \leqslant 1$, then the Zariski decomposition is

$$\underbrace{(3-x-y)\mathbf{f}_1 + (3-x)\mathbf{f}'_1 + \frac{3-x-y}{2}e + e'}_{\text{positive part}} + \underbrace{\frac{x+y-1}{2}e}_{\text{negative part}},$$

- if $1 \leqslant y \leqslant 2 - x$, then the Zariski decomposition is

$$\underbrace{(3-x-y)\mathbf{f}_1 + (4-x-y)\mathbf{f}'_1 + \frac{3-x-y}{2}e + e'}_{\text{positive part}} + \underbrace{\frac{x+y-1}{2}e + (y-1)\mathbf{f}'_1}_{\text{negative part}},$$

- if $2 - x \leqslant y \leqslant 3 - x$, then the Zariski decomposition is

$$\frac{3 - x - y}{2}\underbrace{(2\mathbf{f}_1 + 4\mathbf{f}'_1 + \mathbf{e} + 2\mathbf{e}') + \frac{x + y - 1}{2}\mathbf{e} + (x + 2y - 3)\mathbf{f}'_1 + (x + y - 2)\mathbf{e}'}.$$

positive part   negative part

Similarly, if $1 \leqslant x \leqslant 3$, then $P(x)|_H - y\mathbf{f}_1 \sim_{\mathbb{R}} (3 - x - y)\mathbf{f}_1 + (3 - x)\mathbf{f}'_1 + \frac{3-x}{2}(\mathbf{e} + \mathbf{e}')$. If $1 \leqslant x \leqslant 3$, then the divisor $P(x)|_H - v\mathbf{f}_1$ is pseudo-effective $\iff y \leqslant 3 - x$. If $1 \leqslant x \leqslant 3$ and $0 \leqslant y \leqslant 3 - x$, its Zariski decomposition can be described as follows:

- if $0 \leqslant y \leqslant \frac{3-x}{2}$, then the positive part of the Zariski decomposition is

$$(3 - x - y)\mathbf{f}_1 + (3 - x)\mathbf{f}'_1 + \frac{3 - x - y}{2}\mathbf{e} + \frac{3 - x}{2}\mathbf{e}',$$

and the negative part of the Zariski decomposition is $\frac{y}{2}\mathbf{e}$,
- if $\frac{3-x}{2} \leqslant y \leqslant 3 - x$, then the Zariski decomposition is

$$\frac{3 - x - y}{2}\underbrace{(2\mathbf{f}_1 + 4\mathbf{f}'_1 + \mathbf{e} + 2\mathbf{e}') + \frac{y}{2}\mathbf{e} + \frac{x + 2y - 3}{2}\mathbf{e}' + (x + 2y - 3)\mathbf{f}'_1}.$$

positive part   negative part

Integrating the volume of the divisor $P(x)|_H - y\mathbf{f}_1$, we get $S(W^H_{\bullet,\bullet}; \mathbf{f}_1) = \frac{49}{52}$.

Now, we compute $S(W^{H,\mathbf{f}_1}_{\bullet,\bullet,\bullet}; Z)$. Let $P(x, y)$ be the positive part of the Zariski decomposition of the divisor $P(x)|_H - y\mathbf{f}_1$, and let $N(x, y)$ be its negative part. Recall that

$$S(W^{H,\mathbf{f}_1}_{\bullet,\bullet,\bullet}; Z) = F_Z(W^{H,\mathbf{f}_1}_{\bullet,\bullet}) + \frac{3}{26}\int_0^3\int_0^\infty \left((P(x,y)\cdot\mathbf{f}_1)_H\right)^2 dydx$$

by Theorem 1.112, where

$$F_Z(W^{H,\mathbf{f}_1}_{\bullet,\bullet,\bullet}) = \frac{6}{26}\int_0^3\int_0^\infty (P(x,y)\cdot\mathbf{f}_1)_H \mathrm{ord}_Z\left(N'_H(x)\big|_{\mathbf{f}_1} + N(x,y)\big|_{\mathbf{f}_1}\right) dydx.$$

Recall that $N'_H(x)$ is the part of the divisor $N(x)|_H$ whose support does not contain $\mathbf{f}_1$, so that $N'_H(x) = N(x)|_H$ in our case, which implies that $\mathrm{ord}_Z(N'_H(x)|_{\mathbf{f}_1}) = 0$ for $x \in [0, 3]$. Thus, we have

$$F_Z(W^{H,\mathbf{f}_1}_{\bullet,\bullet,\bullet})$$
$$= \frac{6}{26}\int_0^1\int_1^{2-x}\left(\frac{3 - x - y}{2} + y - (y - 1)\right)(y - 1)dydx$$
$$+ \frac{6}{26}\int_0^1\int_{2-x}^{3-x}\left(\frac{3 - x - y}{2} + y - (x + 2y - 3)\right)(x + 2y - 3)dydx$$

$$+ \frac{6}{26} \int_1^3 \int_{\frac{3-x}{2}}^{3-x} \left(\frac{3-x-y}{2} + y - (x+2y-3) \right) (x+2y-3) dy dx$$

$$= \frac{67}{208}$$

by Theorem 1.112, which also gives

$$S(W_{\bullet,\bullet,\bullet}^{H,f_1}; Z) = \frac{67}{208} + \frac{3}{26} \int_0^1 \int_0^{1-x} (1+y)^2 dy dx$$

$$+ \frac{3}{26} \int_0^1 \int_{1-x}^1 \left(\frac{3-x-y}{2} + y \right)^2 dy dx$$

$$+ \frac{3}{26} \int_0^1 \int_1^{2-x} \left(\frac{3-x-y}{2} + y - (y-1) \right)^2 dy dx$$

$$+ \frac{3}{26} \int_0^1 \int_{2-x}^{3-x} \left(\frac{3-x-y}{2} + y - (x+2y-3) \right)^2 dy dx$$

$$+ \frac{3}{26} \int_1^3 \int_0^{\frac{3-x}{2}} \left(\frac{3-x-y}{2} + y \right)^2 dy dx$$

$$+ \frac{3}{26} \int_1^3 \int_{\frac{3-x}{2}}^{2-x} \left(\frac{3-x-y}{2} + y - (x+2y-3) \right)^2 dy dx = \frac{49}{52}.$$

Since $S_X(H) < 1$ and $S(W_{\bullet,\bullet}^H; f_1) < 1$, we have $\beta(F) > 0$ by Theorem 1.112. □

Let ℓ_1 and ℓ_2 be distinct rulings of the surface $\mathscr{S}$ that pass through the point $\pi \circ \phi(Z)$. Then at least one of these rulings intersects the curve $\mathscr{S}$ transversally. Thus, without loss of generality, we may assume that ℓ_1 intersects the curve $\mathscr{S}$ transversally.

Let $Y = (\pi \circ \phi)^*(\ell_1)$. Then Y is smooth. Let $\mathbf{e} = E|_Y$, let $\mathbf{e}' = E'|_Y$, let $\mathbf{f}$ and $\mathbf{f}'$ be the irreducible components of the fiber $(\pi \circ \phi)^{-1}(Z)$ such that $\mathbf{f}$ intersects the curve $\mathbf{e}$, and $\mathbf{f}'$ intersects the curve $\mathbf{e}'$. Then $Z = \mathbf{f} \cap \mathbf{f}'$ and $\mathbf{e}, \mathbf{e}', \mathbf{f}$ and $\mathbf{f}'$ are (-1)-curves on Y, which is the smooth sextic del Pezzo surface. Let us apply Theorem 1.112 to $Y, \mathbf{f}, Z$. As usual, we will use notations introduced in Section 1.7.

Let x be a non-negative real number, $P(x) = P(-K_X - xY)$ and $N(x) = N(-K_X - xY)$. It follows from the proof of Lemma 4.65 that $-K_X - xY$ is not pseudo-effective for $x > 2$. Moreover, if $0 \leqslant x \leqslant 2$, then

$$P(x)|_Y \sim_{\mathbb{R}} \begin{cases} 2(\mathbf{f} + \mathbf{f}') + \mathbf{e} + \mathbf{e}' & \text{if } 0 \leqslant x \leqslant 1, \\ 2(\mathbf{f} + \mathbf{f}') + (2-x)(\mathbf{e} + \mathbf{e}') & \text{if } 1 \leqslant x \leqslant 2, \end{cases}$$

and

$$N(x)|_Y = \begin{cases} 0 & \text{if } 0 \leqslant x \leqslant 1, \\ (x-1)(\mathbf{e} + \mathbf{e}') & \text{if } 1 \leqslant x \leqslant 2. \end{cases}$$

Thus, it follows from Corollary 1.110 that

$$S(W_{\bullet,\bullet}^Y; \mathbf{f}) = \frac{3}{28} \int_0^1 \int_0^\infty \mathrm{vol}\Big(P(x)\big|_Y - y\mathbf{f}\Big) dy dx$$

$$= \frac{3}{28} \int_0^1 \int_0^\infty \mathrm{vol}\big((2-y)\mathbf{f} + 2\mathbf{f}' + \mathbf{e} + \mathbf{e}'\big) dy dx$$

$$+ \frac{3}{28} \int_1^2 \int_0^\infty \mathrm{vol}\big((2-y)\mathbf{f} + 2\mathbf{f}' + (2-x)(\mathbf{e} + \mathbf{e}')\big) dy dx$$

$$= \frac{3}{28} \int_0^1 \int_0^1 \big((2-y)\mathbf{f} + 2\mathbf{f}' + \mathbf{e} + \mathbf{e}'\big)^2 dy dx$$

$$+ \frac{3}{28} \int_0^1 \int_1^2 \big((2-y)\mathbf{f} + (3-y)\mathbf{f}' + (2-y)\mathbf{e} + \mathbf{e}'\big)^2 dy dx$$

$$+ \frac{3}{28} \int_1^2 \int_0^{2-x} \big((2-y)\mathbf{f} + 2\mathbf{f}' + (2-x)(\mathbf{e} + \mathbf{e}')\big)^2 dy dx$$

$$+ \frac{3}{28} \int_1^2 \int_{2-x}^x \big((2-y)\mathbf{f} + (4-x-y)\mathbf{f}' + (2-x)(\mathbf{e} + \mathbf{e}')\big)^2 dy dx$$

$$+ \frac{3}{28} \int_1^2 \int_x^2 \big((2-y)\mathbf{f} + (4-x-y)\mathbf{f}' + (2-y)\mathbf{e} + (2-x)\mathbf{e}'\big)^2 dy dx$$

$$= \frac{3}{28} \int_0^1 \int_0^1 (6 - 2y - y^2) dy dx + \frac{3}{28} \int_0^1 \int_1^2 (2-y)(4-y) dy dx$$

$$+ \frac{3}{28} \int_1^2 \int_0^{2-x} (2xy - 2x^2 - y^2 - 4y + 8) dy dx$$

$$+ \frac{3}{28} \int_1^2 \int_{2-x}^x (2-x)(6 + x - 4y) dy dx$$

$$+ \frac{3}{28} \int_1^2 \int_x^2 (2-y)(6 - 2x - y) dy dx = 1.$$

Here, we used the Zariski decomposition of $P(x)|_Y - y\mathbf{f}$ that can be described as follows:

- if $0 \leqslant x \leqslant 1$ and $0 \leqslant y \leqslant 1$, then $P(x)|_Y - y\mathbf{f}$ is nef,
- if $0 \leqslant x \leqslant 1$ and $1 \leqslant y \leqslant 2$, then

$$P(x)|_Y - y\mathbf{f} \sim_{\mathbb{R}} \underbrace{(2-y)\mathbf{f} + (3-y)\mathbf{f}' + (2-y)\mathbf{e} + \mathbf{e}'}_{\text{positive part}} + \underbrace{(y-1)\mathbf{e} + (y-1)\mathbf{f}'}_{\text{negative part}},$$

- if $1 \leqslant x \leqslant 2$ and $0 \leqslant y \leqslant 2 - x$, then $P(x)|_Y - y\mathbf{f}$ is nef,

- if $1 \leqslant x \leqslant 2$ and $2 - x \leqslant y \leqslant x$, then

$$P(x)|_Y - y\mathbf{f} \sim_{\mathbb{R}} \underbrace{(2 - y)\mathbf{f} + (4 - x - y)\mathbf{f}' + (2 - x)(\mathbf{e} + \mathbf{e}')}_{\text{positive part}} + \underbrace{(x + y - 2)\mathbf{f}'}_{\text{negative part}},$$

- if $1 \leqslant x \leqslant 2$ and $x \leqslant y \leqslant 2$, then

$$P(x)|_Y - y\mathbf{f} \sim_{\mathbb{R}}$$
$$\underbrace{(2 - y)\mathbf{f} + (4 - x - y)\mathbf{f}' + (2 - y)\mathbf{e} + (2 - x)\mathbf{e}'}_{\text{positive part}} + \underbrace{(y - x)\mathbf{e} + (x + y - 2)\mathbf{f}'}_{\text{negative part}}.$$

Let $P(x, y)$ be the positive part of the Zariski decomposition of the divisor $P(x)|_Y - y\mathbf{f}$, and let $N(x, y)$ be its negative part. Arguing as in the proof of Lemma 4.66, we get

$$S\left(W^{Y,\mathbf{f}}_{\bullet,\bullet,\bullet}; Z\right) = F_Z\left(W^{Y,\mathbf{f}}_{\bullet,\bullet,\bullet}\right) + \frac{3}{28} \int_0^2 \int_0^\infty \left((P(x, y) \cdot \mathbf{f})_Y\right)^2 dy\, dx$$

$$= F_Z\left(W^{Y,\mathbf{f}}_{\bullet,\bullet,\bullet}\right) + \frac{3}{28} \int_0^1 \int_0^1 (y + 1)^2 dy\, dx$$

$$+ \frac{3}{28} \int_0^1 \int_1^2 (3 - y)^2 dy\, dx + \frac{3}{28} \int_1^2 \int_0^{2-x} (2 - x + y)^2 dy\, dx$$

$$+ \frac{3}{28} \int_1^2 \int_{2-x}^x (4 - 2x)^2 dy\, dx + \frac{3}{28} \int_1^2 \int_x^2 (4 - x - y)^2 dy\, dx$$

$$= F_Z\left(W^{Y,\mathbf{f}}_{\bullet,\bullet}\right) + \frac{39}{56}.$$

Recall from Theorem 1.112 that

$$F_Z\left(W^{Y,\mathbf{f}}_{\bullet,\bullet,\bullet}\right) = \frac{6}{28} \int_0^2 \int_0^\infty (P(x, y) \cdot \mathbf{f})_Y \operatorname{ord}_Z\left(N_Y'(x)\big|_{\mathbf{f}} + N(x, y)\big|_{\mathbf{f}}\right) dy\, dx,$$

where $N_Y'(x)$ is the part of the divisor $N(x)|_Y$ whose support does not contain the curve $\mathbf{f}$. In our case, we have $N_Y'(x) = N(x)|_Y$, so that Z is not contained in its support. Then

$$F_Z\left(W^{Y,\mathbf{f}}_{\bullet,\bullet,\bullet}\right)$$
$$= \frac{6}{28} \int_0^1 \int_1^2 (3 - y)(y - 1) dy\, dx + \frac{6}{28} \int_1^2 \int_{2-x}^x (4 - 2x)(x + y - 2) dy\, dx$$
$$+ \frac{6}{28} \int_1^2 \int_x^2 (4 - x - y)(x + y - 2) dy\, dx = \frac{17}{56},$$

which implies that $S(W^{Y,\mathbf{f}}_{\bullet,\bullet,\bullet}; Z) = 1$. Since we also have $S_X(Y) < 1$ by Remark 4.62, we conclude that $\beta(F) > 0$ by Theorem 1.112, which is a contradiction. Thus, we proved that all smooth Fano threefolds №3.9 and 4.2

are K-polystable. Note that the K-polystability of a general member of the deformation family №4.2 has been recently proved in [123].

4.7 Ruled Fano threefolds

There are exactly 21 families of smooth Fano threefolds whose members are $\mathbb{P}^1$-bundles over surfaces. To be precise, we have

Theorem 4.67 ([205]) *Let X be a smooth Fano threefold such that $X = \mathbb{P}(\mathcal{E})$ for some vector bundle $\mathcal{E}$ of rank two on a surface S. Then X can be described as follows:*

(i) $S = \mathbb{P}^2$ *and one of the following holds:*

    (a) *X is a smooth Fano threefold №2.24, and $\mathcal{E}$ is a stable bundle;*

    (b) *X is the unique smooth Fano threefold №2.27, and $\mathcal{E}$ is a stable bundle;*

    (c) *X is the unique smooth Fano threefold №2.31, and $\mathcal{E}$ is a semistable bundle;*

    (d) *X is the unique smooth Fano threefold №2.32, and $\mathcal{E} = \mathcal{T}_{\mathbb{P}^2}$;*

    (e) *$X = \mathbb{P}^1 \times \mathbb{P}^2$, and $\mathcal{E} = O_{\mathbb{P}^2} \oplus O_{\mathbb{P}^2}$;*

    (f) *X is the unique smooth Fano threefold №2.35, and $\mathcal{E} = O_{\mathbb{P}^2} \oplus O_{\mathbb{P}^2}(1)$;*

    (g) *X is the unique smooth Fano threefold №2.36, and $\mathcal{E} = O_{\mathbb{P}^2} \oplus O_{\mathbb{P}^2}(2)$.*

(ii) $S = \mathbb{P}^1 \times \mathbb{P}^1$ *and one of the following holds:*

    (a) *X is the unique smooth Fano threefold №3.17, and $\mathcal{E}$ is a stable bundle;*

    (b) *X is the unique smooth Fano threefold №3.25, and $\mathcal{E} = O_S(\ell_1) \oplus O_S(\ell_2)$;*

    (c) *$X = \mathbb{P}^1 \times \mathbb{P}^1 \times \mathbb{P}^1$ is the unique smooth Fano threefold №3.27, and $\mathcal{E} = O_S \oplus O_S$;*

    (d) *$X = \mathbb{P}^1 \times \mathbb{F}_1$ is the unique smooth Fano threefold №3.28, and $\mathcal{E} = O_S \oplus O_S(\ell_1)$;*

    (e) *X is the unique smooth Fano threefold №3.31, and $\mathcal{E} = O_S \oplus O_S(\ell_1 + \ell_2)$;*

    where ℓ_1 and ℓ_2 are different rulings of the surface S.

(iii) $S = \mathbb{F}_1$ *and one of the following holds:*

    (a) *X is the unique smooth Fano threefold №3.24, and $\mathcal{E} = \pi^*(\mathcal{T}_{\mathbb{P}^2})$;*

    (b) *$X = \mathbb{P}^1 \times \mathbb{F}_1$ and $\mathcal{E} = O_S \oplus O_S$;*

Table 4.1

S	$X = \mathbb{P}(\mathcal{E})$	K-poly-stable	$\mathcal{E}$	$\mathrm{Aut}(X)$	Sections
$\mathbb{P}^2$	Fano threefold №2.27	Yes	stable	reductive	4.2
$\mathbb{P}^2$	Fano threefold №2.31	No	semistable	non-reductive	3.6, 3.7
$\mathbb{P}^2$	Fano threefold №2.32	Yes	stable	reductive	3.2
$\mathbb{P}^2$	$\mathbb{P}^1 \times \mathbb{P}^2$	Yes	semistable	reductive	3.2, 3.1, 3.3
$\mathbb{P}^2$	Fano threefold №2.35	No	unstable	non-reductive	3.6, 3.7
$\mathbb{P}^2$	Fano threefold №2.36	No	unstable	non-reductive	3.6, 3.7
$\mathbb{P}^1 \times \mathbb{P}^1$	Fano threefold №3.17	Yes	stable	reductive	4.2
$\mathbb{F}_1$	Fano threefold №3.24	No	stable	non-reductive	3.6, 3.7
$\mathbb{P}^1 \times \mathbb{P}^1$	Fano threefold №3.25	Yes	semistable	reductive	3.3
$\mathbb{P}^1 \times \mathbb{P}^1$	$\mathbb{P}^1 \times \mathbb{P}^1 \times \mathbb{P}^1$	Yes	semistable	reductive	3.2, 3.1, 3.3
$\mathbb{P}^1 \times \mathbb{P}^1$	$\mathbb{P}^1 \times \mathbb{F}_1$	No	unstable	non-reductive	3.3, 3.6, 3.7
$\mathbb{F}_1$	$\mathbb{P}^1 \times \mathbb{F}_1$	No	semistable	non-reductive	3.3, 3.6, 3.7
$\mathbb{F}_1$	Fano threefold №3.30	No	unstable	non-reductive	3.6, 3.7
$\mathbb{P}^1 \times \mathbb{P}^1$	Fano threefold №3.31	No	unstable	non-reductive	3.6, 3.7
S_7	$\mathbb{P}^1 \times S_7$	No	semistable	non-reductive	3.3, 3.6, 3.7
S_d	$\mathbb{P}^1 \times S_d$ for $d \leqslant 6$	Yes	semistable	reductive	3.1

(c) X is the unique smooth Fano threefold №3.30, and
$$\mathcal{E} = \pi^*(\mathcal{O}_{\mathbb{P}^2} \oplus \mathcal{O}_{\mathbb{P}^2}(1));$$

where $\pi \colon S \to \mathbb{P}^2$ is a blow up of a point.

(iv) S is a smooth del Pezzo surface such that $K_S^2 \leqslant 7$, $X = \mathbb{P}^1 \times S$, and
$\mathcal{E} = \mathcal{O}_S \oplus \mathcal{O}_S$.

From Sections 3.1, 3.3, 3.6, 3.7, 4.2, we know the solution of the Calabi Problem for all smooth Fano threefolds in Theorem 4.67 except for exactly one family: the family №2.24. This is summarized in Table 4.1, where S_d is a smooth del Pezzo surface of degree d.

The goal of this section is to solve the Calabi Problem for the remaining family №2.24.

Let X be a smooth Fano threefold №2.24. Then X is a divisor in $\mathbb{P}^2 \times \mathbb{P}^2$ of degree $(1, 2)$, let $\mathrm{pr}_1 \colon X \to \mathbb{P}^2$ and $\mathrm{pr}_2 \colon X \to \mathbb{P}^2$ be the projections to the first and the second factors, respectively. The morphism pr_1 is a conic bundle, and pr_2 is a $\mathbb{P}^1$-bundle, which is given by the projectivization of a rank two stable vector bundle on $\mathbb{P}^2$ explicitly described in [10]. Let $\mathscr{C}$ be the discriminant curve of the conic bundle pr_1. Then $\mathscr{C}$ is a reduced cubic curve. Moreover, since X is smooth, the curve $\mathscr{C}$ is either smooth or nodal.

By Lemma A.60, we can choose coordinates $([x : y : z], [u : v : w])$ on $\mathbb{P}^2 \times \mathbb{P}^2$ such that one of the following three cases holds:

- The threefold X is given by

$$(\mu v w + u^2)x + (\mu u w + v^2)y + (\mu u v + w^2)z = 0 \qquad (4.13)$$

for some $\mu \in \mathbb{C}$ such that $\mu^3 \neq -1$. In this case, the curve $\mathscr{C}$ is given by

$$\mu^2(x^3 + y^3 + z^3) = (\mu^3 + 4)xyz.$$

It is singular $\iff \mu \in \{0, 2, -1 \pm \sqrt{3}i\} \iff \mathscr{C}$ is a union of three lines.
- The threefold X is given by

$$(vw + u^2)x + (uw + v^2)y + w^2 z = 0. \qquad (4.14)$$

The curve $\mathscr{C}$ is given by $x^3 + y^3 - 4xyz = 0$. It is irreducible and singular.
- The threefold X is given by

$$(vw + u^2)x + v^2 y + w^2 z = 0. \qquad (4.15)$$

The curve $\mathscr{C}$ is given by $x(x^2 - 4yz) = 0$. It is the union of a line and a conic.

In the remaining part of this section, we will show that X is K-polystable in the first case, and X is strictly K-semistable in the other two cases.

Lemma 4.68 *The group* $\mathrm{Aut}(X)$ *is finite except the following cases:*

(i) *X is given by* (4.13) *with* $\mu \in \{0, 2, -1 \pm \sqrt{3}i\}$,
(ii) *X is given by* (4.15).

In the first case, we have $\mathrm{Aut}^0(X) \cong \mathbb{G}_m^2$, *and in the second*, $\mathrm{Aut}^0(X) \cong \mathbb{G}_m$.

Proof The assertion follows from the proof of [45, Lemma 10.2]. $\square$

The four threefolds given by (4.13) with $\mu \in \{0, 2, -1 \pm \sqrt{3}i\}$ are all isomorphic to each other. They are known to be K-polystable [202].

Remark 4.69   If $\mu^3 = -1$, then $\{(\mu vw + u^2)x + (\mu uw + v^2)y + (\mu uv + w^2)z = 0\}$
is a singular threefold in $\mathbb{P}^2 \times \mathbb{P}^2$, isomorphic to the threefold

$$\{xvw + yuw + zuv = 0\} \subset \mathbb{P}^2 \times \mathbb{P}^2,$$

see [221], which has three isolated ordinary double points, and is not $\mathbb{Q}$-factorial.

Lemma 4.70   *Let Y be a divisor in $\mathbb{P}^2 \times \mathbb{P}^2$ that is contained in the pencil*

$$\lambda(xu^2 + yv^2 + zw^2) + \mu(xvw + yuw + zuv) = 0,$$

where $[\lambda : \mu] \in \mathbb{P}^1$. Then Y is a K-polystable Fano threefold.

Proof   Let G be the subgroup in $\mathrm{Aut}(\mathbb{P}^2 \times \mathbb{P}^2)$ generated by α, β and γ defined as follows:

$$\alpha\colon ([x:y:z],[u:v:w]) \mapsto ([y:x:z],[v:u:w]),$$
$$\beta\colon ([x:y:z],[u:v:w]) \mapsto ([y:z:x],[v:w:u]),$$
$$\gamma\colon ([x:y:z],[u:v:w]) \mapsto ([\epsilon x : \epsilon^2 y : z],[\epsilon u : \epsilon^2 v : w]),$$

where ϵ is a primitive cube root of unity. Then $G \cong \mu_3 \rtimes \mathfrak{S}_3$, it preserves Y, and it acts on the threefold Y faithfully, so that we can identify the group G with a subgroup in $\mathrm{Aut}(Y)$.

Let $\pi_1\colon Y \to \mathbb{P}^2$ and $\pi_2\colon Y \to \mathbb{P}^2$ be the projections to the first and the second factors, respectively. Then π_1 and π_2 are G-equivariant, and the induced G-actions on both factors of $\mathbb{P}^2 \times \mathbb{P}^2$ are faithful (cf. [77, Theorem 4.7]).

We claim that $\alpha_G(Y) \geqslant 1$. To prove this claim, let us apply Theorem 1.52 with $\mu = 1$. First, we observe that Y has no G-fixed points because $\mathbb{P}^2$ has no G-fixed points.

Suppose that Y contains a G-invariant irreducible rational curve C. Then $\pi_1(C)$ is not a point and is not a line since $\mathbb{P}^2$ does not have G-fixed points and G-invariant lines. Then $\pi_1(C)$ is an irreducible G-invariant rational curve of degree at least 2, so that G acts faithfully on its normalization, which is isomorphic to $\mathbb{P}^1$. But $\mathrm{Aut}(\mathbb{P}^1)$ does not contain a subgroup that is isomorphic to $\mu_3 \rtimes \mathfrak{S}_3$. This shows that Y does not contain G-invariant irreducible rational curves.

To prove that $\alpha_G(Y) \geqslant 1$ it is enough to show that Theorem 1.52(1) does not hold. Suppose it does. Then the threefold Y contains a G-invariant irreducible surface S such that $-K_Y \sim_{\mathbb{Q}} aS + \Delta$, where $a \in \mathbb{Q}$ such that $a > 1$, and Δ is an effective $\mathbb{Q}$-divisor on Y. If Y is smooth, then there are non-negative integers r and s such that

$$\frac{1}{a}\pi_1^*\big(\mathcal{O}_{\mathbb{P}^2}(2)\big) + \frac{1}{a}\pi_2^*\big(\mathcal{O}_{\mathbb{P}^2}(1)\big) - \frac{1}{a}\Delta \sim_{\mathbb{Q}} \frac{1}{a}\big(-K_Y\big) - \frac{1}{a}\Delta$$
$$\sim_{\mathbb{Q}} S \sim \pi_1^*\big(\mathcal{O}_{\mathbb{P}^2}(r)\big) + \pi_2^*\big(\mathcal{O}_{\mathbb{P}^2}(s)\big),$$

which gives $r = 1$ and $s = 0$ since $a > 1$. But $|\pi_1^*(\mathcal{O}_{\mathbb{P}^2}(1))|$ does not contain G-invariant divisors. Thus, we may assume that Y is given by $xvw + yuw + zuv$.

Let $S_{u,x}$, $S_{v,y}$, $S_{w,z}$ be the surfaces $\{u = x = 0\}$, $\{v = y = 0\}$, $\{w = z = 0\}$, respectively, let $S'_{u,x}$, $S'_{v,y}$, $S'_{w,z}$ be the surfaces

$$\{x = yw + zv = 0\}, \quad \{y = xw + zu\}, \quad \text{and } \{z = xv + yu = 0\},$$

respectively. Then $S_{u,x} \cong S_{v,y} \cong S_{w,z} \cong \mathbb{P}^1 \times \mathbb{P}^1$, $S'_{u,x} \cong S'_{v,y} \cong S'_{w,z} \cong \mathbb{F}_1$, and these six surfaces are contained in Y. But S is not one of them since they are not G-invariant.

Let ℓ be a general ruling of the surface $S_{u,x} \cong \mathbb{P}^1 \times \mathbb{P}^1$ that is contracted by π_1 to a point. Then $\ell \cap \mathrm{Sing}(Y) = \varnothing$ and $1 = -K_Y \cdot \ell = aS \cdot \ell + \Delta \cdot \ell > S \cdot \ell$, so that $S \cdot \ell = 0$, which implies that ℓ and S are disjoint. Similarly, let ℓ' be a general ruling of the surface $S'_{u,x} \cong \mathbb{F}_1$. Then ℓ' and S must also be disjoint. Thus, if C is a general fiber of the conic bundle π_1, then $S \cdot C = S \cdot (\ell + \ell') = 0$, so that S is contracted by π_1.

Since π_1 does not contract surfaces to points, we see that $\pi_1(S)$ is an irreducible curve. Then $\pi_1(S)$ is not the discriminant curve of the conic bundle π_1 because this curve is reducible in this case, and none of its irreducible components is G-invariant (as there is no G-invariant line). This implies that $S \sim \pi_1^*(\mathcal{O}_{\mathbb{P}^2}(t))$ for some $t \in \mathbb{Z}_{>0}$. Arguing as above, we conclude that $t = 1$, which is impossible since $|\pi_1^*(\mathcal{O}_{\mathbb{P}^2}(1))|$ contains no G-invariant surfaces. Thus, we have $\alpha_G(Y) \geqslant 1$, so that Y is K-polystable by Theorem 1.48. $\qquad\square$

Corollary 4.71 *If X is given by (4.14) or (4.15), then X is strictly K-semistable.*

Proof Suppose that X is given by (4.14). Let X_s be the divisor in $\mathbb{P}^2 \times \mathbb{P}^2$ given by

$$\left(svw + u^2\right)x + \left(suw + v^2\right)y + w^2z = 0,$$

where $s \in \mathbb{C}$. Then X_s is smooth for all s. Moreover, scaling coordinates x, y, z, u, v, w, we see that $X_s \cong X$ for every $s \neq 0$. This gives us a test configuration for X, whose special fiber is the threefold X_0, which is a K-polystable smooth Fano threefold by Lemma 4.70. Then X is strictly K-semistable by Corollary 1.13.

In a similar way, we see that the threefold given by (4.15) is also strictly K-semistable. $\qquad\square$

A general threefold in the family №2.24 has finite automorphisms group by Lemma 4.68, so that it is K-stable by Theorem 1.11.

5

Proof of Main Theorem: Remaining Cases

5.1 Family №2.8

Let X be a smooth Fano threefold №2.8. Then there exists a quartic surface $S_4 \subset \mathbb{P}^3$ such that its singular locus consists of one (isolated) ordinary double point O, and the following commutative diagram exists:

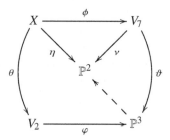

where φ is a double cover branched over S_4, ϑ is a blow up of the point O, θ is the blow up of the preimage of the point O, ϕ is a double cover branched over the proper transform of the surface S_4, ν is a $\mathbb{P}^1$-bundle, η is a (standard) conic bundle, and dashed arrow is a linear projection from the point O.

Without loss of generality, we may assume that $O = [0 : 0 : 0 : 1]$. Then S_4 is given by

$$t^2 f_2(x, y, z) + t f_3(x, y, z) + f_4(x, y, z) = 0,$$

where f_2, f_3, f_4 are homogeneous polynomials of degree 2, 3, 4, respectively, and x, y, z and t are coordinates on $\mathbb{P}^3$. If Δ is the discriminant curve of the standard conic bundle η, then Δ is given by $f_3^2(x, y, z) - 4f_2(x, y, z)f_4(x, y, z) = 0$.

Denote by $\overline{S}_4$ the proper transform on V_7 of the surface S_4, and denote by $\widetilde{S}_4$ its preimage on X. Then $\overline{S}_4$ and $\widetilde{S}_4$ are isomorphic smooth K3 surfaces, and ϑ induces minimal resolutions $\overline{S}_4 \to S_4$. Similarly, we have that the surfaces $\theta(\widetilde{S}_4)$ and S_4 are isomorphic, and that θ induces minimal resolutions $\widetilde{S}_4 \to \theta(\widetilde{S}_4)$.

Let E and E_O be the exceptional divisors of the birational maps θ and ϑ, respectively. Then $E \cong \mathbb{P}^1 \times \mathbb{P}^1$, $E_O \cong \mathbb{P}^2$, and ϕ induces a double cover $E \to E_O$, which is branched over the conic $E_O \cap \overline{S}_4$ in E_O. Let $C = E \cap \overline{S}_4$. Then C is a curve of degree $(1,1)$ on E, which is the preimage of the branching curve $E_O \cap \overline{S}_4$.

Let Q be the cone in $\mathbb{P}^3$ given by $f_2(x, y, z) = 0$, let C_2 be the conic in $\mathbb{P}^2$ given by the same equation, where we consider x, y, z also as coordinates on $\mathbb{P}^2$, let $\overline{Q}$ be the proper transform on V_7 of the surface Q, and let $\widetilde{Q}$ be its preimage on X. Then $\widetilde{Q} \cap E = C$, and $\widetilde{Q}$ is the preimage of C_2 via the conic bundle η. Moreover, this conic bundle induces a double cover $E \to \mathbb{P}^2$, which is branched over the conic C_2. This shows that $\widetilde{Q}|_E = 2C$, so that either $\widetilde{Q}$ is tangent to E along C, or $\widetilde{Q}$ is singular along the curve C, which happens only if f_2 divides f_3.

Let H be a plane in $\mathbb{P}^3$. Then $-K_X \sim \widetilde{S}_4 \sim \widetilde{Q} + E$. Note that $\widetilde{Q}$ and E are G-invariant for every (finite) subgroup $G \subset \mathrm{Aut}(X)$, so that $\alpha_G(X) \leqslant \frac{3}{4}$ since $\mathrm{lct}(X, \widetilde{Q} + E) \leqslant \frac{3}{4}$.

Lemma 5.1 *Let S_4 be the quartic surface in $\mathbb{P}^3$ that is given by*

$$t^2\left(x^2 + y^2 + z^2 + (x + y + z)^2\right) + \mu t\left(x^3 + y^3 + z^3 - (x + y + z)^3\right)$$
$$+ x^4 + y^4 + z^4 + (x + y + z)^4 = 0,$$

where μ is a general complex number, e.g. $\mu = 5$. Then S_4 is smooth away from O, which is an (isolated) ordinary double point of the surface S_4. Moreover, the surface S_4 admits a natural action of the symmetric group $\mathfrak{S}_4$. This action lifts to the threefold X, so that we identify $\mathfrak{S}_4$ with a subgroup in $\mathrm{Aut}(X)$. Let G be the subgroup in $\mathrm{Aut}(X)$ generated by $\mathfrak{S}_4$ and the Galois involution τ of the double cover η. Then $\alpha_G(X) = \frac{3}{4}$.

Proof First, let us describe the action of the group $\mathfrak{S}_4$ on $\mathbb{P}^3$ such that S_4 is $\mathfrak{S}_4$-invariant. To do this, we let $x_0 = x$, $x_1 = y$, $x_2 = z$, $x_3 = -(x + y + z)$ and $x_4 = t$, we consider x_0, x_1, x_2, x_3, x_4 as coordinates on $\mathbb{P}^4$, and we identify our $\mathbb{P}^3$ with $\{x_0 + x_1 + x_2 + x_3 = 0\} \subset \mathbb{P}^4$. Then our surface S_4 is given in $\mathbb{P}^4$ by

$$x_0 + x_1 + x_2 + x_3 = x_4^2\left(x_0^2 + x_1^2 + x_2^2 + x_3^2\right) + \mu x_4\left(x_0^3 + x_1^3 + x_2^3 + x_3^3\right) + x_0^4 + x_1^4 + x_2^4 + x_3^4 = 0,$$

so that it is $\mathfrak{S}_4$-invariant for the $\mathfrak{S}_4$-action on $\mathbb{P}^4$ that permutes the coordinates x_0, x_1, x_2, x_3. Since the hyperplane $\{x_0 + x_1 + x_2 + x_3 = 0\}$ is also $\mathfrak{S}_4$-invariant, this described the desired $\mathfrak{S}_4$-action on $\mathbb{P}^3$ such that S_4 is $\mathfrak{S}_4$-invariant.

Let us say a few words about the generality of $\mu \in \mathbb{C}$. We ask for two natural conditions. First, we want S_4 to be smooth away from O since otherwise X would be singular. Second, we want the following octic curve $C_8 \subset S_4$ to be irreducible and reduced:

$$\begin{cases} x^2 + y^2 + z^2 + (x + y + z)^2 = 0, \\ \mu t\left(x^3 + y^3 + z^3 - (x + y + z)^3\right) + x^4 + y^4 + z^4 + (x + y + z)^4 = 0. \end{cases}$$

We will assume that both conditions are satisfied. For instance, this is true for $\mu = 5$.

Let $\widetilde{C}_8$ be the irreducible curve in $\widetilde{S}_4$ that is a proper transform of the octic curve C_8 via the birational morphism $\widetilde{S}_4 \to S_4$ induced by $\vartheta \circ \phi$. Then $\widetilde{C}_8$ is smooth rational curve, which is (-2)-curve on the K3 surface $\widetilde{S}_4$. $\widetilde{S}_4 \cap \widetilde{Q} = \widetilde{C}_8 \cup C$ and $-K_X \cdot \widetilde{C}_8 = 10$.

We already know that $\alpha_G(X) \leqslant \frac{3}{4}$. Let us apply Lemma A.30 to prove that $\alpha_G(X) = \frac{3}{4}$. We know that X does not have G-fixed points, because η is G-equivariant (we already used this implicitly), and the group $\mathfrak{S}_4$ does not have fixed points in $\mathbb{P}^2$. Thus, we see that the condition Lemma A.30(i) is satisfied.

We claim that the condition Lemma A.30(ii) is also satisfied for $\mu = \frac{3}{4}$. Indeed, suppose that X contains a G-invariant surface S such that $-K_X \sim_{\mathbb{Q}} aS + \Delta$, where a is a positive rational number such that $a > \frac{4}{3}$, and Δ is a G-invariant $\mathbb{Q}$-divisor on the threefold X whose support does not contain S. Now, intersecting $aS + \Delta$ with a general fiber of the conic bundle η, we see that $S \neq E$, so that $\theta(S)$ is also a surface. Since $a\theta(S) + \theta(\Delta) \sim_{\mathbb{Q}} \varphi^*(2H)$, we get $\theta(S) \sim \varphi^*(H)$, so that $\varphi \circ \theta(S)$ is the plane $t = 0$, because this plane is the only $\mathfrak{S}_4$-invariant plane in $\mathbb{P}^3$. Thus, we have $S \sim (\vartheta \circ \phi)^*(H)$. Let ℓ be a general fiber of η. As above, we have $2 = -K_X \cdot \ell = aS \cdot \ell + \Delta \cdot \ell \geqslant aS \cdot \ell = 2a > \frac{8}{3}$, which is absurd. Thus, the condition Lemma A.30(ii) is satisfied for $\mu = \frac{3}{4}$.

Suppose that $\alpha_G(X) < \frac{3}{4}$. Applying Lemma A.30, we see that there is a G-invariant effective $\mathbb{Q}$-divisor D on the threefold X such that $D \sim_{\mathbb{Q}} -K_X$, the pair $(X, \lambda D)$ is strictly log canonical for some $\lambda < \frac{3}{4}$, and the only log canonical center of this log pair is a smooth irreducible rational G-invariant curve Z. Let us seek a contradiction.

By Corollary A.21, we have $-K_X \cdot C \leqslant 8$. Then $Z \neq \widetilde{C}_8$ since $-K_X \cdot \widetilde{C}_8 = 10$.

Now, we observe that $\eta(Z)$ is not a point because $\mathbb{P}^2$ does not have $\mathfrak{S}_4$-invariant points. Similarly, we see that $\eta(Z)$ is not a line. Then $\eta(Z)$ must be a conic by Corollary A.13. In particular, the subgroup $\mathfrak{S}_4$ acts faithfully on the curve Z because it must act faithfully on the curve $\eta(Z)$. Therefore, the Galois involution τ must act trivially on the curve Z because $Z \cong \mathbb{P}^1$ does not admit a faithful G-action. This shows that $Z \subset \widetilde{S}_4$.

We claim that $Z = C$. Indeed, suppose that this is not the case. Then $Z \not\subset E$ because otherwise we would have $Z = E \cap \widetilde{S}_4 = C$. Since $\eta(Z)$ is a conic, we see that $\eta(Z) = C_2$ because C_2 is the only $\mathfrak{S}_4$-invariant conic in $\mathbb{P}^2$. Therefore, we have $Z \subset \widetilde{Q}$, so that $Z = C$, because we know that $Z \neq \widetilde{C}_8$.

Recall that $C = \widetilde{S}_4 \cap E$ and $C = \widetilde{Q} \cap E$. Observe that $(X, \lambda\widetilde{Q} + \lambda E)$ is log canonical at a general point of the curve C. Thus, using Lemma A.34, we may assume that either $\widetilde{Q}$ or E is not contained in the support of the divisor D. Similarly, we may assume that the surface $\widetilde{S}_4$ is not contained in the support of the divisor D.

If $E \not\subset \mathrm{Supp}(D)$, we have $1 = D \cdot L \geqslant \mathrm{mult}_C(D) > \frac{4}{3}$, where L is a general ruling of the surface $E \cong \mathbb{P}^1 \times \mathbb{P}^1$. Therefore, we see that $E \subset \mathrm{Supp}(D)$, so that $\widetilde{Q} \not\subset \mathrm{Supp}(D)$. Let ℓ be a general fiber of η that is contained in $\widetilde{Q}$. Then $\mathrm{mult}_C(D) \leqslant D \cdot \ell = 2$.

Let $f \colon \widehat{X} \to X$ be the blow up of the curve C, and let F be the f-exceptional surface. Then the action of the group G lifts to $\widehat{X}$. Since $C \cong \mathbb{P}^1$ is a complete intersection of the surfaces $\widetilde{S}_4$ and E, its normal bundle in X splits as $\mathcal{O}_{\mathbb{P}^1}(-2) \oplus \mathcal{O}_{\mathbb{P}^1}(2)$, so that $F \cong \mathbb{F}_4$. Let s_F be the (-4)-curve in F, and let l_F be a fiber of the $\mathbb{P}^1$-bundle $F \to C$. Then, since $F^3 = 0$, we have $F|_F = -s_F - 2l_F$. Let $\widehat{E}, \widehat{Q}, \widehat{S}_4$ be proper transforms on $\widehat{X}$ of the surfaces $\widetilde{E}, \widetilde{Q}, \widetilde{S}_4$, respectively. Then $\widehat{E}|_F \sim s_F$, $\widehat{Q}|_F \sim s_F + 6l_F$ and $\widehat{S}_4|_F \sim s_F + 4l_F$. Thus, we see that $\widehat{E} \cap F = s_F$ and $\widehat{S}_4 \cap F$ are disjoint G-invariant smooth irreducible curves, which are both sections of the $\mathbb{P}^1$-bundle $F \to C$. On the other hand, we see that the intersection $\widehat{Q} \cap F$ consists of the curve s_F and 6 fibers of the $\mathbb{P}^1$-bundle $F \to C$, which are mapped to a $\mathfrak{S}_4$-orbit in C of length 6. This shows that the surface $\widetilde{Q}$ is singular at the points of this orbit, which can also be checked explicitly.

Let $\widehat{D}$ be the proper transform of the divisor D on $\widehat{X}$. Since $\lambda\mathrm{mult}_C(D) < 2$, it follows from Lemma A.27 that F contains a G-invariant irreducible curve C such that C is a section of the projection $F \to C$, the curve C is a log canonical center of the log pair $(\widehat{X}, \lambda\widehat{D} + (\lambda\mathrm{mult}_C(D) - 1)F)$, and $\mathrm{mult}_C(D) + \mathrm{mult}_{\mathsf{C}}(\widehat{D}) \geqslant \frac{2}{\lambda} > \frac{8}{3}$. Moreover, using Theorem A.15, we see that the log pair $(F, \lambda\widehat{D}|_F)$ is not log canonical along C. Then $\widehat{D}\big|_F = \delta\mathsf{C} + \Upsilon$, where δ is a rational number such that $\delta > \frac{1}{\lambda} > \frac{4}{3}$, and Υ is an effective $\mathbb{Q}$-divisor whose support does not contain the curve C. But

$$\widehat{D}\big|_F \sim_{\mathbb{Q}} \mathrm{mult}_C(D)s_F + \big(2\mathrm{mult}_C(D) + 2\big)l_F.$$

Since $\mathrm{mult}_C(D) \leqslant 2$ and $\delta > \frac{4}{3}$, this equivalence implies that $\mathsf{C} \sim s_F + nL_F$ for $n \leqslant 4$. Now, using Lemma A.52, we conclude that either $\mathsf{C} = \widehat{E} \cap F$ or $\mathsf{C} = \widehat{S}_4 \cap F$.

Let $\widehat{\ell}$ be the proper transform on $\widehat{X}$ of the general fiber of the conic bundle η that is contained in $\widetilde{Q}$. If $\mathsf{C} = \widehat{E} \cap F$, then $\widehat{\ell}$ intersects C, so that $\mathrm{mult}_{\mathsf{C}}(\widehat{D}) \leqslant \widehat{D} \cdot \widehat{\ell} = 2 - \mathrm{mult}_C(D)$, which contradicts $\mathrm{mult}_C(D) + \mathrm{mult}_{\mathsf{C}}(\widehat{D}) > \frac{8}{3}$. This shows that $\mathsf{C} = \widehat{S}_4 \cap F$.

Observe that $\widehat{S}_4 \cong \widetilde{S}_4$, and $\vartheta \circ \phi \circ f$ induces the minimal resolution $h\colon \widehat{S}_4 \to S_4$, whose exceptional curve is C. Let H_{S_4} be a hyperplane section of the quartic S_4, and let $\widehat{C}_8$ be the proper transform on $\widehat{S}_4$ of the curve C_8. Then $\widehat{C}_8 \sim h^*(2H_{S_4}) - 3C$, so that $\widehat{C}_8$ is a smooth (-2)-curve on the surface $\widehat{S}_4$. In particular, we have $\widehat{C}_8 \cdot C = 6$. On the other hand, we know that $\widehat{S}_4 \not\subset \mathrm{Supp}(\widehat{D})$. Write $\widehat{D}|_{\widehat{S}_4} = bC + c\widehat{C}_8 + \Xi$, where b and c are non-negative numbers, and Ξ is an effective $\mathbb{Q}$-divisor whose support does not contain the curves C and $\widehat{C}_8$. Note that $b \geqslant \mathrm{mult}_C(\widehat{D})$ and $\widehat{D} \sim_{\mathbb{Q}} h^*(2H_{S_4}) - (1 + \mathrm{mult}_C(D))C$. This gives

$$\Xi \sim_{\mathbb{Q}} h^*(2H_{S_4}) - \left(1 + \mathrm{mult}_C(D) + b\right)C \sim_{\mathbb{Q}} \left(1 - b\right)\widehat{C}_8 + \left(2 - \mathrm{mult}_C(D) - b\right)C,$$

where $2 - \mathrm{mult}_C(D) - b \leqslant 2 - \mathrm{mult}_C(D) - \mathrm{mult}_C(\widehat{D}) < 0$. Thus, since Ξ is an effective divisor, we have $b \leqslant 1$, so that

$$0 \leqslant \Xi \cdot \widehat{C}_8 = (1 - b)\widehat{C}_8^2 + (2 - \mathrm{mult}_C(D) - b)C \cdot \widehat{C}_8$$
$$= -2(1 - b) + 6(2 - \mathrm{mult}_C(D) - b) < 0,$$

which is absurd. This contradiction completes the proof of Lemma 5.1.   □

Now, using Theorems 1.11 and 1.51, we see that general smooth Fano three-folds in the family №2.8 are K-stable because their automorphism groups are finite [45].

5.2 Family №2.9

In this section, we present one K-stable smooth Fano threefold №2.9. By Theorem 1.11, this would imply that general Fano threefolds №2.9 are K-stable.

To start with, let $G = \boldsymbol{\mu}_5$ and consider the action of G on $\mathbb{P}^3$ that is given by

$$\left[x : y : z : t\right] \mapsto \left[\omega x : \omega^2 y : \omega^3 z : \omega^4 t\right],$$

where ω is a primitive fifth root of unity. Let us denote by H a general hyperplane in $\mathbb{P}^3$. Let us also introduce the following notations: let $P_x = [1 : 0 : 0 : 0]$, $P_y = [0 : 1 : 0 : 0]$, $P_z = [0 : 0 : 1 : 0]$, $P_t = [0 : 0 : 0 : 1]$, let $L_{xy} = \{x = y = 0\}$, let $L_{xz} = \{x = z = 0\}$, let $L_{xt} = \{x = t = 0\}$, let $L_{yz} = \{y = z = 0\}$, let $L_{yt} = \{y = t = 0\}$, let $L_{zt} = \{z = t = 0\}$, and let H_x, H_y, H_z, H_t be the planes $\{x = 0\}$, $\{y = 0\}$, $\{z = 0\}$, $\{t = 0\}$, respectively.

These points, lines and planes are G-invariant. Moreover, these are all G-invariant points, lines and planes in $\mathbb{P}^3$. Now, we introduce the following three cubic polynomials:

(i) $h(x, y, z, t) = x^2 z + y^2 x + z^2 t + t^2 y$,
(ii) $h'(x, y, z, t) = t^2 x + tyz - x^2 y + z^3$,
(iii) $h''(x, y, z, t) = txy + xz^2 + y^2 z - t^3$.

Let $S_3 = \{h = 0\}$, $S'_3 = \{h' = 0\}$ and $S''_3 = \{h'' = 0\}$. Then S_3 is a smooth cubic surface, which is isomorphic to the Clebsch cubic surface. On the other hand, the surfaces S'_3 and S''_3 are singular: S'_3 has one node (ordinary double point) at the point P_y, and S''_3 has one node at the point P_x. The surfaces S_3, S'_3 and S''_3 are G-invariant.

Remark 5.2 The intersections of G-invariant lines with S_3 can be described as follows: $L_{xy} \cap S_3 = P_z \cup P_t$, and L_{xy} is tangent to S_3 at the point P_t, $L_{xz} \cap S_3 = P_y \cup P_t$, and L_{xy} is tangent to S_3 at the point P_y, L_{xt} is contained in S_3, L_{yz} is contained in S_3, $L_{yt} \cap S_3 = P_x \cup P_z$, and L_{yt} is tangent to S_3 at the point P_z, $L_{zt} \cap S_3 = P_x \cup P_y$, and L_{zt} is tangent to S_3 at the point P_x. The intersections of G-invariant lines with S'_3 can be described as follows: $L_{xy} \cap S'_3 = P_t$, L_{xz} is contained in S'_3, $L_{xt} \cap S'_3 = P_y$, $L_{yz} \cap S'_3 = P_x \cup P_t$, and L_{yz} is tangent to S'_3 at the point P_x, $L_{yt} \cap S'_3 = P_x$, $L_{zt} \cap S'_3 = P_x \cup P_y$, and L_{zt} intersects S'_3 transversally at the point P_x. The intersections of G-invariant lines with S''_3 can be described as follows: $L_{xy} \cap S''_3 = P_z$, $L_{xz} \cap S''_3 = P_y$, $L_{xt} \cap S''_3 = P_y \cup P_z$, and L_{xy} is tangent to S''_3 at the point P_z, $L_{yz} \cap S''_3 = P_x$, $L_{yt} \cap S''_3 = P_x \cup P_z$, and L_{yt} intersects S''_3 transversally at the point P_z, L_{zt} is contained in S''_3.

Let $C = \{h(x, y, z, t) = 0, h'(x, y, z, t) = 0, h''(x, y, z, t) = 0\} \subset \mathbb{P}^3$. Then C is a smooth irreducible curve of genus 5 and degree 7. Note that C is G-invariant. We used the following Magma script to check the smoothness and the genus of this curve:

```
Q:=RationalField();
P<x,y,z,t>:=ProjectiveSpace(Q,3);
X:=Scheme(P,[x^2*z+y^2*x+z^2*t+t^2*y,
            t^2*x+t*y*z-x^2*y+z^3,t*x*y+x*z^2+y^2*z-t^3]);
Degree(X);
IsNonsingular(X);
IsIrreducible(X);
Dimension(X);
IsCurve(X);
C:=Curve(X);
Genus(C);
```

We have $C = S_3 \cap S'_3 \cap S''_3$. On the surface S_3, we have $C \sim 2H\big|_{S_3} + L_{xt}$. This implies that C has no 4-secants. Moreover, this rational equivalence can also be used to compute the genus of the curve C. Observe that C contains P_x and P_y, but it does not contain P_z and P_t. Note also that the quotient C/G is an elliptic curve.

Remark 5.3   The intersections of G-invariant planes with C can be described as follows:

- $H_x|_C = 2P_y +$ the G-orbit of the point $[0 : -1 : 1 : 1]$,
- $H_y|_C = 2P_x +$ the G-orbit of the point $[-1 : 0 : 1 : -1]$,
- $H_z|_C = 4P_x + 3P_y$,
- $H_t|_C = P_x + P_y +$ the G-orbit of the point $[-1 : 1 : 1 : 0]$.

The intersections of G-invariant lines with C can be described as follows: $L_{xy} \cap C = \varnothing$, $L_{xz} \cap C = P_y$, and L_{xz} is tangent to C (ordinary tangency), $L_{xt} \cap C = P_y$, and L_{xt} intersects C transversally at P_y, $L_{yz} \cap C = P_x$, and L_{yz} is tangent to C (ordinary tangency), $L_{yt} \cap C = P_x$, and L_{yt} intersects C transversally at P_x, $L_{zt} \cap C = P_x \cup P_y$, and L_{zt} intersects the curve C transversally at P_x and P_y.

Let us introduce three G-invariant conics in $\mathbb{P}^3$, which will be used later. Observe that the intersection $S_3 \cap S_3'$ consists of the curve C and the conic $C_2' = \{x = 0, yt + z^2 = 0\}$. The intersection $S_3 \cap S_3''$ consists of the curve C and the conic $C_2'' = \{t = 0, xz + y^2 = 0\}$. Therefore, on the surface S_3, we have $C + C_2' \sim C + C_2'' \sim 3H|_{S_3}$. Observe also that the intersection $S_3' \cap S_3''$ consists of the curve C and the conic $C_2''' = \{z = 0, xy - t^2 = 0\}$.

Remark 5.4   The following assertions hold: $P_y \in C_2' \ni P_t$, $P_x \notin C' \not\ni P_z$, $P_x \in C_2'' \ni P_z$, $P_y \notin C_2'' \not\ni P_t$, $P_x \in C_2''' \ni P_y$, $P_z \notin C_2''' \not\ni P_t$, $C \cap C_2'$ consists of the point P_y and the G-orbit of the point $[0 : -1 : 1 : 1]$, $C \cap C_2''$ consists of the point P_x and the G-orbit of the point $[-1 : 1 : 1 : 0]$, $C \cap C_2''' = P_x \cup P_y$, $H_x \cap S_3 = L_{xt} \cup C_2'$, and L_{xt} is tangent to C_2' at the point P_y, $H_t \cap S_3 = L_{xt} \cup C_2''$, and L_{xt} is tangent to C_2'' at the point P_z, $H_x \cap S_3' = L_{xz} \cup C_2'$, and $L_{xz} \cap C_2' = P_y \cup P_t$, $H_z \cap S_3' = L_{xz} \cup C_2'''$, and L_{xz} is tangent to C_2''' at the point P_y, $H_z \cap S_3'' = L_{zt} \cup C_2'''$, and $L_{zt} \cap C_2''' = P_x \cup P_y$, $H_t \cap S_3'' = L_{zt} \cup C_2''$, and L_{zt} is tangent to C_2'' at the point P_x.

Let $\pi : X \to \mathbb{P}^3$ be the blow up of the curve C. Then it follows from [49, Theorem A.1] that X is a smooth Fano threefold №2.9. Since the action of the group G lifts to X, we identify G with a subgroup in $\mathrm{Aut}(X)$. Then there exists the G-equivariant commutative diagram

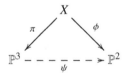

where ϕ is a conic bundle, and ψ is a rational map given by

$$[x : y : z : t] \mapsto [h : h' : h''].$$

The G-action on $\mathbb{P}^2$ has exactly three G-fixed points: $[1:0:0]$, $[0:1:0]$ and $[0:0:1]$. By construction, we have $\psi(C_2') = [0:0:1]$, $\psi(C_2'') = [0:1:0]$ and $\psi(C_2''') = [1:0:0]$. The discriminant curve of the conic bundle ϕ is a quintic curve (we do not need this).

Proposition 5.5 *The Fano threefold X is K-stable.*

We will prove this proposition in several steps in the remaining part of this section.

To start with, let us consider some G-invariant surfaces and curves in the threefold X. Let E be the π-exceptional surface. Denote by $\widetilde{H}$, $\widetilde{H}_x$, $\widetilde{H}_y$, $\widetilde{H}_z$, $\widetilde{H}_t$, $\widetilde{S}_3$, $\widetilde{S}_3'$, $\widetilde{S}_3''$ the proper transforms on the threefold X of the surfaces H, H_x, H_y, H_z, H_t, S_3, S_3', S_3'', respectively. Similarly, we denote by $\widetilde{L}_{xy}$, $\widetilde{L}_{xz}$, $\widetilde{L}_{xt}$, $\widetilde{L}_{yz}$, $\widetilde{L}_{yt}$, $\widetilde{L}_{zt}$, $\widetilde{C}_2'$, $\widetilde{C}_2''$ and $\widetilde{C}_2'''$ the proper transforms on the threefold X of the curves L_{xy}, L_{xz}, L_{xt}, L_{yz}, L_{yt}, L_{zt}, C_2', C_2'' and C_2''', respectively. Let ℓ_x and ℓ_y be the fibers of the natural projection $E \to C$ over P_x and P_y, respectively. Then ℓ_x and ℓ_y are G-invariant curves, and the group G acts faithfully on each of them, because $\pi(\ell_x)$ and $\pi(\ell_y)$ are lines in $\mathbb{P}^2$, and the G-action on $\mathbb{P}^2$ fixes exactly three points.

Remark 5.6 The surfaces $\widetilde{S}_3$, $\widetilde{S}_3'$ and $\widetilde{S}_3''$ are smooth. Moreover, the blow up π induces an isomorphism $\widetilde{S}_3 \cong S_3$, and it induces birational morphisms $\widetilde{S}_3' \to S_3'$ and $\widetilde{S}_3'' \to S_3''$ that contract the curves ℓ_y and ℓ_x, respectively.

Let us introduce three smooth G-invariant curves in E that are sections of the natural projection $E \to C$. First, we let $\widetilde{C} = \widetilde{S}_3|_E$. Second, we observe that $\widetilde{S}_3'|_E = \widetilde{C}' + \ell_y$ for a smooth G-invariant curve $\widetilde{C}'$ that is a section of the natural projection $E \to C$. Similarly, we have $\widetilde{S}_3''|_E = \widetilde{C}'' + \ell_x$ for a smooth G-invariant curve $\widetilde{C}''$ that is a section of the projection $E \to C$. Note that $\widetilde{C}$, $\widetilde{C}'$, $\widetilde{C}''$ are distinct curves (isomorphic to C).

The incidence relation between the curves $\widetilde{L}_{xy}$, $\widetilde{L}_{xz}$, $\widetilde{L}_{xt}$, $\widetilde{L}_{yz}$, $\widetilde{L}_{yt}$, $\widetilde{L}_{zt}$, ℓ_x, ℓ_y, $\widetilde{C}_2'$, $\widetilde{C}_2''$, $\widetilde{C}_2'''$, $\widetilde{C}$, $\widetilde{C}'$, $\widetilde{C}''$ and the surfaces $\widetilde{S}_3$, $\widetilde{S}_3'$, $\widetilde{S}_3''$, $\widetilde{H}_x$, $\widetilde{H}_y$, $\widetilde{H}_z$, $\widetilde{H}_t$ is given in Table 5.1.

Let $\widetilde{P}_x$, $\widetilde{P}_y$, $\widetilde{P}_z$, $\widetilde{P}_t$ be the points in $\widetilde{S}_3$ that are mapped to P_x, P_y, P_z, P_t, respectively. Then $\widetilde{P}_x$ and $\widetilde{P}_y$ are contained in $\widetilde{C}$, the curve ℓ_x contains $\widetilde{P}_x$, the curve ℓ_y contains $\widetilde{P}_y$. Each ℓ_x and ℓ_y has an additional G-fixed point. Denote them by O_x and O_y, respectively.

Corollary 5.7 *The only G-fixed points in X are $\widetilde{P}_x$, $\widetilde{P}_y$, $\widetilde{P}_z$, $\widetilde{P}_t$, O_x and O_y.*

The points O_x and O_y are not contained in $\widetilde{S}_3$, so that they are not contained in $\widetilde{C}$. Observe also that $\widetilde{S}_3'$ contains O_y, O_x, $\widetilde{P}_y$, $\widetilde{P}_t$, the surface $\widetilde{S}_3'$ does not

Table 5.1 *Incidence relations between curves;* • *means that the curve is contained in the corresponding surface;* × *means that the curve is not contained in the corresponding surface.*

⊃	$\tilde{L}_{xy}$	$\tilde{L}_{xz}$	$\tilde{L}_{xt}$	$\tilde{L}_{yz}$	$\tilde{L}_{yt}$	$\tilde{L}_{zt}$	ℓ_x	ℓ_y	$\tilde{C}'_2$	$\tilde{C}''_2$	$\tilde{C}'''_2$	$\tilde{C}$	$\tilde{C}'$	$\tilde{C}''$
$\tilde{S}_3$	×	×	•	•	×	×	×	×	•	•	×	•	×	×
$\tilde{S}'_3$	×	•	×	×	×	×	×	•	•	×	•	×	•	×
$\tilde{S}''_3$	×	×	×	×	×	•	•	×	×	•	•	×	×	•
$\tilde{H}_x$	•	•	•	×	×	×	×	•	•	×	×	×	×	×
$\tilde{H}_y$	•	×	×	•	•	×	•	×	×	×	×	×	×	×
$\tilde{H}_z$	×	•	×	•	×	•	•	•	×	×	•	×	×	×
$\tilde{H}_t$	×	×	•	×	•	•	•	•	×	•	×	×	×	×

Table 5.2 *Incidence relations:* • *means that the point is contained in the corresponding curve, and* × *means that the point is not contained in the corresponding curve.*

∈	$\tilde{L}_{xy}$	$\tilde{L}_{xz}$	$\tilde{L}_{xt}$	$\tilde{L}_{yz}$	$\tilde{L}_{yt}$	$\tilde{L}_{zt}$	ℓ_x	ℓ_y	$\tilde{C}'_2$	$\tilde{C}''_2$	$\tilde{C}'''_2$	$\tilde{C}$	$\tilde{C}'$	$\tilde{C}''$
O_x	×	×	×	×	•	×	•	×	×	×	•	×	•	•
O_y	×	•	×	×	×	•	×	•	×	×	•	×	•	•
$\tilde{P}_x$	×	×	×	•	×	•	•	×	×	•	×	•	×	×
$\tilde{P}_y$	×	×	•	×	×	×	×	•	•	×	×	•	×	×
$\tilde{P}_z$	•	×	•	×	•	×	×	×	×	•	×	×	×	×
$\tilde{P}_t$	•	•	×	•	×	×	×	×	•	×	×	×	×	×

contain $\tilde{P}_x$ and $\tilde{P}_z$, the surface $\tilde{S}''_3$ contains O_x, O_y, $\tilde{P}_x$, $\tilde{P}_z$, and $\tilde{S}''_3$ does not contain $\tilde{P}_y$ and $\tilde{P}_t$.

Lemma 5.8 *The incidence relation between* $\tilde{L}_{xy}$, $\tilde{L}_{xz}$, $\tilde{L}_{xt}$, $\tilde{L}_{yz}$, $\tilde{L}_{yt}$, $\tilde{L}_{zt}$, ℓ_x, ℓ_y, $\tilde{C}'_2$, $\tilde{C}''_2$, $\tilde{C}'''_2$, $\tilde{C}$, $\tilde{C}'$, $\tilde{C}''$ *and the points* O_x, O_y, $\tilde{P}_x$, $\tilde{P}_y$, $\tilde{P}_z$, $\tilde{P}_t$ *is given in Table 5.2.*

Proof Since C does not contain P_z and P_t, the content of the last two rows of the table follows from a corresponding statement about relevant curves in $\mathbb{P}^3$. By the same reason, the content of the second column is obvious since $L_{xy} \cap C = \varnothing$.

Since L_{xt} and L_{yz} are contained in S_3, the curves $\tilde{L}_{xt}$ and $\tilde{L}_{yz}$ do not contain O_x or O_y, which implies the content of the fourth and the fifth columns. Similarly, we see that both curves $\tilde{C}'_2$ and $\tilde{C}''_2$ do not contain O_x or O_y, so that the content of the corresponding columns follows from Remark 5.4.

Recall that L_{xz} and C are contained in S'_3. This surface is smooth at the point P_x, so that $\widetilde{S}'_3$ does not contain $\widetilde{P}_x$. Moreover, by construction, the curve $\widetilde{C}'$ is the proper transform of the curve C on the surface $\widetilde{S}'$ via the birational map $\widetilde{S}' \to S'$ induced by π. This implies that $\widetilde{C}'$ contains O_x, this curve does not contain $\widetilde{P}_x$, and $\widetilde{C}' \cap \ell_y = \widetilde{L}_{xz} \cap \ell_y$ because the line L_{xz} is tangent to the curve C at the point P_y. On the other hand, we know from Remark 5.3 that the line L_{xz} is tangent to the cubic surface S_3 at the point P_y. Moreover, we have $(L_{xz} \cdot S_3)_{P_y} = 2$ and $\pi^*(S_3) = \widetilde{S}_3 + E$. Now, using projection formula we get $\widetilde{S}_3 \cdot \widetilde{L}_{xz} = 1$ since $\widetilde{L}_{xz}$ is tangent to E at $\widetilde{L}_{xz} \cap \ell_y$. Then $\widetilde{S}_3 \cap \widetilde{L}_{xz} = \widetilde{P}_t$, which implies that $\widetilde{L}_{xz}$ does not contain $\widetilde{P}_y$, so that it contains O_y. This implies the content of the third and the fourteenth columns.

The line L_{yt} intersects both C and S_3 transversally at P_x, which implies that $\widetilde{L}_{yt}$ does not contain $\widetilde{P}_x$, so that $O_x \in \widetilde{L}_{yt}$. The remaining content of the sixth column is obvious.

Recall that L_{zt} is contained in S''_3. This surface is smooth at P_y, so that $\widetilde{P}_y \notin \widetilde{S}''_3$. Then $O_y \in \widetilde{L}_{zt}$. We also know that L_{zt} is tangent to S_3 at the point P_x, and it intersects the curve C transversally at this point, which implies that $\widetilde{L}_{zt}$ contains $\widetilde{P}_x$, so that it does not contain O_x. This gives the content of the seventh column.

The contents of the eight and the ninth columns follow from the definition of ℓ_x and ℓ_y.

Recall from Remark 5.4 that both curves L_{xz} and C'''_2 are contained in the surface S'_3, and L_{xz} is tangent to C'''_2 at the point P_y. This gives $\widetilde{C}'''_2 \cap \ell_y = \widetilde{L}_{xz} \cap \ell_y = O_y$. Similarly, both curves L_{zt} and C'''_2 are contained in S''_3, and L_{zt} intersects transversally the conic C'''_2 at the point P_x. Since the induced birational morphism $\widetilde{S}''_3 \to S''_3$ is the blow up of the point P_x, we conclude that $\widetilde{C}'''_2 \cap \ell_x \neq \widetilde{L}_{zt} \cap \ell_x = \widetilde{P}_x$, which implies that $\widetilde{C}'''_2 \cap \ell_x = O_x$ and $\widetilde{P}_x \notin \widetilde{C}'''_2$. These facts can also be shown as follows. The point P_x is a smooth point of the surface S'_3, the point O_x is contained in $\widetilde{S}'_3$, and the curve C'''_2 is contained in S'_3, so that we have $\widetilde{P}_x \notin \widetilde{C}'''_2$, which gives $O_x \in \widetilde{C}'''_2$.

To complete the proof, it is enough to show that $O_x \in \widetilde{C}'' \ni O_y$ and $\widetilde{P}_x \notin \widetilde{C}'' \not\ni \widetilde{P}_y$. Recall that $\widetilde{C}''$ is contained $\widetilde{S}''_3$, which does not contain $\widetilde{P}_y$, so that $\widetilde{P}_y \notin \widetilde{C}''$ and $O_y \in \widetilde{C}''$. Moreover, $\widetilde{S}''_3$ contains $\widetilde{L}_{zt}$ and $\widetilde{C}''_2 + \widetilde{L}_{zt} + \ell_x \sim \pi^*(H)|_{\widetilde{S}''_3}$ because $H_t \cap S''_3 = L_{zt} \cup C''_2$ and $P_x = L_{zt} \cap C''_2$ by Remark 5.4. Furthermore, we have

$$3\pi^*(H)|_{\widetilde{S}''_3} - \widetilde{C}'' - \ell_x \sim \widetilde{S}_3|_{\widetilde{S}''_3} = \widetilde{C}''_2,$$

which implies that $\widetilde{C}'' \sim 2\widetilde{C}''_2 + 3\widetilde{L}_{zt} + 2\ell_x$. This gives $\widetilde{C}'' \cdot \widetilde{L}_{zt} = 1$, so that $\widetilde{C}'' \cap \widetilde{L}_{zt} = O_y$. In particular, the curve $\widetilde{C}''$ does not contain $\widetilde{P}_x$ because $\widetilde{P}_x \in \widetilde{L}_{zt}$. Then $O_x \in \widetilde{C}''$. $\qquad\square$

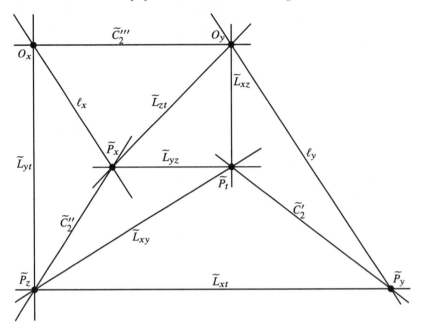

Figure 5.1 The intersection graph of two curves

Observe that every two curves among the smooth curves $\widetilde{L}_{xy}$, $\widetilde{L}_{xz}$, $\widetilde{L}_{xt}$, $\widetilde{L}_{yz}$, $\widetilde{L}_{yt}$, $\widetilde{L}_{zt}$, ℓ_x, ℓ_y, $\widetilde{C}_2'$, $\widetilde{C}_2''$, $\widetilde{C}_2'''$ intersect in at most one point, and if they intersect, then they intersect transversally. Figure 5.1 describes their intersection graph.

Since $\mathrm{Aut}(X)$ is finite [45], to prove the K-stability of the threefold X, it is enough to show that X is K-polystable. Suppose that it is not. Then it follows from Theorem 1.22 that there is a G-invariant prime divisor F over X such that $\beta(F) \leqslant 0$. Let $Z = C_X(F)$. By Theorem 3.17, we conclude that either Z is a G-invariant irreducible curve, or Z is one of the points $\widetilde{P}_x$, $\widetilde{P}_y$, $\widetilde{P}_z$, $\widetilde{P}_t$, O_x, O_y. In both cases, we have $\alpha_{G,Z}(X) < \frac{3}{4}$ by Lemma 1.45. Thus, by Lemma 1.42, there exists a G-invariant effective $\mathbb{Q}$-divisor D on the threefold X such that $D \sim_{\mathbb{Q}} -K_X$ and $Z \subset \mathrm{Nklt}(X, \lambda D)$ for some positive rational number $\lambda < \frac{3}{4}$. Observe that $\mathrm{Nklt}(X, \lambda D)$ contains no surfaces since $\overline{\mathrm{Eff}}(X)$ is generated by E and $\widetilde{S}_3$. Moreover, if Z is a curve, then $Z \cong \mathbb{P}^1$ by Corollary A.14, so that, in particular, it contains a G-fixed point, because the group G is cyclic. Therefore, we see that $\delta_P(X) \leqslant 1$ for some point $P \in \{\widetilde{P}_x, \widetilde{P}_y, \widetilde{P}_z, \widetilde{P}_t, O_x, O_y\}$. Let us show that this is false.

Lemma 5.9 *Let P be one of the points O_x, $\widetilde{P}_x$, $\widetilde{P}_z$ or O_y. Then $\delta_P(X) > 1$.*

Table 5.3

•	H_S	ℓ_x	$\widetilde{L}_{zt}$	$\widetilde{C}_2''$	$\widetilde{C}_2'''$	$\widetilde{C}''$	$\mathcal{L}$
H_S	3	0	1	2	2	7	5
ℓ_x	0	-2	1	1	1	1	5
$\widetilde{L}_{zt}$	1	1	-1	1	1	1	0
$\widetilde{C}_2''$	2	1	1	0	0	5	0
$\widetilde{C}_2'''$	2	1	1	0	0	5	0
$\widetilde{C}''$	7	1	1	5	5	15	10
$\mathcal{L}$	5	5	0	0	0	10	-5

Proof Let $S = \widetilde{S}_3''$. Then S is smooth, it contains the points $O_x, O_y, \widetilde{P}_x, \widetilde{P}_z$, and it contains the curves $\ell_x, \widetilde{L}_{zt}, \widetilde{C}_2'', \widetilde{C}_2'''$ and $\widetilde{C}''$. The intersections of these curves can be described using Remark 5.4 and Lemma 5.8. Namely, we have

$$\ell_x \cap \widetilde{L}_{zt} = \ell_x \cap \widetilde{C}_2'' = \widetilde{L}_{zt} \cap \widetilde{C}_2'' = \widetilde{P}_x, \ell_x \cap \widetilde{C}_2''' = \ell_x \cap \widetilde{C}'' = O_x,$$
$$\widetilde{L}_{zt} \cap \widetilde{C}_2''' = \widetilde{L}_{zt} \cap \widetilde{C}'' = O_y, \widetilde{C}_2'' \cap \widetilde{C}_2''' = \varnothing, \widetilde{C}_2''' \cap \widetilde{C}'' = O_x \cup O_y,$$

and $\widetilde{C}_2''' \cap \widetilde{C}''$ is the preimage of the G-orbit of the point $[-1 : 1 : 1 : 0]$. Note that $\widetilde{P}_z \in \widetilde{C}_2''$.

Let $H_S = \pi^*(H)|_S$. Then $\widetilde{C}_2''' \sim \widetilde{C}_2'', \widetilde{C}_2'' + \widetilde{L}_{zt} + \ell_x \sim H_S, \widetilde{C}'' \sim 2\widetilde{C}_2'' + 3\widetilde{L}_{zt} + 2\ell_x$ on S. We explained these equivalences in the proof of Lemma 5.8. Recall that $E|_{\widetilde{S}_3} = \widetilde{C}'' + \ell_x$.

The cubic surface S_3'' contains 6 lines that pass through P_x, whose union is cut out by the equation $yt + z^2 = 0$. One of these lines is L_{zt}. The remaining lines pass through a point in the G-orbit of the point $[0 : -1 : 1 : 1]$. The proper transforms of these five lines on S are disjoint (-1)-curves that intersect ℓ_x transversally. Let $\mathcal{L}$ be their union. Note that $\mathcal{L}$ is a G-invariant curve in S, which is a disjoint union of five (-1)-curves. On the surface S, we have $\mathcal{L} + \widetilde{L}_{zt} + 3\ell_x \sim 2H_S$. Observe that there is a birational morphism $S \to \mathbb{P}^2$ that contracts the curves $\mathcal{L}$ and $\widetilde{L}_{zt}$, and maps the curve ℓ_x to a conic in $\mathbb{P}^2$ that contains the images of these curves.

The intersections of $H_S, \ell_x, \widetilde{L}_{zt}, \widetilde{C}_2'', \widetilde{C}_2''', \widetilde{C}'', \mathcal{L}$ on S are given in Table 5.3.

Now we are ready to prove that $\delta_P(X) > 1$. We will prove this by applying the results of Section 1.7 to S and a G-invariant curve that contains the point P. As usual, we will use notations introduced in this section.

Take $u \in \mathbb{R}_{\geqslant 0}$. Let $P(u) = P(-K_X - uS)$ and $N(u) = N(-K_X - uS)$. Then

$$-K_X - uS \sim_{\mathbb{R}} (4 - 3u)\pi^*(H) + (u - 1)E \sim_{\mathbb{R}} \pi^*(H) + (1 - u)S,$$

so that $-K_X - uS$ is nef $\iff u \in [0, 1]$, and $-K_X - uS$ is pseudo-effective $\iff u \leqslant \frac{4}{3}$. Moreover, we have

$$P(u) = \begin{cases} -K_X - uS & \text{if } 0 \leqslant u \leqslant 1, \\ (4 - 3u)\pi^*(H) & \text{if } 1 \leqslant u \leqslant \frac{4}{3}, \end{cases}$$

and $N(u) = (u - 1)E$ if $1 \leqslant u \leqslant \frac{4}{3}$. Hence, if $0 \leqslant u \leqslant \frac{4}{3}$, we obtain

$$P(u)\big|_S = \begin{cases} (2 - u)\widetilde{C}_2''' + \widetilde{L}_{zt} + \ell_x & \text{if } 0 \leqslant u \leqslant 1, \\ (4 - 3u)(\widetilde{C}_2''' + \widetilde{L}_{zt} + \ell_x) & \text{if } 1 \leqslant u \leqslant \frac{4}{3}, \end{cases}$$

and $N(u)\big|_S = (u - 1)(\widetilde{C}'' + \ell_x)$ if $1 \leqslant u \leqslant \frac{4}{3}$. Observe that $S_X(S) < 1$ by Theorem 3.17.

Let us compute $S(W^S_{\bullet,\bullet}; \ell_x)$. Take a non-negative real number v. If $0 \leqslant u \leqslant 1$, then

$$P(u)\big|_S - v\ell_x \sim_{\mathbb{R}} (2 - u)\widetilde{C}_2''' + \widetilde{L}_{zt} + (1 - v)\ell_x \sim_{\mathbb{R}} \frac{4 - u - 2v}{2}\ell_x + \frac{2 - u}{2}\mathcal{L} + \frac{u}{2}\widetilde{L}_{zt}.$$

Hence, if $0 \leqslant u \leqslant 1$, the divisor $P(u)\big|_S - v\ell_x$ is pseudo-effective $\iff v \leqslant \frac{4-u}{2}$. If $0 \leqslant u \leqslant 1$ and $0 \leqslant v \leqslant \frac{4-u}{2}$, its Zariski decomposition can be described as follows. If $0 \leqslant v \leqslant 1$, then $P(u)\big|_S - v\ell_x$ is nef. If $1 \leqslant v \leqslant 2 - u$, then the Zariski decomposition is

$$\underbrace{\frac{4 - u - 2v}{2}(\ell_x + \mathcal{L}) + \frac{u}{2}\widetilde{L}_{zt}}_{\text{positive part}} + \underbrace{(v - 1)\mathcal{L}}_{\text{negative part}}.$$

Finally, if $2 - u \leqslant v \leqslant \frac{4-u}{2}$, then the Zariski decomposition is

$$\underbrace{\frac{4 - u - 2v}{2}(\ell_x + \mathcal{L} + \widetilde{L}_{zt})}_{\text{positive part}} + \underbrace{(v - 1)\mathcal{L} + (u + v - 2)\widetilde{L}_{zt}}_{\text{negative part}}.$$

Thus, if $0 \leqslant u \leqslant 1$,

$$\mathrm{vol}\big(P(u)\big|_S - v\ell_x\big) = \begin{cases} 2uv - 2v^2 - 4u - 2v + 7 & \text{if } 0 \leqslant v \leqslant 1, \\ (v - 2)(3v + 2u - 6) & \text{if } 1 \leqslant v \leqslant 2 - u, \\ (u + 2v - 4)^2 & \text{if } 2 - u \leqslant v \leqslant \frac{4 - u}{2}. \end{cases}$$

Similarly, if $1 \leqslant u \leqslant \frac{4}{3}$, then

$$P(u)\big|_S - v\ell_x \sim_{\mathbb{R}} (4 - 3u)(\widetilde{C}_2''' + \widetilde{L}_{zt}) + (4 - 3u - v)\ell_x$$
$$\sim_{\mathbb{R}} \frac{12 - 9u - 2v}{2}\ell_x + \frac{4 - 3u}{2}(\mathcal{L} + \widetilde{L}_{zt}).$$

If $1 \leqslant u \leqslant \frac{4}{3}$, the divisor $P(u)\big|_S - v\ell_x$ is pseudo-effective $\iff v \leqslant \frac{12-9u}{2}$.

If $1 \leqslant u \leqslant \frac{4}{3}$ and $0 \leqslant v \leqslant \frac{12-9u}{2}$, its Zariski decomposition can be described as follows. If $0 \leqslant v \leqslant 4-3u$, then $P(u)|_S - v\ell_x$ is nef. If $4-3u \leqslant v \leqslant \frac{12-9u}{2}$, then the Zariski decomposition is

$$\underbrace{\frac{12-9u-2v}{2}\left(\ell_x + \mathcal{L} + \widetilde{L}_{zt}\right)}_{\text{positive part}} + \underbrace{(3u+v-4)\left(\mathcal{L} + \widetilde{L}_{zt}\right)}_{\text{negative part}}.$$

Thus, if $1 \leqslant u \leqslant \frac{4}{3}$,

$$\mathrm{vol}\left(P(u)|_S - v\ell_x\right) = \begin{cases} 27u^2 - 2v^2 - 72u + 48 & \text{if } 0 \leqslant v \leqslant 4-3u, \\ (12-9u-2v)^2 & \text{if } 4-3u \leqslant v \leqslant \dfrac{12-9u}{2}. \end{cases}$$

Now, using Corollary 1.110, we get

$$\begin{aligned}
S(W_{\bullet,\bullet}^S; \ell_x) &= \frac{3}{16} \int_0^{\frac{4}{3}} \left(P(u) \cdot P(u) \cdot S\right)\mathrm{ord}_{\ell_x}\left(N(u)|_S\right) du \\
&\quad + \frac{3}{16} \int_0^{\frac{4}{3}} \int_0^\infty \mathrm{vol}\left(P(u)|_S - v\ell_x\right) dv\,du \\
&= \frac{3}{16} \int_1^{\frac{4}{3}} (4-3u)^2 \pi^*(H) \cdot \pi^*(H) \cdot \left(3\pi^*(H) - E\right)(u-1) du \\
&\quad + \frac{3}{16} \int_0^{\frac{4}{3}} \int_0^\infty \mathrm{vol}\left(P(u)|_S - v\ell_x\right) dv\,du \\
&= \frac{3}{16} \int_1^{\frac{4}{3}} 3(4-3u)^2(u-1) du + \frac{3}{16} \int_0^{\frac{4}{3}} \int_0^\infty \mathrm{vol}\left(P(u)|_S - v\ell_x\right) dv\,du \\
&= \frac{1}{192} + \frac{3}{16} \int_0^{\frac{4}{3}} \int_0^\infty \mathrm{vol}\left(P(u)|_S - v\ell_x\right) dv\,du \\
&= \frac{1}{192} + \frac{3}{16} \int_0^1 \int_0^1 \left(2uv - 2v^2 - 4u - 2v + 7\right) dv\,du \\
&\quad + \frac{3}{16} \int_0^1 \int_1^{2-u} \left((v-2)(3v+2u-6)\right) dv\,du \\
&\quad + \frac{3}{16} \int_0^1 \int_{2-u}^{\frac{4-u}{2}} (u+2v-4)^2 dv\,du \\
&\quad + \frac{3}{16} \int_1^{\frac{4}{3}} \int_0^{4-3u} \left(27u^2 - 2v^2 - 72u + 48\right) dv\,du \\
&\quad + \frac{3}{16} \int_1^{\frac{4}{3}} \int_{4-3u}^{\frac{12-9u}{2}} (12-9u-2v)^2 dv\,du = \frac{83}{96} < 1.
\end{aligned}$$

Now, we compute $S(W_{\bullet,\bullet}^{S,\ell_x}; O_x)$. Let $P(u,v)$ be the positive part of the Zariski decomposition of the divisor $P(u)|_S - v\ell_x$, and let $N(u,v)$ be its negative part. Recall that

$$F_{O_x}\left(W_{\bullet,\bullet}^{S,\ell_x}\right) = \frac{6}{-K_X^3} \int_0^{\frac{4}{3}} \int_0^\infty \left(P(u,v)\cdot\ell_x\right)_S \mathrm{ord}_{O_x}\left(N_S'(u)\big|_{\ell_x} + N(u,v)\big|_{\ell_x}\right) dvdu.$$

Here, $N_S'(u)$ is the part of the divisor $N(u)|_S$ whose support does not contain ℓ_x. Thus, if $0 \leqslant u \leqslant 1$, then $N_S'(u) = 0$. Similarly, if $1 \leqslant u \leqslant \frac{4}{3}$, then $N_S'(u) = (u-1)\widetilde{C}''$. Note that $\widetilde{C}''|_{\ell_x} = O_x$. On the other hand, the curves $\mathcal{L}$ and $\widetilde{L}_{zt}$ do not contain O_x. Hence, we see that $\mathrm{ord}_{O_x}(N(u,v)|_{\ell_x}) = 0$ in every possible case. Then

$$F_{O_x}\left(W_{\bullet,\bullet}^{S,\ell_x}\right)$$

$$= \frac{6}{16} \int_1^{\frac{4}{3}} \int_0^{4-3u} \left(\left(\frac{12-9u-2v}{2}\ell_x + \frac{4-3u}{2}(\mathcal{L}+\widetilde{L}_{zt})\right)\cdot\ell_x\right)(u-1)dvdu$$

$$+ \frac{6}{16} \int_1^{\frac{4}{3}} \int_{4-3u}^{\frac{12-9u}{2}} \left(\frac{12-9u-2v}{2}(\ell_x + \mathcal{L} + \widetilde{L}_{zt})\cdot\ell_x\right)(u-1)dvdu$$

$$= \frac{6}{16} \int_1^{\frac{4}{3}} \int_0^{4-3u} 2v(u-1)dvdu$$

$$+ \frac{6}{16} \int_1^{\frac{4}{3}} \int_{4-3u}^{\frac{12-9u}{2}} (u-1)(24-18u-4v)dvdu$$

$$= \frac{6}{16} \int_1^{\frac{4}{3}} (u-1)(3u-4)^2 du + \frac{6}{16} \int_1^{\frac{4}{3}} \frac{1}{2}(u-1)(3u-4)^2 du$$

$$= \frac{1}{192}.$$

Thus, it follows from Theorem 1.112 that

$$S\left(W_{\bullet,\bullet}^{S,\ell_x}; O_x\right) = \frac{1}{192} + \frac{3}{-K_X^3} \int_0^{\frac{4}{3}} \int_0^\infty \left((P(u,v)\cdot\ell_x)_S\right)^2 dvdu$$

$$= \frac{1}{192} + \frac{3}{-K_X^3} \int_0^{\frac{4}{3}} \int_0^\infty \left((P(u,v)\cdot\ell_x)_S\right)^2 dvdu$$

$$= \frac{1}{192} + \frac{3}{16} \int_0^1 \int_0^1 (1-u+2v)^2 dvdu$$

$$+ \frac{3}{16} \int_0^1 \int_1^{2-u} (6-u-3v)^2 dvdu$$

$$+ \frac{3}{16} \int_0^1 \int_{2-u}^{\frac{4-u}{2}} (8-2u-4v)^2 dvdu$$

$$+ \frac{3}{16} \int_1^{\frac{4}{3}} \int_0^{4-3u} (2v)^2 dvdu$$

$$+ \frac{3}{16} \int_1^{\frac{4}{3}} \int_{4-3u}^{\frac{12-9u}{2}} (24 - 18u - 4v)^2 dvdu$$

$$= \frac{83}{96} < 1.$$

Since $S_X(S) < 1$ and $S(W_{\bullet,\bullet}^S; \ell_x) < 1$, we see that $\delta_{O_x}(X) > 1$ by Theorem 1.112.

Similarly, we see that $\delta_{\widetilde{P}_x}(X) > 1$ by Theorem 1.112 because

$$S\left(W_{\bullet,\bullet}^{S,\ell_x}; \widetilde{P}_x\right) = \frac{55}{64} + F_{\widetilde{P}_x}\left(W_{\bullet,\bullet}^{S,\ell_x}\right)$$

$$= \frac{55}{64} + \frac{6}{16} \int_0^1 \int_{2-u}^{\frac{4-u}{2}} (u + v - 2)\Big(\big(P(u,v) \cdot \ell_x\big)_S\Big) dvdu$$

$$+ \frac{6}{16} \int_1^{\frac{4}{3}} \int_{4-3u}^{\frac{12-9u}{2}} (3u + v - 4)\Big(\big(P(u,v) \cdot \ell_x\big)_S\Big) dvdu$$

$$= \frac{55}{64} + \frac{6}{16} \int_0^1 \int_{2-u}^{\frac{4-u}{2}} (u + v - 2)(8 - 2u - 4v) dvdu$$

$$+ \frac{6}{16} \int_1^{\frac{4}{3}} \int_{4-3u}^{\frac{12-9u}{2}} (3u + v - 4)(24 - 18u - 4v) dvdu = \frac{167}{192} < 1,$$

since $\mathcal{L}$ and $\widetilde{C}''$ do not pass through $\widetilde{P}_x$, and $\widetilde{L}_{zt}|_{\ell_x} = \widetilde{P}_x$.

Now we will show that $\delta_{\widetilde{P}_z}(X) > 1$. Let us compute $S(W_{\bullet,\bullet}^S; \widetilde{C}_2'')$ and $S(W_{\bullet,\bullet}^{S,\widetilde{C}_2''}; \widetilde{P}_z)$. Take some $v \in \mathbb{R}_{\geqslant 0}$. If $0 \leqslant u \leqslant 1$, then $P(u)|_S - v\widetilde{C}_2'' \sim_{\mathbb{R}} (2 - u - v)\widetilde{C}_2'' + \widetilde{L}_{zt} + \ell_x$, so that $P(u)|_S - v\widetilde{C}_2''$ is pseudo-effective $\iff$ $v \leqslant 2 - u$. If $0 \leqslant u \leqslant 1$ and $0 \leqslant v \leqslant 2 - u$, its Zariski decomposition can be described as follows:

- if $0 \leqslant v \leqslant 1 - u$, then $P(u)|_S - v\widetilde{C}_2''$ is nef,
- if $1 - u \leqslant v \leqslant \frac{5-3u}{3}$, then the Zariski decomposition is

$$\underbrace{(2 - u - v)\widetilde{C}_2'' + \widetilde{L}_{zt} + \frac{3 - u - v}{2}\ell_x}_{\text{positive part}} + \underbrace{\frac{u + v - 1}{2}\ell_x}_{\text{negative part}},$$

- if $\frac{5-3u}{3} \leqslant v \leqslant 2 - u$, then the Zariski decomposition is

$$\underbrace{(2 - u - v)\big(\widetilde{C}_2'' + 3\widetilde{L}_{zt} + 2\ell_x\big)}_{\text{positive part}} + \underbrace{(3u + 3v - 5)\widetilde{L}_{zt} + (2u + 2v - 3)\ell_x}_{\text{negative part}}.$$

Thus, if $0 \leqslant u \leqslant 1$,

$$\mathrm{vol}\big(P(u)\big|_S - v\widetilde{C}_2''\big) = \begin{cases} 7 - 4u - 4v & \text{if } 0 \leqslant v \leqslant 1 - u, \\ \dfrac{15}{2} - 5u - 5v + uv + \dfrac{u^2}{2} + \dfrac{v^2}{2} & \text{if } 1 - u \leqslant v \leqslant \dfrac{5 - 3u}{3}, \\ 5(2 - u - v)^2 & \text{if } \dfrac{5 - 3u}{3} \leqslant v \leqslant 2 - u. \end{cases}$$

Similarly, if $1 \leqslant u \leqslant \frac{4}{3}$,

$$P(u)\big|_S - v\widetilde{C}_2'' \sim_{\mathbb{R}} (4 - 3u - v)\widetilde{C}_2'' + (4 - 3u)\big(\widetilde{L}_{zt} + \ell_x\big).$$

If $1 \leqslant u \leqslant \frac{4}{3}$, the divisor $P(u)|_S - v\widetilde{C}_2''$ is pseudo-effective $\iff v \leqslant 4 - 3u$. If $1 \leqslant u \leqslant \frac{4}{3}$ and $0 \leqslant v \leqslant 4 - 3u$, its Zariski decomposition can be described as follows:

- if $0 \leqslant v \leqslant \frac{8-6u}{3}$, then the positive part of the Zariski decomposition is

$$(4 - 3u - v)\widetilde{C}_2'' + (4 - 3u)\widetilde{L}_{zt} + \left(4 - 3u - \frac{v}{2}\right)\ell_x,$$

and the negative part is $\frac{v}{2}\ell_x$,
- if $\frac{8-6u}{3} \leqslant v \leqslant 4 - 3u$, then the Zariski decomposition is

$$\underbrace{(4 - 3u - v)\widetilde{C}_2'' + (12 - 9u - 3v)\widetilde{L}_{zt} + (8 - 6u - 2v)\ell_x}_{\text{positive part}}$$

$$+ \underbrace{(6u + 3v - 8)\widetilde{L}_{zt} + (3u + 2v - 4)\ell_x}_{\text{negative part}}.$$

Thus, if $1 \leqslant u \leqslant \frac{4}{3}$, then

$$\mathrm{vol}\big(P(u)\big|_S - v\widetilde{C}_2''\big) = \begin{cases} 48 - 72u - 16v + 12uv + 27u^2 + \dfrac{v^2}{2} & \text{if } 0 \leqslant v \leqslant \dfrac{8 - 6u}{3}, \\ 5(4 - 3u - v)^2 & \text{if } \dfrac{8 - 6u}{3} \leqslant v \leqslant 4 - 3u. \end{cases}$$

Using Corollary 1.110 and integrating, we get $S(W_{\bullet,\bullet}^S; \widetilde{C}_2'') = \frac{95}{144} < 1$.

Now we compute $S(W_{\bullet,\bullet}^{S,\widetilde{C}_2''}; \widetilde{P}_z)$. Let $P(u,v)$ be the positive part of the Zariski decomposition of the divisor $P(u)\big|_S - v\widetilde{C}_2''$, and $N(u,v)$ the negative part. Then

$$S(W_{\bullet,\bullet}^{S,\widetilde{C}_2''}; \widetilde{P}_z) = F_{\widetilde{P}_z}(W_{\bullet,\bullet}^{S,\widetilde{C}_2''}) + \frac{3}{16} \int_0^{\frac{4}{3}} \int_0^{\infty} \big((P(u,v) \cdot \widetilde{C}_2'')_S\big)^2 \, dv \, du$$

$$= \frac{3}{16} \int_0^{\frac{4}{3}} \int_0^{\infty} \big((P(u,v) \cdot \widetilde{C}_2'')_S\big)^2 \, dv \, du$$

$$= \frac{3}{16} \int_0^1 \int_0^{1-u} 4 \, dv du + \frac{3}{16} \int_0^1 \int_{1-u}^{\frac{5-3u}{3}} \frac{(5-u-v)^2}{4} \, dv du$$

$$+ \frac{3}{16} \int_0^1 \int_{\frac{5-3u}{3}}^{2-u} (10 - 5u - 5v)^2 dv du$$

$$+ \frac{3}{16} \int_1^{\frac{4}{3}} \int_0^{\frac{8-6u}{3}} \left(8 - 6u - \frac{v}{2} \right)^2 dv du$$

$$+ \frac{3}{16} \int_1^{\frac{4}{3}} \int_{\frac{8-6u}{3}}^{4-3u} (20 - 15u - 5v)^2 dv du = \frac{515}{576} < 1$$

by Theorem 1.112. Here we used the equality $F_{\widetilde{P}_z}\left(W_{\bullet,\bullet}^{S,\widetilde{C}_2''}\right) = 0$. It follows from the fact that ℓ_x, $\widetilde{C}''$ and the support of the divisor $N(u,v)$ do not contain $\widetilde{P}_z$ because

$$F_{\widetilde{P}_z}\left(W_{\bullet,\bullet}^{S,\widetilde{C}_2''}\right) = \frac{6}{-K_X^3} \int_0^{\frac{4}{3}} \int_0^\infty \left(P(u,v)\cdot\widetilde{C}_2''\right)_S \mathrm{ord}_{\widetilde{P}_z}\left(N_S'(u)\big|_{\widetilde{C}_2''} + N(u,v)\big|_{\widetilde{C}_2''}\right) dv du,$$

where

$$N_S'(u) = N(u)\big|_S = \begin{cases} 0 & \text{if } 0 \leqslant u \leqslant 1, \\ (u-1)\ell_x + (u-1)\widetilde{C}'' & \text{if } 1 \leqslant u \leqslant \frac{4}{3}. \end{cases}$$

Since $S(W_{\bullet,\bullet}^S; \widetilde{C}_2'') < 1$ and $S(W_{\bullet,\bullet}^{S,\widetilde{C}_2''}; \widetilde{P}_z) < 1$, we see that $\delta_{\widetilde{P}_z}(X) > 1$ by Theorem 1.112.

Likewise, we can show that $\delta_{O_y}(X) > 1$. Indeed, recall that $O_y \in \widetilde{C}_2'''$ and $\widetilde{C}_2''' \sim \widetilde{C}_2''$. Then $S(W_{\bullet,\bullet}^S; \widetilde{C}_2''') \leqslant S(W_{\bullet,\bullet}^S; \widetilde{C}_2'') < 1$ because $\widetilde{C}_2'''$ is not contained in $\mathrm{Supp}(N(u)|_S)$. Moreover, one can compute $S(W_{\bullet,\bullet}^{S,\widetilde{C}_2'''}, O_y)$ similarly to $S(W_{\bullet,\bullet}^{S,\widetilde{C}_2''}, \widetilde{P}_z)$. Namely, we have

$$S\left(W_{\bullet,\bullet}^{S,\widetilde{C}_2'''}; O_y\right) = \frac{515}{576} + F_{O_y}\left(W_{\bullet,\bullet}^{S,\widetilde{C}_2'''}\right),$$

but now we have $F_{O_y}\left(W_{\bullet,\bullet}^{S,\widetilde{C}_2'''}\right) \neq 0$. On the other hand, we have

$$F_{O_y}\left(W_{\bullet,\bullet}^{S,\widetilde{C}_2'''}\right)$$

$$= \frac{6}{16} \int_0^1 \int_{\frac{5-3u}{3}}^{2-u} (3u + 3v - 5)\left((P(u,v)\cdot\widetilde{C}_2''')_S\right) dv du$$

$$+ \frac{6}{16} \int_1^{\frac{4}{3}} \int_0^{\frac{8-6u}{3}} \left((P(u,v)\cdot\widetilde{C}_2''')_S\right)(u-1)(\widetilde{C}''\cdot\widetilde{C}_2''')_{O_y} dv du$$

$$+ \frac{6}{16} \int_1^{\frac{4}{3}} \int_{\frac{8-6u}{3}}^{4-3u} \left((P(u,v)\cdot\widetilde{C}_2''')_S\right)\left((u-1)(\widetilde{C}''\cdot\widetilde{C}_2''')_{O_y} + 6u + 3v - 8\right) dv du$$

$$= \frac{6}{16} \int_0^1 \int_{\frac{5-3u}{3}}^{2-u} (3u + 3v - 5)(10 - 5v - 5v)\,dv\,du$$

$$+ \frac{6}{16} \int_1^{\frac{4}{3}} \int_0^{\frac{8-6u}{3}} \left(8 - 6u - \frac{v}{2}\right)(u - 1)(\widetilde{C}'' \cdot \widetilde{C}_2''')_{O_y}\,dv\,du$$

$$+ \frac{6}{16} \int_1^{\frac{4}{3}} \int_{\frac{8-6u}{3}}^{4-3u} \left((u - 1)(\widetilde{C}'' \cdot \widetilde{C}_2''')_{O_y} + 6u + 3v - 8\right)(20 - 15u - 5v)\,dv\,du$$

$$= \frac{65}{1728} + \frac{(\widetilde{C}'' \cdot \widetilde{C}_2''')_{O_y}}{192}$$

because $\widetilde{L}_{zt}$ intersects $\widetilde{C}_2'''$ transversally at the point O_y. But $(\widetilde{C}'' \cdot \widetilde{C}_2''')_{O_y} \leqslant \widetilde{C}'' \cdot \widetilde{C}_2''' = 5$. Therefore, we have $S(W_{\bullet,\bullet}^{S,\widetilde{C}_2'''}; O_y) = \frac{515}{576} + \frac{65}{1728} + \frac{(\widetilde{C}'' \cdot \widetilde{C}_2''')_{O_y}}{192} \leqslant \frac{515}{576} + \frac{5}{192} = \frac{1655}{1728} < 1$, which implies that $\delta_{O_y}(X) > 1$ by Theorem 1.112. This completes the proof of the lemma.   □

Finally, we conclude the proof of Proposition 5.5 by the following

Lemma 5.10 $\delta_{\widetilde{P}_y}(X) > 1$ *and* $\delta_{\widetilde{P}_t}(X) > 1$.

Proof Let $S = \widetilde{S}_3'$. Then the surface S is smooth, it contains the points O_x, O_y, $\widetilde{P}_y$, $\widetilde{P}_t$, and it contains ℓ_y, $\widetilde{L}_{xz}$, $\widetilde{C}_2'$, $\widetilde{C}_2'''$, $\widetilde{C}'$. It follows from Remark 5.4 and Lemma 5.8 that

$$\ell_y \cap \widetilde{L}_{xz} = \ell_y \cap \widetilde{C}_2''' = \ell_y \cap \widetilde{C}' = \widetilde{L}_{xz} \cap \widetilde{C}_2''' = \widetilde{L}_{xz} \cap \widetilde{C}' = O_y,$$

$\ell_y \cap \widetilde{C}_2' = \widetilde{P}_y$, $\widetilde{L}_{xz} \cap \widetilde{C}_2' = \widetilde{P}_t$, $\widetilde{C}_2' \cap \widetilde{C}_2''' = \varnothing$, $\widetilde{C}_2''' \cap \widetilde{C}' = O_x \cup O_y$, and $\widetilde{C}_2' \cap \widetilde{C}'$ consists of five points that form the preimage of the G-orbit of the point $[0 : -1 : 1 : 1]$.

The cubic surface S_3' contains 6 lines that pass through P_y. One of them is the line L_{xz}. The remaining five lines pass through a point in the G-orbit of the point $[-1 : 0 : 1 : 1]$. The proper transforms of these five lines on S are disjoint (-1)-curves that intersect the curve ℓ_y transversally. Let $\mathcal{L}$ be their union. Then $\mathcal{L}$ is disjoint from $\widetilde{L}_{xz}$, $\widetilde{C}_2'$ and $\widetilde{C}_2'''$.

On the surface S, the intersection form of the curves ℓ_y, $\widetilde{L}_{xz}$, $\widetilde{C}_2'$, $\widetilde{C}_2'''$, $\widetilde{C}'$ and $\mathcal{L}$ is given in Table 5.4.

To prove that $\delta_{\widetilde{P}_y}(X) > 1$, we will apply Theorem 1.112 to S and the curve ℓ_y. Similarly, to prove that $\delta_{\widetilde{P}_t}(X) > 1$, we will apply Theorem 1.112 to S and $\widetilde{C}_2'$. As usual, we will use notations introduced in Section 1.7.

Take a number $u \geq 0$. Let $P(u) = P(-K_X - uS)$ and $N(u) = N(-K_X - uS)$. As in the proof of Lemma 5.9, we see that $-K_X - uS$ is not pseudo-effective for $u > \frac{4}{3}$. Moreover, if $0 \leqslant u \leqslant \frac{4}{3}$, then

Table 5.4

•	ℓ_y	$\widetilde{L}_{xz}$	$\widetilde{C}_2'$	$\widetilde{C}_2'''$	$\widetilde{C}'$	$\mathcal{L}$
ℓ_y	−2	1	1	1	1	5
$\widetilde{L}_{xz}$	1	−1	1	1	1	0
$\widetilde{C}_2'$	1	1	0	0	5	0
$\widetilde{C}_2'''$	1	1	0	0	5	0
$\widetilde{C}'$	1	1	5	5	15	10
$\mathcal{L}$	5	0	0	0	10	−5

$$P(u)\big|_S = \begin{cases} (2-u)\widetilde{C}_2' + \widetilde{L}_{xz} + \ell_y & \text{if } 0 \leqslant u \leqslant 1, \\ (4-3u)(\widetilde{C}_2' + \widetilde{L}_{xz} + \ell_y) & \text{if } 1 \leqslant u \leqslant \dfrac{4}{3}, \end{cases}$$

and

$$N(u)\big|_S = \begin{cases} 0 & \text{if } 0 \leqslant u \leqslant 1, \\ (u-1)(\widetilde{C}' + \ell_y) & \text{if } 1 \leqslant u \leqslant \dfrac{4}{3}. \end{cases}$$

Observe that $S_X(S) < 1$ by Theorem 3.17.

Let us compute $S(W_{\bullet,\bullet}^S; \ell_y)$. Take a non-negative real number v. If $0 \leqslant u \leqslant 1$, then

$$P(u)\big|_S - v\ell_y \sim_{\mathbb{R}} \frac{4-u-2v}{2}\ell_y + \frac{2-u}{2}\mathcal{L} + \frac{u}{2}\widetilde{L}_{xz}.$$

Therefore, given $0 \leqslant u \leqslant 1$, the divisor $P(u)\big|_S - v\ell_y$ is pseudo-effective $\iff$ $v \leqslant \frac{4-u}{2}$. If $0 \leqslant u \leqslant 1$ and $0 \leqslant v \leqslant \frac{4-u}{2}$, its Zariski decomposition can be described as follows:

- if $0 \leqslant v \leqslant 1$, then $P(u)\big|_S - v\ell_y$ is nef,
- if $1 \leqslant v \leqslant 2-u$, then the Zariski decomposition is

$$\underbrace{\frac{4-u-2v}{2}(\ell_y + \mathcal{L}) + \frac{u}{2}\widetilde{L}_{xz}}_{\text{positive part}} + \underbrace{(v-1)\mathcal{L}}_{\text{negative part}},$$

- if $2-u \leqslant v \leqslant \frac{4-u}{2}$, then the Zariski decomposition is

$$\underbrace{\frac{4-u-2v}{2}(\ell_y + \mathcal{L} + \widetilde{L}_{xz})}_{\text{positive part}} + \underbrace{(v-1)\mathcal{L} + (u+v-2)\widetilde{L}_{xz}}_{\text{negative part}}.$$

Thus, if $0 \leqslant u \leqslant 1$,

$$\text{vol}\big(P(u)\big|_S - v\ell_y\big) = \begin{cases} 2uv - 2v^2 - 4u - 2v + 7 & \text{if } 0 \leqslant v \leqslant 1, \\ (v-2)(3v + 2u - 6) & \text{if } 1 \leqslant v \leqslant 2 - u, \\ (u + 2v - 4)^2 & \text{if } 2 - u \leqslant v \leqslant \dfrac{4-u}{2}. \end{cases}$$

Similarly, if $1 \leqslant u \leqslant \frac{4}{3}$, then

$$P(u)\big|_S - v\ell_y \sim_{\mathbb{R}} \frac{12 - 9u - 2v}{2}\ell_y + \frac{4 - 3u}{2}\big(\mathcal{L} + \widetilde{L}_{xz}\big).$$

If $1 \leqslant u \leqslant \frac{4}{3}$, the divisor $P(u)|_S - v\ell_y$ is pseudo-effective $\Longleftrightarrow v \leqslant \frac{12-9u}{2}$. If $1 \leqslant u \leqslant \frac{4}{3}$ and $0 \leqslant v \leqslant \frac{12-9u}{2}$, its Zariski decomposition can be described as follows:

- if $0 \leqslant v \leqslant 4 - 3u$, then $P(u)|_S - v\ell_y$ is nef,
- if $4 - 3u \leqslant v \leqslant \frac{12-9u}{2}$, then the Zariski decomposition is

$$\underbrace{\frac{12 - 9u - 2v}{2}\big(\ell_y + \mathcal{L} + \widetilde{L}_{xz}\big)}_{\text{positive part}} + \underbrace{(3u + v - 4)\big(\mathcal{L} + \widetilde{L}_{xz}\big)}_{\text{negative part}}.$$

If $1 \leqslant u \leqslant \frac{4}{3}$,

$$\text{vol}\big(P(u)\big|_S - v\ell_y\big) = \begin{cases} 27u^2 - 2v^2 - 72u + 48 & \text{if } 0 \leqslant v \leqslant 4 - 3u, \\ (12 - 9u - 2v)^2 & \text{if } 4 - 3u \leqslant v \leqslant \dfrac{12 - 9u}{2}. \end{cases}$$

Thus, using Corollary 1.110, we get

$$S(W^S_{\bullet,\bullet}; \ell_y)$$

$$= \frac{3}{16}\int_0^{\frac{4}{3}} \big(P(u) \cdot P(u) \cdot S\big)\text{ord}_{\ell_y}\big(N(u)|_S\big)du + \frac{3}{16}\int_0^{\frac{4}{3}}\int_0^{\infty} \text{vol}\big(P(u)\big|_S - v\ell_y\big)dv\,du$$

$$= \frac{3}{16}\int_1^{\frac{4}{3}} 3(4 - 3u)^2(u - 1)du + \frac{3}{16}\int_0^{\frac{4}{3}}\int_0^{\infty} \text{vol}\big(P(u)\big|_S - v\ell_y\big)dv\,du = \frac{83}{96} < 1.$$

Now we compute $S(W^{S,\ell_y}_{\bullet,\bullet}; \widetilde{P}_y)$. Let $P(u,v)$ be the positive part of the Zariski decomposition of the divisor $P(u)\big|_S - v\ell_y$, and let $N(u,v)$ be its negative part. Recall that

$$S(W^{S,\ell_y}_{\bullet,\bullet}; \widetilde{P}_y) = F_{\widetilde{P}_y}\big(W^{S,\ell_y}_{\bullet,\bullet}\big) + \frac{3}{-K_X^3}\int_0^{\frac{4}{3}}\int_0^{\infty} \big((P(u,v) \cdot \ell_y)_S\big)^2 dv\,du,$$

where

$$F_{\widetilde{P}_y}\left(W_{\bullet,\bullet}^{S,\ell_y}\right) = \frac{6}{-K_X^3}\int_0^{\frac{4}{3}}\int_0^\infty \left(P(u,v)\cdot\ell_y\right)_S \mathrm{ord}_{\widetilde{P}_y}\left(N_S'(u)\big|_{\ell_y} + N(u,v)\big|_{\ell_y}\right)dvdu.$$

Here $N_S'(u)$ is the part of the divisor $N(u)\big|_S$ whose support does not contain ℓ_y. Then

$$N_S'(u) = \begin{cases} 0 & \text{if } 0 \leqslant u \leqslant 1, \\ (u-1)\widetilde{C}' & \text{if } 1 \leqslant u \leqslant \dfrac{4}{3}. \end{cases}$$

Thus, since $\widetilde{C}'$, $\mathcal{L}$ and $\widetilde{L}_{xz}$ do not contain the point $\widetilde{P}_y$, we have $F_{\widetilde{P}_y}\left(W_{\bullet,\bullet}^{S,\ell_y}\right) = 0$. Then $S\left(W_{\bullet,\bullet}^{S,\ell_y};\widetilde{P}_y\right) = \frac{55}{64} < 1$, so that $\delta_{\widetilde{P}_y}(X) > 1$ by Theorem 1.112.

Now let us show that $\delta_{\widetilde{P}_t}(X) > 1$. First, we compute $S\left(W_{\bullet,\bullet}^S;\widetilde{C}_2'\right)$. Take some $v \in \mathbb{R}_{\geqslant 0}$. If $0 \leqslant u \leqslant 1$, then $P(u)\big|_S - v\widetilde{C}_2' \sim_{\mathbb{R}} (2-u-v)\widetilde{C}_2' + \widetilde{L}_{xz} + \ell_y$, so that $P(u)\big|_S - v\widetilde{C}_2'$ is pseudo-effective $\iff v \leqslant 2 - u$. If $0 \leqslant u \leqslant 1$ and $0 \leqslant v \leqslant 2 - u$, its Zariski decomposition can be described as follows:

- if $0 \leqslant v \leqslant 1 - u$, then $P(u)\big|_S - v\widetilde{C}_2'$ is nef,
- if $1 \leqslant v \leqslant \frac{5-3u}{3}$, then the Zariski decomposition is

$$\underbrace{(2-u-v)\widetilde{C}_2' + \widetilde{L}_{xz} + \frac{3-u-v}{2}\ell_y}_{\text{positive part}} + \underbrace{\frac{u+v-1}{2}\ell_y}_{\text{negative part}},$$

- if $\frac{5-3u}{3} \leqslant v \leqslant 2 - u$, then the Zariski decomposition is

$$\underbrace{(2-u-v)\left(\widetilde{C}_2' + 3\widetilde{L}_{xz} + 2\ell_y\right)}_{\text{positive part}} + \underbrace{(3u+3v-5)\widetilde{L}_{xz} + (2u+2v-3)\ell_y}_{\text{negative part}}.$$

Thus, if $0 \leqslant u \leqslant 1$,

$$\mathrm{vol}\left(P(u)\big|_S - v\widetilde{C}_2'\right) = \begin{cases} 7 - 4u - 4v & \text{if } 0 \leqslant v \leqslant 1 - u, \\ \dfrac{15}{2} - 5u - 5v + uv + \dfrac{u^2}{2} + \dfrac{v^2}{2} & \text{if } 1 - u \leqslant v \leqslant \dfrac{5-3u}{3}, \\ 5(2-u-v)^2 & \text{if } \dfrac{5-3u}{3} \leqslant v \leqslant 2 - u. \end{cases}$$

Similarly, if $1 \leqslant u \leqslant \frac{4}{3}$, then $P(u)\big|_S - v\widetilde{C}_2' \sim_{\mathbb{R}} (4-3u-v)\widetilde{C}_2' + (4-3u)(\widetilde{L}_{xz} + \ell_y)$, so that $P(u)\big|_S - v\widetilde{C}_2'$ is pseudo-effective $\iff v \leqslant 4 - 3u$. If $1 \leqslant u \leqslant \frac{4}{3}$ and $0 \leqslant v \leqslant 4 - 3u$, its Zariski decomposition can be described as follows:

- if $0 \leqslant v \leqslant \frac{8-6u}{3}$, then the positive part of the Zariski decomposition is

$$(4 - 3u - v)\widetilde{C}'_2 + (4 - 3u)\widetilde{L}_{xz} + \left(4 - 3u - \frac{v}{2}\right)\ell_y,$$

and the negative part is $\frac{v}{2}\ell_y$,
- if $\frac{8-6u}{3} \leqslant v \leqslant 4 - 3u$, then the Zariski decomposition is

$$\underbrace{(4 - 3u - v)\widetilde{C}'_2 + (12 - 9u - 3v)\widetilde{L}_{xz} + (8 - 6u - 2v)\ell_y}_{\text{positive part}}$$

$$+ \underbrace{(6u + 3v - 8)\widetilde{L}_{xz} + (3u + 2v - 4)\ell_y}_{\text{negative part}}.$$

Thus, if $1 \leqslant u \leqslant \frac{4}{3}$,

$$\text{vol}\big(P(u)\big|_S - v\widetilde{C}''_2\big)$$

$$= \begin{cases} 48 - 72u - 16v + 12uv + 27u^2 + \dfrac{v^2}{2} & \text{if } 0 \leqslant v \leqslant \dfrac{8 - 6u}{3}, \\[2mm] 5(4 - 3u - v)^2 & \text{if } \dfrac{8 - 6u}{3} \leqslant v \leqslant 4 - 3u. \end{cases}$$

Using Corollary 1.110 and integrating, we get $S(W^{S}_{\bullet,\bullet}; \widetilde{C}'_2) = \frac{377}{576} < 1$.

Now we compute $S(W^{S,\widetilde{C}'_2}_{\bullet,\bullet}; \widetilde{P}_t)$. Let $P(u,v)$ be the positive part of the Zariski decomposition of the divisor $P(u)\big|_S - v\widetilde{C}'_2$, and $N(u,v)$ the negative part. Then

$$S\big(W^{S,\widetilde{C}'_2}_{\bullet,\bullet}; \widetilde{P}_t\big) = F_{\widetilde{P}_t}\big(W^{S,\widetilde{C}'_2}_{\bullet,\bullet}\big) + \frac{3}{16}\int_0^{\frac{4}{3}}\int_0^\infty \big((P(u,v) \cdot \widetilde{C}'_2)_S\big)^2 dv\, du$$

$$= \text{ord}_{\widetilde{P}_t}\Big(F\big(W^{S,\widetilde{C}'_2}_{\bullet,\bullet}\big)\Big) + \frac{515}{576}$$

by Theorem 1.112. To compute $F_{\widetilde{P}_t}\big(W^{S,\widetilde{C}'_2}_{\bullet,\bullet}\big)$, recall from Theorem 1.112 that

$$F_{\widetilde{P}_t}\big(W^{S,\widetilde{C}'_2}_{\bullet,\bullet}\big) = \frac{6}{-K_X^3}\int_0^{\frac{4}{3}}\int_0^\infty (P(u,v) \cdot \widetilde{C}'_2)_S\, \text{ord}_{\widetilde{P}_t}\big(N'_S(u)\big|_{\widetilde{C}''_2} + N(u,v)\big|_{\widetilde{C}'_2}\big) dv\, du,$$

where, since $\widetilde{C}'_2$ is not contained in the support of the divisor $N(u)\big|_S$, we have

$$N'_S(u) = N(u)\big|_S = \begin{cases} 0 & \text{if } 0 \leqslant u \leqslant 1, \\[2mm] (u-1)\ell_y + (u-1)\widetilde{C}' & \text{if } 1 \leqslant u \leqslant \dfrac{4}{3}. \end{cases}$$

On the other hand, the curves ℓ_y and $\widetilde{C}'$ do not contain the point $\widetilde{P}_t$. Thus, we have

$$F_{\widetilde{P}_t}\left(W_{\bullet,\bullet}^{S,\widetilde{C}_2'}\right) = \frac{6}{16}\int_0^1\int_{\frac{5-3u}{3}}^{2-u}(3u+3v-5)\left(\left(P(u,v)\cdot\widetilde{C}_2'\right)_S\right)dvdu$$

$$+\frac{6}{16}\int_1^{\frac{4}{3}}\int_{\frac{8-6u}{3}}^{4-3u}(6u+3v-8)\left(\left(P(u,v)\cdot\widetilde{C}_2'\right)_S\right)dvdu$$

$$=\frac{6}{16}\int_0^1\int_{\frac{5-3u}{3}}^{2-u}(3u+3v-5)(10-5u-5v)dvdu$$

$$+\frac{6}{16}\int_1^{\frac{4}{3}}\int_{\frac{8-6u}{3}}^{4-3u}(6u+3v-8)(20-15u-5v)dvdu = \frac{65}{1728}.$$

Then $S(W_{\bullet,\bullet}^{S,\widetilde{C}_2'};\widetilde{P}_t) = \frac{805}{864} < 1$. Since we already know that $S(W_{\bullet,\bullet}^S;\widetilde{C}_2') < 1$ and $S_X(S) < 1$, we get $\delta_{\widetilde{P}_t}(X) > 1$ by Theorem 1.112, completing the proof. $\quad\square$

Thus Proposition 5.5 is proved, and general members of the family №2.9 are K-stable.

5.3 Family №2.11

Let V be the cubic threefold in $\mathbb{P}^4$ that is given by

$$xu^2 + 2yuv + zv^2 + 2z^2u + 2x^2v + ay^3 + bxyz = 0,$$

where u, v, x, y, z are homogeneous coordinates on $\mathbb{P}^4$, and a and b are general numbers such that V is smooth, e.g. $a = 5$ and $b = 7$. This threefold has been studied in [111].

Let $G = D_{10} = \langle\alpha,\iota \mid \alpha^5 = 1, \iota^2 = 1, \alpha\cdot\iota = \iota\cdot\alpha^4\rangle$. Then G acts on $\mathbb{P}^4$ via

$$\alpha\big([u:v:x:y:z]\big) = [\omega^2u:\omega^3v:\omega x:y:\omega^4z]$$

and $\iota([u:v:x:y:z]) = [v:u:z:y:x]$, where ω is a primitive fifth root of unity. Moreover, the cubic V is G-invariant, and the only G-invariant linear subspaces in $\mathbb{P}^4$ are the hyperplane $\{y = 0\}$, the plane $\Pi = \{x = z = 0\}$, the plane $\Pi' = \{u = v = 0\}$, the line $L = \{x = y = z = 0\}$, the line $L' = \{v = v = y = 0\}$, and the point $P = [0:0:0:1:0]$. Observe that the point P does not lie on V. Let S_3 be the cubic surface in V that is cut out by the hyperplane $y = 0$. Then S_3 is smooth, it contains the lines L and L', and it is isomorphic to the Clebsch cubic surface [77].

Let $\pi\colon X \to V$ be the blow up of the line L. Then X is a Fano threefold №2.11, and the action of the group G lifts to X, so that we identify G with a subgroup in $\mathrm{Aut}(X)$. Moreover, there exists G-equivariant commutative diagram

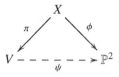

where ϕ is a conic bundle, and ψ is a rational map given by $[u : v : x : y : z] \mapsto [x : y : z]$. Observe that the G-action on $\mathbb{P}^2$ has exactly one G-fixed point: $[0 : 1 : 0]$.

Remark 5.11   The threefold X is given in $\mathbb{P}^4 \times \mathbb{P}^2$ by the equations

$$\begin{cases} sy = tx, \\ sz = rx, \\ ry = tz, \\ u^2x + 2vx^2 + 2uvy + ay^3 + v^2z + bxyz + 2uz^2 = 0, \\ u^2s + 2vxs + 2uvt + ay^2t + v^2r + bxyr + 2uzr = 0, \end{cases}$$

where s, t, r are coordinates on $\mathbb{P}^2$. The surface E is cut out on X by $x = y = z = 0$, so is isomorphic to a surface in $\mathbb{P}^1 \times \mathbb{P}^2$ given by $u^2s + 2uvt + v^2r = 0$, where we consider u and v as coordinates on $\mathbb{P}^1$. The group G acts on $\mathbb{P}^1 \times \mathbb{P}^2$ by

$$\alpha([u : v], [s : t : r]) = ([u : \omega v], [\omega s : t : \omega^4 r])$$

and $\iota([u : v], [s : t : r]) = ([v : u], [r : t : s])$. There are no G-fixed points on E, and there are no G-invariant fibers of the projection $E \to L$ because V does not have G-fixed points. Observe that $E \cong \mathbb{P}^1 \times \mathbb{P}^1$.

Let H be a general hyperplane section of the threefold V, let E be the π-exceptional divisor, and let S be the proper transform of the cubic surface S_3 on the threefold X. Then $-K_X \sim 2S + E$, $S \sim \pi^*(H) - E$, the conic bundle ϕ is given by $|\pi^*(H) - E|$, S is the only G-invariant surface in the linear system $|\pi^*(H) - E|$, the cone of effective divisors on X is generated by E and S, and the cone of nef divisors is generated by $\pi^*(H)$ and S. Note that $\pi^*(H)^3 = 3$, $\pi^*(H) \cdot E^2 = -1$, $\pi^*(H)^2 \cdot E = 0$, $E^3 = 0$, $-K_X^3 = 18$.

Proposition 5.12   *The Fano threefold X is K-stable.*

Thus, since $\mathrm{Aut}(X)$ is finite [45], Theorem 1.11 implies that general smooth Fano threefolds in the family №2.11 are K-stable.

Let us prove Proposition 5.12. By Corollary 1.5, it is enough to show that the threefold X is K-polystable. Suppose that X is not K-polystable. Then, by Theorem 1.22, there are a G-equivariant birational morphism $f : \widetilde{X} \to X$ and a G-invariant dreamy prime divisor $F \subset \widetilde{X}$ such that $\beta(F) = A_X(F) - S_X(F) \leqslant 0$. Let $Z = f(F)$. Then Z is not a surface by Theorem 3.17, so that Z is a

G-invariant irreducible curve, because X does not have G-fixed points. Let us seek a contradiction.

Lemma 5.13 (cf. [92, §4]) $\pi(Z) \subset S_3 \cup \Pi \cup \Pi'$.

Proof Observe that $\pi(Z)$ is a curve. Suppose that this curve is not contained in $S_3 \cup \Pi \cup \Pi'$. By Lemma 1.45, we have $\alpha_{G,Z}(X) < \frac{3}{4}$. Thus, there are a G-invariant effective $\mathbb{Q}$-divisor D on the threefold X and a positive rational number $\lambda < \frac{3}{4}$ such that $Z \subseteq \mathrm{Nklt}(X, \lambda D)$.

Recall that the cone $\overline{\mathrm{Eff}}(X)$ is generated by E and S, and S is the only G-invariant surface in the linear system $|\pi^*(H) - E|$. Thus, since $-K_X \sim 2S + E$, we conclude that the locus $\mathrm{Nklt}(X, \lambda D)$ does not contain surfaces except maybe S. Write $D = aS + \Delta$, where $a \in \mathbb{Q}_{\geqslant 0}$, and Δ is an effective $\mathbb{Q}$-divisor on X whose support does not contain the surface S. Let $\overline{\Delta} = \pi(\Delta)$. Then $Z \subseteq \mathrm{Nklt}(X, \lambda\Delta)$, so that $\pi(Z) \subseteq \mathrm{Nklt}(V, \lambda\overline{\Delta})$. But $\overline{\Delta} \sim_{\mathbb{Q}} (2 - a)H$. Thus, using Corollary A.4, we see that the locus $\mathrm{Nklt}(V, \lambda\overline{\Delta})$ is connected union of finitely many curves. Since V does not have G-fixed points, $\pi(Z)$ is one of these curves.

Choose a positive rational number $\mu \leqslant \lambda$ such that $(V, \mu\overline{\Delta})$ is strictly log canonical. Then $\pi(Z)$ is a minimal log canonical center of the log pair $(V, \mu\overline{\Delta})$ by Corollary A.31. Therefore, the degree of the curve $\pi(Z)$ is at most three by Corollary A.21. On the other hand, the curve $\pi(Z)$ is not a line since L and L' are contained in S. Moreover, since $\pi(Z) \not\subset \Pi$ and $\pi(Z) \not\subset \Pi'$, we see that $\pi(Z)$ is not contained in a plane since Π and Π are the only G-invariant planes in $\mathbb{P}^4$. Thus, we conclude that $\pi(Z)$ is a twisted cubic curve. Then it is contained in the unique G-invariant hyperplane in $\mathbb{P}^4$, which is given by $y = 0$. But $\pi(Z)$ is not contained in S_3, which is a contradiction. $\qquad\square$

Our next step is

Lemma 5.14 $Z \not\subset E$.

Proof Suppose that $Z \subset E$. Let us apply results of Section 1.7 to derive a contradiction. Fix $u \in \mathbb{R}_{\geqslant 0}$. Then $-K_X - uE \sim_{\mathbb{R}} 2\pi^*(H) - (1 + u)E$, so that $-K_X - uE$ is nef if and only if it is pseudo-effective $\iff u \in [0, 1]$. Using this, we see that $S_X(E) = \frac{5}{9} < 1$. Thus, applying Corollary 1.110, we see that $S(W_{\bullet,\bullet}^E; Z) \geqslant 1$.

Recall from Remark 5.11 that $E \cong \mathbb{P}^1 \times \mathbb{P}^1$. We may assume that a fiber of the natural projection $E \to L$ is a divisor of degree $(1, 0)$. Then $\pi^*(H)|_E$ is a divisor of degree $(1, 0)$, the divisor $-E|_E$ has degree $(0, 1)$, and Z has degree (b_1, b_2) for some $b_1 \geqslant 0$ and $b_2 \geqslant 0$. Then $(b_1, b_2) \neq (1, 0)$ since L has no G-fixed points. Therefore, we conclude that $b_2 > 0$. Thus, for a curve $Z_0 \subset E$ of degree $(0, 1)$, we have

$$S\left(W^E_{\bullet,\bullet}; Z\right) \leqslant S\left(W^E_{\bullet,\bullet}; Z_0\right) = \frac{3}{18} \int_0^1 \int_0^{1+u} 4(1 - u + v)dvdu = \frac{7}{9} < 1$$

by Corollary 1.110. The obtained contradiction completes the proof. $\qquad \square$

Observe that $\Pi \cap V = L \cup C$, where C is an irreducible conic. Similarly, $\Pi' \cap V = L' \cup C'$ for an irreducible conic C'. The conics C and C' are G-invariant. Moreover, these are the only G-invariant conics contained in V.

Lemma 5.15 $\pi(Z)$ *is not the conic* C.

Proof Let T_3 be a hyperplane section of V that is cut out by the equation $\lambda x + \mu z = 0$, where λ and μ are complex numbers such that T_3 is smooth. Such numbers always exist, e.g. the cubic surface T_3 is smooth for $a = 5$, $b = 7$, $\lambda = 1$ and $\mu = -1$.

Note that the line L and the conic C are both contained in the surface T_3 by construction. But the surface T_3 is not G-invariant. Let T be its proper transform on X. Then $T \cong T_3$.

Suppose that $\pi(Z) = C$. Then $Z \subset T$. Let us apply results of Section 1.7 to T and Z to derive a contradiction. We will use the notations introduced in this section.

Fix $u \in \mathbb{R}_{\geqslant 0}$. Then $-K_X - uT \sim_{\mathbb{R}} (2 - u)\pi^*(H) + (u - 1)E$, so that $-K_X - uT$ is pseudo-effective if and only if $u \leqslant 2$. Moreover, this divisor is nef if and only if $u \in [0, 1]$. Let $P(u) = P(-K_X - uT)$ and $N(u) = N(-K_X - uT)$. Then

$$P(u) = \begin{cases} -K_X - uT & \text{if } 0 \leqslant u \leqslant 1, \\ (2 - u)\pi^*(H) & \text{if } 1 \leqslant u \leqslant 2, \end{cases}$$

and $N(u) = (u - 1)E$ for $u \in [1, 2]$. Using this, we see that $S_X(T) = \frac{41}{72} < 1$. Now, using Corollary 1.110, we conclude that $S(W^T_{\bullet,\bullet}; C) \geqslant 1$.

Let us compute $S(W^T_{\bullet,\bullet}; C)$. Take $u \in [0, 2]$ and $v \in \mathbb{R}_{\geqslant 0}$. If $u \in [0, 1]$, then

$$P(u)\big|_T - vC \sim_{\mathbb{R}} (2 - u - v)C + L,$$

which easily implies that the divisor $P(u)|_T - vC$ is pseudo-effective if and only if $v \leqslant 2 - u$. If $0 \leqslant u \leqslant 1$ and $0 \leqslant v \leqslant \frac{3}{2} - u$, this divisor is nef and $\text{vol}(P(u)|_T - vC) = 7 - 4u - 4v$. If $0 \leqslant u \leqslant 1$ and $\frac{3}{2} - u \leqslant v \leqslant 2 - u$, the Zariski decomposition of $P(u)|_T - vC$ is

$$\underbrace{(2 - u - v)(C + 2L)}_{\text{positive part}} + \underbrace{(2v + 2u - 3)L}_{\text{negative part}},$$

so that $\text{vol}(P(u)|_T - vC) = 4(v + u - 2)^2$. If $u \in [1, 2]$, then

$$P(u)\big|_T - vC \sim_{\mathbb{R}} (2 - u - v)C + (2 - u)L,$$

so that $P(u)|_T - vC$ is pseudo-effective $\iff v \leqslant 2 - u$. If $1 \leqslant u \leqslant 2$ and $0 \leqslant v \leqslant \frac{2-u}{2}$, this divisor is nef and its volume is $(2 - u)(6 - 3u - 4v)$. If $1 \leqslant u \leqslant 2$ and $\frac{2-u}{2} \leqslant v \leqslant 2 - u$, the Zariski decomposition of the divisor $P(u)|_T - vC$ is

$$\underbrace{(2 - u - v)(C + 2L)}_{\text{positive part}} + \underbrace{(2v + u - 2)L,}_{\text{negative part}}$$

so that $\mathrm{vol}(P(u)|_T - vC) = 4(v + u - 2)^2$. Now, using Corollary 1.110 and integrating, we get $S(W^T_{\bullet,\bullet}; C) = \frac{29}{48} < 1$. The obtained contradiction completes the proof.   □

Lemma 5.16 $\pi(Z) \neq C'$.

Proof Let T_3 be a hyperplane section of V that is cut out by the equation $\lambda u + \mu v = 0$, where λ and μ are complex numbers such that T_3 is smooth. Such numbers always exist, e.g. the cubic surface T_3 is smooth for $a = 5$, $b = 7$, $\lambda = 1$ and $\mu = -1$.

Observe that the line L' and the conic C' are contained in the surface T_3 by construction. But the surface T_3 is not G-invariant. Let T be its proper transform on the threefold X, let $\varpi : T \to T_3$ be the morphism that is induced by π, let $O = T_3 \cap L$, let $E_O = E \cap T$, let R be the hyperplane section of T_3 that is singular at O, and let $\widetilde{R}$ be its proper transform on the surface T. Then ϖ is a blow up of the point O, and E_O is its exceptional curve. Observe that R is irreducible for general a and b, e.g. for $\lambda = -\mu = 1$ and $2b \neq \pm a$. Thus, no line in T_3 passes through O, so that T is a smooth del Pezzo surface of degree 2, and $\widetilde{R}$ is a (-1)-curve in T such that $\widetilde{R} + E_O \sim -K_T$.

Let $\widetilde{L}'$ and $\widetilde{C}'$ be the proper transforms on T of the curves L' and C', respectively. Note that $O \notin L' \cup C'$. Thus, on the surface T, we have

$$\left(\widetilde{L}'\right)^2 = \widetilde{R}^2 = E_O^2 = -1, \left(\widetilde{C}'\right)^2 = \widetilde{C}' \cdot E_O = \widetilde{L}' \cdot E_O = 0,$$
$$\widetilde{C}' \cdot \widetilde{L}' = \widetilde{R} \cdot \widetilde{C}' = \widetilde{R} \cdot E_O = 2, \widetilde{R} \cdot \widetilde{L}' = 1.$$

Suppose that $\pi(Z) = C'$. Then $Z = \widetilde{C}'$. Let us apply results of Section 1.7 to T and $\widetilde{C}'$ to derive a contradiction. Fix $u \in \mathbb{R}_{\geqslant 0}$. Then $-K_X - uT \sim_{\mathbb{R}} (2 - u)\pi^*(H) - E$. Thus, the divisor $-K_X - uT$ is nef $\iff -K_X - uT$ is pseudo-effective $\iff u \in [0, 1]$. Using this, we get $S_X(T) = \frac{3}{8} < 1$, so that $S(W^T_{\bullet,\bullet}; Z) \geqslant 1$ by Corollary 1.110.

Let us compute $S(W^T_{\bullet,\bullet}; Z)$. Take $u \in [0, 1]$ and $v \in \mathbb{R}_{\geqslant 0}$. If $u \in [0, 1]$, then

$$(-K_X - uT)\big|_T - vZ \sim_{\mathbb{R}} (2 - u - v)Z + (2 - u)\widetilde{L}' - E_O$$
$$\sim_{\mathbb{R}} (2 - u - v)\widetilde{R} + v\widetilde{L}' + (3 - 2u - 2v)E_O.$$

This implies that the divisor $(-K_X - uT)|_T - vZ$ is pseudo-effective if and only if $v \leqslant \frac{3-2u}{2}$. If $0 \leqslant u \leqslant 1$ and $0 \leqslant v \leqslant \frac{2-u}{2}$, this divisor is nef and

$$\mathrm{vol}\big((-K_X - uT)\big|_T - vZ\big) = 4v(u-2) + 3(u-2)^2 - 1.$$

If $0 \leqslant u \leqslant 1$ and $\frac{2-u}{2} \leqslant v \leqslant \frac{3-2u}{2}$, the Zariski decomposition of $(-K_X - uT)|_T - vZ$ is

$$\underbrace{(2 - u - v)(\widetilde{R} + \widetilde{L}') + (3 - 2u - 2v)E_O}_{\text{positive part}} + \underbrace{(u + 2v - 2)\widetilde{L}'}_{\text{negative part}},$$

so that $\mathrm{vol}((-K_X - uT)|_T - vZ) = (4 - 2u - 2v)^2 - 1$. Now, using Corollary 1.110 and integrating, we get $S(W^T_{\bullet,\bullet}; C') = \frac{77}{144} < 1$. $\square$

Thus, using Lemmas 5.13 and 5.14, we conclude that Z is contained in the surface S. Observe that $S \cong S_3$, so that we can identify S with the cubic surface S_3 in computations. We also abuse notation and denote by L the curve $E|_S$, and by L' the proper transform on the threefold X of the line L'. Observe that L and L' are G-invariant.

Lemma 5.17 *Either $Z - L$ or $Z - L'$ is pseudo-effective.*

Proof Let $\rho \colon S \dashrightarrow \mathbb{P}^1 \times \mathbb{P}^1$ be the map that is given by $[u : v : x : y : z] \mapsto ([u : v], [x : z])$. Then ρ is a G-equivariant morphism. Moreover, this morphism blows up the points

$$P_0 = \big([-1 : 1], [1 : -1]\big), \quad P_1 = \big([-\omega^3 : 1], [1 : -\omega]\big),$$
$$P_2 = \big([-\omega : 1], [1 : -\omega^2]\big), \quad P_3 = \big([-\omega^4 : 1], [1 : -\omega^3]\big),$$
$$P_4 = \big([-\omega^2 : 1], [1 : -\omega^4]\big),$$

which form one G-orbit in the surface $\mathbb{P}^1 \times \mathbb{P}^1$. We let $A_i = \rho^{-1}(P_i)$ for $i \in \{0, 1, 2, 3, 4\}$. Then each A_i is the line $\{z + \omega^i x = u + \omega^{3i} v = 0\} \subset S$, $\rho(L) \cap \rho(L') = \{P_0, P_1, P_2, P_3, P_4\}$, and the curves $\rho(L)$ and $\rho(L')$ are divisors in $\mathbb{P}^1 \times \mathbb{P}^1$ of degree $(1, 2)$ and $(2, 1)$, respectively. Observe also that $\mathbb{P}^1 \times \mathbb{P}^1$ does not contain G-invariant curves of degree $(1, 0)$, $(0, 1)$, $(1, 1)$.

Let ℓ_1 and ℓ_2 be the fibers of the projections $\mathbb{P}^1 \times \mathbb{P}^1 \to \mathbb{P}^1$ to the first and the second factors, respectively. Then $\rho(Z) \sim b_1 \ell_1 + b_2 \ell_2$ for some positive integers b_1 and b_2. We have

$$L \sim \rho^*(\ell_1 + 2\ell_2) - A_0 - A_1 - A_2 - A_3 - A_4,$$
$$L' \sim \rho^*(2\ell_1 + 1\ell_2) - A_0 - A_1 - A_2 - A_3 - A_4,$$
$$Z \sim \rho^*(b_1 \ell_1 + b_2 \ell_2) - m(A_0 + A_1 + A_2 + A_3 + A_4)$$

for some integer $m \geqslant 0$. If $m = 0$, then we are done. Thus, we assume that $m > 0$.

Intersecting the curve Z with the curve in $|\ell_1|$ that passes through P_0, we obtain $b_2 \geqslant m$. Similarly, intersecting Z with the curve in $|\ell_2|$ that passes through P_0, we obtain $b_1 \geqslant m$. Intersecting Z with the curve in $|\ell_1 + \ell_2|$ that contains P_0, P_1, P_2, we get $b_1 + b_2 \geqslant 3m$. On the other hand, we have

$$Z - mL \sim \rho^*\big((b_1 - m)\ell_1 + (b_2 - 2m)\ell_2\big),$$
$$Z - mL' \sim \rho^*\big((b_1 - 2m)\ell_1 + (b_2 - m)\ell_2\big).$$

Thus, to complete the proof, we may assume that $b_1 < 2m$ and $b_2 < 2m$. Then

$$Z - L \sim (2m - b_1 - 1)L + (b_1 - m)L' + \rho^*\big((b_1 + b_2 - 3m)\ell_2\big),$$

which implies that $Z - L$ is pseudo-effective. $\qquad\square$

Now, we apply the results of Section 1.7 to S and our curve Z to derive a contradiction. Let us use the notations introduced in this section. Fix a non-negative real number u. Let $P(u) = P(-K_X - uS)$ and $N(u) = N(-K_X - uS)$. We have

$$-K_X - uS \sim_{\mathbb{R}} (2-u)\pi^*(H) + (u-1)E \sim_{\mathbb{R}} \pi^*(H) + (1-u)S.$$

Then $-K_X - uS$ is nef $\Longleftrightarrow u \in [0,1]$, and $-K_X - uS$ is pseudo-effective $\Longleftrightarrow u \leqslant 2$. Thus, we have

$$P(u) = \begin{cases} -K_X - uS & \text{if } 0 \leqslant u \leqslant 1, \\ (2-u)\pi^*(H) & \text{if } 1 \leqslant u \leqslant 2, \end{cases}$$

and $N(u) = (u-1)E$ for $u \in [1,2]$. This gives

$$\mathrm{vol}\big(-K_X - uS\big) = \begin{cases} 6u^2 - 21u + 18 & \text{if } 0 \leqslant u \leqslant 1, \\ 3(u-2)^3 & \text{if } 1 \leqslant u \leqslant 2, \end{cases}$$

and

$$P(u)^2 \cdot S = \begin{cases} 7 - 4u & \text{if } 0 \leqslant u \leqslant 1, \\ 3(u-2)^2 & \text{if } 1 \leqslant u \leqslant 2. \end{cases}$$

Integrating, we get $S_X(S) = \frac{41}{72} < 1$. Thus, we have $S(W^S_{\bullet,\bullet}; Z) \geqslant 1$ by Corollary 1.110. Recall from this corollary that

$$S(W^S_{\bullet,\bullet}; Z) = \frac{3}{18} \int_0^2 h(u)\,du,$$

where

$$h(u) = \big(P(u)^2 \cdot S\big)\mathrm{ord}_Z\big(N(u)|_S\big) + \int_0^\infty \mathrm{vol}\big(P(u)\big|_S - vZ\big)\,dv.$$

Observe that $\mathrm{ord}_Z(N(u)|_S) = 0$ unless $Z = L$. Moreover, if $Z = L$, then

$$\mathrm{ord}_Z\left(N(u)|_S\right) = \begin{cases} 0 & \text{if } 0 \leqslant u \leqslant 1, \\ (u-1)L & \text{if } 1 \leqslant u \leqslant 2. \end{cases}$$

Let us show that $S(W^S_{\bullet,\bullet}; Z) < 1$, which would give us a contradiction. We start with

Lemma 5.18 $S(W^S_{\bullet,\bullet}; L) = \frac{5}{9}$.

Proof   Take $v \in \mathbb{R}_{\geqslant 0}$. If $0 \leqslant u \leqslant 1$, then

$$P(u)|_S - vL \sim_{\mathbb{R}} (2-u)\left(-K_S\right) + (u-1-v)L.$$

This divisor is nef $\iff v \leqslant 1$, and it is not pseudo-effective if $v > 1$. So, if $u \in [0,1]$,

$$h(u) = \int_0^1 \left(-v^2 + 2v(2u-3) + (7-4u)\right)dv = \frac{11}{3} - 2u.$$

Similarly, if $1 \leqslant u \leqslant 2$, then

$$h(u) = 3(u-2)^2(u-1) + \int_0^\infty \mathrm{vol}\left(P(u)|_S - vL\right)dv.$$

In this case, we have $P(u)|_S - vL \sim_{\mathbb{R}} (2-u)\left(-K_S\right) + vL$. Observe that this divisor is nef $\iff v \leqslant 2-u$, and it is not pseudo-effective if $v > 2-u$. Thus, if $u \in [1,2]$,

$$\int_0^\infty \mathrm{vol}\left(P(u)|_S - vL\right)dv = \int_0^{2-u} \left(-v^2 - 2v(2-u) + 3(2-u)^2\right)dv = \frac{5}{3}(2-u)^3,$$

so $h(u) = \frac{1}{3}(4u^3 - 15u^2 + 12u + 4)$, with $u \in [1,2]$. Integrating gives us $S(W^S_{\bullet,\bullet}; L) = \frac{5}{9}$. $\square$

Thus we see that $Z \neq L$ and so $Z \not\subseteq N(u)$. If $Z - L$ is pseudo-effective, the proof of Lemma 5.18 therefore gives

$$S(W^S_{\bullet,\bullet}; Z) \leqslant \frac{1}{6}\int_0^2 \int_0^\infty \mathrm{vol}\left(P(u)|_S - vL\right)dv\,du \leqslant S(W^S_{\bullet,\bullet}; L) = \frac{5}{9}.$$

Similarly, if $Z - L'$ is pseudo-effective, then

$$S(W^S_{\bullet,\bullet}; Z) \leqslant \frac{1}{6}\int_0^2 \int_0^\infty \mathrm{vol}\left(P(u)|_S - vL'\right)dv\,du = S(W^S_{\bullet,\bullet}; L') = \frac{179}{288}$$

by the following lemma:

Lemma 5.19 $S(W^S_{\bullet,\bullet}; L') = \frac{179}{288}$.

Proof Take $u \in [0,1]$ and $v \in \mathbb{R}_{\geqslant 0}$. Then

$$P(u)\big|_S - vL' \sim_{\mathbb{R}} (2-u)\big(-K_S\big) - (1-u)L - vL'.$$

This divisor is pseudo-effective $\iff v \leqslant \frac{3-u}{2}$, and it is nef $\iff v \leqslant 1$. So, if $v \leqslant 1$, we have

$$\mathrm{vol}\big(P(u)\big|_S - vL'\big) = -v^2 + 2(u-2)v + (7-4u).$$

Similarly, if $1 \leqslant v \leqslant \frac{3-u}{2}$, then the Zariski decomposition of $P(u)\big|_S - vL'$ is

$$\underbrace{\rho^*\big((3-u-2v)\ell_1 + (2-v)\ell_2\big)}_{\text{positive part}} + \underbrace{(v-1)(A_0 + A_1 + A_2 + A_3 + A_4)}_{\text{negative part}},$$

which implies in this case that $\mathrm{vol}(P(u)\big|_S - vL') = 2(2-v)(3-u-2v)$.

Now we take $u \in [1,2]$ and $v \in \mathbb{R}_{\geqslant 0}$. Then

$$P(u)\big|_S - vL' \sim_{\mathbb{R}} (2-u)\big(-K_S\big) - vL',$$

so that this divisor is pseudo-effective $\iff$ it is nef $\iff v \leqslant 2-u$. Hence

$$\mathrm{vol}\big(P(u)\big|_S - L'\big) = -v^2 - 2(u-2)v + 3(2-u)^2.$$

Using Corollary 1.110 and integrating, gives $S(W^S_{\bullet,\bullet}; L') = \frac{179}{288}$ as desired. $\square$

Since $Z - L$ or $Z - L'$ is pseudo-effective by Lemma 5.17, we see that $S(W^S_{\bullet,\bullet}; Z) < 1$. But we already proved earlier that $S(W^S_{\bullet,\bullet}; Z) \geqslant 1$. The obtained contradiction completes the proof of Proposition 5.12.

5.4 Family №2.12

Let ζ be a primitive seventh root of unity, and let $\widehat{G}$ be the subgroup in $\mathrm{SL}_4(\mathbb{C})$ that is generated by the following matrices:

$$\begin{bmatrix} 1 & 0 & 0 & 0 \\ 0 & \zeta & 0 & 0 \\ 0 & 0 & \zeta^4 & 0 \\ 0 & 0 & 0 & \zeta^2 \end{bmatrix}, \quad \frac{\sqrt{-1}}{\sqrt{7}}\begin{bmatrix} 1 & \sqrt{2} & \sqrt{2} & \sqrt{2} \\ \sqrt{2} & \zeta^2+\zeta^5 & \zeta^3+\zeta^4 & \zeta+\zeta^6 \\ \sqrt{2} & \zeta^3+\zeta^4 & \zeta+\zeta^6 & \zeta^2+\zeta^5 \\ \sqrt{2} & \zeta+\zeta^6 & \zeta^2+\zeta^5 & \zeta^3+\zeta^4 \end{bmatrix}.$$

It follows from [85] that $\widehat{G} \cong \mathrm{SL}_2(\mathbf{F}_7)$, and $\widehat{G}$ gives a subgroup $\mathrm{PSL}_2(\mathbf{F}_7) \subset \mathrm{PGL}_4(\mathbb{C})$ that has no fixed points in $\mathbb{P}^3$. Such a subgroup in $\mathrm{PGL}_4(\mathbb{C})$ is unique up to conjugation [85, 151]. Moreover, it follows from [85, 50] that $\mathbb{P}^3$ contains a unique $\mathrm{PSL}_2(\mathbf{F}_7)$-invariant smooth curve of degree 6 and genus 3. Denote this curve by $\mathscr{C}$. We have the following result:

Proposition 5.20 *Let M be a mobile $\mathrm{PSL}_2(\mathbf{F}_7)$-invariant linear subsystem in $|O_{\mathbb{P}^3}(n)|$, where $n \in \mathbb{Z}_{>0}$. Suppose that $\mathrm{mult}_{\mathscr{C}}(M) \leqslant \frac{n}{4}$. Then $(\mathbb{P}^3, \frac{4}{n}M)$ has canonical singularities.*

Proof This follows from the proof of [50, Theorem 1.9].   □

Let $\pi\colon X \to \mathbb{P}^3$ be a blow up of the curve $\mathscr{C}$. Then X is a Fano threefold №2.12, and there exists $\mathrm{PSL}_2(\mathbf{F}_7)$-equivariant commutative diagram

$$
\begin{array}{ccc}
X & \xrightarrow{\;\sigma\;} & X \\
{\scriptstyle\pi}\downarrow & & \downarrow{\scriptstyle\pi} \\
\mathbb{P}^3 & \dashrightarrow[\tau] & \mathbb{P}^3
\end{array}
$$

where τ is a birational involution given by the linear system of cubic surfaces containing $\mathscr{C}$, and σ is a biregular involution.

Remark 5.21 The involution $\sigma \in \mathrm{Aut}(X)$ can be explicitly constructed as follows. Let

$$
\begin{aligned}
y_0 &= 2\sqrt{2}x_1x_2x_3 - x_0^3, \\
y_1 &= x_0^2 x_1 + \sqrt{2}x_0 x_2^2 + 2x_2 x_3^2, \\
y_2 &= x_0^2 x_2 + \sqrt{2}x_0 x_3^2 + 2x_1^2 x_3, \\
y_3 &= x_0^2 x_3 + \sqrt{2}x_0 x_1^2 + 2x_1 x_2^2.
\end{aligned}
$$

By [85], the ideal sheaf of the curve $\mathscr{C}$ is generated by the cubic polynomials y_0, y_1, y_2, y_3, where x_0, x_1, x_2, x_3 are homogeneous coordinates on $\mathbb{P}^3$. Let $\chi\colon \mathbb{P}^3 \dashrightarrow \mathbb{P}^3$ be the rational map given by $[x_0 : x_1 : x_2 : x_3] \mapsto [y_0 : y_1 : y_2 : y_3]$. Then there is a commutative diagram

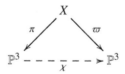

where ϖ is a morphism. Thus, we can consider X as the closure in $\mathbb{P}^3 \times \mathbb{P}^3$ of the graph of the rational map χ, cf. [60, §29]. To be precise, the threefold X is given in $\mathbb{P}^3 \times \mathbb{P}^3$ by

$$x_0y_1+x_1y_0-\sqrt{2}x_2y_2 = x_0y_2+x_2y_0-\sqrt{2}x_3y_3 = x_0y_3+x_3y_0-\sqrt{2}x_1y_1 = 0, \quad (5.1)$$

where we consider y_0, y_1, y_2, y_3 as coordinates on $\mathbb{P}^3$. Indeed, the threefold X is contained in the subset (5.1), so that it should be X, because (5.1) defines a smooth irreducible three-dimensional subvariety in $\mathbb{P}^3 \times \mathbb{P}^3$, which can be checked using Magma:

```
Q:=RationalField();
R<x>:=PolynomialRing(Q);
K<t>:=NumberField(x^2-2);
PxP<x0,x1,x2,x3,y0,y1,y2,y3>:=ProductProjectiveSpace(K,[3,3]);
X:=Scheme(PxP,[x0*y1+x1*y0-t*x2*y2, x0*y2+x2*y0-t*x3*y3,
              x0*y3+x3*y0-t*x1*y1]);
IsNonsingular(X);
IsIrreducible(X);
Dimension(X);
```

Now, we can define $\sigma \in \text{Aut}(\mathbb{P}^3 \times \mathbb{P}^3)$ as follows:

$$\left([x_0 : x_1 : x_2 : x_3], [y_0 : y_1 : y_2 : y_3]\right) \mapsto \left([y_0 : y_1 : y_2 : y_3], [x_0 : x_1 : x_2 : x_3]\right).$$

Then X is σ-invariant, so that we may identify σ with an element in $\text{Aut}(X)$.

Let H be a hyperplane in $\mathbb{P}^3$, and let E be the π-exceptional surface. Then

$$\begin{cases} \sigma^*(E) \sim 8\pi^*(H) - 3E, \\ \sigma^*\left(\pi^*(H)\right) \sim 3\pi^*(H) - E. \end{cases}$$

Using this and Proposition 5.20, we obtain

Theorem 5.22 ([50, Theorem 1.9]) *The threefold $\mathbb{P}^3$ is $\text{PSL}_2(\mathbf{F}_7)$-birationally rigid, and the subgroup of $\text{PSL}_2(\mathbf{F}_7)$-birational selfmaps of $\mathbb{P}^3$ is generated by $\text{PSL}_2(\mathbf{F}_7)$ and τ.*

The involution σ commutes with the $\text{PSL}_2(\mathbf{F}_7)$-action on X. Together, they generate a finite subgroup $G \subset \text{Aut}(X)$ that is isomorphic to $\text{PSL}_2(\mathbf{F}_7) \times \mu_2$, see [50, Lemma 3.8]. Then $\text{Pic}^G(X) = \mathbb{Z}[-K_X]$, so that X is a G-Mori fiber space.

Theorem 5.23 *The threefold X is G-birationally super-rigid.*

Proof Suppose that X is not G-birationally super-rigid. It is well known [53] that there exists a G-invariant mobile linear system $\mathcal{M}$ on the Fano threefold X such that the log pair $(X, \lambda\mathcal{M})$ does not have canonical singularities, where λ is a positive rational number that is defined via $\lambda\mathcal{M} \sim_{\mathbb{Q}} -K_X$. Let us seek a contradiction.

Applying Proposition 5.20 to the log pairs $(\mathbb{P}^3, \lambda\pi(\mathcal{M}))$ and $(\mathbb{P}^3, \lambda\pi \circ \sigma(\mathcal{M}))$, one can easily show that $(X, \lambda\mathcal{M})$ is canonical away from the curve $E \cap \sigma(E)$. However, we would prefer to avoid using Proposition 5.20 because its proof is difficult.

First, we suppose that $(X, \lambda M)$ is not canonical along some G-irreducible curve $C \subset X$. Let M_1 and M_2 be sufficiently general surfaces in M. Then $\mathrm{mult}_C(M_1) = \mathrm{mult}_C(M_2) > \frac{1}{\lambda}$. Thus, intersecting $-K_X$ with the effective one-cycle $M_1 \cdot M_2$, we get

$$\frac{-K_X^3}{\lambda^2} = -K_X \cdot M_1 \cdot M_2 \geqslant \left(-K_X \cdot C\right) \mathrm{mult}_C(M_1) \mathrm{mult}_C(M_2) > \frac{-K_X \cdot C}{\lambda^2},$$

so that $-K_X \cdot C < -K_X^3 = 20$. On the other hand, since $\sigma(C) = C$, we have

$$4\pi^*(H) \cdot C - E \cdot C = \left(4\pi^*(H) - E\right) \cdot C = -K_X \cdot C$$

$$= \left(\pi^*(H) + \sigma^*\left(\pi^*(H)\right)\right) \cdot C = \pi^*(H) \cdot C + \sigma^*\left(\pi^*(H)\right) \cdot C = 2\pi^*(H) \cdot C,$$

so that $E \cdot C = 2\pi^*(H) \cdot C = -K_X \cdot C < 20$. This shows that $C \subset E$ and $\pi(C) = \mathscr{C}$ because the surface E does not contain G-orbits of length less than 24 since $\mathscr{C}$ does not contain $\mathrm{PSL}_2(\mathbf{F}_7)$-orbits of length less than 24 by [50, Lemma 2.16]. Then

$$\frac{1}{\lambda} = M_1 \cdot \ell \geqslant \mathrm{mult}_C(M_1) > \frac{1}{\lambda},$$

where ℓ is a general fiber of the natural projection $E \to \mathscr{C}$. The obtained contradiction shows that the log pair $(X, \lambda M)$ has canonical singularities outside of finitely many points.

Let P be a point in X such that the log pair $(X, \lambda M)$ is not canonical at this point. By [50, Lemma 3.2], one of the following two cases holds:

(i) the $\mathrm{PSL}_2(\mathbf{F}_7)$-orbit of the point $\pi(P)$ is the unique $\mathrm{PSL}_2(\mathbf{F}_7)$-orbit of length 8;
(ii) the length of the $\mathrm{PSL}_2(\mathbf{F}_7)$-orbit of the point $\pi(P)$ is at least 24.

In the first case, the log pair $(\mathbb{P}^3, \lambda\pi(M))$ must be canonical at $\pi(P)$ by [50, Lemma 5.4], so that the log pair $(X, \lambda M)$ must be canonical at P, because $\pi(P) \notin \mathscr{C}$ in this case. Thus, we conclude that the length of the $\mathrm{PSL}_2(\mathbf{F}_7)$-orbit of the point $\pi(P)$ is at least 24. In particular, the G-orbit of the point P consists of at least 24 points.

There is a prime divisor F over X with $C_X(F) = P$ and $\mathrm{ord}_F(\lambda M) > A_X(F) - 1 \geqslant 2$. Thus, we have

$$\mathrm{ord}_F\left(\frac{3}{2}\lambda M\right) = \mathrm{ord}_F(\lambda M) + \frac{\mathrm{ord}_F(\lambda M)}{2} > A_X(F) - 1 + \frac{A_X(F) - 1}{2} \geqslant A_X(F),$$

so that the log pair $(X, \frac{3\lambda}{2}M)$ is not log canonical at P.

We claim that $(X, \frac{3\lambda}{2}M)$ is Kawamata log terminal away from finitely many points. Indeed, if the log pair $(X, \frac{3\lambda}{2}M)$ is not Kawamata log terminal along some G-irreducible curve $C \subset X$, then $(M_1 \cdot M_2)_C \geqslant \frac{16}{9\lambda^2}$ by Theorem A.22, where M_1 and M_2 are general surfaces in $\mathcal{M}$. Using this, we see that

$$\frac{-K_X^3}{\lambda^2} = -K_X \cdot M_1 \cdot M_2 \geqslant \left(-K_X \cdot C \right)\left(M_1 \cdot M_2 \right)_C > \frac{16(-K_X \cdot C)}{9\lambda^2},$$

which gives $2\pi^*(H) \cdot C = -K_X \cdot C \leqslant 11$, so we conclude that $\pi(C)$ is a $\mathrm{PSL}_2(\mathbf{F}_7)$-invariant curve of degree at most 5. But $\mathbb{P}^3$ does not contain $\mathrm{PSL}_2(\mathbf{F}_7)$-invariant curves of degree less than 6 by [50, Lemma 3.7]. This shows that $\mathrm{Nklt}(X, \frac{3\lambda}{2}\mathcal{M})$ consists of finitely many points. Now, applying Corollary A.6, we get $|\mathrm{Nklt}(X, \frac{3\lambda}{2}\mathcal{M})| \leqslant h^0(X, \mathcal{O}_X(-K_X)) = 13$, which is impossible because $P \in \mathrm{Nklt}(X, \frac{3\lambda}{2}\mathcal{M})$, and G-orbit of P consists of at least 24 points. This completes the proof of the theorem. $\qquad\square$

Recall that $\mathrm{Pic}^G(X) = \mathbb{Z}[-K_X]$. Note also that G does not have fixed points on X because $\mathrm{PSL}_2(\mathbf{F}_7)$ has no fixed points in $\mathbb{P}^3$ since the action of this group is given by an irreducible four-dimensional representation of its central extension. This gives

Lemma 5.24 $\alpha_G(X) \geqslant \frac{1}{2}$.

Proof If $\alpha_G(X) < \frac{1}{2}$, then applying Theorem 1.52 with $\mu = \frac{1}{2}$, we see that there exists a G-invariant irreducible rational curve C such that $-K_X \cdot C \leqslant 3$, so that

$$3 \geqslant -K_X \cdot C = \left(\pi^*(H) + \sigma^*\left(\pi^*(H) \right) \right) \cdot C = \pi^*(H) \cdot C + \sigma^*\left(\pi^*(H) \right) \cdot C = 2\pi^*(H) \cdot C,$$

which implies that $\pi^*(H) \cdot C = 1$, so that $\pi(C)$ must be a $\mathrm{PSL}_2(\mathbf{F}_7)$-invariant line in $\mathbb{P}^3$, which does not exist. $\qquad\square$

Thus, applying Corollary 1.81, Theorem 5.23 and Lemma 5.24, we conclude that the threefold X is K-polystable, which also follows from

Lemma 5.25 $\alpha_G(X) \geqslant 1$.

Proof Suppose that $\alpha_G(X) < 1$. Then, applying Theorem 1.52 with $\mu = 1$, we see that the Fano threefold X must contain a G-invariant irreducible smooth rational curve C. But the action of the simple subgroup $\mathrm{PSL}_2(\mathbf{F}_7)$ on the curve C must be trivial, so that the group $G/\mathrm{PSL}_2(\mathbf{F}_7) \cong \mu_2$ has a fixed point in C. Then X contains a G-fixed point, which is not the case. $\qquad\square$

Since $\mathrm{Aut}(X)$ is finite [45], our X is K-stable by Theorem 1.48 and Corollary 1.5. Hence, general Fano threefold №2.12 is K-stable by Theorem 1.11.

5.5 Family №2.13

Consider the group $G \cong 2.\mathfrak{S}_4 \cong \mathrm{GL}_2(\mathbf{F}_3)$. There exists a smooth curve C of genus 2 with a faithful action of G, see e.g. [193, §3.2]. The hyperelliptic double cover $\nu\colon C \to \mathbb{P}^1$ is G-equivariant, where G acts on $\mathbb{P}^1$ via its quotient $\mathfrak{S}_4$. Recall that the group $\mathfrak{S}_4$ has no orbits of length less than 6 on $\mathbb{P}^1$, and it has a unique orbit Σ of length 6. In particular, the hyperelliptic double cover is branched in Σ. So the curve C does not contain G-invariant subsets of cardinality less than 6, and the only G-invariant subset of cardinality 6 is the preimage of Σ on C, which we will also denote by Σ.

By the Riemann–Roch theorem, we know that the linear system $|3K_C|$ has dimension 4. Hence, there exists a faithful action of G on $\mathbb{P}^4$, and a G-equivariant embedding $C \hookrightarrow \mathbb{P}^4$. Observe that $\Sigma \in |3K_C|$, and $|3K_C|$ contains a three-dimensional G-invariant linear subsystem $\nu^*|O_{\mathbb{P}^1}(3)|$. So, we can identify $\mathbb{P}^4 = |3K_C|^\vee = \mathbb{P}(\mathbb{I} \oplus \mathbb{W})$, where $\mathbb{I}$ is the trivial representation of the group G, and $\mathbb{W}$ is its unique irreducible four-dimensional representation. Hence, we conclude that $\mathbb{P}^4$ contains a unique G-invariant hyperplane $H_0 = \mathbb{P}(\mathbb{W})$, the group G acts on H_0 via its quotient $\mathfrak{S}_4$. Similarly, $\mathbb{P}^4$ has a unique G-fixed point P_0, which is not contained in H_0.

Lemma 5.26 *There is a unique G-invariant quadric $Q \subset \mathbb{P}^4$, and this quadric is smooth.*

Proof Let $\rho\colon \mathbb{P}^4 \dashrightarrow H_0$ be the projection from P_0. Recall that $\nu\colon C \to \mathbb{P}^1$ denotes the hyperelliptic double cover. Since the G-invariant linear subsystem $\nu^*|O_{\mathbb{P}^1}(3)| \subsetneq |3K_C|$ defines $H_0 = \mathbb{P}(\mathbb{W})$, we see that $\overline{C} = \rho(C)$ is a twisted cubic and ρ is G-equivariant. Further, the map ρ gives a double cover $C \to \overline{C}$, which is the hyperelliptic double cover ν. We denote by Y the cone in $\mathbb{P}^4$ over the curve $\overline{C}$ with vertex P_0.

Let $\mathcal{Q}$ be the linear system of quadrics in $\mathbb{P}^4$ that pass through C, and let $\overline{\mathcal{Q}}$ be its subsystem that consists of all quadrics that pass through Y. Then $\mathcal{Q}$ is three-dimensional by the Riemann–Roch theorem, and $\overline{\mathcal{Q}}$ is two-dimensional. Note that $\overline{\mathcal{Q}}$ is G-invariant. Thus, by the complete reducibility of the corresponding representation of the group G, there exists a G-invariant quadric $Q \in \mathcal{Q}$ such that $Q \notin \overline{\mathcal{Q}}$.

One can show that the linear system $\overline{\mathcal{Q}}$ is the projectivization of an irreducible three-dimensional representation of the group G, which implies that Q is the unique G-invariant quadric in the linear system $\mathcal{Q}$.

Observe that $C = Y \cap Q$. This implies that Q is smooth. Indeed, if Q were singular, then its vertex would be P_0, which would imply that C is singular. □

Remark 5.27 We can also prove the existence of the G-invariant quadric Q as follows. We have the following exact sequence of G-representations:

$$0 \longrightarrow H^0\big(\mathbb{P}^4, \mathcal{O}_{\mathbb{P}^4}(2) \otimes \mathcal{I}_C\big) \longrightarrow H^0\big(\mathbb{P}^4, \mathcal{O}_{\mathbb{P}^4}(2)\big) \longrightarrow H^0\big(C, \mathcal{O}_{\mathbb{P}^4}(2)\big|_C\big) \longrightarrow 0,$$

where $\mathcal{I}_C$ is the ideal sheaf of the curve C. On the other hand, since C does not contain G-orbits of length 12, we see that 2Σ is the unique G-invariant divisor in $|\mathcal{O}_{\mathbb{P}^4}(2)|_C|$, so that $H^0(C, \mathcal{O}_{\mathbb{P}^4}(2)|_C)$ has unique one-dimensional subrepresentation. On the other hand, the G-representation $H^0(\mathbb{P}^4, \mathcal{O}_{\mathbb{P}^4}(2)) \cong \mathrm{Sym}^2(\mathbb{I} \oplus \mathbb{W})$ contains two trivial one-dimensional subrepresentations of the group G. This can be checked using the following GAP script:

```
G:=SmallGroup(48,29);
T:=CharacterTable(G);
Ir:=Irr(T);
V:=Ir[1]+Ir[8];
S:=SymmetricParts(T,[V],2);
MatScalarProducts(Ir,S);
```

Therefore, we conclude that $H^0(\mathbb{P}^4, \mathcal{O}_{\mathbb{P}^4}(2) \otimes \mathcal{I}_C)$ contains a unique one-dimensional subrepresentation of the group G, so that there exists a unique G-invariant quadric $Q \subset \mathbb{P}^4$.

Let $\pi \colon X \to Q$ be the blow up of the G-invariant quadric Q along the curve C. Since C is an intersection of quadrics [90, Theorem (4.a.1)], the divisor $-K_X$ is ample by Lemma A.56, so that X is a smooth Fano threefold from the family № 2.13. Since the action of the group G lifts to X, we identify G with a subgroup in $\mathrm{Aut}(X)$.

Let H denote the pull back of a hyperplane section of Q, and let E denote the exceptional divisor of π. Then the linear system $|2H - E|$ is basepoint free, and it defines a G-equivariant conic bundle $\psi \colon X \to \mathbb{P}^2$, so that we have a G-equivariant diagram:

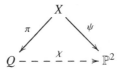

where χ is the map given by the linear system $\overline{Q}$ described in the proof of Lemma 5.26. Since $\overline{Q}$ is the projectivization of an irreducible three-dimensional G-representation, we see that $\mathbb{P}^2$ contains neither G-invariant lines nor G-fixed points, so that X does not contain G-fixed points either.

Lemma 5.28 $\alpha_G(X) \geqslant \frac{3}{4}$.

Proof   First, we claim that there does not exist a G-invariant effective divisor B such that $-K_X \sim_{\mathbb{Q}} bB + \Delta$, where $b > \frac{4}{3}$ and Δ is an effective $\mathbb{Q}$-divisor on X. Indeed, suppose that B is such a divisor, and write $\Delta \sim_{\mathbb{Q}} 3H - E - bB$. If $B = E$, then

$$\Delta \sim_{\mathbb{Q}} 3H - \frac{3}{2}E - \left(b + 1 - \frac{3}{2}\right)E,$$

which is impossible because the cone $\overline{\mathrm{Eff}}(X)$ is generated by E and $2H - E$. Thus, we see that $B \sim mH + kE$ for some $1 \leqslant m \leqslant 2$ and $k \geqslant -\frac{m}{2}$. Moreover, one cannot have $B \sim 2H - E$ because otherwise B is the preimage of a line in $\mathbb{P}^2$ under ψ, while $\mathbb{P}^2$ contains no G-invariant lines; in other words, we have $k > -\frac{m}{2}$. This gives

$$\Delta \sim_{\mathbb{Q}} (3 - bm)H - (1 + bk)E \sim_{\mathbb{Q}} \frac{3 - bm}{2}\left(2H - E\right) + \left(\frac{3 - bm}{2} - bk - 1\right)E,$$

which is a contradiction since $\frac{3-bm}{2} - bk - 1 = \frac{1}{2} - b\left(k + \frac{m}{2}\right) \leqslant \frac{1}{2} - \frac{b}{2} < 0$.

Now assume that $\alpha_G(X) < \frac{3}{4}$. By Lemma A.30, the threefold X contains an effective G-invariant $\mathbb{Q}$-divisor $D \sim_{\mathbb{Q}} -K_X$ and a smooth rational curve Z such that $(X, \lambda D)$ is strictly log canonical for some rational number $\lambda < \frac{3}{4}$, and the curve Z is the unique log canonical center of the log pair $(X, \lambda D)$.

Note that $\pi(Z)$ is not a point since Q does not contain G-invariant points. This implies that Z is not contained in E because $Z \cong \mathbb{P}^1$, but $C = \pi(E)$ is a curve of genus 2.

Using Corollary A.13, we see that $(2H - E) \cdot Z \leqslant 2$, so that $(2H - E) \cdot Z = 2$, because $\mathbb{P}^2$ does not contain G-fixed points and G-invariant lines. If $E \cdot Z = 0$, then we have $H \cdot Z = 1$, so that $\pi(Z)$ must be a G-invariant line in Q, which is impossible since $\mathbb{P}^4$ does not contain G-invariant lines. Then $\pi(Z) \cap C \neq \emptyset$, so that

$$E \cdot Z \geqslant |E \cap Z| \geqslant |C \cap \pi(Z)| \geqslant 6,$$

because the curve C does not contain G-invariant subsets of cardinality less than six. This gives $H \cdot Z = 1 + \frac{E \cdot Z}{2} \geqslant 4$.

The pair $(Q, \lambda\pi(D))$ is not Kawamata log terminal at a general point of the curve $\pi(Z)$. Let μ be a positive rational number such that the pair $(Q, \mu\pi(D))$ is strictly log canonical. Then, since Q does not have G-fixed points, the curve $\pi(Z)$ is a minimal log canonical center of the log pair $(Q, \mu\pi(D))$ by Corollary A.31, so that

$$3H \cdot Z = -K_Q \cdot \pi(Z) \leqslant 7$$

by Corollary A.21. Thus we see that $H \cdot Z \leqslant 2$, which is impossible since $H \cdot Z \geqslant 4$. □

We see that X is K-stable by Theorem 1.51 and Corollary 1.5 since $\mathrm{Aut}(X)$ is finite. Then general Fano threefold №2.13 is K-stable by Theorem 1.11.

5.6 Family №2.16

Let Q_1 be the smooth quadric $\{x_0x_3 + x_1x_4 + x_2x_5 = 0\} \subset \mathbb{P}^5$, and let Q_2 be the quadric $\{x_0^2 + \omega x_1^2 + \omega^2 x_2^2 + x_3^2 + \omega x_4^2 + \omega^2 x_5^2 + x_0x_3 + \omega x_1x_4 + \omega^2 x_2x_5 = 0\} \subset \mathbb{P}^5$, where ω is a primitive cubic root of unity, and $x_0, x_1, x_2, x_3, x_4, x_5$ are coordinates on $\mathbb{P}^5$. Let $V_4 = Q_1 \cap Q_2$. Then V_4 is smooth. Let $G \cong \mu_2^2 \rtimes \mu_3$ be the subgroup in $\mathrm{Aut}(\mathbb{P}^5)$ such that the generator of μ_3 acts by $[x_0 : x_1 : x_2 : x_3 : x_4 : x_5] \mapsto [x_1 : x_2 : x_0 : x_4 : x_5 : x_3]$, the generator of the first factor of μ_2^2 acts by

$$\left[x_0 : x_1 : x_2 : x_3 : x_4 : x_5 \right] \mapsto \left[-x_0 : x_1 : -x_2 : -x_3 : x_4 : -x_5 \right],$$

and the generator of the second factor of μ_2^2 acts by

$$\left[x_0 : x_1 : x_2 : x_3 : x_4 : x_5 \right] \mapsto \left[-x_0 : -x_1 : x_2 : -x_3 : -x_4 : x_5 \right].$$

Then $G \cong \mathfrak{A}_4$, and $\mathbb{P}^5 = \mathbb{P}(\mathbb{U}_3 \oplus \mathbb{U}_3)$, where $\mathbb{U}_3$ is the unique (unimodular) irreducible three-dimensional representation of the group G. Note that Q_1 and Q_2 are G-invariant, so that V_4 is also G-invariant. Thus, we may identify G with a subgroup in $\mathrm{Aut}(V_4)$.

Note that $\mathbb{P}^5$ contains neither G-fixed points nor G-invariant lines, and every G-invariant plane in $\mathbb{P}^5$ is the plane $\{\lambda x_0 + \mu x_3 = \lambda x_1 + \mu x_4 = \lambda x_2 + \mu x_5 = 0\}$ for some $(\lambda, \mu) \neq (0,0)$. Using this, we see that V_4 contains four G-invariant conics: $C_1 = V_4 \cap \{x_0 = x_1 = x_2 = 0\}$, $C_2 = V_4 \cap \{x_3 = x_4 = x_5 = 0\}$, $C_3 = V_4 \cap \{x_0 = \omega x_3, x_1 = \omega x_4, x_2 = \omega x_5\}$, and $C_4 = V_4 \cap \{x_3 = \omega x_0, x_4 = \omega x_1, x_5 = \omega x_2\}$. The conics C_1, C_2, C_3, C_4 are pairwise disjoint.

Let $\pi \colon X \to V_4$ be the blow up of the conic C_1, and let E be the π-exceptional surface. Then X is a smooth Fano threefold №2.16, and the G-action lifts to X, so that we also consider G as a subgroup in $\mathrm{Aut}(X)$. Then there exists a G-equivariant diagram

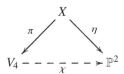

Here, χ is the linear projection from the plane $\{x_0 = x_1 = x_2 = 0\}$, and η is a conic bundle that is given by the net $|\pi^*(H) - E|$, where H is a hyperplane section of the threefold V_4. Note also that $\mathbb{P}^2 = \mathbb{P}(U_3)$, and the discriminant curve of η is a smooth quartic curve.

Lemma 5.29 $E \cong \mathbb{P}^1 \times \mathbb{P}^1$.

Proof We have $E \cong \mathbb{F}_n$ for some non-negative integer n, and $-E|_E \sim s + af$ for some integer a, where s is a section of the projection $E \to C$ with $s^2 = -n$, and f is a fiber of this projection. Then $-2 = E^3 = (s + af)^2 = -n + 2a$. Thus, we see that $a = \frac{n-2}{2}$. But $(\pi^*(H) - E)|_E \sim s + \frac{n+2}{2}f$. Thus, since $|\pi^*(H) - E|$ is basepoint free, we get $n \in \{0, 2\}$. If $n = 2$, then s is contracted by η to a point, which is impossible since G does not have fixed points in $\mathbb{P}^2$. Hence, we see that $n = 0$, so that $E \cong \mathbb{P}^1 \times \mathbb{P}^1$. $\qquad\qquad\square$

Lemma 5.30 *Let C be any G-invariant irreducible smooth rational curve in X such that $C \not\subset E$ and $-K_X \cdot C < 8$. Then $\pi(C)$ is one of the conics C_2, C_3, C_4.*

Proof Let $\overline{C} = \pi(C)$. Suppose that $\overline{C}$ is not one of the conics C_2, C_3, C_4. Then

$$\pi^*(H) \cdot C = H \cdot \overline{C} \geqslant 3$$

since V_4 contains no G-invariant lines, and C_1, C_2, C_3, C_4 are all the G-invariant conics in V_4. Note also that $\eta(C)$ is a curve because G does not have fixed points in $\mathbb{P}^2$. Similarly, we see that $\eta(C)$ is not a line. Hence, we conclude that $(\pi^*(H) - E) \cdot C \geqslant 2$. On the other hand, the number $E \cdot C$ is even since X has no G-orbit of odd length. Moreover, we have

$$7 \geqslant -K_X \cdot C = \pi^*(H) \cdot C + (\pi^*(H) - E) \cdot C \geqslant 5,$$

so that $-K_X \cdot C = 6$, $\pi^*(H) \cdot C = 3$ and $(\pi^*(H) - E) \cdot C = 3$, which gives $E \cdot C = 0$. Hence, we conclude that $\overline{C}$ is a smooth rational cubic curve. Then $\eta(C)$ is a singular cubic curve. This is impossible since G does not have fixed points in $\mathbb{P}^2$. $\qquad\qquad\square$

Now, we are ready to use results described in Section 1.7 to prove that X is K-polystable. Since $\mathrm{Aut}(X)$ is a finite group [45], this would imply that X is K-stable, so that a general member of the deformation family №2.16 is K-stable (see [56] for an alternative proof). We will use notations introduced in Section 1.7.

Lemma 5.31 *If C is a G-invariant irreducible curve in E, then $S(W^E_{\bullet,\bullet}; C) < 1$.*

Proof Let u be any non-negative real number. Then

$$-K_X - uE \sim_{\mathbb{R}} \pi^*(2H) - (1 + u)E,$$

so that $-K_X - uE$ is pseudo-effective $\iff$ $-K_X - uE$ is nef $\iff u \leqslant 1$.

It follows from Lemma 5.29 that $E \cong \mathbb{P}^1 \times \mathbb{P}^1$. Now, using notations introduced in the proof of this lemma, we see that $(-K_X - uE)|_E \sim_{\mathbb{R}} (1+u)s + (3-u)f$.

Observe that $|C - s| \neq \varnothing$ since $C \not\sim f$ as the conic C_1 does not have G-fixed points. Thus, using Corollary 1.110, we get

$$
\begin{aligned}
S(W^E_{\bullet,\bullet}; C) &= \frac{3}{22} \int_0^1 \int_0^\infty \mathrm{vol}\big((-K_X - uE)|_E - vC\big)\,dv\,du \\
&\leqslant \frac{3}{22} \int_0^1 \int_0^\infty \mathrm{vol}\big((-K_X - uE)|_E - vs\big)\,dv\,du \\
&= \frac{3}{22} \int_0^1 \int_0^\infty \mathrm{vol}\big((1 + u - v)s + (3 - u)f\big)\,dv\,du \\
&= \frac{3}{22} \int_0^1 \int_0^{1+u} 2(1 + u - v)(3 - u)\,dv\,du = \frac{67}{88} < 1,
\end{aligned}
$$

as required. $\square$

Let $\widetilde{C}_2$, $\widetilde{C}_3$, $\widetilde{C}_4$ be the proper transforms on X of the conics C_2, C_3, C_4, respectively.

Lemma 5.32 *Let C be one of the curves $\widetilde{C}_2$, $\widetilde{C}_3$, $\widetilde{C}_4$, let $\overline{S}$ be a general hyperplane section of the threefold V_4 that contains $\pi(C)$, and let S be its proper transform on X. Then $S(W^S_{\bullet,\bullet}; C) < 1$.*

Proof Note that the surface $\overline{S}$ is smooth, and it intersects C_1 transversally in two points, so that the surface S is also smooth. Observe also that $-K_S \sim (\pi^*(H) - E)|_S$ and $K_S^2 = 2$, so that S is a weak del Pezzo surface. Then $\eta|_S \colon S \to \mathbb{P}^2$ is the anticanonical map.

Note that $|H|_{\overline{S}} - \pi(C)|$ is a basepoint free pencil. Let C' be the proper transform on the surface S of a general conic in this pencil. On S, we have $(C')^2 = 0$ and $C \cdot C' = 2$. Moreover, we have $\pi^*(H)|_S \sim C + C'$.

Let u be a non-negative real number. Then $-K_X - uS \sim_{\mathbb{R}} (2 - u)\pi^*(H) - E$, implying that $-K_X - uS$ is pseudo-effective $\iff$ $-K_X - uS$ is nef $\iff u \leqslant 1$.

Suppose that $u \in [0, 1]$. Let v be a non-negative real number. Then

$$
(-K_X - uS)|_S - vC \sim_{\mathbb{R}} -K_S + (1 - u - v)C + (1 - u)C',
$$

which implies that $(-K_X - uS)|_S - vC$ is nef for $v \leqslant 1 - u$. On the other hand, we have

$$
\big((-K_X - uS)|_S - vC\big) \cdot C' = \big(-K_S + (1 - u - v)C + (1 - u)C'\big) \cdot C' = 4 - 2u - 2v,
$$

so that $(-K_X - uS)|_S - vC$ is not pseudo-effective for $v > 2 - u$. Moreover,

$$\text{vol}\big((-K_X - uS)\big|_S - vC\big) = \big((-K_X - uS)\big|_S - vC\big)^2 = 6 - 4u.$$

Thus, using Corollary 1.110 and (1.2), we get

$$
\begin{aligned}
S(W^S_{\bullet,\bullet}; C) &= \frac{3}{22} \int_0^1 \int_0^{2-u} \text{vol}\big((-K_X - uS)\big|_S - vC\big)\,dv\,du \\
&\leqslant \frac{3}{22} \int_0^1 \int_0^{1-u} \big((-K_X - uS)\big|_S - vC\big)^2\,dv\,du \\
&\quad + \frac{3}{22} \int_0^1 \frac{2}{3}(6 - 4u)\,du \\
&= \frac{3}{22} \int_0^1 \int_0^{1-u} (14 - 16u - 8v + 4u^2 + 4uv)\,dv\,du + \frac{4}{11} = \frac{37}{44} < 1
\end{aligned}
$$

as claimed.   □

Now, we are ready to show that X is K-polystable. Suppose that it is not K-polystable. By Theorem 1.22, there exists a G-invariant prime divisor F over X such that $\beta(F) \leqslant 0$. Let $Z = C_X(F)$. Then Z is not a surface by Theorem 3.17, so that Z is a G-invariant irreducible curve, because X does not contain G-fixed points.

Using Lemma 1.45, we conclude that $\alpha_{G,Z}(X) < \frac{3}{4}$. Therefore, by Lemma 1.42, there exists a G-invariant effective $\mathbb{Q}$-divisor D on the threefold X such that $D \sim_\mathbb{Q} -K_X$ and the curve Z is contained in $\text{Nklt}(X, \lambda D)$ for some positive rational number $\lambda < \frac{3}{4}$.

Since $|\pi^*(H)|$ does not contain G-invariant surfaces, we see that $\text{Nklt}(X, \lambda D)$ does not contain surfaces. Now, using Corollaries A.31 and A.21, we conclude that Z is a smooth rational curve such that $-K_X \cdot Z < 8$.

By Corollary 1.110 and Lemma 5.31, $Z \not\subset E$, since $S_X(E) < 1$ by Theorem 3.17. Then Z is one of the curves $\widetilde{C}_2, \widetilde{C}_3, \widetilde{C}_4$ by Lemma 5.30. Let S be a general surface in the linear system $|\pi^*(H)|$ that contains the curve C. Then $S_X(S) < 1$ by Theorem 3.17, so that $S(W^S_{\bullet,\bullet}; C) \geqslant 1$ by Corollary 1.110. This contradicts Lemma 5.32.

5.7 Family №2.17

Let C be the harmonic elliptic curve, i.e. the curve $\mathbb{C}/\mathbb{Z}[i]$. Then $\text{Aut}(C)$ has an automorphism θ of order 4 that fixes the zero element $O \in C$, which is induced by the multiplication of $\mathbb{C}$ by i. Let $\mathbb{V}$ be the subgroup in $\text{Aut}(C)$ that consists of the translations by 5-torsion points in C. Then $\mathbb{V} \cong \mu_5^2$, and θ acts on $\mathbb{V}$ by conjugation. If we identify $\mathbb{V}$ with the vector space $\mathbb{F}_5^2$, then this action is given by the linear operator

$$\begin{pmatrix} a \\ b \end{pmatrix} \mapsto \begin{pmatrix} b \\ -a \end{pmatrix}.$$

Note that $(1,2)^T$ is an eigenvector in $\mathbb{F}_5^2$ of this linear operator with eigenvalue 2. Let Γ be the eigenspace with eigenvalue 2. Then Γ is a subgroup in $\mathbb{V}$ that is θ-invariant, so that $\Gamma \cong \mu_5$. Let G be the subgroup in $\mathrm{Aut}(C)$ that is generated by Γ and the automorphism θ. Then $G \cong \mu_5 \rtimes \mu_4$, and Γ is a normal subgroup in G.

Remark 5.33 The group G is known as Frobenius group F_5. In GAP, it can be accessed via `SmallGroup(20,3)`. All irreducible linear representations of the group G can be described as follows: unique four-dimensional representation, and 4 different one-dimensional representations. One also has $H^2(G, \mathbb{G}_m) = 0$.

Let D be the sum of all 5-torsion points in C that corresponds to the subgroup Γ. Then D is a G-invariant divisor by construction. Moreover, since $H^2(G, \mathbb{G}_m)$ is trivial, we see that the line bundle $\mathcal{O}_C(D)$ is G-linearizable [72, Proposition 2.2], so that the action of G on the curve C gives its linear action on $H^0(C, \mathcal{O}_C(D))$, which is faithful, because the divisor D is very ample. By the Riemann–Roch theorem, we have $h^0(C, \mathcal{O}_C(D)) = 5$, so that $|D|$ gives a G-equivariant embedding $C \hookrightarrow \mathbb{P}^4$. By construction, the projective space $\mathbb{P}^4$ contains a G-fixed point because $|D|$ contains G-invariant divisor: the divisor D. Therefore, $H^0(C, \mathcal{O}_C(D))$ is a sum of the four-dimensional irreducible representation and one-dimensional representation. In particular, our $\mathbb{P}^4$ contains a unique G-fixed point.

Let $\phi \colon C \dashrightarrow \mathbb{P}^3$ be the composition of the embedding $C \hookrightarrow \mathbb{P}^4$ and linear projection from the unique G-fixed point. Then ϕ is a morphism since the G-fixed point is not contained in C, because stabilizers in G of every point in C are cyclic since their actions on the Zariski tangent spaces are faithful [53, Theorem 4.4.1]. Moreover, the morphism ϕ is G-equivariant and $\phi(C)$ is G-invariant, where the G-action on $\mathbb{P}^3$ is given by the unique irreducible four-dimensional representation of the group G. This implies that $\phi(C)$ is not contained in a plane in $\mathbb{P}^3$, so that the induced morphism $C \to \phi(C)$ is birational, and $\phi(C)$ is a curve of degree 5. Observe also that $\phi(C)$ cannot have more than 2 singular points because otherwise the curve $\phi(C)$ would be contained in the plane that passes through any 3 of its singular points, which is impossible. Likewise, the curve $\phi(C)$ cannot have 1 or 2 singular points because $\mathbb{P}^3$ does not have G-orbits of length 1 or 2. Therefore, we conclude that $\phi \colon C \to \mathbb{P}^3$ is an embedding. Let us identify C with its image in $\mathbb{P}^3$.

Let $\pi\colon X \to \mathbb{P}^3$ be the blow up of the curve C. Then X is a smooth Fano threefold in the deformation family №2.17 by [17, Theorem 1.1] because $\mathbb{P}^3$ does not have 4-secant lines to C, since otherwise the projection from the 4-secant line would give a birational map $C \dashrightarrow \mathbb{P}^1$. Moreover, the action of the group G lifts to the threefold X, and we have a G-equivariant commutative diagram (see [160, p.117] or [175, §3.1.3]):

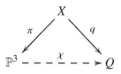

where Q is a smooth quadric surface in $\mathbb{P}^4$, the morphism q is a blow up of a smooth elliptic curve of degree 5, and χ is a rational map that is given by the linear system of cubic surfaces in $\mathbb{P}^3$ that contain the curve C.

Lemma 5.34 $\alpha_G(X) = \frac{3}{4}$.

Proof As we just mentioned, the rational map χ is defined by the four-dimensional linear system of cubic surfaces in $\mathbb{P}^3$ that pass through the curve C. This incomplete linear system is a projectivization of a five-dimensional G-subrepresentation in $H^0(O_{\mathbb{P}^3}(1))$, which contains at least one 1-dimensional subrepresentation of the group G by Remark 5.33. This shows that $\mathbb{P}^3$ contains a G-invariant cubic surface S_3 that passes through C.

Let H be a hyperplane in $\mathbb{P}^3$, let E be the π-exceptional surface, and let $\widetilde{S}_3$ be the proper transform of S_3 on X. Then $-K_X \sim_{\mathbb{Q}} \frac{4}{3}\widetilde{S}_3 + \frac{1}{3}E$, so that $\alpha_G(X) \leqslant \frac{3}{4}$.

To prove that $\alpha_G(X) = \frac{3}{4}$, let us apply Theorem 1.52 with $\mu = \frac{3}{4}$. We see that Theorem 1.52(1) does not hold because the cone of effective divisors on X is generated by E and $\pi^*(5H) - 2E$ (see [175, §3.1.3]), and $\mathbb{P}^3$ does not contain G-invariant planes. Similarly, we see that Theorem 1.52(2) does not hold either, since X does not have G-fixed points because $\mathbb{P}^3$ does not have G-fixed points. Therefore, we have $\alpha_G(X) = \frac{3}{4}$ provided that X does not contain G-invariant smooth rational curves.

Suppose that X contains a G-invariant smooth rational curve $\mathscr{C}$. Then the natural homomorphism $G \to \mathrm{Aut}(\mathscr{C})$ cannot be a monomorphism because $\mathrm{Aut}(\mathbb{P}^1)$ does not contain a subgroup that is isomorphic to $\mu_5 \rtimes \mu_4$. Hence, its kernel is non-trivial, so that it contains the group Γ, because X does not have G-fixed points and every non-trivial normal subgroup of G contains Γ. These means that Γ fixes the curve $\mathscr{C}$ pointwise. Then $\pi(\mathscr{C})$ is an irreducible G-invariant curve in $\mathbb{P}^3$ that is pointwise fixed by Γ, which is impossible because Γ fixes exactly four points in $\mathbb{P}^3$. Hence, we see that X does not contain G-invariant smooth rational curves, so that $\alpha_G(X) = \frac{3}{4}$.   □

Thus, we conclude that X is K-stable by Theorem 1.51 and Corollary 1.5 because the group $\text{Aut}(X)$ is finite [45]. Hence, general Fano threefold №2.17 is also K-stable.

Remark 5.35 In the proof of Lemma 5.34, we mentioned that there is a G-invariant cubic surface $S_3 \subset \mathbb{P}^3$ that passes through the curve C. It is not hard to see that this surface is smooth. Going through the automorphism groups of smooth cubic surfaces [77], we conclude that S_3 is the Clebsch cubic surface. Therefore, we see that $\text{Aut}(S_3) \cong \mathfrak{S}_5$. Moreover, there is a G-equivariant diagram:

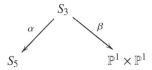

where S_5 is a smooth del Pezzo of degree 5, the morphism α is a blow up of a G-orbit of length 2, and β is a blow up of a G-orbit of length 5. On S_3, we have $C \sim -K_{S_3} + \ell_1 + \ell_2$, where ℓ_1 and ℓ_2 are disjoint lines in S_3 contracted by α. Then $C \cap \ell_1 = \varnothing$, $C \cap \ell_2 = \varnothing$, and $\alpha(C)$ is a G-invariant smooth anticanonical curve in S_5. Therefore, we can construct the curve $C \subset \mathbb{P}^3$ using the quintic del Pezzo surface and its G-equivariant geometry [223].

5.8 Family №2.20

Every smooth Fano threefold №2.20 can be obtained by blowing up the unique smooth Fano threefold №1.15 along a twisted cubic curve. To be more precise, let V_5 be the smooth Fano threefold described in Example 3.2. Then V_5 is a smooth intersection of the Grassmannian $\text{Gr}(2,5) \subset \mathbb{P}^9$ in its Plücker embedding with a linear subspace of codimension 3. Let C be a smooth twisted cubic in V_5, and let X be a blow up of the threefold V_5 along the curve C. Then X is a smooth Fano threefold №2.20, and every smooth threefold in this family can be obtained in this way.

$\text{Aut}(X) \cong \text{Aut}(V_5, C)$, where $\text{Aut}(V_5) \cong \text{PGL}_2(\mathbb{C})$. By [45, Lemma 6.10], there is a unique smooth Fano threefold №2.20 that has an infinite automorphism group. In this case, we have $\text{Aut}(X) \cong \mathbb{G}_m \rtimes \mu_2$. We will prove later in this section that this special smooth Fano threefold №2.20 is K-polystable, which would imply that a general smooth Fano threefold №2.20 is K-stable by Corollary 1.15.

By [164, Lemma 1.5], the stabilizer of a general point in V_5 is a subgroup in $\text{PGL}_2(\mathbb{C})$ isomorphic to $\mathfrak{S}_4$. On the other hand, it follows from [109, Corollary

1.2] that V_5 contains exactly three lines that pass through a general point in V_5. Blowing up the union of these three lines, we obtain a singular Fano threefold №2.20 equipped with a faithful action of the group $\mathfrak{S}_4$, which has a trivial automorphism group. This threefold is a very natural candidate to be tested for K-stability. Unfortunately, it is K-unstable:

Lemma 5.36 *Let O be a point in V_5, let L_1, L_2, L_3 be three lines in V_5 that meet in O, and let $\pi\colon X \to V_5$ be the blow up of the curve $L_1 + L_2 + L_3$. Then X is K-unstable.*

Proof Let $\alpha\colon \widehat{V_5} \to V_5$ be the blow up of the point O, let E_O be the α-exceptional surface, let $\widehat{L}_1, \widehat{L}_2, \widehat{L}_3$ be the proper transforms on $\widehat{V_5}$ of the lines L_1, L_2, L_3, respectively, let $\zeta\colon W \to \widehat{V_5}$ be the blow up of the curve $\widehat{L}_1 + \widehat{L}_2 + \widehat{L}_3$, let E_1, E_2, E_3 be the ζ-exceptional surfaces mapped to $\widehat{L}_1, \widehat{L}_2, \widehat{L}_3$, respectively. Then there exists commutative diagram

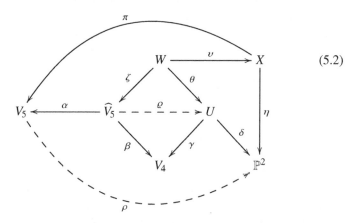

(5.2)

where β is a flopping contraction of the curve $\widehat{L}_1 + \widehat{L}_2 + \widehat{L}_3$, the threefold V_4 is a singular complete intersection of two quadrics in $\mathbb{P}^5$, the map ϱ is a flop of the curve $\widehat{L}_1 + \widehat{L}_2 + \widehat{L}_3$, the morphism θ is a birational contraction of the surfaces E_1, E_2, E_3, the morphism γ is the flopping contraction of the curves $\theta(E_1), \theta(E_2), \theta(E_3)$, the morphism δ is a $\mathbb{P}^1$-bundle, the rational map ρ is a linear projection from the linear span of the curve $L_1 + L_2 + L_3$, the morphism η is a conic bundle, and υ is a small birational map described below. Note also that ρ contracts conics in V_5 that pass through O.

Let S be the proper transform of E_O via ζ, and let $\mathbf{e}_1 = E_1|_S$, $\mathbf{e}_2 = E_2|_S$, $\mathbf{e}_3 = E_3|_S$. Then S is the del Pezzo surface of degree 6, and $\mathbf{e}_1, \mathbf{e}_2, \mathbf{e}_3$ are disjoint (-1)-curves on it. Let ℓ_1, ℓ_2, ℓ_3 be the remaining (-1)-curves in the surface S. Then υ is the flopping contraction of the curves ℓ_1, ℓ_2, ℓ_3. We can flop these curves $\sigma\colon W \dashrightarrow W'$ and obtain the equivariant commutative diagram

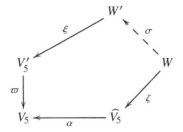

where ξ is the contraction of the surface $\sigma(S) \cong \mathbb{P}^2$ to a singular point of the threefold $\widehat{V}_5$, which is a quotient singularity of type $\frac{1}{2}(1,1,1)$, and ϖ is the $\mathfrak{A}_4$-extremal contraction. Note that ϖ is the symbolic blow up of the curve $L_1 + L_2 + L_3$ (see [186, Example 5.2.3]), which also appears in the proof of [54, Proposition 5.1].

Let us compute $\beta(S)$. Let $H = (\alpha \circ \zeta)^*(H_{V_5})$, where H_{V_5} is a hyperplane section of V_5, and let u be a non-negative real number. Then

$$-K_W - uS \sim_{\mathbb{R}} (\eta \circ \upsilon)^*\big(\mathcal{O}_{\mathbb{P}^2}(2)\big) + (2 - u)S + E_1 + E_2 + E_3.$$

Intersecting $-K_W - uS$ and general fibers of $\eta \circ \upsilon$, we see that $-K_W - uS$ is not pseudo-effective for $u > 2$. Moreover, this divisor is nef for $u \in [0, 1]$. Similarly, if $u \in [1, 2]$, then the Zariski decomposition of the divisor $-K_W - uS$ is

$$-K_W - uS \sim_{\mathbb{R}} \underbrace{(\eta \circ \upsilon)^*\big(\mathcal{O}_{\mathbb{P}^2}(2)\big) + (2 - u)\big(S + E_1 + E_2 + E_3\big)}_{\text{positive part}}$$

$$+ \underbrace{(u - 1)\big(E_1 + E_2 + E_3\big)}_{\text{negative part}}.$$

Hence, in the notations of Section 1.7, we have

$$P\big(-K_W - uS\big) = \begin{cases} 2H - (2 + u)S - (E_1 + E_2 + E_3) & \text{if } u \in [0, 1], \\ 2H - (2 + u)S - u(E_1 + E_2 + E_3) & \text{if } u \in [1, 2], \end{cases}$$

and $N(-K_W - uS) = (u - 1)(E_1 + E_2 + E_3)$ for $u \in [1, 2]$. Therefore, we have

$$S_X(S) = S_W(S) = \frac{1}{26} \int_0^2 \big(P(-K_W - uS)\big)^3 du$$

$$= \frac{1}{26} \int_0^1 \big(9u + 34 - (2 + u)^3\big)du + \frac{1}{26} \int_1^2 \big(3u^3 + 40 - (2 + u)^3\big)du$$

$$= \frac{119}{104} > 1,$$

so that $\beta(S) = -\frac{15}{104}$. Thus, X is not K-semistable by Theorem 1.19. $\qquad\square$

Now, let us prove that the smooth Fano threefold №2.20 with an infinite automorphism group is K-polystable. To do this, we present an explicit construction of this threefold. For an alternative construction, see Section 7.2.

Let Q be the smooth quadric in $\mathbb{P}^4$ given by $xt = yz + w^2$, and let C_3 be the twisted cubic in Q parametrized as $[r^6 : r^4s^2 : r^2s^4 : s^6 : 0]$, where $[r : s] \in \mathbb{P}^1$. Then C_3 is contained in the hyperplane $w = 0$. On Q, this hyperplane cuts out a smooth surface S_2. Let G be the subgroup in $\mathrm{Aut}(\mathbb{P}^4)$ that is generated by the involution τ that acts as $[x : y : z : t : w] \mapsto [t : z : y : x : w]$, and the automorphisms λ_s that act as

$$\left[x : y : z : t : w\right] \mapsto \left[x : s^2y : s^4z : s^6t : s^3w\right],$$

where $s \in \mathbb{G}_m$. Then $G \cong \mathbb{G}_m \rtimes \boldsymbol{\mu}_2$, both Q and C_3 are G-invariant, and G acts faithfully on the quadric Q, so that we identify G with a subgroup in $\mathrm{Aut}(Q)$.

Let $\chi : Q \dashrightarrow \mathbb{P}^6$ be the rational map that is given by

$$\left[x : y : z : t : w\right] \mapsto \left[wx : wy : wz : wt : w^2 : xz - y^2 : yt - z^2\right].$$

Then χ is G-equivariant for the following G-action on $\mathbb{P}^6$: the involution τ acts as

$$\left[x_0 : x_1 : x_2 : x_3 : x_4 : x_5 : x_6\right] \mapsto \left[x_3 : x_2 : x_1 : x_0 : x_4 : x_6 : x_5\right],$$

and the automorphisms λ_s act as

$$\left[x_0 : x_1 : x_2 : x_3 : x_4 : x_5 : x_6\right] \mapsto$$
$$\left[s^3x_0 : s^5x_1 : s^7x_2 : s^9x_3 : s^6x_4 : s^4x_5 : s^8x_6\right].$$

The rational map χ is undefined exactly at the cubic curve C_3, and it contracts the quadric surface S_2 to the G-invariant line $L = \{x_0 = x_1 = x_2 = x_3 = x_4 = 0\} \subset \mathbb{P}^6$.

It is well known that the closure of the image of the map χ is isomorphic to the smooth Fano threefold V_5. In fact, we can find its explicit equations. Namely, observe that the closure of the image of χ is contained in

$$\begin{cases} x_4x_5 - x_0x_2 + x_1^2 = 0, \\ x_4x_6 - x_1x_3 + x_2^2 = 0, \\ x_4^2 - x_0x_3 + x_1x_2 = 0, \\ x_1x_4 - x_0x_6 - x_2x_5 = 0, \\ x_2x_4 - x_3x_5 - x_1x_6 = 0. \end{cases}$$

These equations define a smooth irreducible three-dimensional subscheme of degree 5, which is the closure of the image of χ. Indeed, since V_5 is an intersection of quadrics in $\mathbb{P}^6$, we have

$$h^0\left(\mathbb{P}^6, O_{\mathbb{P}^6}(2) \otimes I_{V_5}\right) = h^0\left(\mathbb{P}^6, O_{\mathbb{P}^6}(2)\right) = h^0\left(V_5, O_{\mathbb{P}^6}(2)\big|_{V_5}\right) = 28 - 23 = 5,$$

where I_{V_5} is the ideal sheaf of the threefold V_5. Thus, the above five linearly independent quadratic equations scheme-theoretically define V_5. Alternatively, we can check this using the following Magma code:

```
Q:=RationalField();
P<x0,x1,x2,x3,x4,x5,x6>:=ProjectiveSpace(Q,6);
X:=Scheme(P,[x4*x5-x0*x2+x1^2,x4*x6-x1*x3+x2^2,
          x4^2-x0*x3+x1*x2,x1*x4-x0*x6-x2*x5,
          x2*x4-x3*x5-x1*x6]);
Degree(X);
IsReduced(X);
IsNonsingular(X);
IsIrreducible(X);
Dimension(X);
```

In the following, we identify V_5 with the closure of the image of the map χ.

Lemma 5.37 *The threefold V_5 does not contain G-fixed points.*

Proof The only G-fixed point in $\mathbb{P}^6$ is $[0:0:0:0:1:0:0] \notin V_5$. □

$\text{Pic}(V_5) = \mathbb{Z}[H_{V_5}]$, where H_{V_5} is a hyperplane section of V_5. Moreover, we have $H_{V_5}^3 = 5$ and $-K_{V_5} \sim 2H_{V_5}$. By construction, the line L is contained in the threefold V_5, and it is also contained in the unique G-invariant hyperplane section of the threefold V_5, which is given by $x_4 = 0$. Let us denote this hyperplane section by $\mathcal{H}$. Then $\mathcal{H}$ is singular along L and $\text{mult}_L(\mathcal{H}) = 2$. Moreover, we have the G-equivariant commutative diagram

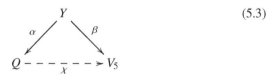

(5.3)

where α is the blow up of the twisted cubic curve C_3, and β is the blow up of the line L, the α-exceptional surface is the proper transform of the surface $\mathcal{H}$, and the β-exceptional surface is the proper transform of the quadric surface S_2.

Let us describe G-invariant irreducible curves in V_5. Observe that such a curve always contains a μ_2-fixed point. Indeed, a G-invariant curve in V_5 admits an effective action of the subgroup $\mathbb{G}_m \subset G$, so it is rational. But an involution acting on an irreducible rational curve always fixes a point in this curve. This shows that every G-invariant irreducible curve in V_5 contains a μ_2-fixed point. Using the defining equation of the threefold V_5, we can find all μ_2-fixed points in V_5 and describe their G-orbits explicitly. In particular, this approach implies that all G-invariant irreducible curves in $\mathcal{H}$ can be described as follows: the line L, the twisted cubic $\mathscr{C}$ given parametrically as $[r^3 : r^2 s : rs^2 : s^3 : 0 : 0 : 0]$ for $[r : s] \in \mathbb{P}^1$, and the smooth rational sextic curve C_γ that is given by the parametric equation

$$[r^6 : -r^4 s^2 : r^2 s^4 : -s^6 : 0 : \gamma r^5 s : -\gamma r s^5],$$

where $\gamma \in \mathbb{C}^*$ and $[r : s] \in \mathbb{P}^1$. $L \cap \mathscr{C} = \varnothing = L \cap C_\gamma$, and the curves $\mathscr{C}$ and C_γ intersect transversally at $[1 : 0 : 0 : 0 : 0 : 0 : 0]$ and $[0 : 0 : 0 : 1 : 0 : 0 : 0]$.

Remark 5.38   Let E_{C_3} be the α-exceptional surface. By [53, Lemma 7.7.3], $E_{C_3} \cong \mathbb{F}_1$. Let $\mathbf{s}$ be the unique (-1)-curve in E_{C_3}, let $\mathbf{f}_x$ and $\mathbf{f}_t$ be the irreducible curves in E_{C_3} that are mapped by the blow up α to the points $[1 : 0 : 0 : 0 : 0]$ and $[0 : 0 : 0 : 1 : 0]$, respectively. Then $\mathbf{s}$ is G-invariant, so that there is a G-equivariant birational morphism $E_{C_3} \to \mathbb{P}^2$ that contracts the curve $\mathbf{s}$. This easily implies the following assertions:

- $|\mathbf{s} + \mathbf{f}_x|$ contains a unique irreducible G-invariant curve C,
- $|2\mathbf{s} + 2\mathbf{f}_x|$ contains a pencil $\mathcal{P}$ generated by the curves $2C$ and $\mathbf{s} + \mathbf{f}_x + \mathbf{f}_t$ such that every other curve in $\mathcal{P}$ is G-invariant, irreducible and smooth.

These are all G-invariant irreducible curves in E_{C_3}. Let F_L be the β-exceptional surface. It follows from the proof of [53, Lemma 13.2.1] that $F_L|_{E_{C_3}} = \mathbf{s}$ and $\beta^*(H_{V_5})|_{E_{C_3}} \sim \mathbf{s} + 3\mathbf{f}_x$. This implies that $\beta(\mathbf{s}) = L$, and $\beta(C)$ is the twisted cubic curve $\mathscr{C}$. Similarly, we see that every smooth curve in $\mathcal{P}$ is mapped by β to the sextic curves C_γ for some $\gamma \in \mathbb{C}^*$.

Similarly, we can describe all G-invariant irreducible curves in V_5. But it is easier to describe G-invariant irreducible curves in the quadric Q, and then use birational map χ. Namely, let P be a μ_2-fixed point in the quadric Q, and let C be the closure of its G-orbit. Then either $P = [a : b : b : a : c]$ for some numbers a, b and c such that $a^2 = b^2 + c^2$, or $P = [a : b : -b : -a : 0]$ for some numbers a and b such that $a^2 = b^2$. In both cases, if $a^2 = b^2$, then either $C = C_3$, or C is another twisted cubic curve in S_2 that is given by the following parametrization:

$$[r^3 : -r^2 s : -rs^2 : s^3 : 0]. \tag{5.4}$$

Since we already described the G-invariant irreducible curves in $\mathcal{H}$, we may assume that $a^2 \neq b^2$. Then the curve C is given by the parametrization: $[ar^6 : br^4s^2 : br^2s^4 : as^6 : cr^3s^3]$, and its image $\chi(C)$ is given by the parametrization:

$$[acr^6 : bcr^4s^2 : bcr^2s^4 : acs^6 : c^2r^3s^3 : b(a-b)r^5s : b(a-b)rs^5]. \quad (5.5)$$

If $a = 0$, then C is the conic $x = t = yz + w^2 = 0$, and $\chi(C)$ is the smooth quartic curve C_4 that is given by the parametrization $[0 : ir^3s : irs^3 : 0 : -r^2s^2 : -r^4 : -s^4]$, where $i = \sqrt{-1}$. Similarly, if $b = 0$, then C is the conic $y = z = xt - w^2 = 0$, and $\chi(C)$ is the conic C_2 given by the parametrization $[r^2 : 0 : 0 : s^2 : rs : 0 : 0]$. If $a \neq 0$ and $b \neq 0$, then $\chi(C)$ is a smooth rational sextic curve in V_5. Since χ induces an isomorphism $Q \setminus S_2 \cong V_5 \setminus \mathcal{H}$, this gives us description of all G-invariant irreducible curves in V_5. These are the curves L, C_2, $\mathscr{C}$, C_4, sextics C_γ, and sextics given by (5.5) with $ab \neq 0$.

Corollary 5.39 *Let C be a G-invariant irreducible curve in V_5 such that $\deg(C) < 6$. Then C is one of the curves L, C_2, $\mathscr{C}$, C_4.*

Remark 5.40 Observe that $C_2 = \{x_1 = x_2 = x_5 = x_6 = x_0x_3 - x_4^2 = 0\} \subset \mathbb{P}^6$. Note also that the curve $\mathscr{C}$ is cut out on V_5 by the equations $x_4 = x_5 = x_6 = 0$, the curve $C_2 + \mathscr{C}$ is cut out by $x_5 = x_6 = 0$, and the curve $L + C_4$ is cut out by $x_0 = x_3 = 0$. As we already mentioned, $L \cap \mathscr{C} = \varnothing$. Similarly, we have $C_2 \cap L = \varnothing$ and $C_2 \cap C_4 = \varnothing$. But

$$L \cap C_4 = [0 : 0 : 0 : 0 : 0 : 1 : 0] \cup [0 : 0 : 0 : 0 : 0 : 0 : 1],$$

and the curves L and C_4 intersect transversally at these points. Similarly, we have

$$C_2 \cap \mathscr{C} = [1 : 0 : 0 : 0 : 0 : 0 : 0] \cup [0 : 0 : 0 : 1 : 0 : 0 : 0],$$

and these curves intersect transversally. Finally, observe that the equations $x_1 = x_2 = 0$ cuts out on V_5 the curve $C_2 + L + \ell + \ell'$, where $\ell = \{x_0 = x_1 = x_2 = x_4 = x_5 = 0\}$ and $\ell' = \{x_1 = x_2 = x_3 = x_4 = x_6 = 0\}$.

Let $\pi \colon X \to V_5$ be the blow up of the curve $\mathscr{C}$, and let $E_{\mathscr{C}}$ be the π-exceptional surface. Then the action of the group G lifts to the threefold X, so that we can identify G with a subgroup of the group $\mathrm{Aut}(X)$.

Lemma 5.41 $\mathrm{Aut}(X) = G$.

Proof Observe that

$$\mathbb{G}_m \rtimes \mu_2 \cong G \subset \mathrm{Aut}(X) \cong \mathrm{Aut}(V_5; \mathscr{C}) \subset \mathrm{Aut}(V_5) \cong \mathrm{PGL}_2(\mathbb{C}).$$

Now, using the classification of algebraic subgroups in $\mathrm{PGL}_2(\mathbb{C})$, see [169], we conclude that either $\mathrm{Aut}(V_5; \mathscr{C}) = G$ or $\mathrm{Aut}(V_5; \mathscr{C}) = \mathrm{Aut}(V_5)$. But $\mathrm{Aut}(V_5; \mathscr{C}) \neq \mathrm{Aut}(V_5)$ since the curve $\mathscr{C}$ is not $\mathrm{Aut}(V_5)$-invariant. □

By [45, Lemma 6.10], the threefold X is the unique smooth Fano threefold №2.20 that has an infinite automorphism group. Observe that $|\pi^*(H_{V_5}) - E_\mathscr{C}|$ is free from basepoints and defines a conic bundle $\eta \colon X \to \mathbb{P}^2$, so that we have the G-equivariant commutative diagram

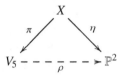

where ρ is the rational map that is given by $[x_0 : x_1 : x_2 : x_3 : x_4 : x_5 : x_6] \mapsto [x_4 : x_5 : x_6]$. Therefore, the composition map $\rho \circ \chi$ is given by $[x : y : z : t : w] \mapsto [w^2 : xz - y^2 : yt - z^2]$.

Let $\widetilde{\mathcal{H}}$ be the proper transform on X of the surface $\mathcal{H}$. We have $\widetilde{\mathcal{H}} \in |\pi^*(H_{V_5}) - E_\mathscr{C}|$, so that $\eta(\widetilde{\mathcal{H}})$ is the unique G-invariant line in $\mathbb{P}^2$. Observe that this line is an irreducible component of the discriminant curve of the conic bundle η. The other irreducible component is a G-invariant irreducible conic that intersects $\eta(\widetilde{\mathcal{H}})$ transversally.

Lemma 5.42 $E_\mathscr{C} \cong \mathbb{P}^1 \times \mathbb{P}^1$.

Proof We have $\mathscr{C} \cong \mathbb{P}^1$ and $N_{\mathscr{C}/V_5} \cong \mathcal{O}_{\mathbb{P}^1}(a) \oplus \mathcal{O}_{\mathbb{P}^1}(b)$ for some integers a and b such that $a + b = 4$ and $a \leqslant b$. Then $E_\mathscr{C} \cong \mathbb{F}_n$ for $n = b - a$. We have to show that $n = 0$.

Let $\mathbf{s}$ be a section of the projection $E_\mathscr{C} \to \mathscr{C}$ such that $\mathbf{s}^2 = -n$, and let $\mathbf{f}$ be a fiber of this projection. Then $-E_\mathscr{C}|_{E_\mathscr{C}} \sim \mathbf{s} + k\mathbf{f}$ for some integer k. Then $-n + 2k = E^3 = -4$, so that $k = \frac{n-4}{2}$. Then $\widetilde{\mathcal{H}}|_{E_\mathscr{C}} \sim \mathbf{s} + (k + 3)\mathbf{f} = \mathbf{s} + \frac{n+2}{2}\mathbf{f}$, which implies that $\widetilde{\mathcal{H}}|_{E_\mathscr{C}} \not\sim \mathbf{s}$. Moreover, we know that $\widetilde{\mathcal{H}}|_{E_\mathscr{C}}$ is a smooth irreducible curve since the surface $\mathcal{H}$ is smooth along the curve $\mathscr{C}$. Thus, we have

$$0 \leqslant \widetilde{\mathcal{H}}\big|_{E_\mathscr{C}} \cdot \mathbf{s} = \left(\mathbf{s} + \frac{n+2}{2}\mathbf{f}\right) \cdot \mathbf{s} = -n + \frac{n+2}{2} = \frac{2-n}{2},$$

so that $n = 0$ or $n = 2$. This can also be deduced from the fact that $\mathscr{C}$ is disjointed from L, so that $N_{\mathscr{C}/V_5} \cong N_{C/Y}$, where $C \cong \mathbb{P}^1$ is the curve in E_{C_3} described in Remark 5.38. Now, using the exact sequence of sheaves

$$0 \longrightarrow N_{C/E_{C_3}} \longrightarrow N_{C/Y} \longrightarrow N_{E_{C_3}/Y}\big|_C \longrightarrow 0,$$

we get $n \in \{0, 2\}$ because $N_{C/E_{C_3}} \cong \mathcal{O}_{\mathbb{P}^1}(1)$ and $N_{E_{C_3}/Y}|_C \cong \mathcal{O}_{\mathbb{P}^1}(3)$.

If $n = 0$, then we are done. If $n = 2$, then $(\pi^*(H_{V_5}) - E_{\mathscr{C}})|_{E_{\mathscr{C}}} \cdot \mathbf{s} = (\mathbf{s} + 2\mathbf{f}) \cdot \mathbf{s} = 0$, so that $\mathbf{s}$ is contracted by η, which is impossible since $-K_X \cdot \mathbf{s} = 3$ in this case. □

The main result of this section is the following proposition, which also implies that general Fano threefolds in the family №2.20 are K-stable by Corollary 1.16.

Proposition 5.43 *The threefold X is K-polystable.*

Let us prove Proposition 5.43. Suppose that the Fano threefold X is not K-polystable. By Theorem 1.22, there is a G-equivariant prime divisor F over X such that $\beta(F) \leqslant 0$. Let $Z = C_X(F)$. Then, using Theorem 3.17 and Lemma 5.37, we see that Z is a curve. Let us use results of Section 1.7. We will use notation introduced in this section.

Lemma 5.44 $Z \not\subset E_{\mathscr{C}}$.

Proof Suppose that $Z \subset E_{\mathscr{C}}$. Recall that $E_{\mathscr{C}} \cong \mathbb{P}^1 \times \mathbb{P}^1$ by Lemma 5.42. Let us use notations introduced in the proof of this lemma. Observe that the pencil $|\mathbf{f}|$ does not contain G-invariant curves because V_5 does not contain G-fixed points by Lemma 5.37. Similarly, the pencil $|\mathbf{s}|$ also does not contain G-invariant curves – otherwise the intersection of such curve with $\widetilde{\mathcal{H}}$ would consist of a single point since we have $\widetilde{\mathcal{H}}|_{E_{\mathscr{C}}} \sim \mathbf{s} + \mathbf{f}$. Hence, we conclude that $Z \sim a\mathbf{s} + b\mathbf{f}$ for some positive integers a and b.

Using Theorem 3.17, we see that $S_X(E_{\mathscr{C}}) < 1$. Using Corollary 1.110, we conclude that $S(W^{E_{\mathscr{C}}}_{\bullet,\bullet}; Z) \geqslant 1$. Let us compute $S(W^{E_{\mathscr{C}}}_{\bullet,\bullet}; Z)$.

Let u be a non-negative real number. Then

$$-K_X - uE_{\mathscr{C}} \sim_{\mathbb{R}} 2\pi^*(H_{V_5}) - (1 + u)E_{\mathscr{C}},$$

so $-K_X - uE_{\mathscr{C}}$ is nef $\iff -K_X - uE_{\mathscr{C}}$ is pseudo-effective $\iff u \in [0, 1]$. Let v be a non-negative real number. Then Corollary 1.110 gives

$$
\begin{aligned}
S(W^{E_{\mathscr{C}}}_{\bullet,\bullet}; Z) &= \frac{3}{26} \int_0^1 \int_0^\infty \mathrm{vol}\Big(\big(-K_X - uE_{\mathscr{C}}\big)\big|_{E_{\mathscr{C}}} - vZ\Big)\,dv\,du \\
&= \frac{3}{26} \int_0^1 \int_0^\infty \mathrm{vol}\Big((1 + u - av)\mathbf{s} + (4 - 2u - bv)\mathbf{f}\Big)\,dv\,du \\
&\leqslant \frac{3}{26} \int_0^1 \int_0^\infty \mathrm{vol}\Big((1 + u - v)\mathbf{s} + (4 - 2u - v)\mathbf{f}\Big)\,dv\,du \\
&= \frac{3}{26} \int_0^1 \int_0^{1+u} 2(1 + u - v)(4 - 2u - v)\,dv\,du = \frac{63}{104} < 1.
\end{aligned}
$$

The obtained contradiction completes the proof of the lemma. □

Thus, we see that $\pi(Z)$ is a G-invariant irreducible curve in V_5 that is different from $\mathscr{C}$. Since we already know the classification of such curves, we can exclude them one by one as in the proof of Lemma 5.44. We start with

Lemma 5.45 $\pi(Z) \neq L$.

Proof   Suppose that $\pi(Z) = L$. Let $\sigma \colon \widehat{X} \to X$ be the blow up along the smooth curve Z, and let $\widehat{S}$ be the σ-exceptional divisor. To start with, let us compute $\beta(\widehat{S})$. Take $u \in \mathbb{R}_{\geqslant 0}$. Let $\widehat{E}_{\mathscr{C}}$ and $\widehat{\mathcal{H}}$ be the proper transforms on $\widehat{X}$ of the surfaces $E_{\mathscr{C}}$ and $\mathcal{H}$, respectively. Then $\sigma^*(-K_X) - u\widehat{S} \sim_{\mathbb{R}} 2\widehat{\mathcal{H}} + \widehat{E}_{\mathscr{C}} + (4-u)\widehat{S}$, so that the divisor $\sigma^*(-K_X) - u\widehat{S}$ is pseudo-effective $\iff u \leqslant 4$. In fact, this divisor is nef if $u \leqslant 1$. Moreover, for $u \in [1,4]$, its Zariski decomposition can be described as follows. If $u \in [1,3]$, then

$$\sigma^*\big(-K_X\big) - u\widehat{S} \sim_{\mathbb{R}} \underbrace{\frac{5-u}{2}\widehat{\mathcal{H}} + \widehat{E}_{\mathscr{C}} + (4-u)\widehat{S}}_{\text{positive part}} + \underbrace{\frac{u-1}{2}\widehat{\mathcal{H}}}_{\text{negative part}}.$$

If $u \in [3,4]$, then

$$\sigma^*\big(-K_X\big) - u\widehat{S} \sim_{\mathbb{R}} \underbrace{(4-u)\big(\widehat{\mathcal{H}} + \widehat{E}_{\mathscr{C}} + \widehat{S}\big)}_{\text{positive part}} + \underbrace{(2-u)\widehat{\mathcal{H}} + (3-u)\widehat{E}_{\mathscr{C}}}_{\text{negative part}}.$$

Therefore we see that

$$\mathrm{vol}\big(\sigma^*\big(-K_X\big) - u\widehat{S}\big) = \begin{cases} 26 - 6u^2 & \text{if } 0 \leqslant u \leqslant 1, \\ \dfrac{67}{2} - 15u + \dfrac{3}{2}u^2 & \text{if } 1 \leqslant u \leqslant 3, \\ 2(4-u)^3 & \text{if } 3 \leqslant u \leqslant 4. \end{cases}$$

Integrating, we get $S_X(\widehat{S}) = \frac{89}{52}$, so that $\beta(\widehat{S}) = A_X(\widehat{S}) - S_X(\widehat{S}) = 2 - \frac{89}{52} = \frac{15}{52} > 0$.

The action of the group G lifts to the threefold $\widehat{X}$, and the surface $\widehat{S}$ is G-invariant. Moreover, since $L \cap \mathscr{C} = \varnothing$, the G-equivariant diagram (5.3) gives a G-equivariant isomorphism $\widehat{S} \cong S_2$. In particular, we see that $\widehat{S} \cong \mathbb{P}^1 \times \mathbb{P}^1$, and $\widehat{S}$ contains exactly two irreducible G-invariant curves, because we proved earlier that S_2 contains exactly two irreducible G-invariant curves: the curve C_3, and the twisted cubic given by (5.4).

Let ℓ_1 and ℓ_2 be two distinct rulings of the surface $\widehat{S} \cong \mathbb{P}^1 \times \mathbb{P}^1$ such that $\sigma(\ell_1) = Z$, and ℓ_2 is a fiber of the projection $\widehat{S} \to Z$. Then $\widehat{\mathcal{H}}|_{\widehat{S}} \sim 2\ell_1 + \ell_2$, and $\widehat{\mathcal{H}}|_{\widehat{S}}$ is a G-invariant irreducible curve in $\widehat{S}$, which is the image of the curve C_3. Similarly, the second irreducible G-invariant curve in $\widehat{S}$ is also contained in $|2\ell_1 + \ell_2|$. In particular, we see that $\widehat{S}$ does not contain irreducible G-invariant curves that are sections of the natural projection $\widehat{S} \to Z$.

Recall that F is a G-invariant prime divisor over X such that $\beta(F) \geqslant 0$ and $Z = C_X(F)$. Thus, using (1.10), we see that $\delta_{G,Z}(X) \leqslant 1$, where $\delta_{G,Z}(X)$ is the number defined in Section 1.5. Let us show that $\widetilde{\delta}_{G,Z}(X) > 1$.

We claim that $\widetilde{\delta}_{G,Z}(X) \geqslant \frac{104}{89}$. Indeed, suppose that $\widetilde{\delta}_{G,Z}(X) < \frac{104}{89}$. Then there exists a G-invariant cool $\mathbb{Q}$-system $\mathcal{D}$ of the divisor $-K_X$ such that

$Z \subseteq \mathrm{Nklt}(X, \lambda\mathcal{D})$ for some rational number $\lambda < \frac{104}{89}$. Let $\widehat{\mathcal{D}}$ be the proper transform on $\widehat{X}$ of the $\mathbb{Q}$-system $\mathcal{D}$. Then

$$K_{\widehat{X}} + \lambda\widehat{\mathcal{D}} + \left(\lambda\mathrm{mult}_Z(\mathcal{D}) - 1\right)\widehat{S} \sim_{\mathbb{Q}} \sigma^*\left(K_X + \lambda\mathcal{D}\right).$$

On the other hand, we know that $\mathrm{mult}_Z(\mathcal{D}) \leqslant S_X(\widehat{S}) = \frac{89}{52}$ since the $\mathbb{Q}$-system $\mathcal{D}$ is cool. Thus, using Lemma A.27, we see that $\widehat{S}$ contains a smooth irreducible G-invariant curve that is a section of the projection $\widehat{S} \to Z$. But, as we explained earlier, such curve does not exist. The obtained contradiction completes the proof of the lemma.

Alternatively, we can find a contradiction using Corollaries 1.102 and 1.109. Namely, it follows from Corollary 1.102 that

$$\frac{A_X(F)}{S_X(F)} \geqslant \delta_Z(X) \geqslant \min\left\{\frac{A_X(\widehat{S})}{S_X(\widehat{S})}, \inf_{\widehat{Z} \subset \widehat{S}} \delta_{\widehat{Z}}\left(\widehat{S}; W_{\bullet,\bullet}^{\widehat{S}}\right)\right\},$$

where the infimum is taken over all irreducible curves $\widehat{Z} \subset \widehat{S}$ that are not contained in the fibers of the projection $\widehat{S} \to Z$. Therefore, since $A_X(F) \leqslant S_X(F)$ and $\frac{A_X(\widehat{S})}{S_X(\widehat{S})} = \frac{104}{89}$, we conclude that $\widehat{S}$ contains an irreducible (horizontal) curve $\widehat{Z}$ such that $\delta_{\widehat{Z}}(\widehat{S}; W_{\bullet,\bullet}^{\widehat{S}}) \leqslant 1$. Using (1.12), we get $S(W_{\bullet,\bullet}^{\widehat{S}}; \widehat{Z}) \geqslant 1$. But we can find $S(W_{\bullet,\bullet}^{\widehat{S}}; \widehat{Z})$ using Corollary 1.109. Namely, we have $\widehat{\mathcal{H}}|_{\widehat{S}} \sim 2\ell_1 + \ell_2$, $\widehat{E}_\mathscr{C} \cap \widehat{S} = \varnothing$ and $\widehat{S}|_{\widehat{S}} \sim -\ell_1$, so that using the Zariski decomposition of the divisor $\sigma^*(-K_X) - u\widehat{S}$ for $u \in [0,4]$ found earlier, we get

$$S\left(W_{\bullet,\bullet}^{\widehat{S}}; \widehat{Z}\right)$$
$$= \frac{3}{26}\int_0^1\int_0^{\frac{u}{2}} 2(u - 2v)(2 - v)\,dv\,du + \frac{3}{26}\int_1^3 \frac{(u-1)(5-u)}{2}\,du$$
$$+ \frac{3}{26}\int_1^3\int_0^{\frac{1}{2}}(1 - 2v)(5 - u - 2v)\,dv\,du + \frac{3}{26}\int_3^4 2(u-2)(4-u)^2\,du$$
$$+ \frac{3}{26}\int_3^4\int_0^{\frac{4-u}{2}} 2(4 - u - 2v)(4 - u - v)\,dv\,du = \frac{63}{104}$$

in the case when $\widehat{Z} = \widehat{\mathcal{H}}|_{\widehat{S}}$. Similarly, if $\widehat{Z} \neq \widehat{\mathcal{H}}|_{\widehat{S}}$, then Corollary 1.109 gives

$$S\left(W_{\bullet,\bullet}^{\widehat{S}}; \widehat{Z}\right) \leqslant S\left(W_{\bullet,\bullet}^{\widehat{S}}; \ell_1\right) = \frac{3}{26}\int_0^1\int_0^u 4(u - v)\,dv\,du$$
$$+ \frac{3}{26}\int_1^3\int_0^1(1 - v)(5 - u)\,dv\,du$$
$$+ \frac{3}{26}\int_3^4\int_0^{4-u} 2(4 - u - v)(4 - u)\,dv\,du = \frac{47}{104}.$$

Thus, we see that $S(W_{\bullet,\bullet}^{\widehat{S}}; \widehat{Z}) < 1$, which is a contradiction. $\square$

The next step is

Lemma 5.46 $\pi(Z) \neq C_2$.

Proof   Suppose that $\pi(Z) = C_2$. Let H be a general hyperplane section of V_5 that contains both curves C_2 and $\mathscr{C}$. By Remark 5.40, the curve $C_2 + \mathscr{C}$ is cut out on V_5 by $x_5 = x_6 = 0$, so that H is cut out on V_5 by $\lambda x_5 + \mu x_6$ for general numbers λ and μ. Then H is smooth. For instance, it is smooth for $\lambda = \mu = 1$. Note that $\pi_*^{-1}(C_2)$ is the fiber of the conic bundle $\eta \colon X \to \mathbb{P}^2$ over the point $[1 : 0 : 0]$, this fiber is smooth, and H is the preimage via the rational map $\eta \circ \pi^{-1}$ of a general line in $\mathbb{P}^2$ that passes through $[1 : 0 : 0]$, which also implies that H is smooth. Then H is a smooth quintic del Pezzo surface.

Let S be the proper transform of the surface H on the threefold X, and let $C = E_{\mathscr{C}}|_S$. Then $S \cong H$ and $Z + C \sim -K_S \sim \pi^*(H_{V_5})|_S$. Observe that $|C|$ is basepoint free and gives a birational morphism $\varpi \colon S \to \mathbb{P}^2$ that contracts four disjoint (-1) curves. Denote these curves by $\ell_1, \ell_2, \ell_3, \ell_4$. Then $2C \sim Z + \ell_1 + \ell_2 + \ell_3 + \ell_4$ because $\varpi(Z)$ is a conic that passes through the points $\varpi(\ell_1), \varpi(\ell_2), \varpi(\ell_3), \varpi(\ell_4)$.

By Corollary 1.110, we have $S(W_{\bullet,\bullet}^S; Z) \geqslant 1$ because $S_X(S) < 1$ by Theorem 3.17. Let us compute $S(W_{\bullet,\bullet}^S; Z)$. Let u be a non-negative real number. Then

$$-K_X - uS \sim_{\mathbb{R}} (2 - u)\pi^*(H_{V_5}) - (1 - u)E_{\mathscr{C}} \sim_{\mathbb{R}} (2 - u)\widetilde{H} + E_{\mathscr{C}}.$$

Then $-K_X - uS$ is nef $\iff u \in [0, 1]$, and $-K_X - uS$ is pseudo-effective $\iff u \in [0, 2]$. If $u \in [1, 2]$, then $P(-K_X - uS) = (2 - u)\pi^*(H_{V_5})$ and $N(-K_X - uS) = (u - 1)E_{\mathscr{C}}$.

First we suppose that $0 \leqslant u \leqslant 1$. Let v be a non-negative real number. Then

$$\left(-K_X - uS\right)\big|_S - vZ \sim_{\mathbb{R}} (2 - u - v)Z + C \sim_{\mathbb{R}} \left(\frac{5}{2} - u - v\right)Z + \frac{1}{2}(\ell_1 + \ell_2 + \ell_3 + \ell_4).$$

Then $(-K_X - uS)|_S - vZ$ is pseudo-effective $\iff v \leqslant \frac{5}{2} - u$. Moreover, if $v \leqslant 2 - u$, then the divisor $(-K_X - uS)|_S - vZ$ is nef. For $2 - u \leqslant v \leqslant \frac{5}{2} - u$, its Zariski decomposition is

$$\underbrace{\left(\frac{5}{2} - u - v\right)(Z + \ell_1 + \ell_2 + \ell_3 + \ell_4)}_{\text{positive part}} + \underbrace{(v + u - 2)(\ell_1 + \ell_2 + \ell_3 + \ell_4)}_{\text{negative part}}.$$

Thus, if $0 \leqslant u \leqslant 1$ and $0 \leqslant v \leqslant \frac{5}{2} - u$,

$$\mathrm{vol}\big((-K_X - uS)|_S - vZ\big) = \begin{cases} 9 - 4u - 4v & \text{if } 0 \leqslant v \leqslant 2 - u, \\ 4\left(\frac{5}{2} - u - v\right)^2 & \text{if } 2 - u \leqslant v \leqslant \frac{5}{2} - u. \end{cases}$$

Now we suppose that $1 \leqslant u \leqslant 2$. Let v be a non-negative real number. Then

$$P\big(-K_X - uS\big)\big|_S - vZ \sim_{\mathbb{R}} (2 - u - v)Z + (2 - u)C$$

$$\sim_{\mathbb{R}} \frac{6 - 3u - 2v}{2}Z + \frac{2 - u}{2}(\ell_1 + \ell_2 + \ell_3 + \ell_4).$$

Then $P(-K_X - uS)|_S - vZ$ is pseudo-effective $\iff v \leqslant \frac{3}{2}(2 - u)$. Moreover, if $v \leqslant 2 - u$, then the divisor $P(-K_X - uS)|_S - vZ$ is nef. Furthermore, if $2 - u \leqslant v \leqslant \frac{3}{2}(2 - u)$, then the Zariski decomposition of this divisor is

$$\underbrace{\frac{6 - 3u - 2v}{2}(Z + \ell_1 + \ell_2 + \ell_3 + \ell_4)}_{\text{positive part}} + \underbrace{(v + u - 2)(\ell_1 + \ell_2 + \ell_3 + \ell_4)}_{\text{negative part}}.$$

Thus, if $1 \leqslant u \leqslant 2$ and $0 \leqslant v \leqslant \frac{3}{2}(2 - u)$,

$$\mathrm{vol}\big(P(-K_X - uS)|_S - vZ\big) = \begin{cases} (2 - u)(10 - 5u - 4v) & \text{if } 0 \leqslant v \leqslant 2 - u, \\ \big(6 - 3u - 2v\big)^2 & \text{if } 2 - u \leqslant v \leqslant \frac{3}{2}(2 - u). \end{cases}$$

Now, using Corollary 1.110, we get $S(W^S_{\bullet,\bullet}; Z) = \frac{171}{208}$, a contradiction. $\qquad\square$

Our next step is the following lemma.

Lemma 5.47 *The curve Z is not contained in $\widetilde{\mathcal{H}}$.*

Proof Suppose that $Z \subset \widetilde{\mathcal{H}}$. Then $\pi(Z)$ is a smooth sextic curve C_γ for some $\gamma \in \mathbb{C}^*$ since $\pi(Z) \neq \mathcal{C}$ and $\pi(Z) \neq L$. In particular, the curve $\pi(Z)$ is disjoint from the line L.

By Corollary 1.110, we have $S(W^{\widetilde{\mathcal{H}}}_{\bullet,\bullet}; Z) \geqslant 1$ since $S_X(S) < 1$ by Theorem 3.17. We compute $S(W^{\widetilde{\mathcal{H}}}_{\bullet,\bullet}; Z)$. Let u be a non-negative real number. Then

$$-K_X - u\widetilde{\mathcal{H}} \sim_{\mathbb{R}} (2 - u)\pi^*(H_{V_5}) - (1 - u)E_{\mathcal{C}} \sim_{\mathbb{R}} (2 - u)\widetilde{\mathcal{H}} + E_{\mathcal{C}}.$$

Then $-K_X - u\widetilde{\mathcal{H}}$ is nef $\iff u \in [0,1]$, and $-K_X - u\widetilde{\mathcal{H}}$ is pseudo-effective $\iff u \in [0,2]$. If $u \in [1,2]$, then $P(-K_X - u\widetilde{\mathcal{H}}) = (2 - u)\pi^*(H_{V_5})$ and $N(-K_X - u\widetilde{\mathcal{H}}) = (u - 1)E_{\mathcal{C}}$.

Recall that $\widetilde{\mathcal{H}} \cong H$, and this surface is non-normal – it is singular along the proper transform of the line L. However, as we already mentioned, the curve Z is contained in its smooth locus. Let $\nu: S \to \widetilde{\mathcal{H}}$ be the normalization, and let $\widetilde{Z} = \nu^{-1}(Z)$. Then

$$S(W^{\widetilde{\mathcal{H}}}_{\bullet,\bullet}; Z) = \frac{3}{26} \int_0^2 \int_0^\infty \mathrm{vol}\Big(\nu^*\big(P(-K_X - u\widetilde{\mathcal{H}})|_{\widetilde{\mathcal{H}}}\big) - v\widetilde{Z}\Big)\,dv\,du$$

by Corollary 1.110 and Remark 1.111. Observe also that the surface S is isomorphic to the surface E_{C_3} described in Remark 5.38. Let us use notations introduced in this remark. Recall that $E_{C_3} \cong \mathbb{F}_1$. As we mentioned in Remark 5.38, we have $v^*(\pi^*(H_{V_5})|_{\widetilde{\mathcal{H}}}) \sim \mathbf{s} + 3\mathbf{f}$ and $v^*(E_{\mathscr{C}}|_{\widetilde{\mathcal{H}}}) \sim \mathbf{s} + \mathbf{f}$. We also observed in Remark 5.38 that $\widetilde{Z} \sim 2(\mathbf{s} + \mathbf{f})$.

Take $v \in \mathbb{R}_{\geqslant 0}$. If $0 \leqslant u \leqslant 1$, then

$$v^*(P(-K_X - u\widetilde{\mathcal{H}})|_{\widetilde{\mathcal{H}}}) - v\widetilde{Z} \sim_\mathbb{R} (1 - 2v)\mathbf{s} + (5 - 2u - 2v)\mathbf{f}.$$

This divisor is pseudo-effective if and only if it is nef, and it is nef if and only if $v \leqslant \frac{1}{2}$. Likewise, if $1 \leqslant u \leqslant 2$, then $v^*(P(-K_X - u\widetilde{\mathcal{H}})|_{\widetilde{\mathcal{H}}}) - v\widetilde{Z} \sim_\mathbb{R} (2 - u - 2v)\mathbf{s} + (6 - 3u - 2v)\mathbf{f}$. This divisor is pseudo-effective $\iff$ it is nef $\iff v \leqslant \frac{2-u}{2}$. Thus we have

$$\mathrm{vol}\left(v^*\left(P(-K_X - u\widetilde{\mathcal{H}})|_{\widetilde{\mathcal{H}}}\right) - v\widetilde{Z}\right)$$
$$= \begin{cases} (1 - 2v)(9 - 4u - 2v) & \text{if } u \in [0,1] \text{ and } 0 \leqslant v \leqslant \frac{1}{2}, \\[2mm] (2 - u - 2v)(10 - 5u - 2v) & \text{if } u \in [1,2] \text{ and } 0 \leqslant v \leqslant \dfrac{2-u}{2}. \end{cases}$$

Now, integrating, we get $S(W^{\widetilde{\mathcal{H}}}_{\bullet,\bullet}; Z) = \frac{47}{208} < 1$, which is a contradiction.   □

By Lemma 1.45, we have $\alpha_{G,Z}(X) < \frac{3}{4}$. Thus, by Lemma 1.42, there is a G-invariant effective $\mathbb{Q}$-divisor D on the threefold X such that $D \sim_\mathbb{Q} -K_X$ and $Z \subseteq \mathrm{Nklt}(X, \lambda D)$ for some positive rational number $\lambda < \frac{3}{4}$.

Lemma 5.48 *If* $\mathrm{Nklt}(X, \lambda D)$ *contains an irreducible surface* S, *then* $S = \widetilde{\mathcal{H}}$.

Proof This follows from the fact that $\overline{\mathrm{Eff}}(X)$ is generated by $\widetilde{\mathcal{H}}$ and $E_{\mathscr{C}}$.   □

Write $D = a\widetilde{\mathcal{H}} + \Delta$, where a is a non-negative rational number, and Δ is an effective $\mathbb{Q}$-divisor whose support does not contain $\widetilde{\mathcal{H}}$. Then $Z \subseteq \mathrm{Nklt}(X, \lambda\Delta)$ by Lemma 5.47. Let $\overline{Z} = \pi(Z)$ and $\overline{\Delta} = \pi(\Delta)$. Then $\overline{Z} \subseteq \mathrm{Nklt}(V_5, \lambda\overline{\Delta})$ and $\overline{\Delta} \sim_\mathbb{Q} (2 - a)H_{V_5}$, so that the locus $\mathrm{Nklt}(V_5, \lambda\overline{\Delta})$ must be connected and one-dimensional by Corollary A.4.

Choose a positive rational number $\mu \leqslant \lambda$, such that $(V_5, \mu\overline{\Delta})$ is strictly log canonical. Then $\overline{Z}$ is a minimal log canonical center of the log pair $(V_5, \mu\overline{\Delta})$ by Corollary A.31 because V_5 does not have G-fixed points. Then Corollary A.21 gives $\deg(Z) = H_{V_5} \cdot \overline{Z} < 4$. Thus, it follows from Corollary 5.39 that $\overline{Z}$ is one of the irreducible curves L, C_2 or $\mathscr{C}$. But $\overline{Z} \neq \mathscr{C}$, $\overline{Z} \neq L$ and $\overline{Z} \neq C_2$ by Lemmas 5.44, 5.45 and 5.46, respectively. The obtained contradiction completes the proof of Proposition 5.43.

5.9 Family №2.21

Smooth Fano threefolds №2.21 are blow ups of the smooth quadric threefold in a twisted quartic curve. It follows from [45] that their automorphism groups are finite with the following exceptions:

(i) a one-dimensional family consisting of threefolds admitting an effective $\mathbb{G}_m$-action,

(ii) a threefold X^a such that $\mathrm{Aut}^0(X^a) \cong \mathbb{G}_a$, it is not K-polystable by Theorem 1.3,

(iii) the K-polystable smooth Fano threefold described in the proof of Lemma 4.15, which admits an effective $\mathrm{PGL}_2(\mathbb{C})$-action.

We already know from Corollary 4.16 that general threefolds in this family are K-stable. In this section, we prove that every smooth Fano threefold №2.21 that admits an effective action of the group $\mathbb{G}_m$ is K-polystable, which would also imply Lemma 4.15.

To describe all smooth Fano threefolds №2.21 that admit an effective $\mathbb{G}_m$-action, we fix the quartic curve $\mathscr{C} \subset \mathbb{P}^4$ given by

$$[u : v] \mapsto [u^4 : u^3v : u^2v^2 : uv^3 : v^4],$$

where $[u : v] \in \mathbb{P}^1$. Let $Q = \{yt - s^2xw + (s^2 - 1)z^2 = 0\} \subset \mathbb{P}^4$, where x, y, z, t, w are coordinates on $\mathbb{P}^4$, and $s \in \mathbb{C} \setminus \{0, \pm 1\}$. Then Q is smooth, and $\mathscr{C} \subset Q$. Fix the $\mathbb{G}_m$-action on $\mathbb{P}^4$ given by

$$[x : y : z : t : w] \mapsto [x : \lambda y : \lambda^2 z : \lambda^3 t : \lambda^4 w], \qquad (5.6)$$

where $\lambda \in \mathbb{G}_m$. Then Q and $\mathscr{C}$ are $\mathbb{G}_m$-invariant, so that we identify $\mathbb{G}_m$ with a subgroup in $\mathrm{Aut}(Q, \mathscr{C})$, which also contains the involution

$$\iota \colon [x : y : z : t : w] \mapsto [w : t : z : y : x].$$

Let Γ be the subgroup in $\mathrm{Aut}(Q, \mathscr{C})$ that is generated by ι and $\mathbb{G}_m$. Then $\Gamma \cong \mathbb{G}_m \rtimes \mu_2$.

Let $\pi \colon X \to Q$ be the blow up of the curve $\mathscr{C}$. Then the $\mathrm{Aut}(Q; \mathscr{C})$-action lifts to X, so that we can identify it with a subgroup in $\mathrm{Aut}(X)$. We see that X admits a $\mathbb{G}_m$-action.

Lemma 5.49 *Every smooth Fano threefold in the family №2.21 that admits an effective action of the group $\mathbb{G}_m$ is isomorphic to X for an appropriate $s \in \mathbb{C} \setminus \{0, \pm 1\}$.*

Proof Let X' be a smooth Fano threefold №2.21 that admits an effective $\mathbb{G}_m$-action. Then X' can be obtained by a $\mathbb{G}_m$-equivariant blow up of a smooth quadric $Q' \subset \mathbb{P}^4$ along a smooth rational quartic curve $\mathscr{C}'$. Now, choosing appropriate coordinates on $\mathbb{P}^4$, we may assume that $\mathscr{C}' = \mathscr{C}$.

The induced $\mathbb{G}_m$-action on the quadric Q is effective. Moreover, this action lifts to an effective action on $\mathbb{P}^4$. Furthermore, keeping in mind that the curve $\mathscr{C}'$ is $\mathbb{G}_m$-invariant, we see that $\mathbb{G}_m$ acts on $\mathbb{P}^4$ as in (5.6). Therefore, since Q' is smooth and $\mathbb{G}_m$-invariant, it is given by $yt - \mu xw + \lambda z^2 = 0$ for some non-zero numbers λ and μ. Since $\mathscr{C}' \subset Q'$, we see that $\lambda = \mu - 1$. Now, letting $\mu = s^2$, we obtain the required assertion. $\qquad\square$

Note that $\mathrm{Aut}^0(X) = \mathrm{Aut}^0(Q; \mathscr{C})$. Moreover, we have the following result:

Lemma 5.50 *If $s \neq \pm\frac{1}{2}$, then $\mathrm{Aut}(Q; \mathscr{C}) = \Gamma$. If $s = \pm\frac{1}{2}$, then $\mathrm{Aut}(Q; \mathscr{C}) \cong$* $\mathrm{PGL}_2(\mathbb{C})$.

Proof Observe that there exists a natural embedding of groups $\mathrm{Aut}(Q; \mathscr{C}) \hookrightarrow$ $\mathrm{Aut}(\mathbb{P}^4; \mathscr{C})$, where the group $\mathrm{Aut}(\mathbb{P}^4; \mathscr{C})$ is isomorphic to $\mathrm{PGL}_2(\mathbb{C})$ and consists of all projective transformations $\phi \colon \mathbb{P}^4 \to \mathbb{P}^4$ given by

$$[x:y:z:t:w] \mapsto [a^4x + 4a^3by + 6a^2b^2z + 4ab^3t + b^4w :$$
$$: a^3cx + (a^3d + 3a^2bc)y + (3a^2bd + 3ab^2c)z + (3ab^2d + b^3c)t + b^3dw :$$
$$: a^2c^2x+(2a^2cd+2abc^2)y+(a^2d^2+4abcd+b^2c^2)z+(2abd^2+2b^2cd)t+b^2d^2w :$$
$$: ac^3x + (3ac^2d + bc^3)y + (3acd^2 + 3bc^2d)z + (ad^3 + 3bcd^2)t + bd^3w :$$
$$c^4x + 4c^3dy + 6c^2d^2z + 4cd^3t + d^4w],$$

where a, b, c and d are some numbers such that $ad - bc = 1$. Hence, to describe $\mathrm{Aut}(Q; \mathscr{C})$, we have to find all such a, b, c, d that $Q = \phi(Q)$. But $\phi^{-1}(Q)$ is given by

$$(1 - 4s^2)a^2c^2xz + (1 - 4s^2)acxt + (abcd - s^2(a^2d^2 + 2abcd + b^2c^2))xw$$
$$+ (4s^2 - 1)a^2c^2y^2 + (a^2d^2 + b^2c^2 - 8abcds^2)yt$$
$$+ (4s^2 - 1)acyz + (1 - 4s^2)bd(ad + bc)yw + (4s^2 - 1)bdzt$$
$$+ ((s^2 - 1)(d^2a^2 + b^2c^2) + (10s^2 - 1)abcd)z^2 - (4s^2 - 1)d^2b^2zw$$
$$+ (4s^2 - 1)b^2d^2t^2 = 0.$$

Keeping in mind that our quadric Q is given by the equation

$$yt - s^2xw + (s^2 - 1)z^2 = 0,$$

we conclude that $Q = \phi(Q)$ if and only if there exists non-zero λ such that

$$ad - bc = 1, \qquad (1 - 4s^2)a^2c^2 = 0, \qquad (1 - 4s^2)ac = 0,$$
$$abcd - s^2(a^2d^2 + 2abcd + b^2c^2) = -\lambda, \qquad (4s^2 - 1)a^2c^2 = 0,$$
$$a^2d^2 + b^2c^2 - 8abcds^2 = \lambda, \qquad (4s^2 - 1)ac = 0, \qquad (1 - 4s^2)bd = 0,$$
$$(4s^2 - 1)bd = 0, \qquad (s^2 - 1)(d^2a^2 + b^2c^2) + (10s^2 - 1)abcd = \lambda(s^2 - 1),$$
$$(4s^2 - 1)d^2b^2 = 0, \qquad (4s^2 - 1)b^2d^2 = 0.$$

Solving this system of equations, we see that one of the following two cases hold:

- $s \neq \pm\frac{1}{2}$ and either $a = d = 0$ or $b = c = 0$,
- $s = \pm\frac{1}{2}$ and a, b, c, d are any numbers with $ad - bc = 1$.

Thus, if $s \neq \pm\frac{1}{2}$, we have $\mathrm{Aut}(Q; \mathscr{C}) = \Gamma$. If $s = \pm\frac{1}{2}$, then $\mathrm{Aut}(Q; \mathscr{C}) = \mathrm{Aut}(\mathbb{P}^4; \mathscr{C})$.   □

Remark 5.51   Let $\epsilon \in \mathbb{C}$, and let Q_ϵ be the quadric threefold in $\mathbb{P}^4$ that is given by

$$\epsilon(t^2 - zw) + 3z^2 - 4yt + xw = 0.$$

Then Q_ϵ is smooth, and Q_ϵ contains $\mathscr{C}$. If $\epsilon \neq 0$, we have $\mathrm{Aut}^0(Q_\epsilon, \mathscr{C}) \cong \mathbb{G}_a$, so that blowing up Q_ϵ along $\mathscr{C}$, we get a threefold X_ϵ in the family №2.21 with $\mathrm{Aut}^0(X_\epsilon) \cong \mathbb{G}_a$. It is easy to see that all threefolds X_ϵ for $\epsilon \neq 0$ are isomorphic to each other (this is the threefold X^a mentioned above). If $\epsilon = 0$, then $Q_\epsilon = Q_0$ is our quadric Q with $s = \pm\frac{1}{2}$, so that blowing up Q_0 along $\mathscr{C}$, we get the unique smooth Fano threefold №2.21 that admits an action of the group $\mathrm{PGL}_2(\mathbb{C})$. We know from Lemma 4.15 that the latter threefold is K-polystable, so that X_ϵ is K-semistable for $\epsilon \neq 0$ by Theorem 1.11.

The group $\mathrm{Aut}(X)$ contains an additional involution $\sigma \notin \mathrm{Aut}(Q; \mathscr{C})$ such that there exists the $\mathrm{Aut}(Q; \mathscr{C})$-equivariant commutative diagram

$$
\begin{array}{ccc}
X & \xrightarrow{\ \sigma\ } & X \\
{\scriptstyle \pi}\big\downarrow & & \big\downarrow{\scriptstyle \pi} \\
Q & \dashrightarrow{\ \tau\ } & Q
\end{array}
$$

and τ is a birational involution that is given by

$$[x : y : z : t : w] \mapsto \left[xz - y^2 : s(xt - yz) : s^2(xw - z^2) : s(yw - zt) : zw - t^2\right].$$

Then $\mathrm{Aut}(X)$ is generated by $\mathrm{Aut}(Q;\mathscr{C})$ and σ, and

$$\begin{cases} \sigma^*(E) \sim 3\pi^*(H) - 2E, \\ \sigma^*(\pi^*(H)) \sim 2\pi^*(H) - E, \end{cases} \tag{5.7}$$

where E is the π-exceptional surface, and H is a hyperplane section of the quadric Q.

Remark 5.52   To see that τ is indeed a birational involution, one can argue as follows. First, substituting $\tau([x : y : z : t : w])$ into the defining equation of the quadric Q, we get

$$s^2\left(yt + (s^2 - 1)wx - s^2z^2\right)\left(yt - s^2xw + (s^2 - 1)z^2\right) = 0,$$

so that $\tau([x : y : z : t : w])$ is contained in Q provided that $[x : y : z : t : w] \in Q \setminus \mathscr{C}$. This shows that τ is a rational selfmap of the quadric Q, which implies that τ is birational. Moreover, let S_6 be the surface in Q cut out by $h = 0$ for $h = xt^2 - 2yzt - xzw + y^2w + z^3$. Then S_6 is singular along the curve $\mathscr{C}$, which implies that τ contracts S_6 to a twisted quartic curve in Q. Now, we observe that S_6 contains $[2 - 4s^2 : 2s : 4s^2 : 2s : 2 - 4s^2]$. If $s \neq \pm\frac{1}{2}$, this point is not contained in $\mathscr{C}$, and it is mapped by τ to $[1 : 1 : 1 : 1 : 1] \in \mathscr{C}$, which implies that $\tau(S_6) = \mathscr{C}$. If $s = \pm\frac{1}{2}$, then S_6 contains $[972 : -189 : 18 : 9 : -8]$, which is mapped by τ to $[2025 : -675 : 225 : -75 : 25] \in \mathscr{C}$, so that $\tau(S_6) = \mathscr{C}$ as well. Moreover, $\tau \circ \tau$ is given by $[x : y : z : t : w] \mapsto [h_0 : h_1 : h_2 : h_3 : h_4]$, where

$$h_0 = -s^2hx,$$

$$h_1 = -s^2hy + s^2\left(yt - s^2xw + (s^2 - 1)z^2\right)(zt - yz),$$

$$h_2 = -s^2hz + s^2\left(yt - s^2xw + (s^2 - 1)z^2\right)(yt + s^2xw - (s^2 + 1)z^2),$$

$$h_3 = -s^2ht - s^2\left(yt - s^2xw + (s^2 - 1)z^2\right)(zt - yw),$$

$$h_4 = -s^2hw.$$

Since $yt - s^2xw + (s^2 - 1)z^2 = 0$ is the defining equation of the quadric threefold Q, this shows that $\tau \circ \tau \colon Q \dashrightarrow Q$ is an identity map, so that τ is a birational involution.

Let $G = \langle \sigma, \Gamma \rangle \subset \mathrm{Aut}(X)$. Then $G \cong \Gamma \times \mu_2 \cong (\mathbb{G}_m \rtimes \mu_2) \times \mu_2$ because σ commutes with the subgroup Γ. In the remaining part of the section, we will show that $\alpha_G(X) \geqslant \frac{3}{4}$, so that X is K-polystable by Theorem 1.51. We start with

Lemma 5.53   *The quadric Q contains neither Γ-invariant lines nor Γ-invariant twisted cubics. Moreover, the only Γ-invariant conics in Q are the conic*

$$\{y = 0, t = 0, (s^2 - 1)z^2 - s^2xw = 0\} \tag{5.8}$$

and the conic

$$\{x = 0, w = 0, yt + (s^2 - 1)z^2 = 0\}. \tag{5.9}$$

Proof All assertions are easy to prove. For instance, if C is a Γ-invariant twisted cubic, then it must be contained in the hyperplane $z = 0$. On the other hand, the smooth quadric surface that is cut out on Q by the equation $z = 0$ does not contain Γ-invariant twisted cubics. We leave the proofs of the remaining assertions to the reader.   □

Let us denote by C_2 and C_2' the irreducible conics (5.8) and (5.9), respectively. Observe that $C_2' \cap \mathscr{C} = \varnothing$, but $C_2 \cap \mathscr{C} = [0:0:0:0:1] \cup [1:0:0:0:0]$, and C_2 intersects the curve $\mathscr{C}$ transversally at these two points. Observe also that the equations

$$\{xt - yz = 0, yw - zt = 0\} \cap Q = \mathscr{C} \cup C_2 \cup \{x = y = z = 0\} \cup \{z = t = w = 0\}.$$

Note also that the lines $\{x = y = z = 0\}$ and $\{z = t = w = 0\}$ are tangent to the curve $\mathscr{C}$ at the points $[0:0:0:0:1]$ and $[1:0:0:0:0]$, respectively.

Let C and C' be the proper transforms on X of the conics C_2 and C_2', respectively. Then the curve C is σ-invariant, while C' is not σ-invariant. Note that Lemma 5.53 implies

Corollary 5.54 *Let C be a G-invariant irreducible curve in X such that $-K_X \cdot C \leqslant 7$. Then C is the conic C, which is given by (5.8).*

Proof We have $\pi^*(H) \cdot C \leqslant 3$ because

$$8 > -K_X \cdot C = \left(\pi^*(H) + \sigma^*(\pi^*(H))\right) \cdot C = \pi^*(H) \cdot C + \sigma^*(\pi^*(H)) \cdot C = 2\pi^*(H) \cdot C,$$

so that $\pi^*(H) \cdot C \leqslant 3$. Thus, we see that either $\pi(C) = C_2$ or $\pi(C) = C_2'$ by Lemma 5.53. On the other hand, we have $0 < -K_X \cdot C = (E + \sigma^*(E)) \cdot C = E \cdot C + \sigma^*(E) \cdot C = 2E \cdot C$, so that $E \cdot C > 0$. Therefore, since $C_2' \cap E = \varnothing$, we have $C = C$, which also follows from the fact that the curve C' is not σ-invariant.   □

Observe that X contains no G-fixed points since Q does not contain Γ-fixed points. Note that $\mathrm{Pic}^G(X) = \mathbb{Z}[-K_X]$, which follows from (5.7). Now, we are ready to prove

Proposition 5.55 $\alpha_G(X) \geqslant \frac{3}{4}$.

Proof Suppose that $\alpha_G(X) < \frac{3}{4}$. Then, arguing as in the proof of Theorem 1.52 and using Lemma 1.42, we see that there exist a rational number $\lambda < \frac{3}{4}$, an irreducible (proper) G-invariant subvariety $Z \subset X$, and a G-invariant effective $\mathbb{Q}$-divisor D on X such that $D \sim_{\mathbb{Q}} -K_X$, the log pair $(X, \lambda D)$ is strictly log canonical, and Z is its unique log canonical center. Then Z is not a point since X has no G-fixed points. Therefore, since $\mathrm{Pic}^G(X) = \mathbb{Z}[-K_X]$, we conclude that Z is a curve.

By Theorem A.20, the curve Z is smooth and rational. Moreover, using Corollary A.21, we see that $-K_X \cdot Z \leqslant 7$, so that $Z = C$ by Corollary 5.54.

We claim that $\mathrm{mult}_C(D) \leqslant 2$. To prove this, let $S = Q \cap \{\alpha(xt - yz) + \beta(yw - tz) = 0\}$, where α and β are general numbers. Then S is a smooth del Pezzo surface of degree 4, so that $|-K_S - C_2|$ is a basepoint free pencil of conics. Let C be a general conic in this pencil, and let $\widetilde{C}$ be its proper transform on X. Then $C \cap C_2$ consists of two distinct points, so that $\widetilde{C} \cap C$ also consists of two distinct points. But $C \not\subset \mathrm{Supp}(D)$, so that we obtain $4 = D \cdot C \geqslant 2\mathrm{mult}_C(D)$, which gives $\mathrm{mult}_C(D) \leqslant 2$ as claimed.

Let $\eta \colon \widehat{X} \to X$ be the blow up of the curve C, and let F be the η-exceptional surface. Then the action of the group G lifts to $\widehat{X}$, and it follows from Lemma A.27 that F has a G-invariant section of the projection $F \to C$. Let us show that G acts on F in such a way that F does not contain any G-invariant sections of the projection $F \to C$.

Let S_y, S_t, S and S' be the surfaces in Q that are cut out by $y = 0$, $t = 0$, $xt - yz = 0$ and $yw - zt = 0$, respectively. Then the following assertions hold:

(i) the surfaces S_y, S_t, S, S' are irreducible;
(ii) the surfaces S_y, S_t, S, S' are $\mathbb{G}_m$-invariant;
(iii) the involution ι swaps the surfaces S_y and S_t;
(iv) the involution ι swaps the surfaces S and S';
(v) $C_2 = S_y \cap S_t \cap S \cap S'$;
(vi) the surfaces S_y, S_t, S, S' are smooth at a general point of the conic C_2;
(vii) any two surfaces among S_y, S_t, S, S' intersect each other transversally at a general point of the conic C_2.

Let $\widetilde{S}_y$, $\widetilde{S}_t$, $\widetilde{S}$, $\widetilde{S}'$ be the proper transforms on X of the surfaces S_y, S_t, S, S', respectively. Then we have $C = \widetilde{S}_y \cap \widetilde{S}_t \cap \widetilde{S} \cap \widetilde{S}'$, the surfaces $\widetilde{S}_y$, $\widetilde{S}_t$, $\widetilde{S}$, $\widetilde{S}'$ are smooth at a general point of the curve C, and any two surfaces among $\widetilde{S}_y$, $\widetilde{S}_t$, $\widetilde{S}$, $\widetilde{S}'$ meet each other transversally at a general point of the curve C. Moreover, we have the following additional two assertions:

(viii) the involution σ swaps the surfaces $\widetilde{S}_y$ and $\widetilde{S}$;
(ix) the involution σ swaps the surfaces $\widetilde{S}_t$ and $\widetilde{S}'$.

Let $\widehat{S}_y$, $\widehat{S}_t$, $\widehat{S}$, $\widehat{S}'$ be the proper transforms on X of the surfaces $\widetilde{S}_y$, $\widetilde{S}_t$, $\widetilde{S}$, $\widetilde{S}'$, respectively. Then each intersection $\widehat{S}_y \cap F$, $\widehat{S}_t \cap F$, $\widehat{S} \cap F$, $\widehat{S}' \cap F$ contains unique irreducible component that is a section of the projection $F \to C$. This gives us 4 sections of the projection $F \to C$, which we denote by $Z_y, Z_t, \mathcal{Z}, \mathcal{Z}'$, respectively. Then $Z_y, Z_t, \mathcal{Z}, \mathcal{Z}'$ are distinct curves because any two surfaces among $\widetilde{S}_y, \widetilde{S}_t, \widetilde{S}, \widetilde{S}'$ intersect each other transversally at a general point of the curve C. Moreover, we have $\iota(Z_y) = Z_t$, $\iota(\mathcal{Z}) = \mathcal{Z}'$, $\sigma(Z_y) = \mathcal{Z}$, $\sigma(Z_t) = \mathcal{Z}'$, and each curve among $Z_y, Z_t, \mathcal{Z}, \mathcal{Z}'$ is $\mathbb{G}_m$-invariant.

Now, using Corollary A.49, we conclude that $F \cong \mathbb{P}^1 \times \mathbb{P}^1$. Then, using Lemma A.48, we conclude that the G-action on F is given by (A.17) for some integers $a > 0$ and b. This implies that F does not contain G-invariant sections, which is a contradiction.

We can prove that F does not contain G-invariant sections without using the explicit description of the G-action on the surface F. Indeed, let $\varrho \colon F \dashrightarrow \mathbb{P}^1$ be the quotient map that is given by the $\mathbb{G}_m$-action on F. Then

- ϱ is G-equivariant,
- $\varrho(Z_y), \varrho(Z_t), \varrho(\mathcal{Z}), \varrho(\mathcal{Z}')$ are four distinct points,
- the group $G/\mathbb{G}_m \cong \mu_2^2$ permutes $\varrho(Z_y), \varrho(Z_t), \varrho(\mathcal{Z}), \varrho(\mathcal{Z}')$ transitively.

Thus, the $G/\mathbb{G}_m$-action on $\mathbb{P}^1$ is effective, which implies that $\mathbb{P}^1$ has no $G/\mathbb{G}_m$-fixed points. Therefore, we conclude that F does not have G-invariant fibers of the rational map ϱ, so that F does not contain G-invariant sections of the projection $F \to C$. $\qquad\square$

Now, using Theorem 1.51, we see that the Fano threefold X is K-polystable, so that a general smooth Fano threefold №2.21 is also K-polystable by Corollary 1.16.

5.10 Family №2.26

Up to isomorphism, there are exactly two smooth Fano threefolds in this family. To describe them, let us recall from [189] the $\mathrm{SL}_2(\mathbb{C})$-action on the unique smooth Fano threefold №1.15, which is described in Example 3.2.

Fix the standard $\mathrm{SL}_2(\mathbb{C})$-action on $W = \mathbb{C}^2$, let $V = \mathrm{Sym}^4(W) \cong \mathbb{C}^5$, and consider the Plücker embedding $\mathrm{Gr}(2, V) \hookrightarrow \mathbb{P}^9 = \mathbb{P}(\bigwedge^2 V)$. As $\mathrm{SL}_2(\mathbb{C})$-representations, we have

$$\bigwedge^2 V^* \cong \mathrm{Sym}^2(W) \oplus \mathrm{Sym}^6(W).$$

We set $A = \mathrm{Sym}^2(W) \subset \bigwedge^2 V^*$ in this decomposition, and note that every non-zero form in A has rank 4. Let $V_5 = \mathrm{Gr}(2, V) \cap \mathbb{P}(A^\perp)$. Then V_5 is the unique smooth Fano threefold in the family №1.15. By construction, this threefold is $\mathrm{SL}_2(\mathbb{C})$-invariant, so that it carries an $\mathrm{SL}_2(\mathbb{C})$-action. In fact, this action is effective, and $\mathrm{Aut}(V_5) \cong \mathrm{PGL}_2(\mathbb{C})$.

Now, let us describe the Hilbert scheme of lines in V_5, see [109, Theorem I], [189, Proposition 2.20], [190, Proposition 3.23]. This scheme can be naturally identified with $\mathbb{P}^2 = \mathbb{P}(A)$ equipped with the induced $\mathrm{SL}_2(\mathbb{C})$-action. Concretely, given a non-zero element $a \in A$, the kernel of a is 1-dimensional, generated by a vector $v_a \in V$. The vector v_a induces a global section of the quotient bundle $V/\mathscr{U}$, where $\mathscr{U}$ is the restriction to V_5 of the tautological vector bundle of the Grassmannian $\mathrm{Gr}(2, V)$. The schematic zero locus of this global section is precisely the line L_a in V_5 associated to a. Using this identification, we can describe the $\mathrm{SL}_2(\mathbb{C})$-orbits in $\mathbb{P}(A)$ as follows:

- the open GIT-polystable orbit,
- the unique invariant conic in $\mathbb{P}(A)$ given by the GIT-unstable orbit in A.

Let L be a line in V_5, then it follows from [109, §1] that there are two possibilities for the normal bundle $\mathcal{N}_{L/V_5}$. Namely, if L is contained in the open $\mathrm{SL}_2(\mathbb{C})$-orbit in $\mathbb{P}(A)$, then $\mathcal{N}_{L/V_5} \cong \mathcal{O}_{\mathbb{P}^1} \oplus \mathcal{O}_{\mathbb{P}^1}$ and we say that L is a *good line*. If L is contained in the invariant conic in $\mathbb{P}(A)$, then we have $\mathcal{N}_{L/V_5} \cong \mathcal{O}_{\mathbb{P}^1}(-1) \oplus \mathcal{O}_{\mathbb{P}^1}(1)$ and we say that L is a *bad line*. Up to the $\mathrm{SL}_2(\mathbb{C})$-action, the threefold V_5 contains exactly one good line and exactly one bad line.

Let $\sigma \colon X \to V_5$ be the blow up of the line L. Then X is one of two smooth Fano threefolds №2.26. In both cases, there exists the commutative diagram

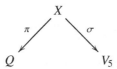

where Q is a smooth quadric in $\mathbb{P}^4$, and π is a blow up of a twisted cubic curve C_3. Let H be the hyperplane section of Q that contains C_3. Then H is smooth if and only if L is a good line. Let $\widetilde{H}$ be the proper transform on X of the surface H, and let F be the π-exceptional divisor. Then $\widetilde{H}$ is the σ-exceptional surface and $2\sigma(F) \sim -K_{V_5}$. Moreover, the surface $\sigma(F)$ is singular along the line L. Furthermore, if L is a bad line, then $\mathrm{Aut}^0(X) \cong \mathbb{G}_a \rtimes \mathbb{G}_m$ by [45, Lemma 6.5], and X is not K-polystable by Theorem 1.3. In fact, we can say more:

Lemma 5.56 *Suppose that L is a bad line. Then X is K-unstable.*

Proof Let Z be the fiber of $F \to C_3$ over the point $\mathrm{Sing}(H)$, let $f\colon \widehat{X} \to X$ be the blow up of the curve Z, and let E be the f-exceptional divisor. Let us show that $\beta(E) < 0$.

Let s_E and l_E be the negative section and a ruling of the surface $E \cong \mathbb{F}_1$, respectively. We denote by $\widehat{H}$ and $\widehat{F}$ the proper transforms on $\widehat{X}$ of the surfaces $\widetilde{H}$ and F, respectively. Then $-E|_E \sim s_E + l_E$, $\widehat{H}|_E \sim s_E + 2l_E$, $\widehat{F}|_E \sim s_E$ and $f^*(-K_X)|_E \sim l_E$.

Now, we observe that $\widetilde{H} \cong \mathbb{F}_2$ and $F \cong \mathbb{F}_3$. Let $s_{\widetilde{H}}$, s_F, $l_{\widetilde{H}}$, l_F be the negative sections and rulings of these surfaces, respectively. Then $F|_{\widetilde{H}} = s_{\widetilde{H}} + \widetilde{C}_3$, where C_3 is the proper transform via the induced birational map $\widetilde{H} \to H$. Moreover, we have $\widetilde{C}_3 \sim s_{\widetilde{H}} + 3l_{\widetilde{H}}$, $-F|_F \sim s_F - 2l_F$, $\widetilde{H}|_F \sim s_F + l_F$, $-\widetilde{H}|_{\widetilde{H}} \sim s_{\widetilde{H}} + l_{\widetilde{H}}$, $-K_X|_{\widetilde{H}} \sim s_{\widetilde{H}} + 3l_{\widetilde{H}}$, $-K_X|_F \sim s_F + 7l_F$. Observe that $Z = s_{\widetilde{H}}$, so that we have $Z \sim l_F$ on the surface F.

Note that $\widehat{H} \cong \widetilde{H}$ and $\widehat{F} \cong F$. Let us denote by $s_{\widehat{H}}$, $s_{\widehat{F}}$, $l_{\widehat{H}}$, $l_{\widehat{F}}$ the negative sections and rulings of the surfaces $\widehat{H} \cong \mathbb{F}_2$ and $\widehat{F} \cong \mathbb{F}_3$, respectively. Then $-\widehat{F}|_{\widehat{F}} \sim s_{\widehat{F}} - l_{\widehat{F}}$, $\widehat{H}|_{\widehat{F}} \sim s_{\widehat{F}}$, $E|_{\widehat{F}} \sim l_{\widehat{F}}$, $-\widehat{H}|_{\widehat{H}} \sim 2s_{\widehat{H}} + l_{\widehat{H}}$, $\widehat{F}|_{\widehat{H}} \sim s_{\widehat{H}} + 3l_{\widehat{H}}$, $E|_{\widehat{H}} \sim s_{\widehat{H}}$. Moreover, we also have $f^*(-K_X)|_{\widehat{H}} \sim s_{\widehat{H}} + 3l_{\widehat{H}}$ and $f^*(-K_X)|_{\widehat{F}} \sim s_{\widehat{F}} + 7l_{\widehat{F}}$.

Take $x \in \mathbb{R}_{\geqslant 0}$. Then $f^*(-K_X) - xE \sim_{\mathbb{R}} 3\widehat{H} + 2\widehat{F} + (5-x)E$, which implies that the divisor $f^*(-K_X) - xE$ is psuedo-effective if and only if $x \leqslant 5$. Moreover, intersecting this divisor with s_E, l_E, $s_{\widehat{H}}$, $s_{\widehat{F}}$, $l_{\widehat{H}}$, $l_{\widehat{F}}$, we see that $f^*(-K_X) - xE$ is nef for $x \in [0,1]$. Thus, if $x \in [0,1]$, then

$$\mathrm{vol}\big(f^*(-K_X) - xE\big) = -K_X^3 - 3x^2\big(-K_X \cdot Z\big) - x^3\big(-\deg(\mathcal{N}_{Z/X})\big) = 34 - 3x^2 - x^3.$$

Similarly, if $x \in [1,3]$, we see that the Zariski decomposition of the divisor $f^*(-K_X) - xE$ is

$$f^*(-K_X) - xE \sim_{\mathbb{R}} \underbrace{\frac{7-x}{2}\widehat{H} + 2\widehat{F} + (5-x)E}_{\text{positive part}} + \underbrace{\frac{1}{2}(x-1)\widehat{H}}_{\text{negative part}}.$$

Thus, if $x \in [1,3]$, then $\mathrm{vol}(f^*(-K_X) - xE) = \frac{1}{4}(x^3 - 9x^2 - 21x + 149)$. Finally, if $x \in [3,5]$, then the Zariski decomposition of the divisor $f^*(-K_X) - xE$ is

$$f^*(-K_X) - xE \sim_{\mathbb{R}} \underbrace{(5-x)\big(\widehat{F} + \widehat{H} + E\big)}_{\text{positive part}} + \underbrace{(x-2)\widehat{H} + (x-3)\widehat{F}}_{\text{negative part}},$$

so that $\mathrm{vol}(f^*(-K_X) - xE) = (5-x)^3$. Now, integrating, we see that $\beta(E) = -\frac{31}{136}$, which implies that X is K-unstable by Theorem 1.19. $\qquad\square$

Now, we suppose that L is a good line. Then it follows from [45] that $\text{Aut}^0(X) \cong \mathbb{G}_m$. Moreover, one can show that $\text{Aut}(X) \cong \mathbb{G}_m \rtimes \mu_2$. In the remaining part of the section, we will show that X is K-semistable and not K-polystable, i.e. X is strictly semistable. To do this, we may assume that $Q = \{x_0 x_3 - x_1 x_2 + x_4^2 = 0\} \subset \mathbb{P}^4$, $H = \{x_4 = 0\} \cap Q$, and

$$C_3 = \left\{ x_0 x_3 - x_1 x_2 = 0, x_0 x_2 - x_1^2 = 0, x_1 x_3 - x_2^2 = 0, x_4 = 0 \right\},$$

where x_0, x_1, x_2, x_3, x_4 are coordinates in $\mathbb{P}^4$. Let Q be the family of quadrics given by

$$x_0 x_3 - x_1 x_2 + t \cdot x_4^2 = 0,$$

where $t \in \mathbb{A}^1$. Let $\widehat{Q}$ be its special member – the singular quadric $x_0 x_3 - x_1 x_2 = 0$. Now, blowing up Q along $C_3 \times \mathbb{A}^1$, we obtain a special test configuration $\mathcal{X} \to \mathbb{A}^1$. Its general fiber is X. Let Y be its special fiber. Then Y is a Fano variety, it has one isolated ordinary double point since Y is the blow up of the quadric $\widehat{Q}$ in the curve C_3, which does not pass through $\text{Sing}(\widehat{Q})$.

Lemma 5.57 *The Fano variety Y is K-polystable.*

Proof Let $f : Y \to \widehat{Q}$ be the blow up of the curve C_3, and let E be its exceptional surface. Observe that $\widehat{Q}$ is a $\mathbb{T}$-variety of complexity one. Namely, the quadric $\widehat{Q}$ admits an effective action of the group $G = \mathbb{G}_m^2 \rtimes \mu_2$, where the $\mathbb{G}_m^2$-action is given by

$$(t_1, t_2).[x_0 : x_1 : x_2 : x_3 : x_4] = [x_0 : t_1 x_1 : t_1^2 x_2 : t_1^3 x_3 : t_2 x_4]$$

and μ_2 acts via the biregular involution $\sigma : [x_0 : x_1 : x_2 : x_3 : x_4] \mapsto [x_3 : x_2 : x_1 : x_0 : x_4]$. Since the curve C_3 is invariant under the G-action, the G-action lifts to the variety Y. Let us use the technique of Section 1.3 and Theorem 1.31 to show that Y is K-polystable. In the following, we will use notations introduced in this section.

Consider the two one-parameter subgroups

$$w_1 : \mathbb{G}_m \to \mathbb{G}_m^2; \quad t \to (t, 1),$$
$$w_2 : \mathbb{G}_m \to \mathbb{G}_m^2; \quad t \to (1, t).$$

These form a basis of N and σ acts on N via $w_1 \mapsto -w_1$ and $w_2 \mapsto w_2$. Let T be the prime divisor in $\widehat{Q}$ given by $x_4 = 0$, and let $\widetilde{T}$ be its strict transform on Y. Then w_2 acts trivially on it, so that T is a horizontal divisor with $w_T = w_2$.

Let $\pi : Y \dashrightarrow \mathbb{P}^1$ be the quotient map by $\mathbb{G}_m^2$. Then $\pi \circ f^{-1}$ is given by

$$[x_0 : x_1 : x_2 : x_3 : x_4] \mapsto [x_0 x_2 : x_1^2].$$

Note that on the quadric $\widehat{Q}$ we have $[x_0x_2 : x_1^2] = [x_2^2 : x_1x_3]$ whenever both are defined. Let F be the fiber of the quotient map over $[1 : 1]$. Then $F = \{x_0x_2 - x_1^2 = x_1x_3 - x_2^2 = 0\}$. Then $C_3 = F \cap T$. Since the domain of $\pi \circ f^{-1}$ intersects C_3, we have $\pi(E) = [1 : 1]$.

The involution σ acts on $\mathbb{P}^1$ by sending $[y_0 : y_1]$ to $[y_1 : y_0]$. There are only two σ-fixed points: $[1 : 1]$ and $[-1 : 1]$. Moreover, the fiber of π over the point $[-1 : 1]$ is integral, and the fiber over $[1 : 1]$ consists of the surfaces E and $\widetilde{F}$. Hence, by Proposition 1.38, it is sufficient to show that $\mathrm{Fut}_Y = 0$ and $\beta(E) > 0$.

Let us compute $\beta(E)$. Take $x \in \mathbb{R}_{\geqslant 0}$. Then $-K_Y - xE \sim_{\mathbb{R}} (2 - x)E + 3\widetilde{T}$, which implies that $-K_Y - xE$ is pseudo-effective $\Longleftrightarrow x \leqslant 2$. Similarly, it is nef $\Longleftrightarrow x \leqslant \frac{1}{2}$. Moreover, if $2 \geqslant x > \frac{1}{2}$, then the ample model of the divisor $-K_Y - xE$ is given by the contraction of the surface $\widetilde{T} \cong \mathbb{P}^1 \times \mathbb{P}^1$ to a curve. Using this, we compute

$$\mathrm{vol}\big(- K_Y - xE\big) = \begin{cases} 7x^3 - 6x^2 + 75x + 34 & \text{if } 0 \leqslant x \leqslant \dfrac{1}{2}, \\ 5(2 - x)^3 & \text{if } \dfrac{1}{2} \leqslant x \leqslant 2. \end{cases}$$

Integrating, we get $S_X(E) = \frac{305}{544}$, so that $\beta(E) = 1 - S_X(E) = \frac{239}{544} > 0$.

Similarly, we see that $\beta(\widetilde{T}) = 0$. Indeed, if $0 \leqslant x \leqslant 1$, then $-K_Y - x\widetilde{T}$ is nef, so that

$$\begin{aligned} \mathrm{vol}\big(- K_Y - x\widetilde{T}\big) &= \big(- K_Y - x\widetilde{T}\big)^3 = \big(f^*((3 - x)T) + (x - 1)E\big)^3 \\ &= 2(3 - x)^3 + 3(3 - x)(x - 1)^2 f^*(T) \cdot E^2 + (x - 1)^3 E^3 \\ &= 2(3 - x)^3 - 9(3 - x)(x - 1)^2 - 7(x - 1)^3 = 34 - 6x^2 - 12x. \end{aligned}$$

Likewise, if $1 \leqslant x \leqslant 3$, then the ample model of this divisor is the quadric $\widehat{Q}$, which implies that $\mathrm{vol}(-K_Y - x\widetilde{T}) = 2(3 - x)^2$ since $f_*(-K_Y - x\widetilde{T}) \sim_{\mathbb{R}} (3 - x)T$. Now, integrating, we get $S_X(\widetilde{T}) = 1$, so that $\beta(\widetilde{T}) = 0$.

The Futaki character of Y is trivial. Indeed, since Fut_Y is σ-invariant, $\mathrm{Fut}_Y(\lambda_{w_1}) = 0$ by Lemma 1.29. Hence, it remains to show that $\mathrm{Fut}_Y(\lambda_{w_2}) = 0$. Since $\widetilde{T}$ is a horizontal divisor with $w_{\widetilde{T}} = w_2$, we have $\mathrm{Fut}_Y(\lambda_{w_2}) = \beta(\widetilde{T}) = 0$ by Corollary 1.37. This shows that Y is K-polystable. $\square$

Now, using Corollary 1.13 and the existence of the test configuration for our smooth Fano threefold X with special K-polystable fiber Y, we obtain

Corollary 5.58 *The Fano threefold X is strictly K-semistable.*

Therefore, the family №2.26 does not contain K-polystable threefolds.

5.11 Family №3.2

Now we construct a special K-stable smooth Fano threefold in family №3.2. By Theorem 1.11, this will imply that general threefolds in this family are K-stable since all smooth threefolds in this family have finite automorphism groups [45].

Let $S = \mathbb{P}^1 \times \mathbb{P}^1$, let H be the divisor of degree $(1,1)$ on S, let

$$\mathbb{P} = \mathbb{P}\big(O_S \oplus O_S(-H) \oplus O_S(-H)\big),$$

let $[s_0 : s_1; t_0 : t_1; u_0 : u_1 : u_2]$ be homogeneous coordinates on the fourfold $\mathbb{P}$ such that $\mathrm{wt}(s_0) = (1,0,0)$, $\mathrm{wt}(s_1) = (1,0,0)$, $\mathrm{wt}(t_0) = (0,1,0)$, $\mathrm{wt}(t_1) = (0,1,0)$, $\mathrm{wt}(u_0) = (0,0,1)$, $\mathrm{wt}(u_1) = (1,1,1)$ and $\mathrm{wt}(u_2) = (1,1,1)$, and let $\pi \colon \mathbb{P} \to S$ be the natural projection. Then the projection π is given by

$$[s_0 : s_1; t_0 : t_1; u_0 : u_1 : u_2] \mapsto [s_0 : s_1; t_0 : t_1],$$

where we consider $[s_0 : s_1; t_0 : t_1]$ as coordinates on S. Let G be the subgroup in $\mathrm{Aut}(\mathbb{P})$ that is generated by the following two transformations:

$$A_1 : [s_0 : s_1; t_0 : t_1; u_0 : u_1 : u_2] \mapsto [s_1 : s_0; t_1 : t_0; u_0 : u_2 : u_1],$$
$$A_2 : [s_0 : s_1; t_0 : t_1; u_0 : u_1 : u_2] \mapsto [s_0 : -is_1; t_0 : -t_1; u_0 : u_1 : iu_2],$$

where $i = \sqrt{-1}$. Observe that G acts naturally and faithfully on S, and that π is G-equivariant. Note also that

(i) S does not contain G-fixed points,

(ii) S does not contain G-invariant curves of degree $(1,0)$, $(0,1)$ or $(1,1)$.

In particular, the fourfold $\mathbb{P}$ does not contain G-fixed points either.

Let L be the tautological line bundle on $\mathbb{P}$ over S, i.e. the line bundle of degree $(2,3,2)$, and let X be the divisor in the linear system $|L^{\otimes 2} \otimes O_S(2,3)|$ that is given by

$$t_0 u_1^2 + t_1 u_2^2 + u_0\big(s_0 t_0^2 u_1 + s_1 t_1^2 u_2 + s_0 t_1^2 u_1 + s_1 t_0^2 u_2\big) + u_0^2\big(s_0^2 t_0^3 + s_1^2 t_1^3 + s_0^2 t_0 t_1^2 + s_1^2 t_0^2 t_1\big)$$
$$= 0.$$

Then X is a smooth Fano threefold №3.2, it is G-invariant, and G acts faithfully on it, so that we can identify G with a subgroup in $\mathrm{Aut}(X)$.

Let $\mathscr{S}$ be the surface cut out by $u_0 = 0$ in X, let $\varpi \colon X \to S$ be the morphism induced by π, let $\mathrm{pr}_1 \colon S \to \mathbb{P}^1$ and $\mathrm{pr}_2 \colon S \to \mathbb{P}^1$ be the projections to the first and the second factors, respectively. Then $\mathscr{S} \cong \mathbb{P}^1 \times \mathbb{P}^1$, ϖ is a conic bundle, and the following commutative diagram is G-equivariant:

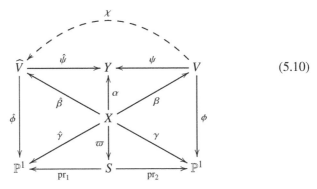

$$(5.10)$$

where Y is a non-$\mathbb{Q}$-factorial Fano threefold with one isolated ordinary double point such that $-K_Y^3 = 16$ and $\mathrm{Pic}(Y) = \mathbb{Z}[-K_Y]$, α is the contraction of the surface $\mathscr{S}$ to the singular point of Y, β and $\hat{\beta}$ are birational morphisms that contract $\mathscr{S}$ to smooth rational curves, ψ and $\hat{\psi}$ are small resolutions of the threefold Y, χ is the Atiyah flop in the curve $\beta(\mathscr{S})$, ϕ is a fibration into quadric surfaces, $\hat{\phi}$ is a fibration into del Pezzo surfaces of degree 4, and γ and $\hat{\gamma}$ are fibrations into del Pezzo surfaces of degree 3 and 6, respectively.

The diagram (5.10) first appeared in [121, Proposition 3.8]. Note that (5.10) extends the diagram (4.1) in Section 4.1 for another singular Fano threefold in the family №1.8.

Lemma 5.59 $\alpha_G(X) \geqslant 1$.

Proof Let us apply Theorem 1.52 with $\mu = 1$. Let F and $\widehat{F}$ be general fibers of the del Pezzo fibrations γ and $\hat{\gamma}$, respectively. Then $-K_X \sim \mathscr{S} + F + 2\widehat{F}$, and it follows from [93, 156] that the cone $\overline{\mathrm{Eff}}(X)$ is generated by the surfaces $\mathscr{S}$, F, $\widehat{F}$. Thus, Theorem 1.52(1) cannot be satisfied because the pencil $|\widehat{F}|$ does not contain G-invariant surfaces since S does not contain G-invariant curves of degree $(1,0)$. Similarly, we see that X does not contain a G-invariant irreducible curve C such that $F \cdot C \leqslant 1$ and $\widehat{F} \cdot C \leqslant 1$, because S does not contain G-fixed points, and S does not contain G-invariant curves of degree $(1,0)$, $(0,1)$ and $(1,1)$. Finally, recall that X does not contain G-fixed points. Thus, we have $\alpha_G(X) \geqslant 1$ by Theorem 1.52. □

Thus, the threefold X is K-stable by Theorem 1.48 and Corollary 1.5. Hence, a general Fano threefold in family №3.2 is also K-stable by Theorem 1.11. In fact, Lemma 5.59 also implies that general Fano threefolds in family №1.8 are K-stable, which we already know from Section 4.1. Indeed, since G acts faithfully on Y, V and $\widehat{V}$, we can identify G with the subgroups in the automorphism groups of these threefolds. Then $\alpha_G(Y) = \alpha_G(V) = \alpha_G(\widehat{V})$ by Lemma 1.47. On the other hand, Lemma 5.59 gives

Corollary 5.60 $\alpha_G(V) \geqslant 1$.

Proof Suppose that $\alpha_G(V) < 1$. Then there is an effective G-invariant $\mathbb{Q}$-divisor D on the threefold V such that $D \sim_\mathbb{Q} -K_V$, and the log pair $(V, \lambda D)$ is not Kawamata log terminal for some positive rational number $\lambda < 1$. Let us seek a contradiction.

Observe that $-K_V \sim \beta(F) + 2\beta(\widehat{F})$, the cone $\overline{\mathrm{Eff}}(X)$ is generated by the surfaces $\beta(F)$ and $\beta(\widehat{F})$, and the pencil $|\beta(\widehat{F})|$ does not contain G-invariant surfaces. This shows that $\mathrm{Nklt}(V, \lambda D)$ does not contain surfaces. Moreover, the pencil $|\beta(F)|$ does not have G-invariant surfaces, so that, in particular, the threefold V does not have G-fixed points. Thus, applying Corollary A.12, we see that the locus $\mathrm{Nklt}(V, \lambda D)$ consists of a smooth rational curve C such that $\beta(F) \cdot C = 1$.

Suppose that $C \neq \beta(\mathscr{S})$. Let $\widehat{C}$ and $\widehat{D}$ be the proper transforms of the curve C and divisor D on the threefold $\widehat{V}$, respectively. Then $\widehat{C}$ is contained in the locus $\mathrm{Nklt}(\widehat{V}, \lambda \widehat{D})$, which does not contain surfaces since χ is a flop. Applying Corollary A.12 again, we see that $\hat{\beta}(F) \cdot \widehat{C} = 1$. Thus, $\varpi \circ \beta^{-1}(C)$ is a G-invariant curve in $S \cong \mathbb{P}^1 \times \mathbb{P}^1$ of degree $(1,1)$, which is impossible since S does not contain G-invariant curves of degree $(1,1)$.

Thus, we see that $C = \beta(\mathscr{S})$. Let $\overline{D}$ be the proper transform of the divisor D on X. Then $\overline{D} + (\mathrm{mult}_C(D) - 1)\mathscr{S} \sim_\mathbb{Q} -K_X$, and the log pair $(X, \overline{D} + (\mathrm{mult}_C(D) - 1)\mathscr{S})$ is not log canonical. Since $\mathrm{mult}_C(D) > 1$ by Lemma A.1, this contradicts Lemma 5.59. $\square$

Thus, we have $\alpha_G(Y) \geqslant 1$, so that it follows from Theorem 1.48 and Corollary 1.5 that Y is K-stable since the group $\mathrm{Aut}(Y) \cong \mathrm{Aut}(X)$ is finite. As noted above, Y has one ordinary double point, and in particular is terminal, hence Y has a smoothing to a smooth Fano threefold №1.8 by [168, Theorem 11] and [122, Theorem 1.4]. Thus, we conclude that general Fano threefold in family №1.8 is K-stable by Theorem 1.11, which we already knew by Example 4.9.

5.12 Family №3.3

Let X be the threefold

$$\{x_1 x_2^2 + y_1 y_2^2 + z_1 z_2^2 + w_1 w_2^2 = 0, \ x_1^2 + y_1^2 + z_1^2 + w_1^2 = 0, \ x_2 + y_2 + z_2 + w_2 = 0\}$$
$$\subset \mathbb{P}^3 \times \mathbb{P}^3,$$

where x_1, y_1, z_1, w_1 are coordinated on the first factor of $\mathbb{P}^3 \times \mathbb{P}^3$, and x_2, y_2, z_2, w_2 are coordinated on the second factor of $\mathbb{P}^3 \times \mathbb{P}^3$. Then X is smooth Fano threefold №3.3. Indeed, the threefold X is a divisor in $\mathbb{P}^1 \times \mathbb{P}^1 \times \mathbb{P}^2$ of degree $(1, 1, 2)$, where we identify

- $\mathbb{P}^1 \times \mathbb{P}^1$ with the quadric $x_1^2 + y_1^2 + z_1^2 + w_1^2 = 0$ in the first factor of $\mathbb{P}^3 \times \mathbb{P}^3$,
- $\mathbb{P}^2$ with the hyperplane $x_2 + y_2 + z_2 + w_2 = 0$ in the second factor of $\mathbb{P}^3 \times \mathbb{P}^3$.

Observe that we have the commutative diagram

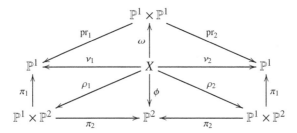

where ρ_1 and ρ_2 are blow ups of smooth curves of genus 3, ϕ is a (non-standard) conic bundle whose discriminant curve is a smooth plane quartic curve, ω is a (standard) conic bundle whose discriminant curve is a smooth curve of bi-degree $(3,3)$, v_1 and v_2 are fibrations into del Pezzo surfaces of degree 5, π_1 and π_2 are natural projections, and pr_1 and pr_2 are projections to the first and second factor, respectively.

Let $G = \mathfrak{S}_4$. Then X admits a natural faithful action of the group G that is given by the (simultaneous) permutations of coordinates on both factors of $\mathbb{P}^3 \times \mathbb{P}^3$. Observe that there are no G-fixed points on X, and that the conic bundles ω and ϕ are G-equivariant. The G-action on $\mathbb{P}^1 \times \mathbb{P}^1$ permutes the two rulings. Thus, we have $\mathrm{Pic}^G(X) \cong \mathbb{Z}^2$. We identify G with a subgroup in $\mathrm{Aut}(X)$.

Lemma 5.61 $\alpha_G(X) \geqslant 1$.

Proof Let S be any G-invariant surface $S \subset X$ such that $-K_X \sim_{\mathbb{Q}} aS + \Delta$, where $a \in \mathbb{Q}_{\geqslant 0}$ and Δ is an effective $\mathbb{Q}$-divisor on X. Then $a \leqslant 1$ because

$$aS + \Delta \sim_{\mathbb{Q}} -K_X \sim v_1^*(\mathcal{O}_{\mathbb{P}^1}(1)) + v_2^*(\mathcal{O}_{\mathbb{P}^1}(1)) + \phi^*(\mathcal{O}_{\mathbb{P}^2}(1)),$$

and $S \sim v_1^*(\mathcal{O}_{\mathbb{P}^1}(m)) + v_2^*(\mathcal{O}_{\mathbb{P}^1}(m)) + \phi^*(\mathcal{O}_{\mathbb{P}^2}(n))$ for some non-negative integers m and n.

Now, we suppose that $\alpha_G(X) < 1$. Since X does not contain G-fixed points, it follows from Lemma A.30 that X contains an effective G-invariant $\mathbb{Q}$-divisor $D \sim_{\mathbb{Q}} -K_X$ and a smooth G-invariant irreducible rational curve Z such that the log pair $(X, \lambda D)$ is strictly log canonical for some positive rational number $\lambda < 1$, and Z is the unique log canonical center of the log pair $(X, \lambda D)$. Applying Corollary A.12 to the del Pezzo fibrations v_1 and v_2, we get $Z \cdot v_1^*(\mathcal{O}_{\mathbb{P}^1}(1)) \leqslant 1$ and $Z \cdot v_2^*(\mathcal{O}_{\mathbb{P}^1}(1)) \leqslant 1$. But $\mathbb{P}^1 \times \mathbb{P}^1$ has no G-fixed points, and $\mathrm{Pic}^G(\mathbb{P}^1 \times \mathbb{P}^1) \cong \mathbb{Z}$, so that $\omega(Z)$ is a curve of degree $(1,1)$. Since $\omega(Z)$ is G-invariant, it is given by

$x_1 + y_1 + z_1 + w_1 = 0$. Likewise, applying Corollary A.13 to the conic bundle ϕ, we see that $\phi(Z)$ is a conic because $\mathbb{P}^2$ does not have G-invariant lines and G-fixed points. Moreover, since $\mathbb{P}^2$ contains a unique G-invariant conic, we see that $\phi(Z)$ is given by $x_2^2 + y_2^2 + z_2^2 + w_2^2 = 0$. Then Z is contained in the support of the subscheme

$$\left\{ \begin{array}{l} x_1 x_2^2 + y_1 y_2^2 + z_1 z_2^2 + w_1 w_2^2 = 0, \\ x_1^2 + y_1^2 + z_1^2 + w_1^2 = 0, x_2^2 + y_2^2 + z_2^2 + w_2^2 = 0, \\ x_2 + y_2 + z_2 + w_2 = 0, x_1 + y_1 + z_1 + w_1 = 0. \end{array} \right\} \subset \mathbb{P}^3 \times \mathbb{P}^3.$$

Denote the later subscheme by C. Using the following Magma code

```
Q:=RationalField();
PxP<x1,y1,z1,x2,y2,z2>:=ProductProjectiveSpace(Q,[2,2]);
C:=Scheme(PxP,[x1*x2^2+y1*y2^2+z1*z2^2-(x1+y1+z1)*(x2+y2+z2)^2,
            x1^2+y1^2+z1^2+(x1+y1+z1)^2,
            x2^2+y2^2+z2^2+(x2+y2+z2)^2]);
IsNonsingular(C);
IsIrreducible(C);
Dimension(C);
```

we conclude that the subscheme C is reduced, irreducible, one-dimensional, and smooth. Then $Z = C$, and C is a smooth (hyperelliptic) curve of genus 3, which is absurd since the curve Z is rational. The obtained contradiction shows that $\alpha_G(X) \geqslant 1$.   □

Therefore, the threefold X is K-stable by Theorem 1.48 and Corollary 1.5 because the group $\mathrm{Aut}(X)$ is finite [45]. The general Fano threefold in family №3.3 is also K-stable.

5.13 Family №3.4

In Section 4.5, we presented one K-stable Fano threefold №3.4, so that general threefolds in this family are K-stable by Theorem 1.11. In this section, we prove the K-stability of another smooth Fano threefold №3.4. The proof is more involved in this case, but we believe that it can be used to prove that all smooth Fano threefolds in the family №3.4 are K-stable.

Using notations of [188, Section 2.2], consider the scroll $\mathbb{F}_1 = \mathbb{F}(0, 1)$ with coordinates t_0 and t_1 of weight $(1, 0)$, and coordinates u_0 and u_1 of weights $(-1, 1)$ and $(0, 1)$, respectively. The blow up morphism $\beta \colon \mathbb{F}_1 \to \mathbb{P}^2$ is given by $[t_0 : t_1; u_0 : u_1] \mapsto [u_1 : t_0 u_0 : t_1 u_0]$, so that it contracts the curve $u_0 = 0$ to the point $[1 : 0 : 0]$, the projection $\upsilon \colon \mathbb{F}_1 \to \mathbb{P}^1$ is given by $[t_0 : t_1; u_0 : u_1] \mapsto [t_0 : t_1]$, and the curve $u_1 = 0$ is the preimage of a line in $\mathbb{P}^2$ that does not contain the point $[1 : 0 : 0]$. We fix coordinates $[s_0 : s_1]$ on the first factor of $\mathbb{P}^1 \times \mathbb{F}_1$. Let $\mathbb{P}^1 \times \mathbb{F}_1 \to \mathbb{P}^1 \times \mathbb{P}^1$ be the morphism $([s_0 : s_1], [t_0 : t_1; u_0 : u_1]) \mapsto ([s_0 : s_1], [t_0 : t_1])$, and we consider $([s_0 : s_1], [t_0 : t_1])$ also as coordinates on $\mathbb{P}^1 \times \mathbb{P}^1$. Let μ_1 be the transformation in $\mathrm{Aut}(\mathbb{P}^1 \times \mathbb{F}_1)$ given by $([s_0 : s_1], [t_0 : t_1; u_0 : u_1]) \mapsto ([s_0 : s_1], [t_1 : t_0; u_0 : -u_1])$, let μ_2 be the transformation $([s_0 : s_1], [t_0 : t_1; u_0 : u_1]) \mapsto ([s_0 : -s_1], [t_0 : t_1; u_0 : u_1])$, and let G' be the subgroup in $\mathrm{Aut}(\mathbb{P}^1 \times \mathbb{F}_1)$ that is generated by μ_1 and μ_2. Then $G' \cong \mu_2^2$, and the morphism $\mathbb{P}^1 \times \mathbb{F}_1 \to \mathbb{P}^1 \times \mathbb{P}^1$ is G'-equivariant. Moreover, one can check that the induced action of the group G' on $\mathbb{P}^1 \times \mathbb{P}^1$ has the following properties: $\mathbb{P}^1 \times \mathbb{P}^1$ does not have G'-fixed points, $\mathbb{P}^1 \times \mathbb{P}^1$ does not contain G'-invariant curves of degree $(1, 0)$, the only G'-invariant curves of degree $(0, 1)$ in $\mathbb{P}^1 \times \mathbb{P}^1$ are $\{t_0 + t_1 = 0\}$ and $\{t_0 - t_1 = 0\}$, and $\mathbb{P}^1 \times \mathbb{P}^1$ does not contain G'-invariant curves of degree $(1, 1)$.

Let B be the surface in $\mathbb{P}^1 \times \mathbb{F}_1$ that is given by

$$\left(s_0^2 + s_1^2\right)\left(t_0^2 + t_1^2\right)u_0^2 + 9\left(s_0^2 + s_1^2\right)u_1^2 + \left(s_0^2 - s_1^2\right)\left(t_0^2 - t_1^2\right)u_0^2$$
$$+ 4\left(s_0^2 + s_1^2\right)\left(t_0 - t_1\right)u_0 u_1 + 8\left(s_0^2 - s_1^2\right)\left(t_0 + t_1\right)u_0 u_1 = 0.$$

Then B is smooth and G'-invariant. Let $\varpi \colon V \to \mathbb{P}^1 \times \mathbb{F}_1$ be the double cover ramified in B. Then V is a smooth Fano threefold №3.4, so that we can use notations introduced in (4.11). Note that the G'-action lifts to V, and we can extend it to a larger subgroup $G \subset \mathrm{Aut}(V)$, which is generated by the subgroup G' and the Galois involution of the double cover τ. Then (4.11) is G-equivariant. In the following, we will use notations used in this diagram.

Let H_s and H_t be general fibers of the del Pezzo fibrations η_1 and ϕ, respectively, and let E be the α-exceptional surface. Then $-K_V \sim H_s + 2H_t + E$, and $\overline{\mathrm{Eff}}(X)$ is generated by H_s, H_t, E. Note that $|H_t|$ contains two G-invariant surfaces. They are the preimages via γ of the two G'-invariant curves in $\mathbb{P}^1 \times \mathbb{P}^1$ of degree $(0, 1)$. Let H_+ and H_- be the preimages via γ of the curves given by $t_0 \pm t_1 = 0$, respectively. Let us apply the results of Section 1.7 to irreducible G-invariant curves in these two surfaces.

Lemma 5.62 *Let Z be an irreducible G-invariant curve in H_+. Then* $S(W_{\bullet,\bullet}^{H_+}; Z) \leqslant \frac{5}{9}$.

Proof The double cover ϖ gives a double cover $H_+ \to \varpi(H_+)$, where $\varpi(H_+) \cong \mathbb{P}^1 \times \mathbb{P}^1$, and we can identify $([s_0 : s_1], [u_0 : u_1])$ with coordinates on $\varpi(H_+)$. Note that this double cover is branched along the curve $\{2(s_0^2 + s_1^2)u_0^2 + 9(s_0^2 + s_1^2)u_1^2 + 16(s_0^2 - s_1^2)u_0 u_1 = 0\}$. This curve is smooth, so that H_+ is a smooth del Pezzo of degree 4.

Fix $u \in \mathbb{R}_{\geqslant 0}$. Let us consider the Zariski decomposition of the divisor $-K_X - uH_+$. For $u > 2$, this divisor is not pseudo-effective. For $u \in [0,2]$, we have

$$P(u) = \begin{cases} H_s + (2-u)H_+ + E & \text{if } 0 \leqslant u \leqslant 1, \\ H_s + (2-u)(H_+ + E) & \text{if } 1 \leqslant u \leqslant 2, \end{cases}$$

and

$$N(u) = \begin{cases} 0 & \text{if } 0 \leqslant u \leqslant 1, \\ (u-1)E & \text{if } 1 \leqslant u \leqslant 2, \end{cases}$$

where $P(u) = P(-K_X - uH_+)$ and $N(u) = N(-K_X - uH_+)$.

Let ℓ_s and ℓ_u be the pull backs on H_+ of the curves in $\varpi(H_+)$ defined by $\{s_0 = 0\}$ and $\{u_0 = 0\}$, respectively. Then $P(u)|_{H_+} \sim_{\mathbb{R}} \ell_s + \ell_u$ for $u \in [0,1]$. Likewise, if $u \in [1,2]$, then we have $P(u)|_{H_+} \sim_{\mathbb{R}} \ell_s + (2-u)\ell_u$ and $N(u)|_{H_+} = (u-1)\ell_u$. If $Z = E|_{H_+}$, then $Z = \ell_u$ and

$$\begin{aligned} S\left(W_{\bullet,\bullet}^{H_+}; Z\right) &= \frac{3}{18} \int_0^1 \int_0^\infty \mathrm{vol}\left(\ell_s + (1-v)\ell_u\right) dv\, du \\ &+ \frac{3}{18} \int_1^2 (u-1)\left(\ell_s + (2-u)\ell_u\right)^2 du \\ &+ \frac{3}{18} \int_1^2 \int_0^\infty \mathrm{vol}\left(\ell_s + (2-u-v)\ell_u\right) dv\, du \\ &= \frac{1}{6} \int_0^1 \int_0^1 4(1-v)\, dv\, du + \frac{1}{6} \int_1^2 4(u-1)(2-u)\, du \\ &+ \frac{1}{6} \int_1^2 \int_0^{2-u} 4(2-u-v)\, dv\, du = \frac{5}{9} < 1. \end{aligned}$$

If $Z \neq E|_{H_+}$, then $Z \sim a\ell_s + b\ell_u$ for some non-negative integers a and b since G contains the Galois involution of the double cover ϖ. Moreover, we have $b \geqslant 1$ because $|\ell_s|$ does not contain G-invariant curves. This gives $S(W_{\bullet,\bullet}^{H_+}; Z) \leqslant S(W_{\bullet,\bullet}^{H_+}; \ell_u) = \frac{5}{9}$ as required.   □

Lemma 5.63   *Let Z be an irreducible G-invariant curve in H_-. Then $S(W_{\bullet,\bullet}^{H_-}; Z) \leqslant \frac{8}{9}$.*

Proof   The double cover ϖ gives a double cover $H_- \to \varpi(H_-)$, where $\varpi(H_-) \cong \mathbb{P}^1 \times \mathbb{P}^1$. We can identify $([s_0 : s_1],[u_0 : u_1])$ with coordinates on $\varpi(H_+)$. Note that this double cover is branched along the curve given by

$$\left(s_0 - is_1\right)\left(s_0 + is_1\right)\left(2u_0 + (4 - \sqrt{2}i)u_1\right)\left(2u_0 + (4 + \sqrt{2}i)u_1\right) = 0.$$

Therefore we see that H_- is the toric del Pezzo of degree 4 that has 4 nodes.

Let ℓ_s and ℓ_u be irreducible curves in H_{-1} that are preimages of the curves in $\varpi(H_-)$ given by $s_0 - is_1 = 0$ and $2u_0 + (4 - \sqrt{2}i)u_1 = 0$, respectively. Then $Z \sim_{\mathbb{Q}} a\ell_s + b\ell_u$ for some integers $a \geqslant 0$ and $b \geqslant 1$ because G contains the involution of the double cover ϖ, and H_- does not contain irreducible G-invariant curves that are $\mathbb{Q}$-rationally equivalent to $n\ell_s$ for $n \in \mathbb{Z}_{>0}$.

Arguing as in the proof of Lemma 5.62, we see that

$$P(-K_X - uH_-)|_{H_-} \sim_{\mathbb{R}} 2\ell_s + 2\ell_u$$

and $N(-K_X - uH_-) = 0$ for $u \in [0,1]$. If $u \in [1,2]$, then

$$N(-K_X - uH_-)|_{H_-} \sim_{\mathbb{R}} (2u - 2)\ell_u \text{ and } P(-K_X - uH_-)|_{H_-} \sim_{\mathbb{R}} 2\ell_s + (4 - 2u)\ell_u.$$

Thus, if $Z = E|_{H_-}$, we compute $S(W^{H_+}_{\bullet,\bullet}; Z) = \frac{5}{9}$ as in the proof of Lemma 5.62. Similarly, if $Z \neq E|_{H_-}$, then

$$
\begin{aligned}
S\left(W^{H_-}_{\bullet,\bullet}; Z\right) &\leqslant S\left(W^{H_-}_{\bullet,\bullet}; \ell_u\right) \\
&= \frac{3}{18} \int_0^1 \int_0^\infty \text{vol}\big(2\ell_s + (2 - \ell_t)\big) dv\, du \\
&\quad + \frac{3}{18} \int_1^2 \int_0^\infty \text{vol}\big(2\ell_s + (4 - 2u - v)\ell_t\big) dv\, du \\
&= \frac{1}{6} \int_0^1 \int_0^2 (4 - 2v) dv\, du + \frac{1}{6} \int_1^2 \int_0^{4-2u} (8 - 4u - 2v) dv\, du \\
&= \frac{8}{9}
\end{aligned}
$$

as required.   □

Now, we are ready to prove

Proposition 5.64 *The threefold V is K-stable.*

Proof Suppose that V is not K-stable. Then V is not K-polystable by Corollary 1.5 because $\text{Aut}(V)$ is finite [45]. Then, by Theorem 1.22, there is a G-invariant prime divisor F over V such that $\beta(F) = A_V(F) - S_V(F) \leqslant 0$. Let $Z = C_V(F)$. Then Z is not a surface by Theorem 3.17, so that Z is a G-invariant irreducible curve, because V does not have G-invariant points.

Applying Corollary 1.110 and Lemma 5.62, we see that $Z \not\subset H_+$ because $S_V(H_+) < 1$ by Theorem 3.17. Similarly, using Lemma 5.63, we see that $Z \not\subset H_-$.

Using Lemma 1.45, we get $\alpha_{G,Z}(V) < \frac{3}{4}$. Now, using Lemma 1.42, we see that there exists a G-invariant effective $\mathbb{Q}$-divisor D on the threefold V such that $D \sim_{\mathbb{Q}} -K_V$ and $\text{Nklt}(V, \lambda D)$ contains Z for some positive rational number $\lambda < \frac{3}{4}$.

Since $-K_V \sim H_s + 2H_t + E$ and $\mathrm{Eff}(V)$ is generated by H_s, H_t and E, the only possible two-dimensional component of $\mathrm{Nklt}(X, \lambda D)$ can be one of the surfaces H_+ and H_-. Since $Z \not\subset H_+ \cup H_-$, we conclude that Z is an irreducible component of the locus $\mathrm{Nklt}(V, \lambda D)$. Now, applying Corollary A.12 to the del Pezzo fibrations η_1 and ϕ, we conclude that $H_s \cdot Z \leqslant 1$ and $H_t \cdot Z \leqslant 1$. On the other hand, we know that $\mathbb{P}^1 \times \mathbb{P}^1$ does not contain G-invariant points, it does not contain G-invariant curves of degree $(1, 0)$, and it does not contain G-invariant curves of degree $(1, 1)$. Hence, we conclude that $\gamma(Z)$ is a G-invariant curve of degree $(0, 1)$, which is impossible since we already proved that $Z \not\subset H_+ \cup H_-$. $\qquad\qquad\square$

5.14 Family №3.5

Let $S = \mathbb{P}^1 \times \mathbb{P}^1$, let C be a prime divisor in S of degree $(1, 5)$, and let $G = \mathrm{Aut}(S, C)$. We can choose coordinates $([u : v], [x : y])$ on the surface S such that the curve C is given by

$$u\left(x^5 + a_1 x^4 y + a_2 x^3 y^2 + a_3 x^2 y^3\right) + v\left(y^5 + b_1 xy^4 + b_2 x^2 y^3 + b_3 x^3 y^2\right) = 0, \quad (5.11)$$

where each $a_i \in \mathbb{C}$ and each $b_j \in \mathbb{C}$. If all numbers a_i and b_j vanish, then $G \cong \mathbb{G}_m \rtimes \mu_2$. In all other cases, the group G is finite by [45, Corollary 2.7].

Consider the G-equivariant embedding $S \hookrightarrow \mathbb{P}^1 \times \mathbb{P}^2$ given by

$$\left([u : v], [x : y]\right) \mapsto \left([u : v], [x^2 : xy : y^2]\right).$$

Identify S and C with their images in $\mathbb{P}^1 \times \mathbb{P}^2$, identify G with a subgroup in $\mathrm{Aut}(\mathbb{P}^1 \times \mathbb{P}^2)$. Let $\mathrm{pr}_1 \colon \mathbb{P}^1 \times \mathbb{P}^2 \to \mathbb{P}^1$ and $\mathrm{pr}_2 \colon \mathbb{P}^1 \times \mathbb{P}^2 \to \mathbb{P}^2$ be the projections to the first and the second factors, respectively. Then C is a G-invariant curve of degree $(5, 2)$ in $\mathbb{P}^1 \times \mathbb{P}^2$, both projections pr_1 and pr_2 are G-equivariant, $\mathrm{pr}_2(S)$ is a G-invariant conic in $\mathbb{P}^2$.

Let $\pi \colon X \to \mathbb{P}^1 \times \mathbb{P}^2$ be the blow up of the curve C. Then X is a Fano threefold №3.5, and the G-action lifts to X. Therefore, we can identify G with a subgroup in $\mathrm{Aut}(X)$. In fact, it follows from the proof of [45, Lemma 8.7] that $\mathrm{Aut}(X) = G$. In this section, we will prove that X is K-stable for a special choice of the curve C, which would imply that general Fano threefolds №3.5 are K-stable by Theorem 1.11.

Let $\widetilde{S}$ be the proper transform on X of the surface S, let E be the π-exceptional surface, let $H_1 = (\mathrm{pr}_1 \circ \pi)^*(\mathcal{O}_{\mathbb{P}^1}(1))$ and let $H_2 = (\mathrm{pr}_2 \circ \pi)^*(\mathcal{O}_{\mathbb{P}^2}(1))$. Then $\widetilde{S} \sim 2H_2 - E$, which implies that

$$-K_X \sim_{\mathbb{Q}} 2H_1 + \frac{3}{2}\widetilde{S} + \frac{1}{2}E,$$

so that $\alpha_G(X) \leqslant \frac{2}{3}$.

Note that $\widetilde{S} \cong \mathbb{P}^1 \times \mathbb{P}^1$ and $\widetilde{S}|_{\widetilde{S}}$ is a line bundle of degree $(-1, -1)$. Therefore, there exists a birational morphism $\varpi \colon X \to Y$ that contracts $\widetilde{S}$ to an ordinary double point of the singular Fano threefold Y such that $-K_Y^3 = 22$ and $\mathrm{Pic}(Y) = \mathbb{Z}[-K_Y]$. Using this, we obtain the G-equivariant commutative diagram

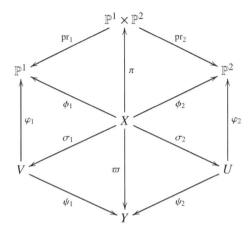

where ϕ_1 is a fibration into quartic del Pezzo surfaces, ϕ_2 is a conic bundle, V and U are smooth weak Fano threefolds, σ_1 and σ_2 are birational contractions of the surface $\widetilde{S}$ to smooth rational curves, ψ_1 and ψ_2 are small resolutions of the threefold Y, ϕ_1 is a fibration into quintic del Pezzo surfaces, and ϕ_2 is a $\mathbb{P}^1$-bundle.

Corollary 5.65 *Suppose that $|H_1|$ contains no G-invariant surfaces. Then $\alpha_G(Y) \geqslant \frac{4}{5}$.*

Proof By Lemma 1.47 and Corollary 1.57, we have $\alpha_G(Y) = \alpha_G(V) \geqslant \frac{4}{5}$. □

Fix an effective $\mathfrak{S}_4$-action on $\mathbb{P}^1$, and consider the corresponding diagonal action on the surface $S = \mathbb{P}^1 \times \mathbb{P}^1$. By Lemma A.54, the surface S contains a unique $\mathfrak{S}_4$-invariant curve of degree $(5, 1)$, and this curve is irreducible and smooth.

Proposition 5.66 *Suppose that C is $\mathfrak{S}_4$-invariant. Then X and Y are K-stable.*

Proof Recall that $G = \mathrm{Aut}(S, C)$. Then $G \cong \mathfrak{S}_4$ by Lemma A.54, and

(i) $\mathbb{P}^1$ does not contain G-invariant points,
(ii) $\mathbb{P}^2$ does not contain G-invariant points,
(iii) $\mathbb{P}^2$ does not contain G-invariant lines,
(iv) $\mathrm{pr}_2(S)$ is the unique G-invariant conic in $\mathbb{P}^2$.

Indeed, the first assertion is obvious. The remaining assertions follow from the fact that the G-action on $\mathbb{P}^2$ is given by an irreducible representation of the group G.

We have $\alpha_G(Y) \geqslant \frac{4}{5}$ by Corollary 5.65, so that Y is K-polystable by Theorem 1.48. Since $\mathrm{Aut}(Y) \cong \mathrm{Aut}(X) = G$, we also conclude that Y is K-stable by Corollary 1.5.

Let us show that X is K-stable. Suppose it is not. By Corollary 1.5 and Theorem 1.22, there are a G-equivariant birational morphism $f \colon \widetilde{X} \to X$ and a G-invariant dreamy prime divisor $F \subset \widetilde{X}$ such that $\beta(F) = A_X(F) - S_X(F) \leqslant 0$. Let $Z = f(F)$. Then Z is not a surface by Theorem 3.17, so that Z is a G-invariant irreducible curve because X has no G-fixed points, since $\mathbb{P}^2$ has no G-fixed points.

Using Lemma 1.45, we get $\alpha_{G,Z}(X) < \frac{3}{4}$. Now, using Lemma 1.42, we see that there are a G-invariant effective $\mathbb{Q}$-divisor D on the threefold X and $\lambda \in \mathbb{Q}_{>0}$ such that $\lambda < \frac{3}{4}$, $D \sim_{\mathbb{Q}} -K_X$ and $\mathrm{Nklt}(X, \lambda D)$ contains Z.

We claim that $\widetilde{S}$ is the only surface that can be contained in the locus $\mathrm{Nklt}(X, \lambda D)$. Indeed, if $\mathrm{Nklt}(X, \lambda D)$ contains a G-invariant surface $\mathcal{S}$, then $-K_X - \mathcal{S}$ is big, so that either $\mathcal{S} \in |2H_2 - E|$, or $\mathcal{S} \in |H_1|$, or $\mathcal{S} \in |H_1 + 2H_2 - E|$. But $\widetilde{S}$ is the only divisor in $|2H_2 - E|$, and $|H_1|$ does not contain G-invariant divisors. Moreover, the surface $\widetilde{S}$ is the fixed locus of the linear system $|H_1 + 2H_2 - E|$, and the pencil $|H_1|$ is its mobile part, so that $|H_1 + 2H_2 - E|$ contains no G-invariant divisors. Thus, if $\mathrm{Nklt}(X, \lambda D)$ contains a G-invariant surface $\mathcal{S}$, then $\mathcal{S} = \widetilde{S}$.

Suppose that $Z \subset \widetilde{S}$. Let us apply results of Section 1.7 to $\widetilde{S}$ and Z. As in Section 1.7, we denote by $V_{\bullet}$ the anticanonical ring of the threefold X with its natural filtration, and we denote by $W_{\bullet,\bullet}^{\widetilde{S}}$ its refinement by the surface $\widetilde{S}$. Using Corollary 1.110, we see that either $S_X(\widetilde{S}) \geqslant 1$ or $S(W_{\bullet,\bullet}^{\widetilde{S}}; Z) \geqslant 1$ (or both). Let us compute $S_X(\widetilde{S})$. Take a positive real number u. If $0 \leqslant u \leqslant 1$, then $-K_X - u\widetilde{S}$ is nef. On the other hand, if $1 \leqslant u \leqslant \frac{3}{2}$, then $P(-K_X - u\widetilde{S}) = 2H_1 + (3 - 2u)H_2$ and $N(-K_X - u\widetilde{S}) = (u - 1)E$. Finally, if $u > \frac{3}{2}$, then $-K_X - u\widetilde{S}$ is not pseudo-effective. This gives

$$S_X(\widetilde{S}) = \frac{1}{20} \int_0^1 \left(-K_X - u\widetilde{S}\right)^3 du + \frac{1}{20} \int_1^{\frac{3}{2}} \left(2H_1 + (3 - 2u)H_2\right)^3 du$$

$$= \frac{1}{20} \int_0^1 \left(20 - 2u^3 - 6u^2 - 6u\right) du + \frac{1}{20} \int_1^{\frac{3}{2}} 6(2u - 3)^2 du = \frac{31}{40},$$

so that $S_X(\widetilde{S}) < 1$, which also follows from Theorem 3.17. Thus, we have $S(W_{\bullet,\bullet}^{\widetilde{S}}; Z) \geqslant 1$. Let us compute $S(W_{\bullet,\bullet}^{\widetilde{S}}; Z)$. Let ℓ_1 and ℓ_2 be the rulings of the surface $\widetilde{S} \cong \mathbb{P}^1 \times \mathbb{P}^1$ that are contracted by $\mathrm{pr}_1 \circ \pi$ and $\mathrm{pr}_2 \circ \pi$, respectively. Then

$-K_X|_{\widetilde{S}} \sim \ell_1 + \ell_2$, $H_1|_{\widetilde{S}} \sim \ell_1$, $H_2|_{\widetilde{S}} \sim \ell_2$, $E|_{\widetilde{S}} \sim \ell_1 + 5\ell_2$, $\widetilde{S}|_{\widetilde{S}} \sim -\ell_1 - \ell_2$. Thus, we have $(-K_X - u\widetilde{S})|_{\widetilde{S}} \sim_{\mathbb{R}} (1+u)(\ell_1 + \ell_2)$. If $1 \leqslant u \leqslant \frac{3}{2}$, then $N(-K_X - u\widetilde{S})|_{\widetilde{S}} = (u-1)E|_{\widetilde{S}}$ and $P(-K_X - u\widetilde{S})|_{\widetilde{S}} \sim_{\mathbb{R}} 2\ell_1 + (6-4u)\ell_2$. Thus, if $Z = E|_{\widetilde{S}}$, then Corollary 1.110 gives

$$
\begin{aligned}
S\big(W_{\bullet,\bullet}^{\widetilde{S}}; Z\big) \\
&= \frac{3}{20} \int_0^1 \int_0^\infty \mathrm{vol}\big((1+u-v)\ell_1 + (1+u-5v)\ell_2\big)dvdu \\
&\quad + \frac{3}{20} \int_1^{\frac{3}{2}} \big(2\ell_1 + (6-4u)\ell_2\big)^2(u-1)du \\
&\quad + \frac{3}{20} \int_1^{\frac{3}{2}} \int_0^\infty \mathrm{vol}\big((2-v)\ell_1 + (6-4u-5v)\ell_2\big)dvdu \\
&= \frac{3}{20} \int_0^1 \int_0^{\frac{1+u}{5}} 2(1+u-v)(1+u-5v)dvdu \\
&\quad + \frac{3}{20} \int_1^{\frac{3}{2}} 4(6-4u)(u-1)du + \frac{3}{20} \int_1^{\frac{3}{2}} \int_0^{\frac{6-4u}{5}} 2(6-4u-5v)(2-v)dvdu \\
&= \frac{193}{1000}.
\end{aligned}
$$

Similarly, if $Z \neq E|_{\widetilde{S}}$, then

$$
\begin{aligned}
S\big(W_{\bullet,\bullet}^{\widetilde{S}}; Z\big) \\
&\leqslant \frac{3}{20} \int_0^1 \int_0^\infty \mathrm{vol}\big((1+u-v)\ell_1 + (1+u-v)\ell_2\big)dvdu \\
&\quad + \frac{3}{20} \int_1^{\frac{3}{2}} \int_0^\infty \mathrm{vol}\big((2-v)\ell_1 + (6-4u-v)\ell_2\big)dvdu \\
&= \frac{3}{20} \int_0^1 \int_0^{1+u} 2(1+u-v)^2 dvdu + \frac{3}{20} \int_1^{\frac{3}{2}} \int_0^{6-4u} 2(6-4u-v)(2-v)dvdu \\
&= \frac{21}{40},
\end{aligned}
$$

because $|Z - \ell_1 - \ell_2|$ is not empty since $|\ell_1|$ and $|\ell_2|$ do not contain G-invariant curves. Hence, we see that $S(W_{\bullet,\bullet}^{\widetilde{S}}; Z) < 1$. The obtained contradiction shows that $Z \not\subset \widetilde{S}$.

Since $Z \not\subset \widetilde{S}$, the curve Z must be an irreducible component of the locus $\mathrm{Nklt}(X, \lambda D)$. Now, applying Corollary A.12 to the del Pezzo fibration $\mathrm{pr}_1 \circ \pi$, we get $H_1 \cdot Z \leqslant 1$, so that $H_1 \cdot Z = 1$ because $|H_1|$ does not have G-invariant surfaces. This gives $Z \not\subset E$. Now, applying Corollary A.13 to the conic bundle $\mathrm{pr}_2 \circ \pi$, we see that $H_2 \cdot Z \leqslant 2$. Then $\mathrm{pr}_2 \circ \pi(Z)$ is either a point, a line, or a conic. Since

$\mathrm{pr}_2 \circ \pi(Z)$ is also G-invariant, we have $\mathrm{pr}_2 \circ \pi(Z) = \mathrm{pr}_2(S)$, so that $Z \subset \widetilde{S}$, which is a contradiction.   □

Thus, we see that a general smooth Fano threefold №3.5 is K-stable by Theorem 1.11.

Remark 5.67 Using [168, Theorem 11] and [122, Theorem 1.4], we see that Y has a smoothing to a Fano threefold №1.10. Using Proposition 5.66 and Theorem 1.11, we conclude (again) that a general smooth Fano threefold №1.10 is K-stable.

Recall from [45] that there is unique smooth Fano threefold №3.5 whose automorphism group is infinite. If $a_1 = a_2 = a_3 = b_1 = b_2 = b_3 = 0$ in (5.11), then X is this threefold. Let us prove that X is K-polystable in this case. To do this, we need two lemmas:

Lemma 5.68 *Let P be a point in $\widetilde{S}$. Then $\delta_P(X) \geqslant \frac{80}{73}$.*

Proof Recall that $\widetilde{S} \cong \mathbb{P}^1 \times \mathbb{P}^1$. Denote by ℓ_1 and ℓ_2 the rulings of this surface that are contracted by $\mathrm{pr}_1 \circ \pi$ and $\mathrm{pr}_2 \circ \pi$, respectively. Let Z be the curve in $|\ell_2|$ such that $P \in Z$. Let us apply Theorem 1.112 $Y = \widetilde{S}$ using notations introduced in this theorem.

Recall from the proof of Proposition 5.66 that $S_X(\widetilde{S}) = \frac{31}{40}$. Moreover, it follows from the proof of Proposition 5.66 that

$$P(u)\big|_{\widetilde{S}} = \begin{cases} (1+u)\ell_1 + (1+u)\ell_2 & \text{if } 0 \leqslant u \leqslant 1, \\ 2\ell_1 + (6-4u)\ell_2 & \text{if } 1 \leqslant u \leqslant \dfrac{3}{2}, \end{cases}$$

and

$$N(u)\big|_{\widetilde{S}} = \begin{cases} 0 & \text{if } 0 \leqslant u \leqslant 1, \\ (u-1)\widetilde{C} & \text{if } 1 \leqslant u \leqslant \dfrac{3}{2}, \end{cases}$$

where $\widetilde{C} = E \cap \widetilde{S}$. Recall that $\widetilde{C} \sim \ell_1 + 5\ell_2$. We have

$$S\left(W^{\widetilde{S}}_{\bullet,\bullet}; Z\right) = \frac{3}{20} \int_0^1 \int_0^{1+u} 2(1+u)(1+u-v)\,dv\,du$$
$$+ \frac{3}{20} \int_0^{\frac{3}{2}} \int_0^{6-4u} 4(6-4u-v)\,dv\,du = \frac{61}{80}$$

and

$$S\left(W^{\widetilde{S},Z}_{\bullet,\bullet,\bullet}; P\right) = F_P + \frac{3}{20} \int_0^1 \int_0^{1+u} (1+u)^2\,dv\,du + \frac{3}{20} \int_1^{\frac{3}{2}} \int_0^{6-4u} 4\,dv\,du$$
$$= F_P + \frac{69}{80},$$

where $F_P = 0$ if $P \notin \widetilde{C}$ and

$$F_P = \frac{6}{20} \int_1^{\frac{3}{2}} \int_0^{6-4u} 2(u-1)dvdu = \frac{1}{20}.$$

Hence, it follows from Theorem 1.112 that

$$\delta_P(X) \geqslant \frac{1}{S(W^{\widetilde{S},Z}_{\bullet,\bullet,\bullet}; P)} \geqslant \frac{80}{73},$$

as required.   □

Lemma 5.69 *Let T be a smooth surface in $|H_1|$, and let P be a point in the surface T. Then $\delta_P(X) > 1$.*

Proof Using Lemma 5.68, we may assume that $P \notin \widetilde{S}$. First, let us compute $S_X(T)$. Let u be a non-negative real number. If $0 \leqslant u \leqslant 1$, then the divisor $-K_X - uT$ is nef. If $1 \leqslant u \leqslant 2$, the positive part of the Zariski decomposition of the divisor $-K_X - uT$ is

$$P(u) = -K_X - uT + (1-u)\widetilde{S} \sim_{\mathbb{R}} (2-u)H_1 + \left(\frac{5}{2} - u\right)\widetilde{S} + \frac{1}{2}E,$$

and its negative part is $N(u) = (u-1)\widetilde{S}$. This gives

$$S_X(T) = \frac{1}{20} \int_0^1 \left(-K_X - uT\right)^3 du + \frac{1}{20} \int_1^2 \left(-K_X - uT + (1-u)\widetilde{S}\right)^3 du$$

$$= \frac{1}{20} \int_0^1 \left(20 - 12u\right) du + \frac{1}{20} \int_1^{\frac{3}{2}} (u-2)(u^2 + 2u - 11)du = \frac{69}{80}$$

because the divisor $-K_X - uT$ is not pseudo-effective for $u > 2$. Thus, we have $S_X(T) < 1$, which also follows from Theorem 3.17.

Since $S_X(T) < 1$, Theorem 1.95 implies that $\delta_P(X) > 1$ if $\delta_P(T; W^T_{\bullet,\bullet}) > 1$. Recall that from (1.12)

$$\delta_P\left(T; W^T_{\bullet,\bullet}\right) = \inf\left\{ \frac{A_T(R)}{S(W^T_{\bullet,\bullet}; R)} \,\middle|\, \begin{array}{l} R \text{ is a prime divisor over } T \\ \text{such that } P \in C_T(R) \end{array} \right\},$$

where $W^T_{\bullet,\bullet}$ and $S(W^T_{\bullet,\bullet}; R)$ are as in Section 1.7. We show that $\delta_P(T; W^T_{\bullet,\bullet}) > 1$. Let $\widetilde{C} = T \cap \widetilde{S}$. Our computations of $S_X(T)$ give

$$P(u)\big|_T = \begin{cases} -K_T & \text{if } 0 \leqslant u \leqslant 1, \\ -K_T + (1-u)\widetilde{C} & \text{if } 1 \leqslant u \leqslant 2, \end{cases}$$

and

$$N(u)\big|_T = \begin{cases} 0 & \text{if } 0 \leqslant u \leqslant 1, \\ (u-1)\widetilde{C} & \text{if } 1 \leqslant u \leqslant 2. \end{cases}$$

Let R be any prime divisor over T. Since $P \notin \widetilde{C}$, Corollary 1.108 implies that

$$S(W^T_{\bullet,\bullet}; R) = \frac{3}{20} \int_0^1 \int_0^\infty \mathrm{vol}(-K_T - vR)dvdu$$

$$+ \frac{3}{20} \int_1^2 \int_0^\infty \mathrm{vol}(-K_T - (u-1)\widetilde{C} - vR)dvdu.$$

But T is a smooth del Pezzo surface of degree 4, so that $\delta(T) = \frac{4}{3}$ by Lemma 2.12. Then

$$\frac{1}{4} \int_0^\infty \mathrm{vol}(-K_T - vR)dv \leqslant \frac{3}{4} A_T(R).$$

Therefore, we have

$$S(W^T_{\bullet,\bullet}; R) \leqslant \frac{3}{20} \int_0^1 3A_T(R)du + \frac{3}{20} \int_1^2 3A_T(R)du = \frac{9}{10} A_T(R),$$

which implies that $\delta_P(T; W^T_{\bullet,\bullet}) \geqslant \frac{10}{9} > 1$. $\qquad\square$

Both Lemmas 5.68 and 5.69 hold for any smooth Fano threefold №3.5. They give

Corollary 5.70   *If $|H_1|$ does not have G-invariant surfaces, then X is K-polystable.*

Proof   Let T be a general surface in $|H_1|$, let F be a G-invariant prime divisor over X, and let $Z = C_X(F)$. If the pencil $|H_1|$ does not contain any G-invariant surfaces, then the restriction of $\phi_1|_Z \colon Z \to \mathbb{P}^1$ is surjective, so that the intersection $T \cap Z$ is not empty. In this case, for every point $P \in T \cap Z$, we have $\frac{A_X(F)}{S_X(F)} \geqslant \delta_P(X) > 1$ by Lemma 5.69, so that X is K-polystable by Theorem 1.22. $\qquad\square$

This corollary implies Proposition 5.66 and the following result:

Corollary 5.71   *Suppose that $\mathrm{Aut}(X)$ is infinite. Then X is K-polystable.*

Proof   We may assume that C is given by $ux^5 + uy^5 = 0$. Then $\mathrm{Aut}(S, C)$ is generated by transformations

$$([u:v],[x:y]) \mapsto ([\lambda^5 u:v],[x:\lambda y])$$

for $\lambda \in \mathbb{C}^*$ and the involution

$$([u:v],[x:y]) \mapsto ([v:u],[y:x]).$$

Then $G = \mathrm{Aut}(S, C) \cong \mathbb{G}_m \rtimes \mu_2$, and the pencil $|H_1|$ does not have G-invariant surfaces, so that X is K-polystable by Corollary 5.70. $\qquad\square$

5.15 Family №3.6

Now, we will construct a K-stable smooth Fano threefold in family №3.6. To do this, let us use the assumptions and notations of Section 4.3 assuming that $d = 8$. Then $V_8 = \mathbb{P}^3$. Let ι and τ be the automorphisms of $\mathbb{P}^3$ given by (4.5) and (4.6), respectively. Let $G = \langle \iota, \tau \rangle$. Then $G \cong D_8$, $\Theta \cong \mu_2$ and $\Gamma \cong \mu_2^2$.

Remark 5.72 Let $\eta\colon \mathrm{GL}_4(\mathbb{C}) \to \mathrm{PGL}_4(\mathbb{C}) = \mathrm{Aut}(\mathbb{P}^3)$ be the natural projection, and

$$A = \begin{pmatrix} 0 & 1 & 0 & 0 \\ 1 & 0 & 0 & 0 \\ 0 & 0 & 0 & 1 \\ 0 & 0 & 1 & 0 \end{pmatrix} \text{ and } B = \begin{pmatrix} 0 & 0 & 1 & 0 \\ 0 & 0 & 0 & i \\ 1 & 0 & 0 & 0 \\ 0 & i & 0 & 0 \end{pmatrix}.$$

Then $A = \eta(\iota)$ and $B = \eta(\tau)$. Note that $\langle A, B \rangle \cong 4.D_8$, and that $H^0(\mathcal{O}_{\mathbb{P}^3}(1))$ splits as a sum of two different two-dimensional representations of the group $\langle A, B \rangle$.

Let $L = \{x_0 - x_2 = x_1 - x_3 = 0\} \subset \mathbb{P}^3$, and $L' = \{x_0 + x_2 = x_1 + x_3 = 0\} \subset \mathbb{P}^3$. Then L and L' are G-invariant. They are the only G-invariant lines in $\mathbb{P}^3$.

Recall from Section 4.3 that X is a blow up of $\mathbb{P}^3$ along the elliptic curve $\mathscr{C} = H_1 \cap H_2$, where $H_1 = \{x_0^2 + x_1^2 + \lambda(x_2^2 + x_3^2) = 0\}$ and $H_2 = \{\lambda(x_0^2 - x_1^2) + x_2^2 - x_3^2 = 0\}$, and λ is a non-zero complex number such that $\lambda^4 \neq 1$. Let $\widetilde{L}$ be the proper transform on X of the line L, and let $\rho\colon \widehat{X} \to X$ be its blow up. Then $\widehat{X}$ is a smooth Fano threefold in family №3.6. Since the action of the group G lifts to $\widehat{X}$, we identify G with a subgroup in $\mathrm{Aut}(\widehat{X})$.

Lemma 5.73 $\alpha_G(\widehat{X}) \geqslant 1$.

Proof Suppose that $\alpha_G(\widehat{X}) < 1$. We will use Theorem 1.52 with $\mu = 1$ to obtain a contradiction. Since X does not have G-fixed points by Lemma 4.29, we see that $\widehat{X}$ does not have G-fixed points, so that Theorem 1.52(2) doesn't hold.

Let $\widehat{F}$ be the proper transform on the threefold $\widehat{X}$ of a general fiber F of the del Pezzo fibration $\phi\colon X \to \mathbb{P}^1$. If Theorem 1.52(3) holds, X contains a G-invariant curve C with $\widehat{F} \cdot C \leqslant 1$. Since $\widehat{X}$ has no G-fixed points, we see that $\rho(C)$ is a G-invariant curve, so that $1 \geqslant \widehat{F} \cdot C = \rho^*(F) \cdot C = F \cdot \rho(C)$, which is impossible by Lemma 4.29.

Let R be the ρ-exceptional surface, and $\widehat{E}$ the proper transform of E, the exceptional surface of $\pi\colon X \to \mathbb{P}^3$. If Theorem 1.52(1) holds, $\widehat{X}$ contains a G-invariant irreducible normal surface S such that $-K_{\widehat{X}} \sim_{\mathbb{Q}} \lambda S + \Delta$ for some rational number $\lambda > 1$ and effective $\mathbb{Q}$-divisor Δ. On the other hand, it follows from [93] that

$$\mathrm{Eff}(\widehat{X}) = \mathbb{R}_{\geqslant 0}\big[\widehat{E}\big] + \mathbb{R}_{\geqslant 0}\big[R\big] + \mathbb{R}_{\geqslant 0}\big[(\pi \circ \rho)^*(O_{\mathbb{P}^3}(2)) - \widehat{E}\big]$$
$$+ \mathbb{R}_{\geqslant 0}\big[(\pi \circ \rho)^*(O_{\mathbb{P}^3}(1)) - R\big],$$

which implies that $S \neq R$, so that $\rho(S)$ is a surface. Write $\widetilde{S} = \rho(S)$ and $\widetilde{\Delta} = \rho(\Delta)$; then, pushing forward to X, $-K_X \sim_{\mathbb{Q}} \lambda \widetilde{S} + \widetilde{\Delta}$. It follows that the surface $\widetilde{S}$ cannot be normal by Lemma 4.29. Since S is normal, we conclude that $\widetilde{S}$ is singular along $\widetilde{L}$. This implies that $\pi(\widetilde{S})$ is a G-invariant cubic surface that contains $\mathscr{C}$ and is singular along L. Then $S \sim (\pi \circ \rho)^*(O_{\mathbb{P}^3}(3)) - \widehat{E} - 2R$, which contradicts the description of the cone $\mathrm{Eff}(\widehat{X})$ given above, and concludes the proof. $\square$

Then $\widehat{X}$ is K-stable by Theorem 1.48 and Corollary 1.5 since $\mathrm{Aut}(X)$ is finite [45]. Therefore, general Fano threefolds in the family №3.6 are K-stable by Theorem 1.11.

5.16 Family №3.8

Let X be a smooth threefold in the family №3.8. Then $X \subset \mathbb{F}_1 \times \mathbb{P}^2$. In fact, the threefold X is a divisor in the linear system $|(\varsigma \circ \mathrm{pr}_1)^*(O_{\mathbb{P}^2}(1)) \otimes \mathrm{pr}_2^*(O_{\mathbb{P}^2}(2))|$, where $\mathrm{pr}_1 : \mathbb{F}_1 \times \mathbb{P}^2 \to \mathbb{F}_1$ and $\mathrm{pr}_2 : \mathbb{F}_1 \times \mathbb{P}^2 \to \mathbb{P}^2$ are projections to the first and the second factors, respectively, and $\varsigma : \mathbb{F}_1 \to \mathbb{P}^2$ is the blow up of a point. Combining $\varsigma \circ \mathrm{pr}_1$ and pr_2, we obtain a morphism $\sigma : X \to Y$ such that Y is a smooth divisor $\mathbb{P}^2 \times \mathbb{P}^2$ of degree $(1, 2)$.

Let $\pi_1 : Y \to \mathbb{P}^2$ and $\pi_2 : Y \to \mathbb{P}^2$ be projections to the first and the second factors, respectively. Then σ is a blow up of a smooth curve C that is a fiber of the morphism π_1. Let $O = \pi_1(C)$. Then ς is a blow up of the point O, and there exists the commutative diagram

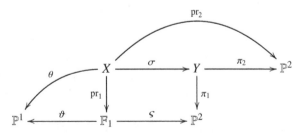

where ϑ is a natural projection and θ is a fibration into del Pezzo surfaces of degree 5.

The threefold Y is a smooth Fano threefold №2.24. By Lemma A.60, we can choose coordinates $([x : y : z], [u : v : w])$ on $\mathbb{P}^2 \times \mathbb{P}^2$ such that Y is given by one of the following three equations:

$$\left(vw + u^2\right)x + v^2y + w^2z = 0, \qquad (5.12)$$

$$\left(vw + u^2\right)x + \left(uw + v^2\right)y + w^2z = 0, \qquad (5.13)$$

$$\left(\mu vw + u^2\right)x + \left(\mu uw + v^2\right)y + \left(\mu uv + w^2\right)z = 0 \qquad (5.14)$$

for some $\mu \in \mathbb{C}$ such that $\mu^3 \neq -1$. Recall also that the morphism π_1 is a conic bundle, whose discriminant curve is a cubic curve, whose equation is given in Lemma A.60. This cubic curve does not contain O since C is smooth. For instance, if Y is given by (5.14) and $O = [1 : 1 : 1]$, then $\mu \neq 2$.

Proposition 5.74 *Suppose that one of the following two cases hold:*

- $O = [1 : 0 : 0]$ *and Y is given by* (5.12),
- $O = [1 : 1 : 1]$ *and Y is given by* (5.14) *with $\mu \neq 2$ and $\mu^3 \neq -1$.*

Then the Fano threefold X is K-polystable.

Remark 5.75 If $O = [1 : 0 : 0]$ and Y is given by (5.12), then

$$\mathrm{Aut}(X) \cong \mathrm{Aut}(Y) \cong \mathrm{Aut}(Y, C) \cong \mathbb{G}_m \rtimes \mu_2,$$

so that X is the unique smooth Fano threefold №3.8 with an infinite automorphism group. Vice versa, if $O = [1 : 1 : 1]$ and Y is given by (5.14) with $\mu \neq 2$ and $\mu^3 \neq -1$, then

$$\mathrm{Aut}(X) \cong \mathrm{Aut}(Y, C) \cong \mathfrak{S}_3,$$

so that the smooth Fano threefold X is K-stable by Proposition 5.74 and Corollary 1.5. Thus, using Theorem 1.11, we conclude that general Fano threefolds №3.8 are K-stable.

Let E be the exceptional surface of the blow up σ, and let E' be the surface $\mathrm{pr}_2^{-1}(\pi_2(C))$. Then $E \cong \mathbb{P}^1 \times \mathbb{P}^1$ and $E' \cong \mathbb{P}^1 \times \mathbb{P}^1$, and $E'|_E$ is a section of the natural projection $E \to C$. Moreover, there exists G-equivariant commutative diagram

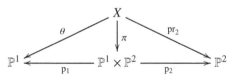

where π is a birational contraction of E' to a curve of degree $(4, 2)$, and p_1 and p_2 are projections to the first and the second factors, respectively. Let $H_1 = (\sigma \circ \pi_1)^*(O_{\mathbb{P}^2}(1))$, $H_2 = (\sigma \circ \pi_2)^*(O_{\mathbb{P}^2}(1))$ and $H_3 = \theta^*(O_{\mathbb{P}^1}(1))$. Then $-K_X \sim H_1 + H_2 + H_3$, $E \sim H_1 - H_3$ and $E' \sim 2H_2 - H_1 + H_3$.

Lemma 5.76   *Let P be a point in E. Then $\delta_P(X) \geqslant \frac{12}{11}$.*

Proof Let ℓ_1 and ℓ_2 be the rulings of $E \cong \mathbb{P}^1 \times \mathbb{P}^1$ contracted by θ and pr_2, respectively. On E, we have $H_1|_E \sim 0$, $H_2|_E \sim 2\ell_2$, $H_3|_E \sim \ell_1$, $E'|_E \sim \ell_1 + 4\ell_2$, $-K_X|_E \sim \ell_1 + 2\ell_2$, $E|_E \sim -\ell_1$. Let C be the curve in $|\ell_2|$ that contains P. By Theorem 1.112, we have

$$\delta_P(X) \geqslant \min\left\{\frac{1}{S_X(E)}, \frac{1}{S(W^E_{\bullet,\bullet}; C)}, \frac{1}{S(W^{E,C}_{\bullet,\bullet,\bullet}; P)}\right\},$$

where $S(W^E_{\bullet,\bullet}; C)$ and $S(W^{E,C}_{\bullet,\bullet,\bullet}; P)$ are defined in Section 1.7. These two numbers can be computed using Corollary 1.110 and Theorem 1.112, respectively.

By Theorem 3.17, we know that $S_X(E) < 1$. Let us compute $S_X(E)$. Take $u \in \mathbb{R}_{\geqslant 0}$. Then $-K_X - uE$ is pseudo-effective $\iff u \leqslant \frac{3}{2}$. For $u \leqslant \frac{3}{2}$, let $P(u)$ be the positive part of the Zariski decomposition of this divisor, and let $N(u)$ be its negative part. Then

$$P(u) = \begin{cases} (1-u)H_1 + H_2 + (1+u)H_3 & \text{if } 0 \leqslant u \leqslant 1, \\ (3-2u)H_2 + 2H_3 & \text{if } 1 \leqslant u \leqslant \dfrac{3}{2}, \end{cases}$$

and

$$N(u) = \begin{cases} 0 & \text{if } 0 \leqslant u \leqslant 1, \\ (u-1)E' & \text{if } 1 \leqslant u \leqslant \dfrac{3}{2}. \end{cases}$$

Hence $S_X(E) = \frac{1}{24}\int_0^1 (24 - 12u - 6u^2)\, du + \frac{1}{24}\int_1^{\frac{3}{2}} 6(3-2u)^2\, du = \frac{17}{24}$.

Now let us compute $S(W^E_{\bullet,\bullet}; C)$. If $u \leqslant 1$, then $N(u)|_E = 0$ and $P(u)|_E \sim (1+u)\ell_1 + 2\ell_2$. Similarly, if $1 \leqslant u \leqslant \frac{3}{2}$, then $N(u)|_E = (u-1)E'|_E$ and $P(u)|_E \sim 2\ell_1 + (6-4u)\ell_2$. Observe that $C \neq E'|_E$. Thus, it follows from Corollary 1.110 that

$$\begin{aligned}
S(W^E_{\bullet,\bullet}; C) &= \frac{3}{(-K_X)^3}\int_0^{\frac{3}{2}}\int_0^{\infty} \mathrm{vol}\big(P(u)|_E - vC\big)\, dv\, du \\
&= \frac{3}{24}\int_0^1\int_0^{\infty} \mathrm{vol}\big((1+u)\ell_1 + (2-v)\ell_2\big)\, dv\, du \\
&\quad + \frac{3}{24}\int_1^{\frac{3}{2}}\int_0^{\infty} \mathrm{vol}\big(2\ell_1 + (6-4u-v)\ell_2\big)\, dv\, du \\
&= \frac{3}{24}\int_0^1\int_0^{2} 2(1+u)(2-v)\, dv\, du \\
&\quad + \frac{3}{24}\int_1^{\frac{3}{2}}\int_0^{6-4u} 4(6-4u-v)\, dv\, du = \frac{11}{12}.
\end{aligned}$$

Finally, let us compute $S(W^{E,C}_{\bullet,\bullet,\bullet};P)$. Using Theorem 1.112, we see that

$$S\left(W^{E,C}_{\bullet,\bullet,\bullet};P\right) = F_P + \frac{3}{24}\int_0^1\int_0^2 (1+u)^2 dvdu + \frac{3}{24}\int_1^{\frac{3}{2}}\int_0^{6-4u} 4dvdu = F_P + \frac{5}{6},$$

where

$$F_P = \begin{cases} 0 & \text{if } P \notin E', \\ \dfrac{6}{24}\displaystyle\int_1^{\frac{3}{2}}\int_0^{6-4u} 2(u-1)dvdu = \dfrac{1}{24} & \text{if } P \notin E'. \end{cases}$$

Thus, we have $S(W^{E,C}_{\bullet,\bullet,\bullet};P) \leqslant \frac{7}{8}$, so that $\delta_P(X) \geqslant \min\left\{\frac{24}{17}, \frac{12}{11}, \frac{8}{7}\right\} = \frac{12}{11}$ as required.   □

Lemma 5.77 *Let P be a point in X, and let S be a surface in the pencil $|H_3|$ that passes through P. Suppose that S is smooth, and $P \notin E$. Then $\delta_P(X) \geqslant \frac{24}{23}$.*

Proof From Theorem 3.17, we know that $S_X(E) < 1$. Let us compute $S_X(E)$ explicitly. Let u be a non-negative real number. Then $-K_X - uS$ is pseudo-effective $\iff u \leqslant 2$. For every $u \leqslant 2$, let $P(u)$ be the positive part of the Zariski decomposition of this divisor, and let $N(u)$ be its negative part. Then

$$P(u) = \begin{cases} H_1 + H_2 + (1-u)H_3 & \text{if } 0 \leqslant u \leqslant 1, \\ (2-u)H_1 + H_2 & \text{if } 1 \leqslant u \leqslant 2, \end{cases}$$

and

$$N(u) = \begin{cases} 0 & \text{if } 0 \leqslant u \leqslant 1, \\ (u-1)E & \text{if } 1 \leqslant u \leqslant 2. \end{cases}$$

Therefore we have $S_X(S) = \frac{1}{24}\int_0^1 (24-15u)\,du + \frac{1}{24}\int_1^2 3(2-u)(5-2u)du = \frac{5}{6}$.

Let $\mathscr{C} = E \cap S$, and let $\varpi\colon S \to \mathbb{P}^2$ be the birational morphism that is induced by pr_1. Then $\mathscr{C}$ is a smooth irreducible curve, S is a smooth del Pezzo surface of degree 5, the morphism φ is a blow up of four distinct points in $\varpi(\mathscr{C})$, and $\varpi(\mathscr{C})$ is a conic, so that $\mathscr{C} \sim 2\ell - e_1 - e_2 - e_3 - e_4$, where e_1, e_2, e_3, e_4 are φ-exceptional curves, and ℓ is the proper transform on S of a general line in $\mathbb{P}^2$. For every $i < j$ in $\{1,2,3,4\}$, let l_{ij} be the proper transform on S of the line in the plane $\mathbb{P}^2$ that passes through the points $\varpi(e_i)$ and $\varpi(e_j)$. Then e_1, e_2, e_3, e_4, l_{12}, l_{13}, l_{14}, l_{23}, l_{24}, l_{34} are all (-1)-curves in the surface S. Observe that $|\mathscr{C}|$ is a basepoint free pencil, which contains exactly three singular curves: the curves $l_{12} + l_{34}$, $l_{13} + l_{24}$, $l_{14} + l_{23}$.

If $P \in l_{12} \cup l_{13} \cup l_{14} \cup l_{23} \cup l_{24} \cup l_{34}$, let C be a curve among l_{12}, l_{13}, l_{14}, l_{23}, l_{24}, l_{34} that contains the point P. Vice versa, if $P \notin l_{12} \cup l_{13} \cup l_{14} \cup l_{23} \cup l_{24} \cup l_{34}$, let C be the unique smooth curve in the pencil $|\mathscr{C}|$ that passes through P. In both cases, we have $C \neq \mathscr{C}$. Moreover, it follows from Theorem 1.112 that

$\delta_P(X) \geqslant \min\left\{\frac{6}{5}, \frac{1}{S(W^S_{\bullet,\bullet};C)}, \frac{1}{S(W^{S,C}_{\bullet,\bullet,\bullet};P)}\right\}$, where $S(W^S_{\bullet,\bullet};C)$ and $S(W^{S,C}_{\bullet,\bullet,\bullet};P)$ are defined in Section 1.7. Let us compute them.

We have $-K_S \sim \mathscr{C} + \ell$ and

$$P(u)\big|_S \sim_{\mathbb{R}} \begin{cases} \mathscr{C} + \ell & \text{if } 0 \leqslant u \leqslant 1, \\ (2-u)\mathscr{C} + \ell & \text{if } 1 \leqslant u \leqslant 2. \end{cases}$$

Let v be a non-negative real number, and let $\tau(u)$ be the largest real number such that the divisor $P(u)|_S - vC$ is pseudo-effective. For $v \in [0, \tau(u)]$, let $P(u, v)$ be the positive part of the Zariski decomposition of the divisor $P(u)|_S - vC$, and let $N(u, v)$ be its negative part. Let us describe $P(u, v)$ and $N(u, v)$.

Suppose that $P \notin l_{12} \cup l_{13} \cup l_{14} \cup l_{23} \cup l_{24} \cup l_{34}$. Then $C \sim \mathscr{C}$. If $u \in [0, 1]$, then

$$P(u)\big|_S - vC \sim_{\mathbb{R}} (1-v)C + \ell \sim_{\mathbb{Q}} \left(\frac{3}{2} - v\right)C + \frac{1}{2}(e_1 + e_2 + e_3 + e_4),$$

so that $\tau(u) = \frac{3}{2}$. Moreover, if $u \in [0, 1]$ and $v \leqslant \frac{3}{2}$, then

$$P(u, v) = \begin{cases} (1-v)C + \ell & \text{if } 0 \leqslant v \leqslant 1, \\ (3-2v)\ell & \text{if } 1 \leqslant v \leqslant \frac{3}{2}, \end{cases}$$

and

$$N(u, v) = \begin{cases} 0 & \text{if } 0 \leqslant v \leqslant 1, \\ (v-1)(e_1 + e_2 + e_3 + e_4) & \text{if } 1 \leqslant v \leqslant \frac{3}{2}. \end{cases}$$

If $u \in [1, 2]$, then

$$P(u)|_S - vC \sim_{\mathbb{R}} (2-u-v)C + \ell \sim_{\mathbb{Q}} \left(\frac{5}{2} - u - v\right)C + \frac{1}{2}(e_1 + e_2 + e_3 + e_4),$$

so that $\tau(u) = \frac{5}{2} - u$. Furthermore, if $u \in [1, 2]$ and $v \leqslant \frac{5}{2} - u$, then

$$P(u, v) = \begin{cases} (2-u-v)C + \ell & \text{if } 0 \leqslant v \leqslant 2 - u, \\ (5-2u-v)\ell & \text{if } 2 - u \leqslant v \leqslant \frac{5}{2} - u, \end{cases}$$

and

$$N(u, v) = \begin{cases} 0 & \text{if } 0 \leqslant v \leqslant 2 - u, \\ (v-2+u)(e_1 + e_2 + e_3 + e_4) & \text{if } 2 - u \leqslant v \leqslant \frac{5}{2} - u. \end{cases}$$

Hence, since $C \neq \mathscr{C}$, it follows from Corollary 1.110 that

$$
\begin{aligned}
S\left(W_{\bullet,\bullet}^S; C\right) &= \frac{3}{24} \int_0^1 \int_0^{\frac{3}{2}} P(u,v) \cdot P(u,v) dv du \\
&+ \frac{3}{24} \int_1^2 \int_0^{\frac{5}{2}-u} P(u,v) \cdot P(u,v) dv du \\
&= \frac{3}{24} \int_0^1 \int_0^1 (5-4v) dv du + \frac{3}{24} \int_0^1 \int_1^{\frac{3}{2}} (2v-3)^2 dv du \\
&+ \frac{3}{24} \int_1^2 \int_0^{2-u} (9-4u-4v) dv du \\
&+ \frac{3}{24} \int_1^2 \int_{2-u}^{\frac{5-2u}{2}} (5-2u-2v)^2 dv du = \frac{9}{16}.
\end{aligned}
$$

Similarly, it follows from Theorem 1.112 that

$$
\begin{aligned}
S\left(W_{\bullet,\bullet,\bullet}^{S,C}; P\right) &= F_P + \frac{3}{24} \int_0^1 \int_0^1 4 dv du \\
&+ \frac{3}{24} \int_0^1 \int_1^{\frac{3}{2}} (6-4v)^2 dv du + \frac{3}{24} \int_1^2 \int_0^{2-u} 4 dv du \\
&+ \frac{3}{24} \int_1^2 \int_{2-u}^{\frac{5-2u}{2}} (10-4u-4v)^2 dv du = F_P + \frac{11}{12},
\end{aligned}
$$

where $F_P = 0$ if $P \notin e_1 \cup e_2 \cup e_3 \cup e_4$, and if $P \in e_1 \cup e_2 \cup e_3 \cup e_4$, then

$$
\begin{aligned}
F_P &= \frac{6}{24} \int_0^1 \int_1^{\frac{3}{2}} (6-4v)(v-1) dv du \\
&+ \frac{6}{24} \int_1^2 \int_{2-u}^{\frac{5-2u}{2}} (10-4u-4v)(v-2+u) dv du = \frac{1}{24}.
\end{aligned}
$$

Thus we have

$$
S\left(W_{\bullet,\bullet,\bullet}^{S,C}; P\right) = \begin{cases} \dfrac{11}{12} & \text{if } P \notin e_1 \cup e_2 \cup e_3 \cup e_4, \\ \dfrac{23}{24} & \text{if } P \in e_1 \cup e_2 \cup e_3 \cup e_4. \end{cases}
$$

Therefore, we see that $\delta_P(X) \geqslant \min\left\{\frac{6}{5}, \frac{16}{9}, \frac{24}{23}\right\} = \frac{24}{23}$ as required. This completes the proof in the case when $P \notin l_{12} \cup l_{13} \cup l_{14} \cup l_{23} \cup l_{24} \cup l_{34}$.

Suppose that $P \notin l_{12} \cup l_{13} \cup l_{14} \cup l_{23} \cup l_{24} \cup l_{34}$. Without loss of generality, we may assume that $P \in l_{12}$ and $C = l_{12}$. If $u \in [0,1]$, then $P(u)|_S - vC \sim_{\mathbb{R}} (2-v)C + l_{34} + e_1 + e_2$, so that $\tau(u) = 2$. If $u \in [0,1]$ and $v \leqslant 2$, then

$$
P(u,v) = \begin{cases} (2-v)C + l_{34} + e_1 + e_2 & \text{if } 0 \leqslant v \leqslant 1, \\ (2-v)(C + l_{34} + e_1 + e_2) & \text{if } 1 \leqslant v \leqslant 2, \end{cases}
$$

and

$$N(u,v) = \begin{cases} 0 & \text{if } 0 \leqslant v \leqslant 1, \\ (v-1)(l_{34} + e_1 + e_2) & \text{if } 1 \leqslant v \leqslant 2. \end{cases}$$

Similarly, if $u \in [1,2]$, then $P(u)|_S - vC \sim_{\mathbb{R}} (3-u-v)C + (2-u)l_{34} + e_1 + e_2$, so that $\tau(u) = 3 - u$. Hence, if $u \in [1,2]$ and $v \leqslant 3 - u$, then

$$P(u,v) = \begin{cases} (3-u-v)C + (2-u)l_{34} + e_1 + e_2 & \text{if } 0 \leqslant v \leqslant 2 - u, \\ (3-u-v)(C + e_1 + e_2) + (2-u)l_{34} & \text{if } 2 - u \leqslant v \leqslant 1, \\ (3-u-v)(C + l_{34} + e_1 + e_2) & \text{if } 1 \leqslant v \leqslant 3 - u, \end{cases}$$

and

$$N(u,v) = \begin{cases} 0 & \text{if } 0 \leqslant v \leqslant 2 - u, \\ (v-2+u)(e_1 + e_2) & \text{if } 2 - u \leqslant v \leqslant 1, \\ (v-1)l_{34} + (v-2+u)(e_1 + e_2) & \text{if } 1 \leqslant v \leqslant 3 - u. \end{cases}$$

Therefore, using Corollary 1.110, we get

$$S(W^S_{\bullet,\bullet}; C) = \frac{3}{24} \int_0^1 \int_0^1 (5 - 2v - v^2) dv du + \frac{3}{24} \int_0^1 \int_1^2 2(v-2)^2 dv du$$

$$+ \frac{3}{24} \int_1^2 \int_0^{2-u} (9 - 4u - 2v - v^2) dv du$$

$$+ \frac{3}{24} \int_1^2 \int_{2-u}^1 (2u^2 + 4uv + v^2 - 12u - 10v + 17) dv du$$

$$+ \frac{3}{24} \int_1^2 \int_1^{3-u} 2(3-u-v)^2 dv du = \frac{19}{24}.$$

Similarly, it follows from Theorem 1.112 that

$$S(W^{S,C}_{\bullet,\bullet,\bullet}; P) = F_P + \frac{3}{24} \int_0^1 \int_0^1 (1+v)^2 dv du + \frac{3}{24} \int_0^1 \int_1^2 (4-2v)^2 dv du$$

$$+ \frac{3}{24} \int_1^2 \int_0^{2-u} (1+v)^2 dv du + \frac{3}{24} \int_1^2 \int_{2-u}^1 (5-2u-v)^2 dv du$$

$$+ \frac{3}{24} \int_1^2 \int_1^{3-u} (6-2u-2v)^2 dv du = F_P + \frac{11}{16},$$

where F_P is calculated as follows. If $P \notin e_1 \cup e_2 \cup l_{34}$, then $F_P = 0$. If $P = C \cap l_{34}$, then

$$F_P = \frac{6}{24} \int_0^1 \int_1^2 (4-2v)(v-1) dv du + \frac{6}{24} \int_1^2 \int_1^{3-u} (6-2u-2v)(v-1) dv du = \frac{5}{48}.$$

Finally, if $P = C \cap e_1$ or $P = C \cap e_2$, then

$$F_P = \frac{6}{24}\int_0^1\int_1^2 (4-2v)(v-1)dvdu + \frac{6}{24}\int_1^2\int_{2-u}^1 (5-2u-v)(v-1)dvdu$$
$$+ \frac{6}{24}\int_1^2\int_1^{3-u}(6-2u-2v)(v-1)dvdu = \frac{5}{48}.$$

Thus we have $S(W_{\bullet,\bullet,\bullet}^{S,C}; P) \leqslant \frac{27}{32}$. Then $\delta_P(X) \geqslant \min\left\{\frac{6}{5}, \frac{24}{19}, \frac{32}{27}\right\} = \frac{32}{27} > \frac{24}{23}$ as required. This completes the proof of the lemma. $\qquad\square$

Using Lemmas 5.76 and 5.77, we obtain

Corollary 5.78 *Let G be a reductive subgroup in $\mathrm{Aut}(X)$ such that the pencil $|H_3|$ does not have G-invariant surfaces. Then X is K-polystable.*

Proof Suppose that X is not K-polystable. Then it follows from Theorem 1.22 that there exists a G-invariant prime divisor F over X such that $\beta(F) = A_X(F) - S_X(F) \leqslant 0$. Let $Z = C_X(F)$. Then the restriction of $\theta|_Z \colon Z \to \mathbb{P}^1$ is surjective because otherwise the pencil $|H_3|$ would contain a G-invariant surface.

Let S be a general surface in $|H_3|$. Then S is smooth, and $S \cap Z \neq \varnothing$. Therefore, for any point $P \in S \cap Z$, we have $\frac{A_X(F)}{S_X(F)} \geqslant \delta_P(X) > 1$ by Lemmas 5.76 and 5.77, which is a contradiction since $A_X(F) \leqslant S_X(F)$. $\qquad\square$

Now we can prove our Proposition 5.74. If $O = [1:0:0]$ and Y is given by (5.12), let G be the subgroup in $\mathrm{Aut}(Y)$ that is generated by the involution

$$([x:y:z],[u:v:w]) \mapsto ([x:z:y],[u:w:u])$$

and the selfmaps $([x:y:z],[u:v:w]) \mapsto ([\lambda^2 x:y:\lambda^4 z],[\lambda u:\lambda^2 v:w])$ for $\lambda \in \mathbb{C}^*$. Likewise, if $O = [1:1:1]$ and Y is given by (5.14) with $\mu \neq 2$ and $\mu^2 \neq -1$, we let G be the subgroup in $\mathrm{Aut}(Y)$ generated by the involution

$$([x:y:z],[u:v:w]) \mapsto ([y:x:z],[v:u:w])$$

and the selfmap $([x:y:z],[u:v:w]) \mapsto ([y:z:x],[v:w:u])$. Then $G \cong \mathbb{G}_m \rtimes \mu_2$ in the former case, and $G \cong \mathfrak{S}_3$ in the latter case. In both cases, the curve C is G-invariant, so that the G-action lifts to the threefold X. Moreover, it is not hard to check that the pencil $|H_3|$ does not contain G-invariant surfaces, so that X is K-polystable in both cases by Corollary 5.78.

5.17 Family №3.10

In this section, we solve the Calabi Problem for all smooth Fano threefolds in family №3.10.

Let Q be a smooth quadric threefold in $\mathbb{P}^4$, let C_1 and C_2 be two disjoint smooth irreducible conics in Q, and let X be the blow up of the quadric Q in these two conics. Then X is a smooth Fano threefold №3.10, and every smooth threefold in this family can be obtained in this way. Moreover, we may assume that $C_1 = \{w^2 + zt = x = y = 0\}$ and $C_2 = \{w^2 + xy = z = t = 0\}$, where x, y, z, t, w are coordinates on $\mathbb{P}^4$. Then, using an appropriate coordinate change, we may assume that the quadric Q is given by one of the following three equations:

(ꓱ) $w^2 + xy + zt + a(xt + yz) + b(xz + yt) = 0$, where $a \in \mathbb{C} \ni b$ such that $a \pm b \pm 1 \ne 0$;

(ꓕ) $w^2 + xy + zt + a(xt + yz) + xz = 0$, where $a \in \mathbb{C}$ such that $a \ne \pm 1$;

(ꓶ) $w^2 + xy + zt + xt + xz = 0$.

The goal of this section is to prove the following result:

Proposition 5.79 *The threefold X is K-polystable $\iff$ Q is given by (ꓱ).*

In all three cases, we have the commutative diagram

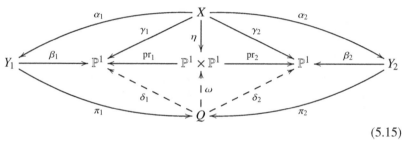

$$(5.15)$$

where δ_1 is a rational map given by $[x : y : z : t : w] \mapsto [x : y]$, the map δ_2 is a rational map given by $[x : y : z : t : w] \mapsto [z : t]$, the map ω is a rational map

$$[x : y : z : t : w] \mapsto ([x : y], [z : t]),$$

the maps π_1 and π_2 are blow ups of the quadric Q at the conics C_1 and C_2, respectively, the maps α_1 and α_2 are blow ups of the proper transforms of these conics, respectively, both β_1 and β_2 are fibrations into quadric surfaces, both γ_1 and γ_2 are fibrations into sextic del Pezzo surfaces, η is a conic bundle, and pr_1 and pr_2 are natural projections. Occasionally, we will consider $[x : y]$ and $[z : t]$ as coordinated on $\mathbb{P}^1 \times \mathbb{P}^1$.

Let $\mathscr{C}$ be the discriminant curve in $\mathbb{P}^1 \times \mathbb{P}^1$ of the conic bundle η. Then $\mathscr{C}$ has at most nodal singularities, and its degree is $(2, 2)$. If Q is given by (ꓱ), then $\mathscr{C}$ is given by

$$a^2(x^2t^2 + y^2z^2) + 2ab(xyz^2 + xyt^2 + ztx^2 + zty^2)$$
$$+ b^2(x^2z^2 + y^2t^2) + 2(a^2 + b^2 - 2)yzxt = 0.$$

If $ab \neq 0$, the curve $\mathscr{C}$ is irreducible and smooth, which also implies that Aut(X) is finite. If $a = 0$ or $b = 0$ (but not both), the curve $\mathscr{C}$ is reducible: it splits as a union of two smooth curves of degree $(1,1)$, which meet at two points. In this case, we have $\mathrm{Aut}^0(X) \cong \mathbb{G}_m$. Similarly, if $a = 0$ and $b = 0$, then $\mathrm{Aut}^0(X) \cong \mathbb{G}_m^2$ and the curve $\mathscr{C}$ is given by $xyzt = 0$, so that X is the unique smooth Fano threefold №3.10 that admits an effective $\mathbb{G}_m^2$-action.

If the quadric Q is given by (⌋), then $\mathscr{C}$ is given by the following equation:

$$a^2t^2x^2 + (2a^2 - 4)xyzt + 2atzx^2 + a^2y^2z^2 + 2ayz^2x + z^2x^2 = 0.$$

If $a \neq 0$, this curve is irreducible and has one node, which implies that Aut(X) is finite. On the other hand, if $a = 0$, then the defining equation simplifies as $zx(zx - 4yt) = 0$, so that the curve $\mathscr{C}$ splits as a union of 3 smooth curves of degree $(0,1)$, $(1,0)$ and $(1,1)$, which meet transversally at 3 distinct points. In this subcase, we have $\mathrm{Aut}^0(X) \cong \mathbb{G}_m$.

Finally, if Q is given by (⌐), then the curve $\mathscr{C}$ is given by

$$2x(t^2x + 2txz - 4tyz + xz^2) = 0,$$

so that $\mathscr{C}$ is a union of a curve of degree $(1,0)$ and a smooth curve of degree $(1,2)$, which implies that Aut(X) is also finite in this case.

Let H be the pull back on X of a general hyperplane section of the quadric threefold Q, let E_1 be the α_1-exceptional surface, and let E_2 be the α_2-exceptional surface. Then

$$\mathrm{Eff}(X) = \mathbb{R}_{\geqslant 0}\big[E_1\big] + \mathbb{R}_{\geqslant 0}\big[E_1\big] + \mathbb{R}_{\geqslant 0}\big[H - E_1\big] + \mathbb{R}_{\geqslant 0}\big[H - E_2\big] + \mathbb{R}_{\geqslant 0}\big[2H - E_1 - E_2\big],$$

the del Pezzo fibration γ_1 is given by $|H - E_1|$, the fibration γ_2 is given by $|H - E_2|$, and the conic bundle η is given by the linear system $|2H - E_1 - E_2|$.

Let us show that X is K-polystable in the case when Q is given by (⌐).

Lemma 5.80 ([202, Theorem 1.1]) *Suppose that Q is given by (⌐) and $a = b = 0$. Then X is K-polystable.*

Proof Let G be the subgroup in Aut$(\mathbb{P}^4)$ generated by

$$[x : y : z : t : w] \mapsto [z : t : x : y : w],$$
$$[x : y : z : t : w] \mapsto [y : x : z : t : w],$$
$$[x : y : z : t : w] \mapsto [x : y : t : z : w],$$
$$[x : y : z : t : w] \mapsto [x : y : rz : t/r : w],$$
$$[x : y : z : t : w] \mapsto [sx : y/s : z : t : w],$$

where $r \in \mathbb{C}^*$ and $s \in \mathbb{C}^*$. Then $G \cong \mathbb{G}_m^2 \rtimes (\mu_2^2 \rtimes \mu_2)$, the quadric Q is G-invariant, and the locus $C_1 \cup C_2$ is G-invariant, so that the action of the group G

lifts to the threefold X. Therefore, we may identify G with a subgroup in $\mathrm{Aut}(X)$. Now, applying Theorem 1.52, we obtain $\alpha_G(X) \geqslant 1$, so that X is K-polystable by Theorem 1.48. $\square$

Lemma 5.81   *If Q is given by $(\square)$ and $ab = 0$, then X is K-polystable.*

Proof   By Lemma 5.80, we may assume that $a \neq 0$ or $b \neq 0$. Without loss of generality, we may assume that $a \neq 0$. Then $b = 0$. Let G be the subgroup in $\mathrm{Aut}(\mathbb{P}^4)$ generated by

$$[x : y : z : t : w] \mapsto [y : x : t : z : w],$$
$$[x : y : z : t : w] \mapsto [z : t : x : y : w],$$
$$[x : y : z : t : w] \mapsto [x/s : ys : z/s : ts : w],$$

where s is any non-zero complex number. Then Q is G-invariant, and $G \cong (\mathbb{G}_m \rtimes \mu_2) \times \mu_2$. Moreover, the locus $C_1 \cup C_2$ is G-invariant, so that the G-action lifts to the threefold X. Therefore, we may identify G with a subgroup in $\mathrm{Aut}(X)$. Note that $\alpha_G(X) \leqslant \frac{2}{3}$.

Observe that X does not have G-fixed points because Q does not have G-fixed points. The conic bundle η in (5.15) is G-equivariant, and G acts on $\mathbb{P}^1 \times \mathbb{P}^1$ such that $\mathbb{P}^1 \times \mathbb{P}^1$ does not contain G-fixed points, $\mathbb{P}^1 \times \mathbb{P}^1$ does not contain G-invariant curves of degree $(1,0)$ or $(0,1)$, and the only G-invariant curves in $\mathbb{P}^1 \times \mathbb{P}^1$ of degree $(1,1)$ are the curves given by $xt + yz = 0$ and $xt - yz = 0$.

Suppose X is not K-polystable. By Theorem 1.22, there exists a G-equivariant birational morphism $f \colon \widetilde{X} \to X$ such that $\beta(F) = A_X(F) - S_X(F) \leqslant 0$ for some G-invariant dreamy prime divisor $F \subset \widetilde{X}$. Let $Z = f(F)$. Then Z is not a surface by Theorem 3.17. Thus, since X does not contain G-fixed points, Z is a G-invariant irreducible curve.

Using Lemma 1.45, we conclude that $\alpha_{G,Z}(X) < \frac{3}{4}$. Thus, by Lemma 1.42, there is a G-invariant effective $\mathbb{Q}$-divisor D on the threefold X such that $D \sim_{\mathbb{Q}} -K_X$ and Z is contained in $\mathrm{Nklt}(X, \lambda D)$ for some positive rational number $\lambda < \frac{3}{4}$. By Theorem A.4, the locus $\mathrm{Nklt}(X, \lambda D)$ is connected. Moreover, since $D \sim_{\mathbb{Q}} 3H - E_1 - E_2$, either the locus $\mathrm{Nklt}(X, \lambda D)$ is one-dimensional, or it contains one G-invariant surface, which is contained in $|2H - E_1 - E_2|$. In the former case, the G-invariant surface in $\mathrm{Nklt}(X, \lambda D)$ is mapped by the conic bundle η to a G-invariant curve in $\mathbb{P}^1 \times \mathbb{P}^1$ of degree $(1,1)$.

If Z is not contained in a two-dimensional component of $\mathrm{Nklt}(X, \lambda D)$, then, applying Corollary A.12 to $(X, \lambda D)$, we find that $(H - E_1) \cdot Z \leqslant 1$ and $(H - E_2) \cdot Z \leqslant 1$, so that either $\eta(Z)$ is a point, or $\eta(Z)$ is a G-invariant irreducible curve of degree $(1,1)$. If Z is contained in a two-dimensional G-irreducible component of the locus $\mathrm{Nklt}(X, \lambda D)$, then this component is mapped by η to a G-invariant curve of degree $(1,1)$ in $\mathbb{P}^1 \times \mathbb{P}^1$. Hence, either $\eta(Z)$ is a G-invariant point, or $\eta(Z)$ is a G-invariant curve of degree $(1,1)$. Since $\mathbb{P}^1 \times \mathbb{P}^1$ contains no G-fixed points, we see that $\eta(Z)$ is a curve given by $xt \pm yz = 0$.

Let S be the unique surface in $|2H - E_1 - E_2|$ that contains Z, let $\overline{S}$ be its image in Q. Then $\overline{S} = \{w^2 + xy + zt + a(xt + yz) = xt \pm yz = 0\}$, so that $\overline{S}$ is a singular quartic del Pezzo surface, whose singular locus consist of 4 points. If $\eta(Z)$ is given by $xt + yz = 0$, these points are $[1 : 0 : -1 : 0 : 0]$, $[1 : 0 : 1 : 0 : 0]$, $[0 : 1 : 0 : -1 : 0]$, $[0 : 1 : 0 : 1 : 0]$. Similarly, if $\eta(Z)$ is given by $xt - yz = 0$, then the surface $\overline{S}$ is singular at the following points: $[-a \pm \sqrt{a^2 - 1} : 0 : 1 : 0 : 0]$ and $[0 : -a \pm \sqrt{a^2 - 1} : 0 : 0 : 1]$. In both cases, the surface $\overline{S}$ contains C_1 and C_2, and $\mathrm{Sing}(\overline{S})$ is disjoint from these conics, so that $S \cong \overline{S}$.

Let $\mathcal{H} = H|_S$, $C_1 = E_1|_S$, $C_2 = E_2|_S$. Then $|C_1|$ and $|\mathcal{H} - C_1|$ are basepoint free pencils, and the surface S contains two curves ℓ and ℓ' such that $C_1 \sim C_2 \sim 2\ell$ and $\mathcal{H} - C_1 \sim 2\ell'$. Then $\ell^2 = (\ell')^2 = 0$ and $\ell \cdot \ell' = \frac{1}{2}$. $\mathcal{H} \sim 2\ell + 2\ell'$. Moreover, there are non-negative integers n and m such that $Z \sim_{\mathbb{Q}} n\ell + m\ell'$. If $n = 0$, then $(2H - E_1 - E_2) \cdot Z = 0$, so that $\eta(Z)$ is a point, which is impossible. Then $n \geqslant 1$, so that $Z - \ell$ is pseudo-effective.

Let us apply results of Section 1.7 to S and Z using notations introduced in this section. First, we note that $S_X(S) < 1$ by Theorem 3.17. Hence, using Corollary 1.110, we conclude that $S(W^S_{\bullet,\bullet}; Z) \geqslant 1$. We show that this is not so.

Let $u \in \mathbb{R}_{\geqslant 0}$, let $v \in \mathbb{R}_{\geqslant 0}$, let $P(u) = P(-K_X - uS)$ and let $N(u) = N(-K_X - uS)$. Then $-K_X - uS$ is not pseudo-effective for $u > \frac{3}{2}$ since $-K_X - uS \sim_{\mathbb{R}} (\frac{3}{2} - u)S + \frac{1}{2}(E_1 + E_2)$. Moreover, if $0 \leqslant u \leqslant 1$, then $P(u)|_S - vZ \sim_{\mathbb{R}} (2 - nv)\ell + (6 - 4u - mv)\ell'$ on the surface S because $N(u) = 0$ and $P(u) = -K_X - uS$ in this case. Similarly, if $1 \leqslant u \leqslant \frac{3}{2}$, then we have $P(u)|_S - vZ \sim_{\mathbb{R}} (6 - 4u - nv)\ell + (6 - 4u - mv)\ell'$ because $N(u) = (u - 1)(E_1 + E_2)$ and $P(u) = (3 - 2u)H$ in this case. Thus, if $Z = C_1$ or $Z = C_2$,

$$
\begin{aligned}
S\big(W^S_{\bullet,\bullet}; Z\big) &= \frac{3}{26} \int_1^{\frac{3}{2}} (6 - 4u)^2 (u - 1) \, du \\
&\quad + \frac{3}{26} \int_0^1 \int_0^\infty \mathrm{vol}\big((2 - 2v)\ell + (6 - 4u)\ell'\big) \, dv \, du \\
&\quad + \frac{3}{26} \int_1^{\frac{3}{2}} \int_0^\infty \mathrm{vol}\big((6 - 4u - 2v)\ell + (6 - 4u)\ell'\big) \, dv \, du \\
&= \frac{1}{104} + \frac{3}{26} \int_0^1 \int_0^1 (2 - 2v)(6 - 4u) \, dv \, du \\
&\quad + \frac{3}{26} \int_1^{\frac{3}{2}} \int_0^{3 - 2u} (6 - 4u - 2v)(6 - 4u) \, dv \, du = \frac{1}{2}.
\end{aligned}
$$

Likewise, if $Z \neq C_1$ and $Z \neq C_2$, then $S(W^S_{\bullet,\bullet}; Z) \leqslant S(W^S_{\bullet,\bullet}; \ell) = \frac{51}{52}$. Thus in every case we have $S(W^S_{\bullet,\bullet}; Z) < 1$, which is a contradiction since we proved earlier that $S(W^S_{\bullet,\bullet}; Z) \geqslant 1$. Therefore X is K-polystable. $\qquad\square$

Lemma 5.82 *Suppose that Q is given by (◲) and $a = b$. Then X is K-polystable.*

Proof By Lemma 5.80, we may assume that $a = b \neq 0$. Then the curve $\mathscr{C}$ is smooth, and the group $\mathrm{Aut}(X)$ is finite. Let G be the finite subgroup in $\mathrm{Aut}(\mathbb{P}^4)$ generated by

$$[x : y : z : t : w] \mapsto [y : x : z : t : w],$$
$$[x : y : z : t : w] \mapsto [x : y : t : z : w],$$
$$[x : y : z : t : w] \mapsto [z : t : x : y : w],$$
$$[x : y : z : t : w] \mapsto [x : y : z : t : -w].$$

Then $G \cong \mu_2 \times (\mu_2^2 \rtimes \mu_2)$, the quadric Q is G-invariant, and $C_1 \cup C_2$ is G-invariant. The action of the group G lifts to X, and we may identify G with a subgroup in $\mathrm{Aut}(X)$. Then X contains no G-fixed points, η is G-equivariant, and G acts on $\mathbb{P}^1 \times \mathbb{P}^1$ such that the only G-fixed points in $\mathbb{P}^1 \times \mathbb{P}^1$ are $([1 : 1], [1 : 1])$ and $([1 : -1], [1 : -1])$, $\mathbb{P}^1 \times \mathbb{P}^1$ does not contain G-invariant curves of degree $(1, 0)$ or $(0, 1)$, and the only G-invariant curves of degree $(1, 1)$ in $\mathbb{P}^1 \times \mathbb{P}^1$ are reducible curves $(x - y)(z - t) = 0$ and $(x + y)(z + t) = 0$.

Suppose X is not K-polystable. By Theorem 1.22, there is a G-invariant prime divisor F over X such that $\beta(F) = A_X(F) - S_X(F) \leqslant 0$. Let $Z = C_X(F)$. Then Z is not a surface by Theorem 3.17, so that Z is a G-invariant curve since X has no G-fixed points.

Arguing as in the proof of Lemma 5.81, we see that either $\eta(Z)$ is a G-invariant point, or $\eta(Z)$ is an irreducible G-invariant curve of degree $(1, 1)$. But $\mathbb{P}^1 \times \mathbb{P}^1$ does not contain irreducible G-invariant curves of degree $(1, 1)$. Thus, we conclude that $\eta(Z)$ is a point. Then either $\eta(Z) = ([1 : 1], [1 : 1])$ or $\eta(Z) = ([1 : -1], [1 : -1])$, so that $\eta(Z) \notin \mathscr{C}$, which implies that Z is a smooth fiber of the conic bundle η.

Let S be the unique surface in the linear system $|H - E_1|$ that contains the curve Z, and let $\overline{S}$ be its image in Q. Then $\overline{S}$ is a smooth quadric surface, $C_1 \subset \overline{S}$, and $\overline{S}$ intersects the conic C_2 transversally in two points, so that S is a smooth sextic del Pezzo surface, and $\pi_1 \circ \alpha_1 = \pi_2 \circ \alpha_2$ induces a birational morphism $\varphi \colon S \to \overline{S}$ that is a blow up of the intersection points $\overline{S} \cap C_2$. We have $E_2|_S = \mathbf{e}_1 + \mathbf{e}_2$, where $\mathbf{e}_1$ and $\mathbf{e}_2$ are (-1)-curves in S contracted by φ. We also have $E_1|_S \sim H|_S \sim \ell_1 + \ell_2 + \mathbf{e}_1 + \mathbf{e}_2$, where ℓ_1 and ℓ_2 are (-1)-curves in S such that $\varphi(\ell_1)$ and $\varphi(\ell_2)$ are intersecting lines that pass through the points $\varphi(\mathbf{e}_1)$ and $\varphi(\mathbf{e}_2)$, respectively. Then $Z \sim \ell_1 + \ell_2$.

As in the proof of Lemma 5.81, we are going to apply results of Section 1.7 to S and Z. By Theorem 3.17, we have $S_X(S) < 1$, so that $S(W^S_{\bullet,\bullet}; Z) \geqslant 1$ by Corollary 1.110.

Let $P(u) = P(-K_X - uS)$ and $N(u) = N(-K_X - uS)$, where u is a non-negative real number. Observe that

$$-K_X - uS \sim_{\mathbb{R}} (2 - u)S + (H - E_2) + E_1 \sim_{\mathbb{R}} (3 - u)H - (1 - u)E_1 - E_2.$$

Then $-K_X - uS$ is nef $\iff u \in [0, 1]$, and $-K_X - uS$ is pseudo-effective $\iff u \in [0, 2]$. Moreover, we have

$$P(u) = \begin{cases} (3 - u)H - (1 - u)E_1 - E_2 & \text{if } 0 \leqslant u \leqslant 1, \\ (3 - u)H - E_2 & \text{if } 1 \leqslant u \leqslant 2, \end{cases}$$

and $N(u) = (u - 1)E_1$ if $1 \leqslant u \leqslant 2$. Let v be a non-negative real number. If $u \in [0, 1]$, then $P(u)|_S - vZ \sim_{\mathbb{R}} (2 - v)(\ell_1 + \ell_2) + \mathbf{e}_1 + \mathbf{e}_2$, so that $P(u)|_S - vZ$ is not pseudo-effective for every $v > 2$. In this case, if $v \in [0, 1]$, then the divisor $P(u)|_S - vZ$ is nef. Furthermore, if $v \in [1, 2]$, then its Zariski decomposition is

$$P(u)\big|_S - vZ \sim_{\mathbb{R}} \underbrace{(2 - v)(\ell_1 + \ell_2 + \mathbf{e}_1 + \mathbf{e}_2)}_{\text{positive part}} + \underbrace{(v - 1)(\mathbf{e}_1 + \mathbf{e}_2)}_{\text{negative part}}.$$

Similarly, if $u \in [1, 2]$, then $P(u)|_S - vZ \sim_{\mathbb{R}} (3 - u - v)(\ell_1 + \ell_2) + (2 - u)(\mathbf{e}_1 + \mathbf{e}_2)$, so that the divisor $P(u)|_S - vZ$ is not pseudo-effective for $v > 3 - u$. Moreover, if $v \in [0, 1]$, then this divisor is nef. Finally, if $1 \leqslant v \leqslant 3 - u$, then its Zariski decomposition is

$$P(u)\big|_S - vZ \sim_{\mathbb{R}} \underbrace{(3 - u - v)(\ell_1 + \ell_2 + \mathbf{e}_1 + \mathbf{e}_2)}_{\text{positive part}} + \underbrace{(v - 1)(\mathbf{e}_1 + \mathbf{e}_2)}_{\text{negative part}}.$$

Thus we have

$$\begin{aligned}
S(W^S_{\bullet, \bullet}; Z) &= \frac{3}{26} \int_0^1 \int_0^1 \left((2 - v)(\ell_1 + \ell_2) + \mathbf{e}_1 + \mathbf{e}_2 \right)^2 dv\, du \\
&\quad + \frac{3}{26} \int_0^1 \int_1^2 \left((2 - v)(\ell_1 + \ell_2 + \mathbf{e}_1 + \mathbf{e}_2) \right)^2 dv\, du \\
&\quad + \frac{3}{26} \int_1^2 \int_0^1 \left((3 - u - v)(\ell_1 + \ell_2) + (2 - u)(\mathbf{e}_1 + \mathbf{e}_2) \right)^2 dv\, du \\
&\quad + \frac{3}{26} \int_1^2 \int_1^{3-u} \left((3 - u - v)(\ell_1 + \ell_2 + \mathbf{e}_1 + \mathbf{e}_2) \right)^2 dv\, du \\
&= \frac{3}{26} \int_0^1 \int_0^1 (6 - 4v)\, dv\, du + \frac{3}{26} \int_0^1 \int_1^2 (2 - v)^2\, dv\, du \\
&\quad + \frac{3}{26} \int_1^2 \int_0^1 2(2 - u)(4 - u - 2v)\, dv\, du \\
&\quad + \frac{3}{26} \int_1^2 \int_1^{3-u} 2(3 - u - v)^2\, dv\, du = \frac{3}{4}.
\end{aligned}$$

The obtained contradiction completes the proof of the lemma.   □

Now, combining the proofs of Lemmas 5.81 and 5.82, we obtain

Lemma 5.83 *Suppose that Q is given by (⊐). Then X is K-polystable.*

Proof By Lemma 5.80, we may assume that $a \neq 0$ and $b \neq 0$. Then $\mathscr{C}$ is smooth, and the group $\mathrm{Aut}(X)$ is finite. Let G be the finite subgroup in $\mathrm{Aut}(\mathbb{P}^4)$ generated by

$$[x : y : z : t : w] \mapsto [y : x : t : z : w],$$
$$[x : y : z : t : w] \mapsto [z : t : x : y : w],$$
$$[x : y : z : t : w] \mapsto [x : y : z : t : -w].$$

Then $G \cong \mu_2^3$, the quadric Q is G-invariant, and the locus $C_1 \cup C_2$ is G-invariant, which implies that the G-action lifts to X, so that we may identify G with a subgroup in $\mathrm{Aut}(X)$. Observe that X does not have G-fixed points because Q does not have G-fixed points.

Recall that the conic bundle η in (5.15) is G-equivariant, and G acts on $\mathbb{P}^1 \times \mathbb{P}^1$ such that $([1 : 1], [1 : 1])$ and $([1 : -1], [1 : -1])$ are the only G-fixed points in $\mathbb{P}^1 \times \mathbb{P}^1$, and $\mathbb{P}^1 \times \mathbb{P}^1$ contains no G-invariant curves of degree $(1, 0)$ or $(0, 1)$. Moreover, the G-invariant curves of degree $(1, 1)$ in $\mathbb{P}^1 \times \mathbb{P}^1$ can be described as follows: $\{xt = yz\}$, $\{xt = yz\}$, and all curves in the pencil $\mathcal{P}$ that is given by $r(xt + yz) = s(xz + yt)$, where $[r : s] \in \mathbb{P}^1$. Note that the pencil $\mathcal{P}$ contains two reducible curves: $\{(x - y)(z - t) = 0\}$ and $\{(x + y)(z + t) = 0\}$, which correspond to $[r : s] = [1 : 1]$ and $[r : s] = [1 : -1]$, respectively.

Suppose X is not K-polystable. By Theorem 1.22, there exists a G-invariant prime divisor F over X with $\beta(F) \leqslant 0$. Let $Z = C_X(F)$. Then $\dim(Z) \leqslant 1$ by Theorem 3.17, so that Z is a G-invariant irreducible curve, because X does not have G-fixed points.

Arguing as in the proof of Lemma 5.81, we see that either $\eta(Z)$ is a G-invariant point, or $\eta(Z)$ is an irreducible G-invariant curve of degree $(1, 1)$. Furthermore, if $\eta(Z)$ is a point, then $\eta(Z) \notin \mathscr{C}$, so that Z is a smooth fiber of the conic bundle η. In this case, for all admissible a and b, the unique surface in $|H - E_1|$ that contains the curve Z is a smooth sextic del Pezzo surface, so that we are exactly in the situation of the proof of Lemma 5.82 and, therefore, we can obtain a contradiction arguing exactly as in this proof. This shows that $\eta(Z)$ is a curve of degree $(1, 1)$.

Let S be the surface in $|2H - E_1 - E_2|$ that contains Z, and let $\overline{S}$ be its image in Q. Then $\overline{S}$ is a quartic del Pezzo surface that contains C_1 and C_2. Since $a \neq 0$ and $b \neq 0$, either the surface $\overline{S}$ is smooth, or $\overline{S}$ has exactly two isolated ordinary double points. Furthermore, if $\overline{S}$ is singular, its singular locus is disjoint from the conics C_1 and C_2. We will provide explicit computations in the end of the proof. In particular, $S \cong \bar{S}$. Now we can proceed as we did in the proof of Lemma 5.81.

Namely, let us apply results of Section 1.7 to S and Z using notations introduced in this section. By Theorem 3.17, we have $S(V_\bullet; S) < 1$. Hence, using Corollary 1.110, we conclude that $S(W^S_{\bullet,\bullet}; Z) \geqslant 1$. Let us show that this is not the case.

Let $\mathcal{H} = H|_S$, $C_1 = E_1|_S$ and $C_2|_S$. Then $C_1 \sim C_2$, both $|C_1|$ and $|\mathcal{H} - C_1|$ are basepoint free pencils. Let C' be a general curve in $|\mathcal{H} - C_1|$. Then $C_1^2 = 0$, $(C')^2 = 0$ and $C_1 \cdot C' = 2$.

Suppose that $Z \sim_\mathbb{Q} \frac{n}{2}C_1 + \frac{m}{2}C'$ for some non-negative integers n and m. Then $n \geqslant 1$ since otherwise $\eta(Z)$ would be a point, which is not so. Thus, if $Z \neq C_1$ and $Z \neq C_2$, to estimate $S(W^S_{\bullet,\bullet}; Z)$ from above we may assume that $n = 1$ and $m = 0$. In this case, arguing as in the proof of Lemma 5.81, we see that

$$
\begin{aligned}
S(W^S_{\bullet,\bullet}; Z) &= \frac{3}{26} \int_0^1 \int_0^\infty \mathrm{vol}\left(\left(1 - \frac{1}{2}v\right)C_1 + (3 - 2u)C'\right) dv du \\
&+ \frac{3}{26} \int_1^{\frac{3}{2}} \int_0^\infty \mathrm{vol}\left(\left(3 - 2u - \frac{1}{2}v\right)C_1 + (3 - 2u)C'\right) dv du \\
&= \frac{3}{26} \int_0^1 \int_0^2 4\left(1 - \frac{1}{2}v\right)(3 - 2u) dv du \\
&+ \frac{3}{26} \int_1^{\frac{3}{2}} \int_0^{6-4u} 4\left(3 - 2u - \frac{1}{2}v\right)(3 - 4u) dv du = \frac{51}{52}.
\end{aligned}
$$

Similarly, if $Z = C_1$ or $Z = C_2$, then arguing as in the end of the proof of Lemma 5.81, we obtain $S(W^S_{\bullet,\bullet}; Z) = \frac{1}{2}$. Thus, we see that $S(W^S_{\bullet,\bullet}; Z) < 1$, so that X is K-polystable.

To complete the proof of the lemma, it is enough to show that every G-invariant curve on the surface S is $\mathbb{Q}$-rationally equivalent to $\frac{1}{2}(nC_1 + mC')$ for some $n \in \mathbb{Z}_{\geqslant 0}$ and $m \in \mathbb{Z}_{\geqslant 0}$. Since $S \cong \bar{S}$, we identify $S = \bar{S}$, so that now S is a quartic del Pezzo surface in $\mathbb{P}^4$.

Suppose that $\eta(Z)$ is given by $xt = yz$. Then $S = Q \cap \{xt = yz\}$. Therefore, the projection $[x : y : z : t : w] \mapsto [x : y : z : t]$ induces a G-equivariant double cover $\varphi \colon S \to Y$ such that Y is the smooth quadric surface in $\mathbb{P}^3$ that is given by $xt = yz$, and the ramification divisor of the double cover φ is the curve $Y \cap \{xy + zt + a(xt + yz) + b(xz + yt) = 0\}$, where we consider x, y, z, t as coordinates on $\mathbb{P}^3$. Explicit computation shows that R is smooth since $a \pm b \neq 1$, $a \pm b \neq -1$ and $b \neq 0$. Then S is also smooth. Since the involution of the double cover φ is contained in G, every G-invariant curve in S is rationally equivalent to $\phi^*(D)$ for some $D \in \mathrm{Pic}(Y)$, which implies the required assertion.

Similarly, we see that the required assertion holds when $\eta(Z)$ is given by $xt = yz$. Therefore, we can proceed to the case when $\eta(Z)$ is an irreducible curve in the pencil $\mathcal{P}$. In this case, we have $S = Q \cap \{r(xt + yz) = s(xz + yt)\}$, where r and s are some numbers such that $(r, s) \neq (0, 0)$, $[r : s] \neq [1 : 1]$, $[r : s] \neq [1 : -1]$. As in the previous case, there exists a G-equivariant double cover $\varphi \colon S \to Y$ such that Y is the quadric in $\mathbb{P}^3$ given by

$$r(xt + yz) = s(xz + yt),$$

and the ramification divisor of φ is the curve $R = Y \cap \{xy + zt + a(xt + yz) + b(xz + yt) = 0\}$. Since $[r : s] \neq [1 : 1]$ and $[r : s] \neq [1 : -1]$, one can check that the quadric Y is smooth. Thus, if the curve R is smooth, we obtain the required assertion as in the previous case. Therefore, we may assume that the curve R is singular.

Since R is singular, explicit computations show that $br + (a \pm 1)s = 0$ or $(b \pm 1)r + as = 0$. In the former case, we have $R = Y \cap \{(x \pm z)(t \pm y) = 0\}$. Similarly, if $(b \pm 1)r + as = 0$, then $R = Y \cap \{(y \pm z)(t \pm x) = 0\}$. In each case, the curve R splits as a union of two smooth conics R_1 and R_2 that intersect transversally at two points, so that S has two isolated ordinary double points, which are disjoint from $C_1 \cup C_2$. As in the previous case, we see that every G-invariant Cartier divisor on S is rationally equivalent to $\phi^*(D)$ for some $D \in \mathrm{Pic}(Y)$. Since any Weil divisor on S becomes Cartier once it is multiplied by 2, the assertion follows. This completes the proof of the lemma.   □

Corollary 5.84 *If Q is given by (ℶ) or (ℸ), then X is strictly K-semistable.*

Proof We only consider the case when Q is given by (ℶ) because the other case is similar. Suppose that Q is given by (ℶ). Let $Q_s = \{w^2 + xy + zt + a(xt + yz) + sxz = 0\} \subset \mathbb{P}^4$, where $s \in \mathbb{C}$. Then the quadric Q_s is smooth, and Q contains both conics C_1 and C_2. Let $X_s \to Q_s$ be the blow up of the conics C_1 and C_2. Scaling coordinates x, y, z, t, w, we see that $X_s \cong X$ for every $s \neq 0$. This gives us a test configuration for X, whose special fiber is X_0, which is a K-polystable smooth Fano threefold №3.10 by Lemma 5.83. Then X is strictly K-semistable by Corollary 1.13.   □

Thus, Proposition 5.79 is completely proved.

5.18 Family №3.12

Let C be a twisted cubic in $\mathbb{P}^3$, L a line in $\mathbb{P}^3$ disjoint from C, and let $\pi \colon X \to \mathbb{P}^3$ be the blow up of the curves L and C. Then X is a Fano threefold №3.12. Moreover, every Fano threefold №3.12 can be obtained this way. Observe that we have the commutative diagram

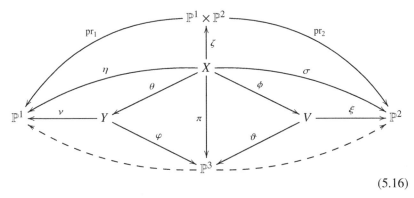

$$(5.16)$$

where φ is the blow up of the line L and ϑ that of C, ϕ is the blow up of the proper transform of the line L, the map θ is the blow up of the proper transform of the curve C, the map ζ is the contraction of the proper transforms of the (quartic) surface in $\mathbb{P}^3$ spanned by the secants of the curve C that intersect L, ξ is a $\mathbb{P}^1$-bundle, ν is a $\mathbb{P}^2$-bundle, σ is a (non-standard) conic bundle, η is a fibration into del Pezzo surfaces of degree 6, the left dashed arrow is the linear projection from the line L, the right dashed arrow is given by the linear system of quadrics that contain C, and pr_1 and pr_2 are projections to the first and the second factors, respectively.

Let H be a plane in $\mathbb{P}^3$, let E_L be the exceptional surface of π that is mapped to L, let E_C be the exceptional surface of π that is mapped to C, and let R be the ζ-exceptional surface. Then $R \sim \pi^*(4H) - 2E_C - E_L$. This gives $-K_X \sim_{\mathbb{Q}} \frac{1}{2}R + 2(\pi^*(H) - E_L) + \frac{3}{2}E_L$, Thus, for every subgroup $G \subset \mathrm{Aut}(X)$, $\alpha_G(X) \leqslant \frac{2}{3}$ because R, E_L and the linear system $|\pi^*(H) - E_L|$ are all G-invariant.

In this section, we prove that one special Fano threefold №3.12 is K-polystable. Namely, starting from now, we assume that L is the line $x_0 = x_3 = 0$, and the twisted cubic C is given by $[s^3 : s^2t : st^2 : t^3]$, where $[s : t] \in \mathbb{P}^1$. Let G be the subgroup in $\mathrm{Aut}(\mathbb{P}^3)$ that is generated by the involution $[x_0 : x_1 : x_2 : x_3] \mapsto [x_3 : x_2 : x_1 : x_0]$, and automorphisms

$$\left[x_0 : x_1 : x_2 : x_3\right] \mapsto \left[x_0 : tx_1 : t^2x_2 : t^3x_3\right],$$

where $t \in \mathbb{C}^*$, and x_0, x_1, x_2, x_3 are coordinates in $\mathbb{P}^3$. Then $G \cong \mathbb{G}_m \rtimes \mu_2$, and the curve C is G-invariant. Thus, the action of the group G lifts to the threefold X, and the diagram (5.16) is G-equivariant. By [45, Lemma 4.6], the threefold X is the unique smooth Fano threefold №3.12 that has an infinite automorphism group.

Proposition 5.85   *The Fano threefold X is K-polystable.*

Thus, by Corollary 1.16, general smooth Fano threefolds №3.12 are K-stable.

The proof of Proposition 5.85 is very similar to the proof of Proposition 4.33 in the case (2.22.D_∞). As in the proof of Proposition 4.33, we first need to collect some information about G-invariant subvarieties in $\mathbb{P}^3$. To do this, we denote by S_2 the quadric surface in $\mathbb{P}^3$ that is given by $x_0 x_3 = x_1 x_2$, and we denote by L' the line in $\mathbb{P}^3$ that is given by $x_1 = x_2 = 0$. For every $q \in \mathbb{C}^*$, we let C_q be the twisted cubic $[s^3 : qs^2 t : qst^2 : t^3]$, where $[s : t] \in \mathbb{P}^1$. Then S_2, L' and C_q are G-invariant, $L \cap L' = \varnothing$ and $C = C_1$. Finally, we let S_4 be the non-normal quartic surface in $\mathbb{P}^3$ that is given by $x_3 x_1^3 = x_0 x_2^3$.

Lemma 5.86   *The following assertions hold:*

 (i) $\mathbb{P}^3$ *contains neither G-fixed points nor G-invariant planes;*
 (ii) S_2 *is the only G-invariant quadric surface in $\mathbb{P}^3$ that contains C;*
 (iii) L *and* L' *are the only G-invariant lines in $\mathbb{P}^3$;*
 (iv) L, L' *and* C_q *are the only G-invariant irreducible curves in $\mathbb{P}^3$;*
 (v) S_4 *contains all G-invariant irreducible curves in $\mathbb{P}^3$;*
 (vi) $L \cap S_2 = [0 : 1 : 0 : 0] \cup [0 : 0 : 1 : 0]$, $L' \cap C = L' \cap Q = [1 : 0 : 0 : 0] \cup [0 : 0 : 0 : 1]$.

Proof   Left to the reader.   □

Corollary 5.87   *The threefold X does not contain G-fixed points.*

We prove Proposition 5.85. Suppose X is not K-polystable. By Theorem 1.22, there exists a G-invariant prime divisor F over X such that $\beta(F) = A_X(F) - S_X(F) \leqslant 0$. Let us seek a contradiction. Let $Z = C_X(F)$. Then Z is not a point by Corollary 5.87, and Z is not a surface by Theorem 3.17, so that Z is a G-invariant irreducible curve.

Lemma 5.88 $Z \not\subset E_L$.

Proof   Suppose that $Z \subset E_L$. Observe that $E_L \cong \mathbb{P}^1 \times \mathbb{P}^1$. Let $\mathbf{s}$ be the section of the natural projection $E_L \to L$ such that $\mathbf{s}^2 = 0$, and $\mathbf{l}$ be the fiber of this projection. Then $E_L|_{E_L} \sim -\mathbf{s} + \mathbf{l}$, $\pi^*(H)|_{E_L} \sim \mathbf{f}$, $R|_{E_L} \sim \mathbf{s} + 3\mathbf{l}$, and E_C and E_L are disjoint. Note that E_L contains exactly two G-invariant irreducible curves. One of them is $R|_{E_L}$, and the other one is cut out on E_L by the proper transform on X of the surface S_4. Thus, we conclude that $Z \sim \mathbf{s} + 3\mathbf{l}$.

Let us use notation introduced in Section 1.7. By Theorem 3.17, we have $S_X(E_L) < 1$. Thus, we conclude that $S(W^{E_L}_{\bullet,\bullet}; Z) \geqslant 1$ by Corollary 1.110. Let us compute $S(W^{E_L}_{\bullet,\bullet}; Z)$. Take $u \in \mathbb{R}_{\geqslant 0}$. Observe that $-K_X - uE_L \sim_{\mathbb{R}} \frac{1}{2}R + 2(\pi^*(H) - E_L) + (\frac{3}{2} - u)E_L$, which implies that $-K_X - uE_L$ is pseudo-effective if and only if $u \leqslant \frac{3}{2}$. Let $P(u) = P(-K_X - uE_L)$ and $N(u) = N(-K_X - uE_L)$. Then

$$P(u) = \begin{cases} -K_X - uE_L & \text{if } 0 \leqslant u \leqslant 1, \\ (8-4u)\pi^*(H) - (3-2u)E_C - 2E_L & \text{if } 1 \leqslant u \leqslant \frac{3}{2}, \end{cases}$$

and

$$N(u) = \begin{cases} 0 & \text{if } 0 \leqslant u \leqslant 1, \\ (u-1)R & \text{if } 1 \leqslant u \leqslant \frac{3}{2}. \end{cases}$$

Take any $v \in \mathbb{R}_{\geqslant 0}$. If $u \in [0, 1]$, we have $P(u)|_{E_L} - vZ \sim_{\mathbb{R}} (1+u-v)\mathbf{s}+(3-u-3v)\mathbf{l}$. Similarly, if $u \in [1, \frac{3}{2}]$ and $v \in \mathbb{R}_{\geqslant 0}$, then $P(u)|_{E_L} - vZ \sim_{\mathbb{R}} (2-v)\mathbf{s}+(6-4u-3v)\mathbf{l}$. Hence, if $Z = R|_{E_L}$, then Corollary 1.110 gives

$$S(W_{\bullet,\bullet}^{E_L}; Z) = \frac{3}{28} \int_1^{\frac{3}{2}} (u-1)E_L \cdot \big((8-4u)\pi^*(H) - (3-2u)E_C - 2E_L\big)^2 du$$

$$+ \frac{3}{28} \int_0^{\frac{3}{2}} \int_0^{\infty} \mathrm{vol}\Big(P(u)\big|_{E_L} - vZ\Big) dv\, du$$

$$= \frac{3}{28} \int_1^{\frac{3}{2}} 4(u-1)(6-4u) du$$

$$+ \frac{3}{28} \int_0^1 \int_0^{\frac{3-u}{3}} 2(1+u-v)(3-u-3v) dv\, du$$

$$+ \frac{3}{28} \int_1^{\frac{3}{2}} \int_0^{\frac{6-4u}{3}} 2(2-v)(6-4u-3v) dv\, du = \frac{19}{56} < 1.$$

Similarly, if $Z \neq R|_{E_L}$, then $S(W_{\bullet,\bullet}^{E_L}; Z) = \frac{3}{28} < 1$. This is a contradiction. □

Let Q be the proper transform of the quadric surface S_2 on the threefold X.

Lemma 5.89 $Z \not\subset Q$.

Proof Suppose that $Z \subset Q$. Let us seek a contradiction. Recall that $\pi(Q) = S_2$ is a smooth quadric surface in $\mathbb{P}^3$ that is given by $x_0 x_3 = x_1 x_2$. It contains the twisted cubic curve C, and it does not contain the lines L and L'. Let us identify $S_2 = \mathbb{P}^1 \times \mathbb{P}^1$ such that C is a curve in S_2 of degree $(1, 2)$. Then π induces a birational morphism $\varpi \colon Q \to \mathbb{P}^1 \times \mathbb{P}^1$ that is a blow up of two intersection points $S_2 \cap L$, which are not contained in the curve C. Moreover, the surface Q is a smooth del Pezzo surface of degree 6 because the points of the intersection $S_2 \cap L$ are not contained in one line in S_2 by Lemma 5.86.

Let us use notation from Section 1.7. By Theorem 3.17, we have $S_X(Q) < 1$. Then $S(W_{\bullet,\bullet}^Q; Z) \geqslant 1$ by Corollary 1.110. We show that $S(W_{\bullet,\bullet}^Q; Z) < 1$.

Take $u \in \mathbb{R}_{\geq 0}$. Then $-K_X - uQ \sim_{\mathbb{R}} 2\pi^*(H) - E_L + (1 - u)(2\pi^*(H) - E_C)$, which implies that $-K_X - uQ$ is nef for every $u \in [0, 1]$. Meanwhile, we have

$$-K_X - uQ \sim_{\mathbb{R}} (4 - 2u)\big(\pi^*(H) - E_L\big) + (3 - 2u)E_L + (u - 1)E_C,$$

so that the divisor $-K_X - uQ$ is pseudo-effective $\iff u \in [0, \frac{3}{2}]$. Moreover,

$$P\big(-K_X - uQ\big) = \begin{cases} -K_X - uQ & \text{if } 0 \leqslant u \leqslant 1, \\ (4 - 2u)\pi^*(H) - E_L & \text{if } 1 \leqslant u \leqslant \frac{3}{2}, \end{cases}$$

and $N(-K_X - uQ) = (u - 1)E_C$ if $1 \leqslant u \leqslant \frac{3}{2}$. For simplicity, we let $P(u) = P(-K_X - uQ)$ and $N(u) = N(-K_X - uQ)$.

Let us introduce some notation on Q. First, we denote by ℓ_1 and ℓ_2 the proper transforms on Q of general curves in $\mathbb{P}^1 \times \mathbb{P}^1$ of degrees $(1, 0)$ and $(0, 1)$, respectively. Second, we denote by e_1 and e_1 the exceptional curves of φ. Third, we let $F_{11}, F_{12}, F_{21}, F_{22}$ be the (-1)-curves on Q such that $F_{11} \sim \ell_1 - e_1$, $F_{12} \sim \ell_1 - e_2, F_{21} \sim \ell_2 - e_1, F_{22} \sim \ell_2 - e_2$. Then $\pi^*(H)|_Q \sim \ell_1 + \ell_2, E_L|_Q = e_1 + e_2$ and $E_C|_Q \sim \ell_1 + 2\ell_2$.

It follows from Lemma 5.86 that either $Z = E_C|_Q$ or $\pi(Z) = C_{-1}$. In both cases, we have $Z \sim \ell_1 + 2\ell_2$. Moreover, if $Z \neq E_C|_Q$, then Corollary 1.110 gives

$$S\big(W^Q_{\bullet,\bullet}; Z\big) = \frac{3}{28} \int_0^{\frac{3}{2}} \int_0^{\infty} \mathrm{vol}\Big(P(u)\big|_Q - v(\ell_1 + 2\ell_2)\Big) dv du$$

$$\leqslant \frac{3}{28} \int_0^{\frac{3}{2}} \int_0^{\infty} \mathrm{vol}\Big(P(u)\big|_Q - v\ell_1\Big) dv du.$$

Similarly, if $Z = E_C|_Q$, then

$$S\big(W^Q_{\bullet,\bullet}; Z\big) = \frac{3}{28} \int_0^{\frac{3}{2}} \Big(P(u) \cdot P(u) \cdot Q\Big) \mathrm{ord}_Z\Big(N(u)\big|_Q\Big) du$$

$$+ \frac{3}{28} \int_0^{\frac{3}{2}} \int_0^{\infty} \mathrm{vol}\Big(P(u)\big|_Q - vZ\Big) dv du$$

$$= \frac{3}{28} \int_1^{\frac{3}{2}} (u - 1)\big((4 - 2u)\pi^*(H) - E_L\big)^2 \cdot \big(2\pi^*(H) - E_C\big) du$$

$$+ \frac{3}{28} \int_0^{\frac{3}{2}} \int_0^{\infty} \mathrm{vol}\Big(P(u)\big|_Q - vZ\Big) dv du$$

$$= \frac{3}{28} \int_1^{\frac{3}{2}} (u - 1)\big(2(4 - 2u)^2 - 2\big) du$$

$$+ \frac{3}{28} \int_0^{\frac{3}{2}} \int_0^{\infty} \mathrm{vol}\Big(P(u)\big|_Q - vZ\Big) dv du$$

$$= \frac{5}{224} + \frac{3}{28} \int_0^{\frac{3}{2}} \int_0^{\infty} \mathrm{vol}\Big(P(u)\big|_Q - v(\ell_1 + 2\ell_2)\Big) dv du$$

$$\leqslant \frac{5}{224} + \frac{3}{28} \int_0^{\frac{3}{2}} \int_0^{\infty} \mathrm{vol}\Big(P(u)\big|_Q - v\ell_1\Big) dv du.$$

Thus, to show that $S(W_{\bullet,\bullet}^Q; Z) < 1$, it is enough to show that the integral in the right hand side of the last formula is less than $\frac{219}{224}$.

Suppose that $u \in [0, 1]$. Take $v \in \mathbb{R}_{\geqslant 0}$. Then

$$P(-K_X - uQ)\big|_Q - vZ \sim_{\mathbb{R}} (3 - u)\ell_1 + 2\ell_2 - e_1 - e_2 \sim_{\mathbb{R}} (3 - u)\ell_1 + F_{21} + F_{22},$$

so that $P(-K_X - uQ)\big|_Q - vZ$ is not pseudo-effective for $v > 3 - u$. Taking intersections with F_{21} and F_{22}, we see that the divisor $P(-K_X - uQ)\big|_Q - vZ$ is nef for $v \leqslant 2 - u$. Similarly, if $2 - u \leqslant v \leqslant 3 - u$, its Zariski decomposition is

$$\underbrace{(3 - u - v)(\ell_1 + F_{21} + F_{22})}_{\text{positive part}} + \underbrace{(v + u - 2)(F_{21} + F_{22})}_{\text{negative part}}.$$

If $u \in [0, 1]$ and $0 \leqslant v \leqslant 3 - u$, then

$$\mathrm{vol}\big(P(-K_X - uQ)\big|_Q - vZ\big) = \begin{cases} 4(3 - u - v) - 2 & \text{if } v \leqslant 2 - u, \\ 2(3 - u - v)^2 & \text{if } 2 - u \leqslant v \leqslant 3 - u. \end{cases}$$

Now we suppose that $u \in [1, 2]$. For $v \in \mathbb{R}_{\geqslant 0}$, we have

$$P(-K_X - uQ)\big|_Q - vZ \sim_{\mathbb{R}} (4 - 2u - v)\ell_1 + (4 - 2u)\ell_2 - e_1 - e_2.$$

Intersecting this divisor with (-1)-curves in Q, we see that it is nef for $v \leqslant 3 - 2u$. Similarly, if $3 - 2u \leqslant v \leqslant 6 - 4u$, its Zariski decomposition is

$$\underbrace{(4 - 2u - v)\ell_1 + (10 - 6u - 2v)\ell_2 - (4 - 2u - v)(e_1 + e_1)}_{\text{positive part}}$$

$$+ \underbrace{(2u + v - 3)(F_{21} + F_{22})}_{\text{negative part}}.$$

Moreover, if $v \geqslant 6 - 4u$, then the divisor $P(-K_X - uQ)\big|_Q - vZ$ is not pseudo-effective. Hence, if $u \in [1, \frac{3}{2}]$ and $0 \leqslant v \leqslant 6 - 4u$, then

$$\mathrm{vol}\big(P(-K_X - uQ)\big|_Q - vZ\big) = \begin{cases} 2(4 - 2u - v)(4 - 2u) - 2 & \text{if } 0 \leqslant v \leqslant 3 - 2u, \\ 2(4 - 2u - v)(6 - 4u - v) & \text{if } 3 - 2u \leqslant v \leqslant 6 - 4u. \end{cases}$$

Now we can compute the required integral as follows:

$$\frac{3}{28} \int_0^{\frac{3}{2}} \int_0^{\infty} \mathrm{vol}\big(P(u)\big|_Q - v f_1\big) \, dv \, du$$

$$= \frac{3}{28} \int_0^1 \int_0^{2-u} \big(4(3 - u - v) - 2\big) \, dv \, dv + \frac{3}{28} \int_0^1 \int_{2-u}^{3-u} 2(3 - u - v)^2 \, dv \, du$$

$$+ \frac{3}{28} \int_1^{\frac{3}{2}} \int_0^{3-2u} \big(2(4 - 2u - v)(4 - 2u) - 2\big) \, dv \, du$$

$$+ \frac{3}{28} \int_1^{\frac{3}{2}} \int_{3-uu}^{6-4u} 2(4 - 2u - v)(6 - 4u - v) \, dv \, du = \frac{109}{112},$$

so that $S(W_{\bullet,\bullet}^Q; Z) \leqslant \frac{5}{224} + \frac{109}{112} = \frac{223}{224} < 1$, and the lemma is proved. □

By Lemma 1.45, $\alpha_{G,Z}(X) < \frac{3}{4}$. Thus, by Lemma 1.42, there is a G-invariant effective $\mathbb{Q}$-divisor D on the threefold X such that $D \sim_{\mathbb{Q}} -K_X$ and $Z \subseteq$ Nklt$(X, \lambda D)$ for some positive rational number $\lambda < \frac{3}{4}$.

Lemma 5.90 *Let S be an irreducible surface in X. Suppose that $S \subset$ Nklt$(X, \lambda D)$. Then either $S = Q$ or $S = E_L$.*

Proof The cone of effective divisors Eff(X) is generated by E_L, E_C, $\pi^*(H) - E_L$, Q, R. Meanwhile we have $D \sim_{\mathbb{Q}} 4\pi^*(H) - E_C - E_L$ and $\lambda < \frac{3}{4}$. Thus, arguing as in the proof of Lemma 4.43, we see that $S = E_L$, $S \sim Q$ or $\pi(S)$ is a plane, so that either $S = E_L$ or $S = Q$ by Lemma 5.86. □

Corollary 5.91 $Z \not\subset E_C$.

Proof Suppose that $Z \subset E_C$. Observe that $\pi(Z)$ is not a point since $\mathbb{P}^3$ does not have G-fixed points by Lemma 5.86. Hence, we see that $\pi(Z)$ is the twisted cubic C.

Let S be a general fiber of η. Then $S \cdot Z = 3$, contradicting Corollary A.12. □

We see that $\pi(Z)$ is a G-invariant curve in $\mathbb{P}^3$ such that $\pi(Z) \not\subset S_2$ and $\pi(Z) \neq L$.

Lemma 5.92 *The curve $\pi(Z)$ is the line L'.*

Proof The proof is essentially the same as the proof of Lemma 4.48. But now we have to use Lemma 5.90. □

Let S be a general surface in the linear system $|2\pi^*(H) - E_C|$ that contains the curve Z. Then $\pi(S)$ is a smooth quadric surface in $\mathbb{P}^3$ that contains C and the line $L' = \pi(Z)$. Note that $\pi(S)$ is not G-invariant. Let us use notation introduced in Section 1.7.

Lemma 5.93 $S(W^S_{\bullet,\bullet}; Z) = \frac{109}{112}$.

Proof Identify $\pi(S) = \mathbb{P}^1 \times \mathbb{P}^1$ such that C is a curve of degree $(1,2)$. Then L' is curve of degree $(1,0)$. Moreover, the morphism π induces a birational morphism $\varpi: S \to \mathbb{P}^1 \times \mathbb{P}^1$ that is a blow up of two intersection points $\pi(S) \cap L$. Observe that these points are not contained in the curves L' and C. Moreover, these two points are not contained in any line in $\pi(S)$ because L is not contained in $\pi(S)$. Hence, we see that S is a smooth del Pezzo surface of degree 6. Thus, the proof of Lemma 5.89 gives $S(W^S_{\bullet,\bullet}; Z) = \frac{109}{112}$. $\qquad\square$

We have $S_X(S) < 1$ by Theorem 3.17. Then $S(W^S_{\bullet,\bullet}; Z) \geqslant 1$ by Corollary 1.110, which contradicts Lemma 5.93. This completes the proof of Proposition 5.85.

5.19 Family №3.13

Let X be a smooth Fano threefold №3.13. Then X is a complete intersection in $\mathbb{P}^2 \times \mathbb{P}^2 \times \mathbb{P}^2$ of 3 divisors of degrees $(1,1,0)$, $(0,1,1)$, $(1,0,1)$, respectively. Therefore, we have $X = \{f = g = h = 0\} \subset \mathbb{P}^2 \times \mathbb{P}^2 \times \mathbb{P}^2$, where

$$f = \begin{bmatrix} x_0 & x_1 & x_2 \end{bmatrix} M_{x,y} \begin{bmatrix} y_0 \\ y_1 \\ y_2 \end{bmatrix}, \quad g = \begin{bmatrix} y_0 & y_1 & y_2 \end{bmatrix} M_{y,z} \begin{bmatrix} z_0 \\ z_1 \\ z_2 \end{bmatrix},$$

$$h = \begin{bmatrix} x_0 & x_1 & x_2 \end{bmatrix} M_{x,z} \begin{bmatrix} z_0 \\ z_1 \\ z_2 \end{bmatrix}$$

for some 3×3 matrices $M_{x,y}, M_{y,z}, M_{x,z}$, and where $[x_0 : x_1 : x_2]$, $[y_0 : y_1 : y_2]$, $[z_0 : z_1 : z_2]$ are coordinates on the first, the second and the third factor of $\mathbb{P}^2 \times \mathbb{P}^2 \times \mathbb{P}^2$, respectively.

Lemma 5.94 $\det(M_{x,y}) \neq 0$, $\det(M_{y,z}) \neq 0$ and $\det(M_{x,z}) \neq 0$.

Proof If $det(M_{x,y}) = 0$, there are $[a_0 : a_1 : a_2]$ and $[b_0 : b_1 : b_2]$ in $\mathbb{P}^2$ such that

$$\begin{bmatrix} a_0 & a_1 & a_2 \end{bmatrix} M_{x,y} = M_{x,y} \begin{bmatrix} b_0 \\ b_1 \\ b_2 \end{bmatrix} = 0,$$

and we can find $[c_1 : c_2 : c_3] \in \mathbb{P}^2$ such that

$$\begin{bmatrix} b_0 & b_1 & b_2 \end{bmatrix} M_{y,z} \begin{bmatrix} c_0 \\ c_1 \\ c_2 \end{bmatrix} = \begin{bmatrix} a_0 & a_1 & a_2 \end{bmatrix} M_{x,z} \begin{bmatrix} c_0 \\ c_1 \\ c_2 \end{bmatrix} = 0,$$

which implies that X is singular at the point $([a_0 : a_1 : a_2], [b_0 : b_1 : b_2], [c_0 : c_1 : c_2])$. This shows that $\det(M_{x,y}) \neq 0$. Similarly, we see that $\det(M_{y,z}) \neq 0 \neq \det(M_{x,z})$.   □

Let $W_{x,y}$, $W_{y,z}$, $W_{x,z}$ be the threefolds in $\mathbb{P}^2 \times \mathbb{P}^2$ that are given by $f = 0$, $g = 0$, $h = 0$, respectively. Then $W_{x,y}$, $W_{y,z}$, $W_{x,z}$ are smooth by Lemma 5.94. Moreover, we have the commutative diagram

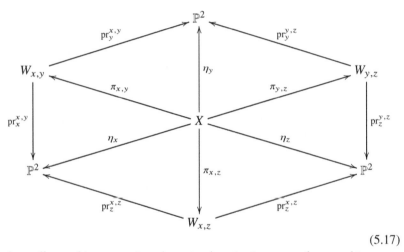

$$(5.17)$$

where all morphisms are given by natural projections, e.g. the morphism $\pi_{x,y}$ is given by

$$\Big([x_0 : x_1 : x_2], [y_0 : y_1 : y_2], [z_0 : z_1 : z_2]\Big) \mapsto \Big([x_0 : x_1 : x_2], [y_0 : y_1 : y_2]\Big),$$

the morphism η_z is given by $([x_0 : x_1 : x_2], [y_0 : y_1 : y_2], [z_0 : z_1 : z_2]) \mapsto [z_0 : z_1 : z_2]$, and the projection $\mathrm{pr}_y^{y,z}$ is given by $([y_0 : y_1 : y_2], [z_0 : z_1 : z_2]) \mapsto [y_0 : y_1 : y_2]$.

Note that the morphisms $\pi_{x,y}, \pi_{y,z}, \pi_{x,z}$ are birational – they blow up smooth rational curves of degree $(2, 2)$. Let $E_{x,y}, E_{y,z}, E_{x,z}$ be their exceptional surfaces, respectively. Then $-K_X \sim E_{x,y} + E_{y,z} + E_{x,z}$. Observe also that η_x, η_y and η_z are (non-standard) conics bundles and $-K_X \sim \eta_x^*(\mathcal{O}_{\mathbb{P}^2}(1)) + \eta_y^*(\mathcal{O}_{\mathbb{P}^2}(1)) + \eta_z^*(\mathcal{O}_{\mathbb{P}^2}(1))$.

Let Δ_x, Δ_y, Δ_z be the discriminant curves of the conic bundles η_x, η_y, η_z, respectively. Then the defining equations of the curves Δ_x, Δ_y, Δ_z are

$$\begin{bmatrix} x_0 & x_1 & x_2 \end{bmatrix} D_x \begin{bmatrix} x_0 \\ x_1 \\ x_2 \end{bmatrix} = 0, \qquad \begin{bmatrix} y_0 & y_1 & y_2 \end{bmatrix} D_y \begin{bmatrix} y_0 \\ y_1 \\ y_2 \end{bmatrix} = 0,$$

$$\begin{bmatrix} z_0 & z_1 & z_2 \end{bmatrix} D_z \begin{bmatrix} z_0 \\ z_1 \\ z_2 \end{bmatrix} = 0$$

for some 3×3 matrices D_x, D_y and D_z.

Lemma 5.95 *The following hold:*

$$D_x = M_{x,z}\left(M_{y,z}^{-1}\right)M_{x,y}^T,$$
$$D_y = M_{y,z}\left(M_{x,z}\right)^{-1}M_{x,y},$$
$$D_z = M_{y,z}^T\left(M_{x,y}\right)^{-1}M_{x,z}.$$

Proof Let $P = [a_0 : a_1 : a_2] \in \mathbb{P}^2$, and let C_P be the fiber of the conic bundle η_x over P. Then there exists a natural embedding $C_P \hookrightarrow \mathbb{P}^1 \times \mathbb{P}^1$ as a curve of degree $(1,1)$, where the first factor of $\mathbb{P}^1 \times \mathbb{P}^1$ is identified with the line in $\mathbb{P}^2$ given by

$$[a_0\ a_1\ a_2]M_{x,y} \begin{bmatrix} y_0 \\ y_1 \\ y_2 \end{bmatrix} = 0, \tag{5.18}$$

and the second factor of $\mathbb{P}^1 \times \mathbb{P}^1$ is identified with the line in $\mathbb{P}^2$ given by

$$[a_0\ a_1\ a_2]M_{x,z} \begin{bmatrix} z_0 \\ z_1 \\ z_2 \end{bmatrix} = 0. \tag{5.19}$$

Moreover, the curve C_P is defined in this $\mathbb{P}^1 \times \mathbb{P}^1$ by the equation

$$[y_0\ y_1\ y_2]M_{y,z} \begin{bmatrix} z_0 \\ z_1 \\ z_2 \end{bmatrix} = 0.$$

So, the curve C_P is singular $\iff$ there is a point $[c_0 : c_1 : c_2]$ in the line (5.19) such that

$$[y_0\ y_1\ y_2]M_{y,z} \begin{bmatrix} c_0 \\ c_1 \\ c_2 \end{bmatrix} = 0$$

for every point $[y_0 : y_1 : y_2] \in \mathbb{P}^2$ that satisfies the condition (5.18). Thus, we conclude that C_P is singular $\iff$ there is a point $[c_0 : c_1 : c_2]$ in the line (5.19) such that

$$\begin{bmatrix} c_0 \\ c_1 \\ c_2 \end{bmatrix} = M_{y,z}^{-1} M_{x,y}^T \begin{bmatrix} a_0 \\ a_1 \\ a_2 \end{bmatrix}.$$

Now, plugging $[z_0 : z_1 : z_2] = [c_0 : c_1 : c_2]$ into (5.19), we see that

the curve C_P is singular $\iff [a_0\ a_1\ a_2] M_{x,z} (M_{y,z}^{-1}) M_{x,y}^T \begin{bmatrix} a_0 \\ a_1 \\ a_2 \end{bmatrix} = 0.$

But $P \in \Delta_{x,y} \iff C_P$ is singular, so that we can let $D_x = M_{x,z}(M_{y,z}^{-1})M_{x,y}^T$ as required. Similarly, we can prove the remaining formulas for D_y and D_z. □

Observe that the conics $\Delta_x, \Delta_y, \Delta_z$ are smooth.

Remark 5.96 Let $C_{x,y}, C_{y,z}, C_{x,z}$ be the curves in $W_{x,y}, W_{y,z}, W_{x,z}$ that are blown up by the morphisms $\pi_{x,y}, \pi_{y,z}, \pi_{x,z}$, respectively. Then $C_{x,y}, C_{y,z}, C_{x,z}$ are given by

$$\begin{bmatrix} x_0 & x_1 & x_2 \end{bmatrix} D_x \begin{bmatrix} x_0 \\ x_1 \\ x_2 \end{bmatrix} = \begin{bmatrix} y_0 & y_1 & y_2 \end{bmatrix} D_y \begin{bmatrix} y_0 \\ y_1 \\ y_2 \end{bmatrix} = f = 0,$$

$$\begin{bmatrix} y_0 & y_1 & y_2 \end{bmatrix} D_y \begin{bmatrix} y_0 \\ y_1 \\ y_2 \end{bmatrix} = \begin{bmatrix} z_0 & z_1 & z_2 \end{bmatrix} D_z \begin{bmatrix} z_0 \\ z_1 \\ z_2 \end{bmatrix} = g = 0,$$

$$\begin{bmatrix} x_0 & x_1 & x_2 \end{bmatrix} D_x \begin{bmatrix} x_0 \\ x_1 \\ x_2 \end{bmatrix} = \begin{bmatrix} z_0 & z_1 & z_2 \end{bmatrix} D_z \begin{bmatrix} z_0 \\ z_1 \\ z_2 \end{bmatrix} = h = 0,$$

respectively.

Linearly changing the coordinates $([x_0 : x_1 : x_2], [y_0 : y_1 : y_2], [z_0 : z_1 : z_2])$, we can simplify the shapes of the polynomials f, g and h. To be precise, we have the following.

Lemma 5.97 (cf. [152, 207]) *One can choose coordinates on $\mathbb{P}^2 \times \mathbb{P}^2 \times \mathbb{P}^2$ such that one of the following two cases holds:*

(★) *the threefold X is given by*

$$\begin{cases} x_0 y_0 + x_1 y_1 + x_2 y_2 = 0, \\ y_0 z_0 + y_1 z_1 + y_2 z_2 = 0, \\ (1+s) x_0 z_1 + (1-s) x_1 z_0 - 2 x_2 z_2 = 0, \end{cases}$$

where $s \in \mathbb{C}$ such that $s \neq \pm 1$.

(♦) *the threefold X is given by*

$$\begin{cases} x_0 y_0 + x_1 y_1 + x_2 y_2 = 0, \\ y_0 z_0 + y_1 z_1 + y_2 z_2 = 0, \\ x_0 z_1 + x_1 z_0 + x_1 z_2 - x_2 z_1 - 2 x_2 z_2 = 0. \end{cases}$$

Proof Linearly changing x_0, x_1, x_2 and y_0, y_1, y_2, we may assume that $M_{x,y} = M_{y,z} = I_3$, so that f and g simplify to $x_0 y_0 + x_1 y_1 + x_2 y_2$ and $y_0 z_0 + y_1 z_1 + y_2 z_2 = 0$, respectively. Then the equations of the curves $\Delta_x, \Delta_y, \Delta_z$ simplify as

$$\begin{bmatrix} x_0 & x_1 & x_2 \end{bmatrix} M_{x,z} \begin{bmatrix} x_0 \\ x_1 \\ x_2 \end{bmatrix} = 0, \qquad \begin{bmatrix} y_0 & y_1 & y_2 \end{bmatrix} \left(M_{x,z}^{-1} \right) \begin{bmatrix} y_0 \\ y_1 \\ y_2 \end{bmatrix} = 0,$$

$$\begin{bmatrix} z_0 & z_1 & z_2 \end{bmatrix} M_{x,z} \begin{bmatrix} z_0 \\ z_1 \\ z_2 \end{bmatrix} = 0,$$

respectively. We can rewrite these equations as

$$\begin{bmatrix} x_0 & x_1 & x_2 \end{bmatrix} \left(\frac{M_{x,z} + M_{x,z}^T}{2} \right) \begin{bmatrix} x_0 \\ x_1 \\ x_2 \end{bmatrix} = 0,$$

$$\begin{bmatrix} y_0 & y_1 & y_2 \end{bmatrix} \left(\frac{M_{x,z}^{-1} + \left(M_{x,z}^{-1} \right)^T}{2} \right) \begin{bmatrix} y_0 \\ y_1 \\ y_2 \end{bmatrix} = 0,$$

$$\begin{bmatrix} z_0 & z_1 & z_2 \end{bmatrix} \left(\frac{M_{x,z} + M_{x,z}^T}{2} \right) \begin{bmatrix} z_0 \\ z_1 \\ z_2 \end{bmatrix} = 0,$$

respectively. In particular, we see that the matrix $\dfrac{M_{x,z} + M_{x,z}^T}{2}$ is not degenerate.

To simplify the bilinear form h, let us consider the coordinate change that corresponds to the automorphism $\phi_A \in \mathrm{Aut}(\mathbb{P}^2 \times \mathbb{P}^2 \times \mathbb{P}^2)$ which is given by the linear transformations

$$\begin{bmatrix} x_0 \\ x_1 \\ x_2 \end{bmatrix} \mapsto A \begin{bmatrix} x_0 \\ x_1 \\ x_2 \end{bmatrix}, \quad \begin{bmatrix} y_0 \\ y_1 \\ y_2 \end{bmatrix} \mapsto \left(A^{-1}\right)^T \begin{bmatrix} y_0 \\ y_1 \\ y_2 \end{bmatrix}, \quad \begin{bmatrix} z_0 \\ z_1 \\ z_2 \end{bmatrix} \mapsto A \begin{bmatrix} z_0 \\ z_1 \\ z_2 \end{bmatrix},$$

where A is some non-degenerate 3×3 matrix. Then h is changed to

$$\begin{bmatrix} x_0 & x_1 & x_2 \end{bmatrix} A^T M_{x,z} A \begin{bmatrix} z_0 \\ z_1 \\ z_2 \end{bmatrix},$$

and the bilinear forms f and g are preserved. We let

$$K = \frac{M_{x,z} + M_{x,z}^T}{2} \quad \text{and} \quad L = \frac{M_{x,z} - M_{x,z}^T}{2}.$$

Since $\det(K) \neq 0$, we can choose A such that $A^T K A$ is any symmetric non-degenerate matrix. In particular, swapping our matrix $M_{x,z}$ with $A^T M_{x,z} A$, we may assume that

$$K = \begin{pmatrix} 0 & 1 & 0 \\ 1 & 0 & 0 \\ 0 & 0 & -2 \end{pmatrix},$$

Then we write

$$L = \begin{pmatrix} 0 & u & v \\ -u & 0 & w \\ -v & -w & 0 \end{pmatrix},$$

so that

$$M_{x,z} = \begin{pmatrix} 0 & 1+u & v \\ 1-u & 0 & w \\ -v & -w & -2 \end{pmatrix}.$$

If $u = 0$, $v = 0$ and $w = 0$, then X is given by ($\bigstar$) with $s = 0$. Thus, we may assume that at least one number among u, v and w is not zero.

Now we choose the matrix A such that $A^T K A = \lambda K$ for some non-zero $\lambda \in \mathbb{C}$, so that our change of coordinates preserve the shape of the matrix $M_{x,z}$ we already achieved. Namely, we take

$$A = \begin{pmatrix} a^2 & b^2 & 2ab \\ c^2 & d^2 & 2cd \\ ac & bd & ad + bc \end{pmatrix}$$

where a, b, c and d are some complex numbers (to be chosen later) such that $ad - bc \neq 0$. Then $\det(A) = (ad - bc)^3 \neq 0$. If $v = 0$ and $u \neq 0$, we let a, $b = \frac{w}{2u}$, $c = 0$ and $d = 1$, which gives

$$A^T M_{x,z} A = \begin{pmatrix} 0 & 1+u & 0 \\ 1-u & 0 & 0 \\ 0 & 0 & -2 \end{pmatrix},$$

so that h becomes $(1+u)x_0z_1 + (1-u)x_1z_0 - 2x_2z_2$ and X is given by (★) with $s = u$. Similarly, if $v = u = 0$, then $w \neq 0$, so that we let $a = \sqrt{w}$, $b = 1$, $c = 0$ and $d = \frac{1}{\sqrt{w}}$, which gives

$$A^T M_{x,z} A = \begin{pmatrix} 0 & 1 & 0 \\ 1 & 0 & 1 \\ 0 & -1 & -2 \end{pmatrix},$$

so that h becomes $x_0z_1 + x_1z_0 + x_1z_2 - x_2z_1 - 2x_2z_2$, which implies that X is given by (♦). Thus, we may assume that $v \neq 0$.

Let $\gamma = \sqrt{4vw + 4u^2}$, so that $w = \frac{\gamma^2-4u^2}{4v}$. If $\gamma \neq 0$, we let $a = -\frac{2u-\gamma}{2\gamma}$, $b = -\frac{2u+\gamma}{2v}$, $c = \frac{v}{\gamma}$ and $d = 1$, which gives

$$A^T M_{x,z} A = \begin{pmatrix} 0 & 1-\frac{\gamma}{2} & 0 \\ 1+\frac{\gamma}{2} & 0 & 0 \\ 0 & 0 & -2 \end{pmatrix},$$

so that h becomes $(1 + \frac{\gamma}{2})x_0z_1 + (1 - \frac{\gamma}{2})x_1z_0 - 2x_2z_2$ and X is given by (★) with $s = -\frac{\gamma}{2}$. Similarly, if $\gamma = 0$, then $4vw + 4u^2 = 0$, so that $w = -\frac{u^2}{v}$ and

$$M_{x,z} = \begin{pmatrix} 0 & 1+u & v \\ 1-u & 0 & -\frac{u^2}{v} \\ -v & \frac{u^2}{v} & -2 \end{pmatrix}.$$

In this case, we let $a = -\frac{u}{v}$, $b = \frac{1-u}{v}$, $c = 1$ and $d = 1$, so that

$$A^T M_{x,z} A = \frac{1}{v^2}\begin{pmatrix} 0 & 1 & 0 \\ 1 & 0 & 1 \\ 0 & -1 & -2 \end{pmatrix},$$

so that our bilinear form h becomes $x_0z_1 + x_1z_0 + x_1z_2 - x_2z_1 - 2x_2z_2$ after scaling by v^2, which implies that X is given by (♦). This completes the proof of the lemma. □

If X is given by (★) with $s = 0$, then X is the unique smooth Fano threefold №3.13 that admits an effective $\mathrm{PGL}_2(\mathbb{C})$-action, and X is K-polystable by Example 1.94 and Lemma 4.18. On the other hand, if X is given by (♦), then X is not K-polystable by

Lemma 5.98 *Suppose that X is given by the equation ($\blacklozenge$). Then $\mathrm{Aut}(X) \cong \mathbb{G}_a \rtimes \mathfrak{S}_3$. Moreover, the threefold X is strictly K-semistable.*

Proof Suppose that the threefold X is given by ($\blacklozenge$). For every $a \in \mathbb{C}$, let us consider the automorphism $\phi_a \in \mathrm{Aut}(\mathbb{P}^2 \times \mathbb{P}^2 \times \mathbb{P}^2)$ given by the following linear transformations:

$$\begin{bmatrix} x_0 \\ x_1 \\ x_2 \end{bmatrix} \mapsto A \begin{bmatrix} x_0 \\ x_1 \\ x_2 \end{bmatrix}, \quad \begin{bmatrix} y_0 \\ y_1 \\ y_2 \end{bmatrix} \mapsto \left(A^{-1}\right)^T \begin{bmatrix} y_0 \\ y_1 \\ y_2 \end{bmatrix}, \quad \begin{bmatrix} z_0 \\ z_1 \\ z_2 \end{bmatrix} \mapsto A \begin{bmatrix} z_0 \\ z_1 \\ z_2 \end{bmatrix},$$

where

$$A = \begin{pmatrix} 1 & a^2 & 2a \\ 0 & 1 & 0 \\ 0 & a & 1 \end{pmatrix}.$$

Each such transformation ϕ_a leaves X invariant, so that we can assume that $\phi_a \in \mathrm{Aut}(X)$. One can check that these transformations form a subgroup in $\mathrm{Aut}(X)$ isomorphic to $\mathbb{G}_a$. Moreover, the group $\mathrm{Aut}(X)$ also contains involutions $\tau_{x,z}, \tau_{x,y}, \tau_{y,z}$ defined as

$$\tau_{x,z} \colon \left([x_0 : x_1 : x_2], [y_0 : y_1 : y_2], [z_0 : z_1 : z_2]\right) \mapsto$$
$$\left([z_0 : z_1 : -z_2], [y_0 : y_1 : -y_2], [x_0 : x_1 : -x_2]\right),$$

$$\tau_{x,y} \colon \left([x_0 : x_1 : x_2], [y_0 : y_1 : y_2], [z_0 : z_1 : z_2]\right) \mapsto$$
$$\left([y_0 + 2y_1 + y_2 : 2y_0 : y_0 + y_2], [x_1 : x_0 - x_2 : 2x_2 - x_1], [z_0 : z_1 : -z_2]\right),$$

$$\tau_{y,z} \colon \left([x_0 : x_1 : x_2], [y_0 : y_1 : y_2], [z_0 : z_1 : z_2]\right) \mapsto$$
$$\left([x_0 : x_1 : -x_2], [z_1 : z_0 + z_2 : z_1 + 2z_2], [y_0 + 2y_1 - y_2 : 2y_0 : y_2 - y_0]\right).$$

One can check that the involution $\tau_{x,z}, \tau_{x,y}, \tau_{y,z}$ together with transformations ϕ_a generate the group $\mathrm{Aut}(X)$. Using this, we conclude that $\mathrm{Aut}(X) \cong \mathbb{G}_a.\mathfrak{S}_3$. This extension of groups splits. To see this, let $\theta = \tau_{x,z} \circ \tau_{x,y} \circ \phi_a$ for $a = \frac{1}{3}$. Then

$$\theta \colon \left([x_0 : x_1 : x_2], [y_0 : y_1 : y_2], [z_0 : z_1 : z_2]\right)$$
$$\mapsto \left([9z_0 + z_1 + 6z_2 : 9z_1 : 9z_2 + 3z_1], [9x_1 : 9x_0 - 2x_1 - 3x_2],\right.$$
$$\left.[5y_0 + 3y_2 + 18y_1 : 18y_0 : -3y_0 - 9y_2]\right).$$

Then $\theta^3 = \mathrm{Id}_X$ and $\tau_{x,z} \circ \theta \circ \tau_{x,z} = \theta^2$, so that $\langle \tau_{x,z}, \theta \rangle \cong \mathfrak{S}_3$. This gives $\mathrm{Aut}(X) \cong \mathbb{G}_a \rtimes \mathfrak{S}_3$.

By Theorem 1.3, the threefold X is not K-polystable. To show that X is K-semistable, observe that X is isomorphic to the threefold given by

$$\begin{cases} x_0 y_0 + x_1 y_1 + x_2 y_2 = 0, \\ y_0 z_0 + y_1 z_1 + y_2 z_2 = 0, \\ x_0 z_1 + x_1 z_0 - 2x_2 z_2 + \epsilon(x_1 z_2 - x_2 z_1) = 0, \end{cases}$$

where ϵ is any non-zero number. As we already mentioned, if $\epsilon = 0$, then these equations define the K-polystable smooth Fano threefold that admits an effective $\mathrm{PGL}_2(\mathbb{C})$-action. Now, arguing as in the proof of Corollaries 4.71 and 5.84, we can construct a test configuration for the threefold X, whose special fiber is a K-polystable Fano threefold, so that X is strictly K-semistable by Corollary 1.13.   □

In the remaining part of this section, we will prove the following result:

Proposition 5.99   *If X is given by ($\bigstar$), then X is K-polystable.*

To prove this result, we suppose that X is given by ($\bigstar$). Then Δ_x is given by $x_0 x_1 = x_2^2$, the curve Δ_y is given by $z_0 z_1 = z_2^2$, and Δ_z is given by $y_0 y_1 = \frac{1-s^2}{4} y_2^2$. Now, let us describe some automorphisms of the threefold X. For every $\lambda \in \mathbb{C}^*$, the group $\mathrm{Aut}(X)$ contains the automorphism $\varphi_\lambda \colon X \to X$ that is given by

$$\left([x_0 : x_1 : x_2], [y_0 : y_1 : y_2], [z_0 : z_1 : z_2]\right)$$
$$\mapsto \left(\left[\lambda x_0 : \frac{x_1}{\lambda} : x_2\right], \left[\frac{y_0}{\lambda} : \lambda y_1 : y_2\right], \left[\lambda z_0 : \frac{z_1}{\lambda} : z_2\right]\right).$$

These automorphisms form a proper subgroup $\Gamma \subsetneq \mathrm{Aut}(X)$, which is isomorphic to $\mathbb{G}_m$. The full automorphism group $\mathrm{Aut}(X)$ also contains the involution $\tau_{x,z}$ that is given by

$$\tau_{x,z} \colon \left([x_0 : x_1 : x_2], [y_0 : y_1 : y_2], [z_0 : z_1 : z_2]\right)$$
$$\mapsto \left([z_1 : z_0 : z_2], [y_1 : y_0 : y_2], [x_1 : x_0 : x_2]\right),$$

the group $\mathrm{Aut}(X)$ also contains the involution $\tau_{x,y}$ given by

$$\tau_{x,y} \colon \left([x_0 : x_1 : x_2], [y_0 : y_1 : y_2], [z_0 : z_1 : z_2]\right)$$
$$\mapsto \left(\left[y_0 : \frac{y_1}{1-s^2} : -\frac{y_2}{2}\right], [x_0 : (1-s^2)x_1 : -2x_2], \left[(s+1)z_1 : \frac{z_0}{s+1} : z_2\right]\right),$$

and it contains the involution $\tau_{y,z}$ which is given by

$$\tau_{y,z} \colon \left([x_0 : x_1 : x_2], [y_0 : y_1 : y_2], [z_0 : z_1 : z_2]\right)$$
$$\mapsto \left([x_1 : x_0 : -x_2], [(1-s)z_0 : (s+1)z_1 : 2z_2], \left[\frac{y_0}{1-s} : \frac{y_1}{s+1} : \frac{y_2}{2}\right]\right).$$

Let G be the subgroup in $\mathrm{Aut}(X)$ generated by $\Gamma \cong \mathbb{G}_m$ and the involutions $\tau_{x,y}, \tau_{x,z}, \tau_{y,z}$. Then Γ is a normal subgroup in G. Note that $G/\Gamma \cong \mathfrak{S}_3$, so that we have $G \cong \mathbb{G}_m.\mathfrak{S}_3$. Actually, this extension of groups splits. To see this, we let $\vartheta = \tau_{x,z} \circ \tau_{x,y}$. Then

$$\vartheta \colon ([x_0 : x_1 : x_2], [y_0 : y_1 : y_2], [z_0 : z_1 : z_2])$$
$$\mapsto \left(\left[\frac{z_0}{s+1} : (s+1)z_1 : z_2\right], \left[(1-s^2)x_1 : x_0 : -2x_2\right], \left[\frac{y_1}{1-s^2} : y_0 : -\frac{y_2}{2}\right]\right).$$

Then $\vartheta \circ \varphi_\lambda = \varphi_\lambda \circ \vartheta$. Now, we let $\vartheta_\lambda = \vartheta \circ \varphi_\lambda$. Then

$$(\vartheta_\lambda)^3 = \mathrm{Id}_X \iff \lambda^3 = (1 - s^2)(1 + s).$$

Moreover, if $\lambda^3 = (1 - s^2)(1 + s)$, then $\tau_{x,z} \circ \vartheta_\lambda \circ \tau_{x,z} = \vartheta_\lambda^2$, which gives $\langle \tau_{x,z} \circ \vartheta_\lambda \rangle \cong \mathfrak{S}_3$. Therefore, choosing $\lambda \in \mathbb{G}_m$ to be one of the three cube roots $\sqrt[3]{(1-s^2)(1+s)}$, we obtain the subgroup $\langle \tau_{x,z}, \vartheta_\lambda \rangle \cong \mathfrak{S}_3$ that gives us a section of the quotient map $G \to G/\Gamma \cong \mathfrak{S}_3$, which defines a splitting $G \cong \mathbb{G}_m \rtimes \mathfrak{S}_3$.

Remark 5.100 If $s = 0$, then $\mathrm{Aut}(X) \cong \mathrm{PGL}_2(\mathbb{C}) \times \mathfrak{S}_3$. If $s \neq 0$, then $\mathrm{Aut}(X) = G$.

To prove the K-polystability of the threefold X, we need to prove one technical lemma. To state it, we find it useful to replace the parameter $s \in \mathbb{C} \setminus \{1, -1\}$ as $s = \frac{r^3 - 1}{r^3 + 1}$ for a non-zero number r such that $r^3 \neq -1$. Then $(1 - s^2)(1 + s) = \frac{8r^6}{(r^3+1)^3}$, so that

$$\sqrt[3]{(1-s^2)(1+s)} = \left\{\frac{2r^2}{r^3+1}, \frac{2\omega r^2}{r^3+1}, \frac{2\omega^2 r^2}{r^3+1}\right\},$$

where ω is a primitive cube root of unity.

Lemma 5.101 *The following assertions hold:*

(i) $\mathrm{Pic}^G(X) = \mathbb{Z}[-K_X]$;
(ii) *the threefold X does not have G-fixed points;*
(iii) *the threefold X contains exactly three distinct G-invariant irreducible curves, which can be parametrically described as follows:*

$$\Big(\big[u^2 : r(r^2 - r + 1)v^2 : ruv\big],$$
$$\big[r(r^2 - r + 1)v^2 : ru^2 : -(r^3 + 1)uv\big],$$
$$\big[ru^2 : (r^2 - r + 1)v^2 : ruv\big]\Big), \quad (5.20)$$

$$\Big(\big[ru^2 : \omega^2(r + 1)(r + \omega^2)v^2 : ruv\big],$$
$$\big[\omega(r + 1)(r + \omega^2)v^2 : \omega r^2 u^2 : -(r^3 + 1)uv\big],$$
$$\big[\omega^2 r^3 u^2 : (r + 1)(r + \omega^2)v^2 : r^2 uv\big]\Big), \quad (5.21)$$

$$\left(\left[ru^2 : \omega(r+1)(r+\omega)v^2 : ruv\right],\right.$$

$$\left[\omega^2(r+1)(r+\omega)v^2 : \omega^2 r^2 u^2 : -(r^3+1)uv\right],$$

$$\left.\left[\omega r^3 u^2 : (r+1)(r+\omega)v^2 : r^2 uv\right]\right), \quad (5.22)$$

where $[u : v] \in \mathbb{P}^1$. *All these three curves are smooth and rational.*

Proof Assertion (i) immediately follows from the description of the action of the group G. If X contains a G-fixed point O, then $\eta_x(O)$ is fixed point by the induced $\langle \Gamma, \tau_{y,z} \rangle$-action, which gives $\eta_x(O) = [0 : 0 : 1]$. Similarly, we get $\eta_y(O) = [0 : 0 : 1]$ and $\eta_z(O) = [0 : 0 : 1]$, so that

$$O = ([0 : 0 : 1], [0 : 0 : 1], [0 : 0 : 1]) \notin X,$$

which is a contradiction. This proves (ii).

Note that the curves (5.20), (5.21) and (5.22) are distinct and G-invariant. Thus to prove assertion (iii), it suffices to show that X contains no other G-invariant irreducible curves. To do this, let C be a G-invariant irreducible curve in the threefold X. We show that C is one of the curves (5.20), (5.21) and (5.22).

To start with, observe that

$$-K_X \cdot C = \left(\eta_x^*(O_{\mathbb{P}^2}(1)) + \eta_y^*(O_{\mathbb{P}^2}(1)) + \eta_z^*(O_{\mathbb{P}^2}(1))\right) \cdot C$$

$$= 3\eta_x^*(O_{\mathbb{P}^2}(1)) \cdot C = 3\eta_y^*(O_{\mathbb{P}^2}(1)) \cdot C = 3\eta_z^*(O_{\mathbb{P}^2}(1)) \cdot C \geqslant 3,$$

so that $\eta_x(C)$, $\eta_y(C)$ and $\eta_z(C)$ are irreducible curves, which are invariant with respect to the induced actions on $\mathbb{P}^2$ of the subgroups $\langle \Gamma, \tau_{y,z} \rangle$, $\langle \Gamma, \tau_{x,z} \rangle$ and $\langle \Gamma, \tau_{x,y} \rangle$, respectively. Thus, if the curves $\eta_x(C)$, $\eta_y(C)$, $\eta_z(C)$ are lines, these are the lines $x_2 = 0$, $y_2 = 0$, $z_2 = 0$. In this case, the curve C must be contained in the subset in $\mathbb{P}^2 \times \mathbb{P}^2 \times \mathbb{P}^2$ given by

$$\begin{cases} x_2 = y_2 = z_2 = 0, \\ x_0 y_0 + x_1 y_1 + x_2 y_2 = 0, \\ y_0 z_0 + y_1 z_1 + y_2 z_2 = 0, \\ (1+s)x_0 z_1 + (1-s)x_1 z_0 - 2x_2 z_2 = 0. \end{cases}$$

But this subset does not contain any curve that is surjectively mapped by η_x, η_y, η_z to the lines $x_2 = 0$, $y_2 = 0$, $z_2 = 0$, respectively. Hence, $\eta_x(C)$, $\eta_y(C)$, $\eta_z(C)$ are not lines.

We see that there are non-zero numbers q_x, q_y, q_z such that $\eta_x(C)$, $\eta_y(C)$, $\eta_z(C)$ are the conics $x_0 x_1 = q_x x_2^2$, $y_0 y_1 = q_y y_2^2$, $z_0 z_1 = q_z z_2^2$, respectively. Therefore, we see that each subgroup $\langle \Gamma, \tau_{y,z} \rangle$, $\langle \Gamma, \tau_{x,z} \rangle$, $\langle \Gamma, \tau_{x,y} \rangle$ acts faithfully on the curve C because they act faithfully on the curves $\eta_x(C)$, $\eta_y(C)$, $\eta_z(C)$, respectively. In particular, C is rational.

The action of the group G on the curve C induces a homomorphism $v \colon G \to$ Aut(C). On the other hand, we have $\langle \Gamma, \vartheta \rangle \cong \mathbb{G}_m \times \mu_3$, and the group Aut$(C)$ does not contain subgroups isomorphic to $\mathbb{G}_m \times \mu_3$ since C is rational [169]. Therefore, since Γ acts on the curve C faithfully, we get $v(\vartheta) \in v(\Gamma)$, so that ker$(v)$ contains ϑ_λ for some $\lambda \in \mathbb{G}_m$.

Let $P = ([x_0 : x_1 : x_2], [y_0 : y_1 : y_2], [z_0 : z_1 : z_2])$ be a sufficiently general point in C. Then $\eta_x(P)$, $\eta_y(P)$, $\eta_z(P)$ is not contained in the lines $x_2 = 0$, $y_2 = 0$, $z_2 = 0$, respectively. Thus, we may assume that $x_2 = y_2 = z_2 = 1$ and $(x_0, x_1, y_0, y_1, z_0, z_1) \neq (0,0,0,0,0,0)$. On the other hand, we have

$$([x_0 : x_1 : 1], [y_0 : y_1 : 1], [z_0 : z_1 : 1]) = P = \vartheta_\lambda(P)$$
$$= \left(\left[\frac{\lambda z_0}{s+1} : \frac{(s+1)z_1}{\lambda} : 1 \right], \left[\frac{(s^2-1)x_1}{2\lambda} : -\frac{\lambda x_0}{2} : 1 \right], \left[\frac{2\lambda y_1}{s^2-1} : -\frac{2y_0}{\lambda} : 1 \right] \right).$$

This gives the following system of linear equations:

$$\begin{pmatrix} 0 & \frac{4r^3}{(r^3+1)^2} & 2\lambda & 0 & 0 & 0 \\ -\frac{2r^3}{r^3+1} & 0 & 0 & 0 & \lambda & 0 \\ 0 & 0 & 0 & 2\lambda & \frac{4r^3}{(r^3+1)^2} & 0 \\ 0 & -\lambda & 0 & 0 & 0 & \frac{2r^3}{r^3+1} \\ \lambda & 0 & 0 & 2 & 0 & 0 \\ 0 & 0 & 2 & 0 & 0 & \lambda \end{pmatrix} \begin{pmatrix} x_0 \\ x_1 \\ y_0 \\ y_1 \\ z_0 \\ z_1 \end{pmatrix} = 0.$$

The determinant of the matrix here is

$$-4 \left(\lambda - \frac{2r^2}{r^3+1} \right)^2 \left(\lambda - \frac{2\omega r^2}{r^3+1} \right)^2 \left(\lambda - \frac{2\omega^2 r^2}{r^3+1} \right)^2.$$

It must vanish since $(x_0, x_1, y_0, y_1, z_0, z_1) \neq (0,0,0,0,0,0)$. Then

$$\lambda \in \left\{ \frac{2r^2}{r^3+1}, \frac{2\omega r^2}{r^3+1}, \frac{2\omega^2 r^2}{r^3+1} \right\}.$$

If $\lambda = \frac{2r^2}{r^3+1}$, then solving the system above, we get

$$P = \left(\left[a : -\frac{r^3+1}{r} b : 1 \right], \left[b : -\frac{r^2}{r^3+1} a : 1 \right], \left[ra : -\frac{r^3+1}{r^2} b : 1 \right] \right)$$

for some $(a,b) \in \mathbb{C} \setminus (0,0)$, so that $f(P) = g(P) = h(P) = 0$ gives $b = -\frac{1}{(r+1)a}$ and

$$P = \left(\left[a : \frac{r^2-r+1}{ra} : 1 \right], \left[-\frac{1}{(r+1)a} : -\frac{ar^2}{r^3+1} : 1 \right], \left[ra : \frac{r^2-r+1}{ar^2} : 1 \right] \right),$$

which implies that P is contained in the curve (5.20), so that C is the curve (5.20). In this case, we have

$$q_x = \frac{r^2 - r + 1}{r}, \quad q_y = \frac{r^2}{(r+1)(r^3+1)}, \quad q_z = \frac{r^2 - r + 1}{r}.$$

Similarly, if $\lambda = \frac{2\omega r^2}{r^3+1}$, then C is the curve (5.21) and

$$q_x = \frac{\omega^2 r^2 - r + \omega}{r}, \quad q_y = \frac{\omega(r\omega + 1)r^2}{(r^2 - r + 1)(r^3 + 1)}, \quad q_z = \frac{\omega^2 r^2 - r + \omega}{r}.$$

Finally, if $\lambda = \frac{2\omega^2 r^2}{r^3+1}$, then C is the curve (5.22) and

$$q_x = \frac{\omega r^2 - r + \omega^2}{r}, \quad q_y = \frac{\omega^2(r\omega^2 + 1)r^2}{(r^2 - r + 1)(r^3 + 1)}, \quad q_z = \frac{\omega r^2 - r + \omega^2}{r}.$$

This completes the proof and also shows that each morphism among η_x, η_y, η_z maps the curves (5.20), (5.21), (5.22) to three different conics in $\mathbb{P}^2$. $\qquad\square$

Now we are ready to prove

Lemma 5.102 *If $s \neq 0$, then $\alpha_G(X) = 1$. If $s = 0$, then $\alpha_G(X) = \frac{2}{3}$.*

Proof First, let us recall that $s = \frac{r^3 - 1}{r^3 + 1}$, where r is a non-zero number such that $r^3 \neq -1$. If $s = 0$, we assume that $r = 1$ to avoid repeating computations.

Since $-K_X \sim E_{x,y} + E_{y,z} + E_{x,z}$, we can conclude that $\alpha_G(X) \leqslant 1$. Moreover, if $s = 0$, then $E_{x,y}$, $E_{y,z}$ and $E_{x,z}$ meet along the curve (5.20), which gives $\alpha_G(X) \leqslant \frac{2}{3}$. Set

$$\mu = \begin{cases} 1 & \text{if } s \neq 0, \\ \dfrac{2}{3} & \text{if } s = 0. \end{cases}$$

We see that $\alpha_G(X) \leqslant \mu$. Suppose that $\alpha_G(X) < \mu$. Let us seek a contradiction.

Recall that $\operatorname{Pic}^G(X) = \mathbb{Z}[-K_X]$ and X has no G-fixed points by Lemma 5.101. Arguing as in the proof of Theorem 1.52 and using Lemma 1.42, we see that there exist an irreducible G-invariant curve $C \subset X$ and a G-invariant effective $\mathbb{Q}$-divisor D on the threefold X such that $D \sim_{\mathbb{Q}} -K_X$, the log pair $(X, \lambda D)$ is strictly log canonical for some rational number $\lambda < \mu$, and C is its unique log canonical center. Then C is one of the curves (5.20), (5.21), (5.22) by Lemma 5.101.

Since $\lambda < 1$ and $C \subseteq \operatorname{Nklt}(X, \lambda D)$, we see that $\operatorname{mult}_C(D) \geqslant \frac{1}{\lambda} > \frac{1}{\mu} \geqslant 1$.

Now let us use assumptions and notations from the proof of Lemma 5.101. Let S_x, S_y, S_z be the surfaces in X that are cut out by $x_0 x_1 = q_x x_2^2$, $y_0 y_1 = q_y y_2^2$, $z_0 z_1 = q_z z_2^2$, respectively. Then $C \subset S_x \cap S_y \cap S_z$, the divisor $S_x + S_y + S_z$ is

G-invariant and $-K_X \sim_\mathbb{Q} \frac{1}{2}(S_x + S_y + S_z)$. Moreover, if $s = 0$ and C is the curve (5.20), then we have $C = E_{x,y} \cap E_{y,z} \cap E_{x,z}$ and we have $S_x = E_{x,y} + E_{x,z}$, $S_y = E_{x,y} + E_{y,z}$, $S_z = E_{x,z} + E_{y,z}$. In all other cases, the surfaces S_x, S_y, S_z are smooth at a general point of the curve C, and they meet each other pairwise transversally at a general point of the curve C.

Indeed, to prove this claim, it is enough to check both assertions for S_x and S_y because the group G acts two-transitively on $\{S_x, S_y, S_z\}$. Let us show that S_x and S_y are smooth at a general point of the curve C, and they meet transversally at a general point of the curve C. This can be explicitly checked at the point $P \in C$ that corresponds to $[u : v] = [1 : 1]$ in the parametrizations (5.20), (5.21) and (5.22). Thus, we can do this in the affine chart $x_2 = y_2 = z_2 = 1$. In this chart, we have

$$X = \{x_0 y_0 + x_1 y_1 + 1 = 0, y_0 z_0 + y_1 z_1 + 1 = 0, (1+s)x_0 z_1 + (1-s)x_1 z_0 - 2 = 0\},$$

the surface S_x is given by $x_0 x_1 = q_x$, and the surface S_y is given by $y_0 y_1 = q_y$, where we consider now $x_0, x_1, y_0, y_1, z_0, z_1$ as coordinates on $\mathbb{A}^6$. If C is the curve (5.20), then

$$P = \left(\frac{1}{r}, r^2 - r + 1, -\frac{r}{r+1}, -\frac{r}{r^3+1}, 1, \frac{r^2 - r + 1}{r}\right),$$

so that the Zariski tangent space to the intersection $S_x \cap S_y$ at the point P is given by

$$\begin{pmatrix}
-\frac{r}{r+1} & -\frac{r}{r^3+1} & \frac{1}{r} & r^2 & r & 1 & 0 & 0 \\
0 & 0 & 1 & \frac{r^2-r+1}{r} & & -\frac{r}{r+1} & -\frac{r}{r^3+1} \\
r^2(r^2-r+1) & 1 & 0 & 0 & & r^2-r+1 & r^2 \\
r^2-r+1 & \frac{1}{r} & 0 & 0 & & 0 & 0 \\
0 & 0 & -\frac{r}{r^3+1} & -\frac{r}{r+1} & & 0 & 0
\end{pmatrix}$$

$$\times \begin{pmatrix}
x_0 - \frac{1}{r} \\
x_1 - r^2 + r - 1 \\
y_0 + \frac{r}{r+1} \\
y_1 + \frac{r}{r^3+1} \\
z_0 - 1 \\
z_1 - \frac{r^2-r+1}{r}
\end{pmatrix} = 0.$$

The determinant of the matrix formed by the first 5 columns of this matrix is $\frac{(r^2-r+1)(r-1)^2}{r+1}$, so that it vanishes if and only if $s = 0$. Thus, if $s \neq 0$ and C is the curve (5.20), then the Zariski tangent space to the intersection $S_x \cap S_y$ at the point P is one-dimensional, so that both surfaces S_x and S_y are smooth at P, and intersect transversally at this point. This proves our claim in the case when C is the curve (5.20).

Similarly, if C is the curve (5.21), then

$$P = \left(1, \frac{\omega^2(r+1)(r+\omega^2)}{r}, -\frac{\omega}{r+\omega}, -\frac{\omega r^2}{r^3+1}, \omega^2 r, \frac{(r+1)(r+\omega^2)}{r^2}\right),$$

and the dimension of the Zariski tangent space to the intersection $S_x \cap S_y$ at this point equals the nullity of the following 5×6 matrix:

$$\begin{pmatrix} -\frac{\omega}{r+\omega} & -\frac{\omega r^2}{r^3+1} & 1 & \frac{\omega^2(r+1)(r+\omega^2)}{r} & 0 & 0 \\ 0 & 0 & w^2 r & \frac{(r+1)(r+w^2)}{r^2} & -\frac{\omega}{r+\omega} & -\frac{\omega}{r^3+1} \\ r(r+1)(r+\omega^2) & w^2 r & 0 & 0 & \frac{\omega^2(r+1)(r+\omega^2)}{r} & r^3 \\ \frac{\omega^2(r+1)(r+\omega^2)}{r} & 1 & 0 & 0 & 0 & 0 \\ 0 & 0 & -\frac{\omega r^2}{r^3+1} & -\frac{\omega}{r+\omega} & 0 & 0 \end{pmatrix}.$$

The determinant of its submatrix formed by the first 5 columns is

$$\frac{\omega(r+1)(r-\omega)^2(r+\omega^2)}{r+w},$$

so that it never vanishes, because $r^3 \neq -1$ and $r \neq \omega$ (if $s = 0$, then $r = 1$ by assumption). Therefore, the Zariski tangent space to $S_x \cap S_y$ at the point P is always one-dimensional, so that both our surfaces S_x and S_y are smooth at P, and intersect transversally at P. This proves our claim in the case when C is the curve (5.21). Now, swapping ω with ω^2, we also obtain the proof of our claim in the case when C is the curve (5.22).

Thus, unless $s = 0$ and C is the curve (5.20), the surfaces S_x, S_y, S_z are smooth at a general point of the curve C, and they meet each other pairwise transversally at a general point of the curve C. In particular, we see that $C \not\subseteq$ Nklt$(X, \frac{\mu}{2}(S_x + S_y + S_z))$. Thus, using Lemma A.34, we may assume that S_x, S_y, S_z are not contained in Supp(D).

If $s = 0$ and C is the curve (5.20), then $1 = D \cdot \ell \geqslant$ mult$_C(D)$, where ℓ is a general fiber of the projection $E_{x,y} \to \pi_{x,y}(E_{x,y})$. But mult$_C(D) > 1$. Therefore, we see that $s \neq 0$ or C is not the curve (5.20). Then $\eta_x(C) \neq \Delta_x$, $\eta_y(C) \neq \Delta_y$ and $\eta_z(C) \neq \Delta_z$.

Let ℓ be a general fiber of the morphism $\eta_x|_{S_x} \colon S_x \to \eta_x(C)$. Then ℓ is not contained in the support of the divisor D since S_x is not contained in its support. On the other hand, the curve ℓ meets the curve C, so that $2 = D \cdot \ell \geqslant$ mult$_C(D)$, which gives mult$_C(D) \leqslant 2$.

Let $\eta \colon \widehat{X} \to X$ be the blow up of the curve C, and let F be the η-exceptional surface. Then the G-action lifts to $\widehat{X}$, and it follows from Lemma A.27 that F

contains a smooth irreducible G-invariant curve $\mathscr{C}$ such that $\mathscr{C}$ is a section of the natural projection $F \to C$. Let us show that such curve does not exist.

Let $\widehat{S}_x$, $\widehat{S}_y$, $\widehat{S}_z$ be the proper transforms on $\widehat{X}$ of the surfaces S_x, S_y, S_z, respectively. Then each intersection among $\widehat{S}_x \cap F$, $\widehat{S}_y \cap F$, $\widehat{S}_z \cap F$ contains a unique component that is a section of the projection $F \to C$. Denote these sections by C_x, C_y, C_z, respectively. Then

- C_x, C_y, C_z are distinct curves,
- C_x, C_y, C_z are Γ-invariant, and Γ acts faithfully on each of these curves,
- the whole group G permutes the curves C_x, C_y, C_z two-transitively.

Thus, using Corollary A.51, we conclude that $F = \mathbb{P}^1 \times \mathbb{P}^1$. Then, using Lemma A.50, we conclude that the G-action on F is given by (A.19) for some integers $a > 0$ and b, which implies that F does not contain G-invariant sections of the projection $F \to C$, which contradicts the existence of the curve $\mathscr{C}$.   □

Now, Proposition 5.99 follows from Theorem 1.51 and Lemma 4.18.

5.20 Family №3.15

Let Q be the quadric $\{x_0^2 + 2x_1x_2 + 2x_1x_4 + 2x_2x_3 = 0\} \subset \mathbb{P}^4$, where x_0, x_1, x_2, x_3, x_4 are homogeneous coordinates on $\mathbb{P}^4$, and note that Q is smooth. Let L be the line $\{x_0 = x_1 = x_2 = 0\}$, let Π be the plane $\{x_3 = x_4 = 0\}$, and let $C = Q \cap \Pi$. Then $L \subset Q$, $L \cap \Pi = \varnothing$, and C is a smooth conic. Let $\pi \colon X \to Q$ be the blow up along the union $L \cup C$. Then X is a smooth Fano threefold from the deformation family №3.15. By [45, Lemma 5.10], the threefold X is the unique smooth member of this family.

Proposition 5.103 *The threefold X is K-polystable.*

Let G be the subgroup in $\mathrm{Aut}(Q)$ generated by the involution ι given by

$$\left[x_0 : x_1 : x_2 : x_3 : x_4\right] \mapsto \left[x_0 : x_2 : x_1 : x_4 : x_3\right]$$

and the transformations $[x_0 : x_1 : x_2 : x_3 : x_4] \mapsto [\lambda x_0 : \lambda^2 x_1 : x_2 : \lambda^2 x_3 : x_4]$ for $\lambda \in \mathbb{C}^*$. Then $G \cong \mathbb{C}^* \rtimes \mu_2$. Since L and C are G-invariant, the action of the group G lifts to X. To prove Proposition 5.103, we will apply Theorem 1.22 to X equipped with this G-action. But first, let us describe the G-equivariant geometry of the threefold X.

Let $\overline{R}$ be the surface $\{x_2x_3 + x_1x_4 = 0\} \cap Q$, and let R be its proper transform on X. Then the surface $\overline{R}$ is irreducible, it is singular along L, and it contains both L and C, but R is smooth, and there is a G-equivariant birational morphism

$\eta\colon X \to \mathbb{P}^1 \times \mathbb{P}^2$ that contracts R to a curve. Thus, we have the G-equivariant commutative diagram

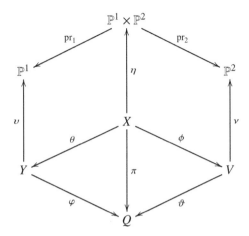

where ϑ is the blow up of the line L, φ is the blow up of the conic C, υ is a fibration into quadric surfaces, ν is a $\mathbb{P}^1$-bundle, pr_1 and pr_2 are projections to the first and the second factors, respectively, θ and ϕ are blow ups of the preimages of L and C, respectively.

Let E_L and E_C be the exceptional surfaces of the morphisms θ and ϕ, respectively. Let $H_Q = \pi^*(O_{\mathbb{P}^4}(1)|_Q)$, let $H_1 = (\mathrm{pr}_1 \circ \eta)^*(O_{\mathbb{P}^1}(1))$ and let $H_2 = (\mathrm{pr}_2 \circ \eta)^*(O_{\mathbb{P}^2}(1))$. Then

$$\mathrm{Pic}(X) = \mathbb{Z}[H_Q] \oplus \mathbb{Z}[E_L] \oplus \mathbb{Z}[E_C],$$
$$\mathrm{Nef}(X) = \mathbb{R}_{\geqslant 0}[H_Q] + \mathbb{R}_{\geqslant 0}[H_1] + \mathbb{R}_{\geqslant 0}[H_2],$$
$$\overline{\mathrm{Eff}}(X) = \mathbb{R}_{\geqslant 0}[E_L] + \mathbb{R}_{\geqslant 0}[E_C] + \mathbb{R}_{\geqslant 0}[R] + \mathbb{R}_{\geqslant 0}[H_1].$$

Note that $H_2 \sim H_Q - E_L$, $H_1 \sim H_Q - E_C$, $R \sim 2H_Q - 2E_L - E_C$, and

$$- K_X \sim 3H_Q - E_L - E_C \sim H_Q + H_1 + H_2 \sim_Q 2E_L + \frac{1}{2}E_C + \frac{3}{2}R, \quad (5.23)$$

so that $\alpha_G(X) = \frac{1}{2}$ by [46, Lemma 8.15]. One can show that $\mathrm{Aut}(X) = G$.

Let L' be the line $\{x_0 = x_1 + 2x_3 = x_2 + 2x_4 = 0\} \subset Q$. Then the line L' is G-invariant. Similarly, for every non-zero $t \in \mathbb{C}$, let

$$C_t = \{(1-t)x_1 - 2tx_3 = (1-t)x_2 - 2tx_4 = 0\} \cap Q.$$

Then C_t is an irreducible G-invariant conic for every non-zero $t \in \mathbb{C}$. Note that $C = C_1$. Note also that $L \cap L' = \varnothing$, $L \cap C_t = \varnothing$ and $L' \cap C_t = \varnothing$ for every $t \neq 0$. Finally, observe that the conics C_{t_1} and C_{t_2} are also disjoint for $t_1 \neq t_2$.

Lemma 5.104 *Let Z be an irreducible G-invariant curve in the quadric hypersurface Q. Then either $Z = L$, or $Z = L'$, or $Z = C_t$ for some non-zero $t \in \mathbb{C}$.*

Proof Observe that the curve Z is rational, so that it contains a ι-fixed point P such that the curve Z is the closure of the $\mathbb{G}_m$-orbit of this point. Thus, looking at the ι-fixed points in Q, we conclude that either $P = [0 : 0 : 0 : 1 : -1]$, or $P = [0 : 2 : -2 : -1 : 1]$, or $P = [4s : 4s^2 : 4s^2 : -2s^2 - 1 : -2s^2 - 1]$ for some non-zero $s \in \mathbb{C}$. Then either $Z = L$, or $Z = L'$, or $Z = C_t$ for $t = -2s^2$. $\qquad\square$

In what follows, we will apply results and notations from Section 1.7 to prove Proposition 5.103. Let Z be an irreducible G-invariant curve in X.

Lemma 5.105 *Suppose that $Z \subset E_C$. Then $S(W_{\bullet,\bullet}^{E_C}; Z) \leqslant \frac{51}{64}$.*

Proof We have $E_C \cong \mathbb{P}^1 \times \mathbb{P}^1$. Let s be a section of the projection $E_C \to C$ such that $s^2 = 0$, and f a fiber of this projection. Then

$$-K_X - uE_C \sim_{\mathbb{R}} 2H_1 + \frac{1}{2}R + \left(\frac{3}{2} - u\right)E_C$$

for $u \in \mathbb{R}_{\geqslant 0}$, so that $-K_X - uE_C$ is pseudo-effective if and only if $u \leqslant \frac{3}{2}$. Moreover, we have

$$P\left(-K_X - uE_C\right) = \begin{cases} -K_X - uE_C & \text{if } 0 \leqslant u \leqslant 1, \\ 2H_1 + (3 - 2u)H_2 & \text{if } 1 \leqslant u \leqslant \frac{3}{2}, \end{cases}$$

and

$$N\left(-K_X - uE_C\right) = \begin{cases} 0 & \text{if } 0 \leqslant u \leqslant 1, \\ (u - 1)R & \text{if } 1 \leqslant u \leqslant \frac{3}{2}. \end{cases}$$

If $u \leqslant 1$, then we have $P(-K_X - uE_C)|_{E_C} \sim (1 + u)s + (4 - 2u)f$. Similarly, if $1 \leqslant u \leqslant \frac{3}{2}$, then $P(-K_X - uE_C)|_{E_C} \sim 2s + (6 - 4u)f$. Note that $R|_{E_C}$ is a smooth curve in $|s + 2f|$. Thus, if $Z = R|_{E_C}$, then Corollary 1.110 gives

$$S\left(W_{\bullet,\bullet}^{E_C}; Z\right) = \frac{3}{32} \int_0^1 \int_0^\infty \text{vol}\big((1 + u - v)s + (4 - 2u - 2v)f\big)\,dv\,du$$

$$+ \frac{3}{32} \int_1^{\frac{3}{2}} 4(u - 1)(6 - 4u)\,du$$

$$+ \frac{3}{32} \int_1^{\frac{3}{2}} \int_0^\infty \text{vol}\big((2 - v)f + (6 - 4u - 2v)s\big)\,dv\,du$$

$$= \frac{3}{32} \int_0^{\frac{1}{2}} \int_0^{1+u} 2(4 - 2u - 2v)(1 + u - v)dvdu$$

$$+ \frac{3}{32} \int_{\frac{1}{2}}^1 \int_0^{2-u} 2(4 - 2u - 2v)(1 + u - v)dvdu$$

$$+ \frac{3}{32} \int_1^{\frac{3}{2}} 4(u - 1)(6 - 4u)du$$

$$+ \frac{3}{32} \int_1^{\frac{3}{2}} \int_0^{3-2u} 2(6 - 4u - 2v)(2 - v)dvdu$$

$$= \frac{15}{32} < \frac{51}{64}.$$

If $Z \neq R|_{E_C}$, then we have $S(W_{\bullet,\bullet}^{E_C}; Z) \leqslant S(W_{\bullet,\bullet}^{E_C}; s)$ because $|Z - s| \neq \emptyset$, since $Z \nsim f$ as the conic C does not have G-fixed points. If $Z \neq R|_{E_C}$, then

$$S(W_{\bullet,\bullet}^{E_C}; Z) \leqslant S(W_{\bullet,\bullet}^{E_C}; s) = \frac{3}{32} \int_0^1 \int_0^\infty \mathrm{vol}\big((1 + u - v)s + (4 - 2u)f\big)dvdu$$

$$+ \frac{3}{32} \int_1^{\frac{3}{2}} \int_0^\infty \mathrm{vol}\big((2 - v)f + (6 - 4u)s\big)dvdu$$

$$= \frac{3}{32} \int_0^1 \int_0^{1+u} 2(4 - 2u)(1 + u - v)dvdu$$

$$+ \frac{3}{32} \int_1^{\frac{3}{2}} \int_0^2 2(6 - 4u)(2 - v)dvdu = \frac{51}{64}$$

by Corollary 1.110. □

Lemma 5.106 *Suppose that $Z \subset E_L$. Then $S(W_{\bullet,\bullet}^{E_L}; Z) \leqslant \frac{29}{32}$.*

Proof First, we observe that $E_L \cong \mathbb{F}_1$. Let f be a fiber of the natural projection $E_L \to L$, and let s be the (-1)-curve in E_L. Then $R|_{E_L}$ is a smooth curve in $|2s + 2f|$.

Take $u \in \mathbb{R}_{\geqslant 0}$. Using (5.23), we see that $-K_X - uE_L$ is pseudo-effective $\iff u \leqslant 2$. Moreover, if $u \leqslant 2$, then

$$P\big(- K_X - uE_L\big) = \begin{cases} - K_X - uE_L & \text{if } 0 \leqslant u \leqslant 1, \\ (2 - u)H_1 + (3 - u)H_2 & \text{if } 1 \leqslant u \leqslant 2, \end{cases}$$

and we have

$$N\big(- K_X - uE_L\big) = \begin{cases} 0 & \text{if } 0 \leqslant u \leqslant 1, \\ (u - 1)R & \text{if } 1 \leqslant u \leqslant 2. \end{cases}$$

If $u \leqslant 1$, then we have $P(-K_X - uE_L)|_{E_L} \sim (1+u)s + 3f$. Similarly, if $1 \leqslant u \leqslant 2$, then we have $P(-K_X - uE_L)|_{E_L} \sim (3-u)s + (5-2u)f$. Thus, if $Z = R|_{E_L}$,

$$
\begin{aligned}
S(W_{\bullet,\bullet}^{E_L}; Z) &= \frac{3}{32} \int_0^1 \int_0^\infty \mathrm{vol}\big((1+u-2v)s + (3-2v)f\big)dvdu \\
&\quad + \frac{3}{32} \int_1^2 (u-1)(3-u)(7-3u)du \\
&\quad + \frac{3}{32} \int_1^2 \int_0^\infty \mathrm{vol}\big((3-u-2v)s + (5-2u-2v)f\big)dvdu \\
&= \frac{3}{32} \int_0^1 \int_0^{\frac{1+u}{2}} (u+1-2v)(5-u-2v)dvdu \\
&\quad + \frac{17}{128} + \frac{3}{32} \int_1^2 \int_0^{\frac{3-u}{2}} (3-u-2v)(7-3u+2v)dvdu
\end{aligned}
$$

by Corollary 1.110, so that $S(W_{\bullet,\bullet}^{E_L}; Z) = \frac{15}{32} < \frac{29}{32}$.

If $Z \neq R|_{E_L}$, then $S(W_{\bullet,\bullet}^{E_L}; Z) \leqslant S(W_{\bullet,\bullet}^{E_L}; s)$ because $|Z - s| \neq \varnothing$, since $Z \not\sim f$ as the line L does not have G-fixed points. Hence, if $Z \neq R|_{E_L}$, Corollary 1.110 gives

$$
\begin{aligned}
S(W_{\bullet,\bullet}^{E_L}; Z) \leqslant S(W_{\bullet,\bullet}^{E_L}; s) &= \frac{3}{32} \int_0^1 \int_0^\infty \mathrm{vol}\big((1+u-v)s + 3f\big)dvdu \\
&\quad + \frac{3}{32} \int_1^2 \int_0^\infty \mathrm{vol}\big((3-u-v)s + (5-2u)f\big)dvdu \\
&\quad - \frac{3}{32} \int_0^1 \int_0^{1+u} (1+u-v)(5-u+v)dvdu \\
&\quad + \frac{3}{32} \int_1^2 \int_0^{3-u} (3-u-v)(7-3u+v)dvdu = \frac{29}{32},
\end{aligned}
$$

as required.   □

Let $\overline{S}$ be the surface $Q \cap \{x_1 x_4 = x_2 x_3\}$. Then $\overline{S}$ is a del Pezzo surface of degree 4 that has four ordinary nodes. It is well-known that $\overline{S}$ is toric, and it contains four lines [65, 42]. Two of them are the lines L and L' described above, and the remaining two lines in $\overline{S}$ are the disjoint lines $\ell = \{x_0 = x_1 = x_3 = 0\}$ and $\ell' = \{x_0 = x_2 = x_4 = 0\}$. Then

$$
\begin{aligned}
L \cap \ell &= [0:0:0:0:1], \\
L' \cap \ell &= [0:0:2:0:-1], \\
L \cap \ell' &= [0:0:0:1:0], \\
L' \cap \ell' &= [0:2:0:-1:0].
\end{aligned}
$$

These are the singular points of $\overline{S}$. By [42, Lemma 2.9], the lines L, L', ℓ, ℓ' generate $\mathrm{Cl}(\overline{S})$, which has rank 2. On the surface $\overline{S}$, we have $2L \sim 2L', 2\ell \sim 2\ell'$ and

$$-K_{\overline{S}} \sim L + L' + \ell + \ell' \sim 2(L + \ell).$$

The surface $\overline{S}$ also contains all conics C_t for $t \in \mathbb{C}^*$ including the conic $C = C_1$, each conic C_t is contained in the smooth locus of the surface $\overline{S}$, and $C_t \sim 2L$ for every $t \in \mathbb{C}^*$. Let us denote by S the proper transforms of the surface $\overline{S}$ on the threefold X.

Lemma 5.107 *Suppose $\pi(Z) = C_t$ for $t \in \mathbb{C} \setminus \{0, 1\}$. Then $S(W_{\bullet,\bullet}^S; Z) = \frac{79}{128}$.*

Proof Take $u \in \mathbb{R}_{\geq 0}$. Observe that

$$-K_X - uS \sim_{\mathbb{R}} \left(\frac{3}{2} - u\right)S + \frac{1}{2}E_L + \frac{1}{3}E_L,$$

which implies that $-K_X - uS$ is pseudo-effective $\iff u \leq \frac{3}{2}$. Moreover, if $u \leq \frac{3}{2}$, then

$$P\big(-K_X - uS\big) = \begin{cases} -K_X - uS & \text{if } 0 \leq u \leq 1, \\ (3 - 2u)H_Q & \text{if } 1 \leq u \leq \frac{3}{2}, \end{cases}$$

and

$$N\big(-K_X - uS\big) = \begin{cases} 0 & \text{if } 0 \leq u \leq 1, \\ (u - 1)(E_L + E_C) & \text{if } 1 \leq u \leq \frac{3}{2}. \end{cases}$$

In particular, we see that Z is not contained in the supports of the divisor $N(-K_X - uS)|_S$. Therefore, using Corollary 1.110, we obtain

$$S\left(W_{\bullet,\bullet}^S; Z\right) = \frac{3}{32}\int_0^1 \int_0^\infty \mathrm{vol}\big((-K_X - uS)|_S - vZ\big)\,du\,dv$$

$$+ \frac{3}{32}\int_1^{\frac{3}{2}} \int_0^\infty \mathrm{vol}\big((3 - 2u)H_Q|_S - vZ\big)\,du\,dv.$$

To compute these integrals, let us say a few words about the geometry of S.

The morphism π induces a birational morphism $\varpi: S \to \overline{S}$, which is the minimal resolution of the two singular points $[0 : 0 : 0 : 1 : 0]$ and $[0 : 0 : 0 : 0 : 1]$ of the surface $\overline{S}$. In particular, the surface S has exactly two singular points, and they are ordinary nodes. Denote the proper transforms on S of the curves L, L', ℓ, ℓ' and C_t by the same symbols, and denote by $\mathbf{e}$ and $\mathbf{e}'$ the two ϖ-exceptional curves such that $\mathbf{e} \cap \ell \neq \varnothing$ and $\mathbf{e}' \cap \ell' \neq \varnothing$. Note that the Mori cone $\overline{\mathrm{NE}}(S)$ is generated by the curves $L, \ell, \ell', \mathbf{e}, \mathbf{e}'$.

Table 5.5

	L	L'	ℓ	ℓ'	$\mathbf{e}$	$\mathbf{e}'$
L	-1	0	0	0	1	1
L'	0	0	$\frac{1}{2}$	$\frac{1}{2}$	0	0
ℓ	0	$\frac{1}{2}$	$-\frac{1}{2}$	0	1	0
ℓ'	0	$\frac{1}{2}$	0	$-\frac{1}{2}$	0	1
$\mathbf{e}$	1	0	1	0	-2	0
$\mathbf{e}'$	1	0	0	1	0	-2

On the surface S, we have $C_t \sim 2L'$, $2L + \mathbf{e} + \mathbf{e}' \sim 2L'$ and $2\ell + \mathbf{e} \sim 2\ell' + \mathbf{e}'$, and the intersections of the curves L, L', ℓ, ℓ', $\mathbf{e}$ and $\mathbf{e}'$ are given in Table 5.5.

Let v be a non-negative real number. If $u \leqslant 1$, then

$$P(-K_X - uS)|_S - vZ \sim_{\mathbb{R}} \left(\frac{3-u}{2} - v\right)Z + (3 - 2u)(\ell + \ell') + \frac{2-u}{2}(\mathbf{e} + \mathbf{e}'),$$

so that the divisor $P(-K_X - uS)|_S - vZ$ is pseudo-effective if and only if $v \leqslant \frac{3-u}{2}$. Moreover, if $u \leqslant 1$ and $v \leqslant \frac{3-u}{2}$, its Zariski decomposition can be described as follows:

- if $0 \leqslant v \leqslant 1$, then $P(-K_X - uS)|_S - vZ$ is nef,
- if $1 \leqslant v \leqslant \frac{3-u}{2}$, then the positive part of the Zariski decomposition is

$$\left(\frac{3-u}{2} - v\right)Z + (5 - 2u - 2v)(\ell + \ell') + \frac{2-u}{2}(\mathbf{e} + \mathbf{e}'),$$

and the negative part is $2(v-1)(\ell + \ell')$.

Similarly, if $1 \leqslant u \leqslant \frac{3}{2}$, then

$$P(-K_X - uS)|_S - vZ \sim_{\mathbb{R}} (3 - 2u - v)Z + (3 - 2u)(\ell + \ell')$$
$$+ \left(\frac{3}{2} - u\right)(\mathbf{e} + \mathbf{e}'),$$

so that this divisor is pseudo-effective $\Longleftrightarrow$ it is nef $\Longleftrightarrow$ $v \leqslant 3 - 2u$. Hence we obtain

$$S(W^S_{\bullet,\bullet}; Z) = \frac{3}{32} \int_0^1 \int_0^1 \left(3u^2 + 8uv - 16u - 12v + 17\right)dudv$$
$$+ \frac{3}{32} \int_0^1 \int_1^{\frac{3-u}{2}} (3 - u - 2v)(7 - 3u - 2v)dudv$$
$$+ \frac{3}{32} \int_1^{\frac{3}{2}} \int_0^{3-2u} 4(3 - 2u - v)(3 - 2u)dudv = \frac{79}{128},$$

as claimed. $\qquad\qquad\qquad\qquad\qquad\qquad\qquad\qquad\qquad\qquad\qquad\qquad\qquad\qquad$ $\square$

Now, let $\overline{H}$ be the hyperplane section of the quadric threefold Q given by $x_0 = 0$, and let H be its proper transform on the threefold X. Then $\overline{H}$ is a smooth quadric surface that contains the lines L and L', and H is a smooth del Pezzo surface of degree six.

Lemma 5.108 *Suppose that $\pi(Z) = L'$. Then $S(W_{\bullet,\bullet}^H; Z) = \frac{49}{64}$.*

Proof Take $u \in \mathbb{R}_{\geqslant 0}$. Note that $-K_X - uH \sim_{\mathbb{R}} (2 - u)H + H_1 + E_L$, which implies that the divisor $-K_X - uH$ is pseudo-effective $\iff u \leqslant 2$. Moreover, if $u \leqslant 2$, then

$$P(-K_X - uS) = \begin{cases} -K_X - uH & \text{if } 0 \leqslant u \leqslant 1, \\ H_1 + (2 - u)H_Q & \text{if } 1 \leqslant u \leqslant 2, \end{cases}$$

and $N(-K_X - uS) = (u - 1)E_L$ for $u \in [1,2]$. Then $Z \not\subset \operatorname{Supp}(N(-K_X - uH)|_H)$, so that

$$S\left(W_{\bullet,\bullet}^H; Z\right) = \frac{3}{32} \int_0^1 \int_0^\infty \operatorname{vol}\big((-K_X - uH)|_H - vZ\big) du\, dv$$
$$+ \frac{3}{32} \int_1^{\frac{3}{2}} \int_0^\infty \operatorname{vol}\big((H_1 + (2 - u)H_Q)|_H - vZ\big) du\, dv$$

by Corollary 1.110.

The conic C intersects $\overline{H}$ transversally at $P_1 = [0 : 1 : 0 : 0 : 0]$ and $P_2 = [0 : 0 : 1 : 0 : 0]$, which are not contained in the lines L and L'. Thus, the morphism π induces a birational morphism $\varpi \colon H \to \overline{H}$ that blows up P_1 and P_2. Let $\mathbf{e}_1$ and $\mathbf{e}_2$ be the ϖ-exceptional curves that are contracted to P_1 and P_2, respectively, let $\mathbf{s}_1$ and $\mathbf{f}_1$ be the proper transform on the surface H of the two rulings of the surface $\overline{H} \cong \mathbb{P}^1 \times \mathbb{P}^1$ that pass through the point P_1, and let $\mathbf{s}_2$ and $\mathbf{f}_2$ be the proper transform on H of the two rulings that pass through P_2. We may assume that $Z \sim \mathbf{s}_1 + \mathbf{e}_1 \sim \mathbf{s}_2 + \mathbf{e}_2$, so that $\mathbf{f}_1 + \mathbf{e}_1 \sim \mathbf{f}_2 + \mathbf{e}_2$ and $\mathbf{f}_1 + \mathbf{s}_2 \sim \mathbf{f}_2 + \mathbf{s}_1$. Observe that $\mathbf{e}_1, \mathbf{e}_2, \mathbf{s}_1, \mathbf{s}_2, \mathbf{f}_1, \mathbf{f}_2$ are all (-1)-curves in H.

Note that $E_L|_H \sim \mathbf{s}_1 + \mathbf{e}_1$, $H_Q|_H \sim \mathbf{f}_1 + \mathbf{s}_1 + 2\mathbf{e}_1$, $H_1|_H \sim \mathbf{f}_1 + \mathbf{s}_2$ and $H|_H \sim \mathbf{f}_1 + \mathbf{e}_1$.

Let v be a non-negative real number. If $u \leqslant 1$, then

$$P(-K_X - uH)|_H - vZ \sim_{\mathbb{R}} (2 - u)\mathbf{f}_1 + \mathbf{f}_2 + (2 - v)\mathbf{s}_1 + (3 - u - v)\mathbf{e}_1,$$

so that this divisor is pseudo-effective if and only if $v \leqslant 2$. Moreover, it is nef for $v \in [0,1]$, and its Zariski decomposition for $v \in [1,2]$ is

$$\underbrace{(3 - u - v)(\mathbf{f}_1 + \mathbf{e}_1) + (2 - v)(\mathbf{s}_1 + \mathbf{f}_2)}_{\text{positive part}} + \underbrace{(v - 1)(\mathbf{f}_1 + \mathbf{f}_2)}_{\text{negative part}}.$$

Similarly, if $1 \leqslant u \leqslant 2$, then

$$P(-K_X - uH)|_H - vZ \sim_{\mathbb{R}} (2 - u)\mathbf{f}_1 + \mathbf{f}_2 + (3 - u - v)\mathbf{s}_1 + (4 - 2u - v)\mathbf{e}_1,$$

so that this divisor is pseudo-effective if and only if $v \leqslant 4 - 2u - v$. Moreover, it is nef for $v \leqslant 2 - u$, and its Zariski decomposition for $v \geqslant 2 - u$ is

$$\underbrace{(4 - 2u - v)(\mathbf{f}_1 + \mathbf{e}_1) + (3 - u - v)(\mathbf{s}_1 + \mathbf{f}_2)}_{\text{positive part}} + \underbrace{(v - 2 + u)(\mathbf{f}_1 + \mathbf{f}_2)}_{\text{negative part}}.$$

Hence, using Corollary 1.110, we obtain

$$
\begin{aligned}
S(W_{\bullet,\bullet}^H; Z) = {} & \frac{3}{32} \int_0^1 \int_0^1 (2uv - 4u - 6v + 10)\,dv\,du \\
& + \frac{3}{32} \int_0^1 \int_1^2 2(2 - v)(3 - u - v)\,dv\,du \\
& + \frac{3}{32} \int_1^2 \int_0^{2-u} (2u^2 + 2uv - 12u - 6v + 16)\,dv\,du \\
& + \frac{3}{32} \int_1^2 \int_{2-u}^{4-2u} 2(3 - u - v)(4 - 2u - v)\,dv\,du = \frac{49}{64}
\end{aligned}
$$

as required.   □

Now, we are ready to prove that X is K-polystable. Suppose that X is not K-polystable. Then, by Theorem 1.22, there is a G-invariant prime divisor F over X such that $\beta(F) \leqslant 0$. Let $Z = C_X(F)$. Then Z is not a surface by Theorem 3.17, so that Z and $\pi(Z)$ are curves since Q has no G-fixed points. Now, applying Lemmas 5.104, 5.105, 5.106, 5.107, 5.108, we get a contradiction with Corollary 1.110 since $S_X(E_C) < 1$, $S_X(E_L) < 1$, $S_X(S) < 1$ and $S_X(H) < 1$ by Theorem 3.17. Therefore, X is K-polystable.

5.21 Family №4.3

Let C be the curve of degree $(1, 1, 2)$ in $\mathbb{P}^1 \times \mathbb{P}^1 \times \mathbb{P}^1$ given by

$$
\begin{cases}
x_0 y_1 - x_1 y_0 = 0, \\
x_0 z_1^2 + x_1 z_0^2 = 0,
\end{cases}
$$

where $[x_0 : x_1]$, $[y_0 : y_1]$ and $[z_0 : z_1]$ are homogeneous coordinates on the first, second and third factors of $\mathbb{P}^1 \times \mathbb{P}^1 \times \mathbb{P}^1$, respectively. Observe that C is smooth and irreducible. Let $\pi \colon X \to \mathbb{P}^1 \times \mathbb{P}^1 \times \mathbb{P}^1$ be the blow up of C. Then X is the smooth Fano threefold №4.3.

Let G be the subgroup of $\mathrm{Aut}(\mathbb{P}^1 \times \mathbb{P}^1 \times \mathbb{P}^1)$ generated by the following transformations:

$$\alpha\colon \big([x_0:x_1],[y_0:y_1],[z_0:z_1]\big) \mapsto \big([x_1:x_0],[y_1:y_0],[z_1:z_0]\big),$$
$$\beta\colon \big([x_0:x_1],[y_0:y_1],[z_0:z_1]\big) \mapsto \big([y_0:y_1],[x_0:x_1],[z_0:z_1]\big),$$
$$\gamma_\epsilon\colon \big([x_0:x_1],[y_0:y_1],[z_0:z_1]\big) \mapsto \big([x_0:\epsilon^2 x_1],[y_0:\epsilon^2 y_1],[z_0:\epsilon z_1]\big),$$

where $\epsilon \in \mathbb{C}^*$. Then $G \cong (\mathbb{G}_m \rtimes \mu_2) \times \mu_2$, and C is G-invariant, so that the G-action lifts to the threefold X. Let R_C be the G-invariant surface $\{x_0 y_1 - x_1 y_0 = 0\} \subset \mathbb{P}^1 \times \mathbb{P}^1 \times \mathbb{P}^1$, let R be its proper transform via π on the threefold X, let E be the π-exceptional surface, and let $H_i = (\mathrm{pr}_i \circ \pi)^*(\mathcal{O}_{\mathbb{P}^1}(1))$, where $\mathrm{pr}_i\colon \mathbb{P}^1 \times \mathbb{P}^1 \times \mathbb{P}^1 \to \mathbb{P}^1$ is the i-th projection. Then

$$-K_X \sim 2H_1 + 2H_2 + 2H_3 - E,$$

and $R \sim H_1 + H_2 - E$, because $C \subset R_C$. Moreover, we have:

Lemma 5.109 *The following assertions hold:*

(i) *both $\mathbb{P}^1 \times \mathbb{P}^1 \times \mathbb{P}^1$ and X do not contain G-fixed points,*
(ii) *if Z is a G-invariant curve in X, then $H_i \cdot Z \geqslant 2$ for every $i \in \{1,2,3\}$,*
(iii) *the linear system $|H_1 + H_2 + H_3|$ contains no G-invariant surfaces,*
(iv) *if D is a non-zero effective G-invariant $\mathbb{Z}$-divisor on X such that $-K_X - D$ is big, then $D = R$.*

Proof The first three assertions follow from the study of the G-action on $\mathbb{P}^1 \times \mathbb{P}^1 \times \mathbb{P}^1$. The remaining assertion immediately follows from the description of the cone of effective divisors of X, which is given in [93]. □

In the remaining part of the section, we will prove that X is K-polystable using results from Section 1.7. As usual, we will use notations introduced in this section. We start with

Lemma 5.110 *If Z is a G-invariant irreducible curve in R, then $S(W^R_{\bullet,\bullet}; Z) < 1$.*

Proof Let us use the descriptions of the cones $\mathrm{Nef}(X)$ and $\mathrm{Eff}(X)$ that are given in [93] to determine the (divisorial) Zariski decomposition of the divisor $-K_X - xR$, where $x \in \mathbb{R}_{\geqslant 0}$. First, if $0 \leqslant x \leqslant 1$, then $-K_X - xR$ is nef. Second, we have

$$-K_X - xR \sim_{\mathbb{R}} (2-x)H_1 + (2-x)H_2 + 2H_3 + (x-1)E,$$

so that $-K_X - xR$ is not pseudo-effective for $x > 2$. Finally, if $1 \leqslant x \leqslant 2$, then

$$P(-K_X - xR) = (2-x)H_1 + (2-x)H_2 + 2H_3$$

and $N(-K_X - xR) = (x-1)E$, where we use notations introduced in Section 1.7.

Let ℓ_1 and ℓ_2 be the rulings of the surface $R \cong \mathbb{P}^1 \times \mathbb{P}^1$ such that ℓ_1 is contracted by both $\mathrm{pr}_1 \circ \pi$ and $\mathrm{pr}_2 \circ \pi$, and ℓ_2 is contracted by $\mathrm{pr}_3 \circ \pi$. Then $(-K_X - xR)|_R \sim_{\mathbb{R}} 2\ell_1 + (x+1)\ell_2$. Let $C = R \cap E$. Then $C \sim 2\ell_1 + \ell_2$. If $1 \leqslant x \leqslant 2$, then $P(-K_X - xR)|_R \sim_{\mathbb{R}} (4 - 2x)\ell_1 + 2\ell_2$ and $N(-K_X - xR)|_R = (x-1)C$. Thus, if $Z = C$, Corollary 1.110 gives

$$S(W^R_{\bullet,\bullet}; Z) = \frac{1}{10} \int_0^1 \int_0^\infty \mathrm{vol}\big(2\ell_1 + (x+1)\ell_2 - yZ\big)\, dy\, dx$$

$$+ \frac{1}{10} \int_1^2 \big((4 - 2x)\ell_1 + 2\ell_2\big)^2 (x-1)\, dx$$

$$+ \frac{1}{10} \int_1^2 \int_0^\infty \mathrm{vol}\big((4 - 2x)\ell_1 + 2\ell_2 - yZ\big)\, dy\, dx$$

$$= \frac{1}{10} \int_0^1 \int_0^1 2(2 - 2y)(x + 1 - y)\, dy\, dx + \frac{1}{10} \int_1^2 4(4 - 2x)(x - 1)\, dx$$

$$+ \frac{1}{10} \int_1^2 \int_0^{2-x} 2(4 - 2x - 2y)(2 - y)\, dy\, dx = \frac{29}{60} < 1.$$

Therefore, to complete the proof, we may assume that $Z \neq C$. Then

$$S(W^R_{\bullet,\bullet}; Z) = \frac{1}{10} \int_0^1 \int_0^\infty \mathrm{vol}\big(2\ell_1 + (x+1)\ell_2 - yZ\big)\, dy\, dx$$

$$+ \frac{1}{10} \int_1^2 \int_0^\infty \mathrm{vol}\big((4 - 2x)\ell_1 + 2\ell_2 - yZ\big)\, dy\, dx$$

$$\leqslant \frac{1}{10} \int_0^1 \int_0^\infty \mathrm{vol}\big(2\ell_1 + (x+1)\ell_2 - y(\ell_1 + \ell_2)\big)\, dy\, dx$$

$$+ \frac{1}{10} \int_1^2 \int_0^\infty \mathrm{vol}\big((4 - 2x)\ell_1 + 2\ell_2 - y(\ell_1 + \ell_2)\big)\, dy\, dx$$

$$= \frac{1}{10} \int_0^1 \int_0^{x+1} 2(2 - y)(x + 1 - y)\, dy\, dx$$

$$+ \int_1^2 \int_0^{4-2x} 2(4 - 2x - y)(2 - y)\, dy\, dx = \frac{13}{24} < 1$$

by Corollary 1.110. $\square$

Now, we are ready to prove that X is K-polystable. Suppose that X is not K-polystable. Then, by Theorem 1.22, there are a G-equivariant birational morphism $f: \widetilde{X} \to X$ and a G-invariant prime divisor $F \subset \widetilde{X}$ such that $\beta(F) = A_X(F) - S_X(F) \leqslant 0$. Let $Z = f(F)$. Then Z is not a surface by Theorem 3.17, so that Z is a G-invariant irreducible curve, because X does not have G-invariant points by Lemma 5.109. Now, using Corollary 1.110 and Lemma 5.110, we see that $Z \not\subset R$ because $S_X(R) < 1$ by Theorem 3.17.

Using Lemma 1.45, we get $\alpha_{G,Z}(X) < \frac{3}{4}$. By Lemma 1.42, there exists a G-invariant effective $\mathbb{Q}$-divisor D on the threefold X such that $D \sim_{\mathbb{Q}} -K_X$ and $Z \subset \mathrm{Nklt}(X, \lambda D)$ for a positive rational number $\lambda < \frac{3}{4}$. By Lemma 5.109, the only possible two-dimensional component of $\mathrm{Nklt}(X, \lambda D)$ is R. Since $Z \not\subset R$, we conclude that Z is an irreducible component of the locus $\mathrm{Nklt}(X, \lambda D)$. Applying Corollary A.12 to $\mathrm{pr}_1 \circ \pi$, $\mathrm{pr}_2 \circ \pi$, $\mathrm{pr}_3 \circ \pi$, we get $H_1 \cdot Z \leqslant 1$, $H_2 \cdot Z \leqslant 1$, $H_3 \cdot Z \leqslant 1$. But this is impossible by Lemma 5.109. The obtained contradiction shows that X is K-polystable.

5.22 Family №4.13

Let X be a smooth Fano threefold №4.13. There is a birational morphism $\pi\colon X \to \mathbb{P}^1 \times \mathbb{P}^1 \times \mathbb{P}^1$ that is a blow up of a smooth curve C of degree $(1,1,3)$. Moreover, one can choose coordinates $([x_0 : x_1], [y_0 : y_1], [z_0 : z_1])$ on $\mathbb{P}^1 \times \mathbb{P}^1 \times \mathbb{P}^1$ such that the curve C is given by one of the following two equations:

$$x_0 y_1 - x_1 y_0 = x_0^3 z_0 + x_1^3 z_1 + \lambda\left(x_0 x_1^2 z_0 + x_0^2 x_1 z_1\right) = 0 \qquad (5.24)$$

for some $\lambda \in \mathbb{C} \setminus \{\pm 1, \pm 3\}$, or

$$x_0 y_1 - x_1 y_0 = x_0^3 z_0 + x_1^3 z_1 + x_0 x_1^2 z_0 = 0. \qquad (5.25)$$

We will prove that X is K-polystable if C is given by (5.24). This would imply

Corollary 5.111 *Suppose that C is given by (5.25). Then X is strictly K-semistable.*

Proof Arguing as in the proof of Corollary 4.71, we construct a test configuration for X, whose special fiber is the threefold X_0, which is the Fano threefold №4.13 that is a blow up of $\mathbb{P}^1 \times \mathbb{P}^1 \times \mathbb{P}^1$ at the smooth curve given by (5.24) with $\lambda = 0$. Assuming that X_0 is K-polystable, we see that X is strictly K-semistable by Corollary 1.13. $\qquad \square$

Suppose that C is given by (5.24). Let $\overline{R} = \{x_0 y_1 - x_1 y_0 = 0\} \subset \mathbb{P}^1 \times \mathbb{P}^1 \times \mathbb{P}^1$, and let $\mathrm{pr}_3\colon \mathbb{P}^1 \times \mathbb{P}^1 \times \mathbb{P}^1 \to \mathbb{P}^1$ be the projection to the third factor. Then we have $\overline{R} \cong \mathbb{P}^1 \times \mathbb{P}^1$, the surface $\overline{R}$ contains the curve C, which is a curve of degree $(3,1)$ on the surface $\overline{R}$, and the projection pr_3 induces a triple cover $C \to \mathbb{P}^1$. If $\lambda = 0$, this triple cover is ramified at exactly 2 points, which implies that $\mathrm{Aut}^0(X) \cong \mathbb{G}_m$ by [45, Corollary 2.7], so that X is the unique smooth Fano threefold in the family №4.13 that has an infinite automorphism group [45].

On the other hand, if $\lambda \neq 0$, the triple cover is ramified at 4 distinct points. Arguing as in the proof of [45, Corollary 8.12], we see that $\mathrm{Aut}(X)$ is finite if $\lambda \neq 0$.

Note, the group $\mathrm{Aut}(X)$ is actually not trivial for every $\lambda \in \mathbb{C} \setminus \{\pm 1, \pm 3\}$. Namely, let A_1, A_2 and A_3 be the automorphisms of $\mathbb{P}^1 \times \mathbb{P}^1 \times \mathbb{P}^1$ defined as follows:

$$A_1 \colon ([x_0 : x_1], [y_0 : y_1], [z_0 : z_1]) \mapsto ([x_0 : -x_1], [y_0 : -y_1], [z_0 : -z_1]),$$
$$A_2 \colon ([x_0 : x_1], [y_0 : y_1], [z_0 : z_1]) \mapsto ([x_1 : x_0], [y_1 : y_0], [z_1 : z_0]),$$
$$A_3 \colon ([x_0 : x_1], [y_0 : y_1], [z_0 : z_1]) \mapsto ([y_0 : y_1], [x_0 : x_1], [z_0 : z_1]).$$

Let G be the subgroup of $\mathrm{Aut}(\mathbb{P}^1 \times \mathbb{P}^1 \times \mathbb{P}^1)$ generated by A_1, A_2 and A_3. Then $|G| = 8$, and the curve C is G-invariant, so that the action of the group G lifts to the threefold X. Thus, we can identify G with a subgroup of the group $\mathrm{Aut}(X)$.

Let us show that X is K-polystable, so that X is K-stable for $\lambda \neq 0$ by Corollary 1.5.

Lemma 5.112   *The following assertions hold:*

(1) $\mathbb{P}^1 \times \mathbb{P}^1 \times \mathbb{P}^1$ *does not contain G-fixed points.*

(2) $\mathbb{P}^1 \times \mathbb{P}^1 \times \mathbb{P}^1$ *does not contain G-invariant irreducible curves of degree* (d_1, d_2, d_3) *such that one of the non-negative integers d_1, d_2 or d_3 is zero.*

(3) $\mathbb{P}^1 \times \mathbb{P}^1 \times \mathbb{P}^1$ *contains sixteen G-invariant irreducible curves of degree* $(1, 1, 1)$. *Four of them lie on $\overline{R}$, and the remaining curves intersect $\overline{R}$ in 2 points.*

(4) *Let Γ be a G-invariant irreducible curve of degree $(1, 1, 1)$ in $\mathbb{P}^1 \times \mathbb{P}^1 \times \mathbb{P}^1$ such that $\Gamma \not\subset \overline{R}$. Then either $\Gamma \cap C = \emptyset$ or $\Gamma \cap C = \Gamma \cap \overline{R}$.*

Proof   Assertions (1) and (2) are obvious. To prove (3) and (4), let $x = \frac{x_1}{x_0}$, $y = \frac{y_1}{y_0}$, $z = \frac{z_1}{z_0}$ be the non-homogeneous coordinates on each factor of $\mathbb{P}^1 \times \mathbb{P}^1 \times \mathbb{P}^1$. There are precisely four irreducible curves of degree $(1, 1)$ on $\mathbb{P}^1_x \times \mathbb{P}^1_y$, which are invariant under the induced action of the group $\langle A_1, A_2 \rangle$. These are the curves given by $y = \pm x^{\pm 1}$. Similarly, there are also 4 irreducible curves of degree $(1, 1)$ on $\mathbb{P}^1_x \times \mathbb{P}^1_z$ invariant under the induced action of the group $\langle A_1, A_3 \rangle$. These are the curves that are given by $z = \pm x^{\pm 1}$. This gives us 16 possibilities for a G-invariant curve in $\mathbb{P}^1 \times \mathbb{P}^1 \times \mathbb{P}^1$ of degree $(1, 1, 1)$. These are the curves given by $(y, z) = (\pm x^{\pm 1}, \pm x^{\pm 1})$. Four of these curves are contained in the surface $\overline{R}$, which is given by $y = x$. On the other hand, each of the remaining twelve curves meets $\overline{R}$ in precisely 2 points. The assertion on the intersection with C is immediate to check. $\square$

Now, let us recall from [93] the descriptions of the Mori cone $\overline{NE}(X)$, the nef cone and the cone of effective divisors of the Fano threefold X. Let l_1, l_2, l_3 be the proper transforms on X of curves of degree $(1,0,0)$, $(0,1,0)$ and $(0,0,1)$ in $\mathbb{P}^1 \times \mathbb{P}^1 \times \mathbb{P}^1$ that meet C. Denote by l_4 the proper transform of a curve of degree $(1,1,0)$ contained in $\overline{R}$, and denote by l_5 a curve contracted by π to a point. Then $NE(X)$ is generated by l_1, l_2, l_3, l_4, l_5. Let H_1, H_2 and H_3 be general fibers of the del Pezzo fibrations $\mathrm{pr}_1 \circ \pi$, $\mathrm{pr}_2 \circ \pi$ and $\mathrm{pr}_3 \circ \pi$, respectively, where $\mathrm{pr}_1 \colon \mathbb{P}^1 \times \mathbb{P}^1 \times \mathbb{P}^1 \to \mathbb{P}^1$ and $\mathrm{pr}_2 \colon \mathbb{P}^1 \times \mathbb{P}^1 \times \mathbb{P}^1 \to \mathbb{P}^1$ are projections to the first and the second factors, respectively. Denote by E_1, E_2, E_3 the exceptional surfaces of the contractions of the extremal rays generated by l_1, l_2, l_3, respectively. Finally, let R be the proper transform on X of the surface $\overline{R}$. Then $E_1 \sim 3H_2 + H_3 - E$, $E_2 \sim 3H_1 + H_3 - E$ and $R = E_3 \sim H_1 + H_2 - E$. Moreover, we have

$$Nef(X) = \mathbb{R}_{\geqslant 0}[H_1] + \mathbb{R}_{\geqslant 0}[H_2] + \mathbb{R}_{\geqslant 0}[H_3]$$
$$+ \mathbb{R}_{\geqslant 0}[2H_1 + H_2 + H_3 - E] + \mathbb{R}_{\geqslant 0}[H_1 + 2H_2 + H_3 - E]$$

and

$$Eff(X) = \mathbb{R}_{\geqslant 0}[H_1] + \mathbb{R}_{\geqslant 0}[H_2] + \mathbb{R}_{\geqslant 0}[H_3] + \mathbb{R}_{\geqslant 0}[H_1 + H_2 - E]$$
$$+ \mathbb{R}_{\geqslant 0}[3H_1 + H_3 - E] + \mathbb{R}_{\geqslant 0}[3H_2 + H_3 - E] + \mathbb{R}_{\geqslant 0}[E].$$

Lemma 5.113 *Let $D \neq 0$ be an effective G-invariant $\mathbb{Z}$-divisor on the threefold X. Suppose that $-K_X - D$ is big. Then $D = R$.*

Proof Since $-K_X \sim 2R + E + 2H_3$, the divisor D must be linearly equivalent to one of the following divisors: H_1, H_2, H_3, $H_1 + H_3$, $H_2 + H_3$, $H_1 + H_2 - E$ or $H_1 + H_2 + H_3 - E$. But the linear systems $|H_1|$, $|H_2|$, $|H_3|$, $|H_1 + H_3|$, $|H_2 + H_3|$, $|H_1 + H_2 + H_3 - E|$ do not contain G-invariant divisors. Thus, we see that $D \sim H_1 + H_2 - E$, so that $D = R$.   □

In the following result and its proof, we use the notations introduced in Section 1.7.

Lemma 5.114 *Let Z be a G-invariant irreducible curve in R. Then $S(W^R_{\bullet,\bullet}; Z) \leqslant \frac{27}{52}$.*

Proof Fix $x \in \mathbb{R}_{\geqslant 0}$. Then the divisor $-K_X - xR$ is pseudo-effective if and only if $x \leqslant 2$. Let $P(x) = P(-K_X - xR)$ and $N(x) = N(-K_X - xR)$. Then

$$P(x) = \begin{cases} -K_X - xR & \text{if } 0 \leqslant x \leqslant 1, \\ (2 - x)(H_1 + H_2) + 2H_3 & \text{if } 1 \leqslant x \leqslant 2, \end{cases}$$

and $N(x) = (x - 1)E$ if $1 \leqslant x \leqslant 2$.

Recall that $R \cong \mathbb{P}^1 \times \mathbb{P}^1$. Let ℓ_1 and ℓ_2 be the rulings of the surface R such that both $\mathrm{pr}_1 \circ \pi$ and $\mathrm{pr}_2 \circ \pi$ contract ℓ_1, and $\mathrm{pr}_3 \circ \pi$ contracts ℓ_2. Then $-K_X|_R \sim -R|_R \sim \ell_1 + \ell_2$. Let $C = R \cap E$. Then $C \sim 3\ell_1 + \ell_2$. If $0 \leqslant x \leqslant 1$, then $P(x)|_R \sim (1+x)(\ell_1 + \ell_2)$. Likewise, if $1 \leqslant x \leqslant 2$, then $P(x)|_R \sim (4-2x)\ell_1 + 2\ell_2$ and $N(x)|_R = (x-1)C$. Thus, if $Z = C$, then

$$
\begin{aligned}
S(W^R_{\bullet,\bullet}; Z) &= \frac{3}{26} \int_0^1 \int_0^\infty \mathrm{vol}\big((1 + x - 3y)\ell_1 + (1 + x - y)\ell_2\big) dy dx \\
&\quad + \frac{3}{26} \int_1^2 \bigg((x-1)\big((4-2x)\ell_1 + 2\ell_2\big)^2 \\
&\qquad + \int_0^\infty \mathrm{vol}\big((4 - 2x - 3y)\ell_1 + (2-y)\ell_2\big) dy \bigg) dx \\
&= \frac{44}{117} < \frac{27}{52}
\end{aligned}
$$

by Corollary 1.110. Thus, to complete the proof, we may assume that $Z \neq C$.

Since the linear systems $|\ell_1|$ and $|\ell_2|$ do not contain G-invariant curves by Lemma 5.112, we have $Z \sim b_1 \ell_1 + b_2 \ell_2$ for some positive integers b_1 and b_2. By Corollary 1.110, we get

$$
\begin{aligned}
S(W^R_{\bullet,\bullet}; Z) &= \frac{3}{26} \int_0^1 \int_0^\infty \mathrm{vol}\big((1 + x - b_1 y)\ell_1 + (1 + x - b_2 y)\ell_2\big) dy dx \\
&\quad + \frac{3}{26} \int_1^2 \int_0^\infty \mathrm{vol}\big((4 - 2x - b_1 y)\ell_1 + (2 - b_2 y)\ell_2\big) dy dx \\
&\leqslant \frac{3}{26} \int_0^1 \int_0^\infty \mathrm{vol}\big((1 + x - y)(\ell_1 + \ell_2)\big) dy dx \\
&\quad + \frac{3}{26} \int_1^2 \int_0^\infty \mathrm{vol}\big((4 - 2x - y)\ell_1 + (2 - y)\ell_2\big) dy dx \\
&= \frac{3}{26} \int_0^1 \int_0^{1+x} 2(1 + x - y)^2 dy dx \\
&\quad + \frac{3}{26} \int_1^2 \int_0^{4-2x} 2(4 - 2x - y)(2 - y) dy dx = \frac{27}{52},
\end{aligned}
$$

which is exactly what we want.   □

Now we are ready to prove

Theorem 5.115 *The threefold X is K-polystable.*

Proof Suppose that X is not K-polystable. By Theorem 1.22, there is a G-invariant prime divisor F over X such that $\beta(F) \leqslant 0$. Let $Z = C_X(F)$. Then Z is not a surface by Theorem 3.17. Thus, since X does not have G-fixed points by Lemma 5.112, we see that Z is a G-invariant irreducible curve. Now, using Lemma 1.45, we get $\alpha_{G,Z}(X) < \frac{3}{4}$. By Lemma 1.42, there are a G-invariant effective $\mathbb{Q}$-divisor D on the threefold X and a positive rational number $\lambda < \frac{3}{4}$ such that $D \sim_{\mathbb{Q}} -K_X$, $Z \subseteq \mathrm{Nklt}(X, \lambda D)$, and $(X, \lambda D)$ is strictly log canonical at a general point of the curve Z. Then $\mathrm{Nklt}(X, \lambda D)$ contains no surfaces except possibly for the surface R by Lemma 5.113.

Using Corollary 1.110, Lemma 5.114 and Theorem 3.17, we see that $Z \not\subset R$. Hence, using Lemma 5.112 and applying Corollary A.12 to $(X, \lambda D)$ and the morphisms $\mathrm{pr}_1 \circ \pi$, $\mathrm{pr}_2 \circ \pi$ and $\mathrm{pr}_3 \circ \pi$, we see that $\pi(Z)$ is a curve of degree $(1, 1, 1)$. Then $\pi(Z)$ is one of the twelve G-invariant curves described in Lemma 5.112.

Let $\psi \colon X \to X'$ be the birational morphism that contracts R to an ordinary double point, let D' be the proper transform of the divisor D on the threefold X', and let $Z' = \psi(Z)$. Then X' is a Fano threefold with terminal Gorenstein singularities, and $D' \sim_{\mathbb{Q}} -K_{X'}$. Moreover, the log pair $(X', \lambda D')$ is strictly log canonical at a general point of the curve Z', and the locus $\mathrm{Nklt}(X', \lambda D')$ is one-dimensional. Then Z' is smooth by Corollary A.14. Thus, using Lemma 5.112, we deduce that $\pi(Z) \cap C$ consists of two points.

Let Y be the unique surface in $|H_1 + H_2|$ that contains Z, let $\overline{Y}$ be its proper transform on $\mathbb{P}^1 \times \mathbb{P}^1 \times \mathbb{P}^1$, and let $\varphi \colon Y \to \overline{Y}$ be the birational morphism that is induced by π. Then φ is the blow up of the intersection $C \cap \overline{Y}$, which consists of two points that are not contained in one ruling of the surface $\overline{Y} \cong \mathbb{P}^1 \times \mathbb{P}^1$. Then Y is a sextic del Pezzo surface. Let us apply results proved in Section 1.7 to Y and Z to derive a contradiction.

Fix $x \geq 0$. Let $P(x) = P(-K_X - xY)$ and $N(x) = N(-K_X - xY)$. Then $-K_X - xY$ is nef $\iff x \leqslant \frac{1}{2}$, and $-K_X - xY$ is pseudo-effective $\iff x \leqslant 2$. Moreover, if $\frac{1}{2} \leqslant x \leqslant 1$, then $N(x) = (2x - 1)R$, so that

$$P(x) = (3 - x)(H_1 + H_2) + 2H_3 + (2x - 2)E.$$

Using Corollary 1.110, we get $S(W^Y_{\bullet, \bullet}; Z) \geqslant 1$ since we have $S_X(Y) < 1$ by Theorem 3.17.

Let e_1 and e_2 be exceptional curves of the morphism φ, let f_1 and f_2 be the proper transforms on Y of the rulings of the surface $\overline{Y}$ that are contracted by both pr_1 and pr_2 and pass through the points $\varphi(e_1)$ and $\varphi(e_2)$, respectively. Then, on the surface Y, we have $E|_Y = e_1 + e_2$, $R|_Y = f_1 + f_2$, $H_1|_Y \sim H_2|_Y \sim f_1 + e_1 \sim f_2 + e_2$.

Let h_1 and h_2 be the proper transform on Y of the rulings of the surface $\overline{Y}$ that are contracted by the projection pr_3 and pass through the points $\varphi(e_1)$ and $\varphi(e_2)$, respectively. Then $H_3|_Y \sim h_1 + e_1 \sim h_2 + e_2$ and $Z \sim f_1 + h_2 \sim f_2 + h_1$. Hence, if $0 \leqslant x \leqslant \frac{1}{2}$, we have $P(x)|_Y \sim_{\mathbb{R}} (2 - 2x)f_1 + 2f_2 + (3 - 2x)e_1 + e_2 + 2h_1$. Similarly, if $\frac{1}{2} \leqslant x \leqslant 1$, then $P(x)|_Y \sim_{\mathbb{R}} (3 - 4x)f_1 + (3 - 2x)f_2 + (3 - 2x)e_1 + e_2 + 2h_1$ and $N(x)|_Y = (2x - 1)(f_1 + f_2)$. Take $y \in \mathbb{R}_{\geqslant 0}$. Then Corollary 1.110 gives

$S(W_{\bullet,\bullet}^{Y};Z)$

$$= \frac{3}{26}\int_0^{\frac{1}{2}}\int_0^\infty \mathrm{vol}\big((2-2x)f_1 + 2f_2 + (3-2x)e_1 + e_2 + 2h_1 - yZ\big)dydx$$

$$+ \frac{3}{26}\int_{\frac{1}{2}}^1\int_0^\infty \mathrm{vol}\big((3-4x)f_1 + (3-2x)f_2 + (3-2x)e_1 + e_2 + 2h_1 - yZ\big)dydx,$$

where e_1, e_2, f_1, f_2, h_1, h_2 are (-1)-curves on the surface Y, and $Z \sim f_1 + h_2 \sim f_2 + h_1$. If $x \leqslant \frac{1}{2}$ and $y \leqslant 1$, then $(2-2x)f_1 + 2f_2 + (3-2x)e_1 + e_2 + 2h_1 - yZ$ is nef, so that

$$\mathrm{vol}\big((2-2x)f_1 + 2f_2 + (3-2x)e_1 + e_2 + 2h_1 - yZ\big) = 4xy - 8x - 8y + 14.$$

If $x \leqslant \frac{1}{2}$ and $1 \leqslant y \leqslant 2$, then the Zariski decomposition of this divisor is

$$\underbrace{(4-2x-y)(f_1+e_1) + (2-y)(h_1+e_1)}_{\text{positive part}} + \underbrace{(y-1)(e_1+e_2)}_{\text{negative part}},$$

so that its volume is $2(4-2x-y)(2-y)$. For $y > 2$, this divisor is not pseudo-effective. Similarly, if $\frac{1}{2} \leqslant x \leqslant 1$ and $0 \leqslant y \leqslant 2-2x$, then

$$\mathrm{vol}\big((3-4x)f_1 + (3-2x)f_2 + (3-2x)e_1 + e_2 + 2h_1\big) = 4xy - 8x^2 - 8x - 8y + 16.$$

If $2-2x \leqslant y \leqslant \min\{2, 6-6x\}$, this divisor has volume $2(6-6x-y)(2-y)$. For $y > \min\{2, 6-6x\}$, this divisor is not pseudo-effective. Now, using Corollary 1.110 and integrating, we get $S(W_{\bullet,\bullet}^{Y};Z) = \frac{257}{312} < 1$. This shows that X is K-polystable. □

If $\lambda \neq 0$, we have that X is K-stable by Corollary 1.5.

Remark 5.116　Let X' be the singular Fano threefold that has been constructed in the proof of Theorem 5.115. One can show that $\mathrm{Aut}(X') \cong \mathrm{Aut}(X)$. Moreover, arguing as in the proof of Theorem 5.115, one can prove that the threefold X' is K-polystable. Furthermore, the threefold X' has a smoothing to a Fano threefold in the family №2.21, so that Theorem 1.11 gives another proof of Corollary 4.16.

5.23 Family №5.1

This family contains a unique smooth threefold. It is K-polystable. To prove this, we have to describe this threefold explicitly and compute its automorphism group. To start with, let Q be a smooth quadric $\{x_1x_2 + x_2x_3 + x_3x_1 + yz = 0\} \subset \mathbb{P}^4$, where x_1, x_2, x_3, y and z are homogeneous coordinates on $\mathbb{P}^4$. Let C be the smooth conic in the quadric Q that is cut out by $y = z = 0$, and let

$P_1 = [1 : 0 : 0 : 0 : 0]$, $P_2 = [0 : 1 : 0 : 0 : 0]$, $P_3 = [0 : 0 : 1 : 0 : 0]$. Then C contains the points P_1, P_2, P_3. Let $\theta \colon Y \to Q$ be the blow up of the points P_1, P_2, P_3, let C be the strict transform on Y of the conic C, and let $\eta \colon X \to Y$ be the blow up of the curve C. Then X is the unique smooth Fano threefold №5.1.

Now let us describe $\mathrm{Aut}(X)$. Let G be a subgroup in $\mathrm{Aut}(Q)$ described as

$$G = \left\{ g \in \mathrm{Aut}(Q) \,\middle|\, g(C) = C \text{ and } g(\{P_1, P_2, P_3\}) = \{P_1, P_2, P_3\} \right\}.$$

Observe that the action of the group G lifts faithfully on the Fano threefold X, so that we can identify G with a subgroup of the automorphism group $\mathrm{Aut}(X)$. Moreover, using the description of the Mori cone $\mathrm{NE}(X)$ given in [93], we conclude that $\mathrm{Aut}(X) = G$. Furthermore, we have $G \cong \mathfrak{S}_3 \times (\mathbb{G}_m \rtimes \mu_2)$ and G acts on Q as follows:

- if $\sigma \in \mathfrak{S}_3$, then σ acts by $[x_1 : x_2 : x_3 : y : z] \mapsto [x_{\sigma(1)} : x_{\sigma(2)} : x_{\sigma(3)} : y : z]$,
- if $\lambda \in \mathbb{G}_m$, then λ acts by $[x_1 : x_2 : x_3 : y : z] \mapsto [\lambda x_1 : \lambda x_2 : \lambda x_3 : \lambda^2 y : z]$,
- if $\iota \in \mu_2$, then ι acts by $[x_1 : x_2 : x_3 : y : z] \mapsto [x_1 : x_2 : x_3 : z : y]$.

Then Q does not contain G-invariant points. Let Z be the smooth conic in Q that is cut out by $x_1 - x_3 = x_2 - x_3 = 0$. Then $C \cap Z = \varnothing$.

Lemma 5.117 *The curves C and Z are the only irreducible G-invariant curves in Q.*

Proof Let $\mathscr{C}$ be a G-invariant irreducible curve in Q that is different from C. Let us show that $\mathscr{C} = Z$. Since $\mathscr{C} \neq C$, it contains a point $P = [x_1 : x_2 : x_3 : y : 1]$ with $y \neq 0$, which implies that $\mathscr{C} = \overline{\mathbb{G}_m.P}$. In particular, for every $\sigma \in \mathfrak{S}_3$, there is $\lambda \in \mathbb{C}^*$ such that

$$\left[x_{\sigma(1)} : x_{\sigma(2)} : x_{\sigma(3)} : y : 1 \right] = \left[x_1 : x_2 : x_3 : \lambda y : \frac{1}{\lambda} \right]$$
$$= \left[\lambda x_1 : \lambda x_2 : \lambda x_3 : \lambda^2 y : 1 \right],$$

so that $\lambda^2 = 1$. Now, using $\sigma = (1, 2)$ and $\sigma = (2, 3)$, we see that $x_1 = x_2 = x_3 \neq 0$, so that $\mathscr{C} = Z$. $\qquad\square$

Let $\phi_C \colon Y_C \to Q$ and $\phi_Z \colon Y_Z \to Q$ be the blow up of the conics C and Z, respectively. Denote by F_C and F_Z the exceptional surfaces of the blow ups ϕ_C and ϕ_Z, respectively. Observe that the action of the group G on the quadric Q lifts to its actions on Y_C and Y_Z, and the surfaces F_C and F_Z are exceptional G-invariant prime divisors over Q.

Lemma 5.118 *The only exceptional G-invariant prime divisors over Q are F_C and F_Z.*

Proof Recall that the center on Q of a G-invariant prime divisor over Q is a G-invariant irreducible subvariety in Q. Therefore, by Lemma 5.117, it is enough to show that the surfaces F_C and F_Z do not contain proper G-invariant irreducible subvarieties.

We start with F_C. Let $\psi_C\colon U_C \to \mathbb{P}^4$ be the blow up of the linear span of the conic C, i.e. the blow up of the plane $y = z = 0$. We have the following G-equivariant diagram:

$$
\begin{array}{ccc}
Y_C & \lhook\joinrel\longrightarrow & U_C \\
{\scriptstyle \phi_C}\big\downarrow & & \big\downarrow{\scriptstyle \psi_C} \\
Q & \lhook\joinrel\longrightarrow & \mathbb{P}^4
\end{array}
$$

Let us describe the G-action on U_C. The fourfold U_C can be covered by two charts. The first one is given in $\mathbb{P}^4 \times \mathbb{A}^1_{y'}$ by $y = y'z$, and the second is given in $\mathbb{P}^4 \times \mathbb{A}^1_{z'}$ by $z = z'y$. Using these charts, the action of the group G can be described as follows:

- if $\sigma \in \mathfrak{S}_3$, then σ acts by $([x_1 : x_2 : x_3 : y : z], y') \mapsto ([x_{\sigma(1)} : x_{\sigma(2)} : x_{\sigma(3)} : y : z], y')$;
- if $\lambda \in \mathbb{G}_m$, then λ acts by

$$
\left([x_1 : x_2 : x_3 : y : z], y'\right) \mapsto \left(\left[x_1 : x_2 : x_3 : \lambda y : \frac{z}{\lambda}\right], \lambda^2 y'\right);
$$

- if $\iota \in \boldsymbol{\mu}_2$, then ι acts by

$$
\left([x_1 : x_2 : x_3 : y : z], y'\right) \mapsto \left([x_1 : x_2 : x_3 : z : y], \frac{1}{y'}\right).
$$

Let E_C be the ψ_C-exceptional divisor. Then E_C can be identified with $\mathbb{P}^2_{x_1,x_2,x_3} \times \mathbb{P}^1_{\mathbf{y},\mathbf{z}}$, and F_C can be identified with its subvariety that is given by $x_1 x_2 + x_2 x_3 + x_3 x_1 = 0$. Moreover, the action of the group G on the threefold E_C can be described as follows:

- if $\sigma \in \mathfrak{S}_3$, then σ acts by

$$
\left([x_1 : x_2 : x_3], [\mathbf{y} : \mathbf{z}]\right) \mapsto \left([x_{\sigma(1)} : x_{\sigma(2)} : x_{\sigma(3)}], [\mathbf{y} : \mathbf{z}]\right);
$$

- if $\lambda \in \mathbb{G}_m$, then λ acts by

$$
\left([x_1 : x_2 : x_3], [\mathbf{y} : \mathbf{z}]\right) \mapsto \left(\left[x_1 : x_2 : x_3\right], \left[\lambda\mathbf{y} : \frac{\mathbf{z}}{\lambda}\right]\right);
$$

- if $\iota \in \boldsymbol{\mu}_2$, then ι acts by

$$
\left([x_1 : x_2 : x_3], [\mathbf{y} : \mathbf{z}]\right) \mapsto \left([x_1 : x_2 : x_3], [\mathbf{z} : \mathbf{y}]\right).
$$

This easily implies that the surface F_C does not contain irreducible G-invariant curves because C does not have $\mathfrak{S}_3$-invariant points. Since F_C does not contain G-invariant points, we see that F_C does not contain proper G-invariant irreducible subvarieties.

Similarly, we see that F_Z does not contain proper G-invariant subvarieties.

$\square$

Now we are ready to prove

Theorem 5.119 *The threefold X is K-polystable.*

Proof Let F be a G-invariant prime divisor over X. By Theorem 1.22, it is enough to prove that $\beta(F) > 0$. If F is a prime divisor on X, then $\beta(F) > 0$ by Theorem 3.17. Therefore, we may assume that F is exceptional over X. Let $\widetilde{Z}$ be the proper transform on X of the curve Z, and let $\sigma \colon \widetilde{X} \to X$ be the blow up of the curve $\widetilde{Z}$. Then F is the σ-exceptional surface by Lemma 5.118.

We claim that $\sigma^*(-K_X) - 2F$ is not big. To prove this fact, observe that there exists the commutative diagram

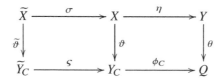

where ϑ is the blow up of the fibers of the projection $F_C \to C$ over the points P_1, P_2, P_3, i.e. the blow up of the preimages of these points via ϕ_C, ς is the blow up of the proper transform of the curve Z, and $\widetilde{\vartheta}$ is the blow up of the preimages of P_1, P_2, P_3 via $\phi_C \circ \varsigma$. Thus, if $\sigma^*(-K_X) - 2F$ is big, then $\varsigma^*(-K_{Y_C}) - 2\widetilde{F}$ is big, where $\widetilde{F}$ is the ς-exceptional surface. But the pseudo-effective cone of the threefold $\widetilde{Y}_C$ is described in [93, Section 10]. Note that $\widetilde{Y}_C$ is a smooth Fano threefold №3.10. Now, using [93, Section 10], we conclude that $\varsigma^*(-K_{Y_C}) - 2\widetilde{F}$ is not big, so that $\sigma^*(-K_X) - 2F$ is not big either.

We see that the pseudo-effective threshold $\tau(F) \leqslant 2$ (see Section 1.2). Thus, it follows from [96, Lemma 2.1] that $S_X(F) \leqslant \frac{3}{4}\tau(F) \leqslant \frac{3}{2} < 2 = A_X(F)$, so that $\beta(F) > 0$. Hence, the threefold X is K-polystable. $\square$

6

The Big Table

In this chapter, we summarize our answers to the Calabi Problem for Fano threefolds. We settle the problem of determining whether the general member of each of the 105 deformation families of Fano threefolds is K-polystable/K-semistable. In some cases, the general member of the family is K-polystable, while there is at least one member that is not K-polystable. A finer problem is to classify, within each family, which smooth Fano threefolds are K-polystable/K-semistable. This is accomplished for 71 of the 105 families. A conjectural picture for each of the remaining cases is then discussed in the final section.

Table 6.1 below contains the list of smooth Fano threefolds. We follow the notation and the numeration of the families in [120]. We also assume the following conventions.

- S_n denotes a smooth del Pezzo surface such that $K_{S_n}^2 = n$ and $S_8 \neq \mathbb{P}^1 \times \mathbb{P}^1$.
- Q denotes a smooth quadric hypersurface in $\mathbb{P}^4$.
- W denotes a divisor in $\mathbb{P}^2 \times \mathbb{P}^2$ of degree $(1,1)$.
- V_n denotes a smooth Fano threefold such that $V_n \neq W$ and

$$-K_{V_n} \sim 2H,$$

where H is a Cartier divisor on V_n such that $H^3 = n \in \{1,2,3,4,5,6,7,8\}$. Note that $V_8 = \mathbb{P}^3$ and V_7 is a blow up of $\mathbb{P}^3$ at a point.

In the first column of Table 6.1, we give the identifier № for a smooth Fano threefold X. The second and the third columns contain the degree $-K_X^3$ and

$$h^{1,2}(X) = \frac{1}{2}h^3(X, \mathbb{Z})$$

of the corresponding Fano threefold X, respectively.

In the fifth column, we present the possibilities for the group $\text{Aut}^0(X)$ within a given deformation class, so that 1 simply means that the group $\text{Aut}(X)$ is finite.

In the sixth column, we put known results about the K-polystability of smooth Fano threefolds, using the following conventions:

| Yes | means that all smooth Fano threefolds in this family are K-polystable;
| Yes★ | means that general Fano threefolds in this family are K-polystable;
| No | means that no smooth Fano threefolds in this family are K-polystable;
| ∃No | means that at least one smooth Fano threefold in this family is not K-polystable.

For instance, the combination of ⟨ Yes★ ⟩ and ⟨ ∃No ⟩ for Fano threefolds №1.10 means that general threefolds in this family are K-polystable but some are not. A priori, we could have a deformation family such that its general member is not K-polystable, but some members are K-polystable. But such a situation is not possible by the Main Theorem.

In the seventh column, we put results about K-semistability of smooth Fano threefolds. Recall that the K-semistability is an open property. We use the following conventions:

| Yes | means that all smooth threefolds in this family are K-semistable;
| Yes★ | means that a general threefold in this family is known to be K-semistable;
| No | means that every smooth Fano threefold in this family is K-unstable;
| ∃No | means that at least one smoooth Fano threefold in this family is K-unstable.

Finally, in the final column of Table 6.1 we put references to the sections of this book or external sources where the corresponding smooth Fano threefolds are discussed in more detail.

Table 6.1 *Smooth Fano threefolds*

№	$-K_X^3$	$h^{1,2}$	Brief description	$\mathrm{Aut}^0(X)$	**K-ps**	**K-ss**	Sections
1.1	2	52	sextic hypersurface in $\mathbb{P}(1,1,1,3)$	1	Yes	Yes	3.5, 4.1
1.2[a]	4	30	quartic threefold in $\mathbb{P}^4$	1	Yes	Yes	3.5, 4.1
1.2[b]	4	30	double cover of smooth quadric threefold	1	Yes	Yes	3.5
1.3	6	20	intersection of quadric and cubic in $\mathbb{P}^5$	1	Yes	Yes	3.5, 4.1
1.4	8	14	complete intersection of three quadrics in $\mathbb{P}^6$	1	Yes	Yes	3.5, 4.1
1.5[a]	10	10	section of $\mathrm{Gr}(2,5) \subset \mathbb{P}^9$ by quadric and linear subspace of dimension 7	1	Yes	Yes	3.5, 4.1
1.5[b]	10	10	double cover of the threefold V_5	1	Yes	Yes	3.5, 4.1
1.6	12	7	section of Hermitian symmetric space $M = G/P \subset \mathbb{P}^{15}$ of type DIII by linear subspace of dimension 8	1	Yes	Yes	4.1
1.7	14	5	section of $\mathrm{Gr}(2,6) \subset \mathbb{P}^{14}$ by linear subspace of codimension 5	1	Yes	Yes	4.1
1.8	16	3	section of Hermitian symmetric space $M = G/P \subset \mathbb{P}^{13}$ of type CI by linear subspace of dimension 10	1	Yes	Yes	4.1, 5.11
1.9	18	2	section of 5-dimensional rational homogeneous contact manifold $G_2/P \subset \mathbb{P}^{13}$ by linear subspace of dimension 11	1	Yes★	Yes★	4.1
1.10	22	0	zero locus of three sections of rank 3 vector bundle $\bigwedge^2 Q$ where Q is universal quotient bundle on $\mathrm{Gr}(7,3)$	1 $\mathbb{G}_a$ $\mathbb{G}_m$ $\mathrm{PGL}_2(\mathbb{C})$	∃No Yes★	Yes★	3.6, 4.1, 5.14

1.11	8	21	V_1 = sextic hypersurface in $\mathbb{P}(1,1,2,3)$		1	Yes	Yes	3.5, 3.4
1.12	16	10	V_2 = quartic hypersurface in $\mathbb{P}(1,1,1,1,2)$		1	Yes	Yes	3.5, 3.4
1.13	24	5	V_3 = cubic hypersurface in $\mathbb{P}^4$		1	Yes	Yes	3.4
1.14	32	2	V_4 = intersection of two quadrics in $\mathbb{P}^5$		1	Yes	Yes	3.4
1.15	40	0	V_5 = linear section of $\mathrm{Gr}(2,5)$ in $\mathbb{P}^9$	$\mathrm{PGL}_2(\mathbb{C})$	1	Yes	Yes	3.4
1.16	54	0	Q = quadric hypersurface in $\mathbb{P}^4$	$\mathrm{PSO}_5(\mathbb{C})$	1	Yes	Yes	3.2, 3.3
1.17	64	0	$V_8 = \mathbb{P}^3$	$\mathrm{PGL}_4(\mathbb{C})$	1	Yes	Yes	3.2, 3.3, 3.4
2.1	4	22	blow up of V_1 in elliptic curve		1	Yes$\star$	Yes$\star$	4.3
2.2	6	20	double cover of $\mathbb{P}^1 \times \mathbb{P}^2$ ramified in surface of degree $(2,4)$		1	Yes$\star$	Yes$\star$	4.5
2.3	8	11	blow up of V_2 in elliptic curve		1	Yes$\star$	Yes$\star$	4.3
2.4	10	10	blow up of $\mathbb{P}^3$ along intersection of two cubics		1	Yes$\star$	Yes$\star$	4.5
2.5	12	6	blow up of V_3 in elliptic curve		1	Yes$\star$	Yes$\star$	4.3
2.6[a]	12	9	divisor on $\mathbb{P}^2 \times \mathbb{P}^2$ of degree $(2,2)$		1	Yes$\star$	Yes$\star$	3.5
2.6[b]	12	9	double cover of W branched in anticanonical surface		1	Yes	Yes	1.5
2.7	14	5	blow up of quadric $Q \subset \mathbb{P}^4$ along intersection of two surfaces in $|O_{\mathbb{P}^4}(2)|_Q|$		1	Yes$\star$	Yes$\star$	4.5
2.8	14	9	double cover of V_7 branched in anticanonical surface		1	Yes	Yes	5.1, [145]
2.9	16	5	blow up of $\mathbb{P}^3$ along curve of degree 7 and genus 5 that is intersection of cubics		1	Yes$\star$	Yes$\star$	5.2

№							
2.10	16	3	blow up of V_4 in elliptic curve	1	Yes★	Yes★	4.3
2.11	18	5	blow up of V_3 along line	1	Yes★	Yes★	5.3
2.12	20	3	blow up of $\mathbb{P}^3$ along curve of degree 6 and genus 3 that is intersection of cubics	1	Yes★	Yes★	5.4
2.13	20	2	blow up of $Q \subset \mathbb{P}^4$ along curve of degree 6 and genus 2	1	Yes★	Yes★	5.5
2.14	20	1	blow up of V_5 in elliptic curve	1	Yes★	Yes★	4.3
2.15	22	4	blow up of $\mathbb{P}^3$ at curve of degree 6 and genus 4 that is intersection of quadric and cubic surfaces	1	Yes★	Yes★	4.4
2.16	22	2	blow up of $V_4 \subset \mathbb{P}^5$ along conic	1	Yes★	Yes★	5.6
2.17	24	1	blow up of quadric $Q \subset \mathbb{P}^4$ along elliptic curve of degree 5	1	Yes★	Yes★	5.7
2.18	24	2	double cover of $\mathbb{P}^1 \times \mathbb{P}^2$ branched in surface of degree (2,2)	1	Yes★	Yes★	4.5
2.19	26	2	blow up of $V_4 \subset \mathbb{P}^5$ along line	1	Yes★	Yes★	4.4
2.20	26	0	blow up of $V_5 \subset \mathbb{P}^6$ along twisted cubic	1, $\mathbb{G}_m$	∃No	Yes★	5.8
2.21	28	0	blow up of $Q \subset \mathbb{P}^4$ along twisted quartic	1, $\mathbb{G}_a$, $\mathbb{G}_m$, $\mathrm{PGL}_2(\mathbb{C})$	∃No	Yes★	4.2, 5.22
2.22	30	0	blow up of $V_5 \subset \mathbb{P}^6$ along conic	1, $\mathbb{G}_m$	∃No	Yes★	1.5, 4.4, [38]

2.23	30	1	blow up of quadric $Q \subset \mathbb{P}^4$ along elliptic curve of degree 4	1	No	No	3.7
2.24	30	0	divisor on $\mathbb{P}^2 \times \mathbb{P}^2$ of degree $(1,2)$	1	∃No	Yes★	4.7
				$\mathbb{G}_m$		Yes★	
				$\mathbb{G}_m^2$	Yes★	Yes★	
2.25	32	1	blow up of $\mathbb{P}^3$ in elliptic curve	1	Yes	Yes	4.3
2.26	34	0	blow up of $V_5 \subset \mathbb{P}^6$ along line	$\mathbb{G}_a \times \mathbb{G}_m$	No	∃No	5.10
				$\mathbb{G}_m$	Yes★	Yes★	
2.27	38	0	blow up of $\mathbb{P}^3$ along twisted cubic	$\mathrm{PGL}_2(\mathbb{C})$	Yes	Yes	4.2
2.28	40	1	blow up of $\mathbb{P}^3$ along plane cubic	$(\mathbb{G}_a)^3 \rtimes \mathbb{G}_m$	No	No	3.6, 3.7
2.29	40	0	blow up of $Q \subset \mathbb{P}^4$ along conic	$\mathbb{G}_m \times \mathrm{PGL}_2(\mathbb{C})$	Yes	Yes	3.3
2.30	46	0	blow up of $\mathbb{P}^3$ along conic	$\mathrm{PSO}_{5;1}(\mathbb{C})$	No	No	3.3, 3.6, 3.7
2.31	46	0	blow up of $Q \subset \mathbb{P}^4$ along line	$\mathrm{PSO}_{5;2}(\mathbb{C})$	No	No	3.3, 3.6, 3.7
2.32	48	0	$W = $ divisor in $\mathbb{P}^2 \times \mathbb{P}^2$ of degree $(1,1)$	$\mathrm{PGL}_3(\mathbb{C})$	Yes	Yes	3.2, 3.3, 3.4
2.33	54	0	blow up of $\mathbb{P}^3$ along line	$\mathrm{PGL}_{4;2}(\mathbb{C})$	No	No	3.3, 3.6, 3.7
2.34	54	0	$\mathbb{P}^1 \times \mathbb{P}^2$	$\mathrm{PGL}_2(\mathbb{C}) \times \mathrm{PGL}_3(\mathbb{C})$	Yes	Yes	3.1, 3.2, 3.3
2.35	56	0	$V_7 = $ blow up of $\mathbb{P}^3$ in one point	$\mathrm{PGL}_{4;1}(\mathbb{C})$	No	No	3.3, 3.6, 3.7
2.36	62	0	$\mathbb{P}(O_{\mathbb{P}^2} \oplus O_{\mathbb{P}^2}(2))$	$\mathrm{Aut}(\mathbb{P}(1,1,1,2))$	No	No	3.3, 3.6, 3.7
3.1	12	8	double cover of $\mathbb{P}^1 \times \mathbb{P}^1 \times \mathbb{P}^1$ branched in surface of degree $(2,2,2)$	1	Yes	Yes	3.5

3.2	14	3	divisor in $\mathbb{P}^2$-bundle	1	Yes★	Yes★	5.11
3.3	18	3	$\mathbb{P}(O_{\mathbb{P}^1\times\mathbb{P}^1} \oplus O_{\mathbb{P}^1\times\mathbb{P}^1}(-1,-1) \oplus O_{\mathbb{P}^1\times\mathbb{P}^1}(2,3))$ such that $X \in |L^{\otimes 2} \otimes O_{\mathbb{P}^1\times\mathbb{P}^1}(-1,-1)|$ where L is tautological line bundle divisor in $\mathbb{P}^1 \times \mathbb{P}^1$ of degree $(1,1,2)$	1	Yes	Yes	5.12, [34]
3.4	18	2	blow up of smooth Fano threefold Y that is contained in family №2.18 along smooth fiber of conic bundle $Y \to \mathbb{P}^2$	1	Yes★	Yes★	4.5, 5.13
3.5	20	0	blow up of $\mathbb{P}^1 \times \mathbb{P}^2$ along curve C of degree $(5,2)$ such that $C \hookrightarrow \mathbb{P}^1 \times \mathbb{P}^2 \to \mathbb{P}^2$ is embedding	$\mathbb{G}_m$	∃No / Yes★	Yes★	5.14
3.6	22	1	blow up of $\mathbb{P}^3$ along disjoint union of line and elliptic curve of degree 4	1	Yes★	Yes★	5.15
3.7	24	1	blow up of W in elliptic curve	1	Yes★	Yes★	4.3
3.8	24	0	blow up of $\mathbb{P}^1 \times \mathbb{P}^2$ along complete intersection of two surfaces that have degree $(0,2)$ and $(1,2)$	1 / $\mathbb{G}_m$	∃No / Yes★	Yes★	5.16
3.9	26	3	blow up of cone $W_4 \subset \mathbb{P}^6$ over Veronese surface $R \subset \mathbb{P}^5$ at its vertex and smooth quartic curve in $R_4 \cong \mathbb{P}^2$	$\mathbb{G}_m$	Yes	Yes	4.6
3.10	26	0	blow up of $Q \subset \mathbb{P}^4$ along disjoint union of two conics	1 / $\mathbb{G}_m$ / $\mathbb{G}_m^2$	∃No / Yes★	Yes	5.17
3.11	28	1	blow up of V_7 in elliptic curve	1	Yes★	Yes★	4.3, [101]
3.12	28	0	blow up of $\mathbb{P}^3$ along disjoint union of line and twisted cubic	1 / $\mathbb{G}_m$	∃No / Yes★	Yes	5.18, [68]

					∃No		
3.13	30	0	intersection of three divisors in $\mathbb{P}^2 \times \mathbb{P}^2 \times \mathbb{P}^2$ that have degree $(1,1,0), (0,1,1)$ and $(1,0,1)$	$\mathbb{G}_a$ $\mathbb{G}_m$ $\mathrm{PGL}_2(\mathbb{C})$	Yes★	Yes	4.2, 5.19
3.14	32	1	blow up of $\mathbb{P}^3$ along plane cubic curve and point that are not coplanar	$\mathbb{G}_m$	No	No	3.7
3.15	32	0	blow up of $Q \subset \mathbb{P}^4$ along disjoint union of line and conic	$\mathbb{G}_m$	Yes	Yes	5.20
3.16	34	0	blow up of V_7 along proper transform via blow up $V_7 \to \mathbb{P}^3$ of twisted cubic passing through blown up point	$\mathbb{G}_a \rtimes \mathbb{G}_m$	No	No	3.6, 3.7
3.17	36	0	divisor on $\mathbb{P}^1 \times \mathbb{P}^1 \times \mathbb{P}^2$ of degree $(1,1,1)$	$\mathrm{PGL}_2(\mathbb{C})$	Yes	Yes	4.2
3.18	36	0	blow up of $\mathbb{P}^3$ along disjoint union of line and conic	$(\mathbb{G}_a \rtimes \mathbb{G}_m) \times \mathbb{G}_m$	No	No	3.3, 3.6, 3.7
3.19	38	0	blow up of $Q \subset \mathbb{P}^4$ at two non-collinear points	$\mathbb{G}_m \times \mathrm{PGL}_2(\mathbb{C})$	Yes	Yes	3.3
3.20	38	0	blow up of $Q \subset \mathbb{P}^4$ along disjoint union of two lines	$\mathbb{G}_m \times \mathrm{PGL}_2(\mathbb{C})$	Yes	Yes	3.3
3.21	38	0	blow up of $\mathbb{P}^1 \times \mathbb{P}^2$ along curve of degree $(2,1)$	$(\mathbb{G}_a)^2 \rtimes (\mathbb{G}_m)^2$	No	No	3.3, 3.6, 3.7
3.22	40	0	blow up of $\mathbb{P}^1 \times \mathbb{P}^2$ along conic in fiber of projection $\mathbb{P}^1 \times \mathbb{P}^2 \to \mathbb{P}^1$	$(\mathbb{G}_a \rtimes \mathbb{G}_m) \times \mathrm{PGL}_2(\mathbb{C})$	No	No	3.3, 3.6, 3.7
3.23	42	0	blow up of V_7 along proper transform via blow up $V_7 \to \mathbb{P}^3$ of irreducible conic passing through blown up point	$(\mathbb{G}_a)^3 \rtimes ((\mathbb{G}_a \rtimes \mathbb{G}_m) \times \mathbb{G}_m)$	No	No	3.3, 3.6, 3.7

3.24	42	0	blow up of W along one fiber of $\mathbb{P}^1$-bundle $W \to \mathbb{P}^2$	$\mathrm{PGL}_{3;1}(\mathbb{C})$	No	No	3.3, 3.6, 3.7
3.25	44	0	blow up of $\mathbb{P}^3$ along two skew lines	$\mathrm{PGL}_{(2,2)}(\mathbb{C})$	Yes	Yes	3.3
3.26	46	0	blow up of $\mathbb{P}^3$ along disjoint union of point and line	$(\mathbb{G}_a)^3 \rtimes (\mathrm{GL}_2(\mathbb{C}) \times \mathbb{G}_m)$	No	No	3.3, 3.6, 3.7
3.27	48	0	$\mathbb{P}^1 \times \mathbb{P}^1 \times \mathbb{P}^1$	$(\mathrm{PGL}_2(\mathbb{C}))^3$	Yes	Yes	3.1, 3.2, 3.3
3.28	48	0	$\mathbb{P}^1 \times S_8 = \mathbb{P}^1 \times \mathbb{F}_1$	$\mathrm{PGL}_2(\mathbb{C}) \times \mathrm{PGL}_{3;1}(\mathbb{C})$	No	No	3.3, 3.6, 3.7
3.29	50	0	blow up of V_7 along line in exceptional surface $E \cong \mathbb{P}^2$ of blow up $V_7 \to \mathbb{P}^3$	$\mathrm{PGL}_{4;3,1}(\mathbb{C})$	No	No	3.3, 3.6, 3.7
3.30	50	0	blow up of V_7 along fiber of $\mathbb{P}^1$-bundle $V_7 \to \mathbb{P}^2$	$\mathrm{PGL}_{4;2,1}(\mathbb{C})$	No	No	3.3, 3.6, 3.7
3.31	52	0	blow up of quadric cone in $\mathbb{P}^4$ with one singular point at vertex	$\mathrm{PSO}_{6;1}(\mathbb{C})$	No	No	3.3, 3.6, 3.7
4.1	24	1	divisor in $(\mathbb{P}^1)^4$ of degree $(1,1,1,1)$	1	Yes	Yes	4.3, [16]
4.2	28	1	blow up of quadric cone in $\mathbb{P}^4$ with one singular point at disjoint union of vertex and elliptic curve of degree 4	$\mathbb{G}_m$	Yes	Yes	4.6
4.3	30	0	blow up of $(\mathbb{P}^1)^3$ at curve of degree $(1,1,2)$	$\mathbb{G}_m$	Yes	Yes	5.21
4.4	32	0	blow up of smooth Fano threefold Y contained in family №3.19 along proper transform of conic on quadric $Q \subset \mathbb{P}^4$ that contains both centers of blow up $Y \to Q$	$\mathbb{G}_m^2$	Yes	Yes	3.3

4.5	32	0	blow up of $\mathbb{P}^1 \times \mathbb{P}^2$ along disjoint union of curves of degree $(2,1)$ and $(1,0)$	$\mathbb{G}_m^2$	No	No	3.3, 3.7
4.6	34	0	blow up of $\mathbb{P}^3$ along three skew lines	$\mathrm{PGL}_2(\mathbb{C})$	Yes	Yes	4.2
4.7	36	0	blow up of $W \subset \mathbb{P}^2 \times \mathbb{P}^2$ along disjoint union of curves of degree $(0,1)$ and $(1,0)$	$\mathrm{GL}_2(\mathbb{C})$	Yes	Yes	3.3
4.8	38	0	blow up of $\mathbb{P}^1 \times \mathbb{P}^1 \times \mathbb{P}^1$ along curve of degree $(0,1,1)$	$(\mathbb{G}_a \rtimes \mathbb{G}_m) \times \mathrm{PGL}_2(\mathbb{C})$	No	No	3.3, 3.6, 3.7
4.9	40	0	blow up of smooth Fano threefold Y contained in family №3.25 along curve $C \cong \mathbb{P}^1$ that is contracted by blow up $Y \to \mathbb{P}^3$	$\mathrm{PGL}_{(2,2);1}(\mathbb{C})$	No	No	3.3, 3.6, 3.7
4.10	42	0	$\mathbb{P}^1 \times S_7$	$\mathrm{PGL}_2(\mathbb{C}) \times (\mathbb{G}_a \rtimes \mathbb{G}_m)^2$	No	No	3.3, 3.6, 3.7
4.11	44	0	blow up of $\mathbb{P}^1 \times \mathbb{F}_1$ along curve $C \cong \mathbb{P}^1$ contained in fiber $F \cong \mathbb{F}_1$ of the projection $\mathbb{P}^1 \times \mathbb{F}_1 \to \mathbb{P}^1$ such that $C^2 = -1$ on F	$(\mathbb{G}_a \rtimes \mathbb{G}_m) \times \mathrm{PGL}_{3;1}(\mathbb{C})$	No	No	3.3, 3.6, 3.7
4.12	46	0	blow up of smooth Fano threefold Y contained in family №2.33 along two curves contracted by blow up $Y \to \mathbb{P}^3$	$(\mathbb{G}_a)^4 \rtimes (\mathrm{GL}_2(\mathbb{C}) \times \mathbb{G}_m)$	No	No	3.3, 3.6, 3.7
4.13	26	0	blow up of $\mathbb{P}^1 \times \mathbb{P}^1 \times \mathbb{P}^1$ along curve of degree $(1,1,3)$	1 / $\mathbb{G}_m$	$\exists$No / Yes★	Yes★	5.22
5.1	28	0	blow up of smooth Fano threefold Y contained in family №2.29 along three curves contracted by blow up $Y \to Q$	$\mathbb{G}_m$	Yes	Yes	5.23

5.2	36	0	blow up of smooth Fano threefold Y contained in family №3.25 along two curves $C_1 \neq C_2$ contracted by blow up $\phi\colon Y \to \mathbb{P}^3$ that are contained in one ϕ-exceptional surface	$\mathrm{GL}_2(\mathbb{C}) \times \mathbb{G}_m$	No	No	3.3, 3.7
5.3	36	0	$\mathbb{P}^1 \times S_6$	$\mathrm{PGL}_2(\mathbb{C}) \times \mathbb{G}_m^2$	Yes	Yes	3.1, 3.3
6.1	30	0	$\mathbb{P}^1 \times S_5$	$\mathrm{PGL}_2(\mathbb{C})$	Yes	Yes	3.1
7.1	24	0	$\mathbb{P}^1 \times S_4$	$\mathrm{PGL}_2(\mathbb{C})$	Yes	Yes	3.1
8.1	18	0	$\mathbb{P}^1 \times S_3$	$\mathrm{PGL}_2(\mathbb{C})$	Yes	Yes	3.1
9.1	12	0	$\mathbb{P}^1 \times S_2$	$\mathrm{PGL}_2(\mathbb{C})$	Yes	Yes	3.1
10.1	6	0	$\mathbb{P}^1 \times S_1$	$\mathrm{PGL}_2(\mathbb{C})$	Yes	Yes	3.1

7

Conclusion

As presented in Table 6.1, we know which smooth Fano threefolds are K-polystable and which are not for 71 of the 105 deformation families. For the remaining 34 families,

№1.9, №1.10, №2.1, №2.2, №2.3, №2.4, №2.5, №2.6, №2.7,
№2.8, №2.9, №2.10, №2.11, №2.12, №2.13, №2.14, №2.15, №2.16,
№2.17, №2.18, №2.19, №2.20, №2.21, №2.22, №3.2, №3.3, №3.4,
№3.5, №3.6, №3.7, №3.8, №3.11, №3.12, №4.1,

the Main Theorem tells us that the general member is K-polystable. In most cases we expect that all smooth members are K-polystable. More precisely, all smooth Fano threefolds in the 27 deformation families

№1.9, №2.1, №2.2, №2.3, №2.4, №2.5, №2.6, №2.7, №2.8,
№2.9, №2.10, №2.11, №2.12, №2.13, №2.14, №2.15, №2.16, №2.17,
№2.18, №2.19, №3.2, №3.3, №3.4, №3.6, №3.7, №3.11, №4.1

have finite automorphism group, and we expect that they are all K-stable. On the other hand, the 7 remaining families

№1.10, №2.20, №2.21, №2.22, №3.5, №3.8, №3.12

contain both K-polystable and non-K-polystable smooth Fano threefolds. In each of these cases, we have a conjectural characterization of K-polystability.

7.1 Family №1.10

Members of the 6-dimensional Family №1.10 are often referred to as *Fano threefolds* V_{22} or *prime Fano threefolds of genus* 12. They can be described as

365

follows. Set $V = \mathbb{C}^7$, and $N = \mathbb{C}^3$. For every smooth prime Fano threefold X of genus 12, there is a net $\eta \colon \bigwedge^2 V \to N$ such that

$$X \simeq \mathrm{Gr}(3, V, \eta) = \{E \in \mathrm{Gr}(3, V) | \bigwedge^2 E \subset \ker \eta\}.$$

The general member of this family has finite automorphism group. In Example 4.12, we exhibited a K-stable Fano threefold in this family, and thus concluded that the general member of the family №1.10 is K-stable, which also follows from [212].

Family №1.10 contains a unique smooth Fano threefold X_{22}^a that has non-reductive automorphism group, namely $\mathbb{G}_a \rtimes \mu_4$ [136]. This special member is not K-polystable by Theorem 1.3, but it is K-semistable by [55, Example 1.4].

There is a 1-parameter subfamily in the family №1.10 consisting of smooth Fano threefolds admitting an effective $\mathbb{G}_m$-action. As explained in Example 4.11, all the threefolds in this subfamily are K-polystable. Together with X_{22}^a, these are all the smooth Fano threefolds №1.10 with infinite automorphism group [182]. Among those, there is one with automorphism group $\mathrm{PGL}_2(\mathbb{C})$, the Mukai–Umemura threefold X_{22}^{MU}, see Example 4.11. It can be constructed as $\mathrm{Gr}(3, V, \eta)$ by taking V to be the irreducible 7-dimensional representation s^6 of $\mathrm{SL}_2(\mathbb{C})$ and N to be the 3-dimensional subspace of $\bigwedge V^*$ that is the image of the Lie algebra under the action; it naturally supports an induced $\mathrm{SL}_2(\mathbb{C})$-action.

We know that the general member of the family №1.10 is K-stable, and there are members of this family that are not K-polystable. The general picture is predicted by the following conjecture by Donaldson, see [79, Section 5.3] for a GIT interpretation of this conjecture.

Conjecture 7.1 (Donaldson) *Let X be a smooth Fano threefold in the family №1.10. Then X is K-polystable if and only if one of the following two conditions is satisfied:*

(i) *either X admits an effective $\mathbb{G}_m$-action,*

(ii) *or no element of $|-K_X|$ has singularities of the form $y^2 = x^3 + t^4 x$ or worse.*

We discuss this conjecture from yet another perspective. In [163, 162], Mukai gives several descriptions of prime Fano threefolds of genus 12, and shows that the moduli space $\mathcal{M}_{22}$ of prime Fano threefolds of genus 12 is birational to the moduli space of plane quartic curves, see also [194]. Namely, for a smooth prime Fano 3-fold X of genus 12, the Hilbert scheme of lines of X is a (possibly singular) quartic curve

$$C_X = \{f(x, y, z) = 0\} \subset \mathbb{P}^2,$$

and the Fano threefold X can be recovered from the quartic curve C_X as the closure of the variety of its polar hexagons:

$$X = \mathrm{VSP}(C_X, 6) = \overline{\left\{ (L_1, \ldots, L_6) \in \mathrm{Hilb}^6(\mathbb{P}^2) \mid f(x, y, z) = l_1^4 + \cdots + l_6^4 \right\}}.$$

Here we write $[x : y : z]$ for coordinates on $\mathbb{P}^2$, $f(x, y, z)$ for the homogeneous quartic polynomial defining the curve C_X, and $l_i = l_i(x, y, z)$ for the linear form defining the line L_i. For instance, if C is the Klein quartic curve, then $\mathrm{VSP}(C, 6)$ is the smooth Fano threefold in the family №1.10 from Example 4.12.

If the threefold X is general, then the quartic curve C_X is irreducible and non-singular. More generally, for every point $P \in C_X$, either P is a smooth point of the curve C_X and the corresponding line $\ell_P \subset X$ has normal bundle $\mathcal{N}_{\ell_P/X} \cong \mathcal{O}_{\mathbb{P}^1} \oplus \mathcal{O}_{\mathbb{P}^1}(-1)$, or P is a singular point of the curve C_X and $\mathcal{N}_{\ell_P/X} \cong \mathcal{O}_{\mathbb{P}^1}(1) \oplus \mathcal{O}_{\mathbb{P}^1}(-2)$.

For the members of the family with infinite automorphism group, we have the following description of the curve C_X.

- If $X = X_{22}^a$, then C_X is a union of two smooth conics that meet at one point.
- If $\mathrm{Aut}^0(X) \cong \mathbb{G}_m$, then the quartic curve C_X is a union of two smooth conics that tangent to each other at two distinct points.
- X is the Mukai–Umemura threefold X_{22}^{MU} if and only if C_X is a double conic [183].

In view of this correspondence between plane quartic curves and smooth members of the family №1.10, it is interesting to compare Donaldson's conjecture with the naive induced correspondence between GIT of plane quartic curves and K-stability of threefolds. Indeed, a plane quartic curve is known to be GIT-stable (respectively, strictly polystable) precisely when it has no worse than $\mathbb{A}_1$ or $\mathbb{A}_2$ singularities (respectively, if it is a double conic or 2 conics tangent at 2 points, at least one of which is smooth). In particular, the members of the family with infinite automorphism group that are K-polystable do have GIT-polystable Hilbert scheme of lines, while the one that is strictly K-semistable has non-GIT-polystable Hilbert scheme of lines.

7.2 Family №2.20

The 3-dimensional family №2.20 contains a unique smooth Fano threefold X with infinite automorphism group [45]. In Proposition 5.43 we showed that this threefold is K-polystable, and proved that the general Fano threefold in the family №2.20 is K-stable. It follows from Remark 1.17 that there is at least one member of the family that is not K-polystable. We explicitly exhibit such a non-K-polystability in Lemma 7.4 below.

Recall that the Fano threefolds in the family №2.20 can be described as blow ups of the unique smooth Fano threefold №1.15 described in Example 3.2, denoted by V_5, along twisted cubic curves. In order to draw the conjectural picture of K-polystability for this family, we describe the Hilbert scheme of twisted cubic curves in V_5 following [116, 189].

The $\mathrm{SL}_2(\mathbb{C})$-action on V_5 has been described in Section 5.10. We use the notation introduced in the very beginning of that section. Recall from [189, Proposition 2.46] that the Hilbert scheme of twisted cubic curves in the threefold V_5 is $\mathrm{SL}_2(\mathbb{C})$-equivariantly isomorphic to $\mathrm{Gr}(2, V)$. To explain this, note that, as $\mathrm{SL}_2(\mathbb{C})$-representations, we have

$$\mathrm{Sym}^2(A) = \mathrm{Sym}^2\big(\mathrm{Sym}^2(W)\big) \cong V \oplus \mathbb{I},$$

where $\mathbb{I}$ is the trivial representation. Composing the Veronese map $A \to \mathrm{Sym}^2(A)$ with the projection $V \oplus \mathbb{I} \to V$ induces an $\mathrm{SL}_2(\mathbb{C})$-equivariant embedding

$$\eta\colon \mathbb{P}^2 = \mathbb{P}(A) \hookrightarrow \mathbb{P}(V) = \mathbb{P}^4.$$

Set $\mathscr{S} = \mathrm{im}(\eta)$. Then $\mathscr{S} \cong \mathbb{P}^2$, and $\mathscr{S} \subset \mathbb{P}^4$ is an $\mathrm{SL}_2(\mathbb{C})$-invariant surface of degree 4. For later use, we also introduce the smooth rational quartic curve $\mathscr{C} \subset \mathscr{S} \subset \mathbb{P}(V)$ that is the image of the unique $\mathrm{SL}_2(\mathbb{C})$-invariant conic in $\mathbb{P}(A)$.

Let $\sigma\colon \mathscr{Y} \to \mathbb{P}^4$ be the blow up of the surface $\mathscr{S}$. By [189, Remark 2.47], there exists an $\mathrm{SL}_2(\mathbb{C})$-equivariant isomorphism $\mathscr{Y} \cong \mathbb{P}(\mathscr{U})$, where $\mathscr{U}$ is the restriction to the threefold V_5 of the tautological vector bundle of the Grassmannian $\mathrm{Gr}(2, V)$. Thus, we obtain the $\mathrm{SL}_2(\mathbb{C})$-equivariant commutative diagram

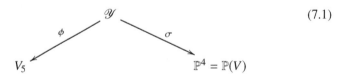

$$(7.1)$$

where $\phi\colon \mathscr{Y} \to V_5$ is the induced $\mathbb{P}^1$-bundle. Let L be a line in $\mathbb{P}^4$ and let $C_L = \phi_*(\sigma^*(L))$. Then C_L is a (possibly singular) twisted cubic curve in V_5. Moreover, one can show that the curve C_L is smooth if and only if $L \cap \mathscr{S} = \varnothing$.

Let X_L be the blow up of the threefold V_5 along the curve C_L. Then X_L is a (possibly singular) Fano threefold №2.20. If the curve C_L is smooth, the threefold X_L is also smooth. In this case, we expect that the smooth Fano threefold X_L is K-polystable if and only if the orbit of the line L considered as a point in $\mathrm{Gr}(2, V)$ is GIT-polystable with respect to the $\mathrm{SL}_2(\mathbb{C})$-action.

Next, we look at the smooth members of the family №2.20 from a slightly different, but more explicit, perspective. We fix the quartic curve $C_4 \subset \mathbb{P}^3$ given by $[r^4 : r^3s : rs^3 : s^4]$ for $[s : r] \in \mathbb{P}^1$. Let $G = \mathrm{Aut}(\mathbb{P}^3, C_4)$. Then G contains transformations

$$[x : y : z : t] \mapsto [x : sy : s^3z : s^4t]$$

for $s \in \mathbb{C}^*$, and G contains the involution $\tau \colon [x : y : z : t] \mapsto [t : z : y : x]$. Since G is naturally embedded to $\mathrm{Aut}(C_4) \cong \mathrm{PGL}_2(\mathbb{C})$, either $G \cong \mathbb{G}_m \rtimes \mu_2$ or $G = \mathrm{Aut}(C_4)$ [169]. The latter is impossible since $H^0(\mathcal{O}_{\mathbb{P}^3}(1)|_{C_4})$ is an irreducible representation of $\mathrm{Aut}(C_4)$, and the embedding $C_4 \hookrightarrow \mathbb{P}^3$ is not linearly normal.

The curve C_4 is contained in the G-invariant smooth surface $S_2 \subset \mathbb{P}^3$ given by $xt = yz$. Let $\chi \colon \mathbb{P}^3 \dashrightarrow \mathbb{P}^6$ be the G-equivariant map given by

$$[x : y : z : t] \mapsto$$
$$[x(xt - yz) : y(xt - yz) : z(xt - yz) : t(xt - yz) : xz^2 - y^2t : x^2z - y^3 : yt^2 - z^3].$$

Then χ is well defined away from C_4, and the closure of its image is isomorphic to V_5. Let $C_2 = \chi(S_2)$. Then C_2 is the unique G-invariant smooth conic in V_5 by Corollary 5.39. Thus, we have the G-equivariant commutative diagram

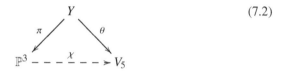

$$(7.2)$$

where π is a blow up of the twisted quartic curve C_4, and θ is a blow up of the conic C_2. Let ℓ be a line in $\mathbb{P}^3$, and let $C_\ell = \theta_*(\pi^*(\ell))$. Then C_ℓ is a (possibly singular) twisted cubic curve in V_5. Moreover, the curve C_ℓ is smooth $\iff$ $\ell \cap C_4 = \varnothing$.

Lemma 7.2 *Every smooth Fano threefold №2.20 can be obtained by blowing up the threefold V_5 along the image of a suitable line $\ell \subset \mathbb{P}^3$ such that $\ell \cap C_4 = \varnothing$.*

Proof Let us recall from [189, Proposition 2.32] the identification of the space of conics in the variety V_5 with $\mathbb{P}(V^*)$. For a hyperplane $H \subset \mathbb{P}(V)$, let $\tilde{H}$ be the proper transform of H in the variety $\mathcal{Y}$, and denote by $\varpi \colon \tilde{H} \to H$ the induced birational morphism. Then ϖ is the blow up of the hyperplane H along a (possibly singular) quartic curve $H \cap \mathcal{S}$, and we can expand diagram (7.1) as follows:

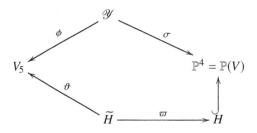

where ϑ is the blow up of a (possibly singular) conic $C_H \subset V_5$. Moreover, one can show that C_H is smooth $\iff H \cap \mathscr{S}$ is smooth, and all conics in V_5 are obtained in this way.

If C_H is smooth, then $\widetilde{H}$ is a smooth Fano threefold №2.22, and $\mathrm{Aut}(\widetilde{H}) \cong \mathrm{Aut}(V_5, C_H)$. In this case, we have the following possibilities:

- The curve $H \cap \mathscr{S}$ is tangent to the curve $\mathscr{C} \subset \mathscr{S}$ at two points, $\mathrm{Aut}(\widetilde{H}) \cong \mathbb{G}_m \rtimes \mu_2$, and $\widetilde{H}$ is the smooth Fano threefold constructed in Example 4.34.
- The curve $H \cap \mathscr{S}$ is smooth, it is tangent to $\mathscr{C} \subset \mathscr{S}$ at one point, and it intersects the curve $\mathscr{C}$ in two extra points. In this case, $\mathrm{Aut}(\widetilde{H})$ is finite, and $\widetilde{H}$ is the non-K-polystable smooth Fano threefold №2.22 explicitly described in Section 7.4.
- The curve $H \cap \mathscr{S}$ is smooth, it intersects $\mathscr{C} \subset \mathscr{S}$ transversally, and $\mathrm{Aut}(\widetilde{H})$ is finite.

Recall that every smooth Fano threefold X in Family №2.20 is isomorphic to X_L for some line $L \subset \mathbb{P}(V)$ such that $L \cap \mathscr{S} = \varnothing$. Let M be the linear subsystem in $|\mathcal{O}_{\mathbb{P}^4}(1)|$ consisting of all hyperplanes that contain L. Suppose that M contains a hyperplane H such that

(i) H is $\mathbb{G}_m$-invariant for some subgroup $\mathbb{G}_m \subset \mathrm{SL}_2(\mathbb{C})$,

(ii) the curve $H \cap \mathscr{S}$ is smooth.

Then we can take $(\mathbb{P}^3, C_4, \ell) = (H, H \cap \mathscr{S}, L)$ in the previous construction, and get that X is the blow up of the threefold V_5 along C_ℓ. Notice that if there is a hyperplane H in M satisfying (1) and (2) above, then it must be tangent to $\mathscr{C}$ at two distinct points. Vice versa, if M contains a hyperplane that is tangent to $\mathscr{C}$ at two distinct points, then this hyperplane is $\mathbb{G}_m$-invariant for the subgroup $\mathbb{G}_m \subset \mathrm{SL}_2(\mathbb{C})$ that fixes these two points, and, moreover, this hyperplane must intersect $\mathscr{S}$ along a smooth curve. So, to complete the proof, it is enough to show that

(★) M contains a hyperplane that is tangent to $\mathscr{C}$ at two distinct points.

Parameter count shows that (★) holds if the line L is general. However, we have to prove this for every line L in $\mathbb{P}^4$ that does not meet the surface $\mathscr{S}$.

The linear system M is a net (a two-dimensional linear system), and L is its base locus. The restriction $M|_{\mathscr{S}}$ is also a net, which does not have base-points since $L \cap \mathscr{S} = \varnothing$. To prove (★), it is enough show that $M|_{\mathscr{S}}$ contains a smooth curve that is tangent to the curve $\mathscr{C}$ at two distinct points. In fact, it is enough to prove that $M|_{\mathscr{S}}$ contains a curve C such that $C|_{\mathscr{C}} = 2P + 2Q$ for two distinct points P and Q in the curve $\mathscr{C}$. If we find such a curve C, then it is automatically smooth since it is cut out by a hyperplane in $\mathbb{P}^4$.

Now, let us explicitly describe the $\mathrm{SL}_2(\mathbb{C})$-action on our $\mathbb{P}^4 = \mathbb{P}(V)$. To do this, we fix the embeddings $\mathbb{P}^1 \hookrightarrow \mathbb{P}^2$ given by $[u : v] \mapsto [u^2 : v^2 : uv]$ and $\mathbb{P}^2 \hookrightarrow \mathbb{P}^5$ given by

$$[x : y : z] \mapsto \left[x^2 : xz : \frac{xy + 2z^2}{3} : yz : y^2 : xy - z^2 \right].$$

Then we equip both $\mathbb{P}^2$ and $\mathbb{P}^5$ with the $\mathrm{SL}_2(\mathbb{C})$-action such that our explicit embeddings are $\mathrm{SL}_2(\mathbb{C})$-equivariant with respect to the standard action of the group $\mathrm{SL}_2(\mathbb{C})$ on $\mathbb{P}^1$. Let $\eta \colon \mathbb{P}^2 \to \mathbb{P}^4$ be the morphism $[x : y : z] \mapsto [x^2 : xz : \frac{xy+2z^2}{3} : yz : y^2]$. Then η is a composition of the embedding $\mathbb{P}^2 \hookrightarrow \mathbb{P}^5$ and projection from the $\mathrm{SL}_2(\mathbb{C})$-fixed point. This gives us the $\mathrm{SL}_2(\mathbb{C})$-action on $\mathbb{P}^4$ such that η is equivariant. This action is given by the monomorphism $\mathrm{SL}_2(\mathbb{C}) \hookrightarrow \mathrm{SL}_5(\mathbb{C})$ given by

$$\begin{pmatrix} a & b \\ c & d \end{pmatrix} \mapsto$$

$$\begin{pmatrix} a^4 & 4a^3b & 6a^2b^2 & 4ab^3 & b^4 \\ a^3c & a^3d + 3a^2bc & 3a^2bd + 3ab^2c & 3ab^2d + b^3c & b^3d \\ a^2c^2 & 2a^2cd + 2abc^2 & a^2d^2 + 4abcd + b^2c^2 & 2abd^2 + 2b^2cd & b^2d^2 \\ c^3a & 3ac^2d + bc^3 & 3acd^2 + 3bc^2d & ad^3 + 3bcd^2 & d^3b \\ c^4 & 4c^3d & 6c^2d^2 & 4cd^3 & d^4 \end{pmatrix}.$$

It should be pointed out that $\mathscr{S} = \mathrm{im}(\eta)$, and the $\mathrm{SL}_2(\mathbb{C})$-invariant curve $\mathscr{C}$ is given by the parametrization $[u^4 : u^3v : u^2v^2 : uv^3 : v^4]$, where $[u : v] \in \mathbb{P}^1$. For simplicity, let us identify $\mathscr{S} = \mathbb{P}^2$ via the embedding η. Then $\mathscr{C}$ is the conic in $\mathbb{P}^2$ given by $xy = z^2$, and the net $M|_{\mathscr{S}}$ is a linear subsystem in $|O_{\mathbb{P}^2}(2)|$ that consists of conics

$$\lambda_1 x^2 + \lambda_2(xy + 2z^2) + \lambda_3 y^2 + \lambda_4 xz + \lambda_5 yz = 0,$$

where $[\lambda_1 : \lambda_2 : \lambda_3 : \lambda_4 : \lambda_5] \in \mathbb{P}^4$. Then $\mathcal{M}|_{\mathscr{C}}$ is also a net since $\mathcal{M}|_{\mathscr{S}}$ does not contain $\mathscr{C}$. Hence, to prove ($\bigstar$), it is enough show that the net $\mathcal{M}|_{\mathscr{C}}$ contains a divisor $2P + 2Q$, where P and Q are two distinct points in $\mathscr{C}$. Suppose that the latter assertion is wrong.

Applying Lemma A.59, we see that the net $\mathcal{M}|_{\mathscr{C}}$ contains divisors $4P$, $4Q$ and $3P + Q$, where P and Q are two distinct points in $\mathscr{C}$. Since $\mathrm{SL}_2(\mathbb{C})$ acts transitively on pairs of distinct points in $\mathscr{C}$, we may assume $P = [0 : 1 : 0]$ and $Q = [1 : 0 : 0]$. Then $4P$ is cut out on $\mathscr{C}$ by $x^2 = 0$, $4Q$ is cut out by $y^2 = 0$, and $3P + Q$ is cut out by $xz = 0$. Since the net $\mathcal{M}|_{\mathscr{S}}$ is uniquely determined by the net $\mathcal{M}|_{\mathscr{C}}$, we see that $\mathcal{M}|_{\mathscr{S}}$ is the net

$$\mu_1 x^2 + \mu_2 y^2 + \mu_3 xz = 0,$$

where $[\mu_1 : \mu_2 : \mu_3] \in \mathbb{P}^2$. But this net contains a basepoint, which is a contradiction.   □

Remark 7.3   The choice of the line ℓ in Lemma 7.2 is not unique even up to G-action. For instance, the following 4 distinct lines in $\mathbb{P}^4$ lie in different G-orbits:

  (i) the line that passes through $[12 : 0 : 3 : -12]$ and $[0 : 3 : 0 : -3 : 12]$,
 (ii) the line that passes through $[-48 : 12{-}3 : 51]$ and $[48 : -12 : 0 : -9 : 21]$,
(iii) the line that passes through the points

$$\left[144 + 96\sqrt{5} : -36 : 21 - 12\sqrt{5} : 63 - 30\sqrt{5}\right],$$
$$\left[240 + 48\sqrt{5} : -12\sqrt{5} + 36 : -21 + 15\sqrt{5} : -39 + 21\sqrt{5}\right],$$

(iv) the line that passes through the points

$$\left[1008 - 480\sqrt{5} : -84 + 48\sqrt{5} : 9 : 9 + 6\sqrt{5}\right],$$
$$\left[624 - 336\sqrt{5} : -84 + 60\sqrt{5} : 9 - 3\sqrt{5} : -15 - 3\sqrt{5}\right].$$

Moreover, they all are disjoint from the curve C_4. However, one can show that the corresponding smooth Fano threefolds №2.20 are isomorphic.

Let X_ℓ be a blow up of the threefold V_5 at the curve C_ℓ. If C_ℓ is smooth, then X_ℓ is a smooth Fano threefold №2.20. In this case, we expect that X_ℓ is K-polystable if and only if the G-orbit of the line ℓ considered as a point in $\mathrm{Gr}(2,4)$ is GIT-polystable with respect to the induced G-action. In Table 7.1, we list all lines that are not GIT-stable. There, we assume that $(a, b) \in \mathbb{C}^2 \setminus (0,0)$ and use conventions from Section 5.10.

Table 7.1

Line ℓ	Equation	GIT-stability	C_ℓ
$L_{x,t}$	$x = t = 0$	polystable	smooth twisted cubic
$L_{y,z}$	$y = z = 0$	polystable	union of a good line and two bad lines such that bad lines intersect the good line
$L_1(a,b)$	$x = t - ay - bz = 0$	strictly semistable	smooth twisted cubic
$L_2(a,b)$	$t = x - ay - bz = 0$	strictly semistable	smooth twisted cubic
$L_3(a,b)$	$y - ax = z - bx = 0$	strictly semistable	union of a conic and a bad line
$L_4(a,b)$	$z - at = y - bt = 0$	strictly semistable	union of a conic and a bad line
$L_{x,z}$	$x = z = 0$	unstable	union of the conic C_2 and a bad line
$L_{y,t}$	$y = t = 0$	unstable	union of the conic C_2 and a bad line
$L_{x,y}$	$x = y = 0$	unstable	triple bad line
$L_{z,t}$	$z = t = 0$	unstable	triple bad line

If $\ell = L_{x,t}$, then X_ℓ is the unique smooth Fano threefold in the family №2.20 that has an infinite automorphism group. This threefold is K-polystable by Proposition 5.43.

Lemma 7.4 *Let $\ell = L_1(a,b)$ or $\ell = L_2(a,b)$. Then X_ℓ is strictly K-semistable.*

Proof This immediately follows from Proposition 5.43 and Corollary 1.13. $\square$

Our conjecture says that all smooth Fano threefolds in Family №2.20 other than the ones from Lemma 7.4 are K-polystable. This conjecture cannot be extended to singular threefolds: if ℓ is given by $t - y = x - z = 0$, then ℓ is GIT-stable, but $\psi(\ell)$ is a point, so that C_ℓ is a union of three lines that meet at $\psi(\ell)$, and X_ℓ is K-unstable by Lemma 5.36.

7.3 Family №2.21

Smooth Fano threefolds of the 2-dimensional family №2.21 can be described as blow ups of the smooth quadric threefold in $\mathbb{P}^4$ along a twisted quartic curve.

By [45, Lemma 9.2], the general member of this family has finite automorphism group, and all smooth members that have infinite automorphism groups can be described as follows.

(i) There is a one-dimensional subfamily in the family №2.21 consisting of smooth threefolds admitting an effective $\mathbb{G}_m$-action, see Section 5.9 for their description.

(ii) There exists a unique smooth Fano threefold X^a in the family with non-reductive automorphism group, and $\mathrm{Aut}^0(X^a) \cong \mathbb{G}_a$.

(iii) There is a unique threefold X in the family with $\mathrm{Aut}^0(X) \cong \mathrm{PGL}_2(\mathbb{C})$.

The threefold X^a in (ii) is not K-polystable by Theorem 1.3. On the other hand, we showed in Section 5.9 that all remaining smooth Fano threefolds №2.21 that have infinite automorphism groups are K-polystable, and concluded in Corollary 4.16 that the general smooth Fano threefold in family №2.21 is K-stable.

In order to draw the conjectural picture of K-polystability for this family, let us fix the standard $\mathrm{SL}_2(\mathbb{C})$-action on $W = \mathbb{C}^2$, set $V = \mathrm{Sym}^4(W)$, let Z be the $\mathrm{SL}_2(\mathbb{C})$-invariant twisted quartic curve in $\mathbb{P}^4 = \mathbb{P}(V)$, which is given by $[u : v] \mapsto [v^4 : uv^3 : u^2v^2 : u^3v : u^4]$. Then Z is given by the vanishing of the following quadratic forms:

$$f_0 = x_3^2 - x_2x_4, f_1 = x_2x_3 - x_1x_4, f_2 = x_2^2 - x_0x_4,$$
$$f_3 = x_1x_2 - x_0x_3, f_4 = x_1^2 - x_0x_2, f_5 = 3x_2^2 - 4x_1x_3 + x_0x_4.$$

Let Q be a (possibly singular) quadric threefold in $\mathbb{P}^4$ that contains the quartic curve Z, and let $\pi \colon X \to Q$ be the blow up of the quadric Q along Z. Then Q is given by

$$s_0f_0 + s_1f_1 + s_2f_2 + s_3f_3 + s_4f_4 + s_5f_5 = 0$$

for some $[s_0 : s_1 : s_2 : s_3 : s_4 : s_5] \in \mathbb{P}(V \oplus \mathbb{C})$. Applying Corollary A.58 to $\mathbb{P}(V \oplus \mathbb{I})$, where $\mathbb{I}$ is the trivial representation of the group $\mathrm{SL}_2(\mathbb{C})$, equipped with a natural action of the group $\mathrm{PGL}_2(\mathbb{C})$, we see that X is GIT-stable except for the seven cases described in Table 7.2.

If Q is smooth (so that X is smooth as well), we expect that X is K-polystable if and only if the threefold X is GIT-polystable. We point out that [172, Theorem 3.4] implies the ($\Rightarrow$)-direction of this conjecture, which also follows from Corollary 1.13.

Table 7.2

Case	Equation of Q	Is X GIT-semistable?	$\mathrm{Aut}^0(X)$	Is Q smooth?
(0)	$f_5 = 0$	GIT-polystable	$\mathrm{PGL}_2(\mathbb{C})$	Yes
(1)	$f_2 + \lambda f_5 = 0$, $\lambda \in \mathbb{C}$	GIT-polystable	$\mathbb{G}_m$	Yes if $\lambda \notin \{0, -1, 3\}$
(2)	$f_0 = 0$	GIT-unstable	$\mathbb{G}_a \rtimes \mathbb{G}_m$	No
(2')	$f_0 + f_5 = 0$	strictly GIT-semistable	$\mathbb{G}_a$	Yes
(3a)	$f_0 + 3f_2 + \lambda f_5 = 0$, $\lambda \in \mathbb{C}$	strictly GIT-semistable	1	Yes if $\lambda \notin \{0, -1, 3\}$
(3b)	$f_1 = 0$	GIT-unstable	$\mathbb{G}_m$	No
(3b')	$f_1 + f_5 = 0$	strictly GIT-semistable	1	Yes

7.4 Family №2.22

Let X be a smooth Fano threefold in the 1-parameter family №2.22. Then X can be described both as the blow up of $\mathbb{P}^3$ along a smooth twisted quartic curve, and the blow up of V_5, the unique smooth threefold №1.15, along a smooth conic. More precisely, there is a smooth twisted quartic curve $C_4 \subset \mathbb{P}^3$, a smooth conic $C \subset V_5$, and a commutative diagram

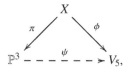

where π is the blow up of $C_4 \subset \mathbb{P}^3$, ϕ is the blow up of $C \subset V_5$, V_5 is embedded in $\mathbb{P}^6$ as described in Section 5.10, and ψ is given by the linear system of cubics containing C_4.

The curve C_4 is contained in a unique smooth quadric surface $S_2 \subset \mathbb{P}^3$, and ϕ contracts the proper transform of this surface. Note that $\mathrm{Aut}(X) \cong \mathrm{Aut}(\mathbb{P}^3, C_4) \cong \mathrm{Aut}(S_2, C_4)$. Choosing appropriate coordinates on $\mathbb{P}^3$, we may assume that S_2 is given by $x_0 x_3 = x_1 x_2$, where x_0, x_1, x_2, x_3 are coordinates on $\mathbb{P}^3$. Fix the isomorphism $S_2 \cong \mathbb{P}^1 \times \mathbb{P}^1$ given by

$$([u : v], [x : y]) \mapsto [xu : xv : yu : yx],$$

where $([u : v], [x : y])$ are coordinates in $\mathbb{P}^1 \times \mathbb{P}^1$. Swapping $[u : v]$ and $[x : y]$ if necessary, we may assume that C_4 is a curve of degree $(1, 3)$ in S_2, so that

$C_4 = \{uf_3(x,y) = vg_3(x,y)\}$, where $f_3(x,y)$ and $g_3(x,y)$ are coprime cubic forms.

The projection $([u : v],[x : y]) \mapsto [u : v]$ gives a triple cover $C_4 \to \mathbb{P}^1$, which is ramified in at least two points. Hence, after an appropriate change of coordinates $[u : v]$, we may assume that this triple cover is ramified over the points $[1 : 0]$ and $[0 : 1]$. This means that both forms $f_3(x,y)$ and $g_3(x,y)$ have multiple roots. Hence, changing coordinates $[x : y]$ if necessary, we may assume that these roots are $[0 : 1]$ and $[1 : 0]$, respectively. Keeping in mind that C_4 is smooth, we see that $C_4 = \{u(x^3+ax^2y) = v(y^3+by^2x)\}$ for some complex numbers a and b, after a suitable scaling of the coordinates. If $a = b = 0$, then the curve C_4 is given by $ux^3 = vy^3$, so that $\mathrm{Aut}(X) \cong \mathbb{G}_m \rtimes \mu_2$, and X is the unique smooth Fano threefold №2.22 with an infinite automorphism group [45]. In this case, we know that X is K-polystable (see Section 4.4).

If $a = 0$ and $b \neq 0$, we can scale the coordinates $([u : v],[x : y])$ further and assume that the curve C_4 is given by

$$ux^3 = v(y^3 + y^2x). \tag{7.3}$$

If $a \neq 0$ and $b = 0$, then we can scale the coordinates and swap them to put the defining equation of the curve C_4 into (7.3). In this case, we have

Lemma 7.5 *If C_4 is given by (7.3), then X is strictly K-semistable.*

Proof This follows from Corollary 1.13, cf. the proof of Corollary 4.71. □

Hence, to solve the Calabi Problem for every smooth threefold in the family №2.22, we may assume that $a \neq 0$ and $b \neq 0$. Therefore, scaling further the coordinates on S_2, we may assume that $C_4 = \{u(x^3 + \lambda x^2y) = v(y^3 + \lambda y^2x)\}$ for some $\lambda \in \mathbb{C}^*$. Then $\lambda \neq \pm 1$ since C_4 is smooth. Moreover, if $\lambda = \pm 3$, then we can change our coordinates such that C_4 is given by (7.3). Hence, we may also assume that $\lambda \neq \pm 3$. We believe that X is K-stable for all remaining values of the parameter λ. By Proposition 4.33, we know that X is K-stable if λ is general. We remark that by taking $\lambda = \pm\sqrt{3}$, we obtain the smooth Fano threefold №2.22 with automorphism group $\mathfrak{A}_4$ described in Example 4.37.

7.5 Family №3.5

Let X be a smooth Fano threefold in the 5-parameter family №3.5. Then X can be described as the blow up of $\mathbb{P}^1 \times \mathbb{P}^2$ along a curve of degree $(5,2)$. To describe X explicitly, let $S = \mathbb{P}^1 \times \mathbb{P}^1$, and let C be a smooth curve in S of degree

(1,5). Arguing as in Section 7.4, we can choose coordinates $([u : v], [x : y])$ on the surface S such that the curve C is given by the following equation:

$$u\left(x^5 + a_1 x^4 y + a_2 x^3 y^2 + a_3 x^2 y^3\right) + v\left(y^5 + b_1 y^4 x + b_2 y^3 x^2 + b_3 y^2 x^3\right) = 0,$$

where $a_1, a_2, a_3, b_1, b_2, b_3$ are some complex numbers. The shape of this equation simply means that the point $([1 : 0], [0 : 1])$ and the point $([0 : 1], [1 : 0])$ are among ramifications points of the finite degree five cover $\eta \colon C \to \mathbb{P}^1$ that is given by $([u : v], [x : y]) \mapsto [u : v]$. Note that the ramification index of the point $([1 : 0], [0 : 1])$ is

$$\begin{cases} 2 \text{ if } a_3 \neq 0, \\ 3 \text{ if } a_3 = 0 \text{ and } a_2 \neq 0, \\ 4 \text{ if } a_3 = a_2 = 0 \text{ and } a_1 \neq 0, \\ 5 \text{ if } a_3 = a_2 = a_1 = 0. \end{cases}$$

Similarly, the ramification index of the point $([0 : 1], [1 : 0])$ is

$$\begin{cases} 2 \text{ if } b_3 \neq 0, \\ 3 \text{ if } b_3 = 0 \text{ and } b_2 \neq 0, \\ 4 \text{ if } b_3 = b_2 = 0 \text{ and } b_1 \neq 0, \\ 5 \text{ if } b_3 = b_2 = b_1 = 0. \end{cases}$$

Without loss of generality, we may assume that $([1 : 0], [0 : 1])$ has the largest ramification index among all ramifications points of the morphism $\eta \colon C \to \mathbb{P}^1$, and the ramification index of the point $([0 : 1], [1 : 0])$ is the second largest index. If both these indices are 5, then $a_1 = a_2 = a_3 = b_1 = b_2 = b_3 = 0$, so that η does not have other ramification points, and the equation of the curve C simplifies as $ux^5 + vy^5 = 0$, so that $\mathrm{Aut}(S, C) \cong \mathbb{G}_m \rtimes \boldsymbol{\mu}_2$. In all other cases, $\mathrm{Aut}(S, C)$ is finite by [45, Corollary 2.7].

Consider the $\mathrm{Aut}(S)$-equivariant embedding $S \hookrightarrow \mathbb{P}^1 \times \mathbb{P}^2$ given by

$$([u : v], [x : y]) \mapsto ([u : v], [x^2 : xy : y^2]),$$

which gives an embedding $\mathrm{Aut}(S) \hookrightarrow \mathrm{Aut}(\mathbb{P}^1 \times \mathbb{P}^2)$. Let us identify S and C with their images in $\mathbb{P}^1 \times \mathbb{P}^2$, and let us identify $\mathrm{Aut}(S)$ with a subgroup of the group $\mathrm{Aut}(\mathbb{P}^1 \times \mathbb{P}^2)$. Then C is a smooth curve of degree $(5, 2)$ in $\mathbb{P}^1 \times \mathbb{P}^2$.

Let $\pi \colon X \to \mathbb{P}^1 \times \mathbb{P}^2$ be the blow up of the curve C. Then X is a Fano threefold №3.5, and every smooth Fano threefold in this deformation family can be obtained in this way. Since the $\mathrm{Aut}(S, C)$-action lifts to X, we identify $\mathrm{Aut}(S, C)$ with a subgroup in $\mathrm{Aut}(X)$. Arguing as in the proof of [45, Lemma 8.7], we get $\mathrm{Aut}(X) = \mathrm{Aut}(S, C)$.

If $a_1 = a_2 = a_3 = b_1 = b_2 = b_3 = 0$, the threefold X is K-polystable by Corollary 5.71. In Section 5.14, we proved that X is K-stable for a general choice of $a_1, a_2, a_3, b_1, b_2, b_3$. On the other hand, arguing as in the proof of Corollary 4.71, we obtain

Lemma 7.6 *Let $(a_1, a_2, a_3) = (0, 0, 0) \neq (b_1, b_2, b_3)$. Then X is strictly K-semistable.*

Proof Take $\lambda \in \mathbb{C}$. Let C_λ be the curve in S given by

$$ux^5 + v\left(y^5 + \lambda b_1 xy^4 + \lambda^2 b_2 x^2 y^3 + \lambda^3 b_3 x^3 y^2\right) = 0,$$

and let X_λ be the Fano threefold №3.5 obtained by blowing up $\mathbb{P}^1 \times \mathbb{P}^2$ along the curve C_λ. We know that X_0 is K-polystable. On the other hand, we have $X_\lambda \cong X$ for every $\lambda \neq 0$. This gives a test configuration for X, whose special fiber is a K-polystable Fano threefold. Then X is strictly K-semistable by Corollary 1.13.   □

If $(a_1, a_2, a_3) \neq (0, 0, 0)$, then we must have $(b_1, b_2, b_3) \neq (0, 0, 0)$ by our assumption on the ramification indices. We believe that X is always K-stable in this case. Let us restate this conjecture in a coordinate-free language.

Let R be the effective divisor on C that is the ramification divisor of the finite cover η, let $P_1 = ([1 : 0], [0 : 1])$ and $P_2 = ([0 : 1], [1 : 0])$. Then $P_1, P_2 \in \mathrm{Supp}(R)$, so that

$$R = n_1 P_1 + n_2 P_2 + \underbrace{n_3 P_3 + n_4 P_4 + n_5 P_5 + \cdots + n_k P_k}_{\text{zero} \iff a_1 = a_2 = a_3 = b_1 = b_2 = b_3 = 0}$$

for some points $P_3, \ldots, P_k$ in the curve C, and some integers $n_1, n_2, \ldots, n_k$ in $\{1, 2, 3, 4\}$. Keeping in mind our assumptions on the ramification indices, we may further assume that $n_1 \geqslant n_2 \geqslant n_3 \geqslant \cdots \geqslant n_k$. By the Riemann–Hurwitz formula, we have $n_1 + \cdots + n_k = 8$. If the curve C is general, then $k = 8$ and $n_1 = n_2 = n_3 = \cdots = n_k = 1$. Note that $n_1 = 4$ if and only if $a_1 = a_2 = a_3 = 0$. Similarly, we have $a_1 = a_2 = a_3 = b_1 = b_2 = b_3 = 0$ if and only if $R = 4(P_1 + P_2)$. If $n_1 = 4$, then the log Fano curve $(C, \frac{1}{5}R)$ is K-polystable if and only if $R = 4(P_1 + P_2)$ by [98, Corollary 1.6]. Likewise, if $n_1 \leqslant 3$, then the log Fano curve $(C, \frac{1}{5}R)$ is K-stable. Thus, we can translate our conjecture as follows:

(i) X is K-polystable $\iff$ the log Fano curve $(C, \frac{1}{5}R)$ is K-polystable;
(ii) X is K-stable $\iff$ the log Fano curve $(C, \frac{1}{5}R)$ is K-stable.

Observe that $p_1 \circ \pi \colon X \to \mathbb{P}^1$ is a fibration by del Pezzo surfaces of degree 4, and each singular fiber of this fibration is a normal del Pezzo surface that has Du Val singularities. We can also restate our conjecture as follows: X is K-stable $\iff$ the singular fibers of $p_1 \circ \pi$ have singular points of type $\mathbb{A}_1$, $\mathbb{A}_2$ or $\mathbb{A}_3$. The ($\Rightarrow$)-direction of this conjecture holds by Lemma 7.6.

7.6 Family №3.8

Let X be a smooth Fano threefold in the 3-parameter family №3.8. Then X can be described as the blow up of $\mathbb{P}^1 \times \mathbb{P}^2$ along a curve of degree $(4, 2)$. The explicit description of X is similar to that of family №3.5, so that we omit details. Let $S = \mathbb{P}^1 \times \mathbb{P}^1$, and let C be a smooth curve in S that is given by

$$u\left(x^4 + a_1 x^3 y + a_2 x^2 y^2\right) + v\left(y^4 + b_1 y^3 x + b_2 y^2 x^2\right) = 0$$

for some complex numbers a_1, a_2, b_1, b_2, where $([u : v], [x : y])$ are coordinates on S. Identify S and C with subvarieties in $\mathbb{P}^1 \times \mathbb{P}^2$ via the embedding $S \hookrightarrow \mathbb{P}^1 \times \mathbb{P}^2$ given by

$$\left([u : v], [x : y]\right) \mapsto \left([u : v], [x^2 : xy : y^2]\right).$$

Let $\pi \colon X \to \mathbb{P}^1 \times \mathbb{P}^2$ be the blow up along the curve C. Then X is a Fano threefold №3.8, and every smooth Fano threefold in this deformation family can be obtained in this way.

If $a_1 = a_2 = b_1 = b_2 = 0$, then $\mathrm{Aut}(X) \cong \mathbb{G}_m \rtimes \boldsymbol{\mu}_2$, so that X is the unique smooth threefold in the deformation family №3.8 that has an infinite automorphism group [45]. In this case, the threefold X is K-polystable by Proposition 5.74 and Remark 5.75. In other cases, the group $\mathrm{Aut}(X)$ is finite, so that X is K-polystable $\iff$ it is K-stable.

Lemma 7.7 *Let $a_1 = a_2 = 0$ and $(b_1, b_2) \neq (0, 0)$. Then X is strictly K-semistable.*

Proof See the proof of Lemma 7.6. $\square$

Let $\eta \colon C \to \mathbb{P}^1$ be the quadruple cover given by the projection $([u : v], [x : y]) \mapsto [u : v]$, and let R be its ramification divisor. Write

$$R = \sum_{i=1}^{k} n_i P_i,$$

where $P_1, P_2, \ldots, P_k$ are points in the curve C, and $n_1, n_2, \ldots, n_k$ are integers in $\{1, 2, 3\}$. Note that $n_1 + \cdots + n_k = 6$, and $\mathrm{Supp}(R)$ contains $([1 : 0], [0 : 1])$ and $([0 : 1], [1 : 0])$. If $k = 2$, then $n_1 = 3$ and $n_2 = 3$, so that $R = 3([1 : 0], [0 : 1]) + 3([0 : 1], [1 : 0])$, which means that $a_1 = a_2 = b_1 = b_2 = 0$. We know that X is K-polystable in this case. Vice versa, if $k > 2$, then $\mathrm{Aut}(X)$ is finite, so that X is K-polystable $\iff$ it is K-stable. Moreover, we expect that the Fano threefold X is K-stable if and only if ramification indices of all ramification points of η are at most 3. We can restate this as follows:

$$X \text{ is K-stable} \iff \text{each } n_i \leqslant 2 \iff \text{the log Fano curve } \left(C, \tfrac{1}{4}R\right) \text{ is K-stable.}$$

Alternatively, we can also restate this as follows: X is K-stable $\iff$ the singular fibers of $p_1 \circ \pi$ have singular points of type $\mathbb{A}_1$ or $\mathbb{A}_2$. Note that $p_1 \circ \pi \colon X \to \mathbb{P}^1$ is a fibration into del Pezzo surfaces of degree 5.

7.7 Family №3.12

Let X be a smooth Fano threefold in the 1-parameter family №3.12. Then X can be described as the blow up of $\mathbb{P}^1 \times \mathbb{P}^2$ along a curve of degree $(3, 2)$. To describe X explicitly, let $S = \mathbb{P}^1 \times \mathbb{P}^1$, and let C be a smooth curve in S of degree $(3, 1)$. In Section 7.4, we showed that we can choose coordinates $([u : v], [x : y])$ on S such that the curve C is given by one of the following three equations:

(i) $ux^3 + vy^3 = 0$,
(ii) $ux^3 + v(y^3 + y^2x) = 0$,
(iii) $u(x^3 + \lambda x^2 y) + v(y^3 + \lambda y^2 x) = 0$, where $\lambda \in \mathbb{C}^*$ such that $\lambda \neq \pm 1$ and $\lambda \neq \pm 3$.

As in Sections 7.5 and 7.6, we identify S and C with subvarieties in $\mathbb{P}^1 \times \mathbb{P}^2$ using the embedding $S \hookrightarrow \mathbb{P}^1 \times \mathbb{P}^2$ given by

$$([u : v], [x : y]) \mapsto ([u : v], [x^2 : xy : y^2]).$$

Let $\pi \colon X \to \mathbb{P}^1 \times \mathbb{P}^2$ be the blow up of the curve C. Then X is a Fano threefold №3.12. Moreover, every smooth Fano threefold in this family can be obtained in this way.

If C is given by $ux^3 + vy^3 = 0$, then $\mathrm{Aut}(X) \cong \mathbb{G}_m \rtimes \mu_2$, so that X is the unique smooth threefold in the deformation family №3.12 that has an infinite automorphism group [45]. In this case, the threefold X is K-polystable by Proposition 5.85.

If C is given by $ux^3 + v(y^3 + y^2x) = 0$, then, arguing as in the proof of Lemma 7.6, we see that X is strictly K-semistable, so that, in particular, X is not K-polystable.

In the remaining case, we believe that the threefold X is K-stable for all $\lambda \notin \{0, \pm 1, \pm 3\}$. In this case, the fibration $p_1 \circ \pi \colon X \to \mathbb{P}^1$ has exactly four singular fibers, and each of them has one singular point, which is an ordinary node.

We can describe X as a blow up of $\mathbb{P}^3$ along the line $x_0 = x_3 = 0$ and the twisted cubic

$$\left\{ x_0 x_2 - x_1^2 + a x_1 x_3 = 0, x_1 x_3 - x_2^2 - a x_3^2 + b x_0 x_2 = 0, x_0 x_3 - x_1 x_2 + b x_0 x_1 = 0 \right\},$$

where a and b are some complex numbers. If $a = 0$ and $b = 0$, then X is the K-polystable threefold with $\mathrm{Aut}(X) \cong \mathbb{G}_m \rtimes \mu_2$. Vice versa if $a = 0$ or $b = 0$ (but not both), then we can scale the coordinates appropriately and assume that $b = 1$ or $a = 1$, respectively. Up to isomorphism, this gives us one special smooth Fano threefold №3.12. This threefold is our strictly K-semistable smooth Fano threefold №3.12 described above. Our conjecture says that all other smooth Fano threefolds in this family are K-stable.

Appendix

Technical Results Used in Proof of Main Theorem

A.1 Nadel's vanishing and Kollár–Shokurov connectedness

In this short section, we present one important result, known as Nadel's vanishing, and some of its corollaries. To state it, we recall basic facts about singularities of pairs following [62, 129, 130, 132].

Let X be a normal variety such that K_X is a $\mathbb{Q}$-Cartier divisor, let $\pi\colon \widehat{X} \to X$ be its resolution of singularities. Denote the π-exceptional divisors by $E_1,\ldots,E_m$. Then

$$K_{\widehat{X}} + \sum_{i=1}^{m} e_i E_i \sim_{\mathbb{Q}} \pi^*(K_X) \qquad (A.1)$$

for some rational numbers $e_1,\ldots,e_m$. For each $i \in \{1,\ldots,m\}$, we let $A_X(E_i) = 1 - e_i$ and say that $A_X(E_i)$ is the log discrepancy of the divisor E_i. We say that

- X has terminal singularities if each $e_i < 0$,
- X has canonical singularities if each $e_i \leqslant 0$,
- X has Kawamata log terminal singularities if each $e_i < 1$,
- X has log canonical singularities if each $e_i \leqslant 1$.

One can show that these definitions do not depend on the choice of the morphism π.

If X is smooth, then its singularities are terminal. Moreover, if X is a surface, then X is smooth if and only if it has terminal singularities. Similarly, if X is a surface, then it has canonical singularities if and only if X has Du Val singularities. Likewise, if X is a surface, then it follows from [130, Theorem 3.6] that X has Kawamata log terminal singularities if and only if X has quotient singularities. In all dimensions, Kawamata log terminal singularities are rational by [130, Theorem 11.1]. Starting from now, we assume that the variety X has Kawamata log terminal singularities.

Let B_X be an effective $\mathbb{Q}$-divisor on X. Then

$$B_X = \sum_{i=1}^{r} a_i B_i, \tag{A.2}$$

where each B_i is a prime Weil divisor on X, and each a_i is a non-negative rational number. We say that (X, B_X) is a log pair, B_X is its boundary, and $K_X + B_X$ is its log canonical divisor. Let us define singularity classes for the log pair (X, B_X) following [130, 132].

Let $\widehat{B}_1, \ldots, \widehat{B}_r$ be the proper transforms on $\widehat{X}$ of the divisors $B_1, \ldots, B_r$, respectively. Let us also replace (if necessary) the resolution of singularities $\pi \colon \widehat{X} \to X$ by a slightly better one such that the divisor

$$\sum_{i=1}^{r} \widehat{B}_i + \sum_{i=1}^{m} E_i$$

has simple normal crossing singularities. Such resolution of singularities exists [115, 131], and it is often called a log resolution of the log pair (X, B_X). Suppose, in addition, that the divisor B_X is a $\mathbb{Q}$-Cartier divisor. Then there are rational numbers $d_1, \ldots, d_m$ such that

$$K_{\widehat{X}} + \sum_{i=1}^{r} a_i \widehat{B}_i + \sum_{i=1}^{m} d_i E_i \sim_{\mathbb{Q}} \pi^*\big(K_X + B_X\big). \tag{A.3}$$

Using this, we define the log pull back of the pair (X, B_X) as follows:

$$\bigg(\widehat{X}, \sum_{i=1}^{r} a_i \widehat{B}_i + \sum_{i=1}^{m} d_i E_i\bigg).$$

This new log pair is often denoted as $(\widehat{X}, B^{\widehat{X}})$. We say that

- (X, B_X) has Kawamata log terminal singularities if each $a_i < 1$ and each $d_j < 1$,
- (X, B_X) has log canonical singularities if each $a_i \leqslant 1$ and each $d_j \leqslant 1$.

Both these definitions do not depend on the choice of the log resolution $\pi \colon \widehat{X} \to X$. Moreover, it is easy to check (using the definition) that (X, B_X) has log canonical singularities if and only if $(\widehat{X}, B^{\widehat{X}})$ has log canonical singularities. Note that $B^{\widehat{X}}$ is not always effective. Nevertheless, our definition still works in this case. Similarly, one can show that the log pair (X, B_X) has Kawamata log terminal singularities if and only if the log pair $(\widehat{X}, B^{\widehat{X}})$ has Kawamata log terminal singularities.

Let P be a point in X. Then we can localize our definitions of singularities at this point. Namely, we say that the pair (X, B_X) has log canonical singularities at P if the following two conditions are satisfied:

- for every $\widehat{B}_i$ in (A.3) such that $P \in B_i$, $a_i \leqslant 1$,
- for every E_i in (A.3) such that $P \in \pi(E_i)$, $d_i \leqslant 1$.

Likewise, we say that the log pair (X, B_X) has Kawamata log terminal singularities at the point P if the following two conditions are satisfied:

- for every $\widehat{B}_i$ in (A.3) such that $P \in B_i$, $a_i < 1$,
- for every E_i in (A.3) such that $P \in \pi(E_i)$, $d_i < 1$.

Lemma A.1   *Suppose that X is smooth at P. Then the following assertions hold:*

(i) *if $\mathrm{mult}_P(B_X) \leqslant 1$, then (X, B_X) is log canonical at P;*
(ii) *if $\mathrm{mult}_P(B_X) < 1$, then (X, B_X) is Kawamata log terminal at P;*
(iii) *if $\mathrm{mult}_P(B_X) > \dim(X)$, then (X, B_X) is not log canonical at P;*
(iv) *if $\mathrm{mult}_P(B_X) \geqslant \dim(X)$, then (X, B_X) is not Kawamata log terminal at P.*

Proof   This is [130, Lemma 8.10] and [62, Exercise 6.18].   □

Example A.2   Suppose that $X = \mathbb{P}^2$. Let ℓ be a line in X. Then $-K_X \sim 3\ell$, so that $\alpha(X) \leqslant \frac{1}{3}$. If $\alpha(X) < \frac{1}{3}$, there is an effective divisor B_X on the surface X such that the log pair (X, B_X) is not log canonical at a point $P \in X$, and $B_X \sim_{\mathbb{Q}} -\lambda K_X$ for some positive rational number $\lambda < \frac{1}{3}$. Now, choosing ℓ to be a general line containing P, we get $1 > 3\lambda = B_X \cdot \ell \geqslant \mathrm{mult}_P(B_X) > 1$ by Lemma A.1. This shows that $\alpha(X) = \frac{1}{3}$.

To measure how far is the log pair (X, B_X) from being log canonical, we can use the following number, which is called the log canonical threshold:

$$\mathrm{lct}(X, B_X) = \sup\{\lambda \in \mathbb{Q}_{>0} \mid (X, \lambda B_X) \text{ has log canonical singularities}\}.$$

We can localize it at the point $P \in X$ as follows:

$$\mathrm{lct}_P(X, B_X) = \sup\{\lambda \in \mathbb{Q}_{>0} \mid (X, \lambda B_X) \text{ has log canonical singularities at } P\}.$$

Similarly, if Z is an irreducible subvariety of the variety X, we let

$$\mathrm{lct}_Z(X, B_X) = \sup\{\lambda \in \mathbb{Q}_{>0} \mid (X, \lambda B_X) \text{ is log canonical at every point in } Z\}.$$

Now let us denote by $\mathrm{Nklt}(X, B_X)$ the subset in X consisting of all points where the singularities of the pair (X, B_X) are not Kawamata log terminal. To be precise, let

$$\mathrm{Nklt}(X, B_X) = \left(\bigcup_{a_i \geqslant 1} B_i\right) \bigcup \left(\bigcup_{d_i \geqslant 1} \pi(E_i)\right) \subsetneq X.$$

This locus has been introduced in [197, Definition 3.14] as the locus of log canonical singularities of the log pair (X, B_X). Because of this, it is often denoted by $\mathrm{LCS}(X, B_X)$. Observe that $\mathrm{Nklt}(X, B_X) = \varnothing \iff (X, B_X)$ has Kawamata log terminal singularities. The locus $\mathrm{Nklt}(X, B_X)$ can be equipped with a subscheme structure as follows: let

$$\mathcal{I}\big(X, B_X\big) = \pi_*\bigg(O_{\widehat{X}}\Big(-\sum_{i=1}^{m}\lfloor d_i\rfloor E_i - \sum_{i=1}^{r}\lfloor a_i\rfloor B_i\Big)\bigg).$$

Since the $\mathbb{Q}$-divisor B_X is assumed to be effective, $\mathcal{I}(X, B_X)$ is an ideal sheaf [137, §9.2], which is commonly known as the multiplier ideal sheaf of the log pair (X, B_X).

Since $\mathcal{I}(X, B_X)$ is an ideal sheaf, it defines some subscheme of the variety X, which we denote by $\mathcal{L}(X, B_X)$. The subscheme $\mathcal{L}(X, B_X)$ is usually called the log canonical singularities subscheme of the log pair (X, B_X). Note that $\mathrm{Supp}(\mathcal{L}(X, B_X)) = \mathrm{Nklt}(X, B_X)$. If (X, B_X) has log canonical singularities, then $\mathcal{L}(X, B_X)$ is reduced (possibly empty).

Theorem A.3 ([137, Theorem 9.4.8]) *Let D be an arbitrary Cartier divisor on X, and let H be some nef and big $\mathbb{Q}$-divisor on the variety X. Suppose that $D \sim_{\mathbb{Q}} K_X + B_X + H$. Then $H^i(O_X(D) \otimes \mathcal{I}(X, B_X)) = 0$ for every $i \geqslant 1$.*

Theorem A.3, known as Nadel's vanishing theorem or simply Nadel's vanishing [165], implies the following result, which is known as the Kollár–Shokurov connectedness theorem or simply Kollár–Shokurov connectedness [197, 129].

Corollary A.4 *If $-(K_X + B_X)$ is big and nef, then $\mathrm{Nklt}(X, B_X)$ is connected.*

Proof See the proof of Corollary A.6 below. □

This result is [197, Connectedness Lemma], [130, Theorem 17.4], [132, Corollary 5.49].

Example A.5 Suppose $X = \mathbb{P}^1 \times \mathbb{P}^1$. Then $\alpha(X) \leqslant \frac{1}{2}$ since $-K_X \sim 2\ell_1 + 2\ell_2$, where ℓ_1 and ℓ_2 are curves in X of degree $(1, 0)$ and $(0, 1)$, respectively. If $\alpha(X) < \frac{1}{2}$, there exists an effective divisor B_X on the surface X such that the pair (X, B_X) is not log canonical at a point $P \in X$, and $B_X \sim_{\mathbb{Q}} -\lambda K_X$ for some positive rational number $\lambda < \frac{1}{2}$. In this case, intersecting B_X with ℓ_1 and ℓ_2, we see that the locus $\mathrm{Nklt}(X, B_X)$ is zero-dimensional, so that $\mathrm{Nklt}(X, B_X) = P$ by Corollary A.4, which implies that $\mathrm{Nklt}(X, \ell_1 + B_X) = \ell_1 \cup P$. But $\mathrm{Nklt}(X, \ell_1 + B_X)$ is connected by Corollary A.4. Thus, choosing ℓ_1 not passing through the point P, we obtain a contradiction. This shows that $\alpha(X) = \frac{1}{2}$.

Let us present more corollaries of Theorem A.3.

Corollary A.6   *Let us use the assumptions and notations introduced in The-orem A.3. Let Σ be the union of zero-dimensional irreducible components of the locus $\mathrm{Nklt}(X, B_X)$. Then Σ contains at most $h^0(O_X(D))$ points of the variety X.*

Proof   Let $\mathcal{L} = \mathcal{L}(X, B_X)$. Using the exact sequence of sheaves

$$0 \longrightarrow O_X(D) \otimes \mathcal{I}(X, B_X) \longrightarrow O_X(D) \longrightarrow O_{\mathcal{L}} \otimes O_X(D) \longrightarrow 0,$$

and applying Theorem A.3, we obtain the surjection

$$H^0\big(O_X(D)\big) \twoheadrightarrow H^0\big(O_{\mathcal{L}} \otimes O_X(D)\big),$$

which gives $|\Sigma| \leqslant h^0(O_{\mathcal{L}} \otimes O_X(D)) \leqslant h^0(O_X(D))$ as required.   □

Corollary A.7   *Let us use the assumptions and notations introduced in Theo-rem A.3. If the locus $\mathrm{Nklt}(X, B_X)$ is a finite set, then $|\mathrm{Nklt}(X, B_X)| \leqslant h^0(O_X(D))$.*

Corollary A.8   *Let $\mathcal{M}$ be a non-empty linear system on X that is basepoint free, let M be a general divisor in $\mathcal{M}$, let Z be a union of some one-dimensional irreducible components of the locus $\mathrm{Nklt}(X, B_X)$, and let D_M be a Cartier divisor on M such that*

$$D_M \sim_{\mathbb{Q}} K_M + B_X\big|_M + H_M$$

for some nef and big $\mathbb{Q}$-divisor H_M on the variety M. Then

$$M \cdot Z \leqslant h^0(M, O_M(D_M)).$$

Proof   First, we observe that M is normal and has Kawamata log terminal singularities. But $(M, B_X|_M)$ is not Kawamata log terminal at every point of the intersection $Z \cap M$. Moreover, these points are isolated components of the locus $\mathrm{Nklt}(M, B_X|_M)$, so that it follows from Corollary A.6 that $M \cdot Z = |M \cap Z| \leqslant h^0(M, O_M(D_M))$.   □

Corollary A.9   *Let $\mathcal{M}$ be a non-empty basepoint free linear system on the variety X, let M be a general divisor in $\mathcal{M}$, let Z be a union of some one-dimensional irreducible components of $\mathrm{Nklt}(X, B_X)$. Suppose that $-K_X$ is nef and big, and $B_X \sim_{\mathbb{Q}} -\lambda K_X$ for some rational number $\lambda < 1$. Then $M \cdot Z \leqslant h^0(M, O_M(\mathcal{M}|_M))$.*

Proof   Apply Corollary A.8 with $H_M = -(1 - \lambda)K_X|_M$.   □

Corollary A.10   *Suppose $X = \mathbb{P}^3$ and $B_X \sim_{\mathbb{Q}} -\lambda K_X$ for some rational number $\lambda < \frac{3}{4}$. Let Z be the union of one-dimensional components of $\mathrm{Nklt}(X, B_X)$. Then $O_{\mathbb{P}^3}(1) \cdot Z \leqslant 1$.*

Proof   Apply Corollary A.8 with $\mathcal{M} = |O_{\mathbb{P}^3}(1)|$ and $D_M = O_M$.   □

Corollary A.11 *Suppose X is a smooth Fano threefold such that $-K_X \sim 2H$ for some ample Cartier divisor H on it such that $H^3 \geqslant 2$, and $B_X \sim_{\mathbb{Q}} -\lambda K_X$ for a positive rational number $\lambda < 1$. Let Z be the union of all one-dimensional components of* $\mathrm{Nklt}(X, B_X)$. *Then $H \cdot Z \leqslant H^3 + 1$.*

Proof Note that $|H|$ is basepoint free, H is a smooth del Pezzo surface, $-K_H \sim H|_H$ and

$$h^0\big(H, O_H(-K_H)\big) = K_H^2 + 1 = H^3 + 1.$$

Thus, we can apply Corollary A.9 with $M = |H|$. □

Corollary A.12 *Suppose that $-K_X$ is nef and big, $B_X \sim_{\mathbb{Q}} -\lambda K_X$ for some rational number $\lambda < 1$, and there exists a surjective morphism with connected fibers $\phi \colon X \to \mathbb{P}^1$. Set $H = \phi^*(O_{\mathbb{P}^1}(1))$. Let Z be the union of one-dimensional components of* $\mathrm{Nklt}(X, \lambda B_X)$. *Then $H \cdot Z \leqslant 1$.*

Proof Apply Corollary A.8 with $M = |H|$ and $D_M = O_M$. □

Corollary A.13 *Suppose that $-K_X$ is nef and big, $B_X \sim_{\mathbb{Q}} -\lambda K_X$ for some rational number $\lambda < 1$, and there exists a surjective morphism with connected fibers $\phi \colon X \to \mathbb{P}^2$. Set $H = \phi^*(O_{\mathbb{P}^2}(1))$. Let Z be the union of one-dimensional components of* $\mathrm{Nklt}(X, \lambda B_X)$. *Then $H \cdot Z \leqslant 2$.*

Proof Apply Corollary A.8 with $M = |H|$ and $D_M = M|_M$. □

Let us conclude this section by the following known application of Theorem A.3.

Corollary A.14 *Suppose that $-(K_X + B_X)$ is nef and big, and* $\mathrm{dimNklt}(X, B_X) = 1$; *then* $\mathrm{Nklt}(X, B_X)$ *has the following properties:*

(o) *the locus* $\mathrm{Nklt}(X, B_X)$ *is connected,*

(i) *each irreducible component is isomorphic to $\mathbb{P}^1$,*

(ii) *any two intersecting irreducible components intersect transversally at one point,*

(iii) *no three irreducible components intersect at one point,*

(iv) *no irreducible components form a cycle.*

Proof Note that assertion (o) follows from Corollary A.4, and all other assertions follow from [165, Theorem 4.1]. For the convenience of the reader, let us prove assertion (i), which also follows from [92, Theorem 6.3.5].

Let C be an irreducible component of the locus $\mathrm{Nklt}(X, B_X)$, let I_C be its ideal sheaf, let $\mathcal{J} = \mathcal{I}(X, B_X)$, and let $\mathcal{L} = \mathcal{L}(X, B_X)$. Then $\mathcal{J} \subseteq I_C$, while

$\mathcal{L}$ is one-dimensional. But $h^1(X, \mathcal{J}) = 0$ and $h^2(X, \mathcal{J}) = 0$ by Theorem A.3. Hence, using the exact sequence

$$0 \longrightarrow \mathcal{J} \longrightarrow O_X \longrightarrow O_{\mathcal{L}} \longrightarrow 0,$$

we get the following exact sequence of cohomology groups:

$$0 = H^1(O_X) \longrightarrow H^1(O_{\mathcal{L}}) \longrightarrow H^2(\mathcal{J}) = 0,$$

which gives $h^1(O_{\mathcal{L}}) = 0$. Now, looking at the exact sequence of sheaves

$$0 \longrightarrow \mathcal{I}_C/\mathcal{J} \longrightarrow O_{\mathcal{L}} \longrightarrow O_C \longrightarrow 0$$

on the subscheme $\mathcal{L}$, we get the following exact sequence of cohomology groups:

$$0 = H^1(O_{\mathcal{L}}) \longrightarrow H^1(O_C) \longrightarrow H^2(\mathcal{I}_C/\mathcal{J}),$$

where $h^2(\mathcal{I}_C/\mathcal{J}) = 0$ because $\mathcal{L}$ is one-dimensional. Thus, we see that $h^1(O_C) = 0$, which implies that C is a smooth rational curve (see [165, Section 4] for details). $\qquad\square$

A.2 Inversion of adjunction and Kawamata's subadjunction

Let X be a normal projective variety that has Kawamata log terminal singularities, and let B_X be an effective $\mathbb{Q}$-divisor on the variety X that is given by (A.2). The following result is commonly known as inversion of adjunction.

Theorem A.15 ([132, Theorem 5.50]) *Suppose that $a_1 = 1$, B_1 is a Cartier divisor, and B_1 has Kawamata log terminal singularities. The following assertions are equivalent:*

- *(X, B_X) is log canonical at every point of the divisor B_1;*
- *the singularities of the log pair $(B_1, \sum_{i=2}^r a_i B_i|_{B_1})$ are log canonical.*

Corollary A.16 *Suppose that X is a surface, (X, B_X) is not log canonical at some point $P \in B_1$, and the curve B_1 is smooth at this point. If $a_1 \leqslant 1$, then*

$$\left(\sum_{i=2}^r a_i B_i\right) \cdot B_1 \geqslant \left(\left(\sum_{i=2}^r a_i B_i\right) \cdot B_1\right)_P > 1.$$

Note that Corollary A.16 can be proved without using the more powerful Theorem A.15. Instead, one can use basics of intersection multiplicities (see the proof of [32, Theorem 7]).

Example A.17 (cf. Example A.5) Suppose $X = \mathbb{P}^1 \times \mathbb{P}^1$. If $\alpha(X) < \frac{1}{2}$, there exists an effective divisor B_X on the surface X such that the pair (X, B_X) is not log canonical at a point $P \in X$, and $B_X \sim_{\mathbb{Q}} -\lambda K_X$ for some rational number $\lambda < \frac{1}{2}$. Write $B_X = a\ell + \Delta$, where a is a non-negative rational number, ℓ is the curve in X of degree $(1, 0)$ that passes through the point P, and Δ is an effective $\mathbb{Q}$-divisor whose support does not contain ℓ. Since B_X is a $\mathbb{Q}$-divisor of degree $(2\lambda, 2\lambda)$ and $\lambda < \frac{1}{2}$, we see that $a < 1$, so that

$$1 > 2\lambda = B_X \cdot \ell = \Delta \cdot \ell \geqslant (\Delta \cdot \ell)_P > 1$$

by Corollary A.16, so that $\alpha(X) \geqslant \frac{1}{2}$.

Let Z be a proper irreducible subvariety of the variety X. Following [125, Definition 1.3], we say that Z is a center of log canonical singularities or a log canonical center of the log pair (X, B_X) if one of the following conditions is satisfied:

- $Z = B_i$ for $\widehat{B}_i$ in (A.3) such that $a_i \geqslant 1$,
- $Z = \pi(E_i)$ for some E_i in (A.3) such that $d_i \geqslant 1$

for *some choice* of the log resolution $\pi \colon \widehat{X} \to X$. If Z is a log canonical center of the log pair (X, B_X), then $Z \subseteq \mathrm{Nklt}(X, B_X)$. Using Lemma A.1, we get

Corollary A.18 *Suppose that X is non-singular at a general point of the subvariety Z. If Z is a center of log canonical singularities of the log pair (X, B_X), then $\mathrm{mult}_Z(B_X) \geqslant 1$.*

From now on and until the end of this section, we assume, additionally, that

($\bigstar$) the pair (X, B_X) has log canonical singularities in every point of the subvariety Z.

We need this additional assumption because centers of log canonical singularities behave much better under it. It can be illustrated by the following result:

Lemma A.19 ([125, Proposition 1.5]) *Let Z' be a proper irreducible subvariety in X. Suppose that Z and Z' are centers of log canonical singularities of the log pair (X, B_X). Then every irreducible component of the intersection $Z \cap Z'$ is a center of log canonical singularities of the log pair (X, B_X).*

If Z is a log canonical center of the log pair (X, B_X), we say that it is a minimal log canonical center if Z does not contain a proper irreducible subvariety that is also a center of log canonical singularities of the log pair (X, B_X).

Theorem A.20 ([126, Theorem 1]) *Suppose that Z is a minimal center of log canonical singularities of the log pair (X, B_X). Then Z is normal and has rational singularities. Let H be an ample $\mathbb{Q}$-Cartier $\mathbb{Q}$-divisor on X. Then $(K_X + B_X + H)|_Z \sim_\mathbb{Q} K_Z + B_Z$ for an effective $\mathbb{Q}$-divisor B_Z on Z such that (Z, B_Z) has Kawamata log terminal singularities.*

This result is Kawamata's subadjunction theorem or Kawamata's subadjunction.

Corollary A.21 *Suppose that $-K_X$ is ample, $B_X \sim_\mathbb{Q} \lambda(-K_X)$ for a rational number λ, and that Z is a curve that is a minimal log canonical center of (X, B_X). Then Z is smooth. Moreover, if $\lambda < 1$, then $-K_X \cdot Z \leqslant \frac{2}{1-\lambda}$ and Z is rational. If $\lambda > 1$, then $-K_X \cdot Z \geqslant \frac{2g-2}{\lambda-1}$, where g is the genus of the curve Z.*

Proof By Theorem A.20, the curve Z is smooth. Let g be its genus. Choose a small rational number $\epsilon > 0$. Set $H = \epsilon(-K_X)$. Then $(\lambda - 1 + \epsilon)(-K_X \cdot Z) = (K_X + B_X + H) \cdot Z \geqslant 2g - 2$ by Theorem A.20. Since ϵ can be arbitrary small, we get $(\lambda - 1)(-K_X \cdot Z) \geqslant 2g - 2$, which implies all required assertions. □

A.3 Mobile log pairs and Corti's inequality

Let us use the assumptions and notations introduced in Section A.1. Recall from A.1 there that X is a normal projective variety with Kawamata log terminal singularities, and B_X is an effective $\mathbb{Q}$-divisor on X. In this book, we occasionally consider log pairs like $(X, \lambda \mathcal{M})$, where $\mathcal{M}$ is a non-empty linear system on X, and λ is a non-negative rational number. For instance, we will use the following result, known as Corti's inequality, in the proof of Theorem 5.23.

Theorem A.22 ([61, Theorem 3.1]) *Let Z be an irreducible subvariety in X such that the variety X is non-singular at its general point, let $\mathcal{M}$ be a mobile linear system on X, and let λ be a positive rational number. If the log pair $(X, \lambda \mathcal{M})$ is not log canonical at a general point of the subvariety Z, then*

$$\mathrm{mult}_Z\left(M \cdot M'\right) \geqslant \frac{4}{\lambda^2}$$

for two general divisors M and M' in the linear system $\mathcal{M}$.

More generally, we consider log pairs $(X, B_X + \mathcal{M}_X)$ with $\mathcal{M}_X$ defined as

$$\mathcal{M}_X = \sum_{i=1}^{s} c_i \mathcal{M}_i, \tag{A.4}$$

where each $\mathcal{M}_i$ is a non-empty mobile linear system on X, i.e. it has no fixed components, and each c_i is a non-negative rational number. For the log pair $(X, B_X + \mathcal{M}_X)$, we say that B_X is the fixed part of its boundary, and $\mathcal{M}_X$ is the mobile part of its boundary.

We can work with the log pair $(X, B_X + \mathcal{M}_X)$ in the same way as with a usual log pair. In fact, replacing each linear system $\mathcal{M}_i$ in (A.4) with its general member, we can handle the mobile part $\mathcal{M}_X$ as a $\mathbb{Q}$-divisor. If $B_X = 0$, then $(X, \mathcal{M}_X)$ is said to be a mobile log pair. Mobile log pairs naturally appear in many problems, see [4, §1.8] and [53, §2.2].

Suppose that the following condition is satisfied: both B_X and $\mathcal{M}_X$ are $\mathbb{Q}$-Cartier. Then we can replace (A.3) by

$$K_{\widehat{X}} + \sum_{i=1}^{r} a_i \widehat{B}_i + \sum_{i=1}^{s} c_i \widehat{\mathcal{M}}_i + \sum_{i=1}^{m} d_i E_i \sim_{\mathbb{Q}} \pi^* \left(K_X + B_X + \mathcal{M}_X \right), \qquad \text{(A.5)}$$

where each $\widehat{\mathcal{M}}_i$ is a proper transform on $\widehat{X}$ of the mobile linear system $\mathcal{M}_i$, and the log resolution $\pi \colon \widehat{X} \to X$ is chosen in such way that each linear system $\widehat{\mathcal{M}}_i$ is basepoint free. Now, following [130, Definition 4.6], we say that the pair $(X, B_X + \mathcal{M}_X)$ is log canonical at the point $P \in X$ if the following two conditions are satisfied:

- $a_i \leqslant 1$ in (A.4) for every $\widehat{B}_i$ such that $P \in B_i$,
- $d_i \leqslant 1$ in (A.4) for every E_i such that $P \in \pi(E_i)$.

Similarly, we say that $(X, B_X + \mathcal{M}_X)$ is Kawamata log terminal at P if the following two conditions are satisfied:

- $a_i < 1$ in (A.4) for every $\widehat{B}_i$ such that $P \in B_i$,
- $d_i < 1$ in (A.4) for every E_i such that $P \in \pi(E_i)$.

These are the same definitions we gave in Section A.1 for (X, B_X) since we do not impose any constraints on the coefficients $c_1, \ldots, c_s$ of the mobile part of the boundary.

Remark A.23   It follows from [130, Theorem 4.8] that the pair $(X, B_X + \mathcal{M}_X)$ has log canonical (Kawamata log terminal, respectively) singularities if and only if the log pair

$$\left(X, B_X + \sum_{i=1}^{s} \sum_{j=1}^{N} \frac{c_i}{N} M_i^j \right)$$

has log canonical (Kawamata log terminal, respectively) singularities for some $N \gg 0$, where each M_i^j is a general divisor in the linear system $\mathcal{M}_i$.

For mobile pairs, we can also define canonical singularities and terminal singularities as it is done in [130, Definition 3.5]. Namely, we say that the log pair $(X, \mathcal{M}_X)$ is canonical (terminal, respectively) at the point P if the following condition is satisfied:

- for every E_i in (A.3) such that $P \in \pi(E_i)$, $d_i \leqslant 0$ ($d_i < 0$, respectively).

Of course, these definitions also make sense for non-mobile pairs, but they behave better for mobile pairs. In this book, we only consider them for mobile log pairs (occasionally).

The following result, known as the Noether–Fano inequality, is used in Example 1.94, and also in the proof of Theorem 5.23.

Theorem A.24 *Suppose that X is a Fano variety with at most terminal singularities, there exists a reductive subgroup $G \subseteq \mathrm{Aut}(X)$ such that $\mathrm{rk}\,\mathrm{Cl}^G(X) = 1$, and for every G-invariant mobile linear system $\mathcal{M}$ on the variety X, the log pair $(X, \lambda \mathcal{M})$ has canonical singularities for $\lambda \in \mathbb{Q}_{>0}$ defined via $\lambda \mathcal{M} \sim_{\mathbb{Q}} -K_X$. Then X is G-birationally super-rigid, i.e. the following two conditions are satisfied:*

(i) *there is no G-equivariant dominant rational map $X \dashrightarrow Y$ such that general fibers of the map $X \dashrightarrow Y$ are rationally connected, and $0 < \dim(Y) < \dim(X)$,*

(ii) *there is no G-equivariant birational non-biregular map $X \dashrightarrow X'$ such that X' is a Fano variety with at most terminal singularities, and $\mathrm{rk}\,\mathrm{Cl}^G(X') = 1$.*

Proof This is well-known. See, for example, [53, Chapter 3.1.1], where this assertion has been proved in the case when G is a finite group. $\square$

Arguing as in Section A.1, we can define the locus $\mathrm{Nklt}(X, B_X + \mathcal{M}_X)$, the multiplier ideal sheaf $\mathcal{I}(X, B_X + \mathcal{M}_X)$ and the log canonical singularities subscheme $\mathcal{L}(X, B_X + \mathcal{M}_X)$. Likewise, we can generalize other notions and results presented in Appendices A.1 and A.2 for log pairs whose boundaries have non-empty mobile parts.

A.4 Equivariant tie breaking and convexity trick

Let X be a projective variety with Kawamata log terminal singularities, and let G be a reductive subgroup in $\mathrm{Aut}(X)$.

Lemma A.25 ([91, Lemma 2.7]) *Let P be a point in X that is fixed by the group G. Then the induced linear G-action on the Zariski tangent space $T_P(X)$ is faithful.*

Corollary A.26 *If X is a curve, and G fixes a smooth point in X, then G is cyclic.*

Let B_X be an effective $\mathbb{Q}$-divisor on X that is given by (A.2), let $\mathcal{M}_X$ be a mobile boundary on X that is given by (A.4), let Z be a proper irreducible subvariety in X. Suppose that both B_X and $\mathcal{M}_X$ are $\mathbb{Q}$-Cartier, and that both B_X and $\mathcal{M}_X$ are G-invariant. The latter condition means that for any $g \in G$, any B_i in (A.2), and any $\mathcal{M}_i$ in (A.4), there are B_j in (A.2) and $\mathcal{M}_k$ in (A.4) such that $g(B_i) = B_j$ and $g(\mathcal{M}_i) = \mathcal{M}_k$.

Lemma A.27 *Suppose that* $\dim(Z) = \dim(X) - 2$, *the variety X is smooth along Z, and the subvariety Z is smooth and G-invariant. Let* $\eta\colon \widetilde{X} \to X$ *be the blow up of Z, let F be the η-exceptional divisor, let $B_{\widetilde{X}}$ and $\mathcal{M}_{\widetilde{X}}$ be the proper transforms on $\widetilde{X}$ of B_X and $\mathcal{M}_X$, respectively. Suppose that* $Z \subseteq \mathrm{Nklt}(X, B_X + \mathcal{M}_X)$, *but* $\mathrm{mult}_Z(B_X) + \mathrm{mult}_Z(\mathcal{M}_X) < 2$. *Then the G-action lifts to $\widetilde{X}$, and F contains a unique G-invariant irreducible proper subvariety $\widetilde{Z}$ such that the induced morphism $\eta|_{\widetilde{Z}}\colon \widetilde{Z} \to Z$ is birational, and the log pair*

$$\left(\widetilde{X}, B_{\widetilde{X}} + \mathcal{M}_{\widetilde{X}} + \left(\mathrm{mult}_Z(B_X) + \mathrm{mult}_Z(\mathcal{M}_X) - 1\right)F\right) \tag{A.6}$$

is not Kawamata log terminal along $\widetilde{Z}$. Moreover,

$$\mathrm{mult}_Z\left(B_X\right) + \mathrm{mult}_Z\left(\mathcal{M}_X\right) + \mathrm{mult}_{\widetilde{Z}}\left(B_{\widetilde{X}}\right) + \mathrm{mult}_{\widetilde{Z}}\left(\mathcal{M}_{\widetilde{X}}\right) \geqslant 2. \tag{A.7}$$

Proof The required assertion follows from [40, Remark 2.5]. Namely, we have

$$K_{\widetilde{X}} + B_{\widetilde{X}} + \mathcal{M}_{\widetilde{X}} + \left(\mathrm{mult}_Z\left(B_X\right) + \mathrm{mult}_Z\left(\mathcal{M}_X\right) - 1\right)$$
$$F \sim_{\mathbb{Q}} \eta^*\left(K_X + B_X + \mathcal{M}_X\right),$$

which implies that the log pair (A.6) is the log pull back of the log pair $(X, B_X + \mathcal{M}_X)$. Thus, since $\mathrm{mult}_Z(B_X) + \mathrm{mult}_Z(\mathcal{M}_X) < 2$, the divisor F contains a proper G-invariant G-irreducible subvariety $\widetilde{Z}$ such that the induced morphism $\eta|_{\widetilde{Z}}\colon \widetilde{Z} \to Z$ is surjective, and the log pair (A.6) is not Kawamata log terminal along $\widetilde{Z}$.

Since $\mathrm{mult}_Z(B_X) + \mathrm{mult}_Z(\mathcal{M}_X) < 2$, the log pair $(\widetilde{X}, B_{\widetilde{X}} + \mathcal{M}_{\widetilde{X}} + F)$ is not log canonical along $\widetilde{Z}$. Now, applying Theorem A.15, we see that $(F, B_{\widetilde{X}}|_F + \mathcal{M}_{\widetilde{X}}|_F)$ is not log canonical along the subvariety $\widetilde{Z}$ either. Since $\widetilde{Z}$ is a divisor in F, we have $\operatorname{ord}_{\tilde{Z}}(B_{\tilde{X}}|_F + M_{\tilde{X}}|_F) > 1$. Let ℓ be a sufficiently general fiber of the natural projection $F \to Z$. Then

$$2 > \operatorname{mult}_Z(B_X) + \operatorname{mult}_Z(M_X) = \left(B_{\tilde{X}}\big|_F + M_{\tilde{X}}\big|_F\right) \cdot \ell$$

$$\geqslant \operatorname{ord}_{\tilde{Z}}\left(B_{\tilde{X}}\big|_F + M_{\tilde{X}}\big|_F\right)|\ell \cap \tilde{Z}| > |\ell \cap \tilde{Z}|$$

by Lemma A.1. Then $|\ell \cap \tilde{Z}| = 1$, and the induced morphism $\eta|_{\tilde{Z}} \colon \tilde{Z} \to Z$ is birational.

Applying Lemma A.1 to the pair $(X, B_X + M_X)$, we get $\operatorname{mult}_Z(B_X) + \operatorname{mult}_Z(M_X) \geqslant 1$. Now, applying Lemma A.1 to (A.6), we obtain (A.7), cf. [30, Corollary 2.7]. $\qquad\qquad\qquad\qquad\qquad\qquad\qquad\qquad\qquad\qquad\qquad\square$

Starting from now and until the end of this section, we suppose, in addition, that

($\bigstar$) $(X, B_X + M_X)$ is log canonical at every point of the subvariety Z.

If Z is a minimal center of log canonical singularities of the log pair $(X, B_X + M_X)$, then the subvariety $g(Z)$ is also a minimal center of log canonical singularities of this log pair for every $g \in G$, so that Lemma A.19 gives $Z \cap g(Z) \neq \varnothing \iff Z = g(Z)$. If the subvariety Z is a divisor in X that is a minimal center of log canonical singularities of the pair $(X, B_X + M_X)$, then X does not contain other log canonical centers of this log pair that meet Z. If $\dim(Z) \leqslant \dim(X) - 2$, this is not always true because Z may be contained in a center of log canonical singularities of larger dimension. In this case, we can often modify the boundary $B_X + M_X$ to obtain a similar assertion.

Lemma A.28 ([53, Lemma 2.4.10]) *Suppose that Z is a minimal center of log canonical singularities of $(X, B_X + M_X)$, $\dim(Z) \leqslant \dim(X) - 2$, and $B_X + M_X \sim_{\mathbb{Q}} H$ for an ample $\mathbb{Q}$-divisor H on the variety X. For a sufficiently divisible $n \gg 0$, let*

$$\mathcal{D} = \big\{D \in |nH| \ : \ g(Z) \subset \operatorname{Supp}(D) \text{ for every } g \in G\big\}.$$

Then $\mathcal{D}$ is a G-invariant linear subsystem in $|nH|$ that does not have fixed components. Fix $\epsilon \in \mathbb{Q}_{>0}$. Then there are rational numbers $1 \gg \epsilon_1 \geqslant 0$ and $1 \gg \epsilon_2 \geqslant 0$ such that

$$(1 - \epsilon_1)\big(B_X + M_X\big) + \epsilon_2 \mathcal{D} \sim_{\mathbb{Q}} (1 + \epsilon)H,$$

the pair $(X, (1 - \epsilon_1)(B_X + M_X) + \epsilon_2 \mathcal{D})$ is log canonical at every point of the subvariety Z, and Z is the only center of log canonical singularities of this log *pair that intersects Z. Moreover, if the original log pair* $(X, B_X + \mathcal{M}_X)$ *has log canonical singularities, then*

$$\text{Nklt}\big(X, (1 - \epsilon_1)(B_X + \mathcal{M}_X) + \epsilon_2 \mathcal{D}\big) = \bigsqcup_{g \in G} \{g(Z)\},$$

so that the new log pair $(X, (1 - \epsilon_1)(B_X + \mathcal{M}_X) + \epsilon_2 \mathcal{D})$ *also has log canonical singularities, and* $\text{Nklt}(X, (1 - \epsilon_1)(B_X + \mathcal{M}_X) + \epsilon_2 \mathcal{D})$ *is a G-irreducible subvariety in X.*

Proof See the proofs of [125, Theorem 1.10] and [126, Theorem 1]. □

This lemma is an equivariant version of the so-called Kawamata–Shokurov trick or tie breaking [125, 126]. Using Lemma A.28 and Corollary A.4, we obtain

Corollary A.29 *Suppose that X is a Fano variety, and* $B_X + \mathcal{M}_X \sim_{\mathbb{Q}} -\nu K_X$ *for some rational number* $\nu < 1$, *and the subvariety Z is a minimal center of log canonical singularities of the log pair* $(X, B_X + \mathcal{M}_X)$. *Then Z is G-invariant.*

This corollary implies the following technical result.

Lemma A.30 *Suppose that* $G = \mathbb{G}_m^r \rtimes B$ *for a finite group B, and X is a Fano threefold such that* $\alpha_G(X) < \mu$ *for some positive rational number* $\mu \leqslant 1$. *Suppose, in addition, that the following two conditions are satisfied:*

(i) *X does not contain G-fixed points,*
(ii) *X does not contain a G-invariant surface S such that* $-K_X \sim_{\mathbb{Q}} aS + \Delta$,
  where $a > \frac{1}{\mu}$ *and* Δ *is an effective* $\mathbb{Q}$-*divisor on X.*

Then X contains an effective G-invariant $\mathbb{Q}$-*divisor* $D \sim_{\mathbb{Q}} -K_X$ *and a smooth G-invariant irreducible rational curve Z such that* $(X, \lambda D)$ *is strictly log canonical for some positive rational number* $\lambda < \mu$, *and Z is the unique log canonical center of the log pair* $(X, \lambda D)$.

Proof By Lemma 1.42, our X contains an effective G-invariant $\mathbb{Q}$-divisor D such that the log pair $(X, \lambda D)$ is strictly log canonical for some positive rational number $\lambda < \mu$. Then $\text{Nklt}(X, \lambda D)$ is at most one-dimensional by (ii).

The locus $\text{Nklt}(X, \lambda D)$ is connected by Corollary A.4. Using Corollary A.29 and (i), we see that this locus is one-dimensional, and there are no points in X that are log canonical centers of the pair $(X, \lambda D)$.

Now, using Lemma A.19, we conclude that $\text{Nklt}(X, \lambda D)$ consists of a single curve Z. By Corollary A.14 or by Theorem A.20, the curve Z is smooth and rational. □

Let us present another application of Lemma A.28 and Theorem A.3.

Corollary A.31   *Suppose that X is a Fano variety that does not contain G-fixed points, the locus $\mathrm{Nklt}(X, B_X)$ is one-dimensional, and $B_X \sim_{\mathbb{Q}} -\lambda K_X$ for some $\lambda \in \mathbb{Q} \cap (0,1)$. Let C be an irreducible G-invariant curve in X that is contained in the locus $\mathrm{Nklt}(X, B_X)$. Choose $\delta \in \mathbb{Q} \cap (0,1]$ such that $(X, \delta B_X)$ is log canonical and not Kawamata log terminal. Then C is a minimal log canonical center of the log pair $(X, \delta B_X)$.*

Proof   Let Z be a minimal log canonical center of the pair $(X, \delta B_X)$. By Corollary A.29, the subvariety Z is G-invariant, so that Z is a curve since X contains no G-fixed points.

If $Z = C$, we are done. Hence, we assume that $Z \neq C$. Let us seek for a contradiction.

We observe that $Z \subset \mathrm{Nklt}(X, B_X)$. But it follows from Corollaries A.4 and A.14 that the locus $\mathrm{Nklt}(X, B_X)$ has the following properties:

- (o) it is connected,
- (i) each irreducible component is isomorphic to $\mathbb{P}^1$,
- (ii) any two intersecting irreducible components intersect transversally at one point,
- (iii) no three irreducible components intersect at one point,
- (iv) no irreducible components form a cycle.

Thus, irreducible curves in $\mathrm{Nklt}(X, B_X)$ form a tree, and Z and C are G-fixed vertices in this tree of curves. But this tree contains a unique path that joins these two vertices, so that this path must be G-invariant, and all its vertices also must be G-invariant, which implies that C contains a G-fixed point, which is a contradiction. □

If the log pairs $(X, \frac{1}{1-\alpha} B_X)$ and $(X, \frac{1}{\alpha} M_X)$ are log canonical at some point $P \in X$ for some $\alpha \in \mathbb{Q} \cap (0,1)$, then $(X, B_X + M_X)$ is also log canonical at this point. This gives

Corollary A.32   *Suppose that $B_X \sim_{\mathbb{Q}} \lambda H$, $M_X \sim_{\mathbb{Q}} \mu H$, $B_X + M_X \sim_{\mathbb{Q}} \nu H$ for some ample $\mathbb{Q}$-Cartier $\mathbb{Q}$-divisor H on the variety X, and rational numbers λ, μ, $\nu = \lambda + \mu$. If $(X, B_X + M_X)$ is not log canonical at a point $P \in X$, then $(X, \frac{\nu}{\lambda} B_X)$ or $(X, \frac{\nu}{\mu} M_X)$ is not log canonical at this point.*

Applying the same idea to the components of the divisor B_X, we obtain

Corollary A.33   *If X is a Fano variety, (X, B_X) is not log canonical at a point $P \in X$, and $\mathrm{rk}\, \mathrm{Cl}^G(X) = 1$, then X contains a G-irreducible effective Weil divisor B such that the log pair (X, bB) is not log canonical at P for $b \in \mathbb{Q}_{>0}$ such that $bB \sim_{\mathbb{Q}} B_X$.*

Now, let us generalize this corollary for arbitrary varieties.

Lemma A.34 *Let D be some G-invariant effective $\mathbb{Q}$-divisor on the variety X such that $D \sim_{\mathbb{Q}} B_X + M_X$ and $\mathrm{Supp}(D) \subseteq \mathrm{Supp}(B_X)$, but $D \neq B_X + M_X$. Then there exists a non-negative rational number μ such that the $\mathbb{Q}$-divisor $(1 + \mu)B_X - \mu D$ is effective, but its support does not contain at least one G-irreducible component of $\mathrm{Supp}(D)$. Moreover, if $(X, B_X + M_X)$ is not log canonical at some point $P \in X$, and (X, D) is log canonical at this point, then $(X, (1 + \mu)(B_X + M_X) - \mu D)$ is also not log canonical at P.*

Proof The proof is essentially the same as the proof of [40, Lemma 2.2]. Namely, we have

$$D = \sum_{i=1}^{r} b_i B_i \sim_{\mathbb{Q}} B_X + M_X \sim_{\mathbb{Q}} \sum_{i=1}^{r} a_i B_i + \sum_{i=1}^{s} c_i M_i, \qquad (A.8)$$

where each b_i is a non-negative number, and each B_i is a prime Weil divisor from (A.2). For every non-negative rational number ϵ, consider the divisor $(1 + \epsilon)B_X - \epsilon D$. Then

$$(1 + \epsilon)B_X - \epsilon D = \sum_{i=1}^{r} \big(\epsilon(a_i - b_i) + a_i\big) B_i,$$

and (A.8) implies that at least one number among $a_1 - b_1, a_2 - b_2, \ldots, a_r - b_r$ is negative. Then we can choose $\epsilon \geqslant 0$ such that $\epsilon(a_i - b_i) + a_i \geqslant 0$ for every $i \in \{1, \ldots, r\}$, but at least one of these numbers is zero. Then we can let μ be this ϵ.

Finally, if both pairs (S, D) and $(S, (1+\mu)(B_X + M_X) - \mu D)$ are log canonical at P, then the log pair $(S, B_X + M_X)$ is also canonical at P because

$$B_X + M_X = \frac{\mu}{1 + \mu} D + \frac{1}{1 + \mu} \Big((1 + \mu)(B_X + M_X) - \mu D\Big)$$

and $\frac{\mu}{1+\mu} + \frac{1}{1+\mu} = 1$. $\qquad \square$

Example A.35 (cf. Examples A.5 and A.17)   Suppose $X = \mathbb{P}^1 \times \mathbb{P}^1$. If $\alpha(X) < \frac{1}{2}$, there exists an effective divisor B_X on the surface X such that the pair (X, B_X) is not log canonical at a point $P \in X$, and $B_X \sim_{\mathbb{Q}} -\lambda K_X$ for some positive rational number $\lambda < \frac{1}{2}$. Let ℓ_1 and ℓ_2 be curves in X of degree $(1, 0)$ and $(0, 1)$ that pass through P, respectively. Then $-K_X \sim 2\ell_1 + 2\ell_2$, but $(X, \ell_1 + \ell_2)$ is log canonical. Thus, if $\alpha(X) < \frac{1}{2}$, it follows from Lemma A.34 that there is an effective divisor B'_X on X such that $B'_X \sim_{\mathbb{Q}} -\lambda K_X$, the log pair (X, B'_X) is not log canonical at some point $P \in X$, but $\mathrm{Supp}(B'_X)$ does not contain one of the curves ℓ_1 or ℓ_2. Without loss of generality, we may assume that $\ell_1 \not\subset \mathrm{Supp}(B'_X)$.

Then it follows from Lemma A.1 that $1 > 2\lambda = B'_X \cdot \ell_1 \geqslant \mathrm{mult}_P(B'_X) > 1$, which is absurd. This shows that $\alpha(X) \geqslant \frac{1}{2}$.

Let us conclude this section by proving one simple result, which is used in Example 4.51.

Lemma A.36 (cf. [177, Theorems 1.6]) *Let X be a del Pezzo surface such that $K^2_X = 2$, and X has one ordinary double point. Then*

$$\alpha(X) = \begin{cases} \dfrac{2}{3} \ if \ |-K_X| \ contains \ a \ tacnodal \ curve \ singular \ at \ \mathrm{Sing}(X), \\ \dfrac{3}{4} \ otherwise. \end{cases}$$

Proof   Recall that $|-K_X|$ gives a double cover $\omega\colon X \to \mathbb{P}^2$ that is branched over a reduced quartic curve R. Since X contains one ordinary double point, the curve R also has one ordinary double point, which implies that R is irreducible. Thus, if C is a singular curve in the linear system $|-K_X|$, then $C = \omega^*(L)$ for a line $L \subset \mathbb{P}^2$ such that either L passes through the point $\mathrm{Sing}(R)$, or L is tangent to R at a smooth point of the curve R. Let

$$\alpha_1(X) = \inf\left\{\mathrm{lct}\left(X, D\right) \mid D \ \text{is a divisor in} \ |-K_X|\right\}.$$

It is not hard to compute $\alpha_1(X)$. Namely, we have

$$\alpha_1(X) = \begin{cases} \dfrac{2}{3} \ \text{if} \ |-K_X| \ \text{contains a tacnodal curve singular at} \ \mathrm{Sing}(X), \\ \dfrac{3}{4} \ \text{otherwise.} \end{cases}$$

Note that [177, Theorems 1.4] claims that $\alpha_1(X) = \frac{2}{3}$, which is wrong in general.

Now, arguing almost as in the proof of [177, Theorems 1.6], we obtain $\alpha(X) = \alpha_1(X)$. Namely, suppose that $\alpha(X) < \alpha_1(X)$. Using Lemma A.34, we see that X contains an effective $\mathbb{Q}$-divisor D such that $D \sim_{\mathbb{Q}} -K_X$, the pair $(X, \lambda D)$ is not log canonical at some point $P \in X$ for some positive rational number $\lambda < \alpha_1(X)$, and $\mathrm{Supp}(D)$ does not contain at least one irreducible component of every curve in $|-K_X|$.

Suppose that $\omega(P)$ is a smooth point of the curve R. Then $|-K_X|$ contains a unique curve T that is singular at $P - \omega(T)$ is the line that is tangent to R at the point $\omega(P)$. If T is irreducible, then $T \not\subset \mathrm{Supp}(D)$, so that Lemma A.1 gives

$$2 = K^2_X = T \cdot D \geqslant \mathrm{mult}_P(T)\mathrm{mult}_P(D) \geqslant 2\mathrm{mult}_P(D) > 2.$$

Therefore, we conclude that $T = T_1 + T_2$, where T_1 and T_2 are two irreducible curves such that $-K_X \cdot T_1 = -K_X \cdot T_2 = 1$ and $T_1 \not\subset \mathrm{Supp}(D)$. Then $1 = T_1 \cdot D \geqslant$

$\mathrm{mult}_P(D)$, which contradicts Lemma A.1. This shows that either $\omega(P) \notin R$ or $P = \mathrm{Sing}(X)$.

Now, let $\eta \colon \widetilde{X} \to X$ be a blow up of the point P, let E be the η-exceptional curve, and let $\widetilde{D}$ be the proper transform on $\widetilde{X}$ of the divisor D. Then $\widetilde{D} \sim_{\mathbb{Q}}$ $\eta^*(D) - mE$ for some rational number $m \geqslant 0$. If $P \neq \mathrm{Sing}(X)$, then $m = \mathrm{mult}_P(D)$, so that $m > \frac{1}{\lambda}$ by Lemma A.1.

If X is smooth at P, we let $\delta = 1$. Likewise, if X is singular at P, we let $\delta = 0$. Then the log pair $(\widetilde{X}, \lambda\widetilde{D} + (\lambda m - \delta)E)$ is not log canonical at some point $Q \in E$. Therefore, applying Lemma A.1 to $(\widetilde{X}, \lambda\widetilde{D} + (\lambda m - \delta)E)$, we get

$$m + \mathrm{mult}_Q(\widetilde{D}) > \frac{1 + \delta}{\lambda}. \tag{A.9}$$

Furthermore, if $\lambda m - \delta \leqslant 1$, applying Corollary A.16 to $(\widetilde{X}, \lambda\widetilde{D} + (\lambda m - \delta)E)$, we get

$$\frac{1}{\lambda} < \left(\widetilde{D} \cdot E\right)_Q \leqslant \widetilde{D} \cdot E = \begin{cases} m & \text{if } P \neq \mathrm{Sing}(X), \\ 2m & \text{if } P = \mathrm{Sing}(X). \end{cases}$$

In particular, if $P = \mathrm{Sing}(X)$, then we have $m > \frac{1}{2\lambda}$ as we mentioned earlier.

Since $\omega(P) \notin R$ or $P = \mathrm{Sing}(X)$, the linear system $|-K_X|$ contains a curve C such that the curve C passes through P, and its proper transform on $\widetilde{X}$ passes through the point Q. Denote by $\widetilde{C}$ the proper transform of the curve C on the surface $\widetilde{X}$. If C is irreducible, then the curve C is not contained in the support of the divisor D, so that

$$\mathrm{mult}_Q(\widetilde{D}) \leqslant \widetilde{D} \cdot \widetilde{C} = \begin{cases} 2 - m & \text{if } P \neq \mathrm{Sing}(X), \\ 2 - 2m & \text{if } P = \mathrm{Sing}(X). \end{cases}$$

If $P \neq \mathrm{Sing}(X)$, this contradicts (A.9). If $P = \mathrm{Sing}(X)$, we get $2 - 2m \geqslant \mathrm{mult}_Q(\widetilde{D})$, but (A.9) gives $m + \mathrm{mult}_Q(\widetilde{D}) > \frac{1}{\lambda}$, so that we have $2 - \frac{1}{\lambda} > m > \frac{1}{2\lambda}$, which gives $\lambda > \frac{4}{3}$. Since $\lambda < \frac{4}{3}$, we see that C is reducible.

Thus, we have $C = C_1 + C_2$, where C_1 and C_2 are smooth irreducible curves such that $-K_X \cdot C_1 = -K_X \cdot C_2 = 1$. If $\mathrm{Sing}(X) \notin C$, then $C_1^2 = C_2^2 = -1$ and $C_1 \cdot C_2 = 2$. Likewise, if $\mathrm{Sing}(X) \in C$, then $\mathrm{Sing}(X) \in C_1 \cap C_2$, so that $C_1^2 = C_2^2 = -\frac{1}{2}$ and $C_1 \cdot C_2 = \frac{3}{2}$. Furthermore, we also know that one of the curves C_1 or C_2 is not contained in $\mathrm{Supp}(D)$. Hence, without loss of generality, we may assume that $C_2 \not\subset \mathrm{Supp}(D)$.

Let $\widetilde{C}_1$ and $\widetilde{C}_2$ be proper transforms on $\widetilde{X}$ via η of the curves C_1 and C_2, respectively. Then both curves $\widetilde{C}_1$ and $\widetilde{C}_2$ are smooth. Moreover, we also know that $Q \in \widetilde{C}_1$ or $Q \in \widetilde{C}_2$. If $Q \in \widetilde{C}_2$, then $\widetilde{C}_2$ intersects E transversally at Q, so that $\mathrm{mult}_Q(\widetilde{D}) \leqslant \widetilde{D} \cdot \widetilde{C}_2 = 1 - m$, which contradicts (A.9) because $\lambda < \alpha_1(X) \leqslant \frac{3}{4}$.

Therefore, we conclude that $Q \in \tilde{C}_1$. Observe that the curve $\tilde{C}_1$ intersects E transversally at the point Q.

Write $D = aC_1 + \Delta$, where a is a non-negative rational number, and Δ is an effective $\mathbb{Q}$-divisor on the surface X whose support does not contain C_1. Then

$$1 = -K_X \cdot C_2 = (aC_1 + \Delta) \cdot C_2 = aC_1 \cdot C_2 + \Delta \cdot C_2 \geq aC_1 \cdot C_2 = \begin{cases} 2a \text{ if } \mathrm{Sing}(X) \notin C, \\ \dfrac{3a}{2} \text{ if } \mathrm{Sing}(X) \in C. \end{cases}$$

Thus, we see that

$$a \leq \begin{cases} \dfrac{1}{2} \text{ if } \mathrm{Sing}(X) \notin C, \\ \dfrac{2}{3} \text{ if } \mathrm{Sing}(X) \in C. \end{cases} \tag{A.10}$$

In particular, we see that $\lambda a < 1$.

Let $\tilde{\Delta}$ be the proper transform on $\tilde{X}$ of Δ. Then $\tilde{\Delta} \sim_{\mathbb{Q}} \eta^*(\Delta) - nE$ for some rational number $n \geq 0$. If $P \neq \mathrm{Sing}(X)$, then $m = n + a$. If $P = \mathrm{Sing}(X)$, then $m = n + \frac{a}{2}$. Note that $(\tilde{X}, \lambda a\tilde{C}_1 + \lambda\tilde{\Delta} + (\lambda m - \delta)E)$ is not log canonical at the point $Q = \tilde{C}_2 \cap E$. Applying Corollary A.16, we obtain $(\lambda m - \delta) + \lambda\tilde{\Delta} \cdot \tilde{C}_1 > 1$, so that $m + \tilde{\Delta} \cdot \tilde{C}_1 > \frac{1+\delta}{\lambda}$. On the other hand, we have $\tilde{\Delta} \cdot \tilde{C}_1 = (\eta^*(\Delta) - nE) \cdot \tilde{C}_1 = \Delta \cdot C_1 - n = 1 - aC_1^2 - n$. Since $C_1^2 < 0$, we get

$$a > \begin{cases} \dfrac{\frac{1+\delta}{\lambda} - 1}{1 - C_1^2} \text{ if } P \neq \mathrm{Sing}(X), \\ \dfrac{\frac{1+\delta}{\lambda} - 1}{\frac{1}{2} - C_1^2} \text{ if } P = \mathrm{Sing}(X). \end{cases} \tag{A.11}$$

If $P \neq \mathrm{Sing}(X)$ and $\mathrm{Sing}(X) \notin C$, then $\delta = 1$ and $C_1^2 = -1$, so that $a > \frac{1}{\lambda} - \frac{1}{2} > \frac{5}{6}$. If $P \neq \mathrm{Sing}(X)$ and $\mathrm{Sing}(X) \in C$, then $\delta = 1$ and $C_1^2 = -\frac{1}{2}$, so that $a > \frac{4}{2\lambda} - \frac{2}{3} > \frac{10}{9}$. In both cases, we get a contradiction with (A.10). Thus, we have $P = \mathrm{Sing}(X)$.

Now, we have $\delta = 0$ and $C_1^2 = -\frac{1}{2}$, so that (A.11) gives $a > \frac{1}{\lambda} - 1 > \frac{1}{3}$, which does not contradict (A.10), but this inequality can still be used to obtain a contradiction. Namely, since $P = \mathrm{Sing}(X)$, the point P is contained in both curves C_1 and C_2, so that

$$0 \leq \tilde{\Delta} \cdot \tilde{C}_2 = (\eta^*(\Delta) - nE) \cdot \tilde{C}_2 = \Delta \cdot C_2 - n = 1 - \frac{3a}{2} - n,$$

which gives $n + \frac{3a}{2} \leq 1$. Thus, since $n + \frac{a}{2} = m > \frac{1}{2\lambda} > \frac{2}{3}$, we get $\frac{2}{3} + a < n + \frac{3a}{2} \leq 1$, which contradicts $a > \frac{1}{3}$ and completes the proof. $\qquad \square$

A.5 α-invariants of del Pezzo surfaces over non-closed fields

Let $\mathbb{F}$ be any field that has characteristic zero, e.g. $\mathbb{F} = \mathbb{Q}$ or $\mathbb{F} = \mathbb{C}(x)$. If C is a smooth conic in $\mathbb{P}^2$ defined over the field $\mathbb{F}$, then

$$\alpha(C) = \begin{cases} 1 & \text{if } C \text{ contains an } \mathbb{F}\text{-point,} \\ \frac{1}{2} & \text{if } S \text{ does not contain } \mathbb{F}\text{-points.} \end{cases}$$

In this section, we will generalize this result for smooth del Pezzo surfaces, i.e. smooth geometrically irreducible surfaces with ample anticanonical divisor.

Namely, let S be a smooth del Pezzo surface defined over $\mathbb{F}$, and let $\overline{\mathbb{F}}$ be the algebraic closure of the field $\mathbb{F}$. Recall from Section 1.4 that

$$\alpha(S) = \inf\big\{\operatorname{lct}(S, D) \,\big|\, D \text{ is an effective } \mathbb{Q}\text{-divisor on } S$$
$$\text{defined over } \mathbb{F} \text{ such that } D \sim_{\mathbb{Q}} -K_S\big\}.$$

If $\mathbb{F} = \overline{\mathbb{F}}$, all possible values of the number $\alpha(S)$ have been computed in [30, 154], see Table 2.1. To summarize these results, let

$$\alpha_n(S) = \inf\left\{\operatorname{lct}\left(S, \frac{1}{n}D\right) \,\Big|\, D \text{ is a divisor in } |-nK_S|\right\}$$

for every $n \in \mathbb{N}$. Clearly, we have $\alpha(S) \leqslant \alpha_n(S)$ for every $n \in \mathbb{N}$ and

$$\alpha(S) = \inf_{n \in \mathbb{N}} \alpha_n(S).$$

Note also that the number $\alpha_1(S)$ is not very hard to compute – to do this, we have to compute log canonical thresholds of all singular curves in $|-K_S|$. Moreover, we have

Theorem A.37 ([176, 30, 154])   *If $\mathbb{F}$ is algebraically closed, then $\alpha(S) = \alpha_1(S)$.*

In general, we may have $\alpha(S) \neq \alpha_1(S)$ if the field $\mathbb{F}$ is not algebraically closed.

Example A.38   Let $f(t)$ be an arbitrary irreducible polynomial in $\mathbb{F}[t]$ that has degree 5, let $\xi_1, \xi_2, \xi_3, \xi_4, \xi_5$ be its roots in $\overline{\mathbb{F}}$, let $\pi\colon S \to \mathbb{P}^2$ be the blow up of the reduced subscheme consisting of the points

$$[\xi_1 : \xi_1^2 : 1], [\xi_2 : \xi_2^2 : 1], [\xi_3 : \xi_3^2 : 1], [\xi_4 : \xi_4^2 : 1], [\xi_5 : \xi_5^2 : 1],$$

let C_2 be the conic in $\mathbb{P}^2$ that is given by $yz = x^2$, and let C be its proper transform on S, where x, y, z are coordinates on $\mathbb{P}^2$. Then S is a quartic del Pezzo surface defined over the field $\mathbb{F}$, and C is a line in S. We will see in Lemma A.41 that $\alpha(S) = \frac{2}{3}$. On the other hand, one can show that $\alpha_1(S) \geqslant \frac{3}{4}$.

In the remaining part of this section, we will find all values of the number $\alpha(S)$ without assuming that the field $\mathbb{F}$ is algebraically closed. Unless it is explicitly stated otherwise, we will assume that everything we deal with is defined over the field $\mathbb{F}$. We will use basic facts about del Pezzo surfaces over non-closed fields, which can be found in [143, 198]. To avoid confusion, let us present the glossary we will use:

- a point is an $\mathbb{F}$-point;
- a curve is a (possibly geometrically reducible) curve defined over $\mathbb{F}$;
- a conic is a (geometrically irreducible) curve isomorphic to a smooth conic in $\mathbb{P}^2$;
- a singular conic is a curve isomorphic to a reduced singular conic in $\mathbb{P}^2$;
- a line in S is a geometrically irreducible curve $C \subset S$ such that $C^2 = -1$;
- a conic in S is a geometrically irreducible curve $C \subset S$ such that $C^2 = 0$;
- a singular conic in S is a singular curve $C \subset S$ such that $-K_S \cdot C = 2$ and $C^2 = 0$;
- if $K_S^2 = 3$, an Eckardt point in S is a point $P \in S$ such that there exists a curve in the linear system $|-K_S|$ that has multiplicity 3 at the point P;
- a divisor on S is a Weil divisor on S defined over $\mathbb{F}$;
- $\mathrm{Pic}(S)$ is a group of divisors on S modulo rational equivalence;
- a $\mathbb{Q}$-divisor on S is a $\mathbb{Q}$-divisor on S defined over $\mathbb{F}$;
- $\overline{\mathbb{F}}$ is the algebraic closure of the field $\mathbb{F}$.

Note that lines in S are isomorphic to $\mathbb{P}^1$. Thus, if S contains a line, it also contains a point. Similarly, conics in S are isomorphic to smooth conics in $\mathbb{P}^2$, and singular conics in S are isomorphic to reduced singular conics in $\mathbb{P}^2$. In particular, if S contains a singular conic, then it contains a point. Recall that $|-K_S|$ gives an embedding $S \hookrightarrow \mathbb{P}^n$ for $K_S^2 \geqslant 3$, where $n = K_S^2$. In this case, lines, conics and singular conics in S are just usual embedded lines, conics and singular conics in $\mathbb{P}^n$, respectively.

First, we present in Table A.1 all possible values of the number $\alpha(S)$.

Now, let us explain in detail how to compute the numbers in this table. To start with, let us compute α-invariants of two-dimensional Severi–Brauer varieties.

Lemma A.39 *Suppose that $K_S^2 = 9$. Then*

$$\alpha(S) = \alpha_1(S) = \begin{cases} 1 & \text{if } S \text{ contains a point,} \\ \dfrac{1}{3} & \text{if } S \text{ does not contain points.} \end{cases}$$

Table A.1

K_S^2	Conditions imposed on the surface S	$\alpha(S)$
9	S contains a point	$\frac{1}{3}$
9	S does not contain points	1
8	S is a blow up of $\mathbb{P}^2$ in one point	$\frac{1}{3}$
8	$S \cong \mathbb{P}^1 \times C$ for a conic C	$\frac{1}{2}$
8	S is a quadric in $\mathbb{P}^3$	$\frac{1}{2}$
8	$S \cong C \times C'$ for two non-isomorphic conics C and C' such that both C and C' do not contain points	1
8	$\mathrm{Pic}(S) = \mathbb{Z}[-K_S]$	1
7	S is a blow up of $\mathbb{P}^2$ in two points	$\frac{1}{3}$
6	S contains a line or a conic	$\frac{1}{2}$
6	S contains no line or conic, but S contains a point	$\frac{2}{3}$
6	S contains no line, conic or point	1
5	S contains a line	$\frac{1}{2}$
5	$\mathrm{Pic}(S) \neq \mathbb{Z}[-K_S]$, but S does not contain lines	$\frac{2}{3}$
5	$\mathrm{Pic}(S) = \mathbb{Z}[-K_S]$	$\frac{4}{5}$
4	S contains a line or a singular conic	$\frac{2}{3}$
4	S contains no line or singular conic, but $\lvert -K_S \rvert$ contains a cuspidal curve	$\frac{3}{4}$
4	S contains no line or singular conic, $\lvert -K_S \rvert$ contains no tacnodal curves, $\lvert -K_S \rvert$ contains cuspidal curves	$\frac{5}{6}$
4	S contains no line or singular conic, and $\lvert -K_S \rvert$ contains no tacnodal and cuspidal curves	1
3	S contains an Eckardt point	$\frac{2}{3}$
3	S contains no Eckardt point, but $\lvert -K_S \rvert$ contains a tacnodal curve	$\frac{3}{4}$
3	S contains no Eckardt point, $\lvert -K_S \rvert$ contains no tacnodal curves, but $\lvert -K_S \rvert$ contains a cuspidal curve	$\frac{5}{6}$
3	S contains no Eckardt point, $\lvert -K_S \rvert$ contains no tacnodal or cuspidal curves	1
2	$\lvert -K_S \rvert$ contains a tacnodal curve	$\frac{3}{4}$
2	$\lvert -K_S \rvert$ contains no tacnodal curve, but $\lvert -K_S \rvert$ contains a cuspidal curve	$\frac{5}{6}$
2	$\lvert -K_S \rvert$ contains no tacnodal or cuspidal curve	1
1	$\lvert -K_S \rvert$ contains a cuspidal curve	$\frac{5}{6}$
1	$\lvert -K_S \rvert$ contains no cuspidal curve	1

Proof If the surface S contains a point, then $S \cong \mathbb{P}^2$, so that $\alpha(S) = \frac{1}{3}$, see Example A.2. Thus, we may assume that S contains no points. Then $\mathrm{Pic}(S) = \mathbb{Z}[-K_X]$ and $\alpha(S) \leqslant 1$.

We claim that $\alpha(S) = 1$. Indeed, suppose that $\alpha(S) < 1$. Then S contains an effective $\mathbb{Q}$-divisor D such that $D \sim_{\mathbb{Q}} -K_S$, and $(S, \lambda D)$ is not log canonical for some $\lambda \in \mathbb{Q} \cap (0,1)$. Since $\mathrm{Pic}(S) = \mathbb{Z}[-K_X]$, we deduce that the locus $\mathrm{Nklt}(S, \lambda D)$ must be zero-dimensional. Then $\mathrm{Nklt}(S, \lambda D)$ must be a point by Corollary A.4. Since $\mathrm{Nklt}(S, \lambda D)$ is defined over $\mathbb{F}$, we see that S contains a point, which is a contradiction. $\qquad\square$

Now, let us consider del Pezzo surfaces of small degree.

Lemma A.40 *Suppose that $K_S^2 \leqslant 3$. Then $\alpha(S) = \alpha_1(S)$.*

Proof The assertion follows from [40, Theorem 1.12]. Indeed, suppose that $\alpha(S) < \alpha_1(S)$. Then there exists an effective $\mathbb{Q}$-divisor D on the surface S such that $D \sim_{\mathbb{Q}} -K_S$, and the log pair $(S, \lambda D)$ is not log canonical for some positive rational number $\lambda < \alpha_1(S)$. Applying Lemma A.34, we may assume that $\mathrm{Supp}(D)$ does not contain at least one irreducible component of every curve in $|-K_S|$. Applying [40, Theorem 1.12], we see that the log pair $(S, \lambda D)$ has log canonical singularities, which is a contradiction. $\qquad\square$

Using Lemma A.40, it is not hard to find all the possible values of the number $\alpha_1(S)$ in the case when $K_S^2 \in \{1, 2, 3\}$, which are presented in Table A.1. See [176] for details. Now, we deal with quartic del Pezzo surfaces.

Lemma A.41 *Suppose that $K_S^2 = 4$. If the surface S contains a line or a singular conic, then $\alpha(S) = \alpha_2(S) = \frac{2}{3}$. Otherwise, we have $\alpha(S) = \alpha_1(S)$.*

Proof Recall that the del Pezzo surface S is a complete intersection of two quadrics in $\mathbb{P}^4$. Note that $\alpha(S) \geqslant \frac{2}{3}$ by [30, Theorem 1.7]. On the other hand, if S contains a line L, then projection from this line $\mathbb{P}^4 \dashrightarrow \mathbb{P}^2$ gives a birational morphism $\pi \colon S \to \mathbb{P}^2$ that contracts a geometrically reducible curve $\mathscr{C}$, which splits over $\overline{\mathbb{F}}$ as a union of five (-1)-curves, so that $3L + \mathscr{C} \sim -2K_S$, which gives $\alpha(S) \leqslant \alpha_2(S) \leqslant \frac{2}{3}$, so that $\alpha(S) = \alpha_2(S) = \frac{2}{3}$. Therefore, to proceed, we may assume that the surface S does not contain lines.

Similarly, if S contains a singular conic C, then $|-K_S - C|$ is a basepoint free pencil, so that it contains a unique curve C' that passes through $\mathrm{Sing}(C)$, so that $\mathrm{lct}(S, C + C') \leqslant \frac{2}{3}$, which gives $\alpha(S) \leqslant \alpha_2(S) \leqslant \alpha_1(S) \leqslant \frac{2}{3}$, which implies that $\alpha(S) = \alpha_2(S) = \alpha_1(S) = \frac{2}{3}$. Hence, to proceed, we may assume that S does not contain singular conics as well.

To complete the proof, we must show that $\alpha(S) = \alpha_1(S)$. Suppose that $\alpha(S) < \alpha_1(S)$. By Lemma A.34, the surface S contains an effective $\mathbb{Q}$-divisor D

such that $D \sim_{\mathbb{Q}} -K_S$, the log pair $(S, \lambda D)$ is not log canonical for some positive rational number $\lambda < \alpha_1(S)$, and the support of the divisor D does not contain at least one irreducible component of every curve in $|-K_S|$. Arguing as in the proof of [30, Lemma 3.4], we see that Nklt$(S, \lambda D)$ does not contain curves because S does not contain lines. Thus, it follows from Corollary A.4 that the locus Nklt$(S, \lambda D)$ is a point. For simplicity, we let $P = $ Nklt$(S, \lambda D)$.

Let $\eta \colon \widetilde{S} \to S$ be a blow up of the point P, and let E be the η-exceptional curve. Then $\widetilde{S}$ is a smooth cubic surface. Let $\widetilde{D}$ be the proper transform on $\widetilde{S}$ of the divisor D. Then $\widetilde{D} + (\text{mult}_P(D) - 1)E \sim_{\mathbb{Q}} -K_{\widetilde{S}}$. Thus, arguing as in the proof of Lemma 2.8 we see that $\text{mult}_P(D) \leqslant 2$. Now, arguing as in Lemma A.27, we see that the curve E contains a point Q such that the log pair $(\widetilde{S}, \lambda \widetilde{D} + (\lambda \text{mult}_P(D) - 1)E)$ is not Kawamata log terminal at Q, so that $(\widetilde{S}, \widetilde{D} + (\text{mult}_P(D) - 1)E)$ is not log canonical at Q.

Observe that $|-K_{\widetilde{S}} - E|$ is a basepoint free pencil. Let $\widetilde{Z}$ be the curve in this pencil that passes through Q. Since the pair $(\widetilde{S}, \widetilde{D} + (\text{mult}_P(D) - 1)E)$ is not log canonical at Q, it follows from [40, Theorem 1.12] that $\widetilde{Z} \cap E = Q$, and Supp$(\widetilde{D})$ contains all irreducible components of the curve $\widetilde{Z}$. Let $Z = \pi(\widetilde{Z})$. Then either Z is a geometrically reducible curve that has a tacnodal singularity at P, or Z is a geometrically irreducible curve that has a cuspidal singularity at P. Therefore, we see that $Z \in |-K_S|$, and the support of the divisor D contains all irreducible components of the curve Z, which contradicts our initial assumption. □

If $K_S^2 = 4$, then using Lemma A.41 and going through all singular curves in $|-K_S|$, we can find all the possibilities for the number $\alpha(S)$ in this case. Note also that the proof of Lemma A.41 implies

Corollary A.42 *If $K_S^2 = 4$ and S does not contain points, then $\alpha(S) = 1$.*

We deal with quintic del Pezzo surfaces in several lemmas. First, we prove

Lemma A.43 *Suppose that $K_S^2 = 5$ and $\text{Pic}(S) = \mathbb{Z}[-K_S]$. Then $\alpha(S) = \alpha_2(S) = \frac{4}{5}$.*

Proof The proof is similar to the proof of [30, Lemma 5.8]. Let us prove that $\alpha(S) \geqslant \frac{4}{5}$. Suppose that $\alpha(S) < \frac{4}{5}$. Then S contains an effective $\mathbb{Q}$-divisor $D \sim_{\mathbb{Q}} -K_S$, and the log pair $(S, \lambda D)$ is not Kawamata log terminal for a positive rational number $\lambda < \frac{4}{5}$. Since $\text{Pic}(S) = \mathbb{Z}[-K_S]$, the locus Nklt$(S, \lambda D)$ is zero-dimensional. By Corollary A.4, the locus Nklt$(S, \lambda D)$ consists of a single point O, which is defined over $\mathbb{F}$.

Over the field $\overline{\mathbb{F}}$, the surface S contains ten (-1)-curves. But none of these ten curves contains O because $\text{Pic}(S) = \mathbb{Z}[-K_S]$. Moreover, over $\overline{\mathbb{F}}$, the surface S contains five smooth curves Z_1, Z_2, Z_3, Z_4, Z_5 such that $-K_S \cdot Z_i = 1$ and

$O = Z_1 \cap Z_2 \cap Z_3 \cap Z_4 \cap Z_5$. These are conics in X defined over $\overline{\mathbb{F}}$ which contain O. Let $\mathscr{C} = Z_1 + Z_2 + Z_3 + Z_4 + Z_5$. Then $\mathscr{C}$ is defined over the field $\mathbb{F}$, the curve $\mathscr{C}$ is irreducible, $\mathscr{C} \sim_{\mathbb{Q}} -2K_S$, and

$$\mathrm{lct}\left(S, \frac{1}{2}\mathscr{C}\right) = \frac{4}{5}. \tag{A.12}$$

Using Lemma A.34, we may assume that $\mathscr{C} \not\subset \mathrm{Supp}(D)$. Then $10 = \mathscr{C} \cdot D \geqslant 5\mathrm{mult}_O(D)$.

Let $\pi \colon \widetilde{S} \to S$ be the blow up of the point O, let E be the π-exceptional curve, and let $\widetilde{D}$ be the proper transform of the divisor D on the surface $\widetilde{S}$. Using Lemma A.27, we see that E contains a point Q such that $(\widetilde{S}, \lambda\widetilde{D} + (\lambda\mathrm{mult}_O(D)-1)E)$ is not Kawamata log terminal at the point Q, which is defined over $\mathbb{F}$. Then $\mathrm{mult}_Q(\widetilde{D}) + \mathrm{mult}_O(D) \geqslant \frac{2}{\lambda} > \frac{5}{2}$ by Lemma A.27. Observe also that $\widetilde{S}$ is a smooth del Pezzo surface of degree 4.

Let $\widetilde{Z}_1, \widetilde{Z}_2, \widetilde{Z}_3, \widetilde{Z}_4, \widetilde{Z}_5$ be the proper transform on $\widetilde{S}$ of the curves Z_1, Z_2, Z_3, Z_4, Z_5, respectively. Note that $\widetilde{Z}_1, \widetilde{Z}_2, \widetilde{Z}_3, \widetilde{Z}_4, \widetilde{Z}_5$ are disjoint (-1)-curves, which (a priori) are defined over $\overline{\mathbb{F}}$. Moreover, since $\mathrm{Pic}(S) = \mathbb{Z}[-K_S]$, we have $Q \notin \widetilde{Z}_1 \cup \widetilde{Z}_2 \cup \widetilde{Z}_3 \cup \widetilde{Z}_4 \cup \widetilde{Z}_5$. Furthermore, we have the following Sarkisov link:

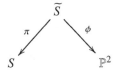

where ϕ is a contraction of the curves $\widetilde{Z}_1, \widetilde{Z}_2, \widetilde{Z}_3, \widetilde{Z}_4$ and $\widetilde{Z}_5$. The curve $\phi(E)$ is the unique conic in $\mathbb{P}^2$ that passes through the points $\phi(\widetilde{Z}_1), \phi(\widetilde{Z}_2), \phi(\widetilde{Z}_3), \phi(\widetilde{Z}_4)$ and $\phi(\widetilde{Z}_5)$.

Over the algebraic closure $\overline{\mathbb{F}}$, the plane $\mathbb{P}^2$ contains five lines L_1, L_2, L_3, L_4, L_5 such that L_i is the line that goes through $\phi(Q)$ and $\phi(\widetilde{Z}_i)$. Let $\widetilde{L}_1, \widetilde{L}_2, \widetilde{L}_3, \widetilde{L}_4, \widetilde{L}_5$ be the proper transforms on $\widetilde{S}$ of the lines L_1, L_2, L_3, L_4, L_5, respectively. Then

$$\pi\left(\widetilde{L}_1\right) + \pi\left(\widetilde{L}_2\right) + \pi\left(\widetilde{L}_3\right) + \pi\left(\widetilde{L}_4\right) + \pi\left(\widetilde{L}_5\right) \sim_{\mathbb{Q}} -3K_S,$$

and $\pi(\widetilde{L}_1) + \pi(\widetilde{L}_2) + \pi(\widetilde{L}_2) + \pi(\widetilde{L}_4) + \pi(\widetilde{L}_5)$ is an irreducible curve defined over the field $\mathbb{F}$. Moreover, the log pair $(S, \frac{4}{15}(\pi(\widetilde{L}_1)+\pi(\widetilde{L}_2)+\pi(\widetilde{L}_2)+\pi(\widetilde{L}_4)+\pi(\widetilde{L}_5)))$ has Kawamata log terminal singularities. Hence, using Lemma A.34, we may assume that $\mathrm{Supp}(D)$ does not contain $\pi(\widetilde{L}_1), \pi(\widetilde{L}_2), \pi(\widetilde{L}_3), \pi(\widetilde{L}_4), \pi(\widetilde{L}_5)$. Then $3 - \mathrm{mult}_O(D) = \widetilde{D} \cdot \widetilde{L}_1 \geqslant \mathrm{mult}_Q(\widetilde{D})$, implying that $\mathrm{mult}_O(D) + \mathrm{mult}_Q(\widetilde{D}) \leqslant 3$.

Let $\xi \colon \widehat{S} \to \widetilde{S}$ be the blow up of the point Q, and let F be the ξ-exceptional divisor. Denote by $\widehat{E}$ and $\widehat{D}$ the proper transforms on $\widehat{S}$ of the divisors E and $\widetilde{D}$, respectively. Using Lemma A.27 again, we see that F contains a unique point P such that the log pair

$$\left(\widehat{S}, \lambda\widehat{D} + \left(\lambda\mathrm{mult}_O(D) - 1\right)\widehat{E} + \left(\lambda\mathrm{mult}_O(D) + \lambda\mathrm{mult}_Q(\widetilde{D}) - 2\right)F\right)$$

is not Kawamata log terminal at P, and

$$\lambda\mathrm{mult}_P(\widehat{D}) + \left(\lambda\mathrm{mult}_O(D) - 1\right)\mathrm{mult}_P(\widehat{E}) + \lambda\mathrm{mult}_O(D) + \lambda\mathrm{mult}_Q(\widetilde{D}) > 3.$$
$$(A.13)$$

Let $\widehat{T}$ be the proper transform on $\widehat{S}$ of the line in $\mathbb{P}^2$ that is tangent to $\phi(E)$ at $\phi(Q)$. Then $\pi \circ \xi(\widehat{T})$ is a cuspidal curve in $|-K_S|$. Thus, using Lemma A.34, we may assume that $\mathrm{Supp}(\widehat{D})$ does not contain $\widehat{T}$. Hence, if $P \in \widehat{E}$, then $P \in \widehat{T}$, so that

$$5 - 2\mathrm{mult}_O(D) - \mathrm{mult}_Q(\widetilde{D}) = \widehat{T}\cdot\widehat{D} \geqslant \mathrm{mult}_P(\widehat{D}) > 5 - 2\mathrm{mult}_O(D) - \mathrm{mult}_Q(\widetilde{D})$$

by (A.13). Then $P \notin \widehat{E}$, so (A.13) gives $\mathrm{mult}_O(D) + \mathrm{mult}_Q(\widetilde{D}) + \mathrm{mult}_P(\widehat{D}) > \frac{15}{4}$.

Observe that $\mathbb{P}^2$ contains a unique line L that passes through $\phi(Q)$ such that its proper transform on $\widehat{S}$ contains the point P. Since the line L is defined over $\mathbb{F}$, it does not contain any of the $\overline{\mathbb{F}}$-points $\phi(\widetilde{Z}_1)$, $\phi(\widetilde{Z}_2)$, $\phi(\widetilde{Z}_3)$, $\phi(\widetilde{Z}_4)$, $\phi(\widetilde{Z}_5)$. Now, we denote by $\widehat{L}$ the proper transform of the line L on the surface $\widehat{S}$. Then $\pi \circ \xi(\widehat{L})$ is a nodal curve in $|-K_S|$, so that, using Lemma A.34, we may assume that $\mathrm{Supp}(\widehat{D})$ does not contain $\widehat{L}$. Then

$$5 - 2\mathrm{mult}_O(D) - \mathrm{mult}_Q(\widetilde{D}) = \widehat{L} \cdot \widehat{D} > \frac{15}{4} - \mathrm{mult}_O(D) - \mathrm{mult}_Q(\widetilde{D}),$$

which gives $\mathrm{mult}_O(D) < \frac{5}{4}$. Then $(S, \lambda D)$ is Kawamata log terminal at O by Lemma A.1, which contradicts our assumption.

We see that $\alpha(S) \geqslant \frac{4}{5}$. To show that $\alpha(S) = \alpha_2(S) = \frac{4}{5}$, recall that S always contains a point [204, 195]. Thus, arguing as above, we can find a curve $\mathscr{C} \in |-2K_S|$ such that the equality (A.12) holds. This gives $\alpha(S) \leqslant \alpha_2(S) \leqslant \frac{4}{5}$. $\quad\square$

Now, we are ready to prove the following result:

Lemma A.44 *Suppose that $K_S^2 = 5$ and $\mathrm{Pic}(S) \neq \mathbb{Z}[-K_S]$. Then*

$$\alpha(S) = \begin{cases} \alpha_1(S) = \dfrac{1}{2} & \text{if } S \text{ contains a line,} \\[2mm] \alpha_2(S) = \dfrac{2}{3} & \text{if } S \text{ does not contain lines.} \end{cases}$$

Proof If the surface S contains a line L, then the linear system $|-K_S - L|$ gives a birational map $\pi\colon S \to Q$ such that Q is a smooth quadric surface in $\mathbb{P}^3$, and $\pi(L)$ is a hyperplane section of the quadric Q. Moreover, the morphism π contracts a curve $\mathscr{E}$, which splits over the algebraic closure $\overline{\mathbb{F}}$ as a disjoint union of three (-1)-curves that intersect the line L. Then $2L + \mathcal{E} \sim -K_S$, so that $\alpha(S) \leqslant \alpha_1(S) \leqslant \frac{1}{2}$, and $\alpha(S) = \frac{1}{2}$ by [30, Theorem 1.7].

To complete the proof of the lemma, we may assume that the surface S contains no lines. Since $\mathrm{Pic}(S) \neq \mathbb{Z}[-K_S]$ and S contains a point, this implies that $\mathrm{Pic}(S) \cong \mathbb{Z}^2$ and there exists the following Sarkisov link:

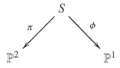

where π is a birational morphism, and ϕ is a conic bundle. Moreover, the morphism π contracts an irreducible curve $\mathcal{E}$ that splits over $\overline{\mathbb{F}}$ as a union of four disjoint (-1)-curves.

Let $\mathscr{C}$ be a fiber of the conic bundle ϕ over a point in $\mathbb{P}^1$. Then $\frac{3}{2}\mathscr{C} + \frac{1}{2}\mathcal{E} \sim_{\mathbb{Q}} -K_S$, so that $\alpha(S) \leqslant \alpha_2(S) \leqslant \frac{2}{3}$. We claim that $\alpha(S) = \frac{2}{3}$. Indeed, suppose that $\alpha(S) < \frac{2}{3}$. Then S contains an effective $\mathbb{Q}$-divisor $D \sim_{\mathbb{Q}} -K_S$, and the pair $(S, \lambda D)$ is strictly log canonical for a positive rational number $\lambda < \frac{2}{3}$.

We claim that $\mathrm{Nklt}(S, \lambda D)$ is zero-dimensional. Indeed, suppose $\mathrm{Nklt}(S, \lambda D)$ contains an irreducible curve C. Then $\frac{3}{2}\mathscr{C} + \frac{1}{2}\mathcal{E} \sim_{\mathbb{Q}} D = \frac{1}{\lambda}C + \Delta$, where Δ is an effective $\mathbb{Q}$-divisor whose support does not contain C. In particular, we see that $C \neq \mathcal{E}$ because $\mathscr{C}$ and $\mathcal{E}$ generate the Mori cone of the surface S. Then $C \sim \pi^*(\mathcal{O}_{\mathbb{P}^2}(d)) - m\mathcal{E}$ for some positive integer d and some non-negative integer m. We have

$$\frac{3}{2}\mathscr{C} + \frac{1}{2}\mathcal{E} \sim_{\mathbb{Q}} \frac{d}{2\lambda}\mathscr{C} + \left(\frac{d}{2\lambda} - \frac{m}{\lambda}\right)\mathcal{E} + \Delta,$$

so that $m \leqslant \frac{d}{2}$, $\frac{d}{2\lambda} \leqslant \frac{3}{2}$ and $\frac{d}{2\lambda} - \frac{m}{\lambda} \leqslant \frac{1}{2}$, which leads to a contradiction since $\lambda < \frac{2}{3}$.

Using Corollary A.4, we see that the locus $\mathrm{Nklt}(S, \lambda D)$ consists of a single point O. Note that $O \notin \mathcal{E}$, so that the log pair $(\mathbb{P}^2, \lambda\pi(D))$ is not Kawamata log terminal at $\pi(O)$. Let L be a line in $\mathbb{P}^2$ that does not contain $\pi(O)$. Then

$$L \cup O \subseteq \mathrm{Nklt}(\mathbb{P}^2, L + \lambda\pi(D)),$$

but $\mathrm{Nklt}(\mathbb{P}^2, L + \lambda\pi(D))$ contains no curves except L. This contradicts Corollary A.4. The obtained contradiction shows that $\alpha(S) = \alpha_2(S) = \frac{2}{3}$. $\square$

We compute α-invariants of sextic del Pezzo surfaces in the following lemma:

Lemma A.45 *Suppose that $K_S^2 = 6$. Then*

$$\alpha(S) = \begin{cases} 1 & \text{if } S \text{ does not contain lines, conics and points,} \\ \dfrac{2}{3} & \text{if } S \text{ does not contain lines and conics, but } S \text{ contains a point,} \\ \dfrac{1}{2} & \text{if } S \text{ contains a line or a conic.} \end{cases}$$

Proof Let us describe the geometry of the del Pezzo surface S over the algebraic closure $\overline{\mathbb{F}}$. Over the field $\overline{\mathbb{F}}$, we have a birational morphism $\varpi \colon S \to \mathbb{P}^2$ that blows up three distinct non-collinear points P_1, P_2, P_2. Let E_1, E_2, E_3 be ϖ-exceptional curves that are mapped to the points P_1, P_2, P_2, respectively. For every i and j in $\{1,2,3\}$, let L_{ij} be the proper transform on S of the line in $\mathbb{P}^2$ that passes through the points P_i and P_j, Then the set

$$\left\{ E_1, E_2, E_3, L_{12}, L_{13}, L_{23} \right\} \tag{A.14}$$

contains all (-1)-curves in S. Moreover, there exists the diagram

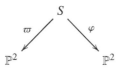

where φ is the contraction of the (-1)-curves L_{12}, L_{13}, L_{23}. In general, this diagram as well as the morphisms ϖ and φ are not defined over $\mathbb{F}$.

The Galois group $\mathrm{Gal}(\overline{\mathbb{F}}/\mathbb{F})$ naturally acts on the set (A.14), and its possible splitting into the $\mathrm{Gal}(\overline{\mathbb{F}}/\mathbb{F})$-orbits can be described as follows:

- (D_{12}) $\{E_1, E_2, E_3, L_{12}, L_{13}, L_{23}\}$,
- $(\mathfrak{S}_3)$ $\{E_1, E_2, E_3\}$ and $\{L_{12}, L_{13}, L_{23}\}$,
- $(\mu_2^2.a)$ $\{E_1, L_{23}\}$ and $\{E_2, E_3, L_{12}, L_{13}\}$,
- $(\mu_2^2.b)$ $\{E_2, L_{13}\}$ and $\{E_1, E_3, L_{12}, L_{23}\}$,
- $(\mu_2^2.c)$ $\{E_3, L_{12}\}$ and $\{E_1, E_2, L_{13}, L_{23}\}$,
- (μ_2) $\{E_1, L_{23}\}$, $\{E_2, L_{13}\}$ and $\{E_3, L_{12}\}$,
- $(\mu_2.a)$ $\{E_1, L_{23}\}$, $\{E_2, L_{12}\}$ and $\{E_3, L_{13}\}$,
- $(\mu_2.b)$ $\{E_2, L_{13}\}$, $\{E_1, L_{12}\}$ and $\{E_3, L_{23}\}$,
- $(\mu_2.c)$ $\{E_3, L_{12}\}$, $\{E_1, L_{13}\}$ and $\{E_2, L_{23}\}$,
- $(\mu_2.a')$ $\{E_1\}$, $\{L_{23}\}$, $\{E_2, E_3\}$ and $\{L_{12}, L_{13}\}$,
- $(\mu_2.b')$ $\{E_2\}$, $\{L_{13}\}$, $\{E_1, E_3\}$ and $\{L_{12}, L_{23}\}$,
- $(\mu_2.c')$ $\{E_3\}$, $\{L_{12}\}$, $\{E_1, E_2\}$ and $\{L_{13}, L_{23}\}$,
- (1) $\{E_1\}$, $\{L_{23}\}$, $\{E_2\}$, $\{L_{13}\}$, $\{E_3\}$, $\{L_{12}\}$.

Suppose that S contains a line L. Then L is a $\mathrm{Gal}(\overline{\mathbb{F}}/\mathbb{F})$-invariant curve in (A.14). We may assume that $L = L_{12}$. Then $2L_{12} + \frac{3}{2}(E_1 + E_2) + \frac{1}{2}(L_{13} + L_{23}) \sim_{\mathbb{Q}} -K_S$, where both curves $E_1 + E_2$ and $L_{13} + L_{23}$ are defined over $\mathbb{F}$. Hence, in this case, we have $\alpha(S) \leqslant \frac{1}{2}$, so that $\alpha(S) = \frac{1}{2}$ by [30, Theorem 1.7].

Similarly, if S contains a conic C, then the linear system $|-K_S - C|$ gives a birational map $\pi \colon S \to Q$ such that Q is a smooth quadric surface in $\mathbb{P}^3$, and $\pi(C)$ is its hyperplane section. In this case, the morphism π contracts a curve $\mathscr{E}$ that splits over $\overline{\mathbb{F}}$ as a disjoint union of two (-1)-curves that intersect C, which gives $2C + \mathscr{E} \sim -K_S$, so that $\alpha(S) \leqslant \frac{1}{2}$, which implies that $\alpha(S) = \frac{1}{2}$ by [30, Theorem 1.7].

Thus, to complete the proof, we may assume that S does not contain lines and conics. This assumption also implies that S does not contain singular conics. Indeed, if S contains a singular conic C, then the linear system $|C|$ gives a conic bundle $S \to \mathbb{P}^1$, so that S also contains a (smooth) conic. Thus, our assumptions impose strong restrictions on the way the group $\mathrm{Gal}(\overline{\mathbb{F}}/\mathbb{F})$ acts on the set (A.14). Namely, we only can have splittings of this set into the orbits described in (D_{12}), $(\mathfrak{S}_3)$, $(\mu_2^2.a)$, $(\mu_2^2.b)$, $(\mu_2^2.c)$, (μ_2).

If S has no points, let $\mu = 1$. If S has a point, we let $\mu = \frac{2}{3}$. We claim that $\alpha(S) \leqslant \mu$. Indeed, if S does not contain points, the claim is obvious. If S contains a point P, then the surface S contains a curve Z that splits over $\overline{\mathbb{F}}$ as $Z = Z_1 + Z_2 + Z_3$, where Z_1, Z_2, Z_3 are smooth rational curves such that $-K_S \cdot Z_1 = -K_S \cdot Z_2 = -K_S \cdot Z_3 = 2$. Indeed, we can let Z_i be the proper transform via ϖ of the line in $\mathbb{P}^2$ that passes through the points $\varpi(P)$ and $\varpi(E_i)$. Thus, in this case, we have $Z_1 + Z_2 + Z_3 \sim -K_S$ and

$$\mathrm{lct}\big(S, Z_1 + Z_2 + Z_3\big) = \mathrm{lct}_P\big(S, Z_1 + Z_2 + Z_3\big) = \mu = \frac{2}{3},$$

so that $\alpha(S) \leqslant \mu$. Note that the curve Z is defined over $\mathbb{F}$.

We claim that $\alpha(S) = \mu$. Indeed, suppose that $\alpha(S) < \mu$. Then S contains an effective $\mathbb{Q}$-divisor D such that $D \sim_{\mathbb{Q}} -K_S$, and the log pair $(S, \lambda D)$ is strictly log canonical for a positive rational number $\lambda < mu$. Let us seek for a contradiction.

If $\mathrm{Nklt}(S, \lambda D)$ is zero-dimensional, then it consists of a single point by Corollary A.4, which must be defined over $\mathbb{F}$, so that $\lambda < \mu = \frac{2}{3}$, and we can obtain a contradiction arguing exactly as in the end of the proof of Lemma A.44. Therefore, we conclude that the locus $\mathrm{Nklt}(S, \lambda D)$ contains an irreducible curve C. Then $D = \frac{1}{\lambda}C + \Delta$, where Δ is an effective $\mathbb{Q}$-divisor whose support does not contain C. We have

$$2 = \big(L_{12} + E_2\big) \cdot D = \frac{1}{\lambda}\big(L_{12} + E_2\big) \cdot C + \big(L_{12} + E_2\big) \cdot \Delta \geqslant \frac{1}{\lambda}\big(L_{12} + E_2\big) \cdot C,$$

$$2 = \left(L_{23} + E_3\right) \cdot D = \frac{1}{\lambda}\left(L_{23} + E_3\right) \cdot C + \left(L_{23} + E_3\right) \cdot \Delta \geqslant \frac{1}{\lambda}\left(L_{23} + E_3\right) \cdot C,$$

$$2 = \left(L_{13} + E_1\right) \cdot D = \frac{1}{\lambda}\left(L_{13} + E_1\right) \cdot C + \left(L_{13} + E_1\right) \cdot \Delta \geqslant \frac{1}{\lambda}\left(L_{13} + E_1\right) \cdot C$$

because divisors $L_{12} + E_2$, $L_{23} + E_3$, $L_{13} + E_1$ are nef. Thus, we see that

$$\begin{cases} (L_{12} + E_2) \cdot C \leqslant 1, \\ (L_{23} + E_3) \cdot C \leqslant 1, \\ (L_{13} + E_1) \cdot C \leqslant 1. \end{cases} \tag{A.15}$$

In particular, this gives $-K_S \cdot C = (L_{12}+E_2) \cdot C + (L_{23}+E_3) \cdot C + (L_{13}+E_1) \cdot C \leqslant 3$. Therefore, keeping in mind that S does not contain lines, conics and singular conics, the curve C is irreducible, and the del Pezzo surface S is an intersection of quadrics in its anticanonical embedding in $\mathbb{P}^6$, we obtain the following cases:

(i) $C = E_1 + L_{23}$,

(ii) $C = E_2 + L_{13}$,

(iii) $C = E_3 + L_{12}$,

(iv) $C = E_1 + E_1 + E_3$,

(v) $C = L_{12} + L_{13} + L_{23}$,

(vi) $C \sim L_{12} + E_1 + E_2$,

(vii) $C \sim L_{12} + L_{13} + E_1$.

The first three cases contradict (A.15). If $C = E_1 + E_1 + E_3$, then

$$3 = \left(L_{12} + L_{13} + E_1\right) \cdot D = \frac{1}{\lambda}\left(L_{12} + L_{13} + E_1\right) \cdot C + \left(L_{12} + L_{13} + E_1\right) \cdot \Delta$$

$$\geqslant \frac{1}{\lambda}\left(L_{12} + L_{13} + E_1\right) \cdot C = \frac{3}{\lambda},$$

which is impossible since $\lambda < 1$. Similarly, if $C = L_{12} + L_{13} + L_{23}$, then

$$3 = \left(L_{12} + E_1 + E_2\right) \cdot D = \frac{1}{\lambda}\left(L_{12} + E_1 + E_2\right) \cdot C + \left(L_{12} + E_1 + E_2\right) \cdot \Delta$$

$$\geqslant \frac{1}{\lambda}\left(L_{12} + E_1 + E_2\right) \cdot C = \frac{3}{\lambda},$$

which is a contradiction. Thus, we see that either $C \sim L_{12} + E_1 + E_2$ or $C \sim L_{12} + L_{13} + E_1$.

If $C \sim L_{12} + E_1 + E_2$, then $|C|$ gives the birational map $\varpi : S \to \mathbb{P}^2$, so that it is defined over $\mathbb{F}$, which implies in particular that S contains a point, so that $\mu = \frac{2}{3}$ and

$$3 = \left(L_{12} + L_{13} + E_1\right) \cdot D \Rightarrow \frac{1}{\lambda}\left(L_{12} + L_{13} + E_1\right) \cdot C = \frac{2}{\lambda} > \frac{2}{\mu} = 3$$

because $L_{12} + L_{13} + E_1$ is nef. Similarly, if $C \sim L_{12} + L_{13} + E_1$, then $\mu = \frac{2}{3}$ and

$$3 = \frac{1}{\lambda}(L_{12}{+}E_1{+}E_2){\cdot}C + (L_{12}{+}E_1{+}E_2){\cdot}\Delta \geqslant \frac{1}{\lambda}(L_{12}{+}E_1{+}E_2){\cdot}C = \frac{2}{\lambda} > \frac{2}{\mu} = 3$$

because $L_{12}{+}E_1{+}E_2$ is nef. The obtained contradiction completes the proof.   □

If $K_S^2 = 7$, then S is a blow up of $\mathbb{P}^2$ in two points, so that $\alpha(S) = \frac{1}{3}$ by [30, Theorem 1.7]. Similarly, if S is a blow up of $\mathbb{P}^2$ in one point, we get $\alpha(S) = \frac{1}{3}$. Finally, we prove

Lemma A.46 *Suppose that $K_S^2 = 8$, and S is not a blow up of $\mathbb{P}^2$ in one point. Then*

$$\alpha(S) = \begin{cases} \dfrac{1}{2} & \text{if S is a smooth quadric surface in $\mathbb{P}^3$ or $S \cong \mathbb{P}^1 \times C$ for a conic C,} \\ 1 & \text{otherwise.} \end{cases}$$

Proof Note that $\alpha(S) \leqslant 1$ since $|-K_S|$ is not empty. Moreover, if S is a smooth quadric surface in $\mathbb{P}^3$, then $\alpha(S) \leqslant \mathrm{lct}(S, 2H) = \frac{1}{2}$ for any hyperplane section H of the surface S, so that $\alpha(S) = \frac{1}{2}$ by [30, Theorem 1.7]. Similarly, we see that $\alpha(S) = \frac{1}{2}$ if $S \cong \mathbb{P}^1 \times C$ for an arbitrary conic C defined over $\mathbb{F}$. Furthermore, if $\mathrm{Pic}(S) = \mathbb{Z}[-K_S]$, then S does not have points. In this case, arguing as in the proof of Lemma A.39, we see that $\alpha(S) = 1$.

Now, we may assume that $\mathrm{Pic}(S) \neq \mathbb{Z}[-K_S]$ and S is not a quadric in $\mathbb{P}^3$. This implies that $\mathrm{rk}\,\mathrm{Pic}(S) = 2$, and $X \not\cong C \times C$ for any conic C. Thus, it follows from [198, Lemma 3.4] that $S = C_1 \times C_2$, where C_1 and C_2 are two non-isomorphic conics such that neither of them contains points. We claim $\alpha(S) = 1$. Indeed, suppose that $\alpha(S) < 1$. Then S contains an effective $\mathbb{Q}$-divisor $D \sim_{\mathbb{Q}} -K_S$, and $(S, \lambda D)$ is not log canonical for some positive rational number $\lambda < 1$. If $\mathrm{Nklt}(S, \lambda D)$ is zero-dimensional, then $\mathrm{Nklt}(S, \lambda D)$ must be a point by Corollary A.4, which contradicts our assumption. Thus, we conclude that $\mathrm{Nklt}(S, \lambda D)$ contains an irreducible curve C. Then $D = aC + \Delta$ for some rational number $a \geqslant \frac{1}{\lambda} > 1$, where Δ is an effective $\mathbb{Q}$-divisor on S. Then $C \sim \mathrm{pr}_1^*(-n_1 K_{C_1}) + \mathrm{pr}_2^*(-n_2 K_{C_2})$ for some non-negative integers n_1 and n_2, where $\mathrm{pr}_1 : S \to C_1$ and $\mathrm{pr}_2 : S \to C_2$ are projections to the first and the second factors, respectively. Then

$$\mathrm{pr}_1^*\left(-K_{C_1}\right) + \mathrm{pr}_2^*\left(-K_{C_2}\right) \sim -K_S \sim_{\mathbb{Q}} D \sim_{\mathbb{Q}} \mathrm{pr}_1^*\left(-an_1 K_{C_1}\right) + \mathrm{pr}_2^*\left(-an_2 K_{C_2}\right) + \Delta,$$

which immediately leads to a contradiction since $a > 1$.   □

A.6 Groups acting on Hirzebruch surfaces

In this section, we describe properties of some groups acting faithfully on Hirzebruch surfaces. Let $X = \mathbb{F}_n$, and let G be a reductive subgroup in $\mathrm{Aut}(X)$.

If $n > 0$, we denote by $\pi \colon X \to \mathbb{P}^1$ the natural G-equivariant projection. In this case, we denote by $\mathbf{s}$ the section of π such that $\mathbf{s}^2 = -n$, and we denote by $\mathbf{f}$ a fiber of this projection. Observe that the curve $\mathbf{s}$ is G-invariant, and $|\mathbf{s} + n\mathbf{f}|$ also contains smooth G-invariant curve, which is disjoint from $\mathbf{s}$.

If $n = 0$, we denote by $\pi_1 \colon X \to \mathbb{P}^1$ and $\pi_2 \colon X \to \mathbb{P}^1$ the projections to the first and the second factors, respectively. Then π_1 and π_2 are G-equivariant $\iff \operatorname{rk} \operatorname{Pic}^G(X) = 2$.

We start with the case $G \cong \mathrm{PGL}_2(\mathbb{C})$.

Lemma A.47 ([149, Theorem 5.1]) *Suppose that* $G \cong \mathrm{PGL}_2(\mathbb{C})$. *If* $X = \mathbb{P}^1 \times \mathbb{P}^1$, *then*

(i) *either G acts trivially on one of the factors of the surface X;*

(ii) *or G acts diagonally on X, and the only proper closed G-invariant subvariety in the surface X is its diagonal.*

Similarly, if $n \geqslant 1$, then X contains exactly two proper closed irreducible G-invariant subvarieties: the section $\mathbf{s}$ and a unique G-invariant curve in $|\mathbf{s} + n\mathbf{f}|$ disjoint from $\mathbf{s}$.

Now, we consider the case when $G \cong (\mathbb{G}_m \rtimes \mu_2) \times \mu_2$.

Lemma A.48 *Suppose that* $X = \mathbb{P}^1 \times \mathbb{P}^1$, $G \cong (\mathbb{G}_m \rtimes \mu_2) \times \mu_2$ *and* $\operatorname{rk} \operatorname{Pic}^G(X) = 2$. *Then G contains two involutions σ and τ such that $G = \langle \mathbb{G}_m, \sigma, \tau \rangle$, $\langle \mathbb{G}_m, \sigma \rangle \cong \mathbb{G}_m \rtimes \mu_2$, and up to conjugation in $\operatorname{Aut}(X)$ the G-action on X can be described as follows: either*

$$
\left.
\begin{aligned}
\lambda &\colon ([x_0 : x_1], [y_0 : y_1]) \mapsto ([\lambda x_0 : x_1], [y_0 : y_1]) \\
\sigma &\colon ([x_0 : x_1], [y_0 : y_1]) \mapsto ([x_1 : x_0], [y_0 : y_1]) \\
\tau &\colon ([x_0 : x_1], [y_0 : y_1]) \mapsto ([x_0 : x_1], [-y_0 : y_1])
\end{aligned}
\right\}, \tag{A.16}
$$

or there are $a \in \mathbb{Z}_{>0}$ and $b \in \mathbb{Z}$ such that $\gcd(a, b) = 1$ and

$$
\left.
\begin{aligned}
\lambda &\colon ([x_0 : x_1], [y_0 : y_1]) \mapsto ((\lambda^a x_0 : x_1], [\lambda^b y_0 : y_1]) \\
\sigma &\colon ([x_0 : x_1], [y_0 : y_1]) \mapsto ([x_1 : x_0], [y_1 : y_0]) \\
\tau &\colon ([x_0 : x_1], [y_0 : y_1]) \mapsto ([x_0 : x_1], [-y_0 : y_1])
\end{aligned}
\right\}, \tag{A.17}
$$

where $\lambda \in \mathbb{G}_m$, and $([x_0 : x_1], [y_0 : y_1])$ are coordinates on $X = \mathbb{P}^1 \times \mathbb{P}^1$.

Proof As π_1 and π_2 are G-equivariant, they induce homomorphisms $\rho_1 \colon G \to \operatorname{Aut}(\mathbb{P}^1)$ and $\rho_2 \colon G \to \operatorname{Aut}(\mathbb{P}^1)$, respectively. Up to a change of coordinates we have, for $\lambda \in \mathbb{G}_m$,

$$\rho_1(\lambda)([x_0 : x_1]) = [\lambda^a x_0 : x_1],$$
$$\rho_2(\lambda)([y_0 : y_1]) = [\lambda^b y_0 : y_1],$$

where $a \in \mathbb{Z}_{>0}$ and $b \in \mathbb{Z}$ such that $\gcd(a, b) = 1$.

Recall that $G \cong (\mathbb{G}_m \rtimes \mu_2) \times \mu_2$. Let σ be the generator of the factor μ_2 in $\mathbb{G}_m \rtimes \mu_2$, and let τ be the generator of the direct factor μ_2. Observe that $\rho_1(\sigma)$ is an involution that normalizes $\rho_1(\mathbb{G}_m)$ but does not commute with it. Then $\rho_1(\sigma)([x_0 : x_1]) = [\alpha x_1 : x_0]$ for some $\alpha \in \mathbb{G}_m$. Rescaling the coordinate x_0 if necessary, we may assume that $\alpha = 1$. Moreover, since $\rho_1(\tau)$ commutes with $\rho_1(\mathbb{G}_m)$, we get $\rho_1(\tau)([x_0 : x_1]) = [\pm x_0 : x_1]$. Replacing τ by $\sqrt[a]{-1}\tau$ if necessary, we may assume that $\rho_1(\tau)$ is trivial.

Suppose that $b \neq 0$. As above, up to a change of coordinates we obtain

$$\rho_2(\sigma)([y_0 : y_1]) = [y_1 : y_0],$$
$$\rho_2(\tau)([y_0 : y_1]) = [\pm y_0 : y_1].$$

However, since $\rho_1(\tau)$ is trivial, $\rho_2(\tau)$ cannot be trivial, so that $\rho_2(\tau)([y_0 : y_1]) = [-y_0 : y_1]$. This gives the action (A.17).

Now, we suppose that $b = 0$. Then $\rho_2(\tau)$ is a non-trivial involution, so that, up to a change of coordinates, we have $\rho_2(\tau)([y_0 : y_1]) = [-y_0 : y_1]$. Since $\rho_2(\sigma)$ commutes with τ, either it is trivial, or $\rho_2(\sigma)([y_0 : y_1]) = [y_1 : y_0]$. In the former case, we get the action (A.16). In the latter case, we get the action (A.17) with $a = 1$ and $b = 0$. □

Corollary A.49 *Suppose $X = \mathbb{F}_n$ with $n > 0$, and $G \cong (\mathbb{G}_m \rtimes \mu_2) \times \mu_2$. Then n is even, and there exists the G-equivariant commutative diagram*

where ψ is a birational map, ϕ is an isomorphism, π_1 is the projection to the first factor, and the G-action on $\mathbb{P}^1 \times \mathbb{P}^1$ is as in (A.16).

Proof As we already mentioned, there exists a smooth G-invariant curve $C \in |s + n\mathbf{f}|$. Since C is G-invariant and $C \cong \mathbb{P}^1$, we conclude that C contains a G-orbit of length 2. Blowing up this G-orbit and contracting the proper transforms of two curves in $|\mathbf{f}|$ that meet this orbit, we obtain the G-equivariant commutative diagram

$$X \dashrightarrow^{\theta} \mathbb{F}_m$$

$$\pi \downarrow \qquad \downarrow$$

$$\mathbb{P}^1 = = = \mathbb{P}^1$$

where θ is the constructed birational map, and $m = n - 2$.

Applying this construction $\lfloor \frac{n-1}{2} \rfloor$ times, we get a G-equivariant commutative diagram

$$X \dashrightarrow^{\psi} \mathbb{F}_r$$

$$\pi \downarrow \qquad \downarrow \varpi$$

$$\mathbb{P}^1 = = = \mathbb{P}^1$$

such that ψ is a birational map, ϖ is a natural projection, $\psi(s)$ and $\psi(C)$ are two disjoint G-invariant sections of the projection ϖ, and

$$r = \begin{cases} 0 & \text{if } n \text{ is even,} \\ 1 & \text{if } n \text{ is odd.} \end{cases}$$

A similar idea was used in the proof of [33, Lemma B.15].

If $r = 1$, then there exists a G-equivariant birational contraction $\mathbb{F}_1 \to \mathbb{P}^2$, which implies that $\mathbb{P}^2$ contains a G-fixed point, which gives an embedding $G \hookrightarrow \mathrm{GL}_2(\mathbb{C})$ by Lemma A.25. However, the group $\mathrm{GL}_2(\mathbb{C})$ does not contain subgroups isomorphic to G, so that $r = 0$. Now, applying Lemma A.48, we obtain the required assertion. □

Now, we consider the case when $G \cong (\mathbb{G}_m \times \pmb{\mu}_3) \rtimes \pmb{\mu}_2 \cong \mathbb{G}_m \rtimes \mathfrak{S}_3$.

Lemma A.50 *Suppose that* $X = \mathbb{P}^1 \times \mathbb{P}^1$, $G \cong (\mathbb{G}_m \times \pmb{\mu}_3) \rtimes \pmb{\mu}_2$ *and* $\mathrm{rk}\,\mathrm{Pic}^G(X) = 2$. *Then there are an involution* $\sigma \in G$ *and an element of order three* $\tau \in G$ *that, together with the subgroup* $\mathbb{G}_m$, *generate the group* G, *and up to conjugation in* $\mathrm{Aut}(X)$ *the* G-action on the surface X can be described as follows: either

$$\left. \begin{aligned} \lambda &\colon ([x_0 : x_1], [y_0 : y_1]) \mapsto ([\lambda x_0 : x_1], [y_0 : y_1]) \\ \sigma &\colon ([x_0 : x_1], [y_0 : y_1]) \mapsto ([x_1 : x_0], [y_0 : y_1]) \\ \tau &\colon ([x_0 : x_1], [y_0 : y_1]) \mapsto ([x_0 : x_1], [\omega y_0 : y_1]) \end{aligned} \right\}, \qquad (A.18)$$

or there are $a \in \mathbb{Z}_{>0}$ *and* $b \in \mathbb{Z}$ *such that* $\gcd(a, b) = 1$ *and*

$$\left. \begin{aligned} \lambda &\colon ([x_0 : x_1], [y_0 : y_1]) \mapsto ([\lambda^a x_0 : x_1], [\lambda^b y_0 : y_1]) \\ \sigma &\colon ([x_0 : x_1], [y_0 : y_1]) \mapsto ([x_1 : x_0], [y_1 : y_0]) \\ \tau &\colon ([x_0 : x_1], [y_0 : y_1]) \mapsto ([x_0 : x_1], [\omega y_0 : y_1]) \end{aligned} \right\} \qquad (A.19)$$

where ω is a primitive cube root, $\lambda \in \mathbb{G}_m$, and $([x_0 : x_1], [y_0 : y_1])$ are coordinates on X.

Proof Arguing as in the proof of Lemma A.48, we see that there are two natural group homomorphisms $\rho_1 \colon G \to \mathrm{Aut}(\mathbb{P}^1)$ and $\rho_2 \colon G \to \mathrm{Aut}(\mathbb{P}^1)$. Up to a change of coordinates, for $\lambda \in \mathbb{G}_m$ we have $\rho_1(\lambda)([x_0 : x_1]) = [\lambda^a x_0 : x_1]$ and $\rho_2(\lambda)([y_0 : y_1]) = [\lambda^b y_0 : y_1]$ for some integers $a > 0$ and b such that $\gcd(a, b) = 1$.

Fix an isomorphism $G \cong (\mathbb{G}_m \times \boldsymbol{\mu}_3) \rtimes \boldsymbol{\mu}_2$. Let τ be a generator of the factor $\boldsymbol{\mu}_3$, and let σ be the generator of the semi-direct factor $\boldsymbol{\mu}_2$. Then $\rho_1(\sigma)([x_0 : x_1]) = [x_1 : x_0]$. Since the centralizer of the torus $\rho_1(\mathbb{G}_m)$ in $\mathrm{Aut}(\mathbb{P}^1)$ coincides with $\rho_1(\mathbb{G}_m)$, we conclude that $\rho_1(\tau)([x_0 : x_1]) = [\gamma x_0 : x_1]$, where γ is a (possibly trivial) cube root of unity. Therefore, replacing τ by $\sqrt[a]{\gamma^2}\tau$, we may assume that $\tau \in \ker(\rho_1)$.

Suppose that $b \neq 0$. Up to a change of coordinates, we have $\rho_2(\sigma)([y_0 : y_1]) = [y_1 : y_0]$ and $\rho_2(\tau)([y_0 : y_1]) = [\omega y_0 : y_1]$, where ω is a cube root of unity. Since $\tau \in \ker(\rho_1)$, we have $\tau \notin \ker(\rho_2)$, so that ω is a primitive cube root of unity. This gives the action (A.19).

Suppose $b = 0$. Up to a change of coordinates, we have $\rho_2(\tau)([y_0 : y_1]) = [\omega y_0 : y_1]$ for a primitive cube root of unity ω. For the element $\rho_2(\sigma)$ we have two options: it is either trivial, or $\rho_2(\sigma)([y_0 : y_1]) = [y_1 : y_0]$. Thus, in the former case, we get the action (A.18). Likewise, in the latter case, we get the action (A.19) with $a = 1$ and $b = 0$. □

Corollary A.51 *Suppose $X = \mathbb{F}_n$ with $n > 0$, and $G \cong (\mathbb{G}_m \times \boldsymbol{\mu}_3) \rtimes \boldsymbol{\mu}_2$. Then n is even, and there exists the G-equivariant commutative diagram*

$$
\begin{array}{ccc}
X & \overset{\psi}{\dashrightarrow} & \mathbb{P}^1 \times \mathbb{P}^1 \\
\pi \downarrow & & \downarrow \pi_1 \\
\mathbb{P}^1 & \overset{\phi}{\longrightarrow} & \mathbb{P}^1
\end{array}
$$

where ψ is a birational map, ϕ is an isomorphism, π_1 is the projection to the first factor, and the G-action on $\mathbb{P}^1 \times \mathbb{P}^1$ is as in (A.18).

Proof The proof is the same as the proof of Corollary A.49. The only difference is that now we should use Lemma A.50 instead of Lemma A.48. □

Now, we present the very result used in the proof of Lemma 5.1.

Lemma A.52 *Suppose that $X = \mathbb{F}_4$ and $G = \mathfrak{S}_4$. Then $|s + k\mathbf{f}|$ does not contain G-irreducible curves for $k \in \{5, 6, 7, 8, 9\}$, and $|s + 4\mathbf{f}|$ contains a unique G-invariant curve.*

Proof If C is a G-irreducible curve in $|\mathbf{s} + k\mathbf{f}|$ for $k \geqslant 5$, then $|C \cap \mathbf{s}| \leqslant C \cdot \mathbf{s} = k - 4$, which gives $k \geqslant 10$ as $\mathbb{P}^1$ does not have $\mathfrak{S}_4$-orbits of length less than 6.

As we already mentioned, the linear system $|\mathbf{s} + 4\mathbf{f}|$ contains an irreducible G-invariant curve C. If $\hat{C}$ is another G-invariant curve in $|\mathbf{s} + 4\mathbf{f}|$, then $|C \cap \hat{C}| \leqslant C \cdot \hat{C} = 4$, which is impossible as well. This shows that C is the only G-invariant curve in $|\mathbf{s} + 4\mathbf{f}|$.   □

The following lemma is used in Example 4.37.

Lemma A.53 *Suppose that* $X = \mathbb{P}^1 \times \mathbb{P}^1$, $G = \mathfrak{A}_4$, *and the G-action on X is diagonal. Then X contains two G-invariant curves of degree* $(1, 3)$, *and both of them are smooth. Moreover, if* $\mathscr{C}$ *is one of these curves, then the group* $\mathrm{Aut}(X, \mathscr{C})$ *is finite.*

Proof To start with, we describe the G-action on the surface X. Let $\widehat{G} = 2.\mathfrak{A}_4 \cong \mathrm{SL}_2(\mathbb{F}_3)$, and let $\mathbb{U}_2$ be a two-dimensional irreducible representation of the group $\widehat{G}$. This gives us a faithful G-action on $\mathbb{P}^1 = \mathbb{P}(\mathbb{U}_2)$, which gives the diagonal G-action on X.

Let Δ be the G-invariant diagonal curve in X, let H be a divisor on X of degree $(1, 3)$. Then $H|_\Delta$ is a divisor on $\Delta \cong \mathbb{P}^1$ of degree 4, and the restriction map gives the following epimorphism of $\widehat{G}$-representations:

$$\mathbb{U}_2 \otimes \mathrm{Sym}^3(\mathbb{U}_2) \cong H^0(O_X(H)) \twoheadrightarrow H^0\big(O_\Delta(H|_\Delta)\big) \cong \mathrm{Sym}^4(\mathbb{U}_2).$$

But $H^0(O_\Delta(H|_\Delta))$ contains two non-isomorphic one-dimensional $\widehat{G}$-subrepresentations because the curve Δ contains exactly two G-orbits of length 4. Therefore, we conclude that $H^0(O_X(H))$ also contains two non-isomorphic one-dimensional $\widehat{G}$-subrepresentations.

Thus, we see that $|H|$ has at least two G-invariant curves. These curves are irreducible and smooth because X does not contain G-invariant curves of degree $(1, 0)$ and $(0, 1)$ since otherwise intersecting them with Δ we would get G-fixed points, which do not exist. This also implies that $|H|$ contains exactly two G-invariant curves.

Let $\mathscr{C}$ be a G-invariant curve in X of degree $(1, 3)$. We claim that $\mathrm{Aut}(X, \mathscr{C})$ is finite. Indeed, if it is not finite, then arguing as in the proof of [45, Corollary 2.7], we see that the projection on the first factor $X \to \mathbb{P}^1$ induces a G-equivariant Galois triple cover $\mathscr{C} \to \mathbb{P}^1$ branched in two points, which must form a G-invariant subset. The latter is impossible since the length of the smallest G-orbit in $\mathbb{P}^1$ is 4. This shows that $\mathrm{Aut}(X, \mathscr{C})$ is finite.   □

Similarly, we obtain the following result, which is used in Section 5.14.

Lemma A.54 *Suppose that $X = \mathbb{P}^1 \times \mathbb{P}^1$, $G = \mathfrak{S}_4$, and the G-action on X is diagonal. Then X contains a unique G-invariant curve of degree $(1,5)$, and this curve is smooth. Moreover, if $\mathscr{C}$ is this curve, then $\mathrm{Aut}(X, \mathscr{C}) \cong G$.*

Proof Let Δ be the diagonal curve in S, and let H be a divisor on X of degree $(1,5)$. Since Δ is G-invariant, the restriction $H^0(\mathcal{O}_X(H)) \twoheadrightarrow H^0(\mathcal{O}_\Delta(H|_\Delta))$ is an epimorphism of two representations of the group $2.\mathfrak{S}_4$. On the other hand, the curve Δ contains a unique G-orbit of length 6, so that $|H|_\Delta|$ contains a unique G-invariant divisor. Therefore, we see that $H^0(\mathcal{O}_\Delta(H|_\Delta))$ contains a unique one-dimensional subrepresentation of $2.\mathfrak{S}_4$, which implies that $H^0(\mathcal{O}_X(H))$ contains one-dimensional subrepresentation of this group. Hence, we conclude that $|H|$ contains a G-invariant divisor $\mathscr{C}$.

We claim that $\mathscr{C}$ is reduced and irreducible. Indeed, otherwise we have $\mathscr{C} = \ell + D$ for some effective G-invariant divisor D on X, and a G-invariant ruling ℓ of the surface X. Then $\ell \cap \Delta$ is a G-invariant point in Δ, which does not exist. Hence, we see that $\mathscr{C}$ is reduced and irreducible. This also implies that $\mathscr{C}$ is the unique G-invariant divisor in $|H|$.

Keeping in mind that $\mathscr{C}$ is a divisor of degree $(5,1)$, we see that $\mathscr{C}$ is a smooth curve.

Arguing as in the very end of the proof of Lemma A.53, we see that $\mathrm{Aut}(X, \mathscr{C})$ is finite, which implies that $\mathrm{Aut}(X, \mathscr{C}) = G$ because the group $G \cong \mathfrak{S}_4$ is not contained in any finite subgroup in $\mathrm{Aut}(\mathscr{C}) \cong \mathrm{PGL}_2(\mathbb{C})$ except itself. $\square$

A.7 Auxiliary results

In this section, we present a few sporadic lemmas.

Lemma A.55 *Let X be an arbitrary normal projective algebraic variety of dimension n, let A and B be Cartier divisors on X such that A is big and nef, and $A + aB$ is nef for some $a \in \mathbb{Z}_{>0}$. Then*

$$\sum_{k=0}^{ma} h^0(X, mA + kB) = \frac{m^{n+1}}{n!} \int_0^a (A + uB)^n \, du + O(m^n).$$

Moreover, for any Cartier divisor D on X and any $i > 0$, we have

$$\sum_{k=0}^{ma} h^i(X, mA + kB + D) = O(m^{n-i}).$$

Proof Let $V = \mathbb{P}(\mathcal{O} \oplus B)$, let $\pi \colon V \to X$ be a $\mathbb{P}^1$-bundle, let H be the tautological line bundle on V, and let $\mathcal{L} = aH + \pi^*(A)$. Then $\mathcal{L}$ is nef by [166, Lemma IV.2.6(2)].

Consider the section $\sigma \colon X \to V$ that corresponds to the embedding $O \hookrightarrow O \oplus B$. Then $\sigma^*(H) = O_X$, and the normal bundle of $\sigma(X)$ in V is $\pi^*(-B)$. Then $\sigma^*(\mathcal{L}) = A$ and $\sigma(X) \sim H - \pi^*(B)$. Let $\mathcal{L} \sim a\sigma(X) + \pi^*(A + aB)$. Then

$$\mathcal{L}^{n+1-i} \cdot \pi^*(A + aB)^i = aA^{n-i} \cdot (A + aB) + \mathcal{L}^{n-i} \cdot \pi^*(A + aB)^{i+1}$$

for every $i \in \{0, \ldots, n\}$. This gives

$$\mathcal{L}^{n+1} = \sum_{j=0}^{n} aA^{n-j} \cdot (A + aB)^j = \sum_{i=0}^{n} a^{i+1} \sum_{j=i}^{n} \binom{j}{i} A^{n-i} \cdot B^i$$

$$= \sum_{i=0}^{n} a^{i+1} \binom{n+1}{i+1} A^{n-i} \cdot B^i.$$

As $\binom{n+1}{i+1} = (n+1)\binom{n}{i} \cdot \frac{1}{i+1}$, we have

$$\mathcal{L}^{n+1} = (n+1) \cdot \sum_{i=0}^{n} \binom{n}{i}(A^{n-i} \cdot B^i)\frac{a^{i+1}}{i+1} = (n+1) \cdot \int_0^a (A + uB)^n du.$$

Thus, to prove the first required equality, it remains to notice that

$$H^0(V, m\mathcal{L}) = H^0\left(X, S^{ma}(O \oplus B) \otimes O_X(mA)\right) \cong \bigoplus_{j=0}^{ma} H^0(X, mA + jB).$$

Since $\mathcal{L}$ is nef, we have by asymptotic Riemann–Roch that

$$h^0(V, m\mathcal{L}) = \frac{m^{n+1}}{(n+1)!}\mathcal{L}^{n+1} + O(m^n),$$

which implies the first required equality.

Now let us prove the second required equality. Using Leray's spectral sequence, we get

$$H^p\left(V, m\mathcal{L} \otimes \pi^*(D)\right) \cong H^p\left(X, \pi_*(m\mathcal{L} \otimes \pi^*D)\right) \cong H^p\left(\bigoplus_{j=0}^{ma} O(mA + jB + D)\right)$$

since $R^q\pi_*(m\mathcal{L} \otimes \pi^*(D)) = 0$ for all $q > 0$. Now, using [106, Corollary 7], we get

$$h^i(V, m\mathcal{L} + \pi^*(D)) \leqslant O(m^{n-i}),$$

which implies the second required equality.   □

Lemma A.56 (cf. [160, Example 1.5]) *Let Q be a smooth quadric hypersurface in $\mathbb{P}^4$, let C be a smooth curve in Q such that C is a scheme-intersection of surfaces in $|O_{\mathbb{P}^4}(2)|_Q|$, and let $\pi \colon X \to Q$ be the blow up of the curve C. Then X is a Fano threefold.*

Proof Let E be the π-exceptional surface. Then $|\pi^*(O_{\mathbb{P}^4}(2)|_Q)-E|$ is basepoint free, which implies that the divisor $K_X \sim \pi^*(O_{\mathbb{P}^4}(3)|_Q) - E$ is ample. $\square$

Lemma A.57 *Let W be the standard two-dimensional $SL_2(\mathbb{C})$-representation equipped with some basis, let W^* be the dual representation, and let u, v be the dual basis in W^*. Consider the representation $\operatorname{Sym}^4(W^*)$ with the basis*

$$(e_0, e_1, e_2, e_3, e_4) = (u^4, u^3v, u^2v^2, uv^3, v^4).$$

Then non-GIT-stable $SL_2(\mathbb{C})$-orbits in $\operatorname{Sym}^4(W^*)$ can be described as follows:

(1.1) a closed 2-dimensional orbit $SL_2(\mathbb{C}).\alpha e_2$ with stabilizer $\mathbb{G}_m$, where $\alpha \in \mathbb{C}^*$,

(1.2) the non-closed 2-dimensional orbit $SL_2(\mathbb{C}).e_0$ with stabilizer $\mathbb{G}_a$ and $0 \in \overline{SL_2(\mathbb{C}).e_0}$,

(1.3.a) a non-closed 3-dimensional orbit $SL_2(\mathbb{C}).(e_0 + \alpha e_2)$ with

$$SL_2(\mathbb{C}).\alpha e_2 \subset \overline{SL_2(\mathbb{C}).(e_0 + \alpha e_2)} \not\ni 0,$$

where $\alpha \in \mathbb{C}^*$,

(1.3.b) the non-closed 3-dimensional orbit $SL_2(\mathbb{C}).e_1$ with

$$SL_2(\mathbb{C}).e_0 \subset \overline{SL_2(\mathbb{C}).e_1} \ni 0.$$

Let $\mathbb{P}^4 = \mathbb{P}(\operatorname{Sym}^4(W^*))$ that is equipped with the induced $PGL_2(\mathbb{C})$-action and coordinates. Then non-GIT-stable $PGL_2(\mathbb{C})$-orbits in $\mathbb{P}^4$ can be described as follows:

(2.1) the polystable 2-dimensional orbit $PGL_2(\mathbb{C}).[0 : 0 : 1 : 0 : 0]$ with stabilizer $\mathbb{G}_m$,

(2.2) the unstable 1-dimensional orbit $PGL_2(\mathbb{C}).[1 : 0 : 0 : 0 : 0]$ with stabilizer $\mathbb{G}_a \rtimes \mathbb{G}_m$,

(2.3.a) the strictly semistable 3-dimensional orbit $PGL_2(\mathbb{C}).[1 : 0 : 1 : 0 : 0]$,

(2.3.b) the unstable 2-dimensional orbit $PGL_2(\mathbb{C}).[0 : 1 : 0 : 0 : 0]$ with stabilizer $\mathbb{G}_m$.

The closure of every non-GIT-stable orbit contains the orbit (2.2).

Proof The description of non-GIT stable $SL_2(\mathbb{C})$-orbits in $\operatorname{Sym}^4(W^*)$ is well-known and can be found in [181, 73]. The remaining assertions follow from this description. $\square$

Corollary A.58 ([45, Lemma 9.1]) *In the assumptions and notations of Lemma A.57, we let $\mathbb{P}^5 = \mathbb{P}(\operatorname{Sym}^4(W^*) \oplus \mathbb{I})$, where $\mathbb{I}$ is the trivial representation of the group $SL_2(\mathbb{C})$. For the induced $PGL_2(\mathbb{C})$-action on $\mathbb{P}^5$, non-GIT-stable orbits can be described as follows:*

(3.0) *the polystable fixed point* $[0:0:0:0:0:1]$ *with stabilizer* $\mathrm{PGL}_2(\mathbb{C})$,

(3.1) *the polystable orbit* $\mathrm{PGL}_2(\mathbb{C}).[0:0:1:0:0:\lambda]$ *with stabilizer* $\mathbb{G}_m$, *where* $\lambda \in \mathbb{C}$,

(3.2) *the unstable orbit* $\mathrm{PGL}_2(\mathbb{C}).[1:0:0:0:0:0]$ *with stabilizer* $\mathbb{G}_a \rtimes \mathbb{G}_m$,

(3.2') *the strictly semistable orbit* $\mathrm{PGL}_2(\mathbb{C}).[1:0:0:0:0:1]$ *with stabilizer* $\mathbb{G}_a$,

(3.3.a) *the strictly semistable orbit* $\mathrm{PGL}_2(\mathbb{C}).[1:0:3:0:0:\lambda]$, *where* $\lambda \in \mathbb{C}$,

(3.3.b) *the unstable orbit* $\mathrm{PGL}_2(\mathbb{C}).[0:1:0:0:0:0]$ *with stabilizer* $\mathbb{G}_m$,

(3.3.b') *the strictly semistable orbit* $\mathrm{PGL}_2(\mathbb{C}).[0:1:0:0:0:1]$.

Proof Observe that set-theoretically we have the following decomposition

$$\mathbb{P}\left(\mathrm{Sym}^4(W^*) \oplus \mathbb{I}\right) = \mathbb{P}\left(\mathrm{Sym}^4(W)\right) \sqcup \mathrm{Sym}^4(W^*),$$

so that the required description follows from Lemma A.57. □

Recall that two-dimensional linear systems are called nets.

Lemma A.59 *Let M be a net in $|\mathcal{O}_{\mathbb{P}^1}(4)|$ that is basepoint free. Then $\mathbb{P}^1$ contains two distinct points P and Q such that one of the following (excluding) possibilities holds:*

 (i) *the net M contains $2P + 2Q$;*

(ii) *the net M contains $4P$, $4Q$ and $3P + Q$.*

In the second case, the net M is uniquely determined up to the action of $\mathrm{PGL}_2(\mathbb{C})$.

Proof Identify $H^0(\mathcal{O}_{\mathbb{P}^1}(4))$ with the vector space of quartic polynomials in variables x and y. Let V be the three-dimensional vector subspace in $H^0(\mathcal{O}_{\mathbb{P}^1}(4))$ that corresponds to M, and let $f(x,y)$, $g(x,y)$, $h(x,y)$ be a basis. Then the system of equations

$$\left.\begin{array}{l} f(x,y) = 0 \\ g(x,y) = 0 \\ h(x,y) = 0 \end{array}\right\} \tag{A.20}$$

has no solution in $\mathbb{P}^1$. We want to show that there are numbers a, b, c, α, β, γ such that

$$\alpha f + \beta g + \gamma h = \left(ax^2 + bxy + cy^2\right)^2 \text{ and } b^2 \neq 4ac \tag{A.21}$$

with one possible exception: when, after an appropriate linear change of variables x and y, the vector space V is generated by x^4, y^4 and $x^3 y$. Moreover, if

$V = \mathrm{span}(x^4, y^4, x^3 y)$, then the condition (A.21) is equivalent to the following system of equations:

$$
\begin{cases}
bc = 0, \\
b^2 = 2ac, \\
\alpha = a^2, \\
\beta = c^2, \\
\gamma = ab, \\
b^2 \neq 4ac,
\end{cases}
$$

which does not have solutions, so that this case is really an exception.

Let Π be the two-dimensional subspace in $\mathbb{P}^4 = \mathbb{P}(H^0(\mathcal{O}_{\mathbb{P}^1}(4)))$ that corresponds to $\mathcal{M}$. Since our $\mathbb{P}^4$ is equipped with the natural action of the group $\mathrm{PGL}_2(\mathbb{C})$, we are in a position to use notations of Lemma A.57. Let $\mathscr{S}$ be the closure of the $\mathrm{PGL}_2(\mathbb{C})$-orbit (2.1), and let $\mathscr{C}$ be the closure of the orbit (2.2). Then $\mathscr{C}$ is a curve, and $\mathscr{S}$ is a surface containing $\mathscr{C}$. We refer the reader to Section 7.2 for an explicit description of this curve and surface. Observe that the condition (A.21) holds $\iff \Pi \cap (\mathscr{S} \setminus \mathscr{C}) \neq \varnothing$. Since $\Pi \cap \mathscr{S} \neq \varnothing$, we see that (A.21) is satisfied if we do not assume that $b^2 \neq 4ac$. The inequality $b^2 \neq 4ac$ simply means that the corresponding point in $\Pi \cap \mathscr{S}$ is not in $\mathscr{C}$. In particular, if $\Pi \cap \mathscr{C} = \varnothing$, then we are done. Hence, we may assume that $\Pi \cap \mathscr{C} \neq \varnothing$. Therefore, applying an appropriate linear change of x and y, we may assume that $f = x^4$.

First, we suppose that $\Pi \cap \mathscr{C}$ contains at least two points. Applying an appropriate linear change of variable, we may assume that $h = y^4$. Now, we can choose $g \in V$ such that

$$
g = a_3 x^3 y + a_2 x^2 y^2 + a_1 x y^3
$$

for some numbers a_3, a_2 and a_1. If $|\Pi \cap \mathscr{C}| \geqslant 3$, then $g = \lambda(4x^3 + 6x^2 y^2 + 4xy^3)$ for $\lambda \in \mathbb{C}^*$, so that V contains $x^4, y^4, (x+y)^4, (x^2 + yx + y^2)^2$, and we are done. Therefore, we may assume that $\Pi \cap \mathscr{C}$ consists of two points, which correspond to x^4 and y^4.

If $a_3 = 0$ and $a_2 \neq 0$, we can scale both g and x to get either $g = x^2 y^2$ or $g = x^2 y^2 + y^4$. In the first case, we are done. In the second case, we have $(x - y)^2(x + y)^2 = f - 2g + 3h \in V$, which is exactly what we want. If $a_3 = 0$ and $a_2 = 0$, then we have $V = \mathrm{span}(x^4, y^4, xy^3)$, which is our exception up to a swap of x and y. Therefore, we may assume that $a_3 \neq 0$, so that we can replace g by g/a_3 and assume that $a_3 = 1$.

If $a_2 = a_1 = 0$, then we get our exceptional case. If $a_2 = 0$ and $a_1 \neq 0$, then scaling x, we may assume that $a_1 = 1$. In this case, we have

$$\left(x^2 + \sqrt{-2}xy + y^2\right)^2 = x^4 + y^4 + 2\sqrt{-2}\left(x^3y + xy^3\right) = f + 2\sqrt{-2}g + h \in V.$$

Thus, we may assume that $a_2 \neq 0$. Then, scaling x, we may assume that $a_2 = 1$. Then

$$\begin{cases} f(x,y) = x^4, \\ g(x,y) = x^3y + x^2y^2 + a_1xy^3, \\ h(x,y) = y^4. \end{cases}$$

In order to verify (A.21), it is enough to find some numbers α, β, γ, a, b and c such that

$$\alpha x^4 + \beta y^4 + \gamma\left(x^3y + x^2y^2 + a_1xy^3\right) = \left(ax^2 + bxy + cy^2\right)^2,$$

where $(a,b) \neq (0,0)$ and $(b,c) \neq (0,0)$, which guarantees that $b^2 \neq 4ac$ since we assume that the intersection $\Pi \cap \mathscr{C}$ consists of exactly two points. This gives

$$\begin{cases} \alpha = a^2, \\ \beta = c^2, \\ \gamma a_1 = 2bc, \\ \gamma = 2ab, \\ \gamma = 2ac + b^2. \end{cases}$$

Eliminating $\alpha = a^2$, $\beta = c^2$ and $\gamma = 2ab$, we obtain $aba_1 = bc$ and $2ab - 2ac - b^2 = 0$. Therefore, we can put $a = 1$, $c = a_1$ and then choose a non-zero b using $b^2 - 2b + 2a_1 = 0$. This gives us the required solution to (A.21) since $b \neq 0$.

To complete the proof, we may assume that the intersection $\Pi \cap \mathscr{C}$ consists of one point, which corresponds to the monomial $x^4 \in V$. Thus, in order to verify (A.21), it is enough to find some numbers α, β, γ, a, b and c such that

$$\alpha f + \beta g + \gamma h = \left(ax^2 + bxy + cy^2\right)^2 \text{ and } (b,c) \neq (0,0). \tag{A.22}$$

As before, we have $f = x^4$. But now we can choose g and h such that

$$\begin{cases} f = x^4 \\ g = a_3x^3y + a_2x^2y^2 + a_1xy^3 + a_0y^4 \\ h = b_2x^2y^2 + b_1xy^3 + b_0y^4 \end{cases}$$

for some numbers a_3, a_2, a_1, a_0, b_2, b_1, b_0.

First, let us consider the subcase $b_2 = 0$. Then $b_1 \neq 0$ since $|\Pi \cap \mathscr{C}| = 1$ by assumption. Therefore, dividing h by b_1, we may assume that $b_1 = 1$. Then, replacing g by $g - a_1h$, we may assume that $a_1 = 0$. If $b_0 = 0$, then $a_0 \neq 0$

since (A.20) has no solutions in $\mathbb{P}^1$, so that scaling x, we may assume $a_0 = 1$, which gives

$$\begin{cases} f = x^4, \\ g = a_3 x^3 y + a_2 x^2 y^2 + y^4, \\ h = xy^3, \end{cases}$$

where $(a_3, a_2) \neq (0,0)$ since $|\Pi \cap \mathscr{C}| = 1$, so that we can find $b \neq 0$ using $b^3 - a_2 b + a_3 = 0$, and let $a = \frac{a_2 - b^2}{2}$, $c = 1$, $\alpha = \frac{(a_2 - b^2)^2}{4}$, $\beta = 1$, $\gamma = 2b$, which gives us a solution to (A.22). Hence, to complete the proof, we may assume that $b_0 \neq 0$. Now, appropriately scaling y, we may also assume that $b_0 = 1$. Then $f = x^4$, $g = a_3 x^3 y + a_2 x^2 y^2 + a_0 y^4$, $h = xy^3 + y^4$. If $a_3 = 0$, then $a_2 \neq 0$, so that we can assume that $a_2 = 1$ by scaling x, which implies that $f = x^4$, $g = x^2 y^2 + a_0 y^4$, $h = xy^3 + y^4$, so that we can find $b \neq 0$ using $a_0 b^2 + 2b - 1 = 0$, and let $a = \alpha = 0$, $c = 1$, $\beta = b^2$, $\gamma = 2b$, which is a required solution to (A.22). Hence, we may assume that $a_3 = 1$. Then $f = x^4$, $g = x^3 y + a_2 x^2 y^2 + a_0 y^4$, $h = xy^3 + y^4$. If $a_0 \neq 0$, then one solution to (A.22) is given by $a = \frac{\xi(\xi - 2)}{2a_0}$, $b = 1$, $c = \xi$, $\alpha = \frac{\xi^2(\xi - 2)^2}{4a_0^2}$, $\beta = \frac{\xi(\xi - 2)}{a_0}$, $\gamma = 2\xi$, where ξ is a root of $x^3 - (a_2 + 2)x^2 + 2a_2 x + a_0$. If $a_0 = 0$ and $a_2 = 0$, then

$$\left(x^2 - 4xy - 8y^2\right)^2 = x^4 - 8x^3 y + 64(xy^3 + y^4) = f - 8g + 64h \in V.$$

If $a_0 = 0$ and $a_2 \neq 0$, then $(a, b, c, \alpha, \beta, \gamma) = (1, 2a_2, 0, 1, 4a_2, 0)$ gives a solution to (A.22). This proves the required assertion in the subcase when $b_2 = 0$.

We may assume that $b_2 \neq 0$, so that replacing h by h/b_2, we may assume that $b_2 = 1$. Then, swapping g with $g - a_2 h$, we may assume that $a_2 = 0$. If $b_1 - b_0 = 0$, then $x^2 y^2 \in V$. Similarly, if $b_1 = 0$ and $b_0 \neq 0$, then $(x^2 + 2b_0 y^2)^2 = x^4 + 4b_0(x^2 y^2 + b_0 y^4) = f + 4b_0 h \in V$. Thus, if $b_1 = 0$, then we are done. Hence, we may assume that $b_1 \neq 0$. Then, scaling x, we may assume that $b_1 = 1$, so that we have $f = x^4$, $g = a_3 x^3 y + a_1 xy^3 + a_0 y^4$, $h = x^2 y^2 + xy^3 + b_0 y^4$. If $b_0 = \frac{1}{4}$, then $4h = (2x + y)^2 y^2$ and we are done. So, we may assume that $b_0 \neq \frac{1}{4}$. Moreover, if $a_3 = 0$, then $a_1 \neq 0$ since otherwise V would contain x^4 and y^4, which is excluded by our assumption that $|\Pi \cap \mathscr{C}| = 1$. Hence, if $a_3 = 0$, then

$$\left(a_1 x^2 + 2(a_1 b_0 - a_0)y^2\right)^2 = a_1^2 f - 4(a_1 b_0 - a_0)g + 4a_1(a_1 b_0 - a_0)h \in V,$$

so that we are done if $a_1 b_0 \neq a_0$. If $a_3 = 0$ and $a_1 b_0 = a_0$, then $a_1 x^2 y^2 = a_1 h - g \in V$, so that we are also done. Therefore, to complete the proof, we may assume that $a_3 \neq 0$. Then, dividing g by a_3, we may also assume that $a_3 = 1$. Thus, we have $f = x^4$, $g = x^3 y + a_1 xy^3 + a_0 y^4$ and $h = x^2 y^2 + xy^3 + b_0 y^4$.

Now let us try to find a solution to (A.22) with $a = 1$, $\alpha = 1$, $\beta = 2b$ and $\gamma = b^2 + 2c$. Then to complete this to a solution to (A.22), we must also have $(b,c) \neq (0,0)$ and

$$\begin{cases} b^2 b_0 + 2c b_0 + 2a_0 b - c^2 = 0, \\ 2a_1 b + b^2 - 2bc + 2c = 0. \end{cases}$$

If $b \notin \{0,1\}$, the second equation gives $c = \frac{(2a_1+b)b}{2(b-1)}$, so that the first equation simplifies as

$$(4b_0 - 1)b^3 + (8a_0 - 4a_1 - 4b_0)b^2 + (8a_1 b_0 - 4a_1^2 - 16a_0)b + (8a_0 - 8a_1 b_0) = 0. \tag{A.23}$$

This polynomial equation in b always has a solution because we assumed that $b_0 \neq \frac{1}{4}$. Moreover, if b is a solution to (A.23) and $b \notin \{0,1\}$, than we can let $a = 1$, $c = \frac{(2a_1+b)b}{2(b-1)}$, $\alpha = 1$, $\beta = 2b$, $\gamma = b^2 + \frac{(2a_1+b)b}{(b-1)}$ to get a solution to (A.22). Hence, if (A.23) has a solution $b \notin \{0,1\}$, then we are done. Observe also that $b = 0$ is a solution to (A.23) if and only if $a_0 = a_1 b_0$. Similarly, $b = 1$ is a solution to (A.23) if and only if $a_1 = -\frac{1}{2}$. Moreover, if $a_1 = -\frac{1}{2}$, then (A.23) simplifies as $(b-1)^2((4b_0 - 1)b + 8a_0 + 4b_0) = 0$. Thus, if $a_1 = -\frac{1}{2}$, $a_0 \neq \frac{1-8b_0}{8}$, $b_0 \neq -2a_0$, then $b = \frac{8a_0 + 4b_0}{1-4b_0}$ satisfies (A.23) and $b \notin \{0,1\}$, so that we are done. On the other hand, if $a_1 = -\frac{1}{2}$ and $a_0 = \frac{1-8b_0}{8}$, then

$$\left(2x^2 + 2xy + y^2\right)^2$$
$$= 4x^4 + \left(8x^3 y - 4xy^3 + (1 - 8b_0)y^4\right) + 8\left(x^2 y^2 + xy^3 + b_0 y^4\right)$$
$$= 4f + 8g + 8h \in V,$$

which is exactly what we need. Similarly, if $a_1 = -\frac{1}{2}$ and $b_0 = -2a_0$, then

$$\left(x^2 + xy - 4a_0 y^2\right)^2 = x^4 + 2\left(x^3 - \frac{xy^3}{2} + a_0\right) + 2\left(x^2 y^2 + xy^3 - 2a_0 y^4\right)$$
$$= f + 2g + (1 - 8a_0)h \in V,$$

which gives a solution to (A.22). Hence, if $a_1 = -\frac{1}{2}$, then the required assertion is proved. Therefore, we may assume that $a_1 \neq -\frac{1}{2}$. Then $b = 1$ is not a solution of (A.23).

If $a_0 = a_1 b_0$, then (A.23) simplifies as $b(b + 2a_1)((1 - 4b_0)b + 2a_1 + 4b_0) = 0$, so that $b = -2a_1$ gives us a solution to (A.23) such that $b \notin \{0,1\}$ provided that $a_1 \neq 0$ Hence, if $a_0 = a_1 b_0$ and $a_1 \neq 0$, then we are done. Similarly, if $a_0 = a_1 = 0$, then $b = \frac{4b_0}{4b_0-1}$ gives a solution to the equation (A.23) such that $b \notin \{0,1\}$, because $b_0 \neq 0$ in this case since (A.20) does not have solutions in $\mathbb{P}^1$. Therefore, we proved that (A.22) has a solution, proving the lemma. □

Let us conclude with the following result (cf. [221] and [45, §10]).

Lemma A.60 ([15, 88])   *Let X be a smooth divisor in $\mathbb{P}^2 \times \mathbb{P}^2$ that has degree $(1,2)$. Then one can choose coordinates $([x : y : z], [u : v : w])$ on $\mathbb{P}^2 \times \mathbb{P}^2$ such that X is given by one of the following three equations:*

(i) $(\mu vw + u^2)x + (\mu uw + v^2)y + (\mu uv + w^2)z = 0$ *for some $\mu \in \mathbb{C}$ such that $\mu^3 \neq -1$,*

(ii) $(vw + u^2)x + (uw + v^2)y + w^2z = 0,$

(iii) $(vw + u^2)x + v^2y + w^2z = 0.$

Proof   To prove the required assertion, it is enough to show that X can be given by

$$\left(a_1vw + a_2u^2\right)x + \left(b_1uw + b_2v^2\right)y + \left(c_1uv + c_2w^2\right)z = 0 \qquad \text{(A.24)}$$

for some numbers a_1, a_2, b_1, b_2, c_1 and c_2. Indeed, suppose that X is given by (A.24). Then $a_2b_2c_2 \neq 0$ because X is smooth. Thus, scaling u, v and w appropriately, we may assume that $a_2 = b_2 = c_2 = 1$. Choose a, b and c such that $a^3 = a_1$, $b^3 = b_1$ and $c^3 = c_1$. If $abc \neq 0$, we scale our coordinates as $x \mapsto x$, $y \mapsto yt^2$, $z \mapsto zs^2$, $u \mapsto u$, $v \mapsto \frac{v}{s}$, $w \mapsto \frac{w}{t}$ for $s = \frac{a}{c}$ and $t = \frac{a}{b}$. Then we are in case (1) with $\mu = abc$, and X is singular if and only if $\mu^3 = -1$, so that the remaining assertions follow from [76]. Similarly, if $abc = 0$, then we can scale and permute the coordinates accordingly to get either case (2) or case (1).

Now we prove that we can choose u, v, w, x, y, z so that X is given by (A.24).

Let $\mathrm{pr}_1 \colon X \to \mathbb{P}^2$ be the projection to the first factor. Then pr_1 is a conic bundle, whose discriminant curve $\mathscr{C}$ is a cubic curve. Since X is smooth, $\mathscr{C}$ is either smooth or nodal. If $\mathscr{C}$ is reducible, the required assertion is well-known (see [221] or [45, §10]). Thus, we may assume that $\mathscr{C}$ is irreducible. Then it follows from [76] that we can choose coordinates x, y and z such that $\mathscr{C}$ is given by

$$\alpha x^3 + \beta y^3 + \gamma z^3 + \delta xyz = 0 \qquad \text{(A.25)}$$

for some α, β, γ and δ such that $\alpha \neq 0$ and $\beta \neq 0$. To prove the required assertion, it is enough to choose the coordinates u, v, w such that X is given by the equation (A.24). In the following, we will not change the coordinates x, y and z except for scaling (once).

Let C_x, C_y, C_z be the fibers of the conic bundle pr_1 over $[1 : 0 : 0]$, $[0 : 1 : 0]$, $[0 : 0 : 1]$, respectively. Since $\mathscr{C}$ contains neither $[1 : 0 : 0]$ nor $[0 : 1 : 0]$, both C_x and C_y are smooth. In particular, we can choose u, v and w such that C_x is given by $vw + u^2 = y = z = 0$. Then X is given by

$$\left(vw + u^2\right)x + f_2(u, v, w)y + f_3(u, v, w)z = 0,$$

where $f_2(u, v, w)$ and $f_3(u, v, w)$ are some quadratic polynomials such that C_y is given by the equation $f_2(u, v, w) = x = z = 0$, and the curve C_z is given by $f_3(u, v, w) = x = y = 0$. Abusing notations, we consider all three curves C_x, C_y and C_z as conics in one plane $\mathbb{P}^2$, which are given by the equations $vw + u^2 = 0$, $f_2(u, v, w) = 0$, $f_3(u, v, w) = 0$, respectively. If $\mathscr{C}$ is singular, then $[0 : 0 : 1] = \mathrm{Sing}(\mathscr{C})$, so that C_z is a double line.

Observe that $C_x \cap C_y \cap C_z = \varnothing$ since X is smooth. But $C_x \cap C_y \neq \varnothing$ and $C_x \cap C_z \neq \varnothing$. Therefore, since $\mathrm{Aut}(\mathbb{P}^2; C_x) \cong \mathrm{PGL}_2(\mathbb{C})$ and this group acts faithfully on $C_x \cong \mathbb{P}^1$, we can choose u, v and w such that $[0 : 0 : 1] \in C_y$ and $[0 : 1 : 0] \in C_z$. Then

$$f_2(u, v, w) = a_1 v^2 + a_2 u^2 + a_3 vu + a_4 vw + a_5 uw,$$
$$f_3(u, v, w) = b_1 w^2 + b_2 u^2 + b_3 vu + b_4 vw + b_5 uw,$$

where a_1, a_2, a_3, a_4, a_5, b_1, b_2, b_3, b_4, b_5 are some numbers. Note that we still have some freedom in changing the coordinates u, v and w. Namely, the subgroup in $\mathrm{Aut}(\mathbb{P}^2; C_x)$ that preserves the subset $\{[0 : 0 : 1], [0 : 1 : 0]\}$ is $\mathbb{G}_m \rtimes \mu_2$, where the $\mathbb{G}_m$-action is just the scaling $u \mapsto u, v \mapsto sv, w \mapsto \frac{w}{s}$ for $s \in \mathbb{C}^*$. Using this scaling, we could get the following new equation for our threefold:

$$(u^2 + vw)x + \left(s^2 a_1 v^2 + a_2 u^2 + a_3 svu + a_4 vw + \frac{a_5}{s} uw\right)y$$
$$+ \left(\frac{b_1}{s^2} w^2 + b_2 u^2 + sb_3 vu + b_4 vw + \frac{b_5}{s} uw\right)z = 0,$$

where $a_1 b_1 \neq 0$ since $C_x \cap C_y \cap C_z = \varnothing$. Thus, if $a_5 \neq 0$, we can scale v, w, x and z such that $a_1 = a_5 = b_1 = 1$. Similarly, if $a_5 = 0$, we can scale y and z to get $a_1 = b_1 = 1$. Therefore, we can assume that $a_1 = b_1 = 1$, and either $a_5 = 0$ or $a_5 = 1$. Note also that

$$2a_3 a_4 a_5 - 2a_5^2 - 2a_2 a_4^2 \neq 0 \tag{A.26}$$

because the conic C_y is smooth.

Now we compute the equation of the curve $\mathscr{C}$ using the equation of the threefold X. Namely, the curve $\mathscr{C}$ is given by

$$x^3 - \left(a_3 a_4 a_5 - a_2 a_4^2 - a_5^2\right)y^3 - \left(b_3 b_4 b_5 - b_2 b_4^2 - b_3^2\right)z^3$$
$$- \left(4 - 2a_2 b_4 + a_3 b_5 - 2a_4 b_2 - 2a_4 b_4 + a_5 b_3\right)xyz + \left(a_2 + 2a_4\right)x^2 y$$
$$- \left(b_2 + 2b_4\right)x^2 z - \left(a_3 a_5 - 2a_2 a_4 - a_4^2\right)xy^2$$
$$- \left(b_3 b_5 - 2b_2 b_4 - b_4^2\right)xz^2$$
$$- \left(4a_2 - 2a_2 a_4 b_4 + a_3 a_4 b_5 + a_3 a_5 b_4 - a_4^2 b_2 + a_4 a_5 b_3 - a_3^2 - 2a_5 b_5\right)y^2 z$$
$$- \left(4b_2 - a_2 b_4^2 + a_3 b_4 b_5 - 2a_4 b_2 b_4 + a_4 b_3 b_5 + a_5 b_3 b_4 - 2a_3 b_3 - b_5^2\right)yz^2 = 0.$$

Thus, since $\mathscr{C}$ is given by (A.25), we obtain the following equations:

$$a_2 + 2a_4 = 0, \quad b_2 + 2b_4 = 0, \quad a_3a_5 - 2a_2a_4 - a_4^2 = 0, \quad b_3b_5 - 2b_2b_4 - b_4^2 = 0,$$

$$4a_2 - 2a_2a_4b_4 + a_3a_4b_5 + a_3a_5b_4 - a_4^2b_2 + a_4a_5b_3 - a_3^2 - 2a_5b_5 = 0,$$

$$4b_2 - a_2b_4^2 + a_3b_4b_5 - 2a_4b_2b_4 + a_4b_3b_5 + a_5b_3b_4 - 2a_3b_3 - b_5^2 = 0.$$

Substituting $a_2 = -2a_4$ and $b_2 = -2b_4$ into the third equation, we get $2a_3a_5 + 6a_4^2 = 0$. Hence, if $a_5 = 0$, then $a_4 = 0$, which contradicts (A.26). Therefore, we see that $a_5 = 1$. Then equations simplify as

$$a_2 = -2a_4, \quad b_2 = -2b_4, \quad 3a_4^2 + a_3 = 0, \quad b_3b_5 + 3b_4^2 = 0,$$

$$a_3a_4b_5 + 6a_4^2b_4 - a_3^2 + a_3b_4 + a_4b_3 - 8a_4 - 2b_5 = 0,$$

$$a_3b_4b_5 + a_4b_3b_5 + 6a_4b_4^2 - 2a_3b_3 + b_3b_4 - b_5^2 - 8b_4 = 0,$$

so that $a_3 = -3a_4^2$. In particular, the threefold X is given by

$$(u^2 + vw)x + (v^2 + uw - 3a_4^2uv - 2a_4u^2 + a_4vw)y$$
$$+ (b_3uv - 2b_4u^2 + b_4vw + b_5uw + w^2)z = 0.$$

Now we change our u, v and w as follows: $u \mapsto w - a_4v$, $v \mapsto v$, $w \mapsto 2a_4w - a_4^2v - u$. Then, in new coordinates, the threefold X is given by the equation:

$$(u^2 + vw)x + (uw + av^2)y + (u^2 + c_1w^2 + c_2v^2 + c_3vu + c_4vw + c_5uw)z = 0,$$

where $a = a_4^3 + 1$, $c_1 = 4a_4^2 + 2a_4b_5 - 2b_4$, $c_2 = a_4^4 + a_4^3b_5 - 3a_4^2b_4 - a_4b_3$, $c_3 = -2a_4^3 - a_4b_5 + b_4$, $c_4 = -4a_4^3 - 3a_4^2b_5 + 6a_4b_4 + b_3$, $c_5 = 4a_4 + b_5$. Now, recomputing again the equation of the cubic curve $\mathscr{C}$ in terms of a, c_1, c_2, c_3, c_4, c_5, we see that $\mathscr{C}$ is given by

$$x^3 + ay^3 + (c_1c_3^2 + c_2c_5^2 - c_3c_4c_5 - 4c_1c_2 + c_4^2)z^3$$
$$- (4ac_1 + c_3)xyz + (2c_4 + 1)x^2z$$
$$+ (2ac_5 + c_2)y^2z - (4c_1c_2 + c_3c_5 - c_4^2 - 2c_4)xz^2$$
$$+ (ac_5^2 - 4ac_1 + 2c_2c_5 - c_3c_4)yz^2 = 0.$$

As above, this gives $2c_4 + 1 = 0$, $2ac_5 + c_2 = 0$, $4c_1c_2 + c_3c_5 - c_4^2 - 2c_4 = 0$, $ac_5^2 - 4ac_1 + 2c_2c_5 - c_3c_4 = 0$, so that $c_4 = -\frac{1}{2}$ and $c_2 = -2ac_5$. This gives $c_4 = -\frac{1}{2}$, $c_2 = -2ac_5$, $-8ac_1c_5 + c_3c_5 + \frac{3}{4} = 0$, $3ac_5^2 + 4ac_1 - \frac{c_3}{2} = 0$. Then $c_3 = 6ac_5^2 + 8ac_1$. Substituting this into $-8ac_1c_5 + c_3c_5 + \frac{3}{4} = 0$, we get $6ac_5^3 + \frac{3}{4} = 0$. In particular, we see that $c_5 \neq 0$. Summarizing, we see that $c_5 \neq 0$, $c_4 = -\frac{1}{2}$, $c_2 = -2ac_5$, $c_3 = 6ac_5^2 + 8ac_1$, $a = -\frac{1}{8c_5^3}$. Therefore, our threefold X is given by

$$\left(u^2+vw\right)x+\left(uw-\frac{v^2}{8c_5^3}\right)y+\left(c_1w^2+\frac{v^2}{4c_5^2}-\frac{3uv}{4c_5}-\frac{uvc_1}{c_5^3}-\frac{vw}{2}+c_5uw+u^2\right)z=0.$$

Now, if we change u, v and w as $u \mapsto c_5(u+v+2w)$, $v \mapsto 4c_5^2u+c_5^2v-4c_5^2w$, $w \mapsto 2u-v+w$, then X would be given by

$$\left(u^2c_5^2+c_5^2vw\right)x+\left(c_5uw-\frac{v^2c_5}{8}\right)y+\left((2c_5^2+c_1)w^2+\frac{(c_5^2-4c_1)vu}{4}\right)z,$$

which is a special case of (A.24). This completes the proof of the lemma.   □

References

[1] A. Adler, *On the automorphism group of a certain cubic threefold*, Am. J. Math. **100** (1978), 1275–1280.

[2] H. Abban, Z. Zhuang, *K-stability of Fano varieties via admissible flags*, Forum Math. Pi **10** (2022), Paper No. e15, 43 pp.

[3] H. Abban, Z. Zhuang, *Seshadri constants and K-stability of Fano manifolds*, preprint, arXiv:2101.09246 (2021).

[4] V. Alexeev, *On general elephants of $\mathbb{Q}$-Fano 3-folds*, Compos. Math. **90** (1994), 91–116.

[5] J. Alper, J. Hall, D. Rydh, *A Luna étale slice theorem for algebraic stacks*, Ann. Math. **191** (2020), 675–738.

[6] J. Alper, H. Blum, D. Halpern-Leistner, C. Xu, *Reductivity of the automorphism group of K-polystable Fano varieties*, Invent. Math. **222** (2020), 995–1032.

[7] C. Arezzo, A. Ghigi, G. Pirola, *Symmetries, quotients and Kähler–Einstein metrics*, J. Reine Angew. Math. **591** (2006), 177–200.

[8] A. Avilov, *Automorphisms of singular three-dimensional cubic hypersurfaces*, Eur. J. Math. **4** (2018), 761–777.

[9] A. Avilov, *Biregular and birational geometry of quartic double solids with 15 nodes*, Izv. Math. **83** (2019), 415–423.

[10] W. Barth, *Moduli of vector bundles on the projective plane*, Invent. Math. **42** (1977), 63–91.

[11] W. Barth, *Two projective surfaces with many nodes, admitting the symmetries of the icosahedron*, J. Algebraic Geom. **5** (1996), 173–186.

[12] V. Batyrev, *Toroidal Fano 3-folds*, Math. USSR, Izv. **19** (1982), 13–25.

[13] V. Batyrev, D. Cox, *On the Hodge structure of projective hypersurfaces in toric varieties*, Duke Math. J. **75** (1994), 293–338.

[14] V. Batyrev, E. Selivanova, *Einstein–Kähler metrics on symmetric toric Fano manifolds*, J. Reine Angew. Math. **512** (1999), 225–236.

[15] A. Beauville, *Variétés de Prym et jacobiennes intermédiaires*, Ann. Sci. Éc. Norm. Supér. **10** (1977), 309–391.

[16] G. Belousov, K. Loginov, *K-stability of Fano threefolds of rank 4 and degree 24*, Eur. J. of Math. **8** (2022), 834–852.

[17] J. Blanc, S. Lamy, *Weak Fano threefolds obtained by blowing-up a space curve and construction of Sarkisov links*, Proc. Lond. Math. Soc. **105** (2012), 1047–1075.

[18] H. Blum, M. Jonsson, *Thresholds, valuations, and K-stability*, Adv. Math. **365** (2020), 107062.

[19] H. Blum, Y. Liu, *Openness of uniform K-stability in families of $\mathbb{Q}$-Fano varieties*, to appear in Ann. Sci. Éc. Norm. Supér.

[20] H. Blum, Y. Liu, C. Xu, *Openness of K-semistability for Fano varieties*, preprint, arXiv:1907.02408 (2019).

[21] H. Blum, C. Xu, *Uniqueness of K-polystable degenerations of Fano varieties*, Ann. Math. **190** (2019), 609–656.

[22] S. Boucksom, *Corps d'Okounkov*, Astérisque **361** (2014), 1–41.

[23] S. Boucksom, H. Chen, *Okounkov bodies of filtered linear series*, Compos. Math. **147** (2011), 1205–1229.

[24] S. Boucksom, T. Hisamoto, M. Jonsson, *Uniform K-stability, Duistermaat–Heckman measures and singularities of pairs*, Ann. Inst. Fourier (Grenoble) **67** (2017), 743–841.

[25] J. Bruce, T. Wall, *On the classification of cubic surfaces*, J. Lond. Math. Soc. (1979) **19**, 245–256.

[26] T. Brönnle, *Deformation constructions of extremal metrics*, Ph.D. Thesis, Imperial College London, 2011.

[27] G. Brown, A. Kasprzyk, *Graded Ring Database*, www.grdb.co.uk

[28] I. Cheltsov, *Log canonical thresholds on hypersurfaces*, Sb. Math. **192** (2001), 1241–1257.

[29] I. Cheltsov, *Birationally rigid Fano varieties*, Russian Math. Surveys **60** (2005), 875–965.

[30] I. Cheltsov, *Log canonical thresholds of del Pezzo surfaces*, Geom. Funct. Anal. **18** (2008), 1118–1144.

[31] I. Cheltsov, *On singular cubic surfaces*, Asian J. Math. **13** (2009), 191–214.

[32] I. Cheltsov, *Del Pezzo surfaces and local inequalities*, Springer Proc. Math. Stat. **79** (2014), 83–101.

[33] I. Cheltsov, *Two local inequalities*, Izv. Math. **78** (2014), 375–426.

[34] I. Cheltsov, K. Fujita, T. Kishimoto, T. Okada, *K-stable divisors in $\mathbb{P}^1 \times \mathbb{P}^1 \times \mathbb{P}^2$ of degree* $(1,1,2)$, to appear in Nagoya Math. J.

[35] I. Cheltsov, D. Kosta, *Computing α-invariants of singular del Pezzo surfaces*, J. Geom. Anal. **24** (2014), 798–842.

[36] I. Cheltsov, A. Kuznetsov, K. Shramov, *Coble fourfold, $\mathfrak{S}_6$-invariant quartic threefolds, and Wiman-Edge sextics*, Algebra Number Theory **14** (2020), 213–274.

[37] I. Cheltsov, J. Park, *Sextic double solids*, Birkhäuser, Progr. Math. **282** (2010), 75–132.

[38] I. Cheltsov, J. Park, *K-stable Fano threefolds of rank 2 and degree 30*, Eur. J. of Math. **8** (2022), 834–852.

[39] I. Cheltsov, J. Park, J. Won, *Log canonical thresholds of certain Fano hypersurfaces*, Math. Z. **276** (2014), 51–79.

[40] I. Cheltsov, J. Park, J. Won, *Affine cones over smooth cubic surfaces*, J. Eur. Math. Soc. **18** (2016), 1537–1564.

[41] I. Cheltsov, J. Park, J. Won, *Cylinders in del Pezzo surfaces*, Int. Math. Res. Not. IMRN **2017** (2017), 1179–1230

[42] I. Cheltsov, Yu. Prokhorov, *Del Pezzo surfaces with infinite automorphism groups*, Algebr. Geom. **8** (2021), 319–357.

[43] I. Cheltsov, V. Przyjalkowski, C. Shramov, *Quartic double solids with icosahedral symmetry*, Eur. J. Math. **2** (2016), 96–119.

[44] I. Cheltsov, V. Przyjalkowski, C. Shramov, *Burkhardt quartic, Barth sextic, and the icosahedron*, Int. Math. Res. Not. IMRN **12** (2019), 3683–3703.

[45] I. Cheltsov, V. Przyjalkowski, C. Shramov, *Fano threefolds with infinite automorphism groups*, Izv. Math. **83** (2019), 860–907.

[46] I. Cheltsov, C. Shramov, *Log canonical thresholds of smooth Fano threefolds*, Russian Math. Surveys **63** (2008), 71–178.

[47] I. Cheltsov, C. Shramov, *Extremal metrics on del Pezzo threefolds*, Proc. Steklov Inst. Math. **264** (2009), 30–44.

[48] I. Cheltsov, C. Shramov, *On exceptional quotient singularities*, Geom. Topol. **15** (2011), 1843–1882.

[49] I. Cheltsov, C. Shramov, *Weakly-exceptional singularities in higher dimensions*, J. Reine Angew. Math. **689** (2014), 201–241.

[50] I. Cheltsov, C. Shramov, *Three embeddings of the Klein simple group into the Cremona group of rank three*, Transform. Groups **17** (2012), 303–350.

[51] I. Cheltsov, C. Shramov, *Five embeddings of one simple group*, Trans. Amer. Math. Soc. **366** (2014), 1289–1331.

[52] I. Cheltsov, C. Shramov, *Two rational nodal quartic 3-folds*, Q. J. Math. **67** (2016), 573–601.

[53] I. Cheltsov, C. Shramov, *Cremona Groups and the Icosahedron*, CRC Press, 2016.

[54] I. Cheltsov, C. Shramov, *Finite collineation groups and birational rigidity*, Selecta Math. (N.S.) **25** (2019), 71.

[55] I. Cheltsov, C. Shramov, *Kaehler–Einstein Fano threefolds of degree 22*, to appear in J. Algebraic Geom.

[56] I. Cheltsov, C. Shramov, *K-polystability of two smooth Fano threefolds*, to appear in Springer Proc. Math. Stat.

[57] I. Cheltsov, A. Wilson, *Del Pezzo surfaces with many symmetries*, J. Geom. Anal. **23** (2013), 1257–1289.

[58] I. Cheltsov, K. Zhang, *Delta invariants of smooth cubic surfaces*, Eur. J. Math. **5** (2019), 729–762.

[59] X.-X. Chen, S. Donaldson, S. Sun, *Kähler-Einstein metrics on Fano manifolds. I, II, III*, J. Amer. Math. Soc. **28** (2015), no. 1, 183–197, 199–234, 235–278.

[60] T. Coates, A. Corti, S. Galkin, A. Kasprzyk, *Quantum periods for 3-dimensional Fano manifolds*, Geom. Topol. **20** (2016), 103–256.

[61] A. Corti, *Singularities of linear systems and 3-fold birational geometry*, London Math. Soc. Lecture Note Ser. **281** (2000), 259–312.

[62] A. Corti, J. Kollár, K. Smith, *Rational and Nearly Rational Varieties*, Cambridge University Press, 2003.

[63] G. Codogni, Z. Patakfalvi, *Positivity of the CM line bundle for families of K-stable klt Fano varieties*, Invent. Math. **223** (2021), 811–894.

[64] J. H. Conway, R. T. Curtis, S. P. Norton, R. A. Parker, R. A. Wilson, *Atlas of Finite Groups*, Oxford University Press, 1985.

[65] D. Coray, M. Tsfasman, *Arithmetic on singular Del Pezzo surfaces*, Proc. Lond. Math. Soc. **57** (1988), 25–87.

[66] I. Coskun, E. Riedl, *Normal bundles of rational curves in projective space*, Math. Z. **288** (2018), 803–827.

[67] V. Datar, G. Székelyhidi, *Kähler–Einstein metrics along the smooth continuity method*, Geom. Funct. Anal. **26** (2016), 975–1010.

[68] E. Denisova, *On K-stability of* $\mathbb{P}^3$ *blown up along the disjoint union of a twisted cubic curve and a line*, preprint, arXiv:2202.04421 (2022).

[69] R. Dervan, *On K-stability of finite covers*, Bull. Lond. Math. Soc. **48** (2016), 717–728.

[70] S. Dinew, G. Kapustka, M. Kapustka, *Remarks on Mukai threefolds admitting* $\mathbb{C}^*$*-action*, Mosc. Math. J. **17** (2017), 15–33.

[71] W. Ding, G. Tian, *Kähler–Einstein metrics and the generalized Futaki invariants*, Invent. Math. **110** (1992), 315–335.

[72] I. Dolgachev, *Invariant stable bundles over modular curves* $X(p)$, in: Recent Progress in Algebra (Taejon/Seoul, 1997), Contemp. Math. **224**, Amer. Math. Soc., 1999, 65–99.

[73] I. Dolgachev, *Lectures on Invariant Theory*, Cambridge University Press, 2003.

[74] I. Dolgachev, *Classical Algebraic Geometry. A Modern View*, Cambridge University Press, 2012.

[75] I. Dolgachev, *Lectures on Cremona transformations*, unpublished lecture notes, https://dept.math.lsa.umich.edu/~idolga/lecturenotes.html

[76] I. Dolgachev, M. Artebani, *The Hesse pencil of plane cubic curves*, Enseign. Math. **55** (2009), 235–273.

[77] I. Dolgachev, V. Iskovskikh, *Finite subgroups of the plane Cremona group*, Progr. Math. **269**, Birkhäuser Boston, 2009, 443–548.

[78] S. Donaldson, *Scalar curvature and stability of toric varieties*, J. Differential Geom. **62** (2002), no. 2, 289–349.

[79] S. Donaldson, *Kähler geometry on toric manifolds, and some other manifolds with large symmetry*, Handbook of Geometric Analysis, Adv. Lect. Math. (ALM) **7** (2008), 29–75.

[80] S. Donaldson, *Algebraic families of constant scalar curvature Kähler metrics*, Surv. Differ. Geom. **19**, Int. Press, 2015.

[81] S. Donaldson, *Stability of algebraic varieties and Kahler geometry*, Proc. Sympos. Pure Math. **97** (2018), 199–221.

[82] S. Donaldson, S. Sun, *Gromov-Hausdorff limits of Kähler manifolds and algebraic geometry*, Acta Math. **213** (2014), 63–106.

[83] R. Dye, *Pencils of elliptic quartics and an identification of Todd's quartic combinant*, Proc. Lond. Math. Soc. **34** (1977), 459–478.

[84] W. Edge, *A plane quartic curve with twelve undulations*, Edinburgh Math. Notes **1945** (1945), 10–13.

[85] W. Edge, *The Klein group in three dimensions*, Acta Math. **79** (1947), 153–223.

[86] W. Edge, *The principal chords of an elliptic quartic*, Proc. Roy. Soc. Edinburgh Sect. A **71** (1972), 43–50.

[87] L. Ein, R. Lazarsfeld, M. Mustata, M. Nakamaye, M. Popa, *Restricted volumes and base loci of linear series*, Amer. J. Math. **131** (2009), 607–651.

[88] J. Emsalem, A. Iarrobino, *Réseaux de coniques et Algébres de longueur sept associés*, preprint, unpublished, 1972.

[89] A. Golota, *Delta-invariants for Fano varieties with large automorphism groups*, Internat. J. Math. **31** no. 10 (2020), 2050077, 31 pp.

[90] M. Green, *Koszul cohomology and the geometry of projective varieties*, J. Differential Geom. **19** (1984), 125–167.

[91] H. Flenner, M. Zaidenberg, *Locally nilpotent derivations on affine surfaces with a $\mathbb{G}_m$-action*, Osaka J. Math. **42** (2005), 931–974.

[92] O. Fujino, *Foundations of the Minimal Model Program*, MSJ Memoirs **35** (2017), Mathematical Society of Japan, Tokyo.

[93] K. Fujita, *On K-stability and the volume functions of $\mathbb{Q}$-Fano varieties*, Proc. Lond. Math. Soc. **113** (2016), 541–582.

[94] K. Fujita, *K-stability of Fano manifolds with not small alpha invariants*, J. Inst. Math. Jussieu **18** (2019), 519–530.

[95] K. Fujita, *A valuative criterion for uniform K-stability of $\mathbb{Q}$-Fano varieties*, J. Reine Angew. Math. **751** (2019), 309–338.

[96] K. Fujita, *Uniform K-stability and plt blow ups of log Fano pairs*, Kyoto J. Math. **59** (2019), 399–418.

[97] K. Fujita, *On K-polystability for log del Pezzo pairs of Maeda type*, Acta Math. Vietnam **45** (2020), 943–965.

[98] K. Fujita, *K-stability of log Fano hyperplane arrangements*, J. Algebraic Geom. **30**, no. 4 (2021), 603–630.

[99] K. Fujita, *Toward criteria for K-stability of log Fano pairs*, to appear in Proceedings of the 64th Algebra Symposium at Tohoku university.

[100] K. Fujita , *On Fano threefolds of degree 22 after Cheltsov and Shramov*, preprint, arXiv:2107.04816 (2021).

[101] K. Fujita, *On K-stability for Fano threefolds of rank 3 and degree 28*, to appear in Int. Math. Res. Notices.

[102] K. Fujita, Y. Odaka, *On the K-stability of Fano varieties and anticanonical divisors*, Tohoku Math. J. **70** (2018), 511–521.

[103] T. Fujita, *On the structure of polarized manifolds with total deficiency one. I*, J. Math. Soc. Japan **32** (1980), 709–725.

[104] T. Fujita, *On the structure of polarized manifolds with total deficiency one. II*, J. Math. Soc. Japan **33** (1981), 415–434.

[105] T. Fujita, *On polarized varieties of small Delta-genera*, Tohoku Math. J. **34** (1982), 319–341.

[106] T. Fujita, *Vanishing theorems for semipositive line bundles*, Lecture Notes in Math. **1016** (1983), 519–528.

[107] T. Fujita, *On the structure of polarized manifolds with total deficiency one. III*, J. Math. Soc. Japan **36** (1984), 75–89.

[108] T. Fujita, *On singular Del Pezzo varieties*, Lecture Notes in Math. **1417** (1990), 117–128.

[109] M. Furushima, N. Nakayama, *The family of lines on the Fano threefold V_5*, Nagoya Math. J. **116** (1989), 111–122.

[110] A. Futaki, *An obstruction to the existence of Einstein–Kähler metrics*, Invent. Math. **73** (1983), 437–443.

[111] B. van Geemen, T. Yamauchi, *On intermediate Jacobians of cubic threefolds admitting an automorphism of order five*, Pure Appl. Math. Q. **12** (2016), 141–164.

[112] G. van der Geer, *On the geometry of a Siegel modular threefold*, Math. Ann. **260** (1982), 317–350.

[113] G. Codogni, A. Fanelli, R. Svaldi, L. Tasin, *Fano varieties in Mori fibre spaces*, Int. Math. Res. Not. IMRN **7** (2016), 2026–2067.

[114] K. Hashimoto, *Period map of a certain K3 family with an S_5-action*, J. Reine Angew. Math. **652** (2011), 1–65.

[115] H. Hironaka, *Resolution of singularities of an algebraic variety over a field of characteristic zero. I, II*, Ann. Math. **79** (1964), 109–203, 205–326.

[116] A. Iliev, *The Fano surface of the Gushel threefold*, Compos. Math. **94** (1994), 81–107.

[117] N. Ilten, H. Süß, *K-stability for Fano manifolds with Torus action of complexity 1*, Duke Math. J. **166** (2017), 177–204.

[118] V. Iskovskikh, *Fano 3-folds I*, Math. USSR, Izv. **11** (1977), 485–527.

[119] V. Iskovskikh, *Fano 3-folds II*, Math. USSR, Izv. **12** (1978), 469–506.

[120] V. Iskovskikh, Yu. Prokhorov, *Fano Varieties*, Encyclopaedia of Mathematical Sciences **47** (1999) Springer.

[121] P. Jahnke, T. Peternell, I. Radloff, *Threefolds with big and nef anticanonical bundles II*, Central Eur. J. Math. **9** (2011), 449–488.

[122] P. Jahnke, I. Radloff, *Terminal Fano threefolds and their smoothings*, Math. Z. **269** (2011), 1129–1136.

[123] A.-S. Kaloghiros, A. Petracci, *On toric geometry and K-stability of Fano varieties*, Trans. Amer. Math. Soc. Ser. B **8** (2021), 548–577.

[124] N. Katz, P. Sarnak, *Random Matrices, Frobenius Eigenvalues, and Monodromy*, American Mathematical Society Colloquium Publications, **45**. American Mathematical Society, 1999, 416 pp.

[125] Y. Kawamata, *On Fujita's freeness conjecture for 3-folds and 4-folds*, Math. Ann. **308** (1997), 491–505.

[126] Y. Kawamata, *Subadjunction of log canonical divisors II*, Amer. J. Math. **120** (1998), 893–899.

[127] S. Keel, Y. Hu, *Mori dream spaces and GIT*, Michigan Math. J. **48** (2000), 331–348.

[128] I.-K. Kim, T. Okada, J. Won, *K-stability of birationally superrigid Fano 3-fold weighted hypersurfaces*, preprint, arXiv:2011.07512 (2020).

[129] J. Kollár et al., *Flips and abundance for algebraic threefold*, Astérisque **211**, 1992.

[130] J. Kollár, *Singularities of pairs*, Proc. Sympos. Pure Math. **62** (1997), 221–287.

[131] J. Kollár, *Lectures on Resolution of Singularities*, Princeton University Press, 2007.

[132] J. Kollár, S. Mori, *Birational Geometry of Algebraic Varieties*, Cambridge University Press, 1998.

[133] A. Kuribayashi, K. Komiya, *On Weierstrass points of non-hyperelliptic compact Riemann surfaces of genus three*, Hiroshima Math. J. **7** (1977), 743–768.

[134] A. Kuribayashi, H. Kimura, *Automorphism groups of compact Riemann surfaces of genus five*, J. Algebra **134** (1990), 80–103.

[135] A. Kuznetsov, Yu. Prokhorov, *Prime Fano threefolds of genus* 12 *with a* $\mathbb{G}_m$-*action*, Épijournal de Géom. Algébrique, EPIGA, **2**, 3 (2018).

[136] A. Kuznetsov, Yu. Prokhorov, C. Shramov, *Hilbert schemes of lines and conics and automorphism groups of Fano threefolds*, Jpn. J. Math. **13** (2018), 109–185.

[137] R. Lazarsfeld, *Positivity in Algebraic Geometry* **II**, Springer-Verlag, 2004.

[138] R. Lazarsfeld, M. Mustata, *Convex bodies associated to linear series*, Ann. Sci. Éc. Norm. Supér. **42** (2009), 783–835.

[139] C. Li, *K-semistability is equivariant volume minimization*, Duke Math. J. **166** (2017), 3147–3218.

[140] C. Li, X. Wang, C. Xu, *Algebraicity of the metric tangent cones and equivariant K-stability*, J. Amer. Math. Soc. **34** (2021), 1175–1214.

[141] C. Li, X. Wang, C. Xu, *On the proper moduli spaces of smoothable Kähler-Einstein Fano varieties*, Duke Math. J. **168** (2019), 1387–1459.

[142] C. Li, C. Xu, *Special test configuration and K-stability of Fano varieties*, Math. Ann. **180** (2014), 197–232.

[143] Ch. Liedtke, *Morphisms to Brauer–Severi varieties, with applications to del Pezzo surfaces*, Geometry over Nonclosed Fields. Simons Symp. Springer (2017), 157–196.

[144] Y. Liu, *The volume of singular Kähler–Einstein Fano varieties*, Compos. Math. **154** (2018), 1131–1158.

[145] Y. Liu, *K-stability of Fano threefolds of rank* 2 *and degree* 14 *as double covers*, Math. Z. **303**, no. 2 (2023), Paper No. 38, 9 pp.

[146] Y. Liu, C. Xu, *K-stability of cubic threefolds*, Duke Math. J. **168** (2019), 2029–2073.

[147] Y. Liu, C. Xu, Z. Zhuang, *Finite generation for valuations computing stability thresholds and applications to K-stability*, Ann. of Math. (2) **196**, no. 2 (2022), 507–566.

[148] Y. Liu, Z. Zhu, *Equivariant K-stability under finite group action*, Internat. J. Math. **33**, no. 1 (2022), Paper No. 2250007, 21 pp.

[149] T. Mabuchi, *On the classification of essentially effective* $\mathrm{SL}_n(\mathbb{C})$-*actions on algebraic n-folds*, Osaka J. Math. **16** (1979), 745–758.

[150] T. Mabuchi, S. Mukai, *Stability and Einstein–Kähler metric of a quartic del Pezzo surface*, Lecture Notes in Pure and Appl. Math. **145** (1993), 133–160.

[151] C. Mallows, N. Sloane, *On the invariants of a linear group of order* 336, Math. Proc. Cambridge Philos. Soc. **74** (1973), 435–440.

[152] A. Maltcev, *Foundations of Linear Algebra*, W. H. Freeman, 1963.

[153] M. Manetti, *Differential graded Lie algebras and formal deformation theory*, Proceedings of the 2005 Summer Research Institute, Seattle, WA, USA, July 25 – August 12, 2005. Proc. Sympos. Pure Math. **80** (2009), 785–810.

[154] J. Martinez-Garcia, *Log canonical thresholds of del Pezzo surfaces in characteristic p*, Manuscripta Math. **145** (2014), 89–110.

[155] Y. Matsushima, *Sur la structure du groupe d'homéomorphismes analytiques d'une certaine variété Kählérienne*, Nagoya Math. J. **11** (1957), 145–150.

[156] K. Matsuki, *Weyl groups and birational transformations among minimal models*, Mem. Amer. Math. Soc. **116** (1995).

[157] S. Mori, *Threefolds whose canonical bundles are not numerically effective*, Ann. Math. **116** (1982), 133–176.

[158] S. Mori, S. Mukai, *Classification of Fano 3-folds with $B_2 \geqslant 2$*, Manuscripta Math. **36** (1981), 147–162.

[159] S. Mori, S. Mukai, *Classification of Fano 3-folds with $B_2 \geqslant 2$. Erratum*, Manuscripta Math. **110** (2003), 407.

[160] S. Mori, S. Mukai, *On Fano 3-folds with $B_2 \geqslant 2$*, Adv. Stud. Pure Math. **1** (1983), 101–129.

[161] S. Mori, S. Mukai, *Classification of Fano 3-folds with $B_2 \geqslant 2$, I*, Algebraic and Topological Theories. Papers from the symposium dedicated to the memory of Dr. Takehiko Miyata held in Kinosaki, October 30-November 9, 1984. Tokyo, Kinokuniya, 1986, 496–545.

[162] S. Mukai, *Biregular classification of Fano 3-folds and Fano manifolds of coindex 3*, Proc. Natl. Acad. Sci. USA **86** (1989), 3000–3002.

[163] S. Mukai, *Fano 3-folds*, London Math. Soc. Lecture Note Ser. **179** (1992), 255–263.

[164] S. Mukai, H. Umemura, *Minimal rational threefolds*, Lecture Notes in Math. **1016** (1983), 490–518.

[165] A. Nadel, *Multiplier ideal sheaves and Kähler–Einstein metrics of positive scalar curvature*, Ann. Math. **132** (1990), 549–596.

[166] N. Nakayama, *Zariski-decomposition and Abundance*, MSJ Memoirs **14** (2004), Mathematical Society of Japan, Tokyo.

[167] N. Nakayama, *Classification of log del Pezzo surfaces of index two*, J. Math. Sci. Univ. Tokyo **14** (2007), 293–498.

[168] Y. Namikawa, *Smoothing Fano 3-folds*, J. Algebraic Geom. **6** (1997), 307–324.

[169] K. Nguyen, M. van der Put, J. Top, *Algebraic subgroups of $GL_2(\mathbb{C})$*, Indag. Math. **19** (2008), 287–297.

[170] Y. Odaka, *On the moduli of Kähler–Einstein Fano manifolds*, Proceeding of Kinosaki Symposium (2013), 112–126.

[171] Y. Odaka, Y. Sano, *Alpha invariant and K-stability of $\mathbb{Q}$-Fano varieties*, Adv. Math. **229** No. 5 (2012), 2818–2834.

[172] Y. Odaka, C. Spotti, S. Sun, *Compact moduli spaces of del Pezzo surfaces and Kähler–Einstein metrics*, J. Differential Geom. **102** (2016), 127–172.

[173] S. Okawa, *On images of Mori dream spaces*, Math. Ann. **364** (2016), 1315–1342.

[174] I. Pan, F. Russo, *Cremona transformations and special double structures*, Manuscripta Math. **117** (2005), 491–510.

[175] I. Pan, *On Cremona transformations of $\mathbb{P}^3$ which factorize in a minimal form*, Rev. Un. Mat. Argentina, **54** (2013), 37–58.

[176] J. Park, *Birational maps of Del Pezzo fibrations* J. Reine Angew. Math. **538** (2001), 213–221.

[177] J. Park, J. Won, *Log-canonical thresholds on del Pezzo surfaces of degrees $\geqslant 2$*, Nagoya Math. J. **200** (2010), 1–26.

[178] J. Park, J. Won, *Log canonical thresholds on Gorenstein canonical del Pezzo surfaces*, Proc. Edinb. Math. Soc. **54** (2011), 187–219.

[179] J. Park, J. Won, *K-stability of smooth del Pezzo surfaces*, Math. Ann. **372** (2018), 1239–1276.

[180] L. Petersen, H. Süß, *Torus invariant divisors*, Israel J. Math. **182** (2011), 481–505.

[181] V. Popov, *Structure of the closure of orbits in spaces of finite-dimensional linear SL(2) representations*, Math. Notes **16** (1974), 1159–1162.

[182] Y. Prokhorov, *Automorphism groups of Fano 3-folds*, Russian Math. Surveys **45** (1990), 222–223.

[183] Y. Prokhorov, *On exotic Fano varieties*, Moscow Univ. Math. Bull. **45** (1990), 36–38.

[184] Yu. Prokhorov, *Simple finite subgroups of the Cremona group of rank 3*, J. Algebraic Geom. **21** (2012), 563–600.

[185] Yu. Prokhorov, *G-Fano threefolds. I., II.*, Adv. Geom. **13** (2013), 389–418, 419–434.

[186] Yu. Prokhorov, *Rationality of Fano threefolds with terminal Gorenstein singularities. I*, Proc. Steklov Inst. Math. **307** (2019), 210–231.

[187] M. Reid, *The complete intersection of two or more quadrics*, PhD thesis, Trinity College, Cambridge, 1972.

[188] M. Reid, *Chapters on algebraic surfaces*, Complex algebraic geometry (Park City, UT, 1993), IAS/Park City Math. Ser. **3**, Amer. Math. Soc., Providence, RI (1997), 3–159.

[189] G. Sanna, *Rational curves and instantons on the Fano threefold Y_5*, preprint, arXiv:1411.7994 (2014).

[190] G. Sanna, *Small charge instantons and jumping lines on the quintic del Pezzo threefold*, Int. Math. Res. Not. IMRN **21** (2017), 6523–6583.

[191] C. Salgado, D. Testa, A. Varilly Alvarado, *On the unirationality of del Pezzo surfaces of degree two*, J. Lond. Math. Soc. **90** (2014), 121–139.

[192] E. Sernesi, *Deformations of Algebraic Schemes*, Grundlehren Math. Wiss., **334**, Springer-Verlag, 2006.

[193] T. Shaska, H. Völklein, *Elliptic subfields and automorphisms of genus 2 function fields*, Algebra, arithmetic and geometry with applications, 703–723, Springer, 2004.

[194] F.-O. Schreyer, *Geometry and algebra of prime Fano 3-folds of genus 12*, Compos. Math. **127** (2001), 297–319.

[195] N. Shepherd-Barron, *The rationality of quintic Del Pezzo surfaces — a short proof*, Bull. Lond. Math. Soc. **24** (1992), 249–250.

[196] Y. Shi, X. Zhu, *Kähler–Ricci solitons on toric Fano orbifolds*, Math. Z. **271** (2012), 1241–1251.

[197] V. Shokurov, *3-fold log flips*, Izv. Math. **40** (1993), 95–202.

[198] C. Shramov, V. Vologodsky, *Automorphisms of pointless surfaces*, preprint, arXiv:1807.06477 (2018).

[199] C. Spotti, S. Sun, *Explicit Gromov-Hausdorff compactifications of moduli spaces of Kähler–Einstein Fano manifolds*, Pure Appl. Math. Q. **13** (2017), 477–515.

[200] T. A. Springer, *Linear algebraic groups*, In Algebraic Geometry IV, 1994. Springer, 1–121.

[201] C. Stibitz, Z. Zhuang, *K-stability of birationally superrigid Fano varieties*, Compos. Math. **155** (2019), 1845–1852.

[202] H. Süß, *Kähler–Einstein metrics on symmetric Fano T-varieties*, Adv. Math. **246** (2013), 100–113.

[203] H. Süß, *Fano threefolds with 2-torus action – a picture book*, Doc. Math. **19** (2014), 905–914.

[204] H.P.F. Swinnerton-Dyer, *Rational points on del Pezzo surfaces of degree 5*, Proceedings of the Fifth Nordic Summer School in Mathematics, Wolters–Noordhoff, 1972, 287–290.

[205] M. Szurek, J. Wiśniewski, *Fano bundles of rank 2 on surfaces*, Compos. Math. **76** (1990), 295–305.

[206] G. Székelyhidi, *The Kähler–Ricci flow and K-polystability*, Amer. J. Math. **132** (2010), 1077–1090.

[207] F. Szechtman, *Equivalence and normal forms of bilinear forms*, Linear Algebra Appl. **443** (2014), 245–259.

[208] K. Takeuchi, *Weak Fano threefolds with del Pezzo fibration*, Eur. J. Math. **8** (2022), no. 3, 1225–1290.

[209] D. Testa, A. Varilly-Alvarado, M. Velasco, *Big rational surfaces*, Math. Ann. **351** (2011), 95–107.

[210] G. Tian, *On Kähler–Einstein metrics on certain Kähler manifolds with $c_1(M) > 0$*, Invent. Math. **89** (1987), 225–246.

[211] G. Tian, *On Calabi's conjecture for complex surfaces with positive first Chern class*, Invent. Math. **101** (1990), 101–172.

[212] G. Tian, *Kähler–Einstein metrics with positive scalar curvature*, Invent. Math. **130** (1997), 1–37.

[213] G. Tian, *Existence of Einstein metrics on Fano manifolds*, In: Metric and Differential Geometry, Progr. Math. **297**, Birkhäuser (2012), 119–162.

[214] G. Tian, *K-stability and Kähler-Einstein metrics*, Comm. Pure Appl. Math. **68** (2015), no. 7, 1085–1156.

[215] G. Tian, S.-T. Yau, *Kähler–Einstein metrics on complex surfaces with $C_1 > 0$*, Comm. Math. Phys. **112**, (1987), 175–203.

[216] D. Timashev, *Homogeneous Spaces and Equivariant Embeddings*, Encyclopaedia of Mathematical Sciences, **138**, Springer, 2011.

[217] C. Xu, *K-stability of Fano varieties: an algebro-geometric approach*, EMS Surv. Math. Sci. **8** (2021), 265–354.

[218] C. Xu, *A minimizing valuation is quasi-monomial*, Ann. of Math. (2), **191** (2020), 1003–1030.

[219] X. Wang, X. Zhu, *Kähler–Ricci solitons on toric manifolds with positive first Chern class*, Adv. Math. **188** (2004), 87–103.

[220] X. Wang, *Height and GIT weight*, Math. Res. Lett. **19** (2012), 909–926.

[221] C.T.C. Wall, *Nets of conics*, Math. Proc. Cambridge Philos. Soc. **81** (1973), 351–364.

[222] K. Watanabe, M. Watanabe, *The classification of Fano 3-folds with torus embeddings*, Tokyo J. Math. **5** (1982), 37–48.

[223] J. Wolter, *Equivariant birational geometry of quintic del Pezzo surface*, Eur. J. Math. **4**, (2018), 1278–1292.

[224] Q. Zhang, *Rational connectedness of log $\mathbb{Q}$-Fano varieties*, J. Reine Angew. Math. **590** (2006), 131–142.

[225] Z. Zhuang, *Product theorem for K-stability*, Adv. Math., **371** (2020), 107250, 18pp.

[226] Z. Zhuang, *Optimal destabilizing centers and equivariant K-stability*, Invent. Math. **226**, no. 1 (2021), 195–223.

[227] Z. Zhuang, *Birational superrigidity and K-stability of Fano complete intersections of index one*, Duke Math. J. **169** (2020), 2205–2229.

Index